AF579990

REDOX CHEMISTRY AND INTERFACIAL BEHAVIOR OF BIOLOGICAL MOLECULES

REDOX CHEMISTRY AND INTERFACIAL BEHAVIOR OF BIOLOGICAL MOLECULES

Edited by

Glenn Dryhurst

The University of Oklahoma
Norman, Oklahoma

and

Katsumi Niki

Yokohama National University
Yokohama, Japan

PLENUM PRESS • NEW YORK AND LONDON

Library of Congress Cataloging in Publication Data

International Symposium on Redox Mechanisms and Interfacial Properties of Molecules of Biological Importance (3rd: 1987: Honolulu, Hawaii)
Redox chemistry and interfacial behavior of biological molecules / edited by Glenn Dryhurst and Katsumi Niki.
p. cm.
"Proceedings of the Third International Symposium on Redox Mechanisms and Interfacial Properties of Molecules of Biological Importance, held October 19–23, 1987, in Honolulu, Hawaii" – T.p. verso.
Includes index.
ISBN 0-306-43038-X
1. Bioelectrochemistry – Congresses. 2. Oxidation-reduction reaction – Congresses. 3. Biomolecules – Surfaces – Electric properties – Congresses. I. Dryhurst, Glenn, 1939- . II. Niki, Katsumi. III. Title.
QP517.B53I58 1987 88-29382
574.19'283 – dc19 CIP

Proceedings of the Third International Symposium on Redox Mechanisms and Interfacial Properties of Molecules of Biological Importance, held October 19–23, 1987, in Honolulu, Hawaii

A Division of Plenum Publishing Corporation
233 Spring Street, New York, N.Y. 10013

Printed in the United States of America

PREFACE

The papers in this book were presented at the Third International Symposium on Redox Mechanisms and Interfacial Properties of Molecules of Biological Importance held in Honolulu, Hawaii between October 19-23, 1987. This Symposium was held as part of the 172nd Meeting of The Electrochemical Society which was cosponsored by The Electrochemical Society of Japan with the cooperation of The Japan Society of Applied Physics.

The aim of the Symposium was to bring together a group of electrochemists and bio-medical scientists with interests in electrochemistry from around the world to present their most current research results and/or to present up-to-date reviews of current areas of research activity.

It is quite clear from the diversity of topics covered in the various papers that electrochemistry and electrochemical techniques and principles have much to contribute to our understanding of many important biochemical phenomena. For example, electrochemical studies are providing important insights into the redox properties of biomolecules ranging from relatively small organic molecules such as indoleamine neurotransmitters to very large organic/organometallic molecules which include various redox enzymes or model enzyme systems. Many of the most powerful analytical techniques are now being coupled to electrodes to monitor potential-controlled behaviors of biological molecules at charged interfaces. Electrochemical techniques are now being developed which permit extraordinarily small electrodes to be inserted into single cells to monitor electroactive biomolecules. Other microelectrodes are being employed to control cell growth and to manipulate single cells. Electrochemical techniques utilizing modified electrodes are now being employed for very sensitive and selective analytical measurements. All of these topics and many more are covered in the forty six papers presented in this book. We hope that there is much here to interest not only electrochemists and electroanalytical chemists but also the biomedical research community in general.

It is always of very great importance to the success of any major symposium to have the participation of established scientists from throughout the world and, equally importantly, young investigators at the beginning of their independent scientific careers. We are thus very pleased to acknowledge the financial support provided by the following organizations which permitted many such people to attend the Symposium:

Office of Naval Research
EE&G Princeton Applied Research Corporation
Eli Lilly Corporation
The Organic and Biological Electrochemistry
Division of The Electrochemical Society

We would also like to express our thanks to Melanie Yelity and Lisa Honski of Plenum Publishing Corporation for help and advice concerning the preparation of the manuscript. Finally, we would like to thank Susan Kyncl who typed the entire manuscript and provided help in numerous other ways.

Glenn Dryhurst
Konstanz

Katsumi Niki
Yokohama

May 1988

CONTENTS

SUPRAMOLECULAR EFFECTS IN THE REDOX AND COORDINATION CHEMISTRY OF SUPERSTRUCTURED IRON PORPHYRINS

D. Lexa and J.M. Savéant

Laboratoire d'Electrochimie Moléculaire
de l'Université de Paris 7
Unité Associée au CNRS No. 438
2 place Jussieu
75251 Paris Cedex 05 / France

INTRODUCTION

Enzymatic systems exhibit quite remarkable properties as far as the efficiency, smoothness and selectivity of the chemistry they can carry out is concerned. These are most probably related to a common feature of their structures, i.e., the existence of a reacting center, the prosthetic group, surrounded by protein chains. The latter offer a microenvironment able to modulate the reactivity of the prosthetic group without participating directly to the reaction. Such organized structures are correlatively expected to function with minimal entropy losses. The concept of supramolecular chemistry, and electrochemistry, derives from attempts to imitate, even remotely, these structures and their functions.

In this context, the purpose of the following discussion is to examine the possibility of modifying the reactivity of a central reactant by relatively simple carbon chains superstructures, some of them containing secondary amide groups, a group known to be an essential component of protein chains. As reacting centers we chose iron porphyrins. These have the advantage of exhibiting properties that are relative to aromatic organic chemistry, coordination chemistry and organometallic chemistry. The influence of the above mentioned superstructures, if significant, will thus provide a good model for the modulation of the reactivity of molecules belonging to various different areas of chemistry. Another interest of iron porphyrins is that they serve as prosthetic groups in a number of natural metalloproteins (hemoglobin, myoglobin, cytochromes, cytochrome oxidases). The superstructured porphyrins investigated in the work described below[1-9] are of the "basket-handle" type as shown in Figures 1, 7 and 9 (for the sake of comparison the Collman's picket fence porphyrin[10] was also investigated). They derive from tetraphenylporphyrins (TPP) by grafting, at the ortho-positions of the phenyl rings, by means of ether or amide linkages simple carbon chains (Figure 1), carbon chains containing additional secondary amide groups (Figure 2), or carbon chains containing a hanging nitrogen base (Figure 3). The basket-handle superstructures may also be varied, not only by the nature of the chains but also by their spatial arrangement (cross-trans "CT", adjacent-trans "AT", adjacent-cis "AC") (Figure 1).

The questions we wish to address upon investigation of the redox and coordination properties of this set of superstructured iron porphyrins are the following. Do the superstructures exert a significant influence on the reactivity of the central iron porphyrin? What is the nature of the reactivity modulation brought about by the presence of the basket handle chains? Does it derive from a single effect or from the superposition of several different effects?. What is the order of magnitude of each of these different effects? We will see that the presence of secondary amide groups in the chain is particularly important leading to a built-in "local solvation". In this connection, does "local solvation" build up when increasing the number of secondary amide groups included in the chains? What is the exact nature of this "local solvation" as revealed by the separation of its energetic and entropic aspects?

The modulation of the reactivity by the basket-handle superstructures was mostly investigated for reactions involving a one-unit increase of the negative charge of the porphyrin complex: Fe(II)(Fe(I)$^-$, Fe(I)$^-$(Fe("O")$^{2-}$ redox reactions, complexation of iron(II) by anionic ligands. However, we will see that significant supramolecular effects also exist for the complexation of iron(II) by neutral nitrogen bases.

EXPERIMENTAL

Chemicals

The synthesis and characterization of the ether-linked and amide-linked basket-handle porphyrins of Figure 1 are described in References 12 and 13, respectively (for the picket fence porphyrin see Reference 13). For the superstructured porphyrins containing an increasing number of NHCO groups (Figure 7) see Reference 14 and for the hanging-base basket-handle (Figure 9) porphyrins References 6, 12, 15.

Reactivity Measurements

They were derived from the electrochemistry of the various investigated porphyrins starting with the iron(III) complex. Iron(III) porphyrins exhibit three successives one-electron reduction waves or groups of waves corresponding to the reduction of Fe(III)$^+$ into Fe(II), Fe(II) into Fe(I)$^-$ and Fe(I)$^-$ into Fe("O")$^{2-}$ (for an overview on the electrochemistry of iron porphyrins see Reference 16). Iron(III) and iron(II) porphyrins are five or six coordination complexes, one of the axial ligands possibly being the solvent. Fe(I)$^-$ is the dominant mesomeric form of the complex resulting from the injection of one electron into an iron(II) porphyrin[17-24]. What we note as Fe("O)$^{2-}$ is probably a combination of two dominant mesomeric forms, the iron(O) porphyrin and the anion radical of the iron(I) porphyrin[7-23]. Both the iron(I) and iron("O") porphyrins bear no axial ligand at room temperature. For this reason the Fe(I)$^-$/Fe("O")$^{2-}$ wave is reversible showing no coupling with homogeneous chemical reactions and thus allowing the simple determination of the standard potential as the middle of the cyclic voltammetric cathodic and anodic peak potentials. The situation is more complex with the Fe(III)/(Fe(II) and Fe(II)/Fe(I) waves which show coupling of the electron transfer with homogeneous ligation/deligation reactions in a number of cases. Cyclic voltammetry and thin-layer spectroelectrochemistry were then used jointly for obtaining the characteristic standard potentials and equilibrium constants as well as the rate constants for the complexation of iron(II) by chloride ions. The procedures employed in this purpose have been described in detail[1,4], requiring in several cases an extension of the available theory[25,26] of the kinetics of electron transfers coupled with homogeneous chemical reactions in the context of cyclic voltammetry[27,28].

RESULTS AND DISCUSSION

1. Protection against solvation, "local solvation" and steric hindrance to ligation by basket-handle superstructures in the reactions: $Fe(I)^- + e^- \rightleftharpoons Fe("0")^{2-}$, $Fe(II) + e^- \rightleftharpoons Fe(I)^-$, $Fe(II) + Cl^- \rightleftharpoons Fe(II)Cl^-$.

The title reactions were investigated for the basket-handle porphyrins shown in Figure 1, in five different solvents, dimethylformamide (DMF), N-methylacetamide (NMA), benzonitrile (PhCN), butyronitrile (PrCN), 1,2-dichloroethane (DCE)[1,4]. The main results are listed in Table 1. They involve the standard potentials of the $Fe(I)^- + e^- \rightleftharpoons Fe("0")^{2-}$ and $Fe(II)+e^- \rightleftharpoons Fe(I)^-$ reactions, $E^o_{Fe(I)^-/Fe("0")2-}$ and $E^o_{Fe(II)/Fe(I)^-}$, the equilibrium constant of the $Fe(II) + Cl^- \rightleftharpoons Fe(II)Cl^-$ reaction, K_A, and the association and dissociation rate constants, k_A and k_D, of the latter reaction.

As a preliminary observation, we note that the three reactions are, from a thermodynamic standpoint, more difficult when passing from TPP to TAP (the variation in free energy is of the order of 100 mV for the $Fe(I)^-/Fe("0")^{2-}$ reaction and less for the two others). This results from the inductive effect exerted by the four electron-donating methoxy groups located in ortho-positions of the phenyl groups. We can thus obtain, with a good approximation, an estimate of the free energies featuring the specific effect of the ether-linked basket handles by substracting from the E^o of the considered porphyrin the E^o of TAP. For the chloride association reaction, the free energy change is given by $(RT/F)Ln(K_A/K_A^{TAP})$. In the amide-linked series, we can take TPP itself as the reference, since the NHCO group has a Hammett σ constant close to zero[29]. The free energy changes thus estimated are given in mV between parenthesis in Table 1.

On these bases, a striking observation is that all three reactions, in all five solvents, are made more difficult by the presence of ether-linked superstructures and easier by the presence of secondary amide-linked superstructures. This can be interpreted as follows. All three reactions have in common that they involve a one-unit increase of the negative charge of the porphyrin complex (from -1 to -2 for the first reaction and from 0 to -1 for the two others). The observed free energy changes thus results from the variations of the free energies of solvation bearing in mind that -1 charged complexes are more strongly solvated than the neutral complexes and less solvated that the -2 charged complexes. In this context, the ether-linked chains offer a steric protection against the approach of the solvent molecules thus decreasing the solvation free energy. The negatively charged complexes are thus destabilized as compared to the unprotected porphyrins, the doubly charged species more than the singly charged species. This renders all three reactions more difficult as observed in the ether-linked series. The same phenomenon ought to occur also in the amide-linked series (in the two series the number of links in the chains was purposely the same so as to results in approximately the same protection against solvation). The reason why this destabilizing factor is largely overcompensated in the amide-linked series, rendering all three reactions easier, resides in the interaction between the negative charge in the porphyrin complex and the four secondary amide dipoles. (Evidence has been provided on NMR grounds that in the Fe(II) complexes, the NHCO dipoles point their positive end towards the porphyrin ring[11,12,15].) Upon negatively charging the porphyrins, this orientation

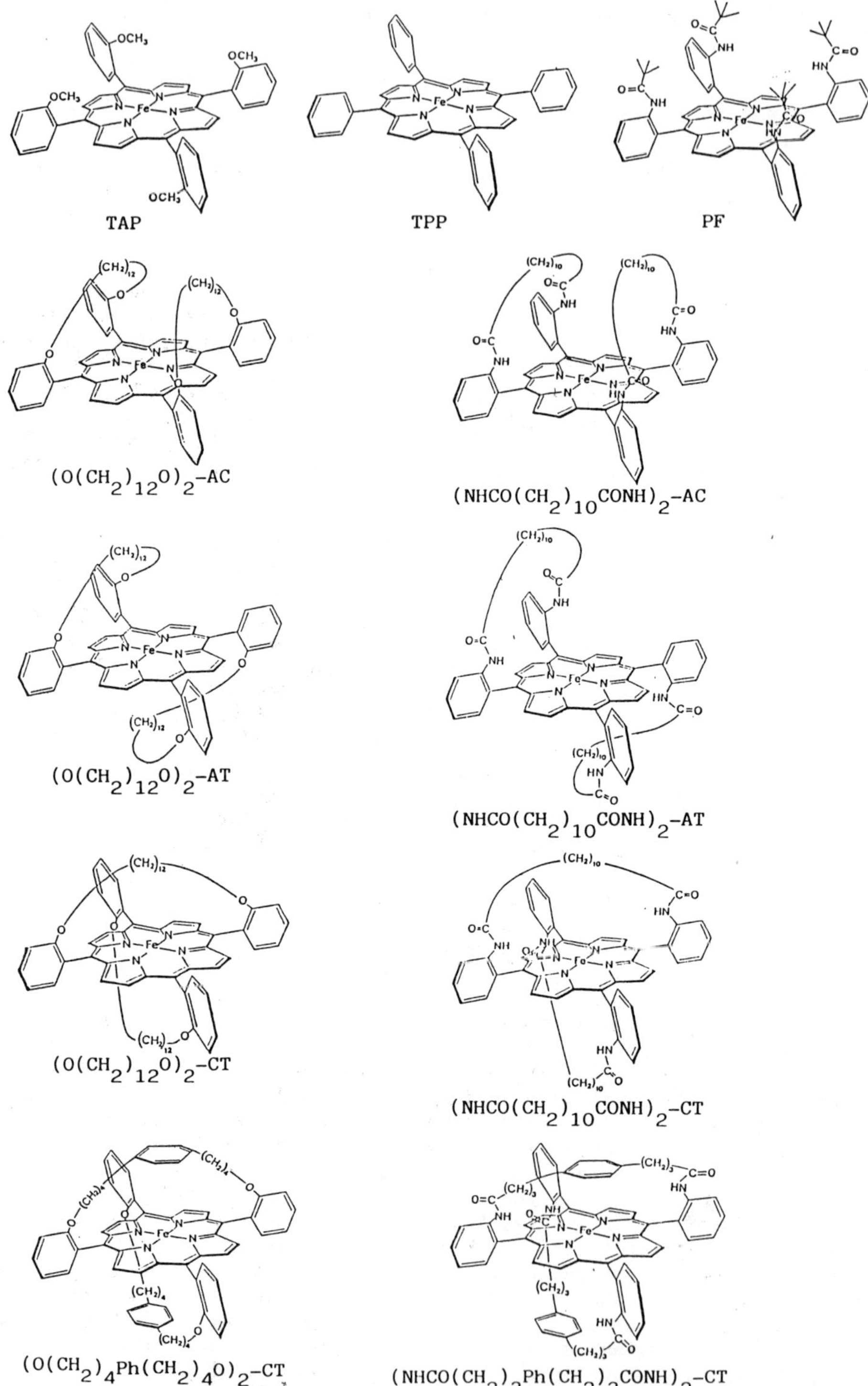

Figure 1. Ether-linked and amide-linked basket-handle porphyrins. TPP: Tetraphenylporphyrin, TAP: tetraanisylporphyrin, PF: Collman's picket fence porphyrin.

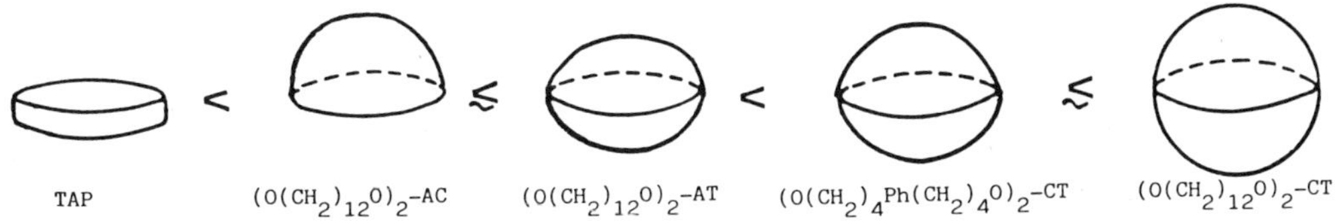

Figure 2. Schematic representation of the protection against solvation offered by the basket-handles.

will obviously remain). In other words, the secondary amide groups play the role of a "local solvent", replacing the first solvation shell of the external solvent. (This "local solvation" phenomenon will be discussed further in the next sections).

Closer inspection of the data in Table 1 shows that, quantitatively, the degree of destabilization (ether-linked basket handles) or stabilization (amide-linked basket handles) is a function of both the nature and arrangement of the chains and of the nature of the solvent.

In this respect, the $Fe(I)^{-}$-$(Fe("O")^{2-}$ reaction is the easiest to analyze since axial coordination is absent in both members of the redox couple. The reaction is thus not influenced by the basic properties of the solvent but rather by its acidic properties (in the Lewis sense). In the ether-linked series, the protection against solvation, as measured by $E^{o}_{BHP} - E^{o}_{TAP}$, varies in DMF and PrCN in the order $(O(CH_2)_{12}O)$-AC $\leq$ $(O(CH_2)_{12}O)$-AT $<$ $(O(CH_2)_4Ph(CH_2)_4O)_2$-CT $\leq$ $(O(CH_2)_{12}O)_2$-CT, suggesting that the degree of protection increases as the volume statistically occupied by the chain increases, taking their random movements into account (Figure 2).

In NMA, the order is the same with the exception of an inversion between $(O(CH_2)_{12}O)_2$-CT and $(O(CH_2)_4Ph(CH_2)_4O)_2$-CT. However, the most striking feature of the data concerning this solvent is that the free energies of protection against solvation are smaller than in the other solvents, particularly with the AC and AT structures. On the other hand, the anion solvating ability of NMA is much larger than that of DMF, PhCN and PrCN (the Gutman acceptor numbers are respectively: 30, 16, 15.5 and 15[29,30]). If the chains would offer a constant volume of protection whatever the solvent, one would thus expect the free energy of protection to be larger in the case of NMA than with the other solvents. The fact that the opposite is observed, particularly with the most flexible AC and AT structures thus points to the concept that the protection is a function of the solvating power of the solvent and of the rigidity of the chains: the more "aggressive" the solvent and the more flexible the chains, the less efficient the protection. In other words, the "aggressive" NMA is, to a larger extent than DMF and PrCN, able to thrust the chains aside or to flatten them out over the porphyrin ring, which is obviously easier with flexible than with rigid chains.

In PhCN, protection against solvation is again strong but varies in a different order (Figure 3). PhCN possesses a property that the other solvents do not, namely, that it lends itself to acceptor $\pi \rightarrow \pi^*$ interactions owing to a defect of electrons in its π orbital caused by the electron-withdrawing character of the cyano group. Since the $Fe(I)^{-}$ and $Fe("O")^{2-}$ complexes possess an excess negative charge in their π^* orbitals it is likely that their solvation by benzonitrile involves, at least partially, $\pi \rightarrow \pi^*$ interactions with parallel aromatic rings (Figure 3). These interactions would be stronger with the $Fe("O")^{-2}$ than with the $Fe(I)^{-}$ complexes. The order of protection efficiency can be explained on

Table 1. Thermodynamic and kinetic data concerning the porphyrins shown in Figure 1.

Solvent[a] (temperature)	Porphyrin	$E^{o}_{Fe(I)^-/Fe("O")^{2-}}$ [c,d]	$E^{o}_{Fe(II)/Fe(I)^-}$ [c,d]	K_A (M^{-1}) [e]	k_A $(M^{-1}.s^{-1})$	k_D (s^{-1})
DMF[b] (20°C)	TAP	-1.767	-1.058	$3.1\ 10^{1}$	$7.2\ 10^{4}$	$2.3\ 10^{3}$
	$(O(CH_2)_{12}O)_2$-AC	-1.842 (-75)	-1.098 (-40)	3.6 (-54)	$1.9\ 10^{4}$	$5.2\ 10^{3}$
	$(O(CH_2)_{12}O)_2$-AT	-1.862 (-95)	-1.093 (-35)	$3.4\ 10^{-1}$ (-114)	$5.3\ 10^{4}$	$1.6\ 10^{5}$
	$(O(CH_2)_{12}O)_2$-CT	-1.947 (-180)	-1.125 (-67)	$6.5\ 10^{-1}$ (-98)	$1.9\ 10^{3}$	$2.9\ 10^{3}$
	$(O(CH_2)_4Ph(CH_2)_4O)_2$-CT	-1.938 (-171)	-1.078 (-29)	$9.6\ 10^{-2}$ (-146)	$2.9\ 10^{2}$	$3.0\ 10^{3}$
	TPP	-1.650	-0.984	$5.6\ 10^{1}$	$1.0\ 10^{5}$	$2.0\ 10^{3}$
	PF	-1.440 (210)	-0.812 (172)	$3.6\ 10^{3}$ (105)	$6.8\ 10^{4}$	$1.9\ 10^{1}$
	$(NHCO(CH_2)_{10}CONH)_2$-AC	-1.455 (195)	-0.850 (134)	$1.8\ 10^{3}$ (88)	$4.6\ 10^{4}$	$2.6\ 10^{1}$
	$(NHCO(CH_2)_{10}CONH)_2$-AT	-1.505 (145)	-0.853 (131)	$3.0\ 10^{3}$ (101)	$3.8\ 10^{4}$	$1.2\ 10^{1}$
	$(NHCO(CH_2)_{10}CONH)_2$-CT	-1.530 (120)	-0.886 (98)	$1.2\ 10^{3}$ (77)	$1.0\ 10^{5}$	$8.6\ 10^{1}$
	$(NHCO(CH_2)_4Ph(CH_2)_4CONH)_2$-CT	-1.505 (145)	-0.900 (84)	$1.1\ 10^{2}$ (17)	$6.8\ 10^{3}$	$6.2\ 10^{1}$
NMA (35°C)	TAP	-1.660	-1.110	f	f	
	$(O(CH_2)_{12}O)_2$-AC	-1.680 (-20)	-1.150 (-40)			
	$(O(CH_2)_{12}O)_2$-AT	-1.695 (-35)	-1.150 (-40)			
	$(O(CH_2)_{12}O)_2$-CT	-1.835 (-175)	-1.200 (-90)			
	$(O(CH_2)_4Ph(CH_2)_4O)_2$-CT	-1.855 (-195)	-1.125 (-15)			
	TPP	-1.535	-1.085	f	f	
	PF	-1.430 (105)	-0.920 (165)			
	$(NHCO(CH_2)_{10}CONH)_2$-AC	-1.430 (105)	-0.987 (100)			
	$(NHCO(CH_2)_{10}CONH)_2$-AT	-1.420 (115)	-0.970 (115)			
	$(NHCO(CH_2)_{10}CONH)_2$-CT	-1.477 (52)	-0.985 (100)			
	$(NHCO(CH_2)_4Ph(CH_2)_4CONH)_2$-CT	-1.443 (92)	-0.945 (140)			
	TAP	-1.734	-1.107	$1.3\ 10^{2}$	$5.6\ 10^{3}$	$4.3\ 10^{1}$
	$(O(CH_2)_{12}O)_2$-AC	-1.958 (-224)	-1.195 (-88)	8.8 (-68)	$4.7\ 10^{2}$	$5.4\ 10^{1}$
	$(O(CH_2)_{12}O)_2$-AT	-1.938 (-204)	-1.195 (-88)	3.9 (-89)	$6.3\ 10^{2}$	$1.6\ 10^{2}$

PhCN (20°C)	$(O(CH_2)_{12}O)_2$-CT	-1.925 (-191)	-1.239 (-132)	2.2 (-103)	$3.3\ 10^{2}$	$1.5\ 10^{2}$
	$(O(CH_2)_4Ph(CH_2)_4O)_2$-CT	-1.988 (-254)	-1.190 (-83)	$4.4\ 10^{-1}$ (-144)	$5.4\ 10^{2}$	$1.3\ 10^{3}$
	TPP	-1.690	-1.029	$1.8\ 10^{2}$	$5.4\ 10^{3}$	$2.9\ 10^{1}$
	PF	-1.536 (154)	-0.777 (253)	$8.5\ 10^{5}$ (214)	$1.6\ 10^{8}$	$2.0\ 10^{2}$
	$(NHCO(CH_2)_{10}CONH)_2$-AC	-1.551 (139)	-0.831 (198)	$5.9\ 10^{5}$ (204)	$2.1\ 10^{8}$	$3.6\ 10^{2}$
	$(NHCO(CH_2)_{10}CONH)_2$-AT	-1.620 (70)	-0.827 (203)	$3.5\ 10^{5}$ (191)	$3.0\ 10^{8}$	$8.7\ 10^{2}$
	$(NHCO(CH_2)_{10}CONH)_2$-CT	-1.630 (60)	-0.895 (134)	$2.5\ 10^{5}$ (183)	$6.5\ 10^{8}$	$2.6\ 10^{3}$
	$(NHCO(CH_2)_4Ph(CH_2)_4CONH)_2$-CT	-1.540 (150)	-0.924 (105)	$9.8\ 10^{2}$ (43)	$1.9\ 10^{6}$	$1.9\ 10^{3}$
PrCN (20°C)	TAP	-1.805	-1.016	$1.0\ 10^{3}$	$1.3\ 10^{6}$	$1.3\ 10^{3}$
	$(O(CH_2)_{12}O)_2$-AC	-1.935 (-130)	-1.183 (-167)	$7.6\ 10^{1}$ (-65)	$4.6\ 10^{3}$	$6.1\ 10^{1}$
	$(O(CH_2)_{12}O)_2$-AT	-1.940 (-135)	-1.191 (-175)	$7.4\ 10^{1}$ (-66)	$2.1\ 10^{3}$	$2.8\ 10^{1}$
	$(O(CH_2)_{12}O)_2$-CT	-2.065 (-260)	-1.197 (-181)	9.7 (-117)	$9.0\ 10^{2}$	$9.0\ 10^{1}$
	$(O(CH_2)_4Ph(CH_2)_4O)_2$-CT	-2.030 (-225)	-1.165 (-149)	2.0 (-157)	$9.4\ 10^{2}$	$4.7\ 10^{2}$
	TPP	-1.682	-1.002	$3.8\ 10^{3}$	$1.0\ 10^{6}$	$2.7\ 10^{2}$
	PF	-1.546 (136)	-0.871 (131)	$3.0\ 10^{5}$ (110)	$1.4\ 10^{8}$	$4.7\ 10^{2}$
	$(NHCO(CH_2)_{10}CONH)_2$-AC	-1.618 (64)	-0.884 (118)	$2.0\ 10^{5}$ (100)	$1.1\ 10^{8}$	$5.2\ 10^{2}$
	$(NHCO(CH_2)_{10}CONH)_2$-AT	-1.622 (60)	-0.889 (113)	$2.5\ 10^{5}$ (104)	$7.5\ 10^{8}$	$3.0\ 10^{3}$
	$(NHCO(CH_2)_{10}CONH)_2$-CT	-1.570 (111)	-0.877 (125)	$5.5\ 10^{3}$ (9)	$7.2\ 10^{6}$	$1.3\ 10^{3}$
DCE (20°C)	TAP	g	-1.048	$4.8\ 10^{3}$	$1.7\ 10^{6}$	$3.5\ 10^{2}$
	$(O(CH_2)_{12}O)_2$-AC		-1.236 (-188)	$2.0\ 10^{2}$ (-80)	$1.2\ 10^{4}$	$6.2\ 10^{1}$
	$(O(CH_2)_{12}O)_2$-AT		-1.275 (-227)	$4.5\ 10^{1}$ (-118)	$7.6\ 10^{2}$	$1.7\ 10^{1}$
	TPP	g	-1.010	$7.2\ 10^{3}$	$1.6\ 10^{6}$	$2.2\ 10^{2}$
	$(NHCO(CH_2)_{10}CONH)_2$-AC		-0.886 (124)	$1.2\ 10^{6}$ (129)	$3.7\ 10^{8}$	$3.1\ 10^{2}$
	$(NHCO(CH_2)_{10}CONH)_2$-AT		-0.879 (131)	$1.0\ 10^{6}$ (125)	$8.8\ 10^{8}$	$8.8\ 10^{2}$

a:supporting electrolyte:0.1M NBu_4ClO_4 unless otherwise stated. b:supporting electrolyte:0.1M $LiClO_4$. c:in V vs the NaCl saturated calomel electrode. d:between parenthesis,in mV, $E^\circ - E^\circ_{TAP}$ for ether-linked basket handles, $E^\circ - E^\circ_{TPP}$ for amide-linked basket handles. e:between parenthesis,in mV, $(RT/F)Ln(K_A/K_A^{TAP})$ for ether-linked basket handles, $(RT/F)Ln(K_A/K_A^{TPP})$ for amide-linked basket handles. f:the association of Cl^- with iron(II) is unmeasurably weak in NMA, owing to the strong solvation of Cl^- by NMA which is a particularly strong acceptor. g: Fe("O") complexes are unstable in DCE catalyzing its reduction into ethylene.

these grounds as pictured in Figure 3. A confirmation of the importance of $\pi \rightarrow \pi^*$ interactions in solvation by benzonitrile is found in the comparison of the $E^\circ_{Fe(I)^-/Fe("O")^{2-}}$s of TAP and TPP in the four solvents:

Solvent	DMF	NMA	PhCN	PrCN
$E^\circ_{TPP} - E^\circ_{TAP}$ (mV)	117	125	44	123

The difference in standard potentials appears to be significantly less in PhCN than in all three other solvents.

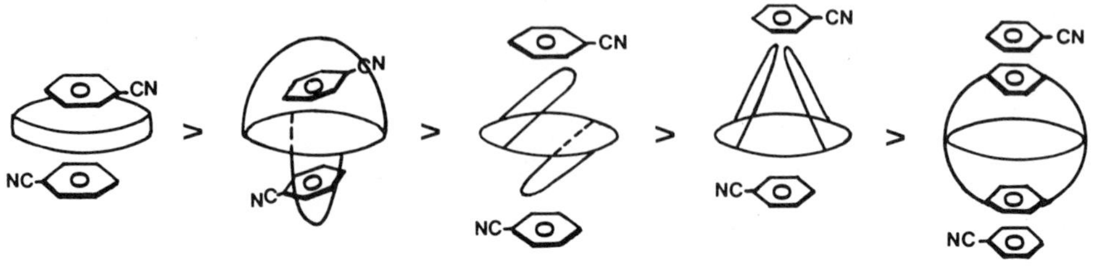

Figure 3. Schematic representation of the $\pi \rightarrow \pi^*$ interactions between benzonitrile molecules and the $Fe("O")^{2-}$-TAP and basket-handle porphyrins.

As discussed above, the most remarkable phenomenon occurs in the amide-linked series where "local solvation" by the NHCO groups largely compensates the destabilization resulting from the protection against solvation which exists here as in the ether-linked series. It is thus interesting to estimate the magnitude of this "local solvation" effect, on the reaction, i.e., the difference of the free energy of interaction of the NHCO dipoles with the $Fe("O")^{2-}$ and $Fe(I)^-$ complexes. A first approach to this problem consists in assuming that the protection against solvation is the same in both the ether-linked and amide-linked series for similar basket-handle structures as would be the case if protection was depending solely on the volume occupied by the chain regardless of the nature of the solvent. Under this assumption, the specific effect of the NHCO groups is measured by $(E^\circ_{amide}-E^\circ_{TPP})-(E^\circ_{ether}-E^\circ_{TAP})$. The results are given in Table 2. This quantity should then be approximately constant in the series of solvent and porphyrins. It is seen in Table 2 that it remains of the same order of magnitude but still varies clearly beyond experimental error. The results obtained in NMA are again of particular interest: the ΔE°'s are smaller than with any other solvent, particularly with the two most flexible AC and AT structures. As seen before, the molecules of NMA, owing to particularly strong anion solvating power, are able to overcome the steric barriers of the chains (especially when these are the most flexible) in the ether-linked series. We can now conclude that this possibility is strongly reduced in the amide-linked series. As sketched in Figure 4, this is due to the fact that in the amide-linked series the NHCO groups play the role of the primary solvation shell, whereas in the ether-linked series NMA tends to build up the primary solvation shell by thrusting the chains apart or flatting them out over the porphyrin ring.

Although weaker, the same effects are observed with the other three solvents: the variations in ΔE° (Table 2) follow the same trend as the protection against solvation in the ether-linked series, itself measured by $E^\circ_{ether}-E^\circ_{TAP}$ (Table 1). This shows the general validity of the concept that protection against solvation is stronger in the amide-linked than in the ether-linked series. It follows that the "local solvation" effect is larger than or equal to the largest value of ΔE° in Table 2, i.e., 0.4 eV

Table 2. Specific effect of the NHCO group ("local solvation") on the $Fe(I)^- + e^- \rightleftharpoons Fe("0")^{2-}$ reaction as measured by: $\Delta E° = (E°_{amide} - E°_{TPP} - (E°_{ether} - E°_{TAP}$ (in mV).

Porphyrin \ Solvent	DMF	NMA	PhCN	PrCN
$(-C_{12}-)_2$-AC	270	125	363	266
$(-C_{12}-)_2$-AT	240	150	274	199
$(-C_{12}-)_2$-CT	300	227	251	320
$(-C_4PhC_4-)_2$-CT	316	287	404	336

(ca. 10^7 in terms of equilibrium constant) showing that we deal with a quite dramatic effect.

The $Fe(II) + e^- \rightleftharpoons Fe(I)^-$ reaction shows the same general trends involving protection against solvation and "local solvation" by secondary amide groups. When analyzing the E° variations with the nature and arrangement of the chains and with the nature of the solvent, an additional factor, namely, axial ligation of iron(II) by solvent molecules acting as electron donors, ought however to be taken into account. This problem does not arise in DCE since it is an extremely poor Lewis base[30]. Quite large protection effects are observed in the ether-linked series in accordance with the good ability of DCE to solvate anions. As expected from the above discussion, the CT isomer offers a significantly better protection than the AC isomer in the $(O(CH_2)_{12}O)_2$ series.

Coming now to the four other solvents which all are able to coordinate iron(II), we note that an increasingly strong complexation of iron(II) would result in a shift of the $Fe(II)/Fe(I)^-$ wave towards negative potentials. Therefore, in the ether-linked series, steric hindrance to axial ligation caused by the basket-handle chains should shift the $Fe(II)/Fe(I)^-$ couple positively by reference to the unencumbered

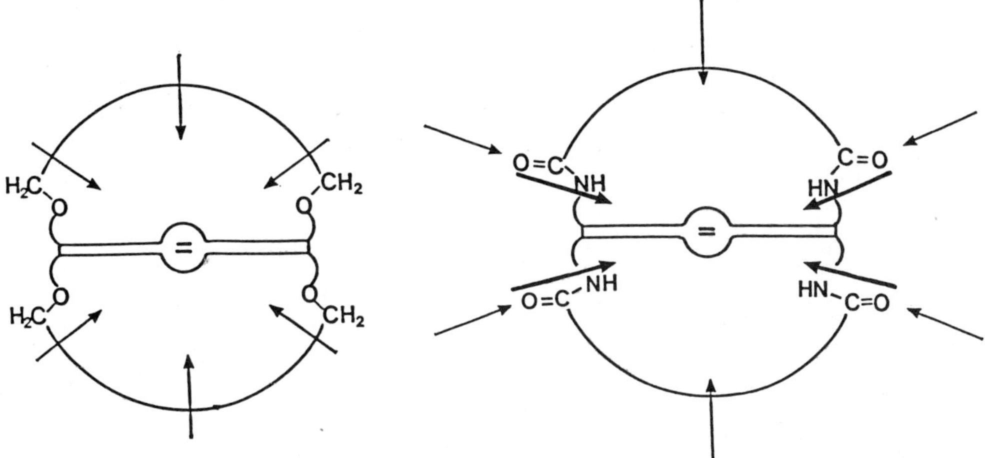

Figure 4. Steric protection against solvation. Contrast between the ether-linked (left) and amide-linked (right) basket-handles.

TAP. This tends to compensate, at least partially, the negative shift due to protection of the negatively charged iron(I) complex from solvation. In agreement with these predictions, we observe (Table 1) that $E^{o}_{ether}-E^{o}_{TAP}$ is less and less negative when passing from DCE to PrCN to PhCN and to DMF, i.e., as the complexing power of the solvent increases. The particularly small values of $|E^{o}_{ether}-E^{o}_{TAP}|$ observed in the case of NMA derive from both its good coordinating power (similar to that of DMF) and the protection against solvation not being very efficient, especially for the most flexible structures, owing to the strong anion-solvating ability of this solvent, as discussed previously.

Another piece of evidence concerns the particularly small values of $|E^{o}_{ether}-E^{o}_{TAP}|$ in the case of the $(O(CH_2)_4Ph(CH_2)_4O)_2$-CT basket-handle structure. Furthermore, it is noticed that $|E^{o}_{ether}-E^{o}_{TAP}|$ decreases as the coordinating power of the solvent increases. In this connection, the case of NMA is particularly striking: whereas the $(O(CH_2)_4Ph(CH_2)_4O)_2$ structure shows, among the various basket-handle structures, the largest value of $|E^{o}_{ether}-E^{o}_{TAP}|$ for the $Fe(I)^- + e^- \rightleftharpoons Fe("O")^{2-}$ reaction, the opposite is true for the $Fe(II) + e^- \rightleftharpoons Fe(I)^-$ reaction emphasizing the role of the freely rotating phenyls in sterically hindering the axial complexation of iron(II) by the solvent molecules.

In the amide-linked series, steric hindrance to axial coordination of solvent molecules should also interfere in the modulation of the $Fe(II) + e^- \rightleftharpoons Fe(I)^-$ reactivity besides protection against solvation and "local solvation". If both protection against solvation and steric hindrance to axial ligation were the same with ether-linked and amide-linked basket-handles of corresponding structure, $(E^{o}_{amide}-E^{o}_{TPP})-(E^{o}_{ether}-E^{o}_{TAP})$ should be a constant when changing superstructure and solvent. As seen in Table 3, this is grossly true, as expected from the fact that protection against solvation should vary less than in the case of the $Fe(I)^- + e^- \rightleftharpoons Fe("O")^{2-}$ reaction where a doubly charged species is involved. However, the observed variations are clearly beyond experimental uncertainty being, therefore, worth commenting on. The ΔE^{o}'s are significantly smaller in the strongly coordinating DMF, not only than in the non-coordinating DCE, but also than in the less coordinating PrCN and PhCN which have comparable anion-solvating abilities. Furthermore, the values obtained with the $(-C_4PhC_4-)_2$-CT structure are significantly smaller than those obtained with the $(-C_{12}-)_2$-CT structure. These observations suggest that not only do the amide-linked structures offer better protection against solvation than the corresponding ether-linked structures, but that in this case axial coordination of solvent molecules resists better to steric hindrance. This implies that NHCO dipoles not only "locally solvate" negative charges but also strengthens the axial coordination by neutral ligands. This will be directly demonstrated in the following with exogeneous and endogeneous nitrogen bases as ligands in a non-coordinating solvent (DCE).

Also with the $Fe(II) + Cl^- \rightleftharpoons Fe(II)Cl^-$ reaction, the main effects of the basket-handle structures derive from protection against solvation and "local solvation" by NHCO groups. In addition, steric hindrance to axial coordination not only by solvent molecules but also by chloride ions should be taken into account too. The number of all these reactivity controlling factors as well as their possible mutual influence (non-additivity) preclude a detailed analysis of the free energy featuring the specific effect of the chains on the thermodynamics of this reaction, namely, $(RT/F)\ln(K_A^{ether}/K_A^{TAP})$ and $RT/F)\ln(K_A^{amide}/K_A^{TPP})$, unlike what was done with the two preceding reactions. We can, however, clearly see how steric hindrance to Cl^- coordination interferes besides solvent coordination upon comparing the $(-C_{12}-)_2$ and $(-C_4PhC_4-)_2$ structures in the ether-linked and amide-linked series (Table 4). It

Table 3. Specific effect of the NHCO groups on the $Fe(II) + e^- \rightleftharpoons Fe(I)^-$ reaction as measured by: $\Delta E^o = (E^o_{amide} - E^o_{TPP}) - (E^o_{ether} - E^o_{TAP})$ (in mV).

Porphyrin \ Solvent	DMF	NMA	PhCN	PrCN	DCE
$(-C_{12}-)_2$-AC	178	140	286	298	312
$(-C_{12}-)_2$-AT	166	155	291	293	--
$(-C_{12}-)_2$-CT	165	190	266	294	358
$(-C_4PhC_4-)_2$-CT	113	155	188	274	--

indeed appears that in all cases, the association with Cl^- is weakened by the presence of a freely rotating phenyl group in the chains. This indicates that the simple C_{12} chains exert a steric discrimination between Cl^- and a solvent molecule as axial ligands, whereas the phenyl containing chains do it to a much lesser extent.

The effect of the basket-handle chains on the kinetics of the reaction:

$$(S)Fe(II) + Cl^- \underset{k_A}{\overset{k_D}{\rightleftharpoons}} Fe(II)Cl^- + (S)$$

(S: solvent molecule) is expected to involve the same factors that were shown to govern its equilibrium constant. That this is indeed the case is best seen representing the data under the form of Bronstedt plots relating the log of the association and dissociation rate constants to the driving force measured by log K_A (Figure 5). In the non-coordination DCE, a good linear correlation is observed. Increasing the association driving force does not result in an increase of k_A and a decrease of k_D but rather in an increase of both rate constants, large for k_A (correlation slope ~ 1.3) and small for k_D (correlation slope ~ 0.3). As sketched in Figure 6, this is compatible with a reaction involving the attack of Cl^- on a four - coordinated iron (II) complex. The transition state then resembles the final state as far as (external and "internal") solvation is concerned. The negative charge is

Table 4. Steric hindrance of Cl^- coordination to Fe(II) by by $(-C_4PhC_4-)_2$ structures.

Solvent	DMF	PhCN	PrCN
$\frac{RT}{F} Ln\left(\frac{K_A^{(O(CH_2)_{12}O)_2-CT}}{K_A^{(O(CH_2)_4Ph(CH_2)_4O)_2-CT}}\right)$ [a]	48	41	40
$\frac{RT}{F} Ln\left(\frac{K_A^{(NHCO(CH_2)_{10}CONH)_2-CT}}{K_A^{(NHCO(CH_2)_3Ph(CH_2)_3CONH)_2-CT}}\right)$ [a]	60	140	95

a : in mV

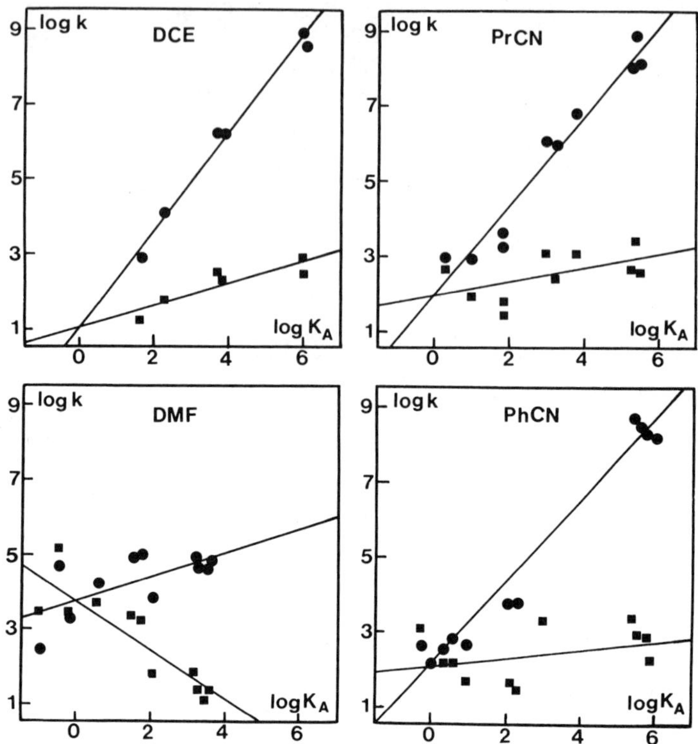

Figure 5. Plots of $\log k_A$ (●) and $\log k_D$ (■) vs $\log K_A$ for the reaction $(S)Fe(II) + Cl^- \rightleftharpoons Fe(II)Cl^- + (S)$ in the four solvents, DCE, PrCN, PhCN and DMF.

even more concentrated in the transition state than in the final state which explains that $\log k_A$ varies more rapidly than $\log K_A$ and that $\log k_D$ increases (slightly) rather than decreases upon increasing $\log K_A$. As the coordinating power of the solvent increases from PrCN to PhCN and DMF, the linearity of the correlation deteriorates, the $\log k$-$\log K_A$ slope decreases, eventually reaching values close to +0.5 and -0.5 in the case of DMF. As sketched in Figure 6, this is compatible with the interference, besides solvation factors, of axial ligation by solvent molecules. If an SN2 type process is assumed, the transition state resembles the final state from the point of view of (external and "internal") solvation and resembles the initial state from the point of view of axial ligation. This explains why the slopes decrease as the coordinating power of the solvent increases. Furthermore, passing from one porphyrin to the other, the SN2 vs SN1 character of the reaction may vary and also the ligation and solvation factors may not be strictly additive resulting in more scatter in the correlation between rate constants and driving force.

2. Supramolecular effects in other reactions involving a one-unit increase of the negative charge

Although analyzed in less details than the previous three reactions, several other reactions involving a one-unit increase of the negative charge on the porphyrin complex have been shown to be the subject of the same supramolecular effects as described above.

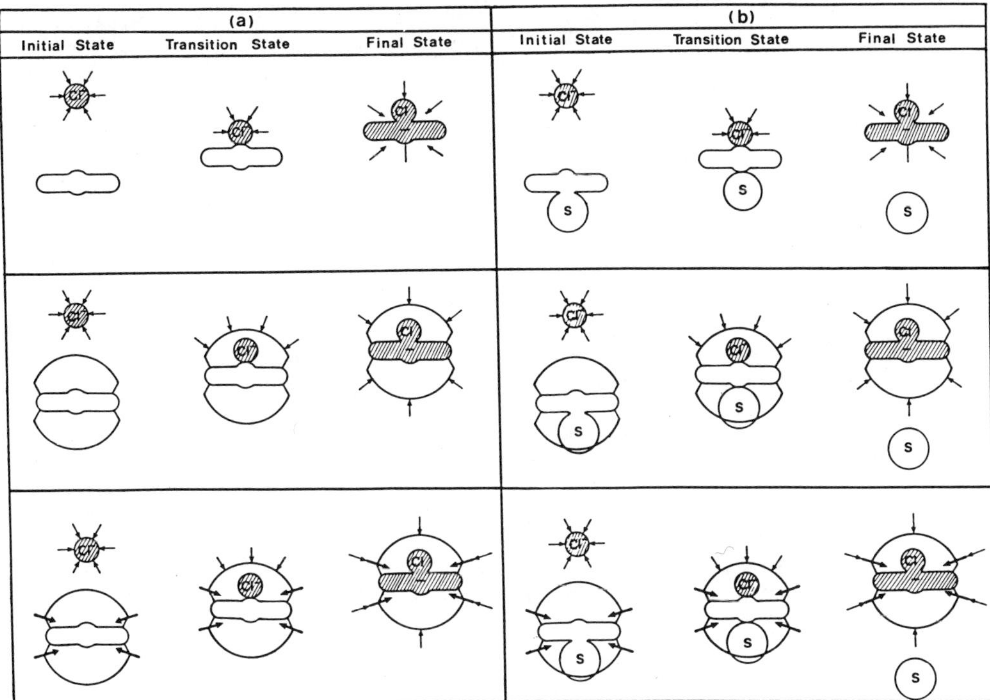

Figure 6. Schematic representation of the initial state, transition state and final state for unprotected, ether-linked and amide-linked superstructured porphyrins for: (a) the reaction of Cl^- with a four-coordinated Fe(II) porphyrin (case of a non-coordinating solvent), and (b) the SN2 displacement of an axially coordinating solvent molecule by Cl^- (case of a coordinating solvent).

This is the case for the reaction[4]:

$$Fe(III)Cl + e^- \rightleftharpoons Fe(II)Cl^-$$

where again the standard potential becomes more negative than in the case of TAP in the presence of ether-linked basket-handle chains and more positive than in the case of TPP in the presence of secondary amide-linked chains. Besides these effects of protection against solvation and of "local solvation", the chains also interfere by means of steric hindrance against axial ligation.

Coordination of iron(II) by bromide ions shows the same general trend as with chloride ions, the association constants being smaller in all cases[9].

Complexation of iron(II) by hydroxyl ions[2]:

$$(S)Fe(II) + OH^- \rightleftharpoons Fe(II)OH^- + (S) \quad \text{(S: solvent molecule)}$$

is another example, similar to the above-described complexation of iron(II) by chloride ions. An important difference, however, is that unprotected iron(III) porphyrins are not stable in the presence of OH^- ions leading to the μ-oxo-dimers. Basket-handle superstructures are a means of preventing the formation of the μ-oxo-dimer thanks to steric hindrance. For these reasons it is not possible to compare the basket-handle porphyrins to the unprotected TAP and TPP. It, however, appears that complexation of iron(II) by OH^- is much weaker in the case of ether-linked basket handles than in the case of amide-linked basket handles which can be rationalized by considering protection against solvation and "local solvation" as described in the preceding section.

Alkyl-iron(III) porphyrins are reduced in two successive reversible one-electron steps[7]:

$$Fe(III)R + e^- \rightleftharpoons Fe(II)R^-$$

$$Fe(II)R^- + e^- \rightleftharpoons Fe(I)R^{2-}$$

In both cases, the standard potential are rendered more negative (as compared to TAP) by the presence of ether-linked basket-handles and more positive (as compared to TPP) by the presence of amide-linked basket handles indicating the occurrence of the same protection against solvation and "local solvation" effects as discussed before[7].

The alkyl-iron(III) complexes can be obtained by reaction of iron(I) with an alkyl halide[7]. It was noted that the reactivity of iron(I) in this respect is enhanced by the presence of ether-linked basket handles and decreased by the presence of amide-linked basket handles, presumably again for the same reasons as before, i.e., in the reaction

$$Fe(I)^- + RX \rightarrow Fe(III)R,$$

the negatively charged iron(I) is destabilized, as compared to the neutral Fe(III)R, by the first type of superstructures and stablized by the second type.

3. Further investigations of the nature of "local solvation" by secondary amide groups included in the superstructures

A first study in this connection consisted in determining the effect of an increasing number of secondary amide groups included in the basket-handle chains[8]. The porphyrins shown in Figure 7 were investigated. For the sake of simplicity, only two reactions, $Fe(I)^- + e^- \rightleftharpoons Fe("0")^{2-}$ in DMF and $Fe(II) + e^- \rightleftharpoons Fe(I)^-$ in 1,2-dichloroethane were considered. In both cases axial coordination and, therefore, steric hindrance to axial coordination does not interfere. In the first case this is because both members of the redox couple are not axially ligated whatever the solvent and in the second because iron(II) is not ligated by DCE. The results are shown in Table 5.

It is observed that the progressive increase of the number of NHCO groups shifts the standard potential of both redox couples toward positive values. The shift is approximately incremental with the number of NHCO groups introduced over the first four of them linking the chains to the phenyl rings. It can, therefore, be concluded that the "local solvation" effect already demonstrated with the four linking secondary amides does build up as the number of NHCO groups increase. The intrinsic effect of the first four NHCO's was estimated to be at least 300-400 mV taking into account that the standard potential variations result both from a positive shift induced by the charge - NHCO interaction and from a negative shift due to a good protection against external solvation. The same two factors interfere in the case of NHCO groups, showing that their intrinsic effect is somewhat larger than 25 mV per group. Even so, however, the effect of the additional NHCO's appears smaller than that of the first four as expected from their larger distance from the central charge. The question

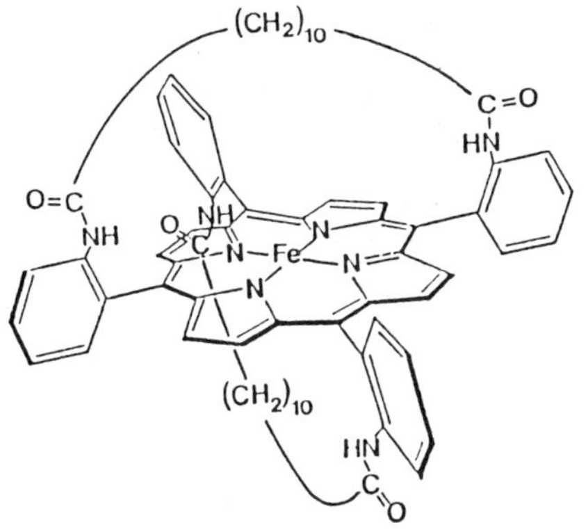

$(NHCO(CH_2)_{10}CONH)_2$-CT

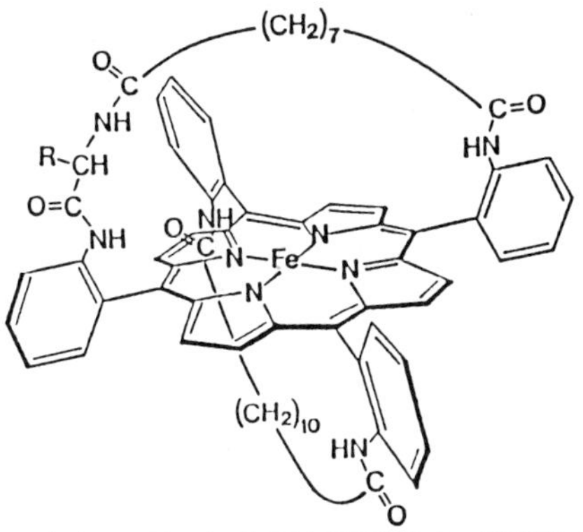

$(NHCOCH(R)NHCO(CH_2)_7CONH)(NHCO(CH_2)_{10}(CONH)$-$CT^{\alpha}$

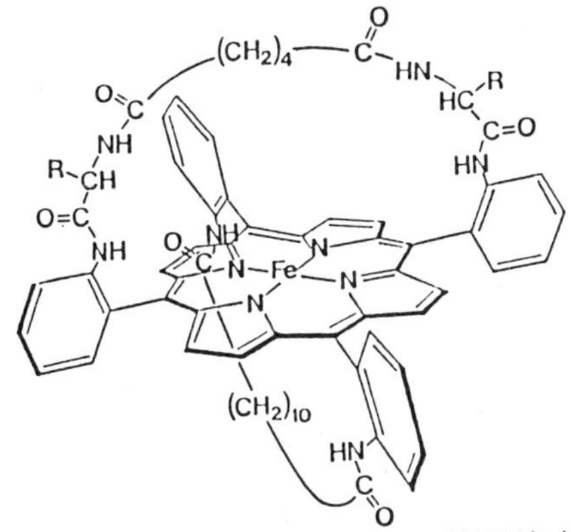

$(NHCOCH(R)NHCO(CH_2)_4CONHCH(R)CONH)(NHCO(CH_2)_{10}CONH)$

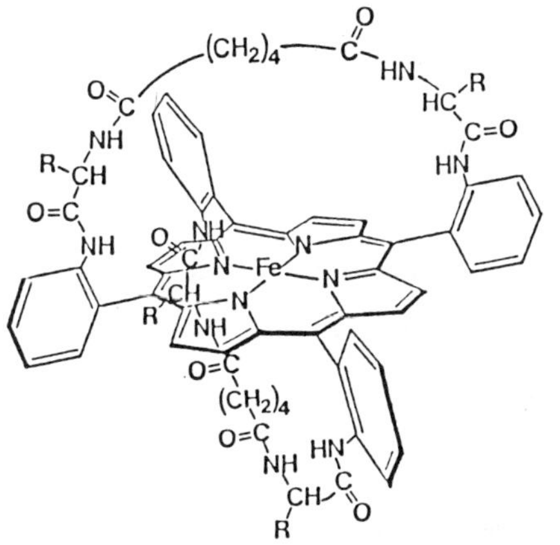

$(NHCOCH(R)NHCO(CH_2)_4CONHCH(R)CONH)_2$-$CT^{\alpha,\beta}$

Figure 7. Basket-handle porphyrins containing an increasing number of secondary amide groups. α:R = phenyl (derived from phenylalanine), β:R = isopropyl (derived from valine).

of their orientation toward the porphyrin ring also arises. In the case of the linking NHCO, there is clear NMR evidence that they already point towards the center of the porphyrin in the Fe(II) complex[12]. In contrast, for the additional NHCO's, although NMR evidence is more sluggish, it seems to indicate that a less favorable orientation occurs[32]. If this is the case, the above results show that the re-orientation toward the porphyrin center takes place upon introduction of negative charges.

Another approach providing deeper insights in the "local solvation" phenomenon consists in investigating the temperature dependence of the reactivity, thus revealing the entropic aspects of the problem. In a preliminary stage, the study was restricted to the Fe(I)/Fe("0") reaction in DMF[33] which allows the investigation of protection against solvation and of "local solvation" without interference of axial ligation. The results obtained by using a non-isothermal cell arrangement are shown in Table 6. The crude entropy variations were corrected from the effect of the interaction between the supporting electrolyte cations (tetrahexylammonium) and the negatively charged porphyrins using the extended Debye-Hückel model.

Table 5. Basket-handle porphyrins containing an increasing number of NHCO groups. Standard potentials of the Fe(I)/Fe("O") and Fe(II)/Fe(I) couples in DMF and DCE, respectively.

Porphyrin	$E°_{Fe(I)^-/Fe("O")^{2-}}$ [a] in DMF	$E°_{Fe(II)/Fe(I)^-}$ [a] in DCE
TPP	- 1.63$_7$	- 1.01$_0$
$(NHCO(CH_2)_{10}CONH)_2$-CT	- 1.54$_0$ (97)	- 0.92$_7$ (83)
$(NHCOCH(R)NHCO(CH_2)_7CONH)(NHCO(CH_2)_{10}(CONH)$-CT [α]	- 1.52$_5$ (112) [b]	- 0.89$_0$ (120)
$(NHCOCH(R)NHCO(CH_2)_4CONHCH(R)CONH)(NHCO(CH_2)_{10}CONH)$-CT[α]	- 1.49$_0$ (147) [c]	- 0.87$_5$ (135)
$(NHCOCH(R)NHCO(CH_2)_4CONHCH(R)CONH)_2$-CT [α,β]	- 1.43$_0$ (207) - 1.42$_0$ (217)	- 0.81$_0$ (200) - 0.80$_5$ (205)

a : standard potential in V vs SCE as determined from the corresponding reversible cyclic voltammetric waves of the Fe(III) hydroxy complex after neutralization by a 0.12 M $HClO_4$ DMF solution in the case of DMF and a 0.12 M $HClO_4$ benzonitrile solution in the case of DCE[2c] unless otherwise stated. Supporting electrolyte : 0.1 M NBu_4PF_6, temp. : 20°C in all cases. Between parenthesis : $E°_{Porph}$ - $E°_{TPP}$ in mV. b : from the cyclic voltammetry of the Fe(III)Cl complex. c : as in a and b, both procedures lead to the same result. d : phenylalanine derived compound. e : valine derived compound.

For the unprotected TPP and TAP porphyrins, the reaction entropy is negative as expected: the solvent is more strongly oriented by the doubly charged iron "O" complex than by the singly charged iron(I) complex.

Due to the inductive electron donating effect of the ortho methoxy group, ΔS is larger in TAP than in TPP (when going from TPP to TAP, solvation of $Fe(I)^-$ increases and thus the entropy decreases owing to the increase of the negative charge on the porphyrin ring and iron atom; the same is true for Fe("O") but to a lesser extent since transfer of charge from the methoxy groups to the doubly charged ring is less.

The entropy of the reaction is clearly smaller with the ether-linked basket-handle porphyrins than with TAP. It is expected that the entropies of both the Fe(I) and Fe("O") complexes increases as a consequence of protection against solvation which leads to a lesser structuration of the solvent around the porphyrins. What the above result shows is that the protection is less at $Fe("O")^{2-}$ than at $Fe(I)^-$. This confirms a previous conclusion, drawn by comparison of the free energies upon changing the solvent, namely, that the solvent is able to overcome the steric barrier offered by the chains to a larger extent in the $Fe("O")^{2-}$ than in the $Fe(I)^-$ complex owing to the larger driving force caused by the doubling of the negative charge. In this respect, the larger value, still smaller than with TAP, found with the $(O(CH_2)_4Ph(CH_2)_4O)_2$-CT structure reflects the stronger steric barrier offered by these phenyl containing chains as compared to the simple C_{12} chains.

Another, even more striking, observation is that, opposite to what happens in the ether-linked series, the secondary amides containing chains give rise to a much larger reaction entropy than TPP. This clearly con-

Table 6. Entropy of the $Fe(I)^- + e^- \rightleftharpoons Fe("O")^{2-}$ reaction in DMF with simple and basket handle porphyrins.

Porphyrin	Entropy (meV/°K)	
	apparent	corrected [a]
TAP	- 0.20	- 0.40
$(O(CH_2)_{12}O)_2$-AC	- 0.71	- 0.91
$(O(CH_2)_{12}O)_2$-AT	- 0.69	- 0.89
$(O(CH_2)_{12}O)_2$-CT	- 0.73	- 0.93
$(O(CH_2)_4Ph(CH_2)_4O)_2$-CT	- 0.60	- 0.80
TPP	- 0.58	- 0.78
$(NHCO(CH_2)_{10}CONH)_2$-AC	- 0.25	- 0.45
$(NHCO(CH_2)_{10}CONH)_2$-AT	- 0.27	- 0.47
$(NHCO(CH_2)_{10}CONH)_2$-CT	0.00	- 0.20
$(NHCO(CH_2)_3Ph(CH_2)_3NHCO)_2$-CT	0.33	0.13
$(NHCOCH(R)NHCO(CH_2)_4CONHCH(R)CONH)_2$-CT	0.17	- 0.03

a : see text.

firms the expectation that the stabilizing interactions between the negative charge and the NHCO dipoles have much less negentropic counterparts than the interactions with the solvent dipoles, owing to the fact that the former dipoles fluctuate much less than the latter. In this connection, we note the following order of entropies in the twelve carbon atom series: AT < AC < CT. It falls in line with a parallel decrease of flexibility of the chains and thus of the fluctuating character of the NHCO dipoles. It should additionally be borne in mind that external solvation still plays a role, even attenuated as compared to unprotected or ether-linked basket-handle porphyrins. This contributes to lowering the entropy although certainly to a lesser extent than in the ether-linked series. This phenomenon appears when comparing the $(NHCO(CH_2)_4Ph(CH_2)_4$-$CONH)_2$-CT and $(NHCO(CH_2)_{10}CONH)_2$-CT structures. The fact that ΔS is larger in the former than in the latter case results from a better protection against external solvation. The value found with the 8 NHCO containing basket-handles, a good solvent protecting structure, probably reflects a greater fluctuating character of the four additional NHCO's, as discussed above.

The two above described studies have allowed a better understanding of the "local solvation" of negatively charged porphyrins by secondary amide groups included in the basket handle chains. On the one hand, it resembles solvation by a solvent in the sense that it builds up as the number of intervening NHCO groups increases, yet, on the other hand, it is different in that the stabilizing interactions have a much lesser negentropic counterpart owing to the scarce fluctuability of the NHCO groups in the chains as compared to solvent molecules.

4. Supramolecular effects in the coordination of iron(II) porphyrins by neutral ligands

The supramolecular effects described so far concern charged molecules thus involving protection against external solvation and strong dipole-charge interactions. What about reactions that do not involve a variation of charge is the question we want to discuss now, taking as an example the complexation of iron(II) porphyrins by neutral exogeneous and endogeneous nitrogen bases. Protection against solvation is expected to play a small role, if any, whereas it is not a priori obvious whether or not interaction with the secondary amide in the chains will interfere significantly. In addition, basket-handle chain steric hindrance to axial ligation is also expected to interfere.

A first investigation concerned the following porphyrins: $(O(CH_2)_{12}O)_2$-CT and AC (with comparison to TAP) and $(NHCO(CH_2)_{10}CONH)_2$-CT and AC (with comparison to TPP)[5] (Figure 1). 1,2-dichloroethane (DCE) was selected as the solvent to avoid complications arising from axial ligation by solvent molecules. Two neutral nitrogen bases, 1,2-dimethylimidazole (1,2-Me_2Im) and 2-methylpyridine (2-MePy) were investigated. Such sterically encumbered ligands were chosen in the aim of easily obtaining the formation constant of the five-coordinated iron(II) porphyrin. The results are shown in Table 7. They suggest the following remarks.

The value of K_A for TPP and 1,2-Me_2Im in DCE is very close to that previously reported in toluene[34] indicating that the influence of the solvent is, as expected, small. The constant for 2-MePy is significantly smaller than for 1,2-Me_2Im paralleling previous observations showing that 1-MeIm is a stronger ligand than pyridine[35]. The electron-donating effect of the four methoxy groups in the ortho position of the phenyl rings in TAP on the association constant is quite small (less than a factor of 2) with both bases. This is also true with T(p-OCH_3)PPFe(II) and 1,2-Me_2Im as reported earlier[34].

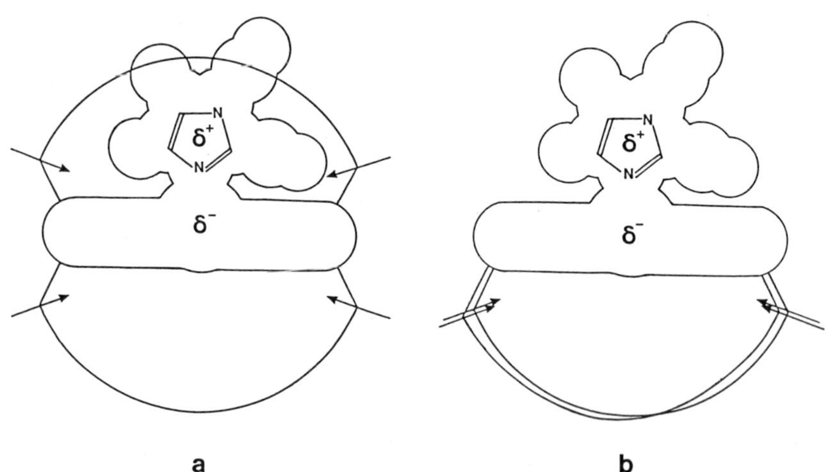

Figure 8. Schematic representation of the stabilization of the Fe(II)-1,2-Me_2Im complex from the dipolar interactions with the NHCO groups in a $(NHCO(CH_2)_{10}CONH)_2$-CT (a) and -AC (b) basket-handle structure. The iron atom is represented at 0.4 Å out of the porphyrin plane.

Table 7. Association constants of 1,2-dimethylimidazole and 2-methylpyridine with unprotected and basket-handle iron(II) porphyrins in 1,2-dichloroethane at 20°C.

Porphyrin	1,2-Me_2Im		2-MePy	
	K_A (M^{-1}) [a]	K_A^{BHP}/K_A^{UNP} [b]	K_A (M^{-1}) [a]	K_A^{BHP}/K_A^{UNP} [b]
TAP	2.6 10^4		7.2 10^1	
$(O(CH_2)_{12}O)_2$-AC	9.5 10^3	0.37 (-25)	4.2 10^1	0.58 (-14)
$(O(CH_2)_{12}O)_2$-CT	1.7 10^3	0.065 (-69)	3.9	0.054 (-74)
TPP	4.1 10^4		1.0 10^2	
$(NHCO(CH_2)_{10}CONH)_2$-AC	1.1 10^6	27 (84)	2.0 10^3	20 (76)
$NHCO(CH_2)_{10}CONH)_2$-CT	4.9 10^4	1.2 (5)	5.0 10^1	0.5 (-18)

a : association equilibrium constant. b : UNP (unprotected) designates TAP in the ether-linked series and TPP in the amide-linked series. Between parenthesis : $(RT/F)Ln(K_A^{BHP}/K_A^{UNP})$ in mV at 20°C.

Concerning the effect of the basket-handle chains on the association constant, the most striking feature of the data listed in Table 7 is the quite significant increase, observed with both bases, when passing from TPP to $(NHCO(CH_2)_{10}CONH)_2$-AC, i.e., when introducing four NHCO groups close to the porphyrin ring without increasing notably steric hindrance against ligation. Since the Hammett σ of an NHCO substituent is practically zero, this cannot be attributed to an electron withdrawing effect through the phenyl rings but rather to a through-space effect of the NHCO dipoles (Figure 8). Another observation, congruent with the latter, is that the association constants are about the same, with both bases, for TPP and the $(NHCO(CH_2)_{10}CONH)_2$-CT porphyrin where steric hindrance against axial ligation certainly exists (Figure 8): the destabilization resulting from steric hindrance thus appears to be compensated by an effect of the NHCO groups of the same kind as in the AC isomer.

On the other hand, the effect of the ether-linked basket-handle structure on the binding of the two bases is best estimated by comparison of the two corresponding porphyrins with TAP since the electron-donating effect of the ether groups through the phenyl rings should be about the same in the series. It is seen (Table 4) that the presence of the chains, in the AC configuration, has a quite small effect on the association constant, i.e., a slight decrease by a factor of 2 to 3. If we assume that the two chains on the same side of the porphyrin ring completely hinder the binding of the base, a decrease of the equilibrium constant by a factor of 2 would results since only one side of the porphyrin is accessible to the axial ligand. It can thus be concluded that protection against solvation has a small effect, if any, on the binding of a neutral axial ligand unlike what was observed with reactions that create a negative charge on the iron porphyrin complex. In the case of the $(O-CH_2)_{12}-O)_2$-CT porphyrin, there is a definite decrease (by a factor of 15 to 20) of the association constant clearly resulting (Figure 8) from steric hindrance by the chain located on the same side as the neutral axial ligand.

The steric hindrance factors are likely to be practically the same in the ether-linked and amide-linked basket-handle series for the same arrangement of the chains. A measure of the specific effect of the NHCO groups on the reactivity of iron(II) towards 1,2-Me_2Im and 2-MePy is, therefore, provided by:

$$\frac{K_A^{amide}}{K_A^{TPP}} : \frac{K_A^{ether}}{K_A^{TAP}}$$ in terms of equilibrium constants and

$$\frac{RT}{F} \, Ln \left(\frac{K_A^{amide}}{K_A^{TPP}} \; \frac{K_A^{TAP}}{K_A^{ether}}\right)$$ in terms of free energy

where "amide" and "ether" designate the amide-linked and ether-linked basket-handle porphyrins, respectively. The values of these parameters are listed in Table 8. It is seen that the presence of the NHCO groups induces a significant increase of the reactivity of iron(II) towards both 1,2-Me_2Im and 2-MePy. This can be explained in terms of stabilizing interactions between the NHCO dipoles and the dipole resulting from the donation of electron density from the nitrogen base to the iron atom and the porphyrin ring. This is similar to the previously evidenced effect of NHCO groups on the reactions that create a negative charge on the iron and the porphyrin ring. There are, however, two important differences between the two cases. One is that the effect is significantly smaller (100 meV at most instead of 300-400 meV) as expected from the fact that dipole-dipole interactions are weaker than dipole-charge interactions. The second derives from the different behaviors towards solvation. It appears that the reactivity of iron(II) towards the two neutral bases is rather insensitive to solvation unlike the case of the reactions creating a negative charge. This points to the concept that the Fe(II)-base dipole is mostly an induced dipole, as such giving rise to significant interactions with the almost fixed NHCO dipoles borne by the basket-handle chains and to weak interactions with the fluctuating solvent dipoles.

Two additional facts are worth noting. The effect of the NHCO groups is somewhat stronger with 1,2-Me_2Im than with 2-MePy (by about the same amount with the AC and CT configurations). This falls in line with the former being a stronger ligand than the latter, hence having a greater propensity to donate more electron density to the iron atom and the porphyrin ring. On the other hand, the effect of the NHCO groups is stronger in the AC configuration than in the CT configuration (by about the same

Table 8. Specific effect of the NHCO groups on the coordination of iron(II) by neutral nitrogen bases.

Basket-Handle	1,2-Me_2Im [a]	2-MePy [a]
$(C12)_2$-AC	73 (109)	34 (89)
$(C12)_2$-CT	18 (73)	10 (56)

a : $(K_A^{amide}/K_A^{TPP})/(K_A^{ether}/K_A^{TAP})$. Between parenthesis, value of (RT/F)Ln of the preceding expression in mV at 20°C, representing the stabilizing effect of the NHCO groups in terms of free energy.

$(O(CH_2)_3(3\text{-}Py\text{-}5)(CH_2)_3O)(O(CH_2)_{12}O)\text{-}CT$

$(O(CH_2)_5CH(1\text{-}Im)(CH_2)_5O)(O(CH_2)_{12}O)\text{-}CT$

$(NHCO(CH_2)_2(3\text{-}Py\text{-}5)(CH_2)_2CONH)(NHCO(CH_2)_{11}CONH)\text{-}CT$

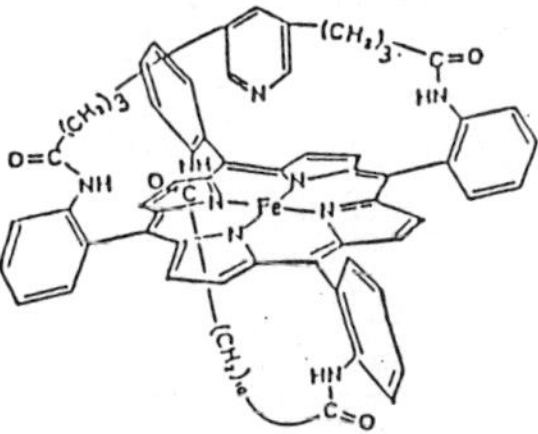

$(NHCO(CH_2)_3(3\text{-}Py\text{-}5)(CH_2)_3CONH)(NHCO(CH_2)_{10}CONH)\text{-}CT$

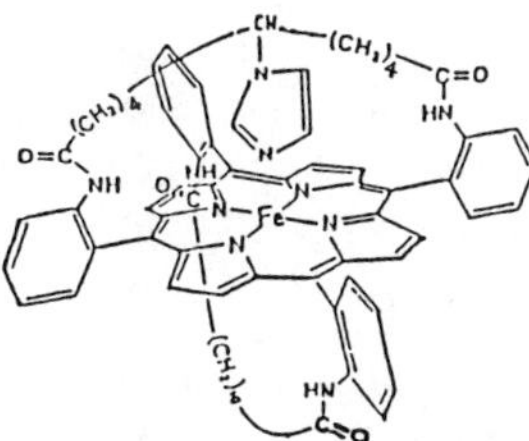

$(NHCO(CH_2)_4CH(1\text{-}Im)(CH_2)_4CONH)(NHCO(CH_2)_{10}CONH)\text{-}CT$

Figure 9. Hanging-base basket-handle porphyrins.

Table 9. Association constant of Fe(II) in hanging base basket-handle porphyrins.

Porphyrin	K_B
$(O(CH_2)_3(3\text{-}Py\text{-}5)(CH_2)_3O)\ (O(CH_2)_{12}O)$-CT	$3.4\ 10^1$
$(O(CH_2)_5CH(1\text{-}Im)(CH_2)_5O)\ (O(CH_2)_{12}O)$-CT	$7.2\ 10^2$
$(NHCO(CH_2)_2\ (3\text{-}Py\text{-}5)\ (CH_2)_2\ CONH)\ (NHCO(CH_2)_{10}CONH)$-CT	$3.9\ 10^3$
$(NHCO(CH_2)_3\ (3\text{-}Py\text{-}5)\ (CH_2)_3\ CONH)\ (NHCO(CH_2)_{10}CONH)$-CT	$2.1\ 10^2$
$(NHCO(CH_2)_4\ CH\ (1\text{-}Im)\ (CH_2)_4\ CONH)\ (NHCO(CH_2)_{10}CONH)$-CT	$8.0\ 10^4$

amount for the two bases). This can be explained by a better geometrical location of the NHCO dipoles for interacting with the induced Fe(II)-base dipole in the former case than in the latter as visible on the schematic representation given in Figure 8.

The same stabilizing effect of NHCO groups on the ligation of iron(II) by neutral nitrogen bases is found in the series of hanging-base basket-handle porphyrins[6] shown in Figure 9. As seen in Table 9, the association constants are significantly larger (by a factor of ca. 100) in the amide-linked series than in the ether-linked series for corresponding structures (compare entries 1 and 3, 2 and 5 in Table 6). We note incidently that the length of the chain bearing the pending base has also a marked effect on the association constant (compare series 3 and 4 in Table 9) for obvious steric reasons.

The same series of hanging-base basket-handle porphyrins has also allowed one to further analyze the factors that control dioxygen binding to iron(II) and O_2 vs CO differentiation in systems modelling hemoglobin and myoglobin (see the detailed discussion of this point in Reference 6). It was indeed found that both O_2 and CO binding correlate quantitatively with the strength of the axial base coordination whatever the origin of its variations (intrinsic coordinating power of the base, spatial arrangement, presence of secondary amide groups).

CONCLUSIONS

Relatively simple supramolecular structures disposed around a reacting center, here an iron porphyrin, are able to strongly influence its reactivity in a manner that is reminiscent of the modulation of the reactivity of a prosthetic group by surrounding protein chains in natural systems.

The largest effects were observed for reactions that involve an increase of the negative charge in the porphyrin complex, redox reactions as well as axial ligand exchange reactions. The main two effects are then: protection of the reacting center from external solvation which renders the reactions more difficult; "local solvation" by secondary amide groups located in the surrounding carbon chains which facilitates the reaction. In addition, the chains are also able to exert a steric discrimination between axial ligands.

"Local solvation" results from the interaction between the NHCO dipoles and the central negative charge. It builds up as the number of NHCO borne by the chains is increased. While "local solvation" resembles solvation by an ordinary solvent in terms of the interaction energies, the two phenomena clearly differ on entropic grounds. The reaction entropies become vanishingly small when interactions with the solvent are replaced by interactions with NHCO groups located in the chain with a fixed geometry. As in natural systems, the building of an ordered structure thus allows a precise control of the reactivity of the reacting center.

The secondary amide groups in the chains not only influence the reactions involving a change of the charge but also reactions taking place between neutral reactants and products such as the axial binding of neutral exogeneous or endogeneous nitrogen bases to iron(II) porphyrins. Quite significant reactivity changes are thus observed, although smaller than with charge-creating reactions (of the order of 100 instead of 10^6-10^7 in terms of equilibrium constants).

ACKNOWLEDGMENT

We acknowledge the contributions of our various coworkers in the above described studies, whose names can be found in the reference list. The collaboration of Dr. M. Momenteau (Institut Curie, Orsay, France) who designed the synthesis of all the basket-handle porphyrins was, of course, essential.

REFERENCES

1. D. Lexa, M. Momenteau, P. Rentien, G. Rytz, J.M. Saveant and F. Xu, J. Am. Chem. Soc., 106, 4755 (1984).

2. D. Lexa, M. Momenteau, J.M. Saveant and F. Xu, Inorg. Chem., 24, 122 (1985).

3. A. Croisy, D. Lexa, M. Momenteau and J.M. Saveant, Organometallics, 4, 1574 (1985).

4. C. Gueutin, D. Lexa, M. Momenteau, J.M. Saveant and F. Xu, Inorg. Chem., 25, 4294 (1986).

5. D. Lexa, M. Momenteau, J.M. Saveant and F. Xu, J. Am. Chem. Soc., 108, 6937 (1986).

6. D. Lexa, M. Momenteau, J.M. Saveant and F. Xu, Inorg. Chem., 25, 4857 (1986).

7. D. Lexa, M.J. Saveant and D.L. Wang, Organometallics, 5, 1428 (1986).

8. D. Lexa, P. Maillard, M. Momenteau and J.M. Saveant, J. Phys. Chem., 91, 1951 (1987).

9. D. Lexa, M. Momenteau, J.M. Saveant and F. Xu, J. Electroanal. Chem., 237, 131 (1987).

10. J.P. Collmann, Acc. Chem. Res., 10, 265 (1977).

11. M. Momenteau, J. Mispelter, B. Loock and E. Bisagni, J. Chem. Soc. Perkins Trans. I, 189 (1983).

12. M. Momenteau, J. Mispelter, B. Loock and J.M. Lhoste, J. Chem. Soc. Perkins Trans. I, 221 (1985).

13. J.P. Collman, R.R. Gagne, T.R. Halper, J.C. Marchon and C.A. Reed, J. Am. Chem. Soc., 95, 7868 (1973).

14. P. Maillard, C. Schaeffer, C. Huel, J.M. Lhoste and M. Momenteau, J. Chem. Soc. Perkins Trans. I, submitted.

15. M. Momenteau, J. Mispelter, B. Loock and J.M. Lhoste, J. Chem. Soc. Perkins Trans. I, 61 (1985).

16. K.M. Kadish, "The Electrochemistry of Metalloporphyrins in Non-Aqueous Media" in Progress of Inorganic Chemistry, S.J. Lippard, Ed., Wiley, New York (1986), Vol. 34, pp. 435-605.

17. I.A. Cohen, D. Ostfeld and B. Lichtenstein, J. Am. Chem. Soc., 74, 4522 (1972).

18. D. Lexa, M. Momenteau and J. Mispelter, Biochim. Biophys. Acta, 338, 151 (1972).

19. K.M. Kadish, G. Larson, D. Lexa and M. Momenteau, J. Am. Chem. Soc., 97, 282 (1975).

20. C.A. Reed, Adv. Chem. Ser., 201, 333 (1982).

21. T. Mashiko, C.A. Reed, K.J. Haller and W.R. Scheidt, Inorg. Chem., 23, 3192 (1984).

22. G.S. Srivatsa, D.T. Sawyer, N.J. Bolt and D.F. Bocian, Inorg. Chem., 24, 2133 (1985).

23. D.L. Hickman, A. Shirazi and H.M. Goff, Inorg. Chem., 24, 563 (1985).

24. D. Lexa, J. Mispelter and J.M. Saveant, J. Am. Chem. Soc., 103, 6806 (1981).

25. J.M. Saveant and E. Vianello, Electrochim. Acta, 8, 905 (1963).

26. C.P. Andrieux and J.M. Saveant, "Electrochemical Reactions" in Investigations of Rates and Mechanisms of Reactions, Vol. 6, 4/E, Part 2, C. Bernasconi, Ed., J. Wiley, (1986), p. 305.

27. D. Lexa, P. Rentien, J.M. Saveant and F. Xu, J. Electroanal. Chem., 191, 253 (1985).

28. J.M. Saveant and F. Xu, J. Electroanal. Chem., 208, 197 (1986).

29. P. Zuman, "Substituent Effects in Organic Polarography", Plenum Press, New York, (1967), p. 46.

30. V. Gutman, "The Donor-Acceptor Approach to Molecular Interactions", Plenum Press, New York, 1978, p. 29.

31. V. Mayer, Stud. Phys. Theor. Chem., 27, 219 (1983).

32. M. Momenteau and P. Maillard, personal communication, 1987.

33. E. Anxolabehere, D. Lexa and J.M. Saveant, in preparation.

34. T. Hashimoto, R.L. Dyer, M.J. Crossler, J.E. Baldwin and F. Basolo, J. Am. Chem. Soc., 104, 2101 (1982).

35. M. Momenteau, B. Loock, E. Bisagni and M. Rougee, Can. J. Chem., 57, 1804 (1979).

ELECTROCHEMISTRY OF METALLOPORPHYRINS CONTAINING A METAL-METAL OR METAL-CARBON BOND

K.M. Kadish

Department of Chemistry
University of Houston
4800 Calhoun Road
Houston, Texas 77004

INTRODUCTION

The synthesis, chemical properties, and electrochemistry of metalloporphyrins with metal-metal and metal-carbon bonds have recently been described in two extensive reviews[1,2]. Porphyrins are known to coordinate with most metallic and pseudo-metallic elements and, in theory, synthesis of numerous organometallic porphyrins is possible. However, to date, the synthesis of metal-carbon σ-bonded complexes has been limited to metalloporphyrins with the following central metals: Fe, Ru, Co, Rh, Ir, Ti, Al, Ga, In, Tl, Si, Ge, Sn and Zn. These organometallic complexes are of importance as model compounds for understanding the functions and relationships of several biologically important macromolecules, as well as for their chemical reactivity. The insertion of small molecules between the metal ion and the carbon atom of a metalloporphyrin may result in activation of the inserted molecule or may generate new monomeric or polymeric materials. In addition, metal-carbon σ-bonded porphyrins can act as precursors in the synthesis of metal-metal bonded derivatives.

This review will cover the general electrochemistry of both transition and non-transition metal organometallic porphyrin complexes. Metalloporphyrin electrochemistry has been reported for σ-bonded complexes containing ten different central metal ions. These include transition metal complexes of iron, cobalt, rhodium, and iridium, as well as metalloporphyrins with aluminum, gallium, indium, thallium (Group 13 metals), silicon, and germanium (Group 14 metals). The first eight types of metalloporphyrins are σ-bonded by one alkyl or one aryl group and have a central metal ion in the +3 oxidation state. The latter two types of compounds contain a central metal ion in the +4 oxidation state and are σ-bonded either with two alkyl or aryl groups or with one σ-bonded ligand and one ionic ligand.

The potentials, mechanisms, and sites of oxidation/reduction in σ-bonded metalloporphyrins depend upon the specific solvent system (all work to date has been carried in nonaqueous media), the macrocycle, and the nature of the specific σ-bonded R group. Most electroreductions are reversible and occur at potentials negatively shifted from those for reduction of the same metalloporphyrin containing coordinated ionic ligands. In contrast, many oxidations of σ-bonded metalloporphyrins are irreversible and are shifted negatively from potentials for oxidation of the same metalloporphyrin containing bound ionic ligands. The magnitude and direction of the potential shift is related to the stability of the singly oxidized species and to the presence or absence of any coupled chemical reactions. For example, (P)Fe(R) (where P = the dianion of a given porphyrin ring and R is the σ-bonded alkyl or aryl group) is oxidized at more negative potentials than (P)FeCl but (P)Co(R) is oxidized at more positive potentials than (P)CoCl. The R group in both iron and the cobalt complexes of (P)M(R) can undergo a metal to nitrogen migration after electrooxidation. Complexes of (P)M(R) and $(P)M(R)_2$ with main group metals may undergo rapid loss of an R group after electrooxidation and when this occurs an irreversible peak is observed at potentials more negative than potentials for the reversible oxidation of (P)MX or $(P)MX_2$. Complexes of (P)In(R) where R = C_6F_5 or C_6F_4H are reversibly oxidized without loss of the σ-bonded ligand and under these conditions, the potentials are only slightly more negative than for oxidation of the same indium porphyrin with ionic ligands. Electrooxidized (P)Tl(R) complexes are also relatively stable and potentials for the reversible abstraction of one electron occur at $E_{1/2}$ values which are negatively shifted by 170 to 420 mV from those of (P)TlCl. This is all discussed in following sections of the manuscript.

Sigma-Bonded Alkyl and Aryl Iron Porphyrins

Sigma-bonded alkyl, aryl, and perfluoroaryliron(III) porphyrins can be obtained by the reaction of (P)FeCl with either Grignard reagents or with alkyl or aryl lithium[2]. The reaction of electrochemically generated iron(I) porphyrins and aryl or vinyl halides also leads to σ-bonded Fe(III) porphyrin complexes as shown in Equation 1[3,4].

$$[(P)Fe^{I}]^- + RX \rightarrow (P)Fe^{III}(R) + X^- \qquad (1)$$

Kinetic measurements of the above substitution reaction indicate that the mechanism is more likely to have SN2 character than to be initiated by an electron-transfer reaction between Fe(I) and the alkyl halide[3].

Many iron-aryl or iron-alkyl σ-bonded porphyrins have been described as low spin complexes on the basis of ^1H NMR data[2,5-8]. However, alkyl and aryl σ-bonded derivatives may exist as a mixture of high and low spin state Fe(III) complexes or, alternatively, (P)Fe(R) can exist as a purely high spin state Fe(III) complex. In general, the spin state of σ-bonded iron(III) porphyrins will vary as a function of the porphyrin ring basicity and/or the ligand field strength of the σ-bonded ligand. The spin state of (P)Fe(R), in turn, seems to directly or indirectly effect the electrochemistry of the complex.

(P)Fe(R) complexes can be divided into groups according to either the spin state of the Fe(III) or to the reactivity of the initial electrooxidation product. (A relationship seems to exist between these two factors.) One group of complexes contains low spin Fe(III) and comprises metalloporphyrins which undergo rapid migration of the σ-bonded R ligand after electrooxidation to generate an N-alkyl or N-aryl derivative as a final species. This reactivity is typified by $(TPP)Fe(C_6H_5)$ and $(OEP)Fe(C_6H_5)$[9]. The second group of metalloporphyrins contain high spin Fe(III). For these complexes, a migration of the electrooxidized species does not occur or is too slow to observe on the electrochemical time scale. This group of complexes shows well defined oxidations and is typified by $(TPP)Fe(C_6H_5)$ and $(TPP)Fe(C_6F_4H)$[8]. A description of the migration reaction and the properties of stable oxidized and reduced σ-bonded iron porphyrins is given in the following sections.

Electrochemically Initiated Migration of Alkyl and Aryl Groups

The reversible one-electron electrooxidation of (TPP)Fe(C_6H_5) and (OEP)Fe(C_6H_5) leads to [(TPP)Fe(C_6H_5)]$^+$ and [(OEP)Fe(C_6H_5)]$^+$. These singly oxidized complexes are not stable and undergo a chemical and electrochemical reaction which leads to the iron(III) N-phenylporphyrins, [(N-C_6H_5TPP)Fe(III)]$^{+2}$ and [(N-C_6H_5OEP)Fe(III)]$^{+2}$. These N-phenyl derivatives can be reversibly reduced by a single electron to quantitatively give [(N-C_6H_5TPP)Fe(II)]$^+$ and [(N-C_6H_5OEP)Fe(II)]$^+$. Further reductions are also possible and give transient species which have tentatively been identified as N-phenyl Fe(I) porphyrins.[9]

The overall two-electron oxidation of (P)Fe(C_6H_5) involves a chemical reaction coupled between two electrochemical reactions (an electrochemical ECE type mechanism) as shown in Equation 2.

$$(P)Fe^{III}(R) \underset{}{\overset{-e}{\rightleftharpoons}} [(P)Fe^{IV}(R)]^+ \rightarrow [(N\text{-}RP)Fe^{II}]^+ \overset{-e}{\rightleftharpoons} [(N\text{-}RP)Fe^{III}]^{+2} \qquad (2)$$

The initial electrooxidized species in Equation 2 was characterized as containing Fe(IV) on the basis of spectroscopic data[9]. A similar assignment of Fe(IV) was also obtained for chemically generated [(TPP)Fe(C_6H_4X)] (where X = CH_3 or H)[10]. The basis of an Fe(IV) assignment in this case was by analysis of ^{1}H NMR data.

The reduction of electrogenerated [(N-C_6H_5P)Fe(II)]$^+$ has also been investigated and was found to proceed via an ECE type mechanism in which the chemical step was a migration of the phenyl group from the nitrogen of the porphyrin ring to the iron center, thus regenerating the initial σ-bonded complex[9]. Thus, the combined oxidation and reoxidation steps of the σ-bonded species and the N-substituted species involve two "reversible" migrations which proceed via ECE type mechanisms.

Reversible migrations of CH=C(C_6H_5)$_2$, C_6H_5, and CH_3 from (TPP)Fe(R) were also reported to occur during chemical oxidation and reduction of the σ-bonded porphyrin[11]. $FeCl_3$ was used as the chemical oxidizing agent and [S_2O_4]$^{-2}$ as the chemical reducing agent. It was initially thought that a similar migration sequence would occur for a large number of σ-bonded alkyl and aryl groups on (P)Fe(R), but this does not seem to be the case. Migrations from (P)Fe(R) to [(N-RP)Fe]$^+$ occur for R = C_6H_5[9-11] and p-$CH_3C_6H_4$[10] or m-$CH_3C_6H_4$[10] but not for R = C_6F_5[8] or C_6F_4H[8]. The (P)Fe/C_6H_5) and (P)Fe(C_6F_4H) complexes contain high spin Fe(III)[8] while (P)Fe(C_6H_5) contains low spin Fe(III) at room temperature. This difference in spin state between the phenyl substituted and the perfluoro-substituted complexes might be one factor related to the lack of (or slow) migration involving σ-bonded perfluoro porphyrins. However, (P)Fe(CH_3) is also low spin at room temperature and no electrochemically initiated migration is observed for this compound[12].

Differences in the donor properties of the σ-bonded ligand may influence the migration reaction of [(P)Fe(R)]$^+$. Alternately, electrochemically initiated migrations may only occur for [(P)Fe(R)]$^+$ complexes which have R groups which contain a certain number of carbon atoms. However, no information is yet available on this subject.

The reactions of σ-bonded cobalt porphyrins are discussed in later sections of this manuscript. However, it can be noted that metal to nitrogen migrations have been reported for σ-bonded cobalt porphyrins with R = CH_3[13], C_2H_5[13], $CO_2C_2H_5$[13], and C_6H_4X[14]. (These reactions differ from those of σ-bonded iron porphyrins in that the initial oxidation proceeds via a cation radical rather than via the oxidized central metal). Metal to nitrogen migrations have also been reported for iridium[15] ($R = CH_3$) and rhodium porphyrins[16-18] ($R = CH_3$ and C_2H_5) but these reactions do not seem to occur via an electrochemical initiation. It thus appears that the iron and cobalt complexes are unique in that they undergo an electrochemically initiated migration reaction of the σ-bonded ligand.

Reversible Electrode Reactions of (P)Fe(R)

The oxidations and reductions of (P)Fe(R) are reversible when the migration step does not occur or is slowed down. In these cases, large shifts of half-wave potentials are observed for both oxidation and reduction of the (P)Fe(R) complexes with respect to the corresponding (P)FeX derivatives (where X is an anionic ligand) in the same media[19].

Half-wave potentials for the reversible oxidation or reduction of (P)Fe(R) are directly influenced by the nature of any electron donating or electron withdrawing groups on the porphyrin ring. There is also a dependence of the R group on reduction potentials of (P)Fe(R). This dependence is small when R is a simple alkyl group[3] but is larger for complexes which contain a σ-bonded aryl group such as C_6H_5, C_6F_5 or C_6F_4H. For example, replacing the protons on C_6H_5 by fluoro groups results in a 70-80 mV positive shift of the half-wave potential for each added F atom[8]. A similar positive shift is observed upon adding electron withdrawing CN groups to the porphyrin ring. For example, the difference between reduction of $(P)Fe(C_6H_5)$ where $P = (CN)_4TPP$ and P = TPP is 670 mV, with the former complex being the most easy to reduce[8].

The formation of neutral $(P)Fe(C_6H_5)(L)$ from $(P)Fe(C_6H_5)$ has been monitored both spectrally[10,20] and electrochemically[20]. Formation constants of $10^{1.6}$ to $10^{2.5}$ were calculated for pyridine binding by the five-coordinate Fe(III) complexes[20] while the six coordinate $(P)Fe(C_6H_5)(Im)$ adducts were characterized by ^1H NMR studies[10]. The reduction of $(P)Fe(C_6H_5)(py)$ is electrochemically reversible and the products of reduction can be spectroscopically identified[10]. Potentials for the reduction of $(P)Fe(C_6H_5)(py)$ are shifted in a negative direction from those of $(P)Fe(C_6H_5)$, consistent with stabilization of the six coordinate species.

Scheme I outlines the possible (simplified) overall electron transfer pathways of a σ-bonded alkyl or aryl iron porphyrin in non-coordinating media. Spin state changes for (P)Fe(R) and chemical reactions involving $[(P)Fe(R)]^+$ are not included in Scheme I. The central Fe(II), Fe(III), Fe(IV) and formally Fe(I) atom may lie in or out of the porphyrin plane depending upon the nature of the porphyrin ring and the R group. The movement of the iron atom is also not shown in the scheme.

Scheme I

$$(P)Fe^{III}(R) \underset{1}{\overset{e}{\rightleftharpoons}} [(P)Fe^{II}(R)]^- \underset{2}{\overset{e}{\rightleftharpoons}} [(P)Fe^{I}(R)]^{-2}$$

$$(P)Fe^{III}(R) \underset{3}{\overset{-e}{\rightleftharpoons}} [(P)Fe^{IV}(R)]^+ \underset{4}{\overset{-e}{\rightleftharpoons}} [(P)Fe^{IV}(R)]^{+2} \underset{5}{\overset{-e}{\rightleftharpoons}} [(P)Fe^{IV}(R)]^{+3}$$

Reactions 1, 3 and 4 have been observed for (TPP)Fe(C_6H_5) and (OEP)Fe(C_6H_5) while reaction 2 has been limited to the more easily reducible ((CN)$_4$TPP)Fe(C_6H_5). Reaction 5 corresponds to generation of an Fe(IV) dication and has yet to be observed. However, this reaction should be accessible within the potential range of many electrochemical solvents.

Although, only five electron transfers are shown in Scheme I, the actual number of electrode reactions which occur for a given σ-bonded iron porphyrin will vary as a function of the porphyrin ring basicity and the specific R ligand of (P)Fe(R). In addition, bonding of a solvent molecule or nitrogeneous base to neutral, oxidized, or reduced (P)Fe(R) will significantly increase the number of possible electron transfers. Under these conditions, the exact nature of the electron transfer involving (P)Fe(R) or (P)Fe(R)(L) as well as the stability of the electrogenerated products will depend upon the donicity of the R group, the basicity of the porphyrin ring, the iron spin state and the nature of any sixth axial ligand.

Sigma-Bonded Cobalt, Rhodium and Iridium Porphyrins

Numerous reports of σ-bonded cobalt, rhodium and iridium porphyrins have appeared in the literature[2] but the electrochemistry of these complexes is limited to only several preliminary studies. Three laboratories have reported the electrochemistry of (TPP)Co(R)[13,14,21,22]. Two laboratories have reported the electrochemistry of (OEP)Ir(R)[23,24], and (OEP)Ir(R)(L)[25] and only a single laboratory has reported the electrochemistry of (TPP)Rh(R)[26-30].

The electroreduction of organocobalt complexes generally leads to cleavage of the cobalt-carbon bond[31-33] and this also seems to be the case for alkylcobalt complexes of (TPP)Co(R)[21]. In contrast, the cobalt-carbon bond cleavage for (TPP)Co(R) species with σ-bonded aryl groups is slower and reversible half-wave potentials have been reported[14].

Little information is available on the stability or spectroscopic properties of reduced aryl-cobalt complexes of the form $[(P)Co(R)]^-$ nor is there much data as to the initial site of electroreduction (i.e., generation of a Co(II) σ-bonded complex or formation of a Co(III) π-anion radical). Values of $E_{1/2}$ for reduction of (TPP)Co(R) vary between -1.30 and -1.32 V vs SCE in Me_2SO[21] and a similar range of potentials (between -1.30 and -1.36 V) in CH_2Cl_2[14] is reported. This fact along with the fact that potentials for reduction of (TPP)Co(R) are similar to those for reduction of (TPP)Rh(R) and vary little with the type of σ-bonded ligand could indicate that the initial reduction of both complexes involves an orbital of the conjugated π system. However, the ultimate product of (TPP)Co(R) reduction is $[(TPP)Co(I)]^-$.

The oxidation of (TPP)Co(R) in Me_2SO was first investigated by rotating disk voltammetry[21]. Potentials in this coordinating solvent are shifted in a negative direction by 200-300 mV from values in CH_2Cl_2. A similar shift of oxidation potentials are observed upon adding Me_2SO to CH_2Cl_2 solutions containing (TPP)Co(R)[34] and this suggests a coordination of Me_2SO by oxidized $[(TPP)Co(R)]^+$. The potentials for reduction of (TPP)Co(R) are not shifted upon going from CH_2Cl_2 to Me_2SO and this suggests only a small interaction of Me_2SO with the neutral form of the complex.

The first mechanistic study involving electrooxidation of σ-bonded cobalt porphyrins was reported for (TPP)Co(R) complexes where R was C_2H_5 and $C_2H_5O_2C$[13]. The reversible formation of $[(TPP)Co(R)]^+$ and

$[(TPP)Co(R)]^{+2}$ was noted at low temperature. However, the singly oxidized $[(TPP)Co(R)]^{+}$ was not stable at room temperature and a metal to nitrogen migration of the R group occurred.

Metal to nitrogen migrations generally seem to occur after chemical[35-38] or electrochemical[13-14] oxidation of σ-bonded cobalt porphyrins. The initial sequence of steps for the electrochemically initiated conversion of (TPP)Co(III)(R) to $[(N\text{-}RTPP)Co(II)]^{+}$ is similar to that reported for the electrochemicaly initiated conversion of (P)Fe(R) to $[(N\text{-}RP)Fe(II)]^{+}$ and is illustrated in Equation 3.

$$(P)Co(R) \overset{-e}{\rightleftharpoons} [(P)Co^{III}(R)] \rightarrow [(N\text{-}P)Co^{II}]^{+} \qquad (3)$$

The above migration can proceed either by a one- or a two-electron intermediate. This was investigated by Callot[37] who synthesized a series of different (TPP)Co(R) derivatives and demonstrated that the reaction proceeded via the one-electron transfer pathway[37-38]. The investigated cobalt metalloporphyrins contained σ-bonded aryl groups of the type C_6H_4X where X = OCH_3, CH_3, H, Br, Cl and NO_2. Oxidation potentials were reported and the kinetics of aryl group migration and/or demetallation after chemical oxidation were related to the nature of the R group on the phenyl ligand[14]. However, no mechanistic details of the actual electrooxidation were provided, nor was the nature of any intermediates in the electrochemical reaction investigated.

Although a good number of similarities exist between the Fe(III) and Co(III) σ-bonded metalloporphyrins, there are sufficient differences between the two series. The cobalt to nitrogen migration of (TPP)Co(R) involves the prior formation of a Co(III) cation radical[13] but the iron to nitrogen migration of (P)Fe(R), involves the initial generation of Fe(IV)[9,10]. The mechanisms for oxidation of (TPP)Co(R) and (TPP)Fe(R) also differ with respect to the fate of the electrogenerated N-substituted porphyrin. Electrogenerated N-alkyl or N-aryl iron porphyrins which have been investigated to date are easier to oxidize than the neutral σ-bonded complexes with the same porphyrin ring and this results in an overall two-electron oxidation of (P)Fe(R), with ultimate formation of an Fe(III) N-alkyl or N-aryl complex. In contrast, the electrogenerated N-alkyl Co(II) complex may or may not be more difficult to oxidize than the neutral σ-bonded cobalt(III) porphyrin and a second electron may or may not be spontaneously abstracted from Co(II) after the migration step.

Crucial to the question of a one- vs a two-electron global oxidation of (P)Co(R) is the nature of the anion associated with $[(N\text{-}RTPP)Co(II)]^{+}$. Potentials for oxidation of $[(N\text{-}C_6H_5TPP)Fe(II)]X$ vary from -0.06 V when X = ClO_4^- to 0.44 V when X = Cl^-. A similar shift in $E_{1/2}$ should occur for electrogenerated N-alkyl or N-aryl cobalt(II) complexes and values of $E_{1/2}$ for this reaction should depend upon the anion of the supporting electrolyte. On the other hand, the oxidation potential of (TPP)Co(R) should also be effected by changes in the anion of the supporting electrolyte so that this variable should also be significant in directing the electrooxidation mechanism. However, this aspect of the σ-bonded cobalt electrochemistry has not been investigated in any great detail to date.

Sigma-Bonded Rh(III) Porphyrins

(P)Rh(R) complexes can be synthesized by chemical[2] or electrochemical methods[27]. The electrochemical method involves an initial generation of (TPP)Rh by the one-electron reduction of $[(TPP)Rh(L)_2]^{+}Cl^{-}$ where L is dimethylamine. The (TPP)Rh monomer is unstable and rapidly dimerizes in solvents such as THF or PhCN[40]. However, the formation of (TPP)Rh(R) is observed[26-29] if (TPP)Rh is generated in the presence of RX, HC≡CR or H_2C=CHR. Twenty one different (TPP)Rh(R) complexes have been synthesized by this method. Thus, electrochemical studies provide a new route to the synthesis of (P)Rh(R) as well as new insights into mechanisms for the reaction of (P)Rh(II) and RX.

The effect of both the R and X group of RX on the overall reaction rate constant for generation of (TPP)Rh(R) from (TPP)Rh and RX have been measured by electrochemical methods[27]. The dependence of the reaction rate constant on X follows the trend I > Br > F ≥ Cl while the dependence of R follows the trend CH_3 > C_2H_5 ≈ C_3H_7 ≈ C_4H_9 ≈ C_5H_{11}. Based on this and other data, an overall reaction scheme was proposed[27] in which the (TPP)Rh radical attacked the carbon in the position α to the halide with consequent loss of either X· or X^-. This proposed mechanism differs, in part, from reactions involving $[(P)Rh]_2$ and RX where formation of (P)Rh(R) and (P)RhX are observed[2].

The radical nature of (P)Rh(R) formation from (P)Rh is evident in the reaction of (P)Rh with alkenes and alkynes. Reaction with HC≡CR or H_2C=CHR where R = n-C_3H_7, n-C_4H_9, n-C_5H_{11} or n-C_6H_{13} results in (TPP)Rh(R) formation with loss of the C_2 fragment[28]. This unexpected product reflects the high stability of the (P)Rh(R) complex.

To date, the electrochemistry of σ-bonded Rh porphyrins has been limited to studies from the group of Kadish[26-30]. (TPP)RH(R) is reduced by one or two one-electron transfer steps in nonaqueous media. Values of $E_{1/2}$ for the first reduction of (TPP)Rh(R) vary between -1.39 and -1.44 V SCE depending on solvent while the second reduction occurs between $E_{1/2}$ = -1.58 V (in THF) and $E_{1/2}$ = -1.87 V vs SCE (in PhCN). Two reductions also occur for (TPP)Rh($COCH_3$)[30] but (TPP)Rh(CH_2Cl)[26] and (OEP)Rh(CH_3) undergo only a single electroreduction within the potential range of the solvent.

(TPP)Rh(CH_2Cl) was the first example of a σ-bonded carbon rhodium(III) species whose synthesis was electrochemically initiated[26]. The infrared, ^1H NMR, UV-visible and voltammetric properties of (TPP)Rh(CH_2Cl) are virtually identical to those of other (TPP)Rh(R) complexes. Potentials for the reversible reduction of 14 different (TPP)Rh(R) complexes where R is an alkyl group also vary little with the nature of R group and ranges between -1.41 and -1.43 V vs SCE[27]. There is no change in $E_{1/2}$ upon going from R = CH_3 to R = C_5H_{11} nor is there a change in $E_{1/2}$ upon going from R = CH_3 to R = CH_2Cl, $CHCl_2$ or CCl_3.

A second reduction of (TPP)Rh(R) is observed in THF or PhCN and for the case of R = CH_3, this reaction occurs at -1.85 V vs SCE. This results in a separation of 420 mV between the first and the second reduction which is consistent with the formation of a π anion radical and a dianion[1]. A similar 410 mV separation is noted for the two reductions of (TPP)Rh($COCH_3$) in THF[30].

The oxidations of (TPP)Rh(CH_3) occur at 0.96 and 1.34 V while those of (TPP)Rh(CH_2Cl) occur at 0.92 and 1.38 V. Both sets of data are consistent with the formation of cation radicals and dications and this is indicated by the UV-visible and ESR data of the oxidation products.

Three oxidation peaks are observed for (TPP)Rh($COCH_3$) in PhCN[30]. The first reversible oxidation occurs at 1.03 V and seems to involve an electron abstraction involving the acetyl group rather than the porphyrin π ring system. The second oxidation is irreversible and occurs at E_{pa} = 1.33 V. A third reversible oxidation is located at $E_{1/2}$ ≈ 1.52 V. It was originally thought that a migration of the $COCH_3$ group occurred but studies of the reaction by thin-layer spectroelectrochemistry indicate that both the singly oxidized and doubly oxidized species can undergo a rhodium-carbon bond cleavage to generate $[(TPP)Rh(III)]^+$ and $[(TPP)Rh(III)]^{+2}$.

Potentials for the first reduction and first oxidation of (TPP)Rh(R) are virtually identical to reversible half-wave potentials for the first reduction and first oxidation of (TPP)Co(R) in CH_2Cl_2. The initial oxidation of (TPP)Co(R) is postulated to occur at the porphyrin π ring system and similar conclusions follow from analysis of the spectral data on $[(TPP)Rh(R)]^+$. As already mentioned, no information is available regarding the site of reduction in (TPP)Co(R), but the initial electroreduction of (TPP)Rh(R) seems to involve electron addition to the σ-bonded group or to the porphyrin π ring system rather than formation of Rh(II).

The addition of a nitrogeneous base to (P)Rh(R) is also possible and six-coordinate $(P)Rh(CH_3)(L)$ complexes with bound amine groups have been spectrally characterized[41]. The effect of a sixth ligand on the electrochemistry of (P)Rh(R)(L) has not been reported.

It is of interest to compare the chemical reactivity of oxidized $(TPP)Rh(COCH_3)$ with singly oxidized (P)Co(R) and (P)Fe(R) on the one hand, and singly oxidized main group σ-bonded complexes of Al, Ga, In, Ge, and Si (which are discussed in later sections) on the other hand. Almost all of the singly oxidized compounds in the above series are unstable and each compound undergoes a rapid cleavage of the metal-carbon bond. They differ, however, in that only the iron and the cobalt derivatives undergo a metal to nitrogen migration. This lack of a migration reaction for the main group metalloporphyrins is presumably due to the absence of an accessible M(II) oxidation. A Rh(II) oxidation state is readily accessible (but highly reactive) and thus one would expect to observe the electrochemically initiated synthesis of $[(N\text{-}RP)Rh(II)]^+$. However, to date, this type of reaction has not been observed.

Especially relevant to studies of (P)Rh(R) are investigations of a possible reverse nitrogen to rhodium migration of the R group. The synthesis of $(OEP)Rh(CH_3)$ is reported to proceed via a chemically generated $(N\text{-}CH_3OEP)Rh^I$ species[41]. A similar migration occurs for Co(I) and Fe(I) porphyrins and thus one might predict an electrochemically initiated migration to occur for the N-substituted complexes of Rh. However, genuine N-substituted Rh(II) porphyrins have never been isolated.

Sigma-Bonded Iridium Porphyrins

Very few σ-bonded Ir(III) porphyrins have been electrochemically investigated. The most detailed study involves complexes of $(OEP)Ir(C_3H_7)(L)$[42] and $(OEP)Ir(C_8H_{13})$[43], the latter of which was also structurally characterized. Unfortunately, the initial analysis[44] of $(OEP)Ir(CH_3)(L)$ electrochemistry where L = CO, CN^-, py, NH_3 and N-MeIm was incorrect. A quasireversible wave was reported at 0.68 V in CH_2Cl_2 for $(OEP)Ir(CH_3)$ while other potentials varied as a function of the bound L ligand in $(OEP)Ir(CH_3)(L)$. The cyclic voltammetric waves were misassigned as due to a reduction of the $(OEP)Ir(CH_3)$ or $(OPE)Ir(CH_3(L)$ complex, despite the positive potential observed. However, more recent studies of the same complexes and of other σ-bonded Ir(III) σ-bonded porphyrins demonstrate that the waves are actually due to an oxidation of the Ir(III) complexes[42,43].

Two reversible one-electron processes are observed for oxidation and reduction of (OEP)Ir(C_8H_{13})[43]. These occur at 0.68 V and -1.79 V vs SCE in THF, 0.1M TBAP. The wave at 0.68 V was shown by rotating disk voltammetry to be an oxidation while the wave at -1.79 V was shown to be a reduction. The oxidation is reversible on both cyclic voltammetric and bulk electrolysis time scales (1-5 min) and involves the abstraction of an electron from the bound axial ligand.

The reduction of (OEP)Ir(C_8H_{13}) proceeds via a reversible one-electron process on the cyclic voltammetric time scale. The generated $[(OEP)Ir(C_8H_{13})]^-$ is not sufficiently stable to be characterized and on longer time scales $[(OEP)Ir]^-$ is formed after a global two-electron reduction and loss of the σ-bonded ligand.

Two reversible waves are observed for oxidation and reduction of (OEP)Ir(C_8H_{13})(CO)[43]. These occur at 0.80 V and -1.65V vs SCE in THF, 0.1 M TBAP. The wave at 0.80 V was ascertained to be an oxidation and the wave at -1.65 V to be a reduction by rotating disk voltammetry. The oxidation is reversible only on cyclic voltammetric time scales and bulk oxidation results in the formation of (OEP)Ir(CO)(ClO_4). In contrast, the reduction of (OEP)Ir(C_8H_{13})(CO) is reversible on bulk electrolysis time scales and leads to the formation of $(OEP)Ir(C_8H_{13})(CO)]^-$, a complex where reduction has occurred at the porphyrin π ring system.

Aluminum, Gallium, Indium, and Thalium σ-Bonded Complexes

Electrochemistry of metalloporphyrins containing Group 13 metals which are σ-bonded to different aryl and alkyl groups has been reported by Kadish and Guilard[46-49]. All of the complexes can be reversibly reduced by one or two single electron transfer reactions.

Values of $E_{1/2}$ for reduction of different (P)M(R) Group 13 complexes containing the same porphyrin ring and the same σ-bonded alkyl or aryl group shift only slightly upon changing the central metal ion. For example, in CH_2Cl_2 the reduction of (OEP)M(C_6H_5) occurs at $E_{1/2}$ = -1.47 V vs SCE for M = Al[46] and In[49] and at $E_{1/2}$ = -1.47 V vs SCE for M = Ga[48] and Tl[46]. The exact values for the first reduction of different (P)M(R) complexes are given in Table 1.

Reduction of the aluminum, gallium and indium σ-bonded complexes involves addition of electrons to the porphyrin π ring system but the site of electron transfer in (P)Tl(R) is not as clear cut. A reduction of Tl(III) to Tl(I) is possible in (P)Tl(R)[46] but the similarity in potentials for reduction of (P)M(R) complexes with all four metals (see Table 1) suggest that reduction of the Tl(III) derivatives also initially occurs at the porphyrin π ring system.

The oxidation of (P)M(R) where M = Al, Ga, In, or Tl proceeds in one to three steps depending on the metal ion, the nature of the R group, and the solvent system. (P)Al(R), (P)Ga(R), and (P)In(R) complexes with σ-bonded alkyl groups all undergo an electrochemically reversible one-electron oxidation which is followed by rapid cleavage of the metal-carbon bond.

$$(P)M(R) \underset{}{\overset{-e}{\rightleftharpoons}} [(P)M(R)]^{\cdot +} \rightarrow [(P)M]^+ + R\cdot \qquad (4)$$

This results in an overall irreversible oxidation peak which occurs at a value of E_p which ranges between 0.64 and 1.05 V depending upon the nature of the alkyl group, the type or porphyrin ring (OEP or TPP) and, of course,

Table 1. Half-Wave Potentials (V vs SCE) for the Reversible First Reduction of Selected (P)M(R) and (P)MCl Complexes in CH_2Cl_2, 0.1 M TBA(PF_6).

R Group on (TPP)M(R)					
Metal	Cl^-	Me	Bu	φ	φOMe
Al[a]	-1.15	-1.18	-1.19	-1.18	-1.14
Ga[b]	-1.16	-1.29	-1.29	-1.22	-*
In[c]	-1.09	-1.24	-1.26	-1.22	-*
Tl[d]	-0.91[a]	-1.27	-*	-1.23	-1.25

R Group on (OEP)M(R)					
Metal	Cl^-	Me	Bu	φ	φOMe
Al[a]	-1.45	-1.47	-1.48	-1.47	-1.47
Ga[b]	-1.41	-1.53	-1.53	-1.49	-*
In[c]	-1.29	-1.50	-1.50	-1.47	-*
Tl[d]	-1.19[a]	-1.52[b]	-*	-1.49	-1.49

[a] Reference 45; [b] Reference 48; [c] Reference 47;

[d] Reference 46; * = no data.

the scan rate. The generated $[(P)Ga]^+$ and $[(P)In]^+$ species which result after loss of the σ-bonded ligand (see Equation 4) are also electroactive and two well-defined oxidations of these porphyrins are observed[48,49].

A reversible (P)M(R) oxidation process may be obtained for complexes with specific σ-bonded ligands. For example, both $(P)Ga(C_2C_6H_5)$[48] and $(P)In(C_2H_2C_6H_5)$[49] are reversibly oxidized. The latter In(III) derivatives are not stable in CH_2Cl_2 but well-defined current-voltage curves are obtained in PhCN. Complexes of (P)In(R) where R = C_6F_4H or C_6F_5 are also stable after electrooxidation and two well-defined oxidations at the porphyrin π ring system are observed[47]. Likewise, all of the investigated (P)Tl(R) complexes are reversibly oxidized by two electrons without cleavage of the Tl-carbon bond on the time scale of cyclic voltammetry[46].

Germanium and Silicon Porphyrins

Sigma-bonded germanium porphyrins were first synthesized for use as NMR shift reagents[50,51]. However, since these early reports there have been very few studies on the chemical properties and reactivity of these type complexes. Pommier and coworkers[52-56] synthesized several (TPP)Ge$(R)_2$ and (OEP)Ge$(R)_2$ compounds which were characterized by UV-visible and ^1H NMR spectroscopy. These porphyrins readily reacted with oxygen or light and the apparent photochemical insertion of O_2 into the germanium carbon bond was compared to similar results for (TPP)Sn$(R)_2$

More recent studies of σ-bonded germanium porphyrins have centered on their possible use as antitumor agents. Miyamoto and coworkers[57] demonstrated that rats fed with dimethyl-5,10,15,20-tetrakis[3',5'-bis(1",1"-dimethylethyl)phenyl]porphinatogermanium(IV) had fewer solid tumors than those not fed with this compound. It was also reported that the compound showed considerable activity toward neoplastic tissues both in vitro and in vivo. This suggests that activation of the methyl groups attached to the germanium atom might play a significant role in the anticancer drug action.

The electrochemistry of Ge(IV) and Si(IV) porphyrins with σ-bonded alkyl and aryl groups has been investigated by Kadish and Guilard[58-61]. All reactions must be carried out both in the dark and in the strict absence of O_2. An insertion of O_2 into the Ge-carbon bond of (P)Ge$(R)_2$ has been postulated[54] but later studies[58-61] show that these reactions are actually more complicated than initially suggested.

The reduction of (P)Fe$(R)_2$[58] and (P)Si$(R)_2$[59] occurs in either one or two single electron transfer steps. $E_{1/2}$ values in the PhCN range between -1.40 and -1.58 V for the first reduction of (OEP)Ge$(R)_2$[58] and (OEP)Si$(R)_2$[59] and no second reduction is observed in this solvent. In contrast, (TPP)Ge$(C_6H_5)_2$ is reduced in two steps which occur at -1.10 and -1.65 V in PhCN. Similar values of -1.17 V and -1.72 V vs SCE are measured for the reduction of (TPP)Ge$(CH_2C_6H_5)_2$ in the same solvent[58].

The oxidation of (P)Ge$(R)_2$ and (P)Si$(R)_2$ involves coupled chemical reactions. These reactions are shown in Scheme II for the case of (OEP)Ge$(C_6H_5)_2$ oxidation in PhCN, 0.1 M TBAP[58].

Scheme II

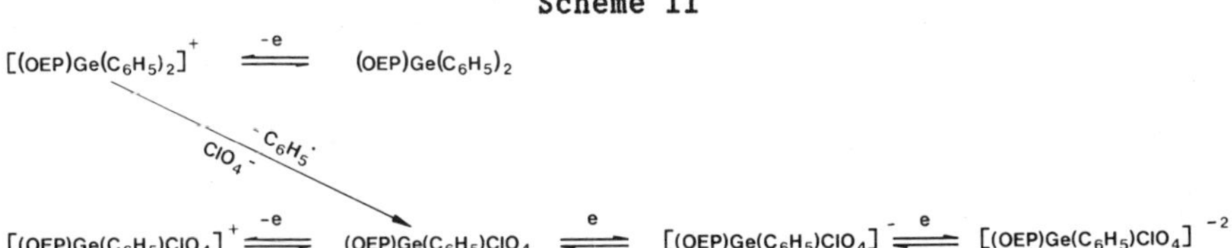

The cleavage of one σ-bonded phenyl group from $[(OEP)Ge(C_6H_5)_2]^+$ is extremely rapid and invariably leads to (OEP)Ge$(C_6H_5)(ClO_4)$ which also undergoes a series of oxidations and reductions given by Reactions 1-3 in Scheme II. Similar reactions occur for other (P)Ge$(R)_2$[58] and (P)Si$(R)_2$[59] derivatives and the final oxidation product after the abstraction of one electron from (P)Ge$(R)_2$ or (P)Si$(R)_2$ is identified as (P)Ge(R)X or (P)Si(R)X where X = ClO_4^-, Cl^- or OH^- depending upon the solvent/supporting electrolyte system. (P)Ge(R)Cl and (P)Si(R)Cl can also be generated from (P)Ge$(R)_2$ or (P)Si$(R)_2$ in solutions of degassed $CHCl_3$ by illumination with visible light[58,59]. For the case of germanium, an electrochemical oxidation of (OEP)Ge$(C_6H_5)_2$, (OEP)Ge(C_6H_5)OH or (OEP)Ge(C_6H_5)Cl gives the same final species which is spectroscopically and electrochemically identified as (OEP)Ge$(C_6H_5)ClO_4$.

The electrochemically initiated conversion of (P)Ge$(R)_2$ to (P)Ge(R)X or (P)Si$(R)_2$ to (P)Si(R)X provides an excellent method for the synthesis of Group 14 metalloporphyrins which contain one σ-bonded and one ionic ligand. In this regard, a number of (P)Ge(R)X[58] and (P)Si(R)X[59] derivatives have already been synthesized and characterized both spectroscopically and electrochemically.

The (OEP)Ge(C_6H_5)ClO_4 complex undergoes two reductions at -0.82 and -1.30 V and a single irreversible oxidation at 1.40 V. (OEP)Ge(C_6H_5)ClO_4 can be converted to (OEP)Ge(C_6H_5)OH by titration with TBA(OH) or by a one-electron electroreduction which is followed by reaction of the electrogenerated porphyrin anion radical with trace H_2O in solution. Likewise, (OEP)Ge(C_6H_5)OH can be converted to (OEP)Ge(C_6H_5)ClO_4 by titration with $HClO_4$ or by a one-electron oxidation of the bound OH^- ligand. These reactions are given in Scheme III.

Scheme III

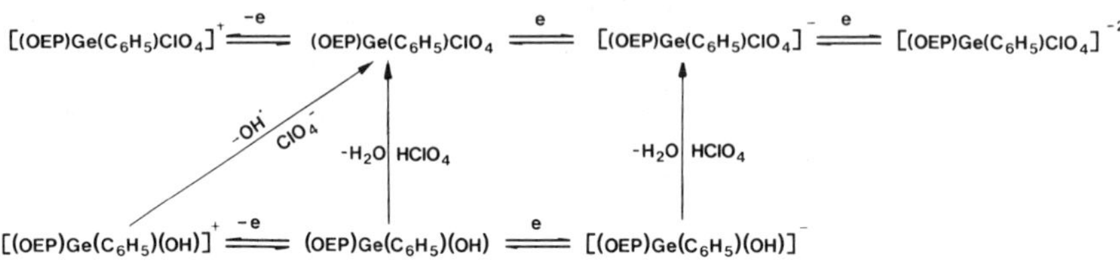

The specific example in Scheme III given is for (OEP)Ge(C_6H_5)X, but the same series of chemical and electrochemical reactions are observed for other (P)Ge(R)X and (P)Si(R)X complexes. Thus, the overall reactions of (P)M$(R)_2$ and (P)M(R)X are given in Scheme IV.

Scheme IV

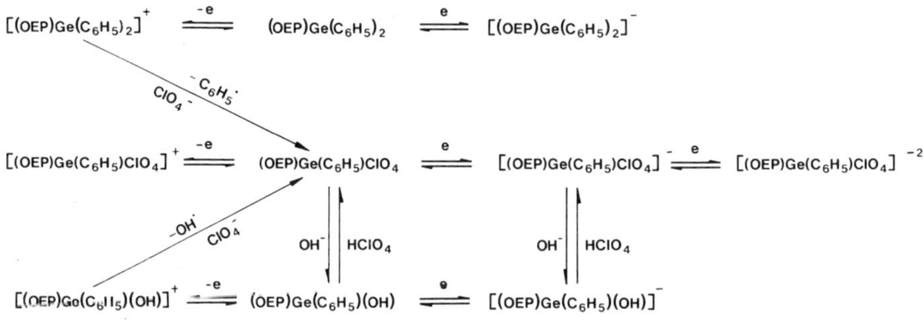

The synthesis, electrochemistry and spectroelectrochemistry of (P)Ge(C_6H_5)(Fc) and (P)Ge$(Fc)_2$ where P is the dianion of octaethylporphyrin or tetraphenylporphyrin and Fc is a ferrocenyl group has also been presented[60,61]. The (P)Ge(C_6H_5)(Fc) complexes provide the first synthetic examples of metalloporphyrins which contain two different σ-bonded axial groups.

The physicochemical properties of neutral (P)Ge$(Fc)_2$ and (P)Ge(C_6H_5)(Fc) are similar to other types of σ-bonded (P)Ge$(R)_2$ derivatives since both groups contain two germanium carbon σ-bonds. However, the Fc derivatives show additional features. The magnitude of the Soret band molar absorptivities for (P)Ge$(Fc)_2$ and (P)Ge(C_6H_5)(Fc) are only half that of other (P)Ge$(R)_2$ complexes. This can be unambiguously attributed to a delocalization of electron density from the porphyrin macrocycle onto the ferrocenyl group. As a consequence, the germanium-carbon σ-bonds in (P)Ge$(Fc)_2$ and (P)Fe(C_6H_5)(Fc) are much more stable than in other (P)Ge$(R)_2$ complexes.

The potentials for oxidation of $(P)Ge(Fc)_2$ and $(P)Ge(C_6H_5)Fc$ are listed in Table 2. The first and second oxidations are assigned as electron abstractions from the σ-bonded Fc ligand. The third oxidation occurs at the porphyrin π ring system. Potential differences between the first and second oxidations of $(P)Ge(Fc)_2$ range between 0.16 to 0.18 V and agree with separations of 0.17 to 0.20 V for $Fc-Si(CH_3)_2-Fe$[62], Fc-Se-Fc[63] and $Fc-CH_2-Fe$[63] in CH_2Cl_2 or CH_3CN. Therefore, based on this similarity between $(P)Ge(Fc)_2$ and other biferrocene derivatives, an interaction between two iron centers of $(P)Ge(Fc)_2$ can be deduced.

Dimeric, Binuclear, and Trinuclear Derivatives

Only a few examples of porphyrin compounds containing metal-metal interactions have been published. In addition to the conducting polymeric porphyrins described by Ibers and Hoffman[64,65], only two classes of metal-metal bonded porphyrin complexes are known. These are porphyrins with donor-acceptor bonds and those with σ-bonds. The latter series consists of both homodinuclear porphyrins such as in $[(P)M]_2$ where M = Ru, and Os, and heterodinuclear derivatives such as in the (P)MM'(L) series where M = In or Tl and M'(L) is one of several different metalate groups.

Binuclear Ru(II) and Os(II) porphyrin dimers of the form $[(P)M]_2$ can be prepared by vacuum pyrolysis of $(P)M(py)_2$ according to Equation 5[66,67].

$$(P)M(py)_2 \xrightarrow[\text{vacuum}]{\Delta} [(P)M]_2 \qquad (5)$$

X-ray and NMR results of $[(OEP)Ru]_2$ are consistent with a qualitative MO diagram that suggests a formal Ru-Ru bond order of 2[68].

The electrochemistry of these two dimers has been reported in dimethoxyethane and CH_2Cl_2. Each compound can undergo up to two reductions and four oxidations. A second reduction is not observed in CH_2Cl_2 due to the proximity of this reaction to the potential for solvent discharge. The

Table 2. Half-Wave Potentials (V vs SCE) for Oxidation of $(P)Ge(Fc)_2$ and $(P)Ge(C_6H_5)Fc$ in CH_2Cl_2, 0.1 M TBAP[a].

Complex	Solvent	Fc ligand(s)		Porphyrin ring	
$(OEP)Ge(Fc)_2$	CH_2Cl_2	0.14	0.32	1.32[b]	1.57
	PhCN	0.17	0.34	1.30[b]	
$(OEP)Ge(C_6H_5)Fc$	PhCN	0.22	-	1.13[b]	
$(TPP)Ge(Fc)_2$	PhCN	0.27	0.43	1.40[b]	
$(TPP)Ge(C_6H_5)Fc$	PhCN	0.29	-	1.24[b]	1.51

[a] Taken from references 59 and 60.

[b] E_{pa} measured at 0.1 Vs^{-1}.

stepwise reduction of $[(P)Ru]_2$ or $[(P)Os]_2$ leads to the formation of anion radicals and dianions with the electrons localized on both of the porphyrin rings. In contrast, the oxidations of $[(OEP)Os]_2$ and $[(OEP)Ru]_2$ are postulated to occur at the central metal ions leading to several dimers with mixed metal oxidation states. This behavior is quite different from that of $[(TPP)Rh]_2$ which undergoes only a single two-electron reduction step and cleaves to generate $[(TPP)Rh]^+$ after electrooxidation[27,40].

Bimetallic σ-Bonded Complexes Containing One Porphyrin Unit and One Metalate Ion

Various indium-metal[69-70] and thallium-metal[71] σ-bonded porphyrin complexes have been synthesized. These metalloporphyrins are represented by (P)InM'(L) and (P)TlM'(L) where P is OEP or TPP and M'(L) is one of the following metallate ions: $Co(CO)_4$, $Cr(CO)_3Cp$, $Mn(CO)_5$, $Mo(CO)_3Cp$ or $W(CO)_3Cp$. Detailed electrochemical studies involving both series of compounds have been carried out[70,71]. The results of these studies indicate that the reduction of binuclear (P)MM'(L) can occur by one or two electrons depending upon the specific porphyrin ring, the temperature, the nature of M and the nature of the bound metallate ion, M'(L). Additional peaks are also observed for oxidation and reduction of (P)M· and $[(P)M]^-$ which are formed by cleavage of the reduced bimetallic species, as well as for $[(P)M]^+$ when some decomposition of the neutral complex occurs. The generated stability of $[(P)MM'(L)]^-$ and $[(P)MM'(L)]^{-2}$ is very low for complexes in both the Tl and In series. A low stability of electrooxidized $[(P)InM'(L)]^+$ is also observed[70] but the singly and doubly oxidized $[(P)TlM'(L)]^+$ and $[(P)TlM'(L)]^{+2}$ derivatives are quite stable and can be spectroscopically characterized[71].

The only known σ-bonded metalloporphyrins containing two porphyrin units are the $[(P)In]_2Fe(CO)_4$ complexes where P is OEP or TPP[69,72]. Three different synthetic methods have been utilized for preparation of these trinuclear complexes. The first method involves nucleophilic substitution of the chloride ion on (P)InCl by $[Fe(CO)_4]^{2-}$ which is generated by Collman's reagent.

$$2\ (P)InCl + [Fe(CO)_4]^{2-} \rightarrow [(P)In]_2Fe(CO)_4 + 2\ Cl^- \qquad (6)$$

The second synthetic method for preparation of $[(P)In]_2Fe(CO)_4$ involves a photochemical reaction of $(P)In(CH_3)$ and $Fe_2(CO)_9$ while the third method proceeds via a reaction involving $[HFe(CO)_4]^-$ and (P)InCl.

The $[(P)In]_2Fe(CO)_4$ complexes can be reduced by two electrons but a rapid cleavage of one or both metal-metal bonds leads to formation of (P)In· and $[Fe(CO)_4]^{2-}$. A further reduction of (P)In· gives the monoanion, $[(P)In]^-$. The $[(P)In]_2(CO)_4$ complexes can also be electrooxidized but the generated product undergoes a rapid cleavage of one or both metal-metal bonds after the first oxidation step which corresponds to a two-electron abstraction.

The electrochemical results[72] show that interactions between the two porphyrin macrocycles of $[(P)In]_2Fe(CO)_4$ are weak and that the two metal-metal bonds are also relatively weak. This agrees with deductions made on the basis of spectrochemical data and demonstrates that the two metalloporphyrin units of $[(P)In]_2Fe(CO)_4$ are equivalent with no interaction occurring across the $Fe(CO)_4$ bridge. Thus, the properties of $[(P)In]_2Fe(CO)_4$ differ from those of the carbenoid bonded $(P)SnFe(CO)_4$ and $(P)GeFe(CO)_4$ complexes[73-75] which are discussed in the following section.

Bimetallic Donor-Acceptor Complexes of the Type (P)MM'(L)

Bimetallic donor-acceptor complexes can be formed with tin and germanium porphyrins as shown in Equation 7[73-75]. The reaction of $Na_2Fe(CO)_4$ with $(P)Sn^{IV}Cl_2$ or $(P)Ge^{IV}Cl_2$ produces $(P)MFe(CO)_4$ where P is OEP, TpTP or TmTP and M is Sn or Ge.

$$(P)MCl_2 \xrightarrow{Na_2Fe(CO)_4} (P)MFe(CO)_4 \qquad (7)$$

Both $(P)GeFe(CO)_4$ and $(P)SnFe(CO)_4$ are extremely stable after electroreduction and two reversible one-electron transfer steps can occur without cleavage of the metal-metal bond[70]. The differences in $E_{1/2}$ between reduction of the (P)SnFe(CO) and $(P)GeFe(CO)_4$ complexes with the same porphyrin ring range between 20 and 70 mV with the former complexes being the easiest to reduce. This is consistent with the slightly higher electronegativity of Ge(II) (2.01) with respect to Sn(II) (1.96) and suggests that the site of electroreduction occurs at the porphyrin π ring system. This conclusion is also suggested by results of thin-layer spectroelectrochemistry and controlled potential electrolysis coupled with ESR.

Each $(P)MFe(CO)_4$ complex undergoes an irreversible oxidation which leads to $[(P)M^{IV}]^{2+}$ and $Fe_3(CO)_{12}$ as the main products. These fragments are formed after the global abstraction of two electrons from $(P)MFe(CO)_4$. Evidence for the above oxidation products also comes from the coulometric value of n. A self-consistent mechanism for the coupled chemical and electrochemical steps in the overall electrooxidation has been presented in the literature[71].

CONCLUSION

This paper has discussed the electrochemistry of metal-carbon and metal-metal bonded metalloporphyrins. The field is quite new and the majority of published studies result from work carried out in our own laboratory during the period 1984 to 1987. To date, relatively few examples of σ-bonded metal-carbon and metal-metal bonded metalloporphyrins are known, but it is theoretically possible to prepare and characterize representative complexes with almost all elements in the periodic table. The field is rapidly expanding and future synthetic developments in porphyrin chemistry should certainly lead to new complexes possessing different metal-metal or metal-carbon interactions and possibly different electrochemistry. These types of complexes are of interest in modelling some biological processes, in synthetic organometallic chemistry and in the preparation of new polymeric materials with conducting properties.

ACKNOWLEDGMENTS

I would like to acknowledge the National Institutes of Health and the National Science Foundation for supporting the work of my research group reported in this review. I would also like to acknowledge joint CNRS/NSF and NATO grants as well as numerous collaborative contributions from Dr. Roger Guilard of the Universite d'Dijon.

REFERENCES

1. K.M. Kadish, Prog. Inorg. Chem., 34, 435-605 (1986).

2. R. Guilard, C. Lecomte and K.M. Kadish, Struct. Bond, 64, 205-268 (1987).

3. D. Lexa, J. Mispelter and J.M. Saveant, J. Am. Chem. Soc., 103, 6806 (1981).

4. D. Lexa and J.M. Saveant, J. Am. Chem. Soc., 104, 3503 (1982).

5. H. Ogoshi, H. Sugimoto, Z. Yoshida, H. Kobayashi, H. Sakai and Y. Maedo, J. Organomet. Chem., 234, 85 (1982).

6. P. Cocolios, G. Lagrange and R. Guilard, J. Organomet. Chem., 253, 65 (1983).

7. J.P. Battioni, D. Lexa, D. Mansuy and J.M. Saveant, J. Am. Chem. Soc., 105, 207 (1983).

8. R. Guilard, B. Boisselier-Cocolios, A. Tabard, P. Cocolios, B. Simonet and K.M. Kadish, Inorg. Chem., 24, 2509 (1985).

9. D. Lançon, P. Cocolios, R. Guilard and K.M. Kadish, J. Am. Chem. Soc., 106, 4472 (1984).

10. (a) A.L. Balch and M.W. Renner, J. Am. Chem. Soc., 108, 2603 (1986); (b) A.L. Balch and M.W. Renner, Inorg. Chem., 25, 303 (1986).

11. D. Mansuy, J.P. Battioni, D. Dupre and E. Sartori, J. Am. Chem. Soc., 104, 6159 (1982).

12. K.M. Kadish, L. Courthaudon and Q.Y. Xu, unpublished results.

13. D. Dolphin, J. Halko and E. Johnson, Inorg. Chem., 20, 4348 (1981).

14. H.J. Callot, R. Cromer, A. Louati and M. Gross, Nouv. J. Chim., 8, 765 (1985).

15. H. Ogoshi, J.I. Setsune and Z. Yoshida, Organomet. Chem., 159, 317 (1978).

16. A.M. Abeysekara, R. Grigg, J. Trocha-Grimshaw and T.J. King, J. Chem. Soc., Perkin I, 2184 (1979).

17. J. Ogoshi, T. Omura and Z. Yoshida, J. Am. Chem. Soc., 95, 1666 (1973).

18. H. Ogoshi, J. Setsune, T. Omura and Z. Yoshida, J. Am. Chem. Soc., 97, 6461 (1975).

19. K.M. Kadish, J. Electroanal. Chem., 168, 261 (1984).

20. D. Lançon, P. Cocolios, R. Guilard and K.M. Kadish, Organometallics, 3, 1164 (1984).

21. M. Perree-Fauvet, A. Gaudemer, P. Boucly and J. Devynk, J. Organomet. Chem., 120, 439 (1976).

22. K.M. Kadish, X.Q. Lin and B.C. Han, Inorg. Chem., 26, 4161 (1987).

23. J.L. Cornillon, J.E. Anderson, C. Swistak and K.M. Kadish, J. Am. Chem. Soc., 108, 7633 (1986).

24. J. Sugimoto, N. Ueda and M. Mori, J. Chem. Soc., Dalton Trans., 1611 (1982).

25. K.M. Kadish, J.L. Cornillon and Y. Deng, submitted to Inorg. Chem.

26. J.E. Anderson, C.-L. Yao and K.M. Kadish, Inorg. Chem., 25, 718 (1986).

27. J.E. Anderson, C.-L. Yao and K.M. Kadish, J. Am. Chem. Soc., 109, 1106 (1987).

28. J.E. Anderson, C.-L. Yao and K.M. Kadish, Organometallics, 6, 706 (1987).

29. J.E. Anderson, Y.H. Liu and K.M. Kadish, Inorg. Chem., 26, 4174 (1987).

30. K.M. Kadish, J.E. Anderson, C.-L. Yao and R. Guilard, Inorg. Chem., 25, 1277 (1986).

31. G. Costa, A. Puxeddu and R. Reisenhofer, J. Chem. Soc., Dalton Trans., 1519 (1972).

32. D. Lexa and J.M. Saveant, J. Am. Chem. Soc., 100, 3220 (1978).

33. D. Lexa and J.M. Saveant, Acc. Chem. Res., 16, 235 (1983).

34. K.M. Kadish and A. Tabard, unpublished results.

35. P.R. Ortiz de Montellano, K.L. Kunze and A. Augusto, J. Am. Chem. Soc., 104, 3345 (1982).

36. H. Ogoshi, E. Watanabe, T. Setsune, N. Koketzu and Z. Yoshida, J. Chem. Soc., Chem. Commun., 943 (1974).

37. H.J. Callot, F. Metz and R. Cromer, Nouv. J. Chim., 82, 759 (1984).

38. H.J. Callot and F. Metz, J. Chem. Soc., Chem. Commun., 947 (1982).

39. Y. Aoymo, T. Yoshida, K. Sakurai and H. Ogoshi, J. Chem. Soc., Chem. Commun., 478 (1983).

40. K.M. Kadish, C.-L. Yao, J.E. Anderson and P. Cocolios, Inorg. Chem., 24, 4515 (1985).

41. H. Ogoshi, J. Setsune, T. Amura and Z. Yoshida, J. Am. Chem. Soc., 97, 6461 (1975).

42. K.M. Kadish, J.L. Cornillon and Y. Deng, submitted for publication.

43. J.L. Cornillon, J.E. Anderson, C. Swistak and K.M. Kadish, J. Am. Chem. Soc., 108, 7633 (1986).

44. H. Sugimoto, N. Ueda and M. Mori, J. Chem. Soc., Dalton Trans., 1611 (1982).

45. K.M. Kadish, A. Endo, A. Zrineh and R. Guilard, manuscript in preparation.

46. A. Tabard, A. Zrineh, R. Guilard and K.M. Kadish, Inorg. Chem., 26, 2459 (1987).

47. K.M. Kadish, B. Boisselier-Cocolios, P. Cocolios and R. Guilard, Inorg. Chem., 24, 2139 (1985).

48. K.M. Kadish, B. Boisselier-Cocolios, A. Coutsolelos, P. Mitaine, R. Guilard, Inorg. Chem., 24, 4521 (1985).

49. A. Tabard, R. Guilard and K.M. Kadish, Inorg. Chem., 25, 4277 (1986).

50. J.E. Maskasky and M.E. Kenney, J. Am. Chem. Soc., 93, 2060 (1971).

51. J.E. Maskasky and M.E. Kenney, J. Am. Chem. Soc., 95, 1443 (1973).

52. C. Cloutour, D. Lafargue, J.A. Richards and J.A. Pommier, J. Organomet. Chem., 137, 157 (1977).

53. C. Cloutour, C. Debaig-Valade, J.C. Pommier, G. Dabosi, M. Martineu, J. Organomet. Chem. 220, 21 (1981).

54. C. Cloutour, D. Lafargue and J.C. Pommier, J. Organomet. Chem., 190, 35 (1983).

55. C. Cloutour, C. Debaig-Valade, C. Gacherieu and J.C. Pommier, J. Organomet. Chem., 269, 239 (1984).

56. C. Cloutour, D. Lafargue and J.C. Pommier, J. Organomet. Chem., 161, 327 (1978).

57. T.K. Miyamoto, N. Sugita, Y. Matsumoto, Y. Sasaski and M. Konno, Chem. Lett., 1695 (1983).

58. K.M. Kadish, Q.Y. Xu, J.-M. Barbe, J.E. Anderson, E. Wang and R. Guilard, J. Am. Chem. Soc., 109, 7705 (1987).

59. K.M. Kadish, Q.Y. Xu, J.-M. Barbe and R. Guilard, Inorg. Chem., in press (1988).

60. K.M. Kadish, Q.Y. Xu, J.-M. Barbe, Inorg. Chem., 26, 2565 (1987).

61. K.M. Kadish, Q.Y. Xu and J.-M. Barbe, Inorg. Chem., in press (1988).

62. A.B. Bocarsly, E.G. Walton, M.G. Bradley and M.S. Wrighton, J. Electroanal. Chem., 100, 283 (1979).

63. J.B. Flanagan, S. Margel, A.J. Bard and F.C. Anson, J. Am. Chem. Soc., 100, 4248 (1978).

64. J.A. Ibers, L.J. Pace, J. Martinsen and B.M. Hoffman, Struct. Bonding, 50, 1 (1982).

65. B.M. Hoffman and J.A. Ibers, Acc. Chem. Res., 16, 15 (1983).

66. J.P. Collman, C.E. Barnes and L.K. Woo, Proc. Natl. Acad. Sci. U.S.A., 80, 7684 (1983).

67. J.P. Collman, C.E. Barnes, T.J. Collins and P.J. Brothers, J. Am. Chem. Soc., 103, 7030 (1981).

68. J.P. Collman, C.E. Barnes, P.N. Swepston and J.A. Ibers, J. Am. Chem. Soc. 106, 3500 (1984).

69. P. Cocolios, C. Moise and R. Guilard, J. Organometal. Chem., 228, C43 (1982).

70. R. Guilard, P. Mitaine, C. Moise, C. Lecomte, A. Boukhris, C. Swistak, A. Tabard, D. Lacombe, J.L. Cornillon and K.M. Kadish, Inorg. Chem., 26, 2467 (1987).

71. R. Guilard, A. Zrineh, M. Ferhat, A. Tabard, P. Mitaine, C. Swistak, P. Richard, C. Lecomte and K.M. Kadish, Inorg. Chem., 27, 697 (1988).

72. R. Guilard, P. Mitaine, C. Moise, P. Cocolios and K.M. Kadish, New Journal of Chemistry, in press (1988).

73. J.-M. Barbe, R. Guilard, C. Lecomte and R. Gerardin, Polyhedron, 3, 889 (1984).

74. K.M. Kadish, B. Boisselier-Cocolios, C. Swistak, J.-M. Barbe and R. Guilard, Inorg. Chem. 25, 121 (1986).

75. K.M. Kadish, C. Swistak, B. Boisselier-Cocolios, J.-M. Barbe and R. Guilard, Inorg. Chem., 25, 4336 (1986).

A STUDY OF THE EFFECTS OF IONIC STRENGTH AND ANION BINDING ON THE REDUCTION POTENTIAL OF CYTOCHROME c USING MICROELECTRODES

Songcheng Sun, David E. Reed and Fred M. Hawkridge*

Department of Chemistry
Virginia Commonwealth University
Box 2006
Richmond, Virginia 23284

INTRODUCTION

Cytochrome c is an important component in the electron transfer chain of the respiratory process of oxidative phosphorylation. It exists in the cytosol between the inner and outer membranes of mitochondria, and functions as an electron carrier between cytochrome c reductase and cytochrome c oxidase[1-3]. The charge transfer process and the effects of ionic strength, anion binding and temperature on the reactivity and conformation of cytochrome c have been widely investigated with electrochemical methods[4-6]. However, most of the measurements in previous studies were carried out in a relatively high concentration of supporting electrolyte due to the need for ionic conductivity in voltammetry. The physiological ionic strength conditions in mitochondria are in a range that is lower than has been used in previous work[4-6]. In order to model the physiological reaction of cytochrome c in mitrochondria, it is informative to conduct experiments under solution conditions that involve low concentrations of supporting electrolytes.

Recently, there has been a surge in the use of very small working electrodes in voltammetry. The use of microelectrodes allows voltammetric measurements to be performed either at very low concentrations or in the absence of supporting electrolytes due to the small iR (ohmic) drop[7-10]. However, previous investigations using microelectrodes have only sparsely described the measurement of formal potentials and other thermodynamic parameters. A majority of the previous studies have been interested in the magnitude of the current responses. In this work, linear sweep voltammetry in a low ionic strength range (1-30 mM) has been used for the measurement of half maximum potentials in order to obtain the formal potential and other thermodynamic information. Using voltammetric measurements of the half maximum potential at microelectrodes, ionic strength effects, anion binding effects and the temperature dependencies of the formal potential of cytochrome c have been studied. In the case of low ionic strength several factors may affect the measurement of formal potentials. Junction potentials, iR drop and other effects have been considered in this work.

Microelectrode arrays greatly enhance the current responses compared with single microelectrodes. Conventional electrochemical equipment can be used for these voltammetric measurements without the need for special current amplification and electrostatic shielding. Microelectrode arrays have not been widely used, probably due to difficulties in fabrication [11-13]. In this work a simple method for making a microelectrode array is described. This array behaves like a single disk microelectrode and it permits voltammetric measurements of cytochrome c at low concentrations.

EXPERIMENTAL

Materials

Horse heart cytochrome c, type VI, Sigma Chemical Co. was purified by chromatography on carboxymethyl cellulose (CM-52, Whatman) according to published procedures[14-15]. To remove phosphate ions from cytochrome c solutions, the central portion of the purified cytochrome c band was first concentrated by pressure dialysis using an Amicon YM5 ultrafiltration membrane. The sample was desalted by chromatography on a desalting column (Bio-Rad No. P-6DG) with purified water as the eluent. A small amount of phosphate ions which bind to, or serve as counter ions for the positively charged cytochrome c, will remain in the desalted sample. To completely remove phosphate ions, the sample was dialyzed against 100 mM tris base (pH = 10) for 24 h at 4°C[16]. Finally, the tris base was removed by passing the sample through the desalting column a second time.

Tris(hydroxymethyl)aminomethane (referred to as tris hereafter) was used as received from Sigma Chemical Co. (Trizma Base, reagent grade). Cacodylic acid, hydroxydimethylarsine oxide (Sigma Chemical Co., 98% pure), was recrystallized twice from 2-propanol. The 4,4'-dithiodipyridine (PySSPy) was purchased from Sigma and used without further purification. Water used in this work was purified with a Milli RO-4 / Milli-Q system (Millipore Corp.) and it exhibited a resistivity of 18 MΩ on delivery. Cytochrome c concentrations were determined by the reduced minus oxidized difference molar absorptivity, $\Delta\varepsilon = 21{,}100\ M^{-1}\ cm^{-1}$ at 550 nm[17]. All other chemicals used in this work were ACS reagent grade.

The gold minigrid used in this work, 333 lines per inch, was purchased from Buckbee-Mears Co. Tin doped indium oxide OTE material was obtained from Donnely Corp., Holland, MI.

Preparation of microelectrode array

PySSPy modified gold microelectrode arrays were used in this work. The microelectrode array was constructed as shown in Figure 1(a), by sandwiching a piece of gold minigrid between two microscope slides which were sealed with epoxy (Dexter Corporation). The minigrid was electrically connected to a metal wire with silver paint (GC Electronics). The whole electrode was then covered with a layer of epoxy to prevent contact with solutions. After the epoxy was completely dried, the electrode was cut with a diamond saw (Buehler Ltd) at a slow rate, ca. 100 r/min. During cutting and polishing the edge was controlled to be perpendicular to the lines of the gold minigrid as shown in Figure 1(b). The electrode was polished successively with No. 600 emery paper (3M) and 0.3, 0.1 and 0.05 µm alumina (Fisher). An electrode array about 5 mm in width will consist of ca. 70 lines. Figure 2 is the picture taken with an electron scanning microscope (JEOL 35C) showing the view from the bottom of the electrode. The shape of the individual electrode disk is prolate rather than a circle with dimensions of 16 µm by 8 µm. The distance between lines, 74 µm, makes each electrode behave as a single disk microelectrode. In this work a conventional potentiostat was used with no Faraday cage (Princeton Applied Research Model 174), and current responses in the range of nA and sub nA were obtained for 20-40 µm cytochrome c solutions.

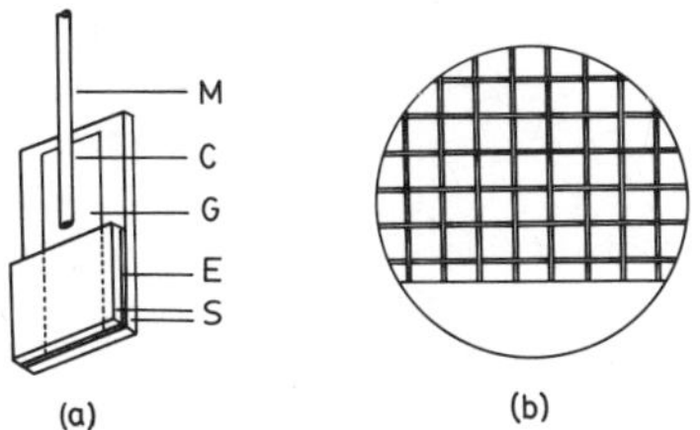

Figure 1. Gold microelectrode array. (a) Construction of gold microelectrode array. M, metal wire; C, silver paint connection; G, gold minigrid; S, microscope slides; E, epoxy resin. (b) side view of microelectrode array. The edge was cut in parallel to lines.

The modification of the Au electrode is similar to the procedure described by Taniguchi et al.[18]. The electrode was dipped into a PySSPy solution (ca. 2 mM) for 10 min under a N_2 atmosphere. The PySSPy modified Au microelectrode arrays are stable at the ionic strengths (1 mM - 200mM) and temperatures (3-55°C) used in this work.

Cell design

In order to eliminate contamination from the reference electrode and to minimize junction potential errors, the cell shown in Figure 3 was used. A Ag/AgCl reference electrode with a relatively high and stable flow rate (ca. 5 ml/24 h) of electrolyte (1 M KCl) was used in order to minimize and stabilize junction potentials. In this work, if the reference electrode was directly in contact with the analyzing solution, the contamination by Cl^- would be a serious problem since it would change the ionic strength at low concentrations of electrolyte (1 mM). Moreover, Cl^- itself could bind

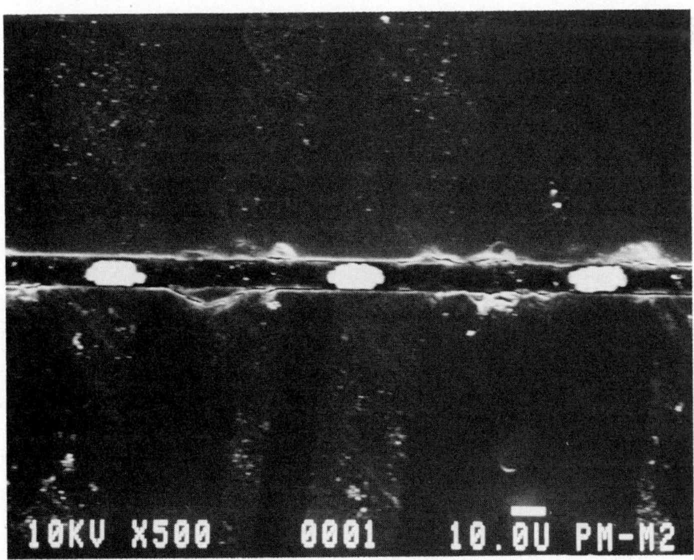

Figure 2. Backscatter electron micrograph of polished microelectrode array.

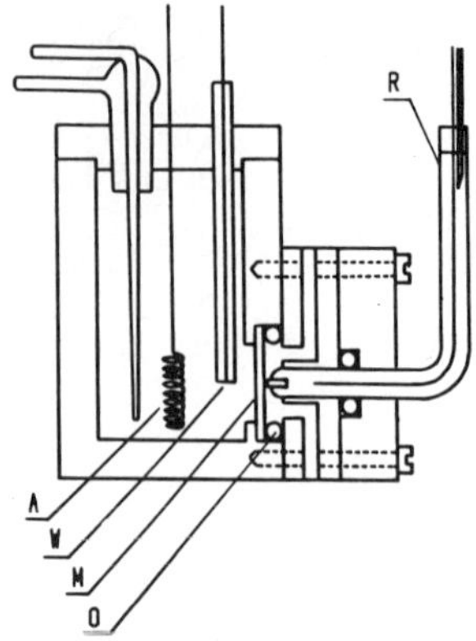

Figure 3. Assembly of electrochemical cell. (W) working microelectrode; (A) Auxiliary electrode; (R) Ag/AgCl reference electrode; (M) YM5 membrane; (O) O-ring seal.

to cytochrome c, causing a change of reduction potentials. To avoid the contamination from the reference electrode, a membrane (YM 5) was used to separate the cell into two parts: a working chamber containing the working electrode and auxiliary electrode, and a reference chamber containing the reference electrode. In the reference chamber, a steady flow of electrolyte (ca. 0.2 ml/min) of the same concentration as that in the working chamber was maintained, preventing contamination by ions from the reference electrode. The large area of the membrane (1.2 cm^2) and the small currents measured with the microelectrode array reduced the iR drop across the membrane to less than 1 mV as determined by measuring the half maximum potential with and without the membrane.

THEORY

Consider the effect of binding by small anions to cytochrome c. The redox reaction of cytochrome c can involve both bound and free cytochrome c, as described by the following equations[19]

$$\begin{array}{ccccc} \text{cyt } c^{o} & + e^- & \rightleftharpoons & \text{cyt } c^{r} & \\ K_o \updownarrow & +pL & & K_r \updownarrow & +qL \\ \text{cyt } c^{o}\text{-}L_p & + e^- & \rightleftharpoons & \text{cyt } c^{r}\text{-}L_q & + (p-q)L \end{array}$$

where L is the binding ion, p and q are the numbers of ions bound and K_o and K_r are the binding constants for the oxidized and reduced forms of cytochrome c, respectively. Assuming the diffusion controlled current arises from both the anion-bound and free cytochrome c, the half-wave potential can be expressed by the following equation

$$E_{1/2} = E^{o\prime} - 0.059 \log \frac{1 + K_o [L]^p}{1 + K_r [L]^q} \tag{1}$$

When the ionic strength is changed, the effect of changes in the activity coefficients of the redox couple must be considered. The formal potential is related to the activity coefficients by[20]:

$$E^{o\prime} = E^{o} + \frac{RT}{nF} \ln \frac{\gamma_{ox}}{\gamma_{red}} \tag{2}$$

where $E^{o\prime}$ is the formal potential, E^{o} is the standard potential extrapolated to zero ionic strength, and γ_{ox} and γ_{red} are the activity coefficients of the oxidized and reduced forms, respectively. The effect of the ionic strength on activity coefficients has been considered by the theoretical work of Kirkwood[21] and Tanford[22] on titration curves of proteins. Several investigations[19,23-25] based on the homogeneous reaction of cytochrome c with small molecules or ions have employed the Debye-Hückel equation for the highly charged and large cytochrome c molecules. The extended Debye-Hückel equation is given by:

$$\log \gamma = \frac{-Z^2 A \mu^{1/2}}{1 + Ba_i \mu^{1/2}} \tag{3}$$

where γ is the activity coefficient of the proteins, Z the charge on the ions in question, and μ is the ionic strength of the solution. A and B equal 0.509 and 0.329 x 10^{-8} $(1/mol)^{1/2}$ cm^{-1} at 25°C, respectively[24], a_i being the ion size parameter in units of angstroms. Substituting the Debye-Hückel expression for activity coefficient into Equation (2)

$$E^{o\prime} = E^{o} - 0.059 \frac{A (Z_{ox}^2 - Z_{red}^2) \mu^{1/2}}{1 + Ba_i \mu^{1/2}} \tag{4}$$

or $$E^{o\prime} = E^{o} - 0.030 (Z_{ox}^2 - Z_{red}^2) f(I) \tag{5}$$

$$\text{where } f(I) = \frac{\mu^{1/2}}{1 + 0.328 a_i \mu^{1/2}} \tag{6}$$

This expression f(I) has been called a steep function[25] since the large radii of proteins tends to "steep" the function.

The half maximum potential for linear sweep voltammograms at very small stationary disk electrodes can be described by the equation[26]

$$E_{m/2} = E^{o\prime} - 0.694 \frac{RT}{nF} \tan^{-1} (0.85 \rho) \tag{7}$$

In this equation ρ is a dimensionless parameter

$$\rho = (nF a^2 \nu / RT D)^{1/2} \tag{8}$$

where a is the radius of the electrode, ν is the potential sweep rate and D is the diffusion coefficient. For simplicity, we use E_s to represent the deviation of the half maximum potential from the formal potential, that is,

$$E_s = - 0.694 \frac{RT}{nF} \tan^{-1} (0.85 \rho) \tag{9}$$

Considering the binding effect, ionic strength effect, size dependence, junction potential and iR drop as well as the overpotential, η, the half maximum potential observed can be expressed as

$$E_{m/2} = E^{\circ} - 0.059 \log \frac{1 + K_o [L]^p}{1 + K_r [L]^q} - 0.030(Z_{ox}^2 - Z_{red}^2) f(I) + E_s + E_j + iR + \eta \qquad (10)$$

RESULTS AND DISCUSSION

The behavior of microelectrode array

Linear sweep voltammograms obtained with a PySSPy modified Au microelectrode array are shown in Figure 4. Well-defined sigmoidal shaped voltammograms were obtained in a purified cytochrome c solution. Cyclic voltammograms using a large PySSPy modified Au electrode (1.4 mm in diameter) are also shown for comparison. The peak separation between the anodic and cathodic peaks of the cyclic voltammograms remains constant, ca. 59 mV at scan rates lower than 50 mVs^{-1}. At these scan rates the electron transfer reactions on the modified electrode are diffusion controlled and the overpotential, η, can be neglected.

The peak currents of reversible cyclic voltammograms using a large electrode are proportional to the square root of the scan rate, but the maximum currents using the microelecrode array are only slightly increased

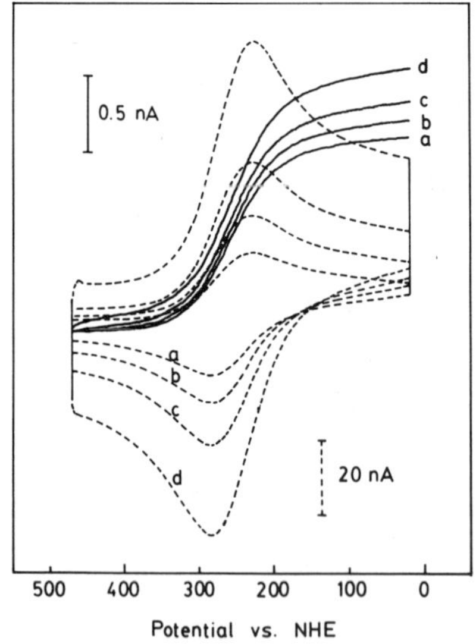

Figure 4. Linear sweep voltammograms (solid lines) at a PySSPy modified gold microelectrode array for the reduction of 44 μM cytochrome c in 0.2 M tris/cacodylic acid buffer at the scan rate of (a) 5 mVs^{-1}, (b) 10 mVs^{-1}, (c) 20 mVs^{-1}, (d) 50 mVs^{-1}. The cyclic voltammogram (dashed lines) at a PySSPy modified gold electrode with a large size (1.4 mm in diameter) was measured in the same solution for comparison.

as the scan rate increases. The dependence of the current on scan rate varies with the size of the microelectrode. This dependence has been described in detail by Osteryoung and coworkers[26]. The maximum current can be calculated according to the following equation[26],

$$I_m/4nFc^*Da = 0.34\exp(-0.66\rho) + 0.66 - 0.13\exp(-11/\rho) + 0.351\rho \quad (11)$$

where all parameters are the same as in Equation (8). We have treated the prolate disk as a circle with a radius of 5.6 μm, so that the areas are the same. The increase of the maximum current calculated from Equation (11) is 9.2% when the scan rate changes from 5 to 50 mVs^{-1} using the diffusion coefficient for cytochrome c of 1.1 x 10^{-6} cm^2s^{-1}. The experimentally determined increase in current is about 10%, quite close to the calculated value, indicating that the array behaves as a single microelectrode and the overlap of the diffusion layer between the individual disk electrodes is negligible.

The half maximum potentials of linear sweep voltammograms using this microelectrode array are constant at scan rates lower than 50 mVs^{-1}. In this work all measurements were carried out with a scan rate of 10 mVs^{-1}. The half maximum potential, $E_{m/2}$, measured with a microelectrode is not equal to $E^{o\prime}$ or the mid-point potential, $(E_{pa}-E_{pc})/2$, measured with a large electrode as shown in Figure 4. Using Equation (9) at a scan rate of 10 mVs^{-1}, with a disk microelectrode of 5.6 μm in radius and a diffusion coefficient of 1.1 x 10^{-6} cm^2s^{-1}, gives an E_s value of 6 mV. This is the same as the experimental value, the difference between 256 mV measured at a large electrode and 262 mV measured at a microelectrode array in 0.2 M tris/cacodylic acid buffer.

The iR drop was estimated in this work by plotting the potential vs. log $(I_m-I)/I$, where I_m is the maximum current. If an iR drop is present, the slope will be lower than the value of 59 mV. Linear sweep voltammograms obtained at these microelectrode arrays have the same slope from high (0.2 M) to low (1 mM) concentrations of supporting electrolyte, indicating that the iR drop is negligible over that concentration range.

Based on the results given above, iR drop and overpotential do not affect the measurement of half maximum potentials in this work. The measured E_s, 6 mV, is treated as a constant and the E_J is assumed to be constant using the cell described earlier. Any change in the half maximum potential measured at different ionic strengths or binding ion concentrations is ascribed to ionic strength or ion binding effects.

Ionic strength effect

The effect of ionic strength on the reduction potential was measured in tris/cacodylic acid buffer, which has been determined to be a weak binding[24] or a non-binding[27] medium for cytochrome c. The ionic strength was varied in the range of 1-30 mM by adding 0.2 M tris/cacodylic acid and accounting for dilution. The measurements were started with solutions containing 1 mM of tris/cacodylic acid in order to fix the pH at 7. The ionic strength contributed by cytochrome c was taken to be 1.5 mM, which is similar to the value estimated in other studies[23-25]. We assume that the weak binding of tris/cacodylic acid does not affect the reduction potential in the concentration range of 1-30 mM of tris/cacodylic acid and that the E_s is a constant which can be subtracted from the half maximum potential. The junction potential, E_J, cannot be measured, but with a high and steady flow rate through the junction of the reference electrode, the change of the junction potential with the change of the concentration of electrolyte in a small range (1-30 mM) is negligible. In this case Equation (10) can be simplified to

$$E_{m/2} = E^\circ - 0.03(Z_{ox}^2 - Z_{red}^2)\ f(I) \qquad (12)$$

where $f(I) = \mu^{1/2} / (1+0.329\ a_1\ \mu^{1/2})$ is the steep function. Figure 5 shows the half maximum potential plotted as a function of f(I), where the a_1 value was taken as 18 Å. A linear relation between the half maximum potential (a value of 6 mV for E_s has been substracted from observed half maximum potentials) and the steep function, f(I), is evident in Figure 5. According to Equation (12), the intercept of the plot as shown in Figure 5 yields a value of 299 mV for the standard potential, E°, extrapolated to zero ionic strength, and the slope -0.278 corresponds to the value of $-0.03(Z_{ox}^2-Z_{red}^2)$. If we assume that the total charge on cytochrome c decreases by 1 unit due to reduction, the total charge on the oxidized and reduced forms can be calculated from the slope as 5.2 and 4.2, respectively. The half maximum potential at zero ionic strength coincides well with the value of 300 mV obtained using a spectrophotometric method where cytochrome c was reduced by the mediator, methyl viologen radical cation[24]. However, the charge calculated from the slope is substantially lower than the reported values, 7.2 and 6.2, for the oxidized and reduced forms, respectively[24]. In the calculation of the net charge, it is assumed that the difference between the total charge of reduced and oxidized forms is one. However, according to the model proposed by Margalit and Schejter[23], the charge on the heme is greatly shielded by the dielectric protein. The contribution of the charge on the heme to the overall charge of cytochrome c is less than 1. We have denoted α ($0 \leq \alpha \leq 1$) as the fraction of the heme charge that contributes to the overall charge. Based on crystallographic data[28], the total charge on the amino acid residues is +8 at pH = 7. Since the charge is +1 on the oxidized heme and 0 on the reduced heme, the total charge will be α+8 and +8 for the oxidized and reduced forms. Substituting these values into the expression for the slope, α is calculated to be 0.6; that is, 40% of the heme charge is shielded. The total charge on the oxidized and reduced forms determined here is therefore +8.6 and +8, respectively, close to the value of +8.3 and +7.6 determined by a spectrophotometric titration with potassium ferrocyanide[23].

The value of a_1, 18 Å, is a best fit value in the steep function f(I) for the plot shown in Figure 5. This is the best value to fit the linear

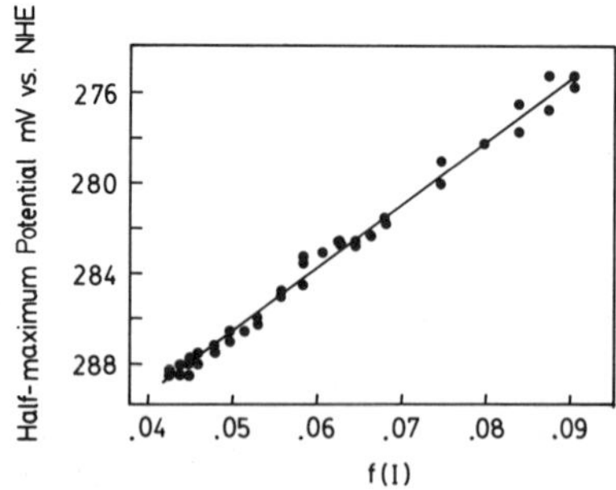

Figure 5. Observed half maximum potential of cytochrome c as a function of ionic strength $f(I) = \mu^{1/2}/(1+0.329a_1\ \mu^{1/2})$, where a_1 = 18 Å.

plot. In the case where the protein reacts with small molecules or ions in solution, a_i is taken as the distance of closest approach[30] $a_i = r_p + r_i$, where r_p is the average radius of the protein and r_i is the radius of a small molecule or ion. Cytochrome c is roughly spherical in shape with dimensions, excluding solvent, of 30 x 34 x 34 Å. The average radius is approximately 16 Å. Electron transfer distances for reactions of cytochrome c have been reported, 18 Å with methyl viologen[24] and 18.5 Å with ferrocyanide[23]. In the case of heterogeneous electron transfer between cytochrome c and a surface modified electrode, the surface modifier can be treated as a small molecule. Surface enhanced Raman spectroscopy[29] indicates that the PySSPy molecules adsorb onto the electrode through the S atoms and the adsorption of PySSPy excludes the cytochrome c molecules from the electrode surface. Electron transfer of cytochrome c takes place through an adsorbed PySSPy layer on the electrode surface. If the electron transfer process at this electrode is treated as a simple redox reaction between cytochrome c and PySSPy, a 2 Å value for half the thickness of the absorbed PySSPy layer is consistent with the value of 18 Å. The ionic strength effect on the redox reactions at this surface modified electrode is also consistent with that observed in homogeneous electron transfer reactions.

Anion binding effect

The binding of small molecules or ions to cytochrome c has been investigated with many different methods such as electrophoresis[27,32], gel filtration[32,33], spectrophotometric titration[23,33], NMR[34,35], ultrafiltration[24], and the direct measurement of formal potentials with cyclic voltammetry[5]. However, the binding constant for chloride, perchlorate and phosphate reported in these studies is varied over a wide range, 1 to 10^4 M^{-1} [33,35-37]. These large differences may be due to a variety of factors. The results, based on the assumption that ligand anions are bound to only one form of cytochrome c, may differ from the value obtained when the effect on both oxidation forms is considered. Moreover, the redox mediator used in previous studies binds tightly to cytochrome c, affecting anion binding[38]. In order to avoid these problems, the redox properties of cytochrome c were studied without mediators on PySSPy modified electrodes.

In this work measurements started with solutions containing 1 mM tris/cacodylic acid at pH = 7.0, and the half maximum potential was measured as a function of the concentration of binding anion in the range of 0.25-35 mM. According to Equation (10) the observed half maximum potential includes both ionic strength and ion binding effects. The net effect of ion binding is obtained by substracting the ionic strength effect which has been measured with tris/cacodylic acid alone. Figure 6 shows the net change of the half maximum potential with respect to the log of the concentration of binding ion. Phosphate has the largest effect on reduction potential compared with perchlorate and sulfate. The common feature of the three curves is that, at higher concentrations of binding anion, the change of the half maximum potentials tends to be linear with respect to the log of the concentration of binding anions. This linear relation can be seen when [L] is large so that Equation (1) can be reduced to

$$E_{m/2} = E^{o\prime} - 0.059 \log \frac{K_o}{K_r} - 0.059\,(p-q) \log [L] \qquad (13)$$

The intercept yields a ratio of K_o/K_r and the slope yields the difference of (p-q). This relation was reported previously[5], where cyclic voltammetry

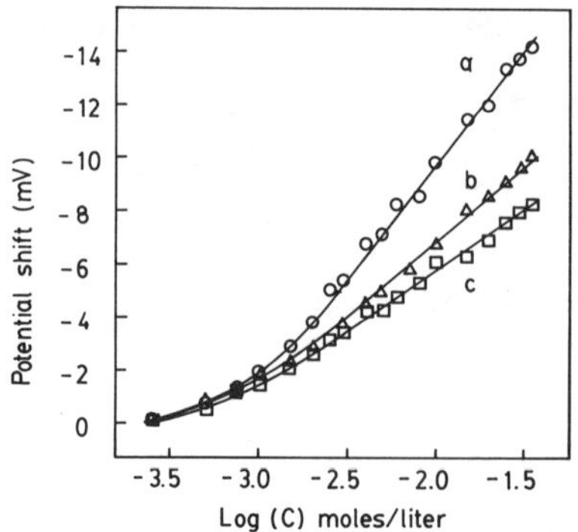

Figure 6. Potential shift as a function of log concentration of binding anions (a) phosphate, (b) perchlorate and (c) sulfate. Solid lines are the simulated curves based on Equation (1) and Reference 13.

was carried out at high concentrations of binding anions. Equation (13) can be derived based on the assumption that the current is only contributed by the cytochrome c bound by anions. At high concentrations of binding ligand or when the binding constant is very large, this assumption is true. The contribution of free cytochrome c to the measured diffusion current can be neglected. However, at very low concentrations of binding ligands the contribution to the diffusion current by free cytochrome c cannot be neglected.

In Equation (1) both K_o and K_r and both p and q are not known. However, the slope and intercept obtained from the linear part of the curve reduces the variables from four to two. Thus, a nonlinear regression can be used to obtain all four variables from Equation (1). Table 1 shows the binding constants and the number of binding sites of both reduced and oxidized cytochrome c. The binding constant of phosphate to ferricytochrome c, 1500 M^{-1}, is close to the result of 1550 M^{-1} obtained using the ultrafiltration technique[24]. There is no previously reported binding constant for sulfate. The numbers of ions bound for all these anions are close to one, smaller than the value of 1.5-2.0 reported previously[24,33]. From Table 1 it can be seen that all three anions bind more strongly to ferricytochrome c than to ferrocytochrome c. This is reflected in the negative shift of the potentials due to the binding of these three anions. It has been suggested that[39,40] anions bind to the positively charged lysine residues surrounding the solvent exposed heme edge. This reduces the charge repulsion between the positively charged heme group of ferricytochrome c and the positively charged lysine residues, thus stabilizing the oxidized form of cytochrome c. However, the heme is neutral in ferrocytochrome c and the stability of ferrocytochrome c will be less affected by the binding of anions. This causes a larger energy difference between ferri- and ferrocytochrome c due to anion binding, and a negative shift of reduction potential will be observed.

The binding parameters for chloride could not be measured since the change of the half maximum potential with change in chloride concentrations is about the same as the change due to the ionic strength effect. This indicates that either the chloride is weakly bound, or it binds with equal strength to the reduced and oxidized forms. Considering a small negative shift by the chloride binding, the sequence of the decrease in binding strength is

Table 1. Binding parameters for some anions

	K_o	K_r	p	q
phosphate	1500	860	1.1	0.99
perchlorate	500	260	0.9	0.82
sulfate	200	112	0.7	0.63

Solution conditions: cytochrome c, 30-40 μM; tris/cacodylic acid,1 mM, pH = 7; the concentrations of anions are 0.25 to 35 mM. The sulfate concentration was varied by adding 0.2M of Na_2SO_4; perchlorate by 0.2 M of $NaClO_4$ and phosphate by 0.2 M of Na_2HPO_4/NaH_2PO_4 buffer at pH = 7. Temperature, 25°C.

phosphate > perchlorate > sulfate >
chloride > tris/cacodylic acid.

It has been reported that redox mediators used in previous studies bind to cytochrome c thus affecting the binding by other small anions[36,41]. In order to determine whether the modifier PySSPy will affect anion binding, the potential shift due to anion binding in the case when the PySSPy modified Au microelectrode array was used has been compared with the case when an In_2O_3 band microelectrode without mediators or promoters was used. The band microelectrode was made using the same procedure as the Au microelectrode array, with the thickness of the In_2O_3 layer being approximately 0.2 - 0.4 μm. Quasi-reversible charge transfer kinetics has been observed on In_2O_3 electrodes without modifiers or mediators[4]. Linear sweep

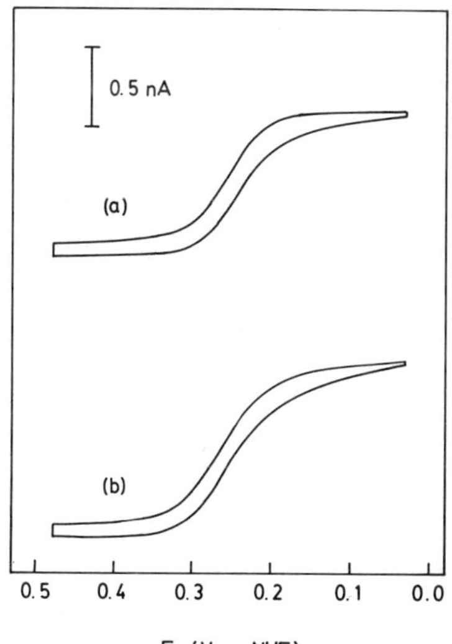

Figure 7. Linear sweep voltammogram of cytochrome c at (a) PySSPy modified gold microelectrode array and (b) Tin doped In_2O_3 band microelectrode in 0.2 M ionic strength of tris/cacodylic acid buffer at scan rate of 10 mVs^{-1}.

voltammograms, using a PySSPy modified Au microelectrode array and an In_2O_3 band electrode, are compared in Figure 7. Nernst responses have been observed on both electrodes at scan rates lower than 50 mVs^{-1}. The same shift of the half maximum potential was observed from tris/cacodylic acid to phosphate on both electrodes at ionic strengths higher than 10 mM. This indicates no effect of PySSPy on the anion binding measurement. However, at low ionic strength the reduction became irreversible on In_2O_3 electrodes and at 1 mM buffer the current response totally disappeared. This may be due to the irreversible adsorption of cytochrome c at low ionic strength. This did not happen on PySSPy modified electrode surfaces even when no supporting electrolyte was present. A Nernst response of cytochrome c could still be observed and was stable for hours, indicating that the strongly adsorbed PySSPy prevents the irreversible adsorption of cytochrome c on the electrode surface.

Temperature dependence

The temperature dependence of the formal potential was studied at different concentrations of electrolytes. A linear relation was obtained at pH = 7.0 in the temperature range of 3-55°C. Figure 8 shows the temperature dependence of the half maximum potentials of cytochrome c in tris/cacodylic acid buffer at ionic strengths of 200, 20 and 5 mM. Since a nonisothermal cell was used, the temperature of the reference electrode was held constant. In this case, the entropy change ($\Delta S_{rc}°$) for the reduction of ferricytochrome c can be given by the difference in entropy between ferro- and ferricytochrome c

$$\Delta S_{rc}° = S_{red}° - S_{ox}° = nF(dE°'/dT) \qquad (14)$$

where F is the Faraday constant and n equals one for the reduction of ferri- to ferrocytochrome c. The slope of the temperature dependence of half maximum potential yields the reaction center entropies, $\Delta S_{rc}°$, of -13.6, -14.5 and -15.4 eu. at ionic strengths of 200, 20 and 5 mM tris/cacodylic acid, respectively. The standard deviation of $\Delta S_{rc}°$ is within ±0.5 eu. When the ionic strength is lower than 5 mM, poor reproducibility made the measurement of the slope difficult. However, the result of -13.6 eu. compares well with the value of -13.4 eu. reported previously using cyclic voltammetry at a conventional electrode (In_2O_3

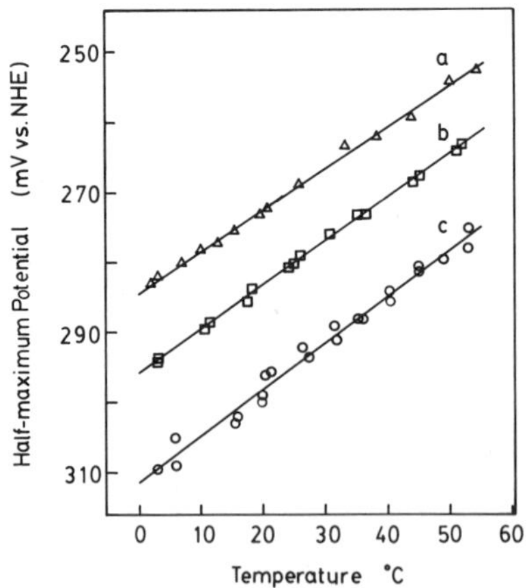

Figure 8. Observed half maximum potential of cytochrome c as a function of temperature at the ionic strengths of tris/cacodylic acid buffer (a) 200 mM, (b) 20 mM and (c) 5 mM.

electrode with an area of ca. 1.2 cm^2) at the same ionic strength of tris/cacodylic acid (0.2 M at pH = 7.0)[4]. Compared with the result, $\Delta S_{rc}°$ = -12.4 eu., reported by Taniguchi et al.[6] using cyclic voltammetry on a PySSPy modified Au electrode of conventional size, this result is slightly higher. The difference is probably due to the binding effect since in their work the solution consisted of phosphate buffer (μ = 0.1 M) and 0.1 M NaCl. The reaction entropy at lower ionic strength has been studied previously. Using the same expression as described earlier for the Debye-Hückel equation, $f(I) = \mu^{1/2}/(1+6\ \mu^{1/2})$, an extrapolation of the entropy value to zero ionic strength yields a value of $\Delta S_{rc}°$ = -16.5 eu.

The reaction center entropy changes for cytochrome c are related to both conformational change and solvent reorganization upon reduction. The conformational change includes closing of the crevice about the solvent-exposed heme edge leading to a more compact structure of cytochrome c when it is reduced. A negative entropy change will be expected for the reduction of cytochrome c. However, the loss of charge on the heme upon reduction will be accompanied by a decrease of ordered solvation layers, causing a positive change of entropy. The net change of entropy depends on a balance between these two factors. The effect of ionic strength on the entropy change seems to be more complicated. The increase in magnitude of entropy change as the ionic strength decreases is probably due to the larger effect of conformation at lower ionic strength compared with the solvent reorganization. A more detailed study of the conformation change is presently under investigation using CD/MCD and other methods.

ACKNOWLEDGEMENT

The authors gratefully acknowledge the support of this work by the National Science Foundation, CHE85-20270. We thank Charles E. Thomas and Karen Sanders for experimental assistance in electron micrographs.

REFERENCES

1. F.S. Mathews, Prog. Biophys. Molecul. Biol.,45, 1 (1985).

2. T.E. Meyer and M.D. Kamen, Adv. Prot. Chem., 35, 105 (1982).

3. F.R. Salemme, Rev. Biochem., 46, 299 (1977).

4. K.B. Koller and F.M. Hawkridge, J. Am. Chem. Soc., 107, 7412 (1985).

5. D.J. Cohen, F.M. Hawkridge, H.N. Blount and C.R. Hartzell, J. Electrochem. Soc., 129, C130 (1982).

6. I. Taniguchi, T. Funatsu, M. Iseki, H. Yamaguchi and K. Yasukouchi, J. Electroanal. Chem., 193, 295 (1985).

7. J.O. Howell and R.M. Wightman, Anal. Chem., 56, 524 (1984).

8. A.M. Bond and P.A. Lay, J. Electroanal. Chem., 199, 285 (1986).

9. M.J. Pena, M. Fleischmann and N. Garrard, J. Electroanal. Chem., 220, 31 (1987).

10. A.M. Bond, M. Fleischmann and J. Robinson, J. Electroanal. Chem., 168, 299 (1984).

11. N. Sleszynski and J. Osteryoung, Anal. Chem., 56, 130 (1984).

12. W. Thormann, P. van den Bosch and A.M. Bond, Anal. Chem., 57, 2764 (1985).

13. M. Ciszkowska and Z. Stojek, J. Electroanal. Chem., 191, 101 (1985).

14. D.L. Brautigan, S. Ferguson-Miller and E. Margoliash, Methods Enzymol., 53D, 131 (1978).

15. E.F. Bowden, F.M. Hawkridge, J.F. Chlebowski, E.E. Bancroft, C. Thorpe and H.N. Blount, J. Am. Chem. Soc., 104, 7641 (1982).

16. B. Errede, G.P. Baight and M.D. Kamen, Proc. Natl. Acad. Sci. U.S.A., 73, 113 (1976).

17. K.J.H. Van Buuren, B.F. Van Gelder, J. Wilting and R. Braams, Biochim. Biophys. Acta, 333, 421 (1974).

18. I. Taniguchi, K. Toyosawa, H. Yamaguchi and K. Yasukouchi, J. Electroanal. Chem., 140, 187 (1982).

19. P.L. Dutton and D.F. Wilson, Biochim. Biophys. Acta, 346, 165 (1974).

20. A.J. Bard and L.R. Faulkner, "Electrochemical Methods", Wiley, New York (1980), p. 52.

21. J.G. Kirkwood, J. Chem. Phys., 2, 351 (1934).

22. C. Tanford and J.G. Kirkwood, J. Am. Chem. Soc., 79, 5333 (1957).

23. R. Margalit and A. Schejter, Eur. J. Biochem., 32, 492 (1973).

24. D. Gopal, G.S. Wilson, R.A. Earl and M.A. Cusanovich, J. Biol. Chem., submitted.

25. T. Goldkorn and A. Schejter, Arch. Biochem. Biophys., 177, 39 (1976).

26. K. Aoki, K. Akimoto, K. Tokuda, H. Matsuda and J. Osteryoung, J. Electroanal. Chem., 171, 219 (1984).

27. G.H. Barlow and E. Margolish, J. Biol. Chem., 241, 1473 (1966).

28. R.E. Dickerson, M.L. Kopka, J. Mienzierl, D.C. Eisenberg and E. Margoliash, J. Biol. Chem., 242, 3015 (1967).

29. I. Taniguchi, M. Iseki, H. Yamaguchi and K. Yasukouchi, J. Electroanal. Chem., 175, 341 (1984).

30. J.G. Beetlestone and D.H. Irvine, Proc. Roy. Soc. Ser., A 122, 401 (1964).

31. E. Margoliash, G.H. Barlow and V. Byers, Nature, 228, 723 (1970).

32. R. Margalit and A. Schejter, Eur. J. Biochem., 46, 387 (1974).

33. R. Margalit and A. Schejter, Eur. J. Biochem., 32, 500 (1980).

34. T. Andersson, J. Angstrom, K.E. Falk and S. Forsen, Eur. J. Biochem., 110, 363 (1980).

35. T. Andersson, E. Thulin and S. Forsen, Biochem., 18, 2487 (1979).

36. B.F. Peterman and R.A. Morton, Can. J. Biochem., 57, 372 (1979).

37. E. Stellwagen and R.G. Schulman, J. Mol. Biol., 75, 683 (1973).

38. G.R. Moore, C.G.S. Eley and G. Williams in Advances in Inorganic and Bioinorganic Mechanism, A.G. Sykes ed., Vol. 3, Academic Press, New York (1984), pp 1-96.

39. G.W. Pettigrew, I. Aviram and A. Schejter, Biochem. Biophys. Res. Commun., 68, 807 (1976).

40. Y.P. Myer, A.F. Salurno, B.C. Verma and A. Pande, J. Biol. Chem., 254, 11202 (1979).

41. W.P. Vorkink and M.A. Cusanovich, Photochem. Photobiol., 19, 205 (1974).

ELECTROCHEMICAL REACTIVITY OF STRONGLY ADSORBED CYTOCHROME c

James L. Willit and Edmond F. Bowden*

Department of Chemistry
North Carolina State University
Raleigh, North Carolina 27695-8204

INTRODUCTION

A major bioelectrochemical research area during the last fifteen years has been the investigation of direct electron transfer reactions of redox proteins at electrode surfaces. Mercury electrode studies[1-4] pioneered the early years of this period during the 1970's. The last ten years have witnessed a substantial emphasis on reactions at solid electrodes since key reports appeared in 1977[5-7]. Several reviews of the field have been published[8-10].

In this report we address cytochrome c, which is the most well-understood electron transfer protein[11,12]. It has occupied a prominent role in interfacial electrochemical investigations due to its high degree of structural and reactivity characterization and its ready availability and purification. Cytochrome c has been found to react in a reproducible, quasi-reversible manner at a number of solid electrode surfaces. Electrode surfaces which have been most successful in this regard are metal oxides[5,13,14] and chemically modified metal electrodes[3,6,15]. Cytochrome c also has the potential to react readily at unmodified metal electrodes, as exemplified by the recent report of a stable, quasi-reversible reaction at bare silver[16].

In the electrode studies of cytochrome c and other proteins, the role of adsorption has generally been recognized as being important. The importance of reversible adsorption in facilitating the electrode reactions of diffusing proteins was recognized[17] and has received much attention with respect to the reactions of diffusing proteins. The present paper is concerned with strong (irreversible) adsorption, for which desorption of the protein adsorbate is negligible on the time scale of the electrochemical experiment. The presence of irreversible adsorption has been recognized as occurring at many electrodes, especially at several bare metal electrodes. In early studies at mercury, it was found that cytochrome c and other proteins adsorbed irreversibly[2-4]. The resultant blocking of the electrode surface by strongly adsorbed proteins resulted in the irreversible electron transfer kinetics which are typically observed for cytochrome c in polarographic experiments. Strong adsorption has also been observed at solid metal electrodes such as silver[18] and gold[19]. At metal electrodes, strongly adsorbed cytochrome c appears to be denatured as evidenced by negative redox potential shifts from native solution values. In contrast, it has been found that cytochrome c can strongly adsorb on at least two metal oxide electrodes, tin oxide[20] and indium oxide[13], while retaining its native redox potential for extended periods of time.

In a recent publication[20] we reported initial measurements of unimolecular electron transfer rate constants (k_{et}) for the reaction of native, strongly absorbed cytochrome c on fluoride-doped tin oxide electrodes in tris/cacodylate media. Unimolecular electron transfer rate constants have also been recently reported by Yokota et al.[21] for cytochrome c immobilized at tin oxide by a Langmuir-Blodgett phospholipid overlayer. Rate constants such as these are unimolecular, having units of s^{-1}, since there is no translational diffusion component to the reaction rate[22]. Weaver and coworkers[23] have treated fundamental aspects of such "intramolecular" electrochemical reactions, so-named since one is in essence measuring the internal electron transfer rate for an electrode-reactant complex. Analogously, one may treat strongly absorbed cytochrome c on tin oxide as an electrode/metalloprotein complex. This concept also has attractive parallels to biological electron transfer reactions, in which two protein redox partners first associate into a complex with the subsequent electron transfer occurring intramolecularly. The formation of protein/protein complexes under low ionic strength conditions has been exploited successfully over the last several years to measure rates of long range electron transfer in proteins[24-26].

In this report, we extend our measurements to phosphate buffer systems, where significantly faster electron transfer rates are observed. The effect of sample lyophilization on cytochrome c adsorption and electron transfer kinetics at tin oxide is also evaluated. We also discuss our current thinking on the electron transfer mechanism, the nature of cytochrome c binding to tin oxide, and the gradual deactivation of adsorbed electroactive cytochrome c that is typically observed.

EXPERIMENTAL

Reagents and Materials

Water was purified using a Milli-Q system with an Organex-Q final stage (Millipore, Bedford, MA). Tris(hydroxymethyl)aminomethane, referred to as tris, was reagent grade (Sigma Chemical Co., St. Louis, MO). Cacodylic acid (98%, Sigma) was recrystallized twice from isopropanol + hexane (50/50 by volume). Phosphate buffers were prepared from monobasic and dibasic potassium salts (ACS certified, Fisher, Fair Lawn, NJ) and ionic strengths were calculated from solution volumes and concentrations. Values of pH were measured to $\pm$0.01 units. Horse heart cytochrome c (Type VI, Sigma) was chromatographically purified[27] using carboxymethylcellulose (CM52, Whatman) equilibrated with 40 mM phosphate and eluted with 70 mM phosphate, concentrated by ultrafiltration using an Amicon ultrafiltration cell (Amicon, Danvers, MA) with an Amicon YM5 membrane, and desalted with Bio-Rad P6DG desalting gel (Bio-Rad, Richmond, CA). Milli-Q water was used to both equilibrate the desalting gel and to elute the purified cytochrome c. Fluoride-doped tin oxide was an 8 μm layer deposited on a 5 mm glass substrate (donated by PPG Industries, Pittsburgh, PA).

Apparatus

The two-chamber voltammetry cell was fabricated from acrylic plastic and featured a side-mounted working electrode (WE). The area of the WE was 0.71 cm^2. The reference electrode, a small Ag/AgCl/1.00 M KCl electrode, was calibrated against an SCE, whose potential was taken to be ±242 mV vs

NHE. All potentials reported in this paper are versus NHE. A platinum wire was used as the auxiliary electrode. The apparatus used for the CV measurements consisted of a Princeton Applied Research Model 362 potentiostat, a Nicolet 3091 digital oscilloscope, and an XY recorder.

X-ray photoelectron spectra (XPS) were acquired with a Perkin Elmer-Physical Electronics Division Model 550 spectrometer with an Mg anode operated at 15.6 kV and 20.0 mA to acquire survey spectra.

Procedures

The following procedures are similar to those previously described[20]. Prior to adsorption and voltammetric measurements, the tin oxide electrodes were cleaned by sequential ten-minute sonications in Alconox and isopropanol followed by two ten-minute sonications in Milli-Q water. The electrodes then were heated in an evacuated Vycor tube for two hours at 525°C. After cooling, the electrodes were transferred to a pH 7, 10 mM ionic strength tris/cacodylic buffer and allowed to equilibrate for at least 12 hours.

Adsorption of the protein on the electrode was accomplished by filling the WE chamber with a dilute solution of cytochrome c (10 μM for tris/cacodylate experiments and 17 μm for phosphate experiments) in pH 7, tris/cacodylate buffer, 10 mM ionic strength for one hour. Since the adsorption is strong and irreversible, concentrations of cytochrome c ranging from 10 μM to 30 μM during the adsorption step have no observable effect on the concentration of strongly adsorbed protein. Following this adsorption step, the cytochrome c containing buffer was removed, the cell was thoroughly rinsed with Milli-Q water and refilled with the desired buffer solution for electrochemistry. Cyclic voltammograms were acquired with the digital oscilloscope and then transferred to the XY recorder. Peak potentials were determined to ±5-15 mV from the oscilloscope screen using scale expansion and baseline estimation. Laviron's method[28] was used to determine quasi-reversible unimolecular electron transfer rate constants from cyclic voltammetric peak separations[20]. An α value of 0.5 was assumed in all calculations[28,20].

RESULTS AND DISCUSSION

Effect of Cytochrome c Lyophilization on Electron Transfer Kinetics in Tris/Cacodylate Media

In a prior publication, initial measurements of k_{et} were reported for tris/cacodylate buffers of pH 6-8.5 and ionic strengths of 1-100 mM. The cytochrome c samples in those experiments had been chromatographically purified according to established procedures[27] and then lyophilized for subsequent storage at -18°C. Solutions were then prepared directly from the lyophilized material. Since lyophilization has been shown to have very deleterious effects on the silver electrochemistry of cytochrome c[16] and also results in the appearance of new chromatographic bands[14], we have repeated our earlier experiments using purified, non-lyophilized samples.

For the electrochemistry of strongly adsorbed cytochrome c on fluoride-doped tin oxide electrodes, sample lyophilization has not been found to have a major negative impact, at least for the solution conditions that have been investigated. Using non-lyophilized samples, we have obtained rate constants (vide infra) very similar to those previously reported[20]. However, the use of non-lyophilized samples appears to result in somewhat higher surface coverage of electroactive cytochrome c, although the change is not dramatic. Despite the lack of a significant kinetic effect in the tin oxide electrochemistry of cytochrome c, the use of freshly chromatographed, non-lyophilized samples is advisable since surface coverage appears to increase and a source of contamination is removed which may become important at other conditions or when investigating more subtle behavior.

Figure 1 shows some typical cyclic voltammograms (CV) recently obtained with non-lyophilized cytochrome c in pH 8.0, 10 mM ionic strength tris/cacodylate buffer. Peaks due to the one-electron oxidation and reduction of strongly adsorbed cytochrome c appear superimposed on the tin oxide background current. The solution cytochrome c concentration in these and all other experiments is zero. The close correspondence between the surface formal potential, taken to be $(E_{p,a} + E_{p,c})/2$, and the solution formal potential, +0.26 V vs NHE[11], is strong evidence that the absorbed cytochrome c evident in Figure 1 is native. The rising background on the cathodic sweep is due to trace oxygen. These CV's are similar in morphology and behavior to those previously described[20] with the exception that coverage is higher. The surface coverage of electroactive cytochrome c in Figure 1 is 0.6 monolayers, as determined by integrating charge under the anodic peak and assuming a geometric electrode area. Because of surface roughness, the microscopic coverage would be somewhat less than this value. Further discussion of responses similar to these can be found in reference 20.

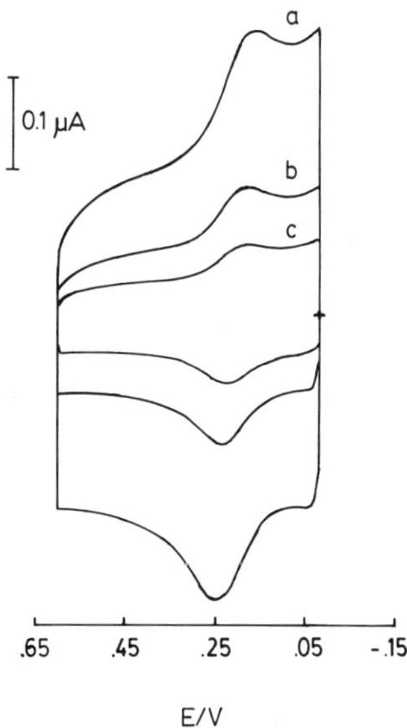

Figure 1. Cyclic voltammograms in tris/cacodylate media of non-lyophilized cytochrome c adsorbed on tin oxide. See Experimental Section for adsorption details. Solution conditions: tris/cacodylate, pH 8, 100 mM ionic strength. Scan rates (mV/s): (a) 50, (b) 20, (c) 10.

Values of k_{et} obtained for non-lyophilized cytochrome c samples are plotted versus scan rate in Figure 2 for pH 6 and pH 8 tris/cacodylate. Because of the greater coverages obtained with non-lyophilized samples, more accurate and complete data have been obtained at pH 6. Previously, only the lowest ionic strength used, 1.0 mM, resulted in surface coverages sufficient for extracting rate parameters[20]. With lyophilized cytochrome c, well-defined voltammetric peaks were obtained up to the highest ionic strength used, 100 mM.

Trends that can be deduced from the results in Figure 2 are identical to those previously found with lyophilized samples. The rate constant magnitudes are relatively small and show a trend of increasing value with increasing scan rate. For either pH 6 or pH 8, ionic strength has little apparent effect on the value of k_{et}. However, at constant ionic strength the value of k_{et} increases with decreasing pH, with the rates at pH 6 being approximately a factor of 2 greater than the pH 8 rates. Mechanistic interpretation of these data is discussed below after first presenting kinetic results obtained with phosphate buffers.

Electron Transfer Kinetics of Adsorbed Cytochrome c in Phosphate Media

Tris/cacodylate buffers were used initially because their components do not specifically bind to cytochrome c[29,30]. Phosphate, on the other hand, is known to bind specifically at certain sites on the surface of cytochrome c. Phosphate Site I, the high affinity site, is located in the vicinity of lysine 87 and has a dissociation constant of 2×10^{-4} M[31]. Phosphate Site II is a lower affinity site located in the vicinity of residues 25, 26, and 27[31]. Both of these sites are located near the front face of the cytochrome c molecule.

In phosphate buffers, well-defined cyclic voltammetric waves due to the one-electron oxidation and reduction of adsorbed cytochrome c were observed. Peak areas were initially larger and the adsorbed protein remained electroactive longer for pH 8 phosphate buffers than for pH 6 buffers. Adsorption was also favored at lower ionic strength, at least in the case of pH 6. These effects of pH on apparent coverage parallel those

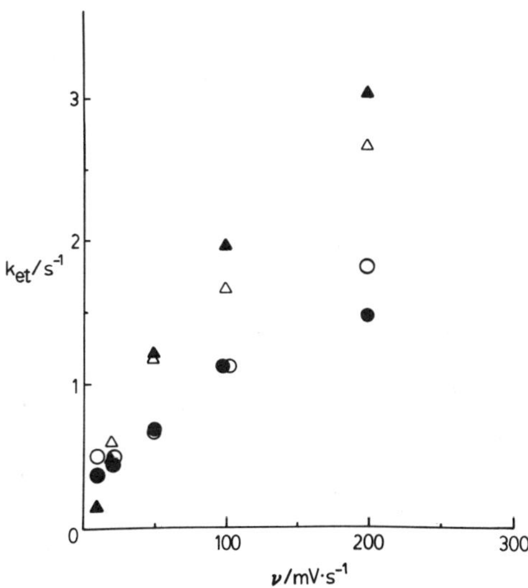

Figure 2. Plot of k_{et} vs ν for non-lyophilized cytochrome c at various tris/cacodylate solution conditions. See Experimental Section for adsorption conditions. Symbols represent an average of measurements taken at two separate electrodes under the same solution conditions. Relative standard deviations ranged from 0.31 to 0.02. Solution conditions: (▲) pH 6, 1.0 mM; (Δ) pH 6, 100 mM; (●) pH 8, 1.0 mM; (O) pH 8, 100 mM. Values of k_{et} were calculated from anodic and cathodic peak potentials which have been corrected for uncompensated iR drops; see reference 20.

seen previously with tris/cacodylate buffers[20]. However, relative to tris/cacodylate, phosphate buffers, especially of pH 8, result in a more robust and persistent voltammetric response from native adsorbed cytochrome c. An excellent example of this enhanced stability is shown in Figure 3. These CV's were obtained after first adsorbing the cytochrome c, obtaining an initial set of voltammetric data in pH 8, 34 mM ionic strength phosphate buffer, and then allowing the electrode to remain at open circuit in the same buffer for 16 hours. The surface coverage calculated from these voltammetric waves is 0.8 monolayers on a geometric electrode area basis. There was little difference between the CV's in Figure 3 and those obtained 18 hours earlier with the same electrode. We tentatively attribute the enhanced stability seen in phosphate to binding of this anion(s) to cytochrome c, since this is known to stabilize its native heme crevice structure[31].

Such long-term adsorbate stability is not seen in pH 6 phosphate buffer. For example, when the electrode in Figure 3 was transferred to a pH 6, 20 mM ionic strength phosphate buffer, the voltammetrically indicated surface coverage decreased almost immediately by roughly one-half. As discussed below in conjunction with Table 2, this lost electroactivity can be substantially restored by reimmersion in pH 8 buffer. However, long-term immersion (several hours) in pH 6, 20 mM ionic strength phosphate buffer leads to a total loss of signal. In a separate experiment, in which a freshly adsorbed electrode was transferred into a higher ionic strength (82 mM) pH 6 phosphate buffer, no voltammetric signals indicative of adsorbed cytochrome c were observed.

Unimolecular electron transfer kinetics for adsorbed cytochrome c in various phosphate media are reported in Table 1. Several points can be made. First, at constant pH, these rate constants are, without exception, larger than those obtained in the corresponding tris/cacodylate non-binding

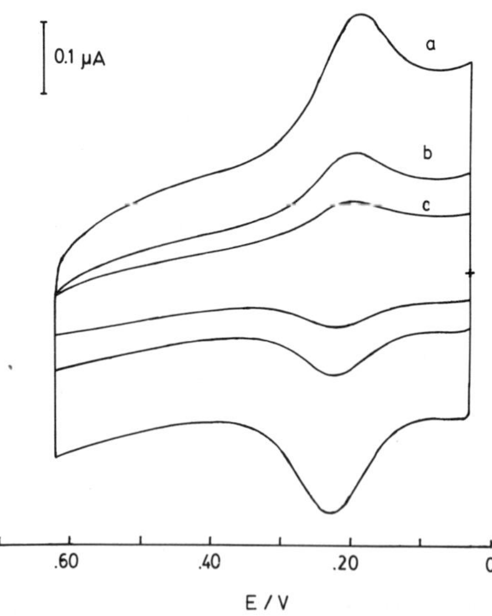

Figure 3. Cyclic voltammograms of non-lyophilized cytochrome c adsorbed on tin oxide after 16 h open-circuit immersion. See Experimental Section for adsorption details. Solution conditions for immersion and CV: potassium phosphate, pH 8, 34 mM ionic strength. Scan rates (mV/s): (a) 50, (b) 20, (c) 10. See text for further details on immersion and electrochemical procedures.

buffers. This result is attributed to an electrostatic effect. The small values of k_{et} observed in tris/cacodylate media have previously been attributed to non-optimal orientation resulting from electrostatic repulsion of the exposed heme edge by a positively charged tin oxide surface[20]. The specific binding of phosphate to the front face of cytochrome c would lower the postulated repulsion, and this is believed to be the cause of the observed rate enhancement. Phosphate adsorption on the tin oxide surface, if present, would be expected to give a similar qualitative effect, and thus cannot be ruled out at present as a contributing cause. Experiments are planned to clarify this situation.

A second observation with regard to Table 1 is that k_{et} is larger at pH 6 than pH 8. This effect is also observed using tris/cacodylate buffers (Figure 2). As discussed previously[20], this result is puzzling and we do not yet have an explanation for it. The opposite result, in fact, would be predicted from simply considering the effect of pH on tin oxide electrostatics, since a pH 6 interface would be more highly protonated. The possibility of an offsetting electrostatic contribution from protonation of cytochrome c was previously described. Another potential factor may be the effect which pH exerts on the flatband potential of tin oxide[32-34]. Ionic strength appears to be much more important in phosphate buffers than

Table 1. Unimolecular electron transfer kinetics for strongly adsorbed cytochrome c on tin oxide in phosphate buffers.[a]

Scan rate (mV/s)	k_{et}/s^{-1} [b,c]		
	pH 8 113 mM[d]	pH 8 34 mM[d]	pH 6 20 mM[d]
500	17.3 (2.5)[e]	4.3 (1.5)[f]	10.6 (3.6)[g]
200	16.1 (2.6)	3.1 (0.7)	6.1 (1.9)
100	14.2 (4.6)	2.4 (0.5)	4.7 (1.4)
50	7.4 (1.9)	2.0 (0.7)	3.0 (0.8)
20	7.5 (5.8)	1.3 (0.7)	1.9 (0.3)
10	4.7 (2.2)	1.1 (0.7)	1.0 (0.6)

[a] See Experimental Section for adsorption procedure.

[b] Calculated per reference 28 for $\alpha = 0.5$ and a Langmuir isotherm.

[c] Uncompensated iR corrections were negligible for these systems and were not made.

[d] Ionic strength of phosphate buffers.

[e] Standard deviation (S.D.) shown in parentheses is based on 3 experiments.

[f] S.D. in parentheses is based on 7 experiments.

[g] S.D. in parentheses is based on 4 experiments.

in tris/cacodylate. At pH 8, Table 1 shows a significant increase in electron transfer rate at a higher ionic strength. Such behavior was not seen in tris/cacodylate, although Figure 2 gives the suggestion of a similar, although much smaller, effect.

The last point to be made from Table 1 is that the values of k_{et} increase with increasing scan rate. This phenomenon has been consistently observed in all experiments to date, regardless of the buffer system used, and has been attributed to adsorbate reorientation[20] or, more generally, adsorbate adjustment, during the potential scan. This point and other aspects of the electron transfer mechanism are discussed in the subsequent section.

As mentioned above, partial loss of apparent electroactive coverage which occurred in pH 6, 20 mM, phosphate could be restored by prompt reimmersion in pH 8, 34 mM, phosphate. Table 2 shows electron transfer kinetic data for a single tin oxide electrode which was evaluated

Table 2. Effect on electron transfer kinetics of repeated alternations of phosphate buffers.[a,b,c]

Scan rate (mV/s)	k_{et}/s^{-1} [d]				
	(1) pH 8 34 mM[e]	(2) pH 6 20 mM	(3) pH 8 34 mM	(4)[f] pH 8 34 mM	(5) pH 6 20 mM
500	3.8	8.8	4.4	3.1	12.0
200	3.1	5.6	3.0	2.9	5.9
100	2.5	3.1	1.8	2.0	4.5
50	2.0	1.8	1.4	1.2	3.6
20	-	1.4	0.7	1.0	2.0
10	-	-	0.3	0.6	1.5

[a] Initial adsorption was performed at pH 7 per Experimental Section. This one tin oxide electrode was then used for the 5 sequential experiments shown, proceeding from the left column to the right. No additional cytochrome adsorption steps were performed during this sequence.

[b] Electrode was rinsed with Milli-Q water between buffer changes.

[c] Experiments 1-3 were performed sequentially with 5-10 minutes of buffer equilibrium following each change. Experiments 4-5 were performed 16 hours later after leaving the electrode immersed in pH 8, 34 mM, buffer at open-circuit.

[d] See Footnote b, Table 1.

[e] Ionic strength.

[f] Figure 3 shows CV's from this experiment.

alternatively in these two low-ionic-strength phosphate buffer systems. It is clear from these data that pH 8 phosphate stabilizes the native form of adsorbed cytochrome c, but results in slower kinetics for electron transfer. In pH 6, the kinetics are reproducibly faster, but the apparent coverage was always reduced by approximately one-half. Significantly, these different adsorption and kinetic behaviors are reversible as seen by comparison of appropriate columns in Table 2. This experiment has been repeated with similar results on a different electrode. However, we caution that in other cases, such reversibility may not be obtained. We have seen some evidence for a kinetic memory effect after immersion in higher ionic strength (113 mM), pH 8, phosphate.

Mechanistic Aspects of the Electron Transfer Reaction

In our initial publication on strongly adsorbed cytochrome c, a model was proposed to account for the slow electron transfer kinetics observed in tris/cacodylate media and the anomalous scan rate dependence of k_{et}[20]. The proposed model invoked facile electron transfer as occurring at the positively charged exposed heme edge site of cytochrome c, which is believed to be the physiologically active site[12]. A second feature of the model was that a net positive excess surface charge resides on the tin oxide at the potentials investigated, as concluded from reports of flat-band potential measurements[32-34]. With this model, the slow electron transfer kinetics were understood in terms of repulsive elecrostatic interactions between the exposed heme edge site and the electrode surface, presumably leading to a less-than-optimum orientation of the adsorbed molecule. To account for the dependence of k_{et} on scan rate, it was further proposed that the cytochrome c adsorbate could undergo some re-orientation in response to the positive charge buildup which occurred during the CV scan. The convolvement of the proposed reorientation kinetics with the electron transfer kinetics could then result in the scan rate dependence observed. It should also be emphasized that the proposed re-orientation of the cytochrome c adsorbate would not have to be drastic in order to cause the observed kinetic effects. For non-adiabatic reactions, only a 2-3 Å change in electron transfer distance would be predicted to result in an order-of-magnitude change in rate constant for typical β-values of ca. 1 $Å^{-1}$ [35]. In the following paragraph, this model is reconsidered in light of the phosphate kinetics. Alternatives to the reorientation model are then considered in the subsequent paragraph.

The phosphate kinetic results are consistent with the proposed model, as mentioned above. Specific binding of phosphate to the front face of cytochrome c would reduce repulsion and promote more favorable orientation of the exposed heme edge. Such rate enhancement effects due to phosphate have been observed before in homogeneous electron transfer reactions of cytochrome c with protein reaction partners which have positively charged docking sites, e.g., cytochrome c self-exchange[12]. Thus, the kinetic effect of phosphate supports the role of interfacial electrostatics as a major rate-determining factor in these reactions. With regard to the proposed adsorbate reorientation capability of strongly adsorbed cytochrome c, the kinetic results obtained with phosphate neither provide new support for nor detract from the proposed model.

The tin oxide/cytochrome c interface is quite complex and there are alternatives to adsorbate reorientation which could also conceivably account for the observed dependence of k_{et} on scan rate. The changing potential which occurs during the CV scan could result in chemical changes to the tin oxide itself[33] or could induce conformational changes or slight denaturation in the adsorbed cytochrome c molecules. An alter-native with a non-dynamic origin is an equilibrium distribution of electron transfer rate constants, which could arise from a distribution of orientations and/or heterogeneous surface effects. Future experimental work will be directed towards differentiating among these alternatives.

Binding of Cytochrome c to Tin Oxide

The results presented clearly show that cytochrome c can bind very strongly to tin oxide, especially as the pH becomes more basic. The detailed nature of the binding is not known at present due to lack of knowledge regarding the complex tin oxide surface. Past[32-34] and more recent[36,37] work points to the heterogeneous nature of the tin oxide aqueous interface. Furthermore, protein binding at solid/aqueous interfaces is in general a complex and rather poorly understood phenomenon due in part largely to the numerous possible interactions which may occur[38].

With these caveats in mind, some aspects of the binding situation are considered. With regard to the electrostatic situation, the charge distribution on tin oxide will be composed of discrete charge sites due to acid/base surface chemistry as well as delocalized charge due to the high concentration of charge carriers. Although the overall net charge on the electrode is believed to be positive at cytochrome c potentials (vide supra), the acid/base surface sites will be largely anionic in character since the solution pH's are basic of the isoelectric point, 5.5[39]. This picture not only provides a reasonable basis for interpreting electron transfer kinetic results, as discussed above, but is also consistent with strong persistent adsorption such as seen in Figure 3. Thus, optimum orientation of cytochrome c for rapid electron transfer would be hindered by the overall positive charge on the electrode, thus leading to slow kinetics, as observed in non-binding tris/cacodylate buffers. On the other hand, cytochrome c does not readily desorb from this positively charged surface even though it is a polycation. We believe, as suggested before[20], that specific interfacial binding interactions, namely, localized ionic interactions and hydrogen bonding, are important in inhibiting desorption of the protein. The statistical nature of polyion desorption is undoubtedly another important factor in this regard[38]. Our proposal is consistent with other reports of protein adsorption at solid/aqueous interfaces which have found that the overall charges on proteins and solid surfaces are often not the major determinant of adsorption/desorption behavior; instead, specific interactions frequently dominate[38].

Another aspect of binding interactions concerns the question of the gradual loss of electroactivity observed for adsorbed cytochrome c. In earlier reports this was attributed to slow desorption[13,20]. However, at pH 8 we have found from initial XPS results that the loss of electroactivity appears to be due to slow denaturation of the adsorbate. Figure 4 shows the N_{1s} portion of XPS spectra for two tin oxide electrodes which had simultaneously undergone the cytochrome c adsorption procedure described in the Experimental Section. Both electrodes were immediately evaluated using CV and gave similar responses indicative of strongly adsorbed cytochrome c at 0.1-0.2 monolayer coverage. One electrode was immediately removed from its cell, rinsed with Milli-Q water, air-dried, and stored in a polypropylene wafer case; its N_{1s} spectrum, shown in the top panel of Figure 4, indicates the presence of protein on the electrode surface. The other electrode was immersed in pH 8, 100 mM ionic strength, tris/cacodylate buffer for 14 hours, after which the voltammetric response due to adsorbed cytochrome c had disappeared. This electrode was then removed from its cell and treated in the same fashion as the first electrode; its spectrum, shown in the bottom panel of Figure 4, reveals an amount of nitrogen comparable to that observed on the freshly adsorbed electrode. A blank electrode, carried through the adsorption procedure in

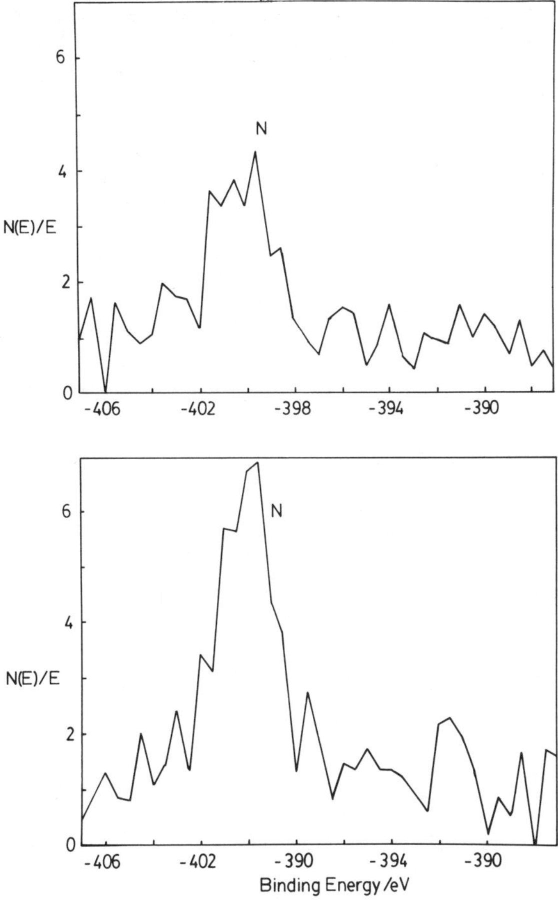

Figure 4. X-ray photoelectron spectroscopy of adsorbed cytochrome c on tin oxide. Enlargements of the N(1s) region of XPS survey spectra are shown. TOP: Electrode with electroactive adsorbed cytochrome c. BOTTOM: Electrode for which voltammetric peaks due to electroactive adsorbed cytochrome c have disappeared due to immersion in buffer. See text for details.

the absence of cytochrome c, showed no evidence of nitrogen. We conclude that the gradual loss of electroactivity of adsorbed cytochrome c on tin oxide at pH 8 is not due to desorption. Further XPS experiments at various pH values are planned.

CONCLUSIONS

Cytochrome c can be strongly adsorbed in its native state on tin oxide electrodes at coverages of several tenths of a monolayer. Measurements of unimolecular electron transfer rate constants for adsorbed cytochrome c can be readily made in the absence of solution cytochrome c using cyclic voltammetry. The kinetic results are consistent with an interfacial model involving electrostatic interaction between the tin oxide and the exposed heme edge of the cytochrome as well as an electrostatically driven adsorbate reorientation capability to account for the anomalous dependence of rate constant on scan rate. Other possible explanations for the kinetic results cannot be ruled out at present. Replacement of non-binding tris/cacodylate buffers with specifically binding phosphate buffers has two important effects. The voltammetrically measured surface coverage increases, attributed in part to stabilization of the native protein structure, and electron transfer rates increase, attributed to reduction of interfacial electrostatic repulsion. The irreversible nature of the cytochrome c adsorption, especially with increasing pH, is consistent with a number of other reports from the protein adsorption literature.

ACKNOWLEDGMENTS

We are very grateful for helpful discussions with Professors V.I. Birss, J.F. Evans, F.M. Hawkridge, R.A. MacPhail, K. Tokuda, and M.J. Weaver, and Dr. M.J. Tarlov. We thank the North Carolina Biotechnology Center (Grant No. 85-G-00705) and the North Carolina Board of Science and Technology (Grant No. 86-0056) for financial support, and Dr. John Sopko for providing tin oxide.

REFERENCES

1. L. Griggio and S. Pinamonti, Atti Ist. Veneto Sci. Lett. Arti. Cl. Sci. Mat. Natur., 124, 15 (1965-66).

2. S.R. Betso, M.H. Klapper and L.B. Anderson, J. Am. Chem. Soc. 94, 8197 (1972).

3. F. Scheller, M. Janchen, J. Lampe, H.-J. Prumke, J. Blanck and E. Palecek, Biochim. Biophys. Acta, 412, 157 (1975).

4. B.A. Kuznetsov, N.M. Mestechkina and G.P. Shumakovich, Bioelectrochem. Bioenerg., 4, 1 (1977).

5. P. Yeh and T. Kuwana, Chem. Lett., 1145 (1977).

6. M.J. Eddowes and H.A.O. Hill, J. Chem. Soc., Chem. Commun., 771 (1977).

7. H.L. Landrum, R.T. Salmon and F.M. Hawkridge, J. Am. Chem. Soc., 99, 3154 (1977).

8. G. Dryhurst, K.M. Kadish, F. Scheller and R. Renneberg, "Biological Electrochemistry", Academic Press, New York (1982), p. 398.

9. E.F. Bowden, F.M. Hawkridge and H.N. Blount, in: "Comprehensive Treatise of Electrochemistry", Volume 10, S. Srinivason, Yu. A. Chizmadzhev, J. O'M. Bockris, B.W. Conway and E. Yeager (Eds.), Plenum Press, New York (1985), p. 297.

10. F.A. Armstrong, H.A.O. Hill and N.J. Walton, Quart. Rev. Biophys. 18, 261 (1986).

11. R.E. Dickerson and R. Timkovich, in "The Enzymes", Volume 11A, P.D. Boyer (Ed.), Academic Press, New York (1975), p. 397.

12. S. Ferguson-Miller, D.L. Brautigan and E. Margoliash, in: "The Porphyrins", Volume 7B, D. Dolphin (Ed.), Academic Press, New York (1979), p. 149.

13. E.F. Bowden, F.M. Hawkridge and H.N. Blount, J. Electroanal. Chem., 161, 355 (1984).

14. K.B. Koller and F.M. Hawkridge, J. Am. Chem. Soc., 107, 7412 (1985).

15. I. Taniguchi, T. Funatsu, M. Iseki, H. Yamaguchi and K. Yasukouchi, J. Electroanal. Chem., 193, 295 (1985).

16. D.E. Reed and F.M. Hawkridge, Anal. Chem., 59, 2334 (1987).

17. M.J. Eddowes, H.A.O. Hill and K. Uosaki, J. Am. Chem. Soc., 101, 7114 (1979).

18. T. Cotton, S.G. Schultz and R.P. van Duyne, J. Am. Chem. Soc., 102, 7960 (1980).

19. C. Hinnen, R. Parsons and K. Niki, J. Electroanal. Chem., 147, 329 (1983).

20. J.L. Willit and E.F. Bowden, J. Electroanal. Chem., 221, 265 (1987).

21. T. Yokota, K. Itoh and A. Fujishima, J. Electroanal. Chem., 216, 289 (1987).

22. E. Laviron, J. Electroanal. Chem., 100, 263 (1979).

23. (a) S.W. Barr, K.L. Guyer, T. T.-T. Li, H.Y. Liu and M.J. Weaver, J. Electrochem. Soc., 131, 1626 (1984). (b) T. T.-T. Li and M.J. Weaver, J. Am. Chem. Soc., 106, 6107 (1984) and preceding papers in this series.

24. S.L. Mayo, W.R. Ellis, R.J. Crutchley and H.B. Gray, Science, 233, 948 (1986).

25. S.E. Peterson-Kennedy, J.L. McGourty, P.S. Ho, C.J. Sutoris, N. Liang, H. Zemel, N.V. Blough, E. Margoliash and B.M. Hoffman, Coord. Chem. Rev., 64, 125 (1985).

26. E. Cheung, K. Taylor, J.A. Kornblatt, A.M. English, G. McLendon and J.R. Miller, Proc. Natl. Acad. Sci. USA, 83, 1330 (1986).

27. D.L. Brautigan, S. Ferguson-Miller and E. Margoliash, Methods Enzymol., 53D, 128 (1978).

28. E. Laviron, J. Electroanal. Chem. 101, 19 (1979).

29. G.H. Barlow and E. Margoliash, J. Biol. Chem., 241, 1473 (1966).

30. R. Margalit and A. Schejter, Eur. J. Biochem., 32, 492 (1973).

31. N. Osheroff, D.L. Brautigan and E. Margoliash, Proc. Natl. Acad. Sci. USA, 77, 4439 (1980).

32. D. Elliott, D.L. Zellmer and H.A. Laitinen, J. Electrochem. Soc., 117, 1343 (1970).

33. N.R. Armstrong and V.R. Shepard, J. Phys. Chem., 85, 2965 (1981).

34. F. Mollers and R. Memming, Ber. Bunsenges. Phys. Chem., 76, 475 (1972).

35. R.A. Marcus and N. Sutin, Biochim. Biophys. Acta, 811, 265 (1985).

36. R.H. Colton and H.E. Hager, J. Electrochem. Soc., 133, 2530 (1986).

37. M.J. Tarlov and J.F. Evans, J. Vac. Sci. Technol., A5, 941 (1987).

38. J. Lyklema, Colloids Surfaces, 10, 33 (1984).

39. S.M. Ahmed, in: "Oxides and Oxide Films", Volume 1, J.W. Diggle (Ed.), Marcel Dekker, New York (1972), p. 319.

REDOX REACTIVITY OF MANGANESE "CAPPED" METALLOPORPHYRINS

Lawrence A. Bottomley*, Jeffrey F. Luck,
Frank L. Neely and James S. Quick

School of Chemistry
Georgia Institute of Technology
Atlanta, Georgia 30332-0400

INTRODUCTION

For the past three years, we have been in pursuit of the redox properties of various pentacoordinate metalloporphyrins which remain coordinated to the same axial ligand upon electron transfer. Porphyrins with a high degree of steric hindrance built onto only one side of the porphyrin seemingly provide a straightforward means for preparing pentacoordinate metalloporphyrin complexes. However, ligand addition studies on the "picket fence", "strapped", "pocket" and "bis-pocket" porphyrins have indicated that these materials are unsuitable for this type of study because diadduct formation can occur[1-4]. Likewise, the various "strapped" metalloporphyrins are not appropriate because of the accessibility of the metal centers to coordination by components of the supporting electrolyte. Metallo derivatives of Baldwin's "capped" porphyrin[5] have been selected (see Figure 1 for the structure of the Mn derivative). We have shown that the presence of the pyrromellitoyl cap impeded formation of Co capped porphyrin-nitrogenous base diadducts, even when a very strong thermodynamic driving force for formation of a hexacoordinate metal center was present[6].

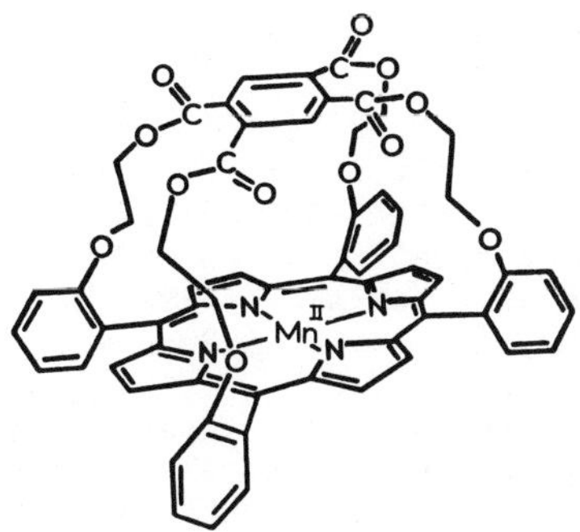

Figure 1. Structure of the Mn(II) derivative of Baldwin's "capped" porphyrin.

High valent metalloporphyrins have been and are presently the focus of considerable effort by several research teams[7-13]. The chemistry of oxygen atom transfer from high valent oxo-Fe and oxo-Mn derivatives of synthetic porphyrins to olefins and paraffins (in a limited number of cases) has been well documented. Recently, nitrido manganese porphyrins have been prepared by Buchler[14], Hill[15] and Groves[16]. Although the N≡ Mn core is isoelectronic with O=Fe, the reactive moiety involved in the oxygen atom transfer chemistry, its reactivity is somewhat limited. Buchler has shown that the nitrido group can be converted to a phosphine imide at reflux in the presence of triphenylphosphine. A potentially more useful reaction pathway was discovered by Groves and Takahashi[16]. Treatment of the nitrido manganese(V) derivative of 5,10,15,20-tetra-mesitylporphyrin with trifluoroacetic anhydride in the presence of an excess of cyclooctene resulted in the transfer of the nitrogen atom to the olefin. Conversion of cyclooctene to the corresponding aziridine proceeded via nitrenoid Mn porphyrin intermediate. During this process the nitrido Mn(V) porphyrin was reduced to the trifluoroacetato Mn(III) porphyrin. This reaction, if generalized, would find widespread synthetic utility especially in the design of new chemotherapeutic agents[17].

The impetus for this study was threefold. First, we sought to place the Mn capped porphyrin within the general framework of metalloporphyrin electrochemistry as recently reviewed by Kadish[18]. Second, we hoped to discover and exploit the unique reactivity of the Mn capped porphyrin afforded by the presence of the nondisplaceable steric blocking group on one side of the macrocycle. Third, we wished to explore the chemistry of high valent Mn capped porphyrin.

The results presented herein show that the redox reactivity of Mn(III) capped porphyrin is comparable to that previously observed for the unhindered analogues. All charge transfer processes involve a penta-coordinate metal center and, in some instances, involve ligand substitution reactions. The presence of the cap does not impede the generation of the high valent nitrido Mn(V) porphyrin. When compared to the reactivity of sterically unencumbered porphyrins, the presence of the cap actually facilitates the production of the nitrenoid complex, the key intermediate in the nitrogen transfer chemistry leading to aziridines.

EXPERIMENTAL

Chemicals

All sterically unencumbered Mn porphyrins used herein were prepared by the method of Adler[19]. The following nitrido Mn(V) porphyrins were prepared from the corresponding Mn(III) porphyrins by either the method of Buchler[14] or Takahashi[20]: nitrido meso-tetrakis(2-methoxytetraphenyl)-porphinatomanganese(V), NMn(o-OCH_3)TPP; nitrido meso-tetrakis(4-methoxy-tetraphenyl)porphinatomanganese(V), NMn(p-OCH_3)TPP, nitrido meso-5,10,15,20-tetraphenylporphinatomanganese(V), NMnTPP. The free base form of Baldwin's capped porphyrins, 5,10,15,20-[pyromellitoyltetrakis(o-(oxyethoxy)]-phenylporphyrin, H_2C_2Cap, was synthesized and purified by the method of Almog et al.[5] Manganese was inserted into the free base by the method of Bruice[7]. The purity of all porphyrinic materials was verified by UV-Vis, NMR and mass spectral measurements. All reagents were obtained from Aldrich Chemical Co. and purified in the manner previously described[21].

Apparatus

The spectroelectrochemical instrumentation and techniques were identical to those previously reported[21]. Experiments were carried out at ambient temperatures (21±1 °C). Unless otherwise noted, all solutions were deoxygenated with solvent-saturated nitrogen and blanketed with nitrogen during all experiments. All potentials are referenced to the SCE.

RESULTS AND DISCUSSION

Electrochemistry of Mn Capped Porphyrin

Voltammograms obtained on (acetato)MnC_2Cap in either 1,2-dichloroethane or pyridine were similar in appearance to those obtained for the sterically hindered Mn tetraphenylporphyrins[22]. In the noncoordinating solvent, three quasireversible charge transfer reactions were observed at $E_{1/2}$ values of 0.96, -0.41 and -1.34 V corresponding to the Mn(III) porphyrin cation radical/Mn(III) porphyrin, the Mn(III)/Mn(II) porphyrin and the Mn(II)/Mn(II) porphyrin anion radical charge transfer reactions, respectively. In pyridine, a competitive equilibrium between the acetate ion and pyridine for the accessible axial position on both the Mn(III) and Mn(II) centers was observed. Half wave potential values of -0.17 and -1.14 were measured for the successive reduction of the monopyridine adduct.

Synthesis and Electrochemistry of Nitrido Mn Capped Porphyrin

Sterically unhindered nitrido Mn(V) porphyrins have been prepared by either the photochemical decomposition of the azido Mn(III) porphyrin[16] or by treatment of either the hydroxo or acetato Mn(III) porphyrin with sodium hypochlorite and aqueous ammonia[14]. We found that $NMnC_2Cap$ could be prepared quantitatively from (acetato)MnC_2Cap by the latter route. Despite the rather basic conditions, the ester linkages holding the cap above the metalloporphyrin core were not cleaved. Presumably, this is because the reaction occurs at the interface between two immiscible phases (CH_2Cl_2 and the aqueous ammonia/sodium hypochlorite phases). The extent of this reaction is easily monitored spectroscopically, as depicted in Figure 2.

Figure 3 depicts cyclic and differential pulse voltammograms typical of $NMnC_2Cap$ in 1,2-dichloroethane. Analysis of data from variable scan rate experiments (over the range 20 to 10,000 mVs^{-1}) and variable pulse amplitude differential pulse experiments (over the range 5 to 100 mV) revealed that all four electrode reactions are electrochemically reversible one-electron transfers. This conclusion is based on the following observed trends. For all reactions, $E_{1/2}$ values were invariant with potential scan rate. Forward peak currents were proportional to the square root of the potential sweep rate and current ratios were equal to unity at all scan rates. At 20 mV/s, each process gave a peak potential separation of 60±2 mV, which increased slightly at higher scan rates. $E_{1/2}-E_p$ and $E_{1/2}-E_{p/2}$ values were 31±2 and 29±2 mV, respectively at a potential sweep rate of 20 mV/s and increased slightly at higher sweep rates. The trends observed for the variable pulse amplitude differential pulse data were also those expected for reversible charge transfers.

Similar current-voltage-potential sweep rate trends were observed for NMnTPP, NMn(o-OCH_3)TPP and NMn(p-OCH_3)TPP. Table 1 lists the potentials observed for the electrode reactions of the nitrido Mn porphyrin complexes investigated. The increased potential separation between the $E_{1/2}$ values of the first and second oxidations (as compared to the unhindered porphyrins) reflects the influence of the cap in inhibiting the formation

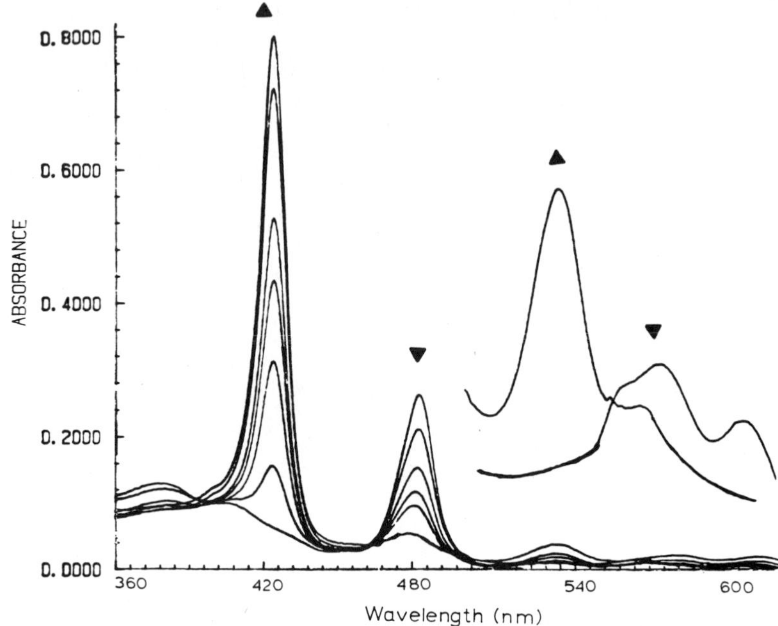

Figure 2. Visible spectra acquired as a function of time during the conversion of acetato MnC_2Cap to $NMnC_2Cap$.

of an associated ion pair between the anion of the supporting electrolyte and the positively charged metalloporphyrin. A similar increase in potential separation was observed for the Co capped porphyrin[6].

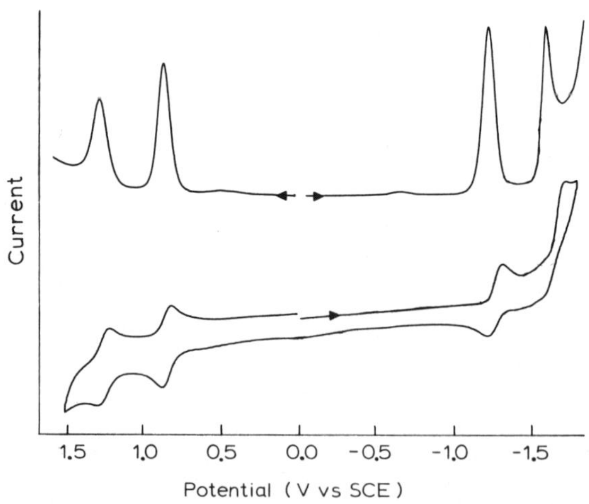

Figure 3. Voltammograms obtained on a millimolar solution of $NMnC_2Cap$ in 1,2-dichloroethane. The lower trace is the steady state cyclic voltammogram recorded at a potential sweep rate of 200 mVs^{-1}. The upper trace is the differential pulse voltammogram acquired at a pulse amplitude of 25 mV, a pulse interval of 0.1 s and a potential sweep rate of 30 mVs^{-1}.

Table 1. Potential Data[a] For the Redox Reactions of Selected Nitrido Mn Porphyrins in $EtCl_2$

Porphyrin	Redox Reaction			
	2nd oxidn	1st oxidn	1st redn	2nd redn
$NMnC_2Cap$	1.24	0.86	-1.23	-1.71
$NMn(o\text{-}OCH_3)TPP$	1.11	0.94	-1.14	-1.56
$NMn(p\text{-}OCH_3)TPP$	1.12	0.93	-1.13	-1.49
NMnTPP	1.20	1.00	-1.09	-1.46

[a]Potentials were measured in 0.1 M tetrabutylammonium perchlorate, referenced against the SCE and were uncorrected for liquid junction potentials.

Voltammograms observed for $NMnC_2Cap$ dissolved in CH_2Cl_2, $EtCl_2$ and pyridine[16] were similar in appearance and gave comparable trends as a function of potential sweep rate and pulse amplitude. The spectrum of $NMnC_2Cap$ in each of these solvents was essentially unchanged. This implies that the Mn core in $NMnC_2Cap$ remains five coordinate even in the presence of a large excess of strong Lewis bases.

Various spectroelectrochemical experiments were carried out on each of the nitrido Mn porphyrins to verify the chemical and electrochemical reversibility of each charge transfer process. Constant potential chronoabsorptometric experiments showed that the products of the first or second reduction processes were stable for at least an hour under strict exclusion of oxygen. Multiple potential step chronoabsorptometric experiments produced comparable results. Analysis of the change in absorbance as a function of applied potential yielded $E^{o\prime}$ values equal to the $E_{1/2}$ values obtained by cyclic voltammetry and confirmed the passage of one electron for each charge transfer process. Constant potential chronoabsorptometric experiments showed that the products of the first oxidation degraded with time. At oxidative electrolysis times longer than five minutes, mixtures of the nitrido Mn porphyrin cation and either the corresponding perchlorato Mn(III) complex or the perchlorato Mn(III) cation radical (depending upon the applied potential) were produced. Details of the results of the spectroelectrochemical experiments will be published elsewhere[23].

The lifetime of the nitrido Mn porphyrin cation is, however, long enough to allow isolation. Takahashi[20] has generated the oxidized form of the nitrido meso-tetrakis(4-methyltetraphenyl)porphinatomanganese(V) by chemical oxidation with ferric perchlorate. The spectrum of this green complex is identical to the spectrum of the oxidized material generated electrochemically. The infrared spectrum of this material exhibited a minor shift in the N=Mn stretch from 1048 cm^{-1} to 1044 cm^{-1} upon oxidation, indicating that the nitrogen moiety remains intact. A strong band at 1282 cm^{-1}, indicative of porphyrin ring oxidation, was observed. An EPR spectrum of the green complex confirms the presence of a d^2 low-spin electronic configuration for the Mn center upon oxidation.

Based on the combined spectral and electrochemical results, we have proposed that all four charge transfer processes of nitrido Mn porphyrins occur at the porphyrin ring and do not involve a formal change in the valence of the central metal atom. The nitrido Mn porphyrins are the first examples in which an electroactive Mn core is deactivated by axial ligation. Although other examples of the modification of the site of charge transfer induced by changes in metalloporphyrin axial ligands are known[24,25], this marks the first time this phenomenon has been observed for a high valent metalloporphyrin.

Reactivity of Nitrido Mn Capped Porphyrin

The evidence presented above demonstrates that nitrido Mn capped porphyrins are remarkably stable, are not activated electrochemically and are not subject to the ligand addition reactions normally encountered with Mn(III) porphyrins. We were, however, able to activate the N≡Mn core by reaction with trifluoroacetic anhydride, TFAA, and produce the Mn(V) nitrenoid species: $\{CF_3CON{=}Mn{-}C_2Cap\}^{+}\{OOCCF_3\}^{-}$. The progress of this reaction was monitored both spectrally and electrochemically. Figure 4 depicts the spectral transition observed for $NMnC_2Cap$ when treated with an excess of TFAA. The final spectrum compares favorably with that previously published for the nitrenoid Mn tetramesitylporphyrin[16a]. When this reaction was monitored electrochemically, the currents associated with the two oxidation processes at $E_{1/2}$ values of 0.86 and 1.24 V decreased in amplitude with a corresponding increase in current for two new processes at $E_{1/2}$ values of 1.11 and 1.52 V. The electroreduction of TFAA masked the two reduction processes of nitrenoid MnC_2Cap.

The kinetics of the nitrenoid formation reaction were evaluated spectrally under pseudo first-order conditions. The rate of the reaction was independent of porphyrin concentration (over the range 10 to 100 μM) and directly dependent upon the TFAA concentration (over the range of 100- to 2000-fold excess). When trifluoracetic acid or acetic anhydride was substituted for TFAA, no reaction was observed. When the reaction was carried out in an excess of either cyclooctene or norbornylene, the apparent rate constant was slightly larger. This observation suggested that the formation of the nitrene is the rate determining step. An apparent rate constant of $14.0 \times 10^{-3}\ s^{-1}$ was computed from the spectral data at a

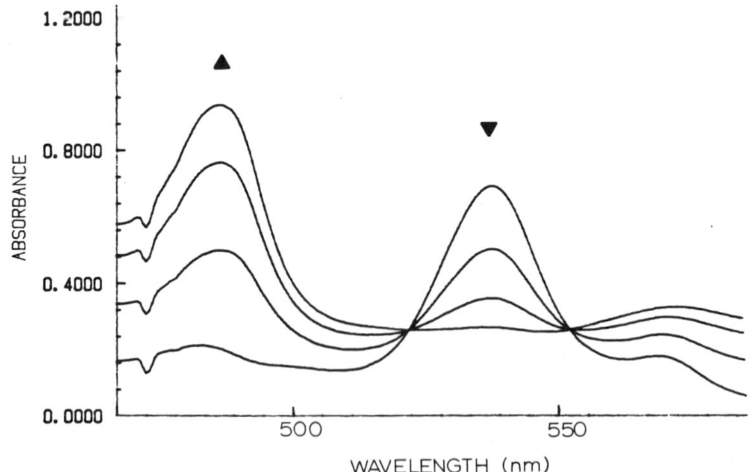

Figure 4. Visible spectra acquired as a function of time during the conversion of $NMnC_2Cap$ to $\{CF_3CON{=}MnC_2Cap\}^{+}\{OOCCF_3\}^{-}$.

1000-fold excess of TFAA. To evaluate the impact of the cap on this reaction, kinetic evaluations identical to that described above for $NMnC_2Cap$ were performed on NMnTPP, $NMn(o\text{-}OCH_3)TPP$ and $NMnp\text{-}OCH_3)TPP$. The apparent rate constants obtained are listed in Table 2.

From the rate data, several inferences can be drawn. First, the enhanced rate of nitrenoid formation for $NMn(p\text{-}OCH_3)TPP$ as compared to NMnTPP suggests that the basicity of the porphyrin ring has a direct effect on the rate of formation of the nitrenoid. This is in accord with expectations. The nitrenoid formation reaction represents nucleophilic attack of the N≡ Mn moiety on the electron deficient carbonyl carbon of TFAA. Thus, increasing the basicity of the porphyrin ring should increase the nucleophilic character of the N≡ Mn core and increase the rate of the reaction.

Secondly, the nitrido moiety on $NMnC_2Cap$ can potentially be located under the cap or opposite the cap relative to the Mn core. In the former case, the reaction rate should be significantly decreased relative to that of the unhindered porphyrins. In the latter case, the observed reaction rate should be comparable to that observed for porphyrins with similar phenyl ring substituents. Since the observed reaction rate for $NMnC_2Cap$ is one order of magnitude greater than $NMn(o\text{-}OCH_3)TPP$, we conclude that the nitrido moiety on $NMnC_2Cap$ is located opposite the cap.

Thirdly, the decrease in rate for TFAA addition to $NMn(o\text{-}OCH_3)TPP$ as compared to $NMn(p\text{-}OCH_3)TPP$ suggests that steric effects are also important in this reaction. Ortho-substituted nitrido Mn tetraphenylporphyrins possess six interconvertible atropisomers. The least sterically hindered atropisomer would have all four $o\text{-}OCH_3$ groups oriented anti to the nitride. Using the nomenclature usually employed for porphyrins which exhibit atropisomerism[1], this would be the β,β,β,β isomer. The most stericaly hindered atropisomer would have all four $o\text{-}OCH_3$ groups oriented syn to the nitride, i.e. the α,α,α,α isomer. The anticipated rate of nitrenoid formation for $NMn(o\text{-}OCH_3)TPP$ would be less than that of another porphyrin with comparable basicity but without ortho-substituents. The decrease reflects the additional time required for the rotation of one or more phenyl rings of the molecule to achieve the favored conformation.

Fourthly, substituted tetraphenylporphyrins with relatively small ortho-substituents are known to isomerize at ambient temperature and at

Table 2. Pseudo First-Order Rate Constants[a] for the Conversion of Nitrido Mn(V) Porphyrins to Nitrenoid Mn(V) Porphyrin in $EtCl_2$

Porphyrin	k_{app} (s^{-1})
$NMnC_2Cap$	4.0×10^{-3}
$NMn(o\text{-}OCH_3)TPP$	1.1×10^{-4}
$NMn(p\text{-}OCH_3)TPP$	2.7×10^{-4}
NMnTPP	1.1×10^{-4}

[a]Rate constants were determined from absorption data acquired at 21±1 °C on 50 μM solutions of the nitrido Mn porphyrin under a 1000-fold excess of TFAA.

rates at least comparable to the observed rates for the nitrenoid formation[1,5]. Statistically, 6.25 % of the molecules of NMn(o-OCH_3)TPP should be in the favored conformation, the β,β,β,β atropisomer, at any point in time. If one views the cap functionality on NMnC_2Cap strictly as a molecular device to prevent isomerization of the β,β,β,β atropisomer, then the observed reaction rate for NMn(o-OCH_3)TPP should be 6.25 % of that observed for NMnC_2Cap. If more than one atropisomer is capable of reaction, then the relative reaction rate of NMn(o-OCH_3)TPP should increase. The observed reaction rate for NMn(o-OCH_3)TPP is 2.7 % of that observed for NMnC_2Cap. From single crystal X-ray diffraction studies on ClFeC_2Cap, the presence of the cap is known to ruffle the porphyrin[26]. Since the observed rate for NMnC_2Cap is even faster than predicted from statistical considerations alone, the enhancement is probably caused by the greater accessibility of the N≡ Mn core due to its position at the apex of the domed porphyrin.

Thus, the rate of nitrenoid formation and, consequently, the rate of nitrogen atom transfer from nitrido Mn porphyrins to olefins seems dependent upon the basicity of the porphyrin ring and upon the steric environment of the nitride. To clarify the relative importance of these factors, we are currently investigating the reactivity of other nitrido Mn porphyrins which possess a greater variation in porphyrin ring basicity and steric bulk about the N≡ Mn moiety. These findings should permit us to select the "optimum" nitrido Mn porphyrin for generalization of nitrogen atom transfer chemistry.

ACKNOWLEDGMENTS

Support of this research by the DHHS through Grant No. HL-33734-02 and by the NIH Biomedical Research Program at Georgia Tech is gratefully acknowledged. We wish to thank Professor Craig Hill for helpful discussions during the early stages of this work.

REFERENCES

1. J.P. Collman, R.R. Gagne, C. Reed, T.R. Halbert, G. Lang and W.T. Robinson, J. Am. Chem. Soc., 97, 1427 (1975).

2. J.E. Baldwin, M.J. Crossley, T. Klose, E.A. O'Rear and M.K. Peters, Tetrahedron, 38, 27 (1982).

3. J.P. Collman, J.I. Brauman, T.J. Collins, B.L. Iverson, G. Lang, R.B. Pettman, J.L. Sessler and M.A. Walters, J. Am. Chem. Soc., 105, 3038 (1983).

4. a) K.S. Suslick and M.M. Fox, J. Am. Chem. Soc., 105, 3507 (1983). b) K.S. Suslick, M.M. Fox and T.J. Reinert, ibid., 106, 4522 (1984).

5. J. Almog, J.E. Baldwin, M.J. Crossley, J.F. DeBernardis, R.L. Dyer, J.R. Huff and M.K. Peters, Tetrahedron, 38, 3587 (1981).

6. L.A. Bottomley, J.-N. Gorce and W.M. Davis, J. Electroanal. Chem., 202, 111 (1986).

7. a) D. Ostovic, C.B. Knobler and T.C. Bruice, J. Am. Chem. Soc., 109, 3444 (1987). b) T.C. Bruice, C.M. Dicken, P.N. Balasubramanian, T.C. Woon, F.-L. Lu, J. Am. Chem. Soc., 109, 3436 (1987). c) W.-H. Wong, D. Ostovic and T.C. Bruice, J. Am. Chem. Soc., 109, 3428 (1987), and references therein.

8. a) C.L. Hill and B.C. Schardt, J. Am. Chem. Soc.,102, 6374 (1980). b) J.A. Smegal, B.C. Schardt and C.L. Hill, J. Am. Chem. Soc., 105, 3510 (1983). c) J.A. Smegal and C.L. Hill, J. Am. Chem. Soc., 105, 351 (1983). d) C.L. Hill, J.A. Smegal and T.J. Henly, J. Org. Chem., 18, 3277 (1983). e) M.J. Camenzind, F.J. Hollander and C.L. Hill, Inorg. Chem., 21, 4301 (1982). f) J.A. Smegal and C.L. Hill, J. Am. Chem. Soc., 105, 2920 (1983).

9. a) T.G. Taylor, T. Nakano, A.R. Miksztal and B.E. Dunlap, J. Am. Chem. Soc., 109, 3625 (1987). b) T.G. Traylor and A.R. Miksztal, J. Am. Chem. Soc., 109, 2770 (1987), and references therein. c) P.S. Traylor, D. Dolphin and T.G. Traylor, J. Chem. Soc. Chem. Commun., 279 (1984).

10. a) J.T. Groves, R.C. Haushalter, M. Nakamura, T.E. Nemo and B.J. Evans, J. Am. Chem. Soc., 103, 2884 (1981). b) B. Boso, G. Lang, T.J. McMurry and J.T. Groves, J. Chem. Phys., 79, 1122 (1983). c) J.T. Groves, R. Quinn, T.J. McMurry, M. Nakamura, G. Lang and B. Boso, J. Am. Chem. Soc., 27, 102 (1985). d) J.T. Groves and Y. Watanabe, Inorg. Chem., 26, 785 (1987). e) J.T. Groves and M.K. Stern, J. Am. Chem. Soc., 109, 3812 (1987), and references therein.

11. B.R. Cook, T.J. Reinert, K.S. Suslick, J. Am. Chem. Soc., 108, 7281 (1986).

12. K. Shin and H.M. Goff, J. Am. Chem. Soc., 109, 3140 (1987), and references therein.

13. S.E. Creager, S.A. Raybuck and R.W. Murray, J. Am. Chem. Soc., 108, 4225 (1986).

14. a) J.W. Buchler, C. Dreher and K.L. Lay, Z. Naturforsch., 37b, 1155 (1982). b) J.W. Buchler, C. Dreher, K.L. Lay, A. Raap and K. Gersonde, Inorg. Chem., 22, 879 (1983).

15. C.L. Hill and F.J. Hollander, J. Am. Chem. Soc., 104, 7318 (1982).

16. a) J.T. Groves and T. Takahashi, J. Am. Chem. Soc., 105, 2073 (1983). b) J.T. Groves, T. Takahashi and W.M. Butler, Inorg. Chem., 22, 884 (1983).

17. N.R. Lomax and V.L. Narayanan, "Chemical Structures of Interest to the Division of Cancer Treatment", Vol. V, National Cancer Institute, Washington, D.C. (1986).

18. K.M. Kadish, Prog. Inorg. Chem., 34, 435 (1987).

19. A.D. Adler, F.R. Longo, J.D. Finarelli, J. Goldmacher, J. Assour and L. Korsakoff, J. Org. Chem., 32, 476 (1967).

20. T. Takahashi, Ph.D. Dissertation, University of Michigan, (1985).

21. L.A. Bottomley, M.R. Deakin and J.-N. Gorce, Inorg. Chem., 23, 3563 (1984).

22. a) K.M. Kadish and S.L. Kelly, Inorg. Chem., 18, 2968 (1979). b) S.L. Kelly and K.M. Kadish, ibid., 21, 3631 (1982).

23. L.A. Bottomley, F.L. Neely and J.-N. Gorce, Inorg. Chem., 27, 0000 (1988).

24. a) G.M. Brown, F.R. Hopf, J.A. Ferguson, T.J. Meyer and D.J. Whitten, J. Am. Chem. Soc., 95, 5939 (1973). b) G.M. Brown, F.R. Hopf, T.J. Meyer and D.J. Whitten, J. Am. Chem. Soc., 197, 5385 (1975). c) P.D. Smith, D. Dolphin and B.R. James, J. Organomet. Chem., 208, 239 (1981). d) B.R. James, D. Dolphin, T.W. Leung, F.W.B. Einstein and A.C. Willis, Can. J. Chem., 62, 1238 (1984).

25. K.M. Kadish, L.A. Bottomley and J.S. Cheng, J. Am. Chem. Soc., 100, 2731 (1978).

26. M. Sabat and J.A. Ibers, J. Am. Chem. Soc., 104, 3715 (1982).

METALLOPORPHYRIN FILMS ON SOLID ELECTRODES: REFLECTANCE SPECTROSCOPY AND AC IMPEDANCE STUDIES IN AQUEOUS AND NONAQUEOUS MEDIA

Tadeusz Malinski* and Janet E. Bennett

Department of Chemistry
Oakland University
Rochester, Michigan 48309-4401

INTRODUCTION

In recent years, much attention has been focused on electrochemical studies of metalloporphyrins, not only as mimetic compounds of the iron porphyrin unit in heme proteins but also as potential electrocatalysts[1-8]. Metalloporphyrins have been found to be applicable in both homogeneous and heterogeneous catalysis[9-11]; and, because oxygen can be reduced directly through a 4-electron pathway on some transition metal porphyrins, catalysis in the heterogeneous electrochemical oxygen reduction reaction has received particular attention[3,7,8,12]. The application of metalloporphyrins to heterogeneous electrocatalysis requires their attachment to solid electrodes which can be realized based on chemisorption, chemical reactions with previously functionalized electrodes, chemical reactions with a functionalized polymer, incorporation of the porphyrin with the polymer film and electrochemical polymerization.

Electrochemical polymerization offers particular advantages in that polymerized porphyrins can form electroactive, adherent and stable films on solid electrodes. Oxidative electropolymerization of several porphyrins and metalloporphyrins have been reported[6,8,13]. Special focus has been placed on amino-substituted porphyrins due to the propensity of aniline to form electroactive polymers[14]. Murray et al. reported on the electropolymerization of tetrakis(o-aminophenyl)porphyrin[6] and several para-, ortho-, and meta-substituted tetrakis(aminophenyl)porphyrins with Co as a central metal[7]. They found that poly-Co(o-NH_2)TPP films are effective catalysts for the electroreduction of oxygen in aqueous solution. Metalloporphyrin films on solid electrodes have been mainly characterized by voltammetry and resonsance Raman spectroscopy. The electrochemistry of ruthenium paradiethylamino substituted tetraphenylporphyrins recently have been investigated[15]. This study reports the ac impedance and UV-visible reflectance spectroscopic studies of paradiethylamino substituted tetraphenylporphyrin films formed via an oxidative electropolymerization process.

The porphyrins used in these studies are tetrakis(p-diethylaminophenyl)porphyrins,(p-Et_2N)TPP) with Ru, Pd, Co, and Mn as central metals (general formula M(p-Et_2N)TPP) (Figure 1a). Preliminary data of the novel meso-tetrakis[2.2]paracyclophanylporphyrin, H_2T([2.2]PCP)P and meso-

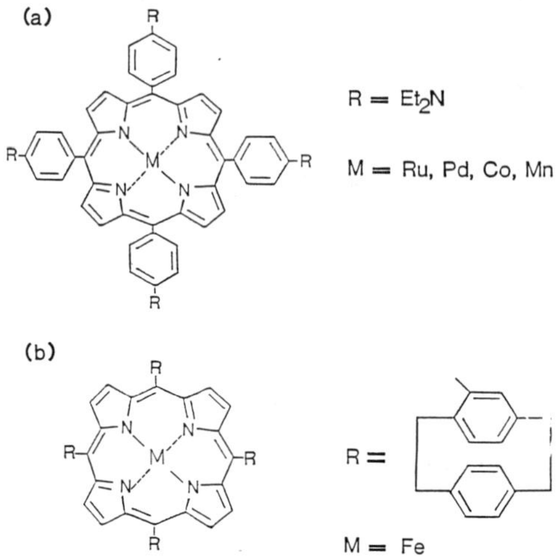

(a)

R = Et_2N

M = Ru, Pd, Co, Mn

(b)

R =

M = Fe

Figure 1. Structures of a) M(p-Et_2)TPP and b) MT([2.2]PCP)P complexes.

tetrakis[2.2]paracyclophanylporphyrin iron chloride, FeT([2.2]PCP)P(Cl) (Figure 1b) are also included in this report.

The impedance method allows polymer films which exhibit a distribution of relaxation times to be studied. Using this method, the interfacial charge injection, diffusional charge transport and charge saturation can be monitored in the film during a single experiment. Although any of several transient techniques could, in principle, be used to study polymeric films, the ac impedance method greatly simplifies data analysis because semi-infinite diffusion conditions cannot be assumed in thin film electrodes. Theoretical analysis of complex impedance data for processes involving layers of finite thickness in the study of solid state WO_3 films was developed by Ho et al.[16].

EXPERIMENTAL

Materials

The syntheses of (p-Et_2N)TPPRu(CO)(t-Bupy) and (p-Et_2N)TPP with Pd, Co, Mn and Fe were according to the procedure of Eaton and Eaton[17]. H_2T([2.2]PCP)P was obtained from the reaction of [2.2]paracyclophane-carbaldehyde with pyrrole[18]. Tetraethylammonium perchlorate(TMAP) and tetrabutylammonium perchlorate (TBAP) (both from Kodak) were recrystallized from water and methanol, respectively, and dried in vacuo. Spectroquality dichloromethane ($MeCl_2$), dichloroethane ($EtCl_2$), and acetonitrile (AN) (Aldrich) were stored over 4-Å molecular sieves. All other reagents were used as received.

Instrumentation

Cyclic and differential pulse voltammetric measurements were obtained from either 0.1 M TBAP/$MeCl_2$ or 0.1 M TBAP/$EtCl_2$ using a conventional three-electrode configuration and an IBM 225 voltammetric analyzer. A saturated calomel electrode (SCE) was used as a reference electrode in all measurements. Platinum disk working electrodes of area 0.03 cm^2 and 0.50 cm^2, used in voltammetric measurements and in reflectance spectroscopy measurements, respectively, were polished sequentially with 1.0, 0.3 and 0.05 µm alumina before use. UV-visible spectroelectrochemistry was performed in a thin-layer cell following the design of Lin et al.[19] with a calculated path length of 0.5 mm and platinum mesh as a working electrode. In-situ diffuse reflectance spectroscopy was accomplished in a special cell with two quartz windows and a platinum disk electrode of area 0.50 cm^2. Both cells were coupled with a Tracor Northern 1710 optical spectrometer multichannel analyzer to obtain time resolved spectra. One hundred rapid scan spectra (2.5 ms) were taken and averaged to give a final spectrum recorded as log R_o/R where R_o is the electrolyte/solvent without the sample or film and R is the electrolyte/solvent system with the sample or film. The chemical changes occurring in the bulk as well as in the films were monitored spectroelectrochemically from 300-900 nm at a potential scan rate from 2-5 mV/s or at a constant potential.

The ac impedance measurements of the porphyrin films were performed using a Solartron 1250 Frequency Response Analyzer and a PAR Model 173 potentiostat in the frequency range from 25.6 KHz to 20.0 mHz at constant dc potential, chosen based on voltammetric measurements of the film in the range between 0.0 and 1.45 V. All ac impedance experiments were run in 0.1 M TBAP/$MeCl_2$ solutions, purged with nitrogen before each run and blanketed with nitrogen during the course of the experiment. The data from each experiment was stored and plotted on a Hewlett-Packard 7470A plotter in the form of a Nyquist plot, i.e., complex impedance plot of $Z_{imaginary}$ vs Z_{real}.

The polymeric films on Pt electrodes were formed by oxidative polymerization of metalloporphyrins (from $EtCl_2$ or $MeCl_2$, 0.1 M TBAP solution) at a constant potential between 1.3 and 1.65 V for varying time periods from 1 to 15 minutes. A PAR Model 173 potentiostat and PAR Model 179 coulometer were used for the electrodeposition process and control of film thickness. Very thin films were also formed by scanning the potential through the oxidation waves of the given metalloporphyrin from 0.0 to 1.65 V at a scan rate of 100 mVs^{-1}.

For morphology studies, the films were coated with ca. 10 nm of gold and examined with a Philips PSEM 500 scanning electron microscope.

RESULTS AND DISCUSSION

Voltammetry and Film Formation

Differential pulse voltammograms for the oxidation of the M(p-Et_2N)-TPP complexes at a Pt electrode in $MeCl_2$, 0.1 M TBAP are shown in Figure 2. The species undergo up to four oxidations (peaks I - IV). Peak potentials for these processes are presented in Table 1. The potentials for the first and second oxidation of Ru(CO)(p-Et_2N)TPP(t-Bupy) at E_{pa} = 0.50 V and E_{pa} = 0.72 V, respectively, are the most negative in the group and have been assigned previously to the formation of the cation radical and dication, respectively[15], based on data from coulometry, ESR and thin-layer spectroelectrochemistry. Peaks I and II for the manganese and palladium complexes, although overlapping, are clearly distinctive while only one, broad, peak with a shoulder is observed for the Co complex at E_{pa} = 0.62 V. Coulometric measurements show a transfer of two faradays per mole of Co(p-Et_2N)TPP in this process. The same amount of charge is also transferred during the first and second oxidation for the Mn(p-Et_2N)TPP and Pd(p-Et_2N)TPP complexes. An analysis of the peak potentials as well as the peak currents of the third wave, appearing around 1.10 V, show them to

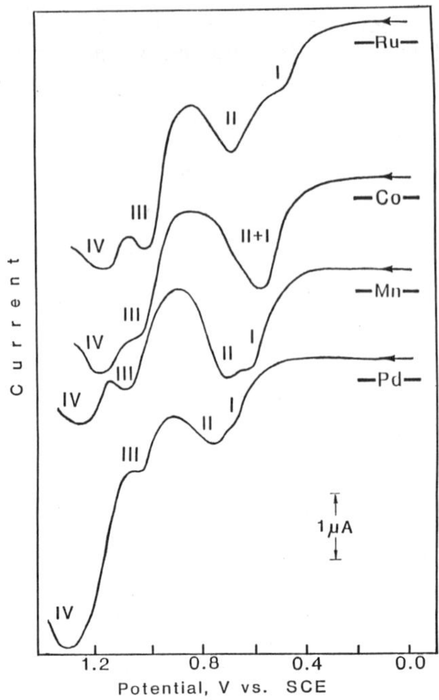

Figure 2. Differential pulse voltammograms of M(m-Et_2N)TPP complexes at a Pt electrode in $MeCl_2$, 0.1 M TBAP.

depend to a large extent on the conditions of the experiment, i.e., solvent, supporting electrolyte, and potential scan rate. At a potential about 1.20 V, a very broad fourth peak is evident. Analysis of the

Table 1. Potentials of Differential Pulse Voltammetric Peaks for Oxidiation of Various Metalloporphyrins in $MeCl_2$, 0.1 M TBAP Solution (scan rate 5 mVs^{-1}, amplitude 10 mV and t_p= 2s)

Compound	Peak Potential, V vs SCE			
	Peak I	Peak II	Peak III	Peak IV
Ru(CO)(p-Et_2N)TPP(t-Bupy)	0.50	0.72	1.04	1.18
Co(p-Et_2N)TPP[c]	0.62[a]	-	1.10	1.26
Mn(p-Et_2N)TPP[c]	0.73	0.80	1.20	1.35
Pd(p-Et_2N)TPP	0.74	0.80	1.09	1.40
H_2([2.2]PCP)P	0.54	0.87	1.08	1.28
Fe([2.2]PCP)P(Cl)	0.96	1.23	1.50[b]	-

[a]peak I and II
[b]peak III and IV
[c]ClO_4^- - axial ligand for oxidation state (III) of metal

dependence of peak current of peaks III and IV and peak width at half-peak current on concentration and scan rate indicated that these peaks are strongly influenced by adsorption. The potential for the third oxidation of $M(p\text{-}Et_2N)TPP$, about 1.10 V, is very similar to the potential for the oxidation of aniline in Cl_2 solution (1.06 V) which leads to polymerization and film formation[14] and, therefore, peaks III and IV can be assigned to the oxidation of the diethylamino substituents.

Figures 3 and 4 show the growth patterns for Ru, Co, Pd and Mn (p-diethylaminophenyl) porphyrins in continuous scan cyclic voltammetry. The cyclic voltammograms of $M(p\text{-}Et_2N)TPP$ are complex and clearly indicate film formation on the surface of the platinum electrode. No polymer films were formed on successive cycling if the potential sweep was not extended beyond 0.9 V. However, if the sweep was extended beyond 0.9 V, rapid polymerization and film formation of $M(Et_2N)TPP$ were seen. In the range of potential scanned between 0.0-0.9 V, two reversible couples (peaks I and II) are observed due to the formation of the cation and dication, respectively. In continuous scanning of the potential in this range, over the time period of at least 15 min, no film formation was noted which demonstrates that the formation ofthe porphyrin radical or dication is generally insufficient to induce polymerization of the $p\text{-}Et_2N$ groups. This is in direct contrast with the data obtained for tetrakis(o-, m- and p-aminophenyl) porphyrins with cobalt as a central metal[7], where polymerization was observed during the first oxidation process at about 0.6 V. However, in the case of the $M(p\text{-}Et_2N)TPP$ complexes, scanning potentials between 0.0 and 0.6 V, for a time period longer than 30 min, produces a small amount of polymeric species. The yield of this film formation process was minimal in comparison with the yield obtained during processes III and IV; and, therefore, can be attributed to a secondary chemical reaction(s) which follow a cation radical decay. When $M(p\text{-}Et_2N)TPP$ is polymerized, the cyclic voltammograms show both anodic and cathodic peaks III and IV growing with successive potential scan. The most rapid increase of the film thickness

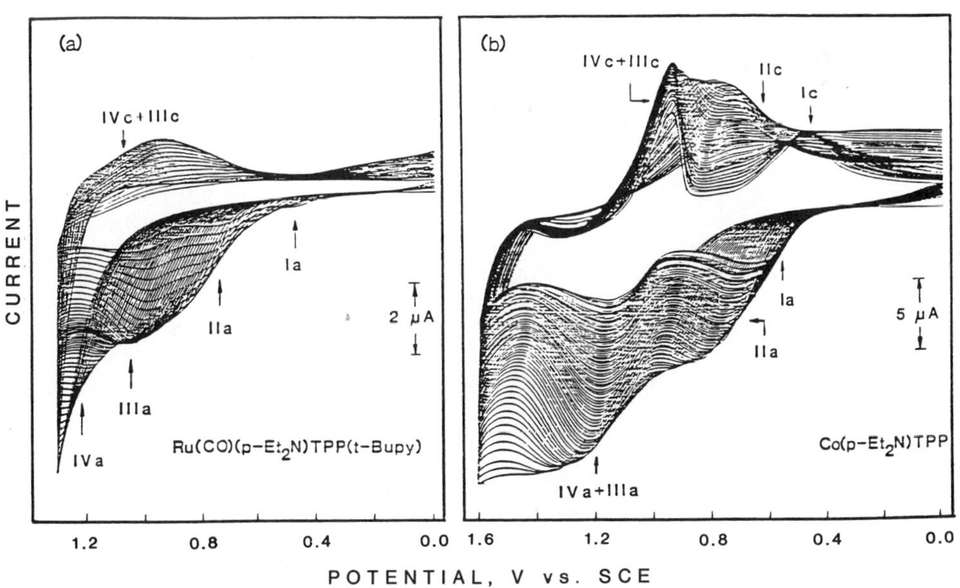

Figure 3. Cyclic voltammetric scans (100 mVs^{-1}) at a Pt electrode in $MeCl_2$, 0.1 M TBAP solution: a) 0.15 mM $Ru(CO)(p\text{-}Et_2N)TPP$-(t-Bupy); b) 0.15 mM $Co(p\text{-}Et_2N)TPP$.

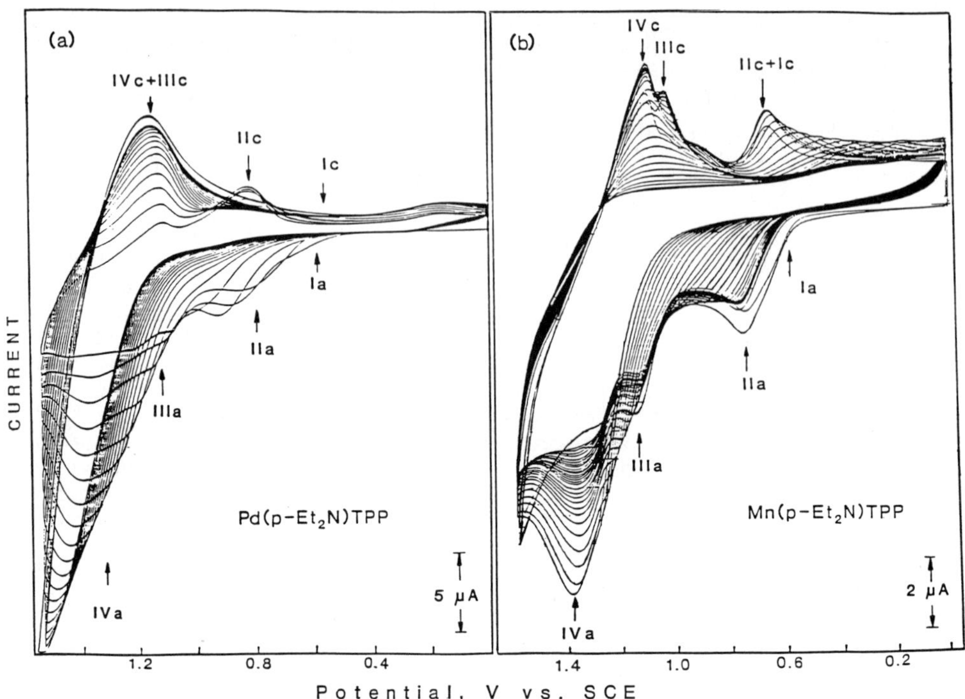

Figure 4. Cyclic voltammetric scans (100 mVs^{-1}) at a Pt electrode in $MeCl_2$, 0.1 M TBAP solution: a) 0.20 mM $Pd(p-Et_2N)TPP$; b) 0.20 mM $Mn(p-Et_2N)TPP$.

was observed for $Pd(p-Et_2N)TPP$, and the relative current increase (i(5) - i(1)/i(1), where 1 and 5 indicate the number of scans) of peaks III and IV, in four consecutive scans following the first scan (at a scan rate of 100 mVs^{-1}) was 235%, Figure 4a. The relative current increase of 145% for $Ru(p-Et_2N)TPP(t-Bupy)$ was smaller than that observed for $Pd(p-Et_2N)TPP$, Figure 3a. A significally slower increase of polymer growth was observed for $Mn(p-Et_2N)TPP$, which showed only a 30% increase of current for peak IV after five scans, in comparison to the current recorded on the first scan, Figure 4b. The slowest process of film formation was observed for $Co(p-Et_2N)TPP$ which was manifested by only a 20% increase of current after the fifth scan, Figure 3b. In thin polymer films, the original four redox couples I-IV can be detected in $TBAP/MeCl_2$, no doubt reflecting significant interactions among the various redox sites in the polymer. When the thickness of the polymer film increases, a positive shift of anodic potential and negative shift of cathodic potential with a simultaneous decrease in peak current is observed for peaks I and II. This is typical behavior indicating increasing resistance to charge propagation in the film. Peaks I and II are no longer observable after approximately 20 scans for the Ru, Co, Pd and Mn polymeric porphyrins.

In thick films, only the anodic process, IVa, with a very broad peak for the reduction process, IVc, are observed on the cyclic voltammetric time scale of 100 mVs^{-1}. During film formation on continuous scanning, the anodic peak IVa of $Co(p-Et_2N)TPP$ moves from its original position of 1.26 to 1.35 V, while the anodic peak IV of $Mn(p-Et_2N)TPP$ moves from 1.35 to 1.40 V. Similarly, peak IVa of $Pd(p = Et_2N)TPP$ moves from 1.40 to 1.58 V, while the anodic peak of $Ru(CO)(p-Et_2N)TPP(t-Bupy)$ moves from 1.04 to 1.35 V. It seems logical to assume the polymerization mechanism in

M(p-Et_2N)TPP complexes, which occurs under processes III and IV, is similar to the aniline-like polymerization mechanism suggested previously[7]. The formation of a nitrogen base salt rather than a potentially highly reactive carbonium ion is believed to be responsible for the high chemical stability of the polymeric aniline film[14]. Although, aniline oxidative polymerization proceeds via para-substitution, in the case of M(p-Et_2N)TPP complexes, para polymerization is impossible since the para position is occupied by the porphyrin. Therefore, the polymerization of M(p-Et_2N)TPP, most likely, proceeds via ortho or meta substitution to produce a polymer with a positive charge located on the nitrogen atom of the Et_2N group. This positive charge is neutralized by the presence of perchlorate ions from the supporting electrolyte. Peaks III and IV are due to the oxidation of diethylamino substituents which is followed by polymerization and film formation. The oxidation which occurs under process IV is more difficult than process III. This may be due to one or both of the following: a shift of the oxidation potential of the remaining diethylamino substituents after the first one is oxidized or a dramatic change of polymeric film structure during the deposition process. In the case of a dramatic change in the film structure, a calculation of the "critical" number of layers would be about 40, assuming four Et_2N are oxidized, or about 160 assuming only one is oxidized. Since the charge transferred under peak III is similar to that observed not only for peak I but also for peak II, and both have been shown to be one-electron transfer processes, process III probably involves an oxidation of only one Et_2N substituent in the porphyrin molecule. Detailed studies of the structure of these polymeric films as well as optimization of the deposition process and role of the electrolyte anion are underway and will be published separately[20].

The oxidation of the novel paracyclophanyl porphyrin-H_2T([2.2]PCP)P also showed four peaks. The differential pulse voltammogram of this compound, obtained in $EtCl_2$, 0.1 M TBAP solution, is shown in Figure 5, while peak potentials for both H_2T([2.2]PCP)P and T([2.2]PCP)PFeCl are listed in Table 1. Figure 5 shows three one-electron transfer processes (peaks I - III) which are all reversible on the cyclic voltammetric time scale. Although peak IV is also a one-electron process, it is irreversible. While peaks I and II are due to the formation of the cation radical and dication, respectively, processes III and IV are due to the oxidation

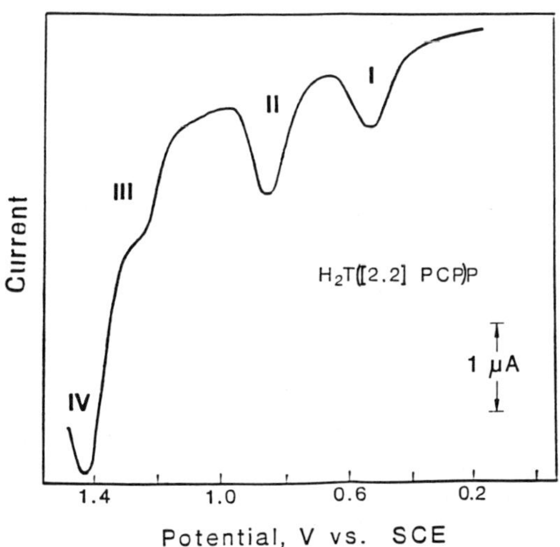

Figure 5. Differential pulse voltammogram of H_2T([2.2]PCP)P at a Pt electrode in $EtCl_2$, 0.1 M TBAP solution.

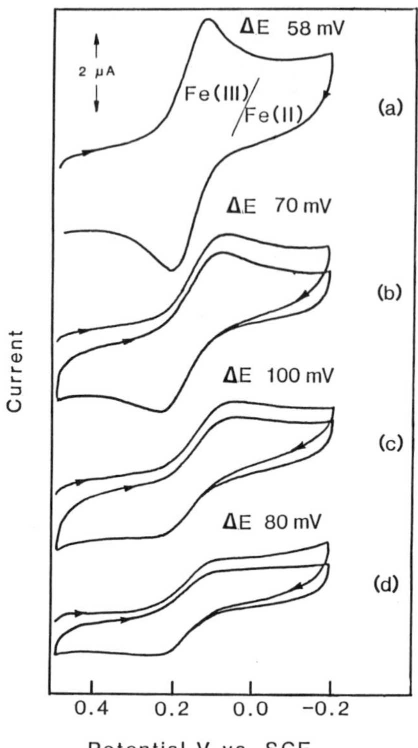

Figure 6. Cyclic voltammograms (100 mVs^{-1}) for $Fe(CN)_6^{3-}/Fe(CN)_6^{4-}$ at at Pt/porphyrin film electrodes in an aqueous 0.1 M KCl solution containing 5 x 10^{-4} M $K_3Fe(CN)_6$: a) Pt electrode; b) Pt/30 monolayers of H_2T([2.2]PCP)PCl; c) Pt/40 monolayers of FeT([2.2]PCP)P; d) Pt/267 monolayers of Ru(CO)(p-Et_2N)-TPP(t-Bupy).

of two paracyclophane substituents. Polymerization and film formation follow the oxidation of the second paracyclophane substituent. A similar film formation was observed for FeT([2.2]PCP)P(Cl). The mechanism of polymerization as well as a detailed electrochemical characterization of these porphyrins are being published separately[21].

The cyclic voltammograms of the redox couple $Fe(CN)_6^{4-/3-}$ obtained on platinum electrodes modified with Ru(CO)(p-Et_2N)TPP(t-Bupy), H_2T([2.2]PCP)P and FeT([2.2]PCP)P(Cl) polymeric films are shown in Figure 6. The peak separations E_{pa}-E_{pc} obtained ranged from 70-100 mV for those porphyrins but this factor depends to a greater extent on the polymer thickness as well as on the type of porphyrin and central metal. Electrodes modified with Pd(p-Et_2N)TPP and Mn(p-Et_2N)TPP show the cyclic voltammograms of the $Fe(CN)_6^{4-/3-}$ couple were similar to that obtained on Pt with peak separations E_{pa}-E_{pc} approaching 58 mV for very thin films.

Preliminary studies of the catalytic activity of these films in the reduction of oxygen were performed in both basic and acidic aqueous solutions. All polymeric porphyrins studied, catalytically reduced O_2 to H_2O_2 or to water. The best catalytic activity for the reduction of O_2 directly to water was shown by Co(p-Et_2N)TPP, Mn(p-Et_2N)TPP, and Pd(p-Et_2N)TPP polymers in 0.1 M H_2SO_4 and the Fe([2.2]PCP)PCl polymeric

Table 2. Peak Potentials for Electroreduction of Dioxygen (O_2-Saturated Solution of 0.5 M H_2SO_4 or 1 M NaOH) on Various Metalloporphyrin Films

Electropolymerized Monomer	Electrolyte	E_{pc} V vs SCE
Co(p-Et_2N)TPP	H_2SO_4	0.38
Co(p-NH_2)TPP	H_2SO_4	0.18[a]
Co(o-NH_2)TPP	H_2SO_4	0.36[a]
Ru(CO)(p-Et_2N)TPP(t-Bupy)	H_2SO_4	0.32
Pd(p-Et_2N)TPP	H_2SO_4	0.26
Fe([2.2]PCP)P(Cl)	NaOH	-0.30

[a]Taken from Reference 7

film in 0.5 M NaOH. The peak potentials for the reduction of dioxygen from O_2 saturated 0.5 M H_2SO_4 in 1 M NaOH solution are shown in Table 2. The data obtained and presented in Table 2 show the shift of the potential for oxygen reduction on polymeric Co(p-Et_2N)TPP to be comparable to that obtained for both the Co(m-NH_2)TPP and Co(p-NH_2)TPP systems[7].

The films formed from M(p-Et_2N)TPP as well as from [2.2]paracyclophanylporphyrins solutions were stable in $MeCl_2$, Et_2Cl_2, CH_3CN, $(CH_3)_2CO$, 0.5 M H_2SO_4 and 1 M NaOH. The cyclic voltammograms obtained in these solvents showed the films to be conductive throughout the potential range of the porphyrin in solution and depending on the film thickness, could provide redox potentials similar to those seen in the bulk solutions. Films are insoluble in all common organic solvents.

UV-Visible Spectroscopy-Absorption and Diffuse Reflectance

Thin-layer spectroelectrochemistry was carried out on both the (p-diethylaminophenyl)porphyrins and the [2.2]paracyclophanylporphyrins prior to diffuse reflectance spectroscopy. Spectral changes accompanying redox reactions of M(p-Et_2N)TPP are abbreviated in Figures 7, 8, and 9 as [MP], $[MP]^{+}$, $[MP]^{2+}$ and $[MP]^{n+}$ where [MP] denotes the metalloporphyrin, $[MP]^{+}$ and $[MP]^{2+}$ denote the ring cation radical and dication, respectively, and $[MP]^{n+}$ denotes the porphyrin film with oxidized diethylamine groups. These examples were chosen to illustrate the variances in the spectral changes occurring within this homologous series of metalloporphyrins. For clarity, the reactions are shown beside each spectrum.

Thin-layer spectroelectrochemistry of M(p-Et_2N)TPP shows typical changes of spectra expected for the generation of the cation radical and dication in the first and second oxidation step (process I and II), i.e., decrease of the Soret peak along with the appearance of a new band between 600-800 nm, in the formation of the ring cation radical; a further decrease in the Soret peak, along with broadening and a bathochromic shift in the formation of the ring dication. The continued decrease in the Soret

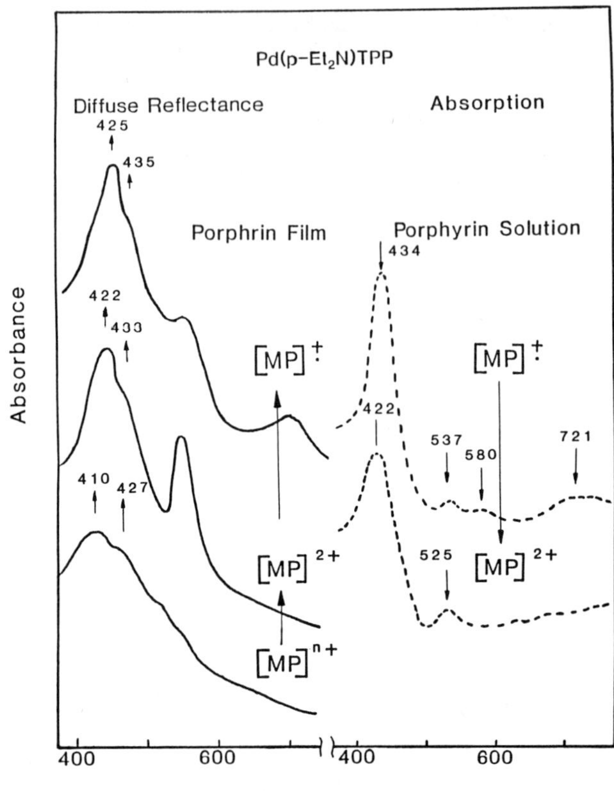

Figure 7. UV-visible diffuse reflectance and absorption spectra of Pd(p-Et_2N)TPP in CH_3CN, 0.1 M TBAP solution. (Controlled-potential electroreduction of the film at 0.7 V and controlled-potential electrooxidation of solution at 0.9 V.)

peak and a noticeable shift to shorter wavelengths are evident for oxidation of the diethylamine group.

Films for diffuse reflectance spectroelectrochemistry were formed on Pt electrodes of area 0.50 cm^2 by applying a constant potential of 1.30 - 1.45 V. After film formation, the electrode was rinsed in solvent and placed in the corresponding electrolyte/solvent system before placement in the cell. The thickness of the film for these experiments ranged from approximately 100-200 monolayers. A comparison of the absorption versus the reflectance spectra of both Pd(p-Et_2N)TPP and Ru(CO)(p-Et_2N)TPP-(t-Bupy) are shown in Figures 7 and 8, respectively.

A primary goal of this experiment was to use diffuse reflectance for evaluation of thicker porphyrin films which cannot be studied by using transparent conductive electrodes and absorption spectroscopy. The main concern was whether the integrity of the porphyrin system had been retained in the electropolymerization scheme and whether the reactions taking place in the film would be comparable to those taking place in the bulk. The cyclic voltammograms of the films in solution suggested that both of these facts were true.

Comparing the results obtained from diffuse reflectance of Pd(p-Et_2N)TPP and Ru(CO)(p-Et_2N)TPP(t-Bupy) films with absorption spectroelectrochemistry of these porphyrins in solution show a relationship

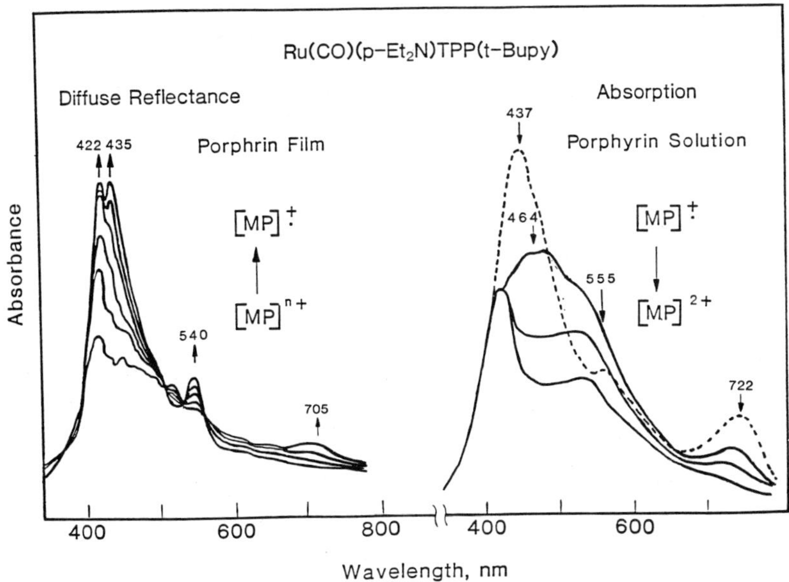

Figure 8. UV-visible diffuse reflectance and absorption spectra of $Ru(CO)(p-Et_2N)TPP(t-Bupy)$ in CH_3CN, 0.1 M TBAP. (Controlled-potential electroreduction of the film at 0.6 V and controlled-potential electrooxidation of the solution at 0.84 V.)

between the two systems, not only in the shape of their peaks but in their position shifts as oxidation is occurring.

Diffuse reflectance spectra obtained for polymeric $Pd(p-Et_2N)TPP$ as well as for $Ru(CO)(p-Et_2N)TPP(t-Bupy)$ show a broad peak in the range 380-580 nm without major features except small shoulders at 410, 427 and 415, 435 nm for the Pd and Ru film, respectively (Figure 7 and 8).

A potential step reduction of the completely oxidized polymeric film first produced a spectrum charcteristic of the dication and, with further reduction, a spectrum of the cation radical with a typical broad peak around 600-800 nm. A split of the Soret peak is observed for both the $Pd(p-Et_2N)TPP$ and $Ru(CO)(p-Et_2N)TPP(t-Bupy)$ films. When observed for porphyrins in solution, such splitting is usually an indication of porphyrin aggregation, therefore, it is not unexpected that a similar effect be seen in polymeric film. Spectral changes of oxidized/reduced film obtained in acetonitrile were very similar to that observed in $EtCl_2$

Diffuse reflectance spectra of $Pd(p-Et_2N)TPP$ and $Ru(CO)(p-Et_2N)TPP$-(t-Bupy) in aqueous solution and 0.1 M TEAP are presented in Figure 9. These spectra indicate changes from the polymeric dication to the polymeric radical of $M(p-Et_2N)TPP$ when the potential is stepped down from 0.90 V to 0.64 V. Radical formation as well as dication formation in polymeric films in aqueous solutions are reversible not only in the cyclic voltammetric time scale but also in the diffuse reflectance spectroelectrochemical time scale. The spectral position of the Soret band in aqueous solution is shifted to longer wavelengths due to the change of axial coordination in the presence of water. No passivation of the modified electrode was observed in the potential range 1.1 and -1.0 V.

Diffuse reflectance FTIR spectra of $M(Et_2N)TPP$ polymeric films closely resemble those of corresponding monomers and are influenced by

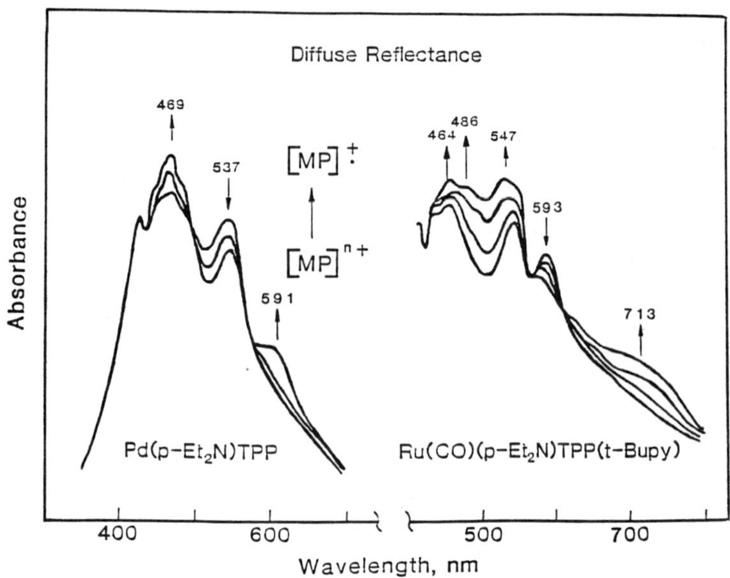

Figure 9. UV-visible diffuse reflectance spectra of polymeric Pd(p-Et_2N)TPP and Ru(CO)(p-Et_2N)TPP(t-Bupy) in aqueous 0.1 M TEAP solution. (Electroreduction at 0.6 V.)

bands of adsorbed (occluded) tetrabutylammonium cations which are difficult to remove from the surface of the polymeric films. In the region below 700 cm^{-1}, bands which could be attributed to the ClO_4^- counter ions were observed. Diffuse reflectance UV-visible and FTIR of T([2.2]paracyclophane) porphyrin films were not obtainable due to high specular reflectance. This was an indication that the surface of these films were very smooth which was later confirmed by scanning electron microscopy.

Morphology of Films

A scanning electron micrograph (Plate 1a) of polymeric Ru(CO)(p-Et_2N)-TPP(t-Bupy) grown electrochemically on a platinum electrode at a constant potential of 1.45 V for 10 min. showed a relatively smooth surface with many craters. A closer look at the surface inside the crater showed a relatively even compact microspheroid surface morphology (Plate 1b). The surfaces of Co(p-Et_2N)TPP and Pd(p-Et_2N)TPP films obtained under similar conditions were rougher than observed for Ru(CO)(p-Et_2N)TPP(t-Bupy) but without craters (Plate 2a and b). Microspheroids are larger resulting in a less compact morphology. The surface of the H_2([2.2]PCP)P film, obtained at 1.60 V, is very smooth with no microspheroidical features visible under the identical electron microscopic conditions used for the observation of M(p-Et_2N)TPP surfaces.

AC Impedance

The methodology of ac impedance for the analysis of charge propagation processes involving layers of finite thickness was comparable to that proposed by Ho et al.[16], and later modified by White et al.[22]. The electrochemical reaction scheme for the reduction of oxidized polymeric film of M (p-Et_2N)TPP can be written as

$$\underset{\text{polymer}}{[M(p\text{-}Et_2N)TPP]^{n+}nClO_4^-} + 1e^- \rightleftharpoons \underset{\text{polymer}}{[M(p\text{-}Et_2N)TPP]^{(n-1)}} + (n-1)ClO_4 + ClO_4^- \quad (1)$$

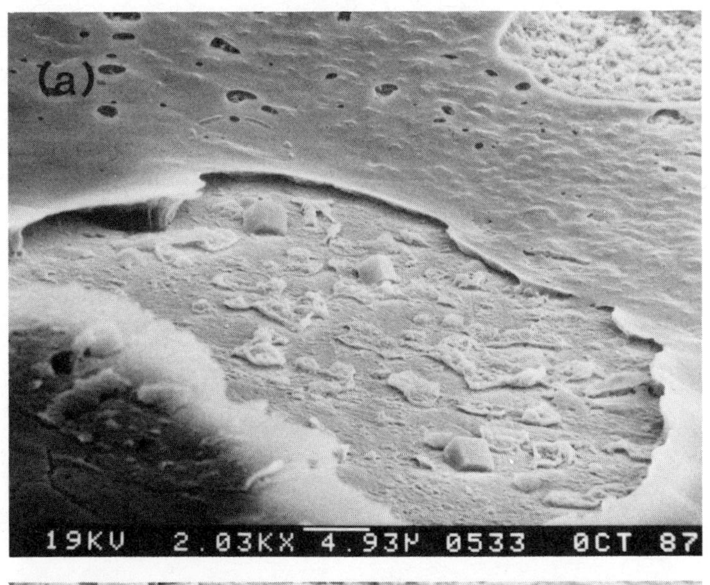

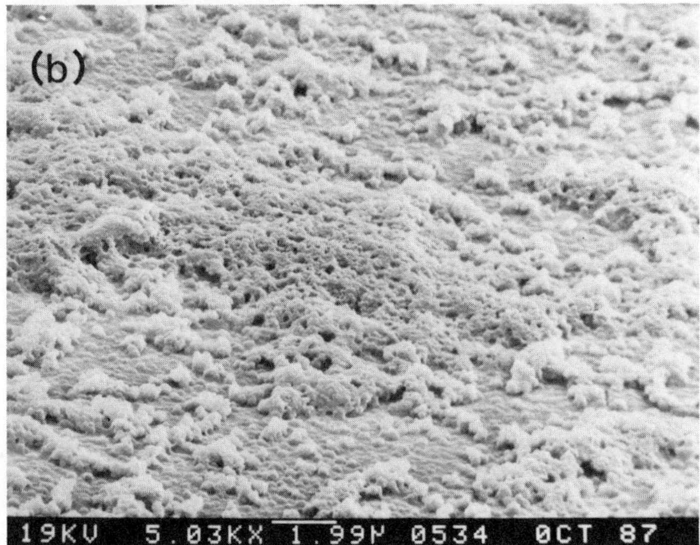

Plate 1. Scanning electron micrograph of the film of Ru(CO)(p-Et_2N)TPP-(t-Bupy) grown electrochemically on a platinum electrode at constant potential 1.45 V for 10 min.: a) 2000:1; b) 5000:1; from inside the upper right crater which appears in (a).

The overall process of the reduction of this polymeric film will involve an electron injection from the platinum electrode to the film, charge propagation through the film and counterion-perchlorate ion diffusion to maintain film neutrality.

Typically, the data obtained from impedance measurements is plotted out in a complex plane diagram of capacitive impedance ($Z_{imaginary}$) vs faradaic impedance (Z_{real}), Figure 10. At high frequency, the rate of charge injection to the polymeric film may be a controlling factor for charge progagation through the film in which case the complex plane plot

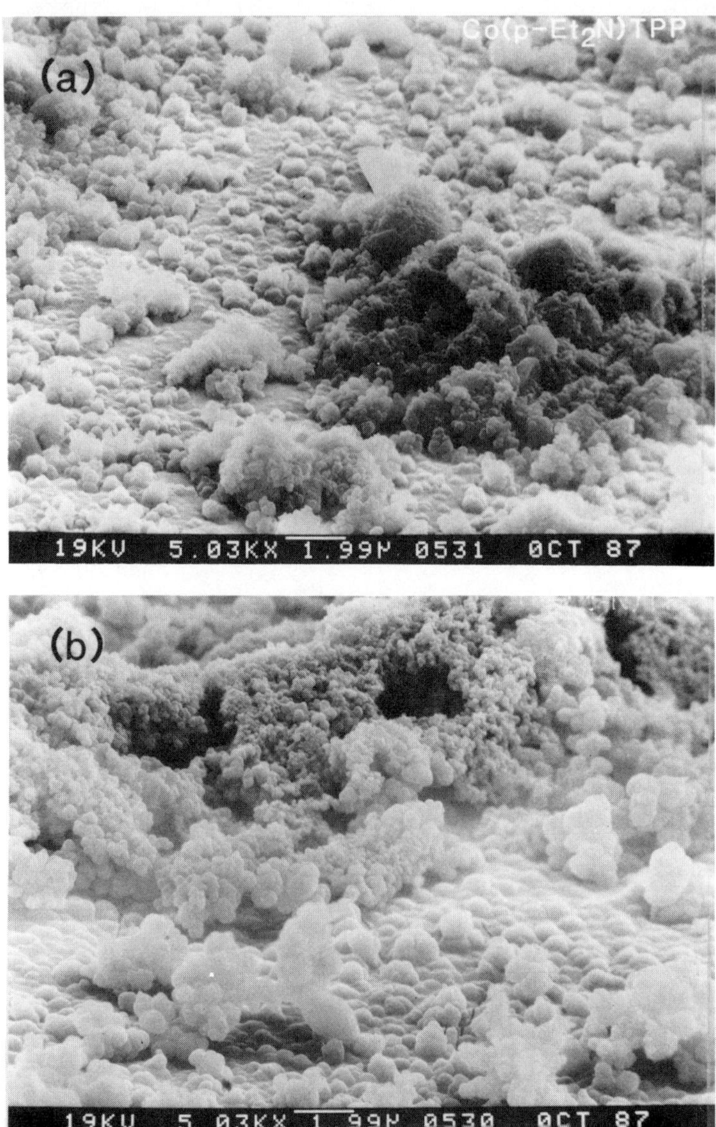

Plate 2. Scanning electrode micrograph of the film of a) CO(p-Et_2N)TPP and b) Pd(p-Et_2N)TPP obtained at constant potential 1.45 V for 10 min.

will be a semicircle in the high frequency region. The injection-charge transfer resistance R_{ct} was determined from the diameter of the semicircle; but if only a part of the semicircle was observed, the diameter was obtained by extrapolation of the circular part of the diagram to the axis, Z_{re} (Figure 10). The resistance of the electrode-film-electrolyte system R_s, is shown in the complex plane as a difference between the zero origin of the plot and intercept of the semicircle with the axis, Z_{re} at high frequencies. The double layer capacitance was calculated from the frequency at which impedance is a maximum on the axis, Z_{im} by using the relation

$$\omega = 1/R_{ct}C_{dl} \tag{2}$$

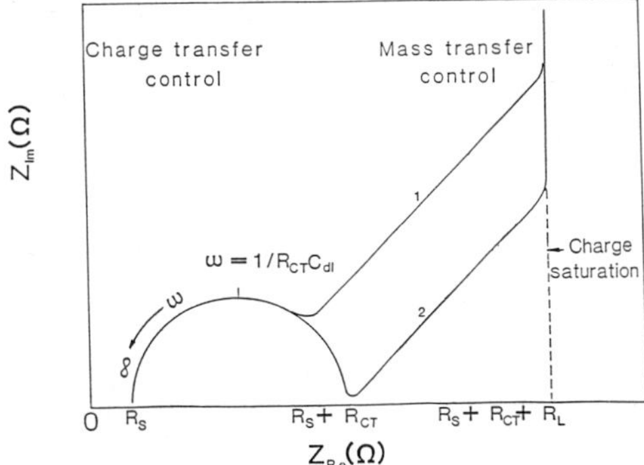

Figure 10. Complex impedance plot expected for thin layer redox polymeric film on solid electrode; 1 and 2 represent mass transfer controlled region for different thickness of the film.

where $\omega = 2\pi f$. At lower frequencies, the charge propagation becomes diffusion controlled and the complex plane plot becomes linear with a slope of $\pi/4$. The diffusion coefficient for charge transfer, D_{ct} was determined by the method described by Ho et al.[16] and compared with the results obtained through the relation

$$D_{ct} = L^2/2R_{ct}C_{dl} \tag{3}$$

where L is a thickness of polymer film. At very low frequency, ($\omega < 1$ Hz), diffusion of the charge across the finite film may be accomplished during one-half cycle of the perturbing signal. At this point, the real component of the impedance will reach a maximum and the phase angle approches $\pi/2$. This leads to a capacitive-like behavior (charge saturation) characterized by a polymer redox capacitance C_L.

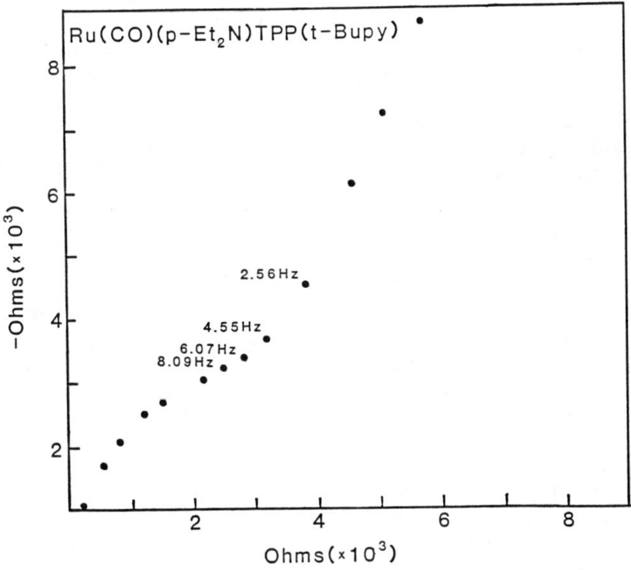

Figure 11. Complex impedance plot for Pt/Ru(CO)(p-Et_2N)TPP(t-Bupy) electrode in $MeCl_2$ containing 0.1 M TBAP; potential 1.16 V, surface coverage 7.1×10^{-8} mol cm^{-2}.

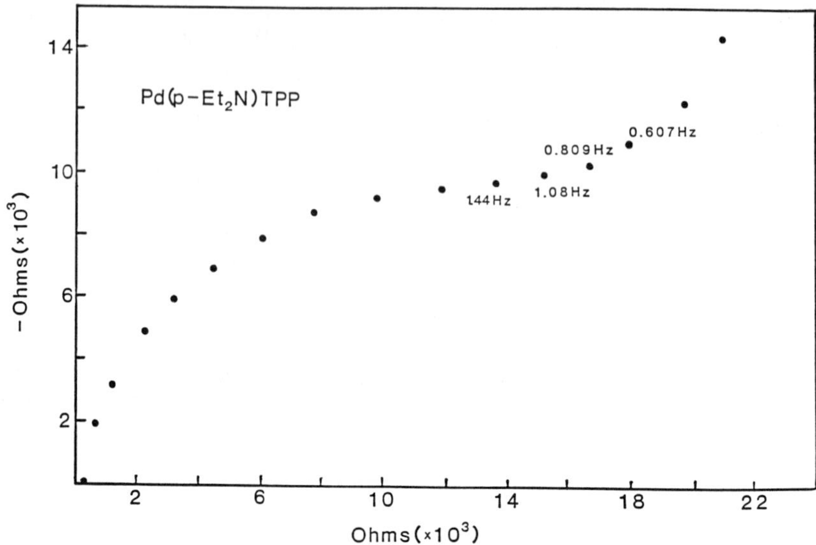

Figure 12. Complex impedance plot for Pt/Pd(p-Et_2n)TPP in $MeCl_2$ containing 0.1 M TBAP; potential 1.30 V, surface coverage 4.7×10^{-8} mol cm^{-2}.

The complex impedance plots of Ru(CO)(p-Et_2N)TPP(t-Bupy) and Pd(p-Et_2N)TPP films are shown in Figures 11 and 12, respectively. The impedance measurements were taken in 0.1 M TBAP/$MeCl_2$ over the frequency range 20 mHz to 25.6 KHz at a constant applied potential E = 1.13 V and 1.40 V, respectively. (E was determined from the cyclic voltammogram of the film.) Thickness of the films studied varied from 296 to 479 monolayers.

Both the film thickness and the charge-injection process effectively limit the range of frequencies over which charge transfer kinetics, diffusion control or charge saturation are observed. For thin films, the kinetic control region may overlap the charge saturation region; for thick films, charge saturation is observed only at very low frequencies and diffusional control dominates over a large frequency range[16,22]. The complex impedance plots obtained for relatively thick films (730 and 479 monolayers for Ru(CO)(p-Et_2N)TPP(t-Bupy) and Pd(p-Et_2N)TPP, respectively) exhibited the two regions of impedance behavior predicted: extensive diffusion control region and kinetic control region. For thinner films (below 80 layers) a charge saturation region was observed which was overlapped with a charge transfer control region. The impedance analysis of a number of polymeric Ru(CO)(p-Et_2N)TPP(t-Bupy) and Pd(p-Et_2N)TPP films with different thicknesses was performed and the typical examples are presented in Table 3. The polymeric film oxidation state was controlled by varying the potential, E, to the Pt/M(p-Et_2N)TPP electrode over the range 0.90-1.40 V. The diffusion coefficient for charge transfer is higher for the Ru(CO)(p-Et_2N)TPP(t-Bupy) film than that for the Pd(p-Et_2N)TPP film. D_{ct} in the range of 0.31 - 1.90×10^{-9} cm^2 s^{-1} calculated for the later one is similar to that observed for the poly(vinylferrocene) film coated on a Pt electrode (0.1 - 1.80×10^{-9} cm^2 s^{-1})[22]. However, a charge transfer resistance of 4 - 29×10^3 ohms was found to be higher than that found for the poly(vinylferrocene) film (300-2600 ohms). The redox capacitance C_L estimated based on ac impedance data was 3.25×10^{-4} F and 3.40×10^{-4} F for polymeric Ru(CO)(p-Et_2N)TPP(t-Bupy) (730 monolayers) and Pd(p-Et_2N)TPP (479 monolayers), respectively.

Table 3. Impedance Data for Pt/Polymeric Porphyrin Electrodes in $MeCl_2$, 0.1 M TBAP Solution

Electropolymerized Monomer	Potential V vs SCE	(x 10^8 mol cm^{-2})
Ru(CO)(p-Et_2N)TPP-(t-Bupy)	0.90	2.89
	1.16	7.13
Pd(p-Et_2N)TPP	1.30	3.12
	1.30	4.68
	1.40	3.12
	1.40	4.68

Electropolymerized Monomer	Number of Monolayers	io (μAcm^{-2})	R_{ct} (x 10)	D_{ct} (x $10^9 cm^2 s^{-1}$)	C_{dl} (uFcm^{-2})
Ru(CO)(p-Et_2N)TPP-(t-Bupy)	296	86.6	9.46	8.90	4.70
	730	130.0	6.27	7.35	5.20
Pd(p-Et_2N)TPP	319	47.5	17.24	0.31	8.50
	479	27.9	29.32	0.82	4.30
	319	210.0	3.87	1.54	7.70
	479	90.5	9.04	1.90	5.90

CONCLUSIONS

This study presents evidence to show that M(p-Et_2N)TPP and MT([2.2]PCP)P can undergo oxidative polymerization leading to film formation on platinum electrodes. Stable adherent metalloporphyrin films thus formed show good electroactivity over substantial thicknesses (over 800 monlayer equivalents) and are stable in most common nonaqueous solvents as well as in aqueous solution. UV-visible diffuse reflectance spectroscopy shows M(p-Et_2N)TPP polymeric films can be reversibly reduced to form dications and cation radicals.

All of the films studied show catalytic activity in the reduction of dioxygen to H_2O_2 or H_2O. Impedance analysis was found to be applicable in the determination of the charge transfer diffusion coefficient, charge transfer resistance, and in the double layer capacity of M(p-Et_2N)TPP films.

REFERENCES

1. T. Kuwana, M. Fujihira, K. Sunakawa and T. Osa, J. Electroanal. Chem. Intertacial. Electrochem., 83, 207 (1978).
2. A. Bettelheim, R.J.H. Chan and T. Kuwana, ibid., 99, 391 (1979).
3. A. Bettelheim, R.J.H. Chan and T. Kuwana, ibid., 110, 93 (1980).
4. D.A. Buttry and F.C. Anson, J. Am. Chem. Soc., 106, 59 (1983).

5. K.A. Macor and T.G. Spiro, ibid., 105, 5601 (1983).

6. B.A. White and R. Murray, J. Electroanal. Chem. Intertacial. Electrochem., 189, 345 (1985).

7. A. Bettelheim, B.A. White, S.A. Raybuck and R.W. Murray, Inorg. Chem., 26, 1009 (1987).

8. K.A. Macor, Y.O. Su, L.A. Miller and T.G. Spiro, ibid., 26, 2594 (1987).

9. R.J. Brodd, V.Z. Leger, R.F. Scarr and A. Kozawa, National Bureau of Standards Special Publications, 455, Proceedings of Workshop held at NBS, Gaithersburg, MD, 253 (1976).

10. C.P. Andrieux, J.M. Dumas-Bouchiat and J.M. Saveant, J. Electroanal. Chem. Intertacial. Electrochem., 114, 159 (1980).

11. J.T. Groves, in: "Catalysis of Organic Reactions"; W.R. Moser, Ed., Marcel Dekker: New York, 1981; pp. 131-138.

12. T.G. Spiro, Ed., "Metal Ion Activation of Dioxygen", Wiley-Interscience, New York, 1980.

13. K.A. Macor and T.G. Spiro, J. Am. Chem. Soc., 105, 5601 (1983).

14. W.S. Huang, B.D. Humphrey and A.G. MacDiarmid, J. Chem. Soc., Faraday Trans. I, 82, 2385 (1985).

15. T. Malinski, D. Chang, L.A. Bottomley and K.M. Kadish, Inorg. Chem., 21, 4248 (1982).

16. C. Ho, I.D. Raistrick and R.A. Haggins, J. Electrochem. Soc., 127, 343 (1980).

17. S. Eaton and G.R. Eaton, Inorg. Chem., 16, 72 (1977).

18. L. Czuchajowski and M. Lozynski, J. Heterocycl. Chem., 25, 349 (1988).

19. X.Q. Lin and K.M. Kadish, Anal. Chem., 58, 1493 (1986).

20. J.E. Bennett and T. Malinski, to be published.

21. J.E. Bennett, A.K. Wisor, M. Lozynski, L. Czuchajowski and T. Malinski, J. Am. Chem. Soc., in press.

22. T.B. Hunter, P.S. Tyler, W.H. Smyrl and H.S. White, J. Electrochem. Soc., 134, 2198 (1987).

BACTERIAL CYTOCHROME P-450 ENZYMES AND REACTIONS ON IMMOBILIZED ELECTRODES. I. PRELIMINARY STUDIES

V.L. Vilker*, F. Khan, D. Shen,
M.M. Baizer and K. Nobe

Department of Chemical Engineering
University of California
Los Angeles, California 90024

INTRODUCTION

Cytochrome P-450 is a hydroxylase enzyme system that can catalyze several commercially important regio- and stereospecific hydroxylation and epoxidation reactions. These typical monooxygenase enzymes require the transfer of two electrons to the heme active center per molecule of substrate converted. Biochemically, electron transfer to or from the enzyme is often linked to the electron flow in the physiological cycle of a cell or microorganism through expensive co-factors like NAD^+ and $NADP^+$.

In electrochemistry, research is focused on the development of chemically modified electrodes that permit direct charge transfer to an active redox enzyme without need for expensive co-factors or linkage to physiological cycles[1-6]. Recently, electrodes which have been modified by attaching organic mediators to the surface have been used to study electron transfer reactions of proteins and other macromolecules[7-16]. In one case, a glucose sensor was developed by immobilizing glucose oxidase and derivatives of ferrocene on graphite electrodes[15]. The ferrocene/-ferricinium couple acts as a mediator between the glucose and the graphite electrode. Immobilized derivatives of ferrocene have also been successfully coupled with other enzymes[16].

Our work is directed at development of biocatalytic processes using the P-450 enzyme system for electrochemical conversion of selected organic compounds. Processes of interest include the synthesis of fine chemicals and biochemicals, and the detection of new biologically-significant compounds in either clinical or biochemical manufacturing settings. Recent monographs on organic synthesis [17] and biosensor development [18] indicate the need for process research and development in these two applications.

Current interest in the use of cytochrome P-450 enzyme systems for commercial syntheses or biosensor development is restrained not only by the need for expensive cofactors, but also by the limited supplies of isolated and purified enzyme which until a few years ago was known to be produced by only mammalian cells. Now, bacterial sources of P-450 offer the possibility to produce larger quantities of the active enzyme. The known bacterial sources of cytochrome P-450 has been summarized by Sligar and Murray [19] and are shown in Table 1. Three interacting protein components are thought to make up bacterial-derived cytochrome P-450 systems. These generic protein components of systems grown in *Pseudomonas putida* by hydroxylation of 1-bornanone (camphor) have been characterized by spectrophotometry, and by their function during substrate conversion [20]: (a) an FAD-coupled flavoprotein (putidaredoxin reductase); (b) an iron sulfide redox (putidaredoxin); (c) a b-type heme protein of the P-450 class (cytochrome m). Only protein m interacts directly with a substrate molecule at the porphyrin heme center during the catalytic reaction cycle. In whole cells, the flavoprotein and iron sulfide redox protein serve as electron transfer shuttles between the cytochrome enzymes and cofactors NADPH or NADH.

The goals of our research are: i) growth of large quantities of cytochrome P-450 in several bacterial systems; ii) investigation by cyclic voltammetry of the electrochemical activity of these P-450 systems for catalyzing organic hydroxylations, iii) immobilization of electrochemically active P-450 enzymes on electrodes which have been previously modified by attaching electron transfer mediators (e.g. ferrocene) to the surface, iv) establishment of the lifetime of the immobilized P-450 electrochemical activity and v) development of potential commercial-scale processes using flow-through porous electrode reactors containing immobilized P-450 enzyme. In this first report about our progress toward these goals, we give account of our efforts to grow large quantities of P-450_{cam} from *P. putida* that have been cultured with camphor as the carbon source. Key features include the use of difference spectrophotometry to follow P-450 production in whole cell suspensions and in suspen-

Table 1. Bacterial P-450 Systems (from Reference 19)

Bacterium	Function
Acinetobacter calcoaceticus (EB104)	ω-Alkane hydroxylase
Bacillus megaterium (ATCC 13368)	15β-Hydroxylase of 3-oxo-Δ-steroids
B. megaterium (ATCC 14581)	ω-2-Fatty acid hydroxylase-epoxidase
Nocardia sp. (NH1)	*p*-Alkylphenyl ether dealkylase
Pseudomonas incognita	Linalool-10-methyl-hydroxlase Linalool-8-methyl-hydroxlase
P. putida (ATCC 17453)	Camphor-5-*exo*-hydroxylase
P. putida (JT810)	*p*-Cymene-7-hydroxylase
P. putida (PL-W)	*p*-Cymene-7-hydroxylase
Rhizobium japonicum (CC 705)	Unknown, but see Ref. 20
Rhodococcus rhodochrous (ATCC 19067)	ω-Alkane hydroxlase
Rhodococcus sp.	Camphor-6-*endo*-hydroxylase

sions of partially purified enzyme. This technique is also used to investigate enzyme stability by measuring 450 nm absorbance conversion to 420 nm. This absorbance conversion indicates active P-450 conversion to denaturated, noncatalytic P-420. We also report on cyclic voltammetry experiments demonstrating the demethylation of anisole by the P-450_{cam} enzyme system with the ferrocene-graphite electrode. These preliminary results indicate direct electron transfer between the ferrocene-graphite electrode and cytochrome m, which suggests that a number of different reactions may be effected. These represent major steps in unlocking the great potential for performing regioselective catalysis with this enzyme system.

EXPERIMENTAL

Chemicals and Biochemicals

Chemicals used in the bacteria growth media were obtained from the following sources: bactotryptone, yeast extract, ascorbic acid (Difco Laboratories, Detroit, MI); D(+)-camphor (Sigma Chemical Company, St. Louis, MO); methylene chloride N,N-dimethylformamide (Burdick and Jackson Chemical Company, Muskegon, MI); Antifoam B. Silicone defoamer (Union Carbide Chemical Company, Phillipsburg, NJ); and ACS-grade mineral salts (Mallinckrodt Chemical Company, Paris, KY).

All the chemicals for the cyclic voltammetry experiments were used as received. The sources of these chemicals are ferrocene and 2-methyltetrahydrofuran (Aldrich); anisole (MCB); t-butyl alcohol (Fischer).

The water used for all solutions and cultures was from a Milli-Q system fitted with an organic scavenging resin bed and 0.22 µm final filter. Argon and carbon monoxide gases used in the enzyme assay were ultrapure grades.

Bacterial Growth and Enzyme Isolation

The *Pseudomonas putida* strain PpG-786 (ATCC 29607) used to generate the P-450_{cam} enzyme system was maintained by weekly transfer to minimum agar plates with D(+)-camphor. Procedures and media used are described in reference [21]. The growth was performed in 2.8-liter Fernback flasks at 30°C using the method of Peterson [22]. Cells were harvested at the late-log-growth phase by centrifugation at 4000 xg for 20 min at 4°C. The cells were then washed three times in 50 mM potassium phosphate buffer (pH 7.4). Following the last wash, the cells were resuspended in the phosphate buffer at concentration of about 25 mg wet weight per mL and stored at 4°C until used.

A crude preparation of the P-450_{cam} proteins was prepared by freezing cell paste at -20°C for about 12 h and then thawing at room temperature to lyse the cells. The paste was next resuspended in phosphate buffer at about 250 mg/mL. Cytochrome m and other cell protein was then isolated in the supernatant after centrifugation of 4500 xg for 30 min at 4°C. This procedure releases about 40-50 % of the P-450_{cam} enzymes [20].

Cytochrome m content of either the whole cell preparations or crude enzyme preparations were determined by comparing the change in absorption at 446 nm relative to 490 nm before and after the suspensions were i) deoxygenated by stripping with argon gas, ii) reduced by adding about 3 mg $Na_2S_2O_4$ per 10 mL suspension, and iii) saturated with carbon monoxide gas by bubbling the gas through the suspension. An extinction coefficient of 91 (ℓ/(mmol-cm) was used to calculate the cytochrome m content [23]. The difference spectra for the treated samples minus untreated were obtained with a Beckman DU-65 scanning spectrophotometer, except for an initial study in which a Bausch and Lomb Model 2000 was used.

Cyclic Voltammetry Experiments

The 3-electrode electrochemical cell was a two-neck round bottom flask (5 mL). The counter electrode was a Pt coil electrode made from Pt wire (0.0508 cm diam., 25.4 cm length). The reference electrode was a saturated Ag/AgCl electrode. The working electrode was a pyrolytic rod (0.635 cm diam). The cross-sectional surface of the graphite electrode exposed to the electrolyte was polished with No. 600 SiC paper before each cyclic voltammetry experiment. After polishing, the surface was covered with 20 μL of a 0.1 M ferrocene solution of THF. After the THF was completely evaporated, the electrode was immersed in the electrolyte, a 0.1 M phosphate buffer solution.

The buffer solution containing anisole (20 mM) was prepared by first dissolving anisole in t-butyl alcohol (0.5 mL) before addition to the electrolyte. The cytochrome m suspension in the phosphate solution (9.5 μM) was maintained at 4°C until use. For each cyclic voltammetry experiment in both the absence and presence of the enzyme, 4 mL of

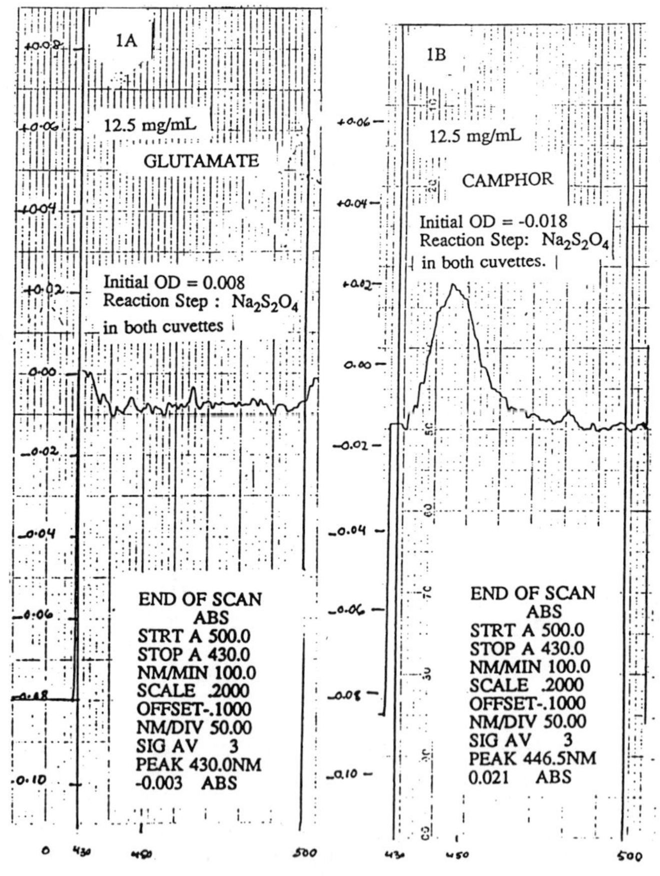

Figure 1. P-450 Production with Camphor Substrate.

electrolyte was used. Cyclic voltammetry was performed with a Solartron Electrochemical Inter- face (Model 1286). Sweep rates were varied between 5-100 mV/s.

RESULTS AND DISCUSSION

Enzyme Production and Stability

The production of P-450 enzyme in *P. putida* when it is cultured on camphor as the sole carbon source is illustrated by the difference spectra shown in Figure 1. These spectra were obtained using the Bausch and Lomb Spectrophotometer with suspension concentrations at 12.5 mg wet cell mass/mL. For cells cultured with glutamate as the carbon source, the difference spectra of Figure 1A indicate no P-450 content in these cells due to the lack of a Soret absorption maximum at 446 nm wavelength. The strong absorption maximum for the cells grown on camphor shown in Figure 1B indicate the presence of $P\text{-}450_{cam}$.

We have also studied the stability of P-450 in whole cells, and in isolated form using this difference spectrophotometry technique. Figure 2 reveals that P-450 in whole cells has limited stability at room temperature. The conversion of P-450 to the denatured, catalytically inactive form, P-420, occurs as the cell samples were aged after harvesting through 1 hour (2A), 20 hours (2B), and 40 hours (2C). This denaturation reaction has been reported by others[23]. We have recently shown that this denaturaation corresponds to a loss of dehalogenation activity by the whole cells, and this loss correlates very well against the conversion to P-420 shown in Figure 2. The rate of denaturation can be significantly reduced for freeze/thawed samples held at 4°C for up to 3 weeks.

Cyclic Voltammetry

Preliminary experiments were first performed between -0.5 to 1.0 V vs Ag/AgCl. These results indicated that ferrocene/ferricinium oxidation-reduction and anisole demethylation were obtained in the 0 to 0.5 V range.

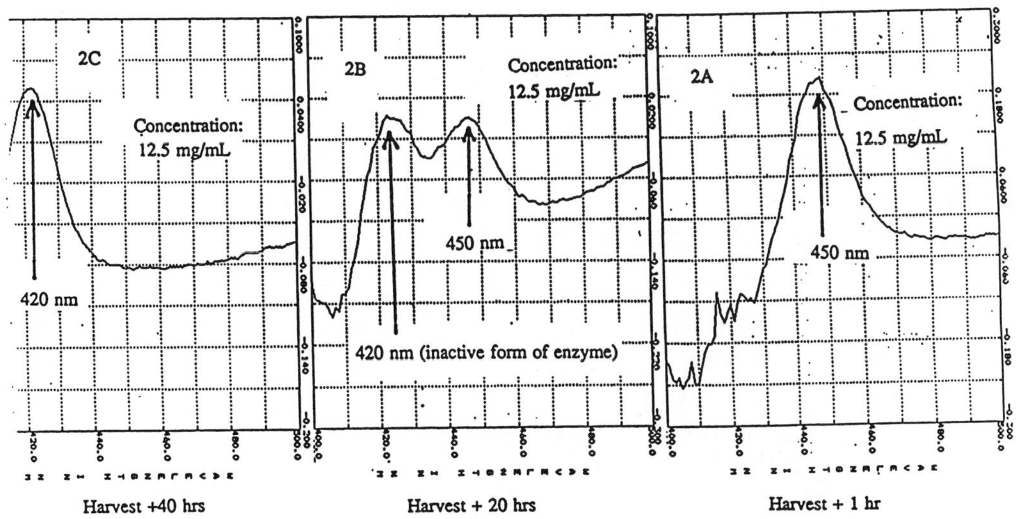

Figure 2. Stability of P-450 Enzyme in *P. putida*.

Figure 3a is a typical cyclic voltammogram (0 - 0.5 V, 5 mV/s) obtained for the graphite-ferrocene electrode in the 0.1 M phosphate buffer solution containing either cytochrome m or anisole. Only ferrocene oxidation and ferricinium ion reduction is indicated. Neither anisole nor cytchrome m alone has any affect on the response of this redox couple.

The presence of both anisole and cytochrome m in the buffer solution gives a voltammogram with a significant oxidation response, as shown in Figure 3b, indicating demethylation of anisole to phenol and formaldehyde. This voltammogram was obtained after about one-half hour exposure of the buffer solution containing both anisole and enzyme to room temperature. A formaldehyde assay after passage of about 20 mC during a potentiostatic experiment at 0.4 V indicated an approximate equivalent formation of formaldehyde confirming the electroenzymological demethylation of anisole.

Figures 3b-3d show the progressive decrease in the enzyme activity with time at room temperature. The two hour voltammogram (Figure 3d) approaches that obtained for the ferrocene/ferricinium couple on the

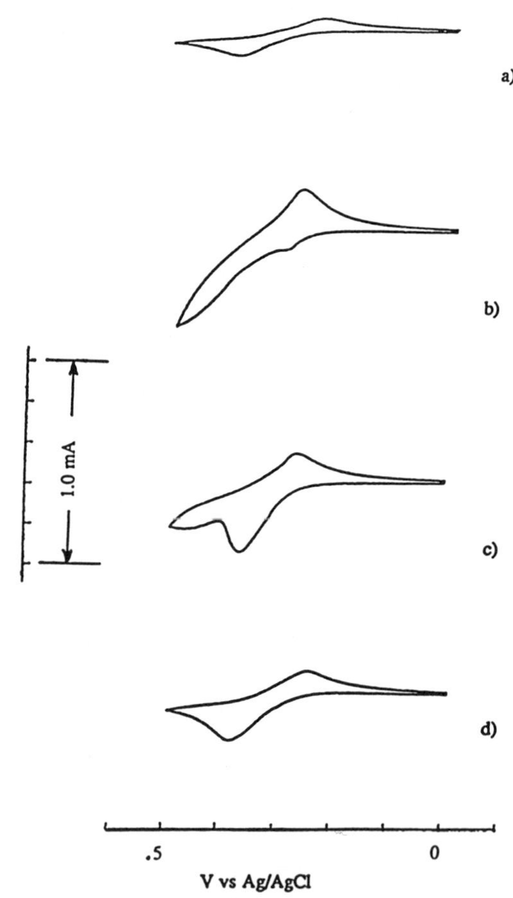

Figure 3. Voltammograms (5 mV/s) of graphite-ferrocene in 0.1 M phosphate buffer solution (pH 7.4): a) containing either anisole (20 mM) or cytochrome; b), c), d) containing both anisole and cytochrome m. Activity decrease with time at room temperature: b) 1/2 h, c) ~ 1 h, d) 2 h.

graphite electrode (Figure 3a). This decline in the demethylation activity of cytochrome m indicates that at room temperature the enzyme is gradually denaturing with time. This result is consistent with our previous observation of the aging of P-450 to the denatured P-420 as shown by the spectrophotometric measurements.

ACKNOWLEDGEMENT

This work was supported by a grant from the UC Energy Research Group (UERG). Dr. A.K. Cho kindly provided the formaldehyde assay.

REFERENCES

1. P.R. Moses, L. Weir and R.W. Murray, Anal. Chem., 47, 1882 (1975).

2. D.S.C. Tse and T. Kuwana, Anal. Chem., 50, 1315 (1978).

3. J.P. Collman, M. Marrocco, P. Denisevich, C. Koval and F.C. Anson, J. Electroanal. Chem., 101, 117 (1979).

4. J.B. Kerr, L.L. Miller and M.R. Van de Mark, J. Am. Chem. Soc., 102, 3383 (1980).

5. A.B. Bocarsly, E.G. Walton and M.S. Wrighton, J. Am. Chem. Soc., 102, 3390 (1980).

6. R.W. Murray, Acc. Chem. Res., 13, 135 (1980).

7. A.W.C. Lin, P. Yeh, A.M. Yacynych and T. Kuwana, J. Electroanal. Chem., 84, 411 (1977).

8. W.J. Albery, M.J. Eddowes, H.A.O. Hill and A.R. Hillman, J. Am. Chem. Soc., 103, 3904 (1981).

9. N.S. Lewis and M.S. Wrighton, Science, 211, 944 (1981).

10. J.J. Kulys and A.S. Samalius, Bioelectrochem. Bioengr., 10, 385 (1983).

11. C.W. Lee, H.B. Gray and F.C. Anson, J. Electroanal. Chem., 172, 289 (1984).

12. P. Yeh and T. Kuwana, Chem. Lett., 1145 (1977).

13. I.V. Berezin, S.D. Varfolomeew and M.V. Lomonosov, Enzyme Eng., 5, 95 (1980).

14. H.A.O. Hill and I.J. Higgins, Phil. Trans. R. Soc. Lond., A302, 267 (1981).

15. A.E.G. Cass, G. Davis, G.D. Francis, H.A.O. Hill, W.J. Aston, I.J. Higgins, E.V. Plotkin, L.D.L. Scott and A.P.F. Turner, Anal. Chem., 56, 667 (1984).

16. A.E.G. Cass, G. Davis, J.J. Green and H.A.O. Hill, J. Electroanal. Chem., 190, 117 (1985).

17. CIBA Foundation Symposium 111, Enzymes in Organic Synthesis, Pitman, London (1985).

18. A.P.F. Turner, I. Karube and G.S. Wilson, Eds., Biosensors: Fundamentals and Applications, Oxford University Press, NY, 750 pages (1987).

19. S.G. Sligar and R.I. Murray, in Cytochrome P-450: Structure, Mechanism and Biochemistry, P.R. Ortiz de Montellano, Ed., Plenum Press, NY, chapter 12 (1986).

20. I.C. Gunsalus and G.C. Wagner, in Methods in Enzymology, 51, 166-168 (1978).

21. T. Lam and V.L. Vilker, Biotechnology and Bioengineering, 29, 151-159 (1987).

22. J.A. Peterson, J. Bacteriology, 103, 714 (1970).

23. D.H. O'Keefe, R.E. Ebel and J.A. Petersen, in Methods in Enzymology, Vol. 52, 151-157 (1978).

INTERFACIAL ELECTROCHEMISTRY OF PROMOTER MODIFIED ELECTRODES FOR THE RAPID ELECTRON-TRANSFER OF CYTOCHROME c

Isao Taniguchi

Department of Applied Chemistry
Faculty of Engineering
Kumamoto University
Kurokami, Kumamoto 860 / Japan

INTRODUCTION

Recently, rapid direct electron-transfer of metalloproteins has been a subject of extensive study of various research groups. Since the pioneering works by Eddowes and Hill[1] and by Yeh and Kuwana[2] for cytochrome c and by Niki et al.[3] for cytochrome c_3, many investigations have been done to date, especially on cytochrome c. At present various functional electrodes, at which rapid electron-transfer of cytochrome c takes place, are known[4-23]. These electrodes so far developed can be classified into several groups[20]: Use of so-called promoters, surface modifiers of the electrode having a function of acceleration of electron-transfer kinetics but the modifiers themselves are electroinactive at the potentials of interest, in the solution[1,5,7,10-13]; use of promoter immobilized electrodes (no promoter in the solution)[4-9]; metal oxides[14-15] and pyrolytic graphite (edge plane)[16] electrodes; other functional electrodes such as surface pretreated bare electrodes[17,18], negatively charged protein (or poly-peptide) coated electrodes[19,20] and so on[21-23]. However, surface functions of individual electrodes have not yet been fully characterized.

Studies from this laboratory[4-6,24-27] suggest that a sulfur atom is effective to immobilize promoters by irreversible adsorption onto gold and silver electrodes. To date, various sulfur containing promoters are known[4-9] to give effective promoter-modified electrodes for cytochrome c. In the present paper, surface structures of such promoter-modified electrodes and electrochemistry of cytochrome c at these electrodes have been investigated to better understand the surface functions of such electrodes.

EXPERIMENTAL

Horse heart cytochrome c (Sigma, type VI) was purified by chromatography on a Whatman CM-32 column at 4°C as described previously[5,24] according to the literatures[14,28]. Pyridine aldehyde thiosemicarbazones (PATS) and benzaldehyde thiosemicarbazone (BATS) were synthesized from thiosemicarbazide and corresponding aldehydes[8], and recrystallized from ethanol. All other reagents were of analytical grade, and used as received.

Cyclic voltammograms of cytochrome c (ca., 0.1-0.3 mM) were measured under various conditions at gold and silver wire electrodes (ca., 5 mm x 0.5 mm diameter) of which surfaces had been modified with various promoters by dipping the electrodes for ca. 10 min into the solutions of promoters of interest.

Surface enhanced Raman scattering (SERS) spectra were measured using a JASCO R-800 laser Raman spectrometer with a He-Ne laser (632.8 nm) or an Ar^+ laser (514.5 and 457.9 nm were used) in a manner similar to those described previously[6,26,27,29].

RESULTS AND DISCUSSION

General features of the surfaces of functional electrodes

From the various electrodes so far developed, the essential requirements for the electrode surface for rapid electron transfer of cytochrome c seem to be summarized roughly as follows: The surface is hydrophilic and has attractive (or at least no repulsive) interaction site(s) with the positively charged surface near the heme edge of cytochrome c; pyridyl and purinyl nitrogen[1,4-13] and/or negatively charged surfaces[2,14,15,19,20] have been reported to be effective. Another requirement is that the electrode surface can prevent cytochrome c from strong irreversible and degradative adsorption onto the electrode; cytochrome c at the electrode surface is preferable to be the native form for rapid electron-transfer, while denatured cytochrome c shows considerable difference in the redox potential[20,30,31] from that of the native form and is usually difficult to be detached from the electrode surface. Figure 1 shows a schematic representation of the various surfaces for the rapid electron-transfer

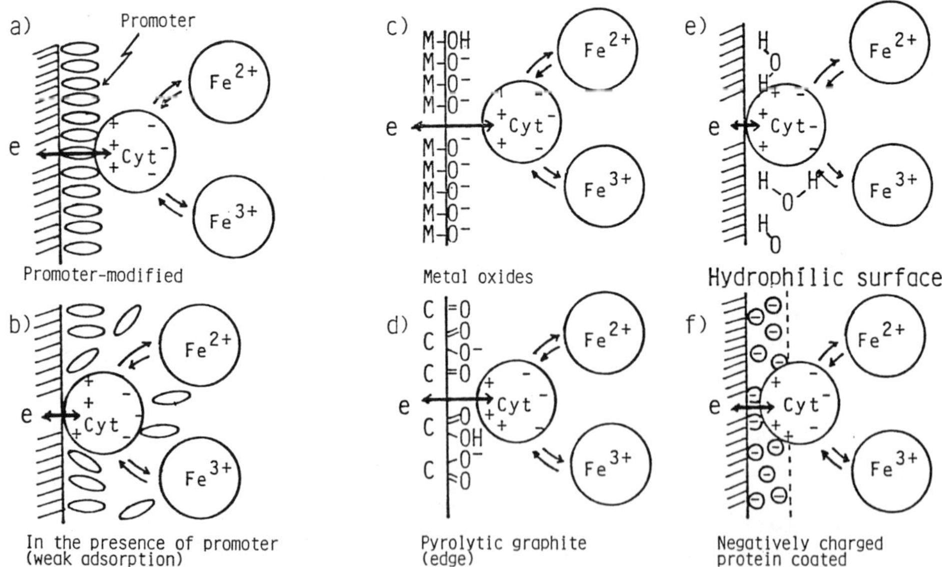

Figure 1. Schematic representation of various surfaces so far developed at which rapid direct electron-transfer of cytochrome c can take place.

of cytochrome c, where the above two requirements would be successfully satisfied.

Among various functional electrodes, in the present work, attention will be focused on functions of the surfaces modified with promoters. At first, note that, as is shown in Figure 1a and 1b, use of a promoter gives two distinguishable groups of surface structures: When a promoter adsorbs more strongly onto the electrode than cytochrome c, the promoter displaces the adsorbed cytochrome c from the electrode surface, where apparent electron-transfer of cytochrome c takes place on the promoter-modified surface (Figure 1a). On the other hand, some promoters do not adsorb so strongly as to exclude cytochrome c completely from the electrode surface but still have a role to prevent strong adsorption of cytochrome c onto the electrode surface; electron-transfer of cytochrome c may occur both at very close to the bare electrode surface and on the promoter-modified surface. Among promoters known to date, bis(4-pyridyl)disulfide (PySSPy) is one of typical compounds for the former[26], and purine as well as 4,4'-bipyridine (PyPy) is the latter[26,30,32]. These phenomena can be distinguishable, for example, by monitoring SERS spectra on an Ag electrode, because SERS signals come only from species absorbed on the electrode surface. SERS signals due only to promoters are observed for the former surface, while the latter case gives SERS signals of both promoter and cytochrome c.

In the present paper, promoters which adsorb on the electrode more strongly than cytochrome c (see Figure 1a) are discussed.

Role of pyridine ring of sulfur containing pyridyl promoters

Sulfur is well known to have a strong interaction with Au and Ag, and sulfur-containing compounds adsorb onto these metals irreversibly to give modified electrodes. PySSPy modified gold (PySSPy-Au) and silver (PySSPy-Ag) electrodes showed well-defined redox waves of cytochrome c with the heterogeneous electron-transfer rate constant (k_{sh}) of ca. $> 5 \times 10^{-3}$ cm s^{-1} (Figure 2). Also, 4-mercaptopyridine (PySH)-modified gold and silver electrodes showed almost reversible electrochemical response of cytochrome c, which was indistinguishable with those observed at PySSPy-Au and PySSPy-Ag electrodes, suggesting that the surface structures of these electrodes are similar to each other.

SERS spectra of these electrodes also support this conclusion; SERS spectra of PySSPy-Au and PySH-Au electrodes and of PySSPy-Ag and PySH-Ag electrodes are very similar to each other and no S-S bond stretching vibration at ca. 550 cm^{-1} was observed for PySSPy-Au and PySSPy-Ag electrodes, suggesting the S-S bond of PySSPy would be cleaved on these electrode surfaces[26,27]. Bis(4-phenyl)disulfide (PhSSPh) and thiophenol (PhSH) also adsorb strongly on Ag and Au electrode surfaces and can exclude preadsorbed cytochrome c from the electrode surfaces. However, PhSSPh- and PhSH-Au electrodes were ineffective for the rapid electron-transfer of cytochrome c; no appreciable redox wave of cytochrome c was observed. SERS measurements of a PhSSPh-Au electrode gave similar Raman signals to those of a PySSPy-Au electrode, suggesting a similar orientation for PySSPy and PhSSPh at the electrode surface (Figure 3); again no signal due to the S-S bond stretching vibration at ca. 550 cm^{-1} was observed. Also, interestingly, a bis(2-pyridyl)disulfide modified gold electrode gave rather poor electrochemistry of cytochrome c ($k_{sh} < 1 \times 10^{-4}$ cm s^{-1}), although this compound adsorbed strongly onto the gold surface and similar orientation for the adsorbed species to a PySSPy-Au electrode was expected from its SERS spectrum (Figure 3). This indicates an important role of the pyridine ring and the position of the

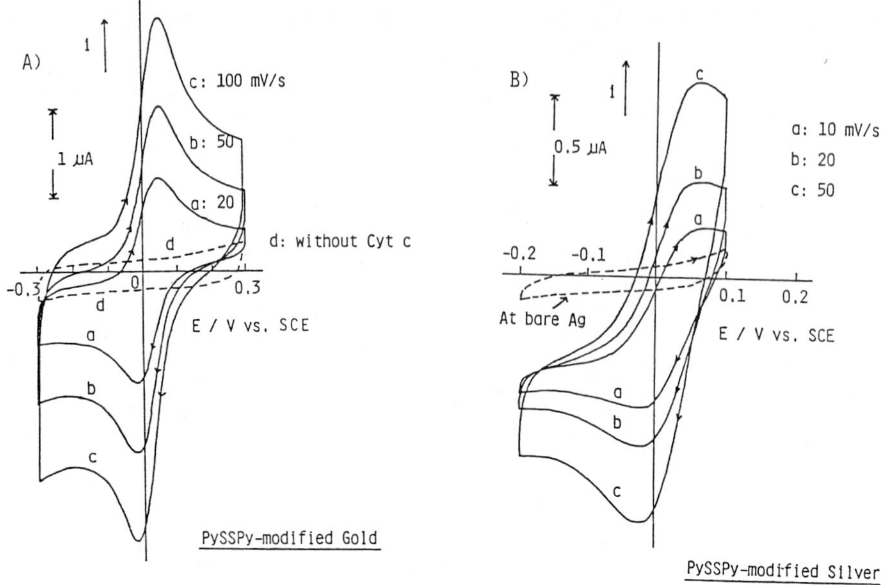

Figure 2. Cyclic voltammograms of cytochrome c (ca. 0.37 mM) in a phosphate buffer solution with 0.1 M $NaClO_4$ (pH 7) at (A) PySSPy-Au and (B) PySSPy-Ag electrodes at 25°C.

nitrogen atom in the pyridine ring of the promoter molecule. Hinnen and and Niki[32] have recently reported that the bis(4-pyridyl)disulfide modified surface has a strong interaction with cytochrome c and cytochrome c can be immobilized on the surface, while the bis(2-pyridyl)disulfide modified surface has weak interaction with cytochrome c.

Similar results were again obtained for pyridine aldehyde thiosemicarbazones (PATS)[8] and benzaldehyde thiosemicarbazone (BATS). From SERS spectra of these compounds on Au (Figure 4) and Ag electrodes no significant difference in the surface structures of these modifiers was expected. However, PATS's acted as promoters for cytochrome c, while BATS did not. Moreover, again, the position of pyridyl nitrogen of PATS affected the reversibility in electrochemical response of cytochrome c (Figure 5), like the cases of bis(4-pyridyl)disulfide and of bis(2-pyridyl)disulfide. This can be explained in terms of the extent of the hydrophilicity of the promoter modified surfaces of the electrodes; PATS modified surfaces would become similar to that of a BATS modified electrode in the order of 4- > 3- > 2-PATS, when we simply assume that PATS adsorbed on the electrode surface through the sulfur atom (Figure 4). Phenyl ring would provide much more hydrophobic surface than pyridine ring. The importance of such hydrophilicity of the electrode surface for the rapid electron-transfer of cytochrome c has also been demonstrated[33] recently using pyrolytic graphite electrodes.

Pyridyl nitrogen becomes ineffective for the rapid electron-transfer of cytochrome c, when it is protonated and positively charged. A PySSPy-Au electrode became ineffective at pH <5, although no significant change in SERS spectra of a PySSPy-Au electrode was observed even in an acidic solution of pH 3. Also, similar results were obtained for PATS-Au electrodes; at PATS-Au electrodes no voltammetric response of cytochrome c

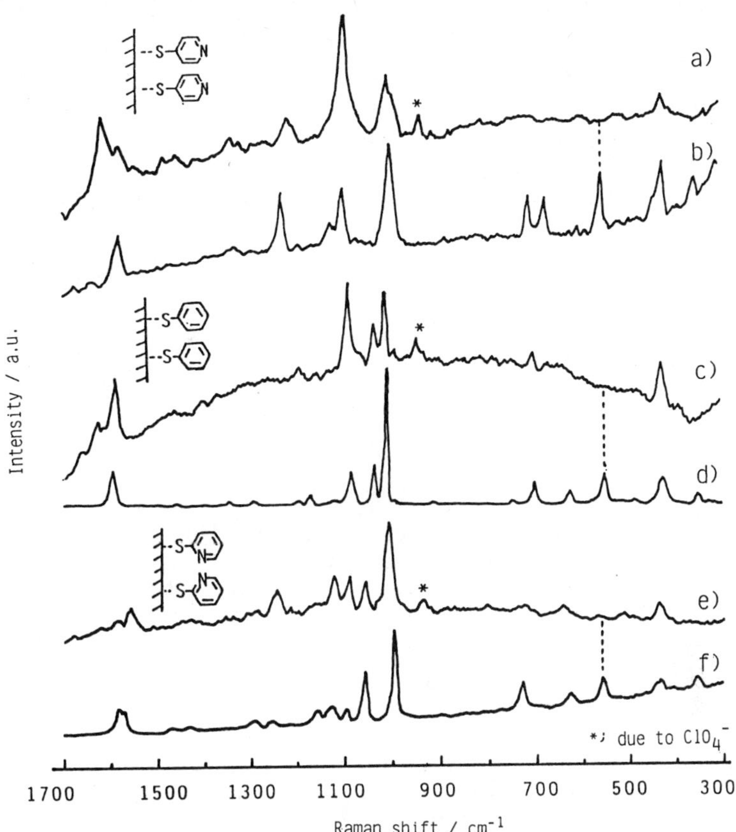

Figure 3. SERS spectra of Au electrodes modified with (a) bis(4-pyridyl)-disulfide (PySSPy), (c) PhSSPh, and (e) bis(2-pyridyl)disulfide in a phosphate buffer solution with 0.1 M $NaClO_4$ (pH 7) at 0 V vs SCE, together with Raman spectra of these modifiers in the solid state (powder, b, d, and f). He-Ne laser (632.8 nm, 30 mV at the cell) was used.

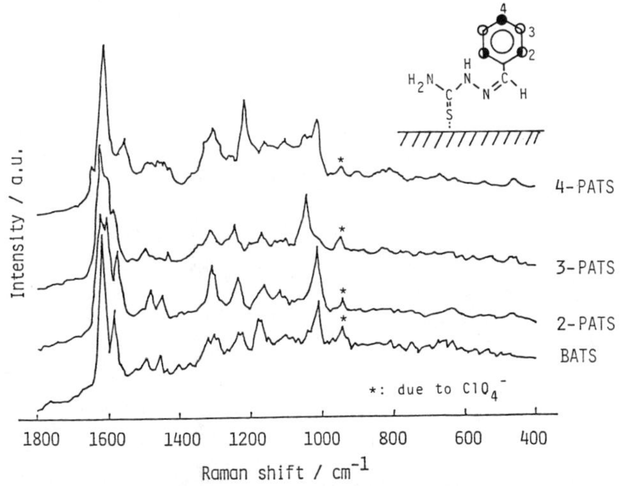

Figure 4. SERS spectra of PATS and BATS modified Au electrodes at 0 V vs SCE in a phosphate buffer solution with 0.1 M $NaClO_4$ (pH 7). He-Ne laser (632.8 nm, 40 mW at the cell) was used.

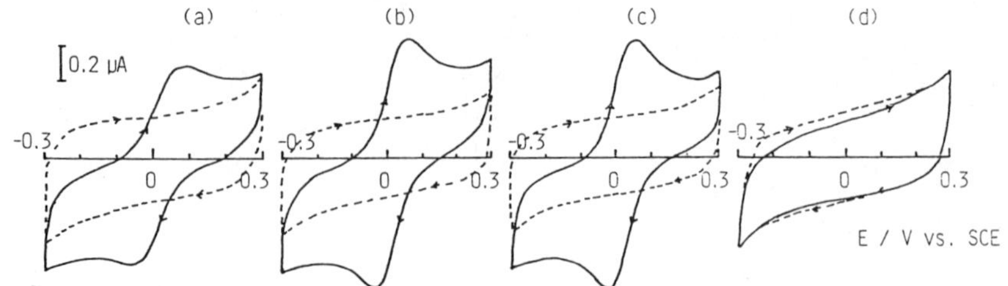

Figure 5. Cyclic voltammograms of cytochrome c (ca. 0.3 mM) in a phosphate buffer solution with 0.1 M $NaClO_4$ (pH 7) at Au electrodes modified with (a) 2-PATS, (b) 3-PATS, (c) 4-PATS and (d) BATS at 25°C. ----: Background currents with no cytochrome c at corresponding modified electrodes. Scan rate: 20 mV/s.

was observed in the solution of pH ca. 4, but SERS signals of the modified electrodes did not change significantly in this solution. 6-Mercaptopurine (PuSH), which also adsorbed strongly through its sulfur atom on an Au electrode in an alkaline form from its SERS spectrum[6], has smaller pK_a value for the nitrogen exposed to the solution, and thus a PuSH-Au electrode works effectively for cytochrome c even in acid solutions of pH above 2.8 (Figure 6).

Sulfur containing amino acids as promoters

L-Cysteine, an amino acid having a sulfur atom, can also be immobilized on Au[9] and Ag surfaces to give promoter modified electrodes; at these electrodes a well-defined redox wave of cytochrome c was obtained.

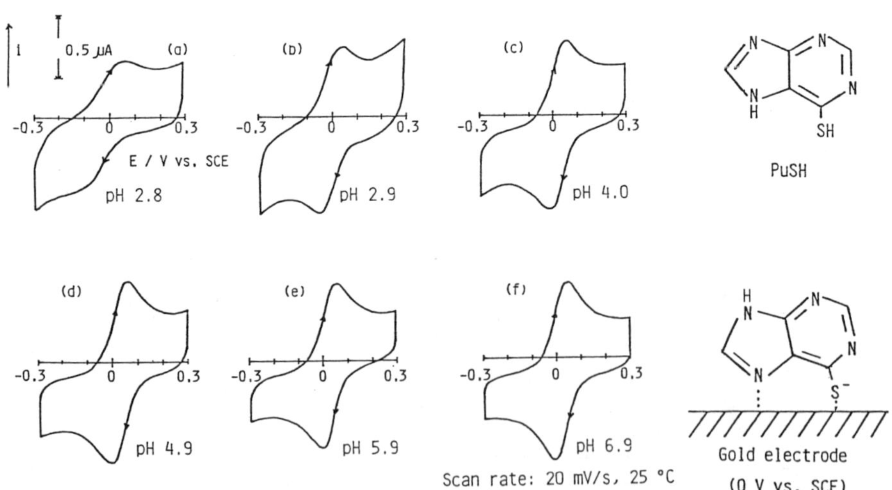

Figure 6. Cyclic voltammograms of 0.2 mM cytochrome c at a PuSH-Au electrode in acidic Britton-Robinson buffer solutions with 0.1 M $NaClO_4$, together with a possible surface structure of the electrode.

SERS spectra of an Ag electrode in a cytochrome c solution showed that added L-cysteine excluded the adsorbed cytochrome c from the electrode surface (Figure 7). Also, no SERS signal due to cytochrome c appeared from an L-cysteine pre-modified Ag electrode in a phosphate buffer solution even when cytochrome c was added to the solution, indicating that L-cysteine adsorbed more strongly than cytochrome c, as were the cases of PySSPy- and PuSH-Au electrodes. D-Cysteine also gave effective modified gold and silver electrodes at which reversible electrochemistry of cytochrome c was observed, indicating D- and L-stereochemistry is not important to provide the suitable electrode surface.

L-cystine, the disulfide of L-cysteine, also acted as a promoter, but showed slower electron-transfer kinetics of cytochrome c than at a cysteine-Au electrode (Figure 8), unlike the cases of PySSPy-Au and PySH-Au electrodes (where voltammograms obtained were indistinguishable from each other). The reason for the difference between cysteine and cystine is not yet clear, but the S-S bond of cystine may not be cleaved, resulting in some difference in the surface structures of these electrodes; unfortunately, however, these compounds did not give strong SERS signals even at an Ag electrode under the present experimental conditions, and more precise information about these electrode surfaces from SERS measurements has been difficult to obtain. A L-methionine modified Au electrode showed very poor electrochemistry of cytochrome c (Figure 8), although L-methionine adsorbed rather strongly onto the gold surface. This may be due to the difference in the length of the carbon chain, which may keep cytochrome c somewhat far from the electrode surface. More detailed experiments on this interesting fact are now under way.

For cysteine as a promoter, the active site facing to the solution would be the dissociated carboxyl group from the following facts. Since the pK_a value of the carboxyl group of cysteine is rather small (< ca. 2),

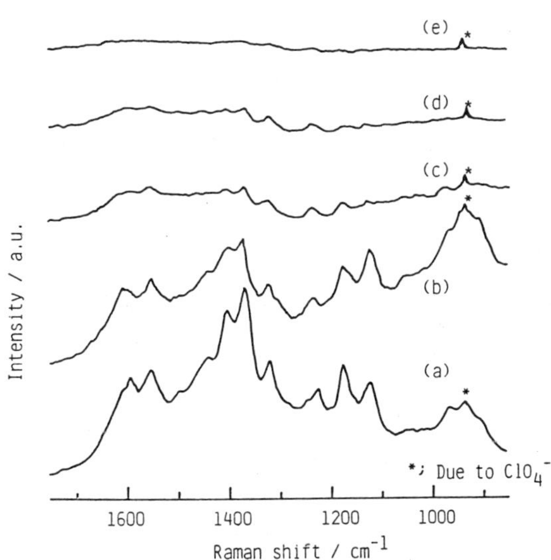

Figure 7. SERS spectra of an Ag electrode at -0.5 V vs SCE in a phosphate buffer solution with 0.1 M $NaClO_4$ (pH 7) containing 90 μM cytochrome c with (a) 0, (b) 10, (c) 87, (d) 100, and (e) 300 μM L-cysteine. Ar^+ laser (514.5 nm, 45 mW at the cell) was used.

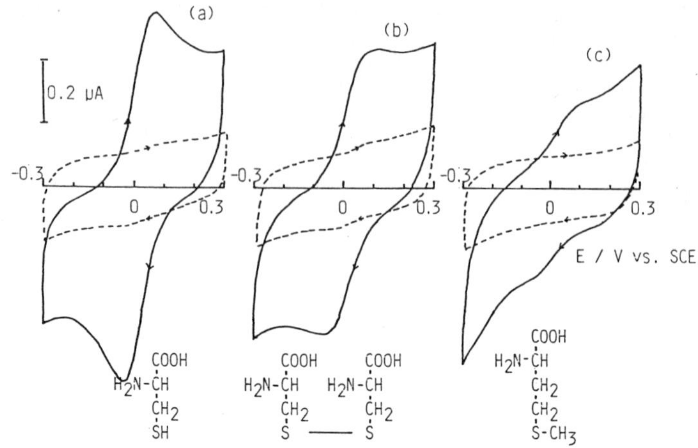

Figure 8. Cyclic voltammograms of cytochrome c (ca. 0.2 mM) in a phosphate buffer solution with 0.1 M $NaClO_4$ (pH 7) at Au electrodes modified with (a) L-cysteine (same as for D-cysteine), (b) L-cysteine, and (c) L-methonine at 25°C. ----: Background currents with no cytochrome c at corresponding modified electrodes. Scan rate: 20 mV/s.

this promoter is expected to act even in acid solutions, like a PuSH-Au electrode[6], and this was demonstrated in an acidic solution of pH 3, although in acidic solutions peak currents of cytochrome c decreased gradually and small additional redox waves appeared at more positive potentials (around 0.2 V vs SCE at pH 3); no such wave at positive potentials was observed at a PuSH-Au electrode. Furthermore, the effect of poly-L-lysine (p-L-Lys, MW 23,000) addition on the electrochemistry of cytochrome c showed marked differences between electrodes modified with pyridyl promoters and with cysteine. At a cysteine-Au electrode, addition of positively charged p-L-Lys remarkably inhibited the electrochemical response of cytochrome c (Figure 9), like at a pyrolytic graphite (edge) electrode[34] and at an In_2O_3 electrode (see Figure 9), as was the case of the physiological reaction between cytochrome c and its oxidase. This can be explained reasonably that the negatively charged surface of a cysteine-Au electrode was blocked by positively charged p-L-Lys, and thus cytochrome c became difficult to approach to the electrode surface. At the concentrations of p-L-Lys used, no significant replacement of cysteine as promoter from the electrode surface by p-L-Lys took place. On the other hand, at electrodes modified with pyridyl promoters such as PySSPy-, PuSH-, and PATS-Au electrodes, addition of p-L-Lys showed much less inhibition in cytochrome c electrochemistry[34,35] (see Figure 9), indicating that p-L-Lys has no significant interaction with the surfaces modified with pyridyl promoters.

CONCLUSIONS

In summary: (1) a sulfur atom is effective to immobilize promoters onto a Au (or Ag) electrode. (2) Pyridyl (or purinyl) nitrogens of promoter molecules have an important role to act as effective promoters, while the phenyl ring inhibits the rapid electron-transfer of cytochrome c, and thus, the position of nitrogen in pyridine ring affects the electrochemistry of cytochrome c. This would be explained in terms of the

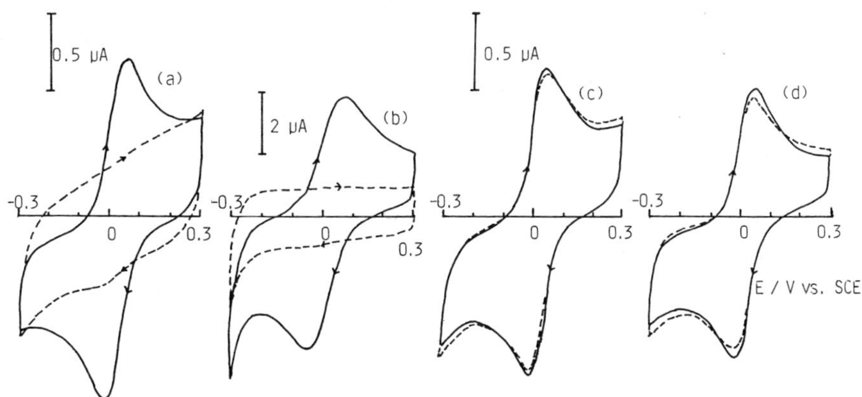

Figure 9. Effect of p-L-Lys addition on cyclic voltammograms of cytochrome c (ca. 0.3 mM for a, c and d, and 0.1 mM for b) in a phosphate buffer solution with 0.1 M $NaClO_4$ (pH 7) at various electrodes at 25°C. (a) L-cysteine-Au, (b) In_2O_3 (ca. 1.2 cm^2), (c) 4-PATS-Au and (d) PySSPy-Au. ----: after 0.15 mg/ml p-L-Lys (MW 23,000) was added to the solution. Scan rate: 20 mV/s.

difference in hydrophilicity of the electrode surface. (3) L- and D-Cysteine can be modified onto Au and Ag electrodes and work as effective promoters, while cystine is less effective and methionine works poorly. (4) For an L-cysteine-Au electrode, the dissociated carboxyl group has an important role for the rapid electron-transfer; this electrode was effective even in acid solutions (due to a small pK_a value for the carboxyl group, ca. 2), but p-L-Lys addition inhibited significantly the electron-transfer of cytochrome c, whereas pyridyl promoters were ineffective in acid solution of pH ca. 4. However, p-L-Lys addition affected much less the electrochemistry of cytochrome c. (5) Surface structures of individual electrodes are necessary to be clarified to understand the electrochemical behavior of cytochrome c in more detail at sulfur containing amino acid modified electrode surfaces. Finally, (6) in general, although the mechanism for rapid electron-transfer of cytochrome c at individual electrodes may be different to each other depending on the type of functional electrodes, the following requirements seem to be still valid for a suitable electrode surface; (a) the hydrophilic surface with a weak (or at least no repulsive) interaction site(s) and (b) a surface at which no significant degradative adsorption of cytochrome c occurs.

ACKNOWLEDGMENTS

The author appreciates Messrs. Nobuhiro Higo and Hirofumi Kurihara, graduate students of the author's research group, for their experimental assistance. The partial financial support of this work by a Grant-in-Aid for Scientific Research from the Ministry of Education, Science and Culture (Nos. 61550602 and 62303011) is also gratefully acknowledged.

REFERENCES

1. M.J. Eddowes and H.A.O. Hill, J. Chem. Soc., Chem. Commun., 771 (1977).

2. P. Yeh and T. Kuwana, Chem. Lett., 1145 (1977).

3. K. Niki, T. Yagi, H. Inokuchi and K. Kimura, J. Electrochem. Soc., 124, 1889 (1977).

4. I. Taniguchi, K. Toyosawa, H. Yamaguchi and K. Yasukouchi, J. Chem. Soc., Chem. Commun., 1032 (1982); J. Electroanal. Chem., 140, 187 (1982).

5. I. Taniguchi, M. Iseki, K. Toyosawa, H. Yamaguchi and K. Yasukouchi, J. Electroanal. Chem., 164, 385 (1984).

6. I. Taniguchi, N. Higo, K. Umekita and K. Yasukouchi, J. Electroanal. Chem., 206, 341 (1986).

7. P.M. Allen, H.A.O. Hill and N.J. Walton, J. Electroanal. Chem., 178, 69 (1984).

8. H.A.O. Hill, D.J. Page, N.J. Walton, and D. Whitford, J. Electroanal. Chem., 187, 315 (1985); H.A.O. Hill, D.J. Page, and N.J. Walton, ibid., 208, 395 (1986).

9. K. Di Gleria, H.A.O. Hill, V.J. Lowe and D.J. Page, J. Electroanal. Chem., 213, 333 (1986).

10. J. Haladjian, P. Bianco, R. Pilard, Electrochim. Acta, 28, 1823 (1983); J. Haladjian, R. Pilard, P. Bianco, L. Asso, J.P. Galy and R. Vidal, ibid., 30, 695 (1985).

11. I. Taniguchi, T. Murakami, K. Toyosawa, H. Yamaguchi and K. Yasukouchi, J. Electroanal. Chem., 131, 397 (1982).

12. Y. Gui and T. Kuwana, J. Electroanal. Chem., 222, 321 (1987); ibid., 226, 199 (1987).

13. R. Dhesi, T.M. Cotton and R. Timkovich, J. Electroanal. Chem., 154, 129 (1983); T.M. Cotton, D. Kaddi and D. Lorga, J. Am. Chem. Soc., 105, 7462 (1983).

14. E.F. Bowden, F.M. Hawkridge, J.F. Chlebowski, E.E. Bancroft, T. Thorpe and H.N. Blount, J. Am. Chem. Soc., 104, 7641 (1982).

15. M.A. Harmer and H.A.O. Hill, J. Electroanal. Chem., 170, 369 (1984).

16. F.A. Armstrong, H.A.O. Hill and B.N. Oliver, J. Chem. Soc., Chem. Commun., 976 (1984).

17. E.F. Bowden, F.M. Hawkridge and H.N. Blount, J. Electroanal. Chem., 161, 355 (1984).

18. H. Durliat and M. Comtat, Anal. Chem., 54, 856 (1982).

19. T. Akaike, BEM (Biomedical Electronics), 1, 86 (1987) (in Japanese).

20. I. Taniguchi and K. Yasukouchi, Hyomen (Surface), 23, 597 (1985) (in Japanese).

21. V.J. Razumas, A.S. Samalius and J.J. Kulys, J. Electroanal. Chem., 164, 195 (1984).

22. D.J. Arnold, K.A. Gerchario and C.W. Anderson, J. Electroanal. Chem., 172, 379 (1984).

23. D.F. Reed and F.M. Hawkridge, Anal. Chem., 59, 2334 (1987).

24. I. Taniguchi, M. Iseki, T. Eto, K. Toyosawa, H. Yamaguchi and K. Yasukouchi, Bioelectrochem. Bioenerg., 13, 373 (1984).

25. I. Taniguchi, T. Funatsu, M. Iseki, H. Yamaguchi and K. Yasukouchi, J. Electroanal. Chem., 193, 295 (1985).

26. I. Taniguchi, M. Iseki, H. Yamaguchi and K. Yasukouchi, J. Electroanal. Chem., 175, 341 (1984).

27. I. Taniguchi, M. Iseki, H. Yamaguchi and K. Yasukouchi, J. Electroanal. Chem., 186, 299 (1985).

28. D.L. Brautigan, S. Ferguson-Miller and E. Margoliash, Methods Enzymol., 53, 128 (1978).

29. I. Taniguchi, K. Umekita and K. Yasukouchi, J. Electroanal. Chem., 202, 315 (1986).

30. C. Hinnen, R. Parsons and K. Niki, J. Electroanal. Chem., 147, 329 (1983).

31. I. Taniguchi and K. Yasukouchi, in: "Study in Organic Chemistry 30 (Recent Adv. Electroorg. Synth.)", S. Torii, Ed., Kodansha-Elsevier, Amsterdam-Tokyo (1987), p. 393.

32. K. Niki, Rev. Polarography, 31, 85 (1985); C. Hinnen and K. Niki; ibid., 33, 64 (1987).

33. F.A. Armstrong and K.J. Brown, J. Electroanal. Chem., 219, 319 (1987).

34. I. Taniguchi, T. Funatsu, K. Umekita, H. Yamaguchi and K. Yasukouchi, J. Electroanal. Chem., 199, 455 (1986).

35. H.A.O. Hill, D.J. Page and N.J. Walton, J. Electroanal. Chem., 217, 141 (1987).

SERS STUDIES ON THE ELECTRODE REACTION OF CYTOCHROME c AT SILVER ELECTRODE IN THE PRESENCE OF PYRIDINE DERIVATIVES

Fan Ke-Jun, I. Satake, K. Ueda,
H. Akutsu and K. Niki*

Electrochemistry Laboratory
Department of Physical Chemistry
Yokohama National University
Tokiwadai 156, Hodogaya-ku
Yokohama 240 / Japan

INTRODUCTION

Heterogeneous electron transfer reactions between electrodes and c-type cytochromes in the bulk of solution have been extensively studied. Cytochrome c exhibits voltammetric responses ranging from reversible to kinetically irreversible at various electrodes[1,2]. The electrode reactions of cytochrome c_3 have been found to exhibit reversible heterogeneous electron transfer at various electrodes without mediators[3]. It has been shown that both cytochrome c and cytochrome c_3 adsorb strongly on various electrode surfaces from aqueous solutions. The voltammetric behavior of these cytochromes in solution is strongly influenced by the nature of the adsorbed films on the electrode[4,5].

Since Hill and his collaborators have found that horse heart cytochrome c undergoes a rapid electron transfer reaction at gold electrode in the presence of 4,4'-bipyridyl[6,7], various surface-modifiers (electron promoters) have been proposed for electron transfer proteins in biological systems[8-12]. Allen et al.[13] have investigated various surface-modifiers for cytochrome c at gold electrodes and postulated a structure for the surface-modifiers at the electrode surface. The surface modifiers should contain a functional group X, which adsorbs or binds to the gold electrode surface through nitrogen, phosphorous, or sulfur, and another functional group Y, anionic or weakly basic, which interacts favorably with the positively charged electron transfer domain (lysine residues) of cytochrome c. Among the isomers of pyridine derivatives, the compounds containing 4-pyridyl facilitate a rapid electron transfer reaction between cytochrome c and gold electrodes. Compounds containing 2-pyridyl do not accelerate the electrode reaction rate of cytochrome c, and its voltammetric response remains the same as that without 2-bipyridyl derivatives[6,13]. These explanations are based only on the structure of surface-modifiers and on the voltammetric responses. Taniguchi et al.[14-16] have used the surface enhanced Raman scattering technique to elucidate the conformation of 4,4'-bipyridyl and bis(4-pyridyl) disulfide, purine, and 6-mercaptopurine at both gold and silver electrodes. The results showed that both 4,4'-bipyridyl and bis(4-pyridyl) disulfide molecules are oriented perpendicularly toward the gold electrode surface with nitrogen or sulfur as a binding site. The 4-isomers direct the pyridyl nitrogen out toward the solution and cytochrome c binds to the pyridyl ring through nitrogen at the 4-position. On the contrary, the 2-isomers point nitrogen more back towards the electrode and cytochrome c is unfavorable to bind nitrogen at the 2-position.

Hinnen and Niki[17] have studied the evolution of the redox properties of an absorbed cytochrome c on gold electrode in the presence of surface-modifiers by using a UV-vis electroreflectance spectroscopic technique coupled with voltammetric (dc and ac) measurements. It was found that there are distinct differences between 4,4'-bipyridyl and bis(4-pyridyl) disulfide as a surface-modifier for cytochrome c. Cytochrome c is maintained adsorbed on the bis(4-pyridyl) disulfide modified gold electrode and its formal potential is 0.00 V (vs SCE), which is the same as that of the native one. In the case of 4,4'-bipyridyl, there is an adsorption equilibrium between 4,4'-bipyridyl and cytochrome c. The electrode is covered by both 4,4'-bipyridyl and cytochrome c when 4,4'-bipyridyl is present in the solution. The formal potential of cytochrome c co-adsorbed with 4,4'-bipyridyl on gold electrode is -0.25 V (vs SCE), which is 0.25 V more negative than that of the native protein and is 0.25 V more positive than that of adsorbed cytochrome c on a bare gold electrode. The electrode reaction of cytochrome c in the bulk takes places through the immobilized layers of cytochrome c and bis(4-pyridyl) disulfide on the gold electrode. The electrode reaction through the cytochrome c co-adsorbed with 4,4'-bipyridyl is very improbable due to its negative formal potential. It is reasonable to assume that the displacement of 4,4'-bipyridyl molecules by cytochrome c from the bulk takes place at the electrode surface. 4,4'-Bipyridyl acts as a surface-active agent and provides a hydrophobic electrode surface where unfolding of cytochrome c is prevented.

In UV-vis electroreflectance spectroscopy, an average optical property of the adsorbed species on the gold electrode is obtained but enough information is not obtainable to elucidate the conformation of both cytochrome c and surface-modifiers at the electrode surface. However, the conformation of biomolecules and surface-modifiers can be exploited to clarify the SERS (surface enhanced Raman scattering) sensitivity dependence on distance from the electrode surface because the SERS responses decay very rapidly with increasing distance[18,19]. In small molecules with dimensions of approximately 0.6 nm (pyridine) all vibrations of the molecule can be enhanced. In large molecules with diameters of about 4 nm (cytochrome c), on the other hand, only groups which are attached directly to the surface give rise to SERS responses. It has been shown that the Raman enhancement is limited to functional groups in biological macromolecules at distances ranging up to 0.5 nm from the electrode surface[20-22].

In the present work, the interfacial behavior of cytochrome c at a silver electrode in the presence of both 4-isomers and 2-isomers of pyridine derivatives are investigated by using voltammetric and SERS techniques. The 4-isomers used are 4,4'-bipyridyl, bis(4-pyridyl) disulfide, and 1,2-bis(4-pyridyl)ethylene and the 2-isomers are 2,2'-bipyridyl, bis(2-pyridyl) disulfide, and 1,2-bis(2-pyridyl) ethylene.

EXPERIMENTAL

Horse heart cytochrome c (Type VI from Sigma Chemical Co.) was purified chromatographically as described previously[23]. 4,4'-Bipyridyl (reagent grade from Tokyo Kasei); 4-BIPY, 2,2'-bipyridyl (reagent grade from Wako Chemical Co.); 2-BIPY, bis(4-pyridyl) disulfide (reagent grade from Sigma Chemical Co.); 4-PYS, bis(2-pyridyl) disulfide (from Dojin Chemical Co.); 2-PYS, trans-1,2-bis(4-pyridyl)ethylene (from Aldrich Chemical Co.); 4-PYCH and 1,2-bis(2-pyridyl)ethylene (from Tokyo Kasei); 2-PYCH were used without further purification. Phase sensitive ac polarographic measurements were made with a Fuso potentiostat Model 311 and a Fuso phase-sensitive detector Model 316. The SERS spectra were excited by the 514.5-nm line of an argon ion laser (Lexel Model 95-4) and recorded with a JASCO R-800 Raman Spectrophotometer in the spectral range 1000 to 1700 cm^{-1} with a slit width of 0.1-0.2 mm (resolution: 1-2 cm^{-1}). The laser power at the light source was approximately 60 mW. A ca. 5 mL glass spectroelectrochemical cell was used with a platinum wire counter electrode and a silver/silver chloride reference electrode in saturated potassium chloride. The working electrode was a polycrystalline silver rod (8 m/m) sheathed in a Teflon holder. Prior to each experiment the working electrode was mechanically polished to a 1- μm finish with alumina and then ultrasonicaly cleaned in Milli-Q water until alumina embedded in the electrode was not detectable. The working electrode was roughened by scanning two oxidation/reduction cycles (ORC) in the base solution. During ORC the potential was scanned between -0.2 and +0.5 V at a scan rate of 50 mV/s. The supporting electrolyte was 0.03 M phosphate buffer at pH 7.0. Voltammetric measurements were carried out at 25.0°C, and SERS measurements were made at 23-25°C. Electrode potentials are given with respect to a silver/silver chloride in saturated potassium chloride unless otherwise stated.

RESULTS

Voltammetric Response of Cytochrome c on Silver Electrode

Cytochrome c exhibits an irreversible voltammetric response in (0.03 M phosphate buffer + 0.12 mM cytochrome c) solution at pH 7.0 as shown in Figure 1. The reduction peak is observed at 0.3 V but the reoxidation peak is hardly observable. The silver electrode, on which cytochrome c is immobilized, exhibits a pseudo-capacitance peak at -0.3 V by phase sensitive ac polarography. Since this pseudo-capacitance peak is independent of the modulation frequency, it corresponds very likely to the redox reaction of adsorbed cytochrome c on the silver electrode. Similar redox behavior has been reported for cytochrome c adsorbed on a gold electrode[5]. That is, the formal potential of the adsorbed cytochrome c on a silver electrode is -0.3 V, which is different from that of the native species in the bulk at +0.06 V. The formal potential of cytochrome c adsorbed on a

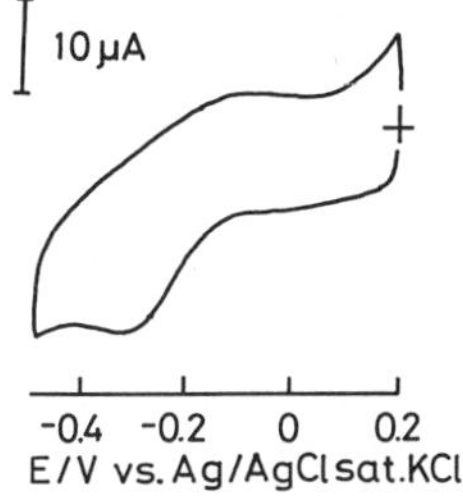

Figure 1. Cyclic voltammogram of 0.12 mM cytochrome c at a silver electrode in 0.03 M phosphate buffer at pH 7.0 and at 25°C.

silver electrode can also be determined by monitoring the shift of the oxidation-state-marker band (Band IV) from the ferri-form at 1375 cm^{-1} to the ferro-form at 1366 cm^{-1} in the SERS spectrum and is estimated to be about -0.35 V, which agrees with the pseudo-capacitance peak potential.

Voltammetric Response of Cytochrome c in the Presence of Surface Modifiers

After the ORC treatment of the silver electrode, cyclic voltammetric measurements of cytochrome c in the presence of surface modifiers were carried out. The results are shown in Figure 2. In the potential range between -0.1 and -0.2V, cytochrome c exhibits no voltammetric response on a bare silver electrode. In the presence of the surface modifiers, cyto-

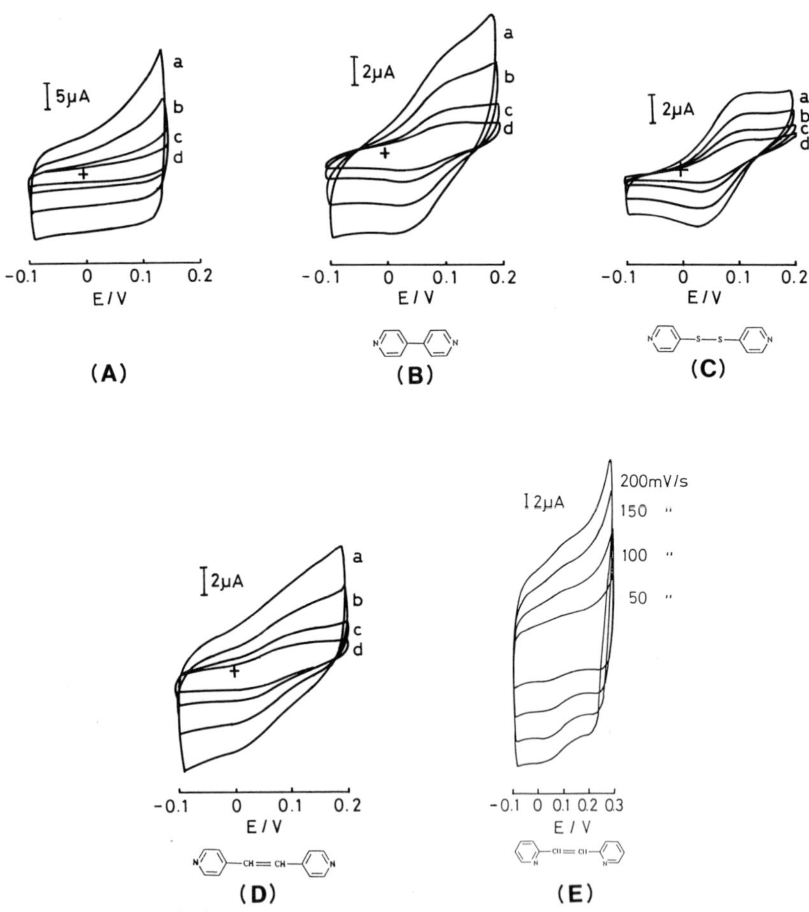

Figure 2. Cyclic voltammograms of 0.10 mM cytochrome c at a silver electrode in 0.05 M phosphate buffer in the presence of surface modifiers. Scan rate: 10, 20, 50, and 100 mV/s. A: no surface modifier; B: 10 mM 4-BIPY; C: 5 mM 4-PYS; D: 4 mM 4-PYCH; E: 0.037 mM cyt. c + 1 mM 2-PYCH on a silver electrode.

chrome c exhibits a quasi-reversible voltammetric response with the peak potential separation of about 60 mV and a formal potential of about +0.1 V, which is the same as that of the native protein (+0.06 V). Their peak heights, however, vary with the surface modifier. The highest peak was observed with 4-PYS and the lowest peak with 4-PYCH. The variation of the peak heights with the surface modifier is probably due to differences in the electrode reaction mechanism of cytochrome c through the layers of the surface modifiers.

SERS of Cytochrome c

It has been reported that horse cytochrome c in the bulk of the solution exists in the six-coordinated low spin state (LS) and that adsorption on silver sol induces a partial transformation from the low spin state to the five-coordinated high spin state (HS)[24,25]. The present SERS results reveal that cytochrome c adsorbed on a silver electrode exists mainly in the high spin state at -0.5 V (ferrous-form), which is characterized by the Raman line at 1546 cm^{-1}, as shown in Figure 3 (curve C).

Effect of 4,4'-Bipyridyl

The SERS spectra of 0.01 mM cytochrome c + 10 mM 4-BIPY and 10 mM 4-BIPY solutions are shown in Figure 3 (curves A and B), in which the SERS

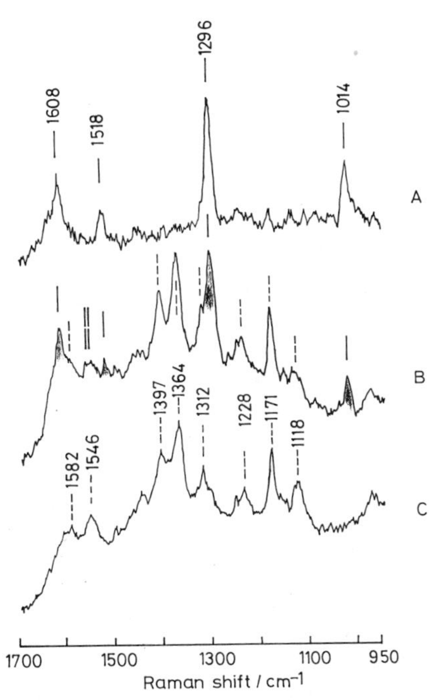

Figure 3. SERS spectra at a silver electrode in the solutions of (A) 10 mM 4-BIPY, (B) 0.01 mM cyt. c+ 10 mM BIPY), and (C) 0.01 mM cyt. c at -0.50 V. Shadow indicates the peaks for 4-BIPY.

signals of both species are observable. That is, 4,4'-bipyridyl is co-adsorbed with cytochrome c on the silver electrode. The relative intensities and the peak positions for cytochrome c are the same both in the absence and the presence of 4-BIPY in the solution except for the peak at 1546 cm^{-1}, which corresponds to the HS state. In the presence of the adsorbed 4-BIPY on the electrode the spin-state-marker band at 1537 cm^{-1} for LS state is induced. That is, the co-adsorption of 4-BIPY with cytochrome c gives rise to the partial transformation from the HS state to the LS state. The redox behavior of cytochrome c co-adsorbed with 4-BIPY on a silver electrode can be estimated by monitoring the potential dependence of the oxidation-state marker band (Band IV) at 1375 cm^{-1} for the ferri-form and 1366 cm^{-1} for the ferro-form. The formal potential of cytochrome c co-adsorbed with 4-BIPY is estimated to be about -0.15 V, which is 0.21 V more negative than that of the native protein and is 0.20 V more positive than that of cytochrome c adsorbed on a bare silver electrode.

Figure 4 compares the normal Raman spectrum of a solid 4-BIPY with SERS of 4-BIPY at three different electrode potentials (+0.1, -0.2, and -0.5 V) on a silver electrode. It is quite obvious that 4-BIPY remains adsorbed on the silver electrode throughout this potential range as was shown by Cotton et al.[27]. No shift in peak positions was observed. When the silver electrode[4b] is transferred into the cytochrome c solution, 4-BIPY molecules adsorbed on the silver electrode are partially displaced by cytochrome c molecules. That is, adsorption of cytochrome c and 4-BIPY on a silver electrode are competitive and the adsorption of 4-BIPY on a silver electrode is considered to be physical in nature rather than chemisorption. Both cytochrome c and 4-BIPY molecules are located within the range of ~ 0.5 nm from the electrode surface, since the SERS signals of both molecules are observable at a silver electrode.

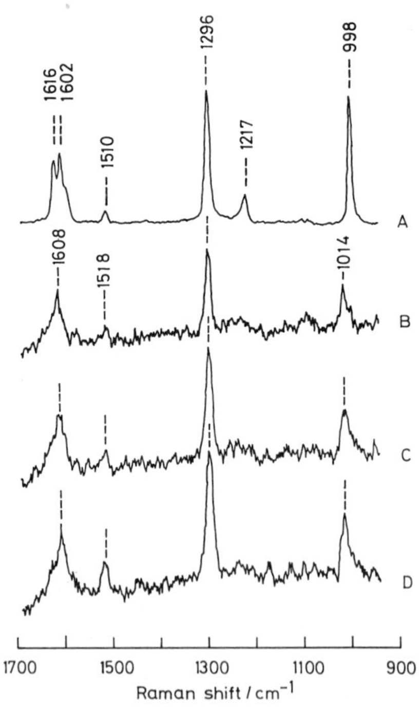

Figure 4. NRS spectrum of 4-BIPY powder and SERS spectra at a silver electrode in 10 mM 4-BIPY at various electrode potentials.

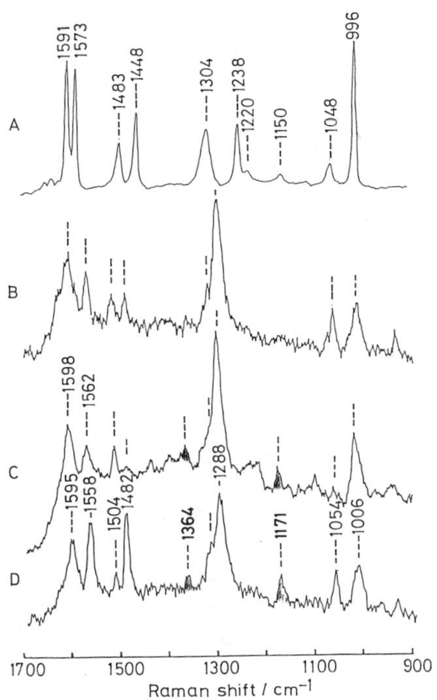

Figure 5. SERS spectra at a silver electrode in the solutions of (A) NRS spectrum of 2-BIPY powder, (B) 5 mM 2-BIPY, (C) 0.05 mM cyt. c (2-BIPY is pre-adsorbed on the silver electrode), and (D) (0.05 mM cyt. c + 5 mM 2-BIPY). Shadow indicates the peaks for cyt. c.

Effect of 2,2'-Bipyridyl

The SERS studies on 2-BIPY at silver electrode by Kim and Itoh[27] revealed that the conformation of 2-BIPY at a silver electrode is a nearly planar cis-conformation. 2,2'-Bipyridyl forms an Ag(I)-N coordination bond similar to that of $Ag(2\text{-BIPY})_2$ at +0.05 V and this coordination bond becomes chemisorption bonding assigned to a strongly bound Lewis acid coordination structure at potentials more negative than +0.05 V. The adsorption of 2-BIPY is so strong that 2-BIPY molecules pre-adsorbed on the silver electrode are not displaced by cytochrome c as shown in Figure 5 (curve C). Small Raman peaks at 1171 and 1364 cm^{-1} for cytochrome c are observable and suggest that the amount of cytochrome c co-adsorbed with 2-BIPY is much less than that co-adsorbed with 4-BIPY shown in Figure 3 (curve B). 2-BIPY molecules are preferentially adsorbed on a silver electrode from the solution of 0.05 mM cytochrome c + 5 mM 2-BIPY as shown in Figure 5 (curve D) and form a stable chemisorbed film on the silver electrode because the concentration of 2-BIPY is much higher than that of cytochrome c and its diffusivity is also higher. The displacement of cytochrome c pre-adsorbed on the silver electrode by 2-BIPY is very slow at -0.5 V. On shifting the electrode potential from -0.5 V to 0.0 V, 2-BIPY partially displaces cytochrome c and both molecules can coexist at the silver electrode.

The voltammetric response of cytochrome c at the 2-BIPY adsorbed silver electrode is irreversible (no enhancement of the electrode reaction rate) probably because the chemisorbed 2-BIPY layer on the silver electrode inhibits the electron transfer reaction of cytochrome c. After washing this silver electrode very thoroughly with the base solution, it was subjected to voltammetry to detect cytochrome c on the 2-BIPY modified surface. There was no indication that cytochrome c was immobilized on this modified surface. That is, the chemisorbed 2-BIPY layer on the silver electrode inhibits both the adsorption and electron transfer reaction to cytochrome c on the electrode.

Effect of Bis(4-pyridyl) disulfide

The SERS signals of 4-PYS are almost identical to the normal Raman spectrum of a solid 4-PYS. That is, the conformation of 4-PYS on the silver electrode is almost the same as that in the bulk. Since no SERS signals for cytochrome c are observed, as shown in Figure 6 (curve A), cytochrome c molecules are considered to be located at least more than 0.5 nm from the electrode surface. The adsorption of 4-PYS is so strong that cytochrome c pre adsorbed on the silver electrode may be completely displaced by 4-PYS and the adsorbed layer of 4-PYS is hardly displaced by cytochrome c (only small Raman peaks are observable when cytochrome c is present in the solution)[4]. This electrode exhibits a quasi-reversible voltammetric response at the potential corresponding to the formal potential of native cytochrome c. That is, cytochrome c remains immobilized on the 4-PYS modified electrode. The electron transfer reaction of

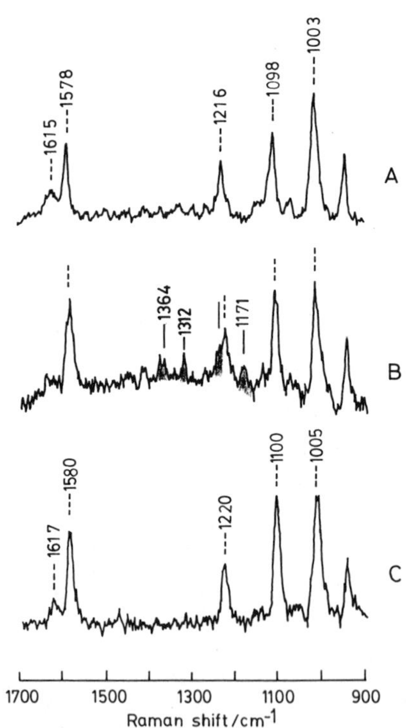

Figure 6. SERS spectra at (A) cyt. c adsorbed silver electrode in 5 mM 4-PYS, (B) 4-PYS adsorbed silver electrode in 0.01 mM cyt. c, and (C) silver electrode in 5 mM 4-PYS. Shadow indicates the peaks for cyt. c.

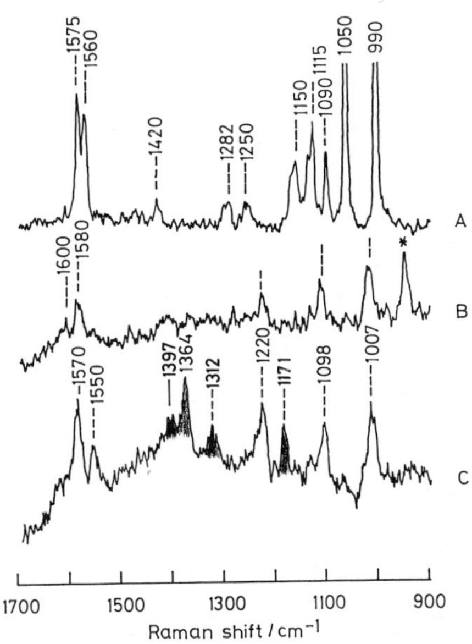

Figure 7. (A) NRS spectrum of 2-PYS powder, SERS spectrum (B) in 10 mM 2-PYS, and (C) in (0.01 mM cyt. c + 10 mM 2-PYS). Shadow indicates the peaks for cyt. c.

cytochrome c immobilized on the 4-PYS modified electrode takes place through the 4-PYS layer and the electrode reaction of cytochrome c in the bulk may take place through the immobilized layers of both 4-PYS and cytochrome c.

Effect of Bis(2-pyridyl) disulfide

There is a marked difference between the normal Raman signals of 2-PYS and those of a solid 2-PYS as shown in Figure 7. The SERS signals of 2-PYS are more like those of 4-PYS. The adsorption behavior of 2-PYS, however, is very different from that of 4-PYS. The adsorption of cytochrome c and 2-PYS on the silver electrode is competitive as is observed in the adsorption from the solution of cytochrome c + 4-BIPY. The SERS spectra shown in Figure 7 reveal that both 2-PYS and cytochrome c adsorb directly onto the silver electrode.

The ac voltammetric response obtained in cytochrome c + 2-PYS solution exhibited two peaks at -0.15 and +0.09 V as shown in Figure 8. This voltammetric behavior is similar to that observed on a gold electrode in a solution of cytochrome c + 4-BIPY[17]. The peak at -0.15 V corresponds to the redox reaction of the adsorbed cytochrome c, which is partially unfolded at the electrode surface, and the peak at +0.09 V corresponds to the redox reaction of native cytochrome c diffusing from the bulk of the solution.

Effect of 1,2-Bis(4-pyridyl)ethylene

Both normal Raman and SERS spectra of 4-PYCH at the silver electrode are basically the same. Cytochrome c cannot displace 4-PYCH at the silver electrode (curve B in Figure 9) but 4-PYCH nearly displaces cytochrome c

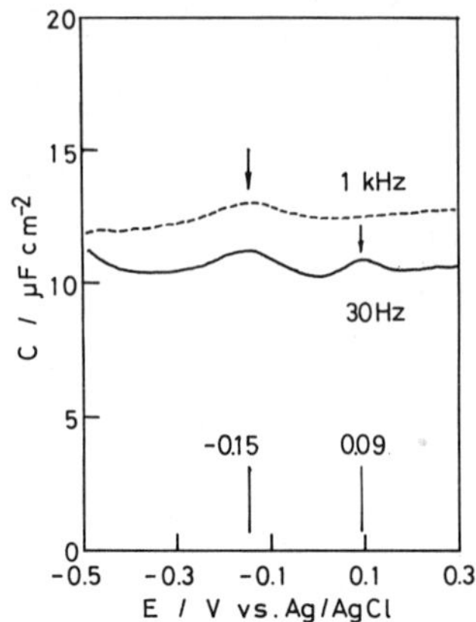

Figure 8. Ac voltammetric response of a silver electrode in (0.04 mM cyt. c + 10 mM 2-PYS). Scan rate: 10 mV/s; modulation: 5 mV.

which is pre-adsorbed on the silver electrode as shown in Figure 9 (curve A). The voltammetric measurements reveal that the immobilized cytochrome c is not detected on the 4-PYCH modified electrode. That is, the electrode

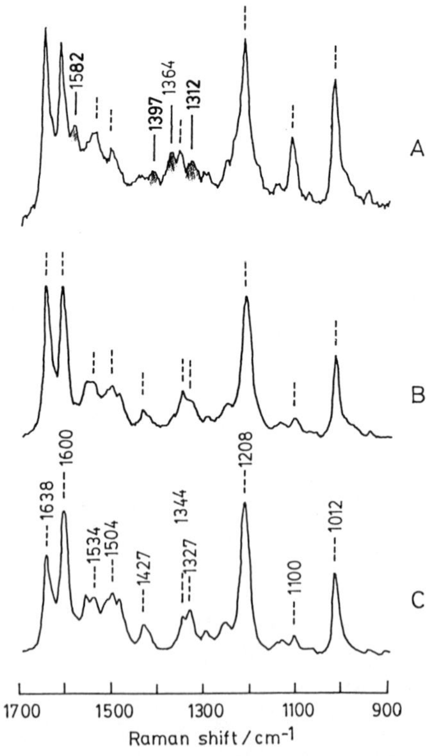

Figure 9. SERS spectra at a silver electrode (A) in 4 mM 4-PYCH (cyt. c is pre-adsorbed on the silver electrode), and (B) in 0.01 mM cyt. c (4-PYCH is pre-adsorbed on the silver electrode), and (C) in 4 mM 4-PYCH.

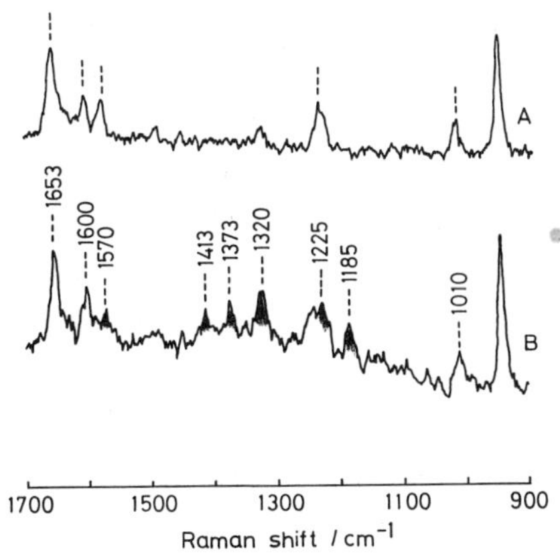

Figure 10. SERS spectra of a silver electrode (A) in 5 mM 2-PYCH and (B) in (0.01 mM cyt. c + 5 mM 2-PYCH).

reaction of cytochrome c in the bulk may take place either through the adsorbed layer of 4-PYCH or its cavity at the formal potential of the native protein. The peak potential separation of the cyclic voltammogram is about 60 mV but the peak heights are markedly suppressed.

Effect of 1,2-Bis(2-pyridyl)ethylene

The SERS spectrum of 2-PYCH at a silver electrode is shown in Figure 10 (curve A) and is basically the same as the normal Raman spectrum of a solid 2-PYCH. The adsorption of cytochrome c and 2-PYCH are competitive from the solution of (0.01 mM cytochrome c + 10 mM 2-PYCH) as shown in Figure 10 (curve B). Cytochrome c in the presence of 2-PYCH shows a voltammetric response visible at 0.08 V in cyclic voltammograms. In contrast to the results reported by Haladjian et al.[11], similar voltammetric behavior of cytochrome c is observed at gold electrode in the presence of 2-PYCH. The role of 2-PYCH in the enhancement of the electrode reaction of cytochrome c is probably the same as that of 4-BIPY and 2-PYS.

DISCUSSION

The present results clearly show that only 2-BIPY among the six pyridine derivatives used inhibits the electrode reaction of cytochrome c in the bulk of the solution. This is probably due to the formation of either stable $Ag(2\text{-}BIPY)_2$ or a chemisorbed film of 2-BIPY on the silver electrode and these films inhibit both the adsorption of cytochrome c on the silver electrode and electron exchange between the electrode and cytochrome c in the bulk.

4-BIPY, 2-PYS and 2-PYCH behave as surface active agents, which provide a hydrophobic electrode surface and prevent partial unfolding of cytochrome c at the silver electrode. It is also interesting to note that the spin state of cytochrome c co-adsorbed with 4-BIPY is the mixture of the HS state, which corresponds to cytochrome c adsorbed on a bare silver electrode, and the LS state, which corresponds to the native protein. The formal potential of cytochrome c co-adsorbed with these surface modifiers is aboiut -0.15 V. The shift of the formal potential of adsorbed cytochrome c on a silver electrode at -0.3 V to that of cytochrome c co-adsorbed with either 4-BIPY or 2-PYS at -0.15 V can be explained in terms of a partial unfolding of cytochrome c. The partial unfolding of cytochrome c at a silver electrode may give rise to an increase in the degree of exposure of the heme of the molecule and accordingly the formal potential shifts toward negative[28]. Since the formal potential of cytochrome c co-adsorbed with either 4-BIPY or 2-PYS is -0.15 V, which is 0.21 V more negative than the formal potential of the native protein, it is very improbable that the electrode reaction of cytochrome c in the bulk take place through this adsorbed cytochrome c layer. On the other hand, the exchange process of adsorbed 4-BIPY (or 2-PYS) on the silver electrode with cytochrome c diffusing from the bulk of the solution would not be difficult because adsorption of both cytochrome c and 4-BIPY (or 2-PYS) is in a dynamic equilibrium. Thus, the electrode reaction of the cytochrome c in the solution becomes possible at the electrode surface covered by the surface modifiers.

On the other hand, 4-PYS is immobilized on the silver electrode, on which cytochrome c is further immobilized, and the electrode reaction of cytochrome c takes place through the adsorbed layers of these compounds.

The variation of the peak heights in cyclic voltammetry can be explained in terms of the structure of the surface modifiers at the electrode surface.

By means of the present results, it has been demonstrated that the surface modifiers for the electrode reaction of cytochrome c either act as surface active agents with which cytochrome c is co-adsorbed on the electrode surface, or form a strongly adsorbed layer through which the electron exchange reaction takes place between cytochrome c (both immobilized species on the surface modifiers and the species in the bulk) and the electrode. The model proposed by Allen et al.[13] for the structure and conformation of the surface modifier cannot explain the present results.

ACKNOWLEDGMENT

The support of a Grant-in-Aid of Scientific Research on Priority Area "Macromolecular Complexes (No. 62612003)" from the Ministry of Education, Science and Culture, Japan, Shimadzu Science Foundation, Kyoto, Japan, and Ajinomoto Co., Toyko, Japan, are gratefully acknowledged.

REFERENCES

1. E.F. Bowden, F.M. Hawkridge and H.N. Blount, in: "Comprehensive Treatise of Electrochemistry", S. Srinivasan, Y.A. Chizmadzhev, J. O'M. Bockris, E.B. Conway and E. Yeager, Eds., Plenum Press, New York, 1985, pp. 297.

2. a) E.F. Bowden, F.M. Hawkridge and H.N. Blount, J. Electroanal. Chem., 161, 355 (1984); b) D.E. Reed and F.M. Hawkridge, Anal. Chem., 59, 2334 (1987).

3. a) K. Niki, T. Yagi, H. Inokuchi and K. Kimura, J. Electrochem. Soc., 124, 1889 (1977); b) K. Niki, T. Yagi, H. Inokuchi and K. Kimura, J. Am. Chem. Soc., 101, 3335 (1979).

4. a) F. Scheller, Bioelectrochem. Bioenerg., 4, 490 (1977); b) B.A. Kuznetsov, G.P. Schumakovich and N.M. Mestechkina, Bioelectrochem. Bioenerg., 4, 512 (1977); c) J. Haladjian, P. Bianco and P.-A. Serre, Bioelectrochem. Bioenerg., 6, 555 (1979).

5. C. Hinnen, R. Parsons and K. Niki, J. Electroanal. Chem., 147, 329 (1983).

6. M.J. Eddowes and H.A.O. Hill, J. Chem. Soc., Chem. Commun., 771 (1977).

7. M.J. Eddowes and H.A.O. Hill, J. Am. Chem. Soc., 101, 4461 (1979).

8. M.J. Eddowes and H.A.O. Hill, Bioelectrochem. Bioenerg., 7, 527 (1980).

9. I. Taniguchi, K. Toyosawa, H. Yamaguchi and K. Yasukouchi, J. Chem. Soc., Chem. Commun., 1032 (1982).

10. I. Taniguchi, K. Toyosawa, H. Yamaguchi and K. Yasukouchi, J. Electroanal. Chem., 140, 187 (1982).

11. J. Haladjian, P. Bianco and R. Pilard, Electrochim. Acta, 28, 1823 (1983).

12. I. Taniguchi, M Iseki, K. Toyosawa, H. Yamaguchi and K. Yasukouchi, J. Electroanal. Chem., 164, 384 (1984).

13. P.M. Allen, H.A.O. Hill, N.J. Walton, J. Electroanal. Chem., 178, 69 (1984).

14. I. Taniguchi, M. Iseki, H. Yamaguchi and K. Yasukouchi, J. Electroanal. Chem., 175, 341 (1984).

15. I. Taniguchi, M. Iseki, H. Yamaguchi and K. Yasukouchi, J. Electroanal. Chem., 186, 299 (1985).

16. I. Taniguchi, N. Higo, K. Umekita and K. Yasukouchi, J. Electroanal. Chem., 206, 341 (1986).

17. C. Hinnen und K. Niki, submitted to J. Electroanal. Chem.

18. M. Kerker, D.S. Wang and H.Chew, Appl. Opt.,19, 4159 (1980).

19. J.I. Gersten and A. Nitzan, J. Chem. Phys., 73, 2023 (1980).

20. E. Koglin and J.M. Sequaris, J. Phys. (Les Ulis, Fr.), 44, C-10,489 (1983).

21. V. Brabec and K. Niki, Biophys. Chem., 23, 63 (1985).

22. K. Niki, Y. Kawasaki, Y. Kimura, Y. Higuchi and N. Yasuoka, Langmuir, 3, 982 (1987).

23. D.L. Brautigan, S. Fergason-Miller and E. Margoliash, "Methods of Enzymology", 33D, 128 (1978)

24. K. Niki, Y. Kawasaki, C. Hinnen, Y. Higuchi and N. Yasuoka, in "Frontiers in Bioinorganic Chemistry", A.V. Xavier, Ed., VCH Verlagsgesellschaft mbH, Weinheim, FRG (1986), pp. 622.

25. P. Hildebrandt and M. Stockburger, in "Spectroscopy of Biological Molecules", A.J.P. Alix, L. Bernard and M. Manfait, Eds., John Wiley and Sons, New York (1985), pp. 25.

26. P. Hildebrandt and M. Stockburger, J. Phys. Chem., 190, 6017 (1986).

27. M. Kim and K. Itoh, Chem. Letters, 357 (1984).

28. E. Stellwagen, Nature, 275, 73 (1978).

THE ELECTROCHEMICAL CHARACTERIZATION OF THE OXENE ADDUCTS OF $PFe(ClO_4)$, PFe^{II}, AND $PFe^{II}(SR)$ [P = TETRAKIS(2,6-DICHLOROPHENYL)PORPHINATO DIANION] IN ACETONITRILE: MODELS FOR COMPOUNDS I AND II OF HORSERADISH PEROXIDASE AND FOR ACTIVATED CYTOCHROME P-450

Donald T. Sawyer*, Hiroshi Sugimoto
Hui-Chan Tung and Paul K.S. Tsang

Department of Chemistry
Texas A&M University
College Station, Texas 77843

INTRODUCTION

The interaction of horseradish peroxidase (an iron(III) heme that has a proximal imidazole) with HOOH results in the formation of a green reactive intermediate known as Compound I. The latter is reduced by one electron to give a red reactive intermediate, Compound II[1]. Both of these intermediates contain a single oxygen atom from HOOH, and Compound I is two oxidizing equivalents above the iron(III)-heme state with a magnetic moment equivalent to three unpaired electrons (S = 3/2). A recent EXAFS study[2] summarizes the physical data in support of formulations of $[(Por^{-}\cdot)Fe^{IV}(O^{2-})]^{+}$ for Compound I, and $[(Por^{2-})Fe^{IV}(O^{2-})]$ for Compound II; and concludes that both species contain an oxo-ferryl group (Fe = O) with a bond length of 1.64 Å.

A recent summary[3] of the activation of O_2 by cytochrome P-450 (an iron(III)-heme protein with a proximal cysteine thiol) concludes that the reactive form of this monooxygenase also contains an oxo-ferryl group $[(RS^{-})(Por^{-}\cdot)Fe^{IV}(O^{2-})]$. The monooxygenase chemistry of cytochrome P-450 has been modeled via the use of $(TPP)Fe^{III}Cl$ (TPP = tetraphenylporphyrin dianion) and $(OEP)Fe^{III}Cl$ (OEP = octaethylporphyrin dianion) with peracids[4,5], iodoso-benzene[4,5], 4-cyano-N,N-dimethyl aniline-N-oxide[6], and hypochlorite[7] to oxygenate model substrates. On the basis of the close parallel with the products from the cytochrome P-450-catalyzed reactions and the net two-oxidizing equivalents of the catalytic cycles for cyt P-450/$(O_2 + 2H^{+} + 2e^{-})$ and HRP/H_2O_2, a general consensus has developed that the reactive intermediate of cytochrome P-450 is analogous to Compound I with an $Fe^{IV}(O^{2-})$ group.

All contemporary work indicates that the reactive intermediate for HRP-I and cytochrome P-450 is an oxygen-atom adduct of $(imid)(Por^{2-})Fe^{III}$ and $(RS^{-})(Por^{2-})Fe^{III}$ [2,8]. The common belief is that atomic oxygen invariably removes two electrons from iron(III) and/or (Por^{2-}) to achieve an $oxo(O^{2-})$ state. Although this misconception is general for the oxygen compounds of transition metals, there is no thermodynamic, electronegativity, or theoretical basis to exclude stable $M(O^{-}\cdot)$ and M(O) species[9]. Thus, the atomic-oxygen adduct of $(Por^{2-})Fe^{III}(B)^{+}$ should be viewed as the resonance hybrid of several valence-bond formulations.

$$[Por^{2-}Fe^{V}(O^{2-})^{+} \leftrightarrow (Por^{-}\cdot)Fe^{IV}(O^{2-})^{+} \leftrightarrow (Por^{o})Fe^{III}(O^{2-})^{+} \leftrightarrow$$

$$[(Por^{2-})Fe^{IV}(O^{-}\cdot)^{+} \leftrightarrow (Por^{-}\cdot)Fe^{III}(O^{-}\cdot)^{+} \leftrightarrow (Por^{o})Fe^{II}(O^{-}\cdot)^{+} \leftrightarrow$$

$$(Por^{2-})Fe^{III}(O)^{+} \leftrightarrow (Por^{-}\cdot)Fe^{II}(O)^{+} \leftrightarrow (Por^{+}\cdot)Fe^{o}(O)^{+}]$$

The last three of these are simple O-atom adducts without intramolecular electron transfer to oxygen, but stablized by d-p orbital overlap [similar to the addition of [O] to CO to give O=C=O or O_2 to heme-Fe(II) to give heme-Fe(II)(O_2)][10].

A recent discussion[9] concludes that high-valent transition metal ions such as iron(IV) are thermodynamically incompatible with the strongly electronegative oxo dianion, particularly when electron transfer results in unpaired valence electrons that can stabilize each other via covalent bond formation. Such stabilization of the two unpaired electrons of ground-state atomic oxygen attenuate its redox potential and it reactivity to an extent proportional to the energy of the metal-oxygen covalent bond.

The results of a recent investigation[11] of model systems provide compelling evidence that stabilized atomic oxygen is present in Compound I and Compound II of horseradish peroxidase and the reactive form of cytochrome P-450. Thus, the combination of tetrakis(2,6-dichlorophenyl)-porphinato iron(III) perchlorate (1, Scheme 1) with pentafluoro-iodoso-benzene, m-chloroperbenzoic acid, or ozone in acetonitrile at -35°C yields a green porphyrin-oxene adduct (2). This species, which has been characterized by spectroscopic, magnetic and electrochemical methods, cleanly and stereospecifically epoxidizes olefins (>99% exo-norbornene-oxide), oxidizes alcohols to aldehydes, oxidatively cleaves diols, and demethylates dimethyl aniline (Scheme 1).

The reaction chemistry and electronic characterization of the green adduct (2) are consistent with an oxygen atom covalently bound to an iron(II)-porphyrin radical center $[P^{-}\cdot)Fe^{II}(O)^{+}]$. The latter has the spectral, magnetic, and redox characteristics of Compound I of horseradish peroxidase (HRP), and the selective stereospecific oxygenase character of the reactive intermediate for cytochrome P-450. Reduction of the green species by one-electron equivalent yields a red species (3, Scheme 1), which has the spectral characteristics and reactivity of Compound II of HRP.

ELECTROCHEMISTRY

The combination of $(Cl_8TPP)Fe^{III}(ClO_4)$ (1) with m-ClPhC(O)OOH in acetonitrile at -35°C results in the rapid and stoichiometric formation of a green product species (2), which is illustrated by the cyclic voltammograms of Figure 1. The same product with identical spectroscopy and electrochemistry results when a 20-fold excess of F_5PhIO is added to the iron(III) porphyrin (only small amounts of 2 are formed for a 1:1 combination). Figure 2 illustrates the electrochemistry of the adduct (2) when formed from the brief exposure of the iron porphyrin to ozone. The electrochemical rest potential of 2 is shifted to +1.35 V vs SCE [from +0.6V for $(Cl_8TPP)Fe^{III}(ClO_4)$ with a new reduction peak at +1.25 V] (curve b, Figure 1). At -35°C the green species (2) has an approximate half-life of

1 h; combination of the reagents at room temperature results in a transient green color prior to the rapid degradation of the porphyrin.

The addition of excess $(H_3O)ClO_4$ to **2** (at -35°C in MeCN) yields an intense blue species (**5**) that exhibits the electrochemistry (curve c, Figure 1) of oxidized iron(III)-porphyrin $[Fe^{III}(Cl_8TPP^{\circ})^{3+}]$. The new redox couples of **2** are absent from the cyclic voltammogram for **5**, which is equivalent to that for $(Cl_8TPP)Fe^{III}(ClO_4)$ (curve a, Figure 1).

The addition of one equivalent of $(Bu_4N)OH$ to a solution of **2** in MeCN at -35°C results in the formation of a red species (**3**) with spectroscopy and electrochemistry (curve d, Figure 1) that are identical with that observed for the product from the controlled potential one-electron reduction of the green species (**2**) in MeCN at -35°C. After the preparation of a pure solution of species **3**, subsequent addition of one equivalent of $(H_3O)ClO_4$ results in the formation of a blue-green species (**4**) that has spectroscopy and electrochemistry that are the same as the product $[(Cl_8TPP^{-}\cdot)Fe^{III}ClO_4)^{+}]$ from the one-electron oxidation of $(Cl_8TPP)Fe^{III}$-(ClO_4).

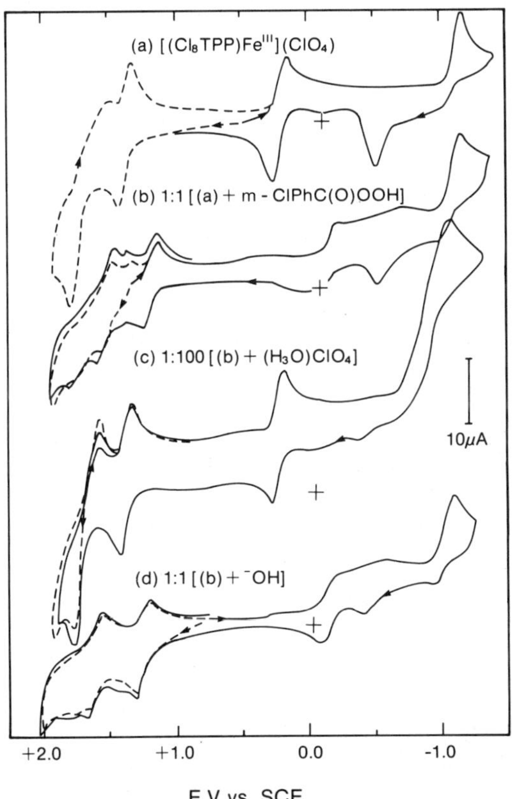

Figure 1. Cyclic voltammograms at a glassy carbon electrode in acetonitrile (0.1 M tetraethylammonium perchlorate) at -35°C for (a) 1 mM $[(Cl_8TPP)Fe^{III}](ClO_4)$, (b) the product from its 1:1 combination with m-ClPhC(O)OOH, (c) the product from the addition of excess protons to the green species (2) of solution (b), and (d) the product from the 1:1 addition of ^{-}OH to the green species (2) of solution (b). Scan rate, 0.1 Vs^{-1}.

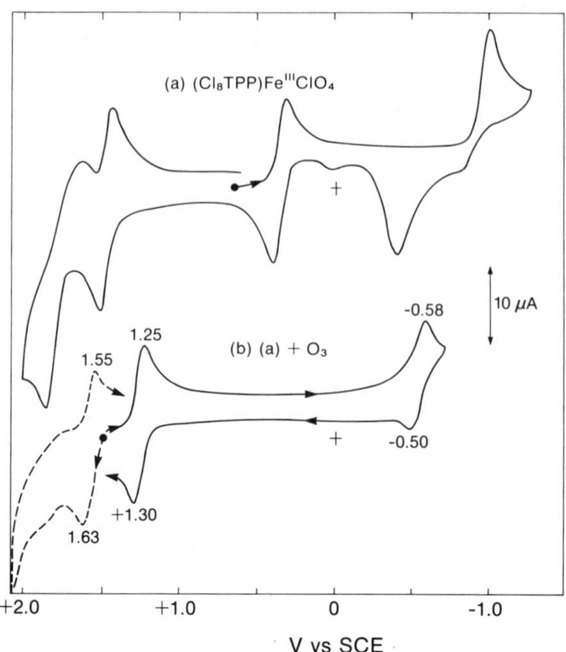

Figure 2. Cyclic voltammograms at a glassy-carbon electrode in acetonitrile (0.1 M tetraethylammonium perchlorate) at -35°C for (a) 1mM $[(Cl_8TPP)Fe^{III}](ClO_4)$, and (b) the product from its exposure for 50s to 0.03 atm of O_3 in O_2 (followed by a purge of the solution with Ar). Scan rate, 0.1 Vs^{-1}.

The electrochemical oxidation of -OH in the presence of (Cl_8TPP^{2-})-$Fe^{III}(ClO_4)$ is illustrated in Figure 3[12]. The addition of one equivalent of -OH causes the Fe(III)/Fe(II) couple to shift to a more negative potential and results in the appearance of a new irreversible oxidation ($E_{1/2}$, +1.09 V vs SCE). With the addition of excess ^{-}OH the anodic voltammogram exhibits two new peaks ($E_{1/2}$, -0.10 V and +0.40 V). Controlled-potential electrolysis at +0.25 V vs SCE $(Cl_8TPP^{2-})Fe^{III}(ClO_4)$ with 3 equivalents of -OH in MeCN at -35°C results in a one-electron oxidation to give species **3** (Scheme 1). The product has a spectrum and a cyclic voltammogram (Figure 2b) that are identical to those for **3**.

Figure 4 illustrates the electrochemistry for $(Cl_8TPP)Fe^{III}(ClO_4)$ and for the product from its 1:1 combination with $PhCH_2S^-$. When O_3 is added to the latter, a new species is formed with a reversible reduction at -0.02 V vs SCE (Figure 4). Consideration of the electrochemical and spectral data prompts the conclusion that $(P^{2-})Fe^{II}(\cdot SCH_2Ph)$ (**6**) is formed from the initial combination (Scheme 2)[13].

The transient exposure of **6** to O_3 appears to produce an ozone adduct (**7**) that is reversibly reduced by one electron to **8**. Species **6** is reduced to form **9**, whose spectrum is transformed upon addition of CO to give **10**.

Table 1 summarizes the reduction potentials in MeCN for atomic oxygen[12], the various iron-oxene species of this model system, the oxene adducts of other metalloporphyrins[14], and the oxidation potentials for ^{-}OH adducts of metal porphyrins[14]. The shift in the two-electron reduction potential for O(g) (from +0.63 V vs NHE in a neutral unbuffered solution to +2.94 V in acidic media) is analogous to that observed when protons are

Scheme 1.

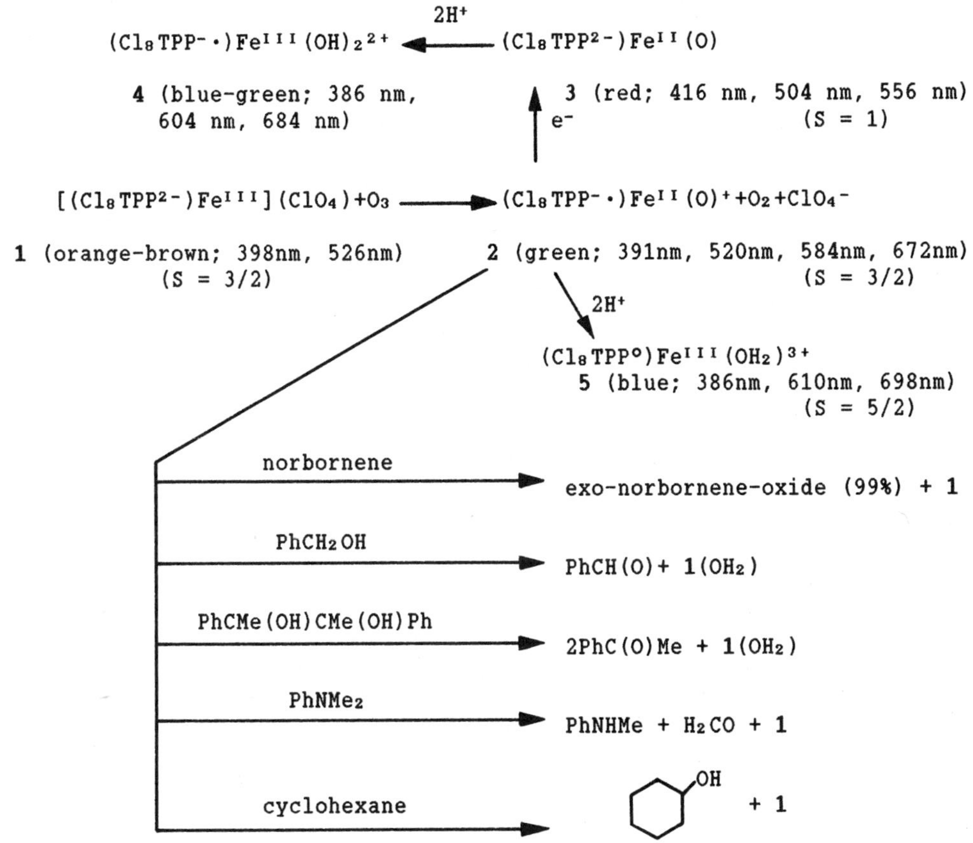

added to the green iron-oxene species (**2**, Scheme 1). Because the addition of an equivalent of ^{-}OH to **2** produces the same red species (**3**) as the addition of an electron to **2**, species **3** is formulated as an iron(II)-oxene. Production of this species by the combination $(Cl_8TPP)Fe^{III}(ClO_4)$, Me_3Py, and HOOH; and its limited epoxidation of olefins are consistent with an iron(II)-oxene (versus iron(III)-$O^{-}\cdot$) formulation. Addition of protons to **3** promotes an intramolecular two-electron transfer to the oxene oxygen [one from the porphyrin ring and one from iron(II)] to give $(Cl_8TPP^{-}\cdot)Fe^{III}(OH_2)^{2+}$ (**4**, Scheme 1). If species **3** contained hypervalent iron or oxidized porphyrin, such a transformation with proton addition would not be expected.

The magnetic moments of **2** (S = 3/2) and for **3** (S = 1) indicate extensive coupling between the ground state triplet p-orbitals of atomic oxygen and the half filled d-orbitals of iron(II). In terms of valence-bond considerations overlap by the metal-d and oxygen-p orbitals will result in the formation of a metal-oxygen σ-bond and a metal-oxygen π-bond. The two-electron reduction potentials under acidic conditions (Table 1) for **2** (+1.94 V vs NHE) and O(g) (+2.94 V) provide an approximate measure of the bond energy for the $(P^{-}\cdot)Fe^{II}{=}O$ covalent double bond; B.E. = ΔE x n x 23.1 kcal = 46.2 kcal (Table 1). Likewise, the two-electron reduction potential for **3** (+1.30 V vs NHE) relative to that for O(g) (+2.94 V) provides an indication of the bond energy for the $(P^{2-})Fe^{II} = O$ covalent double bond; B.E. = +1.43 x 2 x 23.1 = 66 kcal. Thus, the much lower reactivity of **3** with olefins is consistent with the

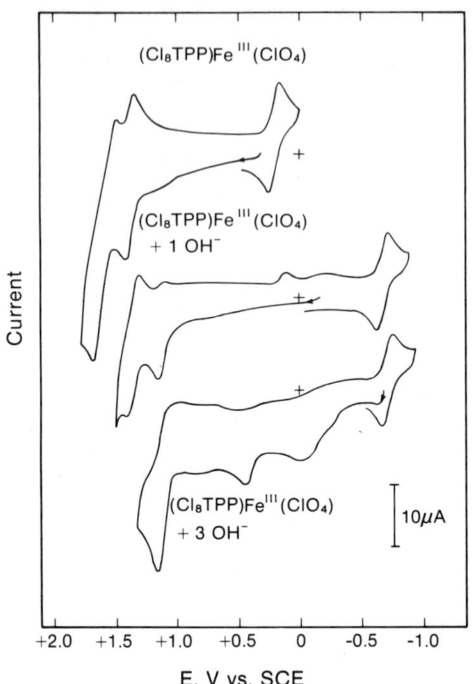

Figure 3. Cyclic voltammograms in MeCN (0.1 M TEAP) of (a) 0.6 mM $(Cl_8TPP)Fe^{III}(ClO_4)$, (b) 0.6mM $(Cl_8TPP)Fe^{III}(ClO_4)$ + 1 equiv TBAOH, and (c) 0.6 mM $(Cl_8TPP)Fe^{III}(ClO_4)$ + 3 equiv TBAOH. Scan rate, 0.1 Vs^{-1}; GCE (area, 0.062 cm^2).

greater stabilization of (O) by the iron(II) center. The redox data for the oxene adducts of $(P^{2-})Cr^{III}(O)^+$ and $(P^{-}\cdot)Mn^{II}(O)^+$ indicate that the oxygen atom is stabilized with covalent bond energies of 66 kcal and 106 kcal, respectively (Table 1).

Scheme 2.

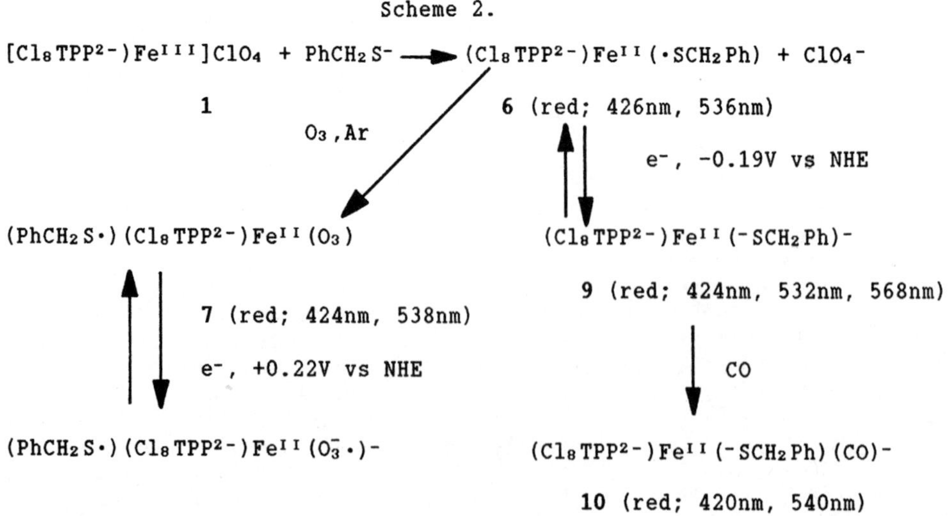

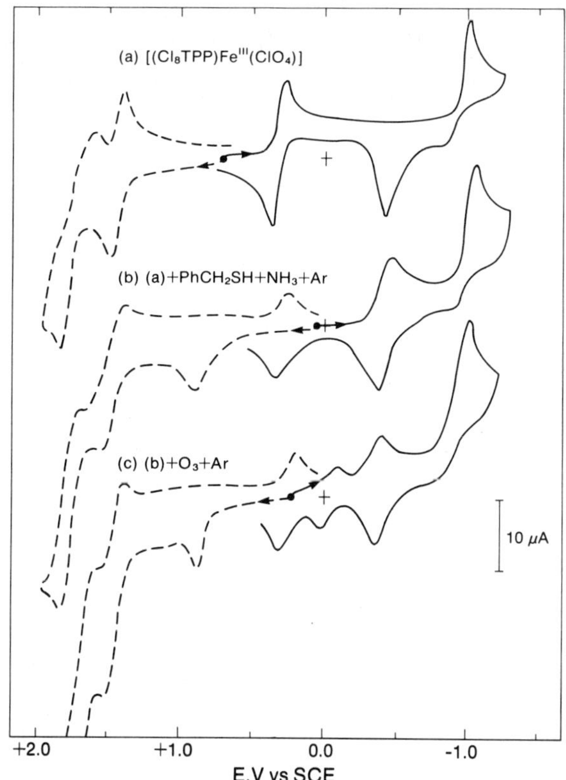

Figure 4. Cylic voltammograms at -35°C for (a) 1 mM $[(Cl_8TPP)Fe^{III}]$-ClO_4, (b) solution (a) plus 1 mM $PhCH_2S^-$, and (c) solution (b) plus 0.03 atm O_3 in O_2 (1 atm) for 50s (purged with Ar). Scan rate, 0.1 Vs^{-1}.

The spectroscopy, electrochemistry, and magnetic properties of 2 indicate that its iron center is equivalent to that of Compound I of HRP. Recent EXAFS studies[2,15] of Compound I confirm that it conctains an Fe=O double bond (bond distance, 1.64 Å), and that its conversion to Compound II (via one-electron reduction) gives a species with an Fe=O group that has the same iron-oxygen bond distance. Again, the spectroscopic and electrochemical properties of 3, and its reduced reactivity with olefins, indicate that the electronic structure of its iron-oxygen center is analogous to that of Compound II of HRP.

The present results indicate that 2 contains a stabilized oxygen atom, and the parallel chemistry with the active form of cytochrome P-450 prompts the conclusion that it also contains stabilized atomic oxygen. We have argued elsewhere[9-16] that the most reasonable electronic formulation for the active form of cytochrome P-450 is $(RS\cdot)(Por^{2-})Fe^{II}(O)$ with an $(RS\cdot)$-Fe(II) covalent bond and an Fe(II)=O covalent double bond. The inability to form 2 with HOOH as the oxidant and the inefficient formation of the active form of cytochrome P-450 via the peroxide shunt[8,17] may mean that HOOH is not formed as an intermediate during the cytochrome P-450

Table 1. Reduction potentials for stabilized zero-valent atomic oxygen (oxene) in acetonitrile at -35°C ($P^{2-} \equiv Cl_8TPP^{2-}$), and oxidation potentials for ^-OH adducts of metal porphyrins.

	E°' V vs NHE	σ-π Double Bond Energy, kcal
A. Double-bond Stabilization [1M(H_3O)(ClO_4]		
$O(g) + 2H^+ + 2e^- \rightarrow H_2O$	+2.94	-
$O_3(g) + 2H^+ + 2e^- \rightarrow O_2(g) + H_2O$	+2.59	16
$(P^-\cdot)Fe^{II}(O)^+ + 2H^+ + 2e^- \rightarrow (P^{2-})Fe^{III}(OH_2)^+$	+1.94	46
$(P^-\cdot)Mn^{II}(O)^+ + 2H^+ + 2e^- \rightarrow (P^-\cdot)Mn^{II}(OH_2)^+$	+0.65	106
$(P^{2-})Cr^{III}(O)^+ + 2H^+ + 2e^- \rightarrow (P^{2-})Cr^{III}(OH_2)^+$	+1.50	66
$(P^{2-})Fe^{II}(O) + 2H^+ + 2e^- \rightarrow (P^{2-})Fe^{II}(OH_2)$	+1.51	66
		π-Single Bond Energy, kcal
B. Pi-bond Stabilization		
$O(g) + e^- \rightarrow O^-\cdot$	+0.67	-
$O_3(g) + e^- \rightarrow O_3^-\cdot$	+0.35	7
$(P^{2-})Fe^{II}(O) + e^- \rightarrow (P^{2-})Fe^{II}(\cdot O^-)^-$	-0.30	22
$(P^{2-})Mn^{II}(O) + e^- \rightarrow (P^{2-})Mn^{II}(\cdot O^-)^-$	-0.85	35
$(P^{2-})Cr^{II}(O) + e^- \rightarrow (P^{2-})Co^{II}(\cdot O^-)$	-0.69	31
$(P^{2-})Co^{II}(O) + e^- \rightarrow (P^{2-})Co^{II}(\cdot O^-)$	+0.30	9
$(RS\cdot)(P^{2-})Fe^{II}(O_3) + e^- \rightarrow (RS\cdot)(P^{2-})Fe^{II}(O_3^-\cdot)$	+0.22	10
C. Redox Reactions of Oxene Adducts		
$(P^-\cdot)Fe^{II}(O)^+ + e^- \rightarrow (P^{2-})Fe^{II}(O)$	+1.49	
$(P^-\cdot)Fe^{III}(O)^{2+} + e^- \rightarrow (P^-\cdot)Fe^{II}(O)^+$	+1.83	
$(P^o)Fe^{III}(O)^{3+} + e^- \rightarrow (P^-\cdot)Fe^{III}(O)^{2+}$	+1.96	
$(P^{2-})Cr^{III}(O)^+ + e^- \rightarrow (P^{2-})Cr^{II}(O)$	+1.16	
D. Oxidation of ^-OH and Its Metal-porphyrin Adducts		
$^-OH \rightarrow \cdot OH + e^-$	+0.92	
$2\ ^-OH \rightarrow \cdot O^- + H_2O + e^-$	+0.59	
$(TPP)Zn(^-OH)^- + {}^-OH \rightarrow (TPP)Zn(O^-\cdot)^- + e^-$	+0.73	
$(TPP)Fe(^-OH)^- \rightarrow (TPP)Fe(\cdot OH) + e^-$	-0.50	
$(Cl_8TPP)Fe(^-OH)^- \rightarrow (Cl_8TPP)Fe(\cdot OH) + e^-$	-0.45	
$(Cl_8TPP)Fe(\cdot OH) + {}^-OH \rightarrow (Cl_8TPP)Fe(\cdot OH)_2 + e^-$; $\xrightarrow{-H_2O} Cl_8TPPFe(O)$	+0.14	
$(Cl_8TPP)Fe^{II}(\cdot OH) \rightarrow (Cl_8TPP)Fe^{III}(\cdot OH)^+ + e^-$	+1.33	

activation cycle. A direct reduction cycle to give the active species seems likely

$$(RSH)(Por)Fe^{III} \xrightarrow{e^-, O_2} (RSH)(Por)Fe^{II}(O_2) \xrightarrow{e^-, H^+} [(RS\cdot)(Por)Fe^{II}(O) + H_2O] \quad (1)$$

Scheme 3.

Catalase Redox Cycle:

$$(PhO^-)PFe^{III}(H_2O) + H_2O_2 \rightarrow (PhO\cdot)PFe^{II}(O) + H_2O$$

Compound I (catalase)
$S = 1/2$

$$(PhO\cdot)PFe^{II}(O) + H_2O_2 \rightarrow (PhO^-)PFe^{III}(H_2O) + O_2$$

Peroxidase Redox Cycle:

$$(Imid)(P^{2-})Fe^{III}(H_2O)^+ + H_2O_2 \rightarrow (Imid)(P^-\cdot)Fe^{II}(O)^+ + 2H_2O$$

$S = 5/2$; Compound I (peroxidase) $S = 3/2$

$$(Imid)(P^-\cdot)Fe^{II}(O)^+ + RH \rightarrow [(ImidH)(P^{2-})Fe^{II}(O)]^+ + R\cdot$$

Compound II

$$\downarrow RH$$

$$[(Imid)(P^{2-})Fe^{III}(OH_2)] + R\cdot$$

Scheme 4.

Cytochrome P-450 Redox Cycle:

$$(RS\cdot)PFe^{II}(OH_2)_{LS} + R'H \xrightarrow{H^+} (RSH)PFe^{III}(OH_2)(R'H)_{HS}^+$$

$S = 1/2$; $S = 5/2$

$$\xrightarrow{e^-}$$

$$(RSH)PFe^{II}(OH_2)(R'H) \xrightarrow{O_2} (RSH)PFe^{II}(O_2)(R'H) + H_2O$$

$S = 4/2$; $S = 0$

$$\xrightarrow{e^-, H^+}$$

$$(RS\cdot)PFe^{II}(O)(OH_2)(R'H) \longrightarrow (RS\cdot)PFe^{II}(OH_2) + R'OH$$

$S = 1/2$; $S = 1/2$

On the basis of the preceding results and arguments reasonable reaction cycles are proposed for the activation of HOOH by the catalase and peroxidase proteins (Scheme 3), and for the activation of O_2 by the cytochrome P-450 protein (Scheme 4).

ACKNOWLEDGEMENT

This work was supported by the National Science Foundation under grant No. CHE-8516247.

REFERENCES

1. (a) P. George, Adv. Catal., 4, 367 (1952); (b) P. George, Biochim. J., 54, 267 (1953); (c) P. George, Biochim. J., 55, 220 (1953).

2. J.E. Denner-Hahn, K.S. Eble, T.J. MacMurray, M. Renner, A.L. Balch, J.T. Groves, J.H. Dawson and K.O. Hodgson, J. Am. Chem. Soc., 108, 7819 (1986).

3. P.F. Guengerich and T.L. McDonald, Acc. Chem. Res., 17, 9 (1986).

4. (a) J.T. Groves, R.C. Haushalter, M. Nakamura, T.E. Nemo and B.J. Evans, J. Am. Chem. Soc., 103, 2884-2886 (1981); (b) J.T. Groves and Y. Watanabe, J. Am. Chem. Soc., 108, 7834-7836 (1986).

5. (a) P.S. Taylor, D. Dolphin and T.G. Traylor, J. Chem. Soc. Chem. Commun. 279-280 (1984); (b) T.G. Traylor, T. Nakano, B.E. Dunlap, P.S. Traylor and D. Dolphin, J. Am. Chem. Soc., 108, 2782-2784 (1986).

6. (a) C.M. Dicken, T.C. Woon and T.C. Bruice, J. Am. Chem. Soc., 108, 1636-1643 (1986); (b) T.S. Calderwood and T.C. Bruice, Inorg. Chem., 25, 3722-3724 (1986); (c) T.S. Calderwood, W.A. Lee and T.C. Bruice, J. Am. Chem. Soc., 107, 8272-8273 (1985).

7. J.P. Collman, T. Kodadek and J.I. Brauman, J. Am. Chem. Soc., 108, 2588-2594 (1986).

8. P.R. Ortiz de Montellano, Ed., "Cytochrome P-450", Plenum Press, New York (1986).

9. D.T. Sawyer, Comments Inorg. Chem. 6, 103-121 (1987).

10. W.A. Goddard, III and B.D. Olafson, Proc. Natl. Acad. Sci., USA 72, 2335-2339 (1975).

11. H. Sugimoto, H.-C. Tung and D.T. Sawyer, J. Am. Chem. Soc., 110, 0000 (1987).

12. P.K.S. Tsang, P. Cofré and D.T. Sawyer, Inorg. Chem., 26, 3604-3609 (1987).

13. H.-C. Tung and D.T. Sawyer, J. Am. Chem. Soc., submitted (1988).

14. H. Sugimoto and D.T. Sawyer, Inorg. Chem., submitted (1988).

15. M. Chance, L. Powers, T. Poulos and B. Chance, Biochemistry, 25, 1266 (1986).

16. H. Sugimoto, L. Spencer and D.T. Sawyer, Proc. Natl. Acad. Sci., USA 84, 1731-1733 (1987).

17. M.B. MacCarthy and R.E. White, J. Biol. Chem., 258, 9153-9158 (1983).

DIRECT ELECTRICAL COMMUNICATION BETWEEN CHEMICALLY MODIFIED ENZYMES AND METAL ELECTRODES: III. ELECTRON-TRANSFER RELAY MODIFIED GLUCOSE OXIDASE AND D-AMINO-ACID OXIDASE

Adam Heller* and Yinon Degani

AT&T Bell Laboratories
Murray Hill, New Jersey 07974

INTRODUCTION

Electrical communications between enzymes and metal electrodes is an element in the bridging of electronics and biochemistry and is of specific relevance to the electrochemical assay of biochemicals. In such assays, the enzyme is usually first reduced by the substrate, then is reoxidized either directly at an electrode[1], or indirectly by oxygen or a diffusing redox mediator[2-30]. Electrochemical or chemical assays of the oxygen consumed, or of the hydrogen peroxide or reduced mediator products, serve in amperometric[2-30] or colorimetric[31,32] assays of substrates such as glucose and cholesterol.

In homogeneous solutions, redox enzymes, such as glucose oxidase, accept electrons from and transfer electrons to small oxidizable/reducible ions or molecules, but do not exchange electrons with simple metal electrodes. In the absence of adequately fast electron transfer, the current flowing through a gold, platinum or carbon electrode immersed in a glucose oxidase containing electrolytic solution does not vary with the concentation of glucose, even though glucose reduces the FAD (oxidized flavin adenine dinucleotide), and the reduced enzyme diffuses to the surface of the electrode. $FADH_2$ is not oxidized electrochemically through the electrode reaction

$$FADH_2 \rightarrow FAD + 2H^+ + 2e^- \qquad E^\circ = -0.05 \text{ V (SHE) at pH 7}$$

because the $FAD/FADH_2$ redox centers are located deep in the enzyme, and even if the enzyme is adsorbed on the electrode, the distance between either of the enzyme's two $FAD/FADH_2$ centers and the surface of the electrode exceeds the distance across which electrons are transferred at measurable rates.

To enable electron-transfer from redox centers of enzymes to metal electrodes, the common practice of the past 60 years has been to add to the enzyme solution a redox couple having an appropriate redox potential, or to introduce oxygen, which is reduced by $FADH_2$ to hydrogen peroxide, a species that can be colorimetrically or amperometrically assayed. Both oxidized electron-shuttling species and oxygen penetrate the enzyme, and upon approaching its $FADH_2$ centers, accept electrons. In amperometric assays the reduced shuttles or H_2O_2 diffuse to an electrode, where they are reoxidized. In the case of glucose oxidase, the ferrocene carboxylate/-ferrocinium carboxylate couple is a particularly effective electron-shuttle[18]. Electrochemical glucose sensors based on this couple, on the O_2/H_2O_2[2-7,9-13,17,20-21,28,30] couple and on electrodes consisting of ions that when dissolved can form fast redox couples, such as TTF/TCNQ, have been reported[8,15,22,23].

Here we discuss chemical modification of redox enzymes. We show that by bonding electron-transfer relays to their protein one can establish direct electrical communication between their redox centers and metal or carbon electrodes.

EXPERIMENTAL

Chemicals

Glucose oxidase [E.C. 1.1.3.4.] type X and D-amino-acid-oxidase [E.C. 1.4.3.3.] type I were purchased from Sigma. Na-HEPES (sodium 4-(2-hydroxy-ethyl)-1-piperazinethane sulfonate), DEC [1-(3-dimethylaminopropyl)-3-ethylcarboiimide.HCl], ferrocenecarboxylic acid, isonicotinoyl chloride. HCl and 4-aminopyridine were purchased from Aldrich, and Sephadex G-15 and Sephadex C-25 from Pharmacia Fine Chemicals. $[Ru(NH_3)_5Cl]Cl_2$ was prepared from $[Ru(NH_3)_6]Cl_3$ (Johnson Matthey), following a reported procedure[33]. Ferrocylacetic acid was made from N,N-dimethylaminoethylferrocene (Aldrich) by a reported procedure[34].

Preparation of Modified Enzymes

Amides Between Ferrocylacetic Acid and Glucose-Oxidase Amines

The procedure followed is outlined in Scheme 1. Ferrocylacetic acid (60 mg), Na-HEPES (200 mg) and urea (480 mg) were dissolved in water (4mL).

Scheme 1

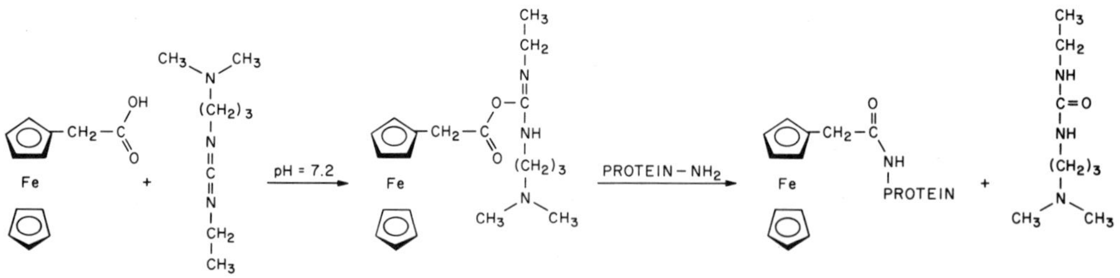

The pH of the solution was adjusted to 7.0 by adding 0.1 M HCl dropwise. Glucose oxidase (100 mg) was then added and dissolved, followed by DEC (50 mg). The pH was readjusted to 7.0, and the solution left to react overnight (16 h) in an ice bath. The chemically modified enzyme was separated from the reaction mixture by gel filtration chromatography on Sephadex G-15.

Amides Between Ferrocenecarboxylic Acid and Glucose Oxidase

Amines or D-Amino-Acid-Oxidase Amines

These were prepared in a procedure similar to that of Scheme 1, except that the ferrocylacetic was replaced by ferrocenecarboxylic acid. All reactions were carried out in an ice bath. 80 mg of ferrocenecarboxylic acid were dissolved in 4 mL of a 0.15 M Na-HEPES solution, to form a slightly turbid pH 7.3 ± 0.1 solution. (When higher, the pH was lowered to this value by adding dropwise, with stirring, 0.1 M HCl.) 100 mg of DEC was then added and dissolved, followed by 810 mg of urea. After the urea dissolved the pH was again adjusted to 7.2-7.3 and 60 mg of glucose oxidase were added and dissolved. When out of the 7.2 to 7.3 range, the pH was again adjusted by adding either 0.15 M Na-HEPES or 0.1 M HCl. The solution was then placed in a glass vial, sealing with a paraffin foil, and left immersed overnight in an ice and water containing dewar (~15 h). The turbid solution was centrifuged, and the supernatant liquid was filtered under mild (~2 atm) pressure through a 0.2 μm pore filter. The modified enzyme was separated from the unreacted ferrocenecarboxylate and from the reaction product of ferrocenecarboxylate and DEC by gel filtration through a 2 cm diameter, 22 cm long column of Sephadex G-15. Prior to the preparation of the column, the Sephadex was soaked for >2 h in a pH 7.0 buffer, prepared by adding to a 0.085 M solution of Na_2HPO_4 drops of concentrated HCl. This buffer was used as the eluent in the separation. The enzyme was eluted in the first orange-colored fraction of a volume of 4 mL or less.

Amides Between the Ruthenium Pentaammine Complex of Isonicotinic

Acid and Glucose Oxidase Amines

The modified enzyme was formed by the reaction sequence outlined in Scheme 2. Glucose oxidase (100 mg) Na-HEPES (500 mg) and urea (240 mg)

Scheme 2

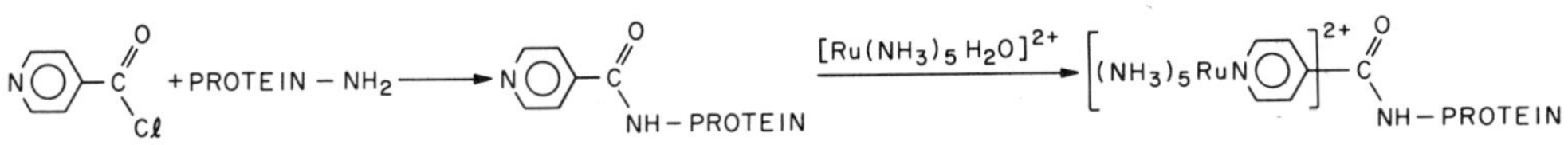

N⟨py⟩–C(=O)Cl + PROTEIN – NH_2 ⟶ N⟨py⟩–C(=O)NH – PROTEIN $\xrightarrow{[Ru(NH_3)_5H_2O]^{2+}}$ $[(NH_3)_5RuN⟨py⟩]^{2+}$–C(=O)NH – PROTEIN

were dissolved in 2 mL of water. Solid isonicotinoyyl chloride.HCl (50 mg) was added to the stirred solution. The resulting solution was kept at room temperature for 1 h. The isonicotinoyl-modified glucose oxidase was separated from the reaction mixture by gel-filtration chromatography (Sephadex G-15). $[Ru(NH_3)_5Cl]Cl_2$ (20 mg) was added to the yellow-enzyme containing fraction, followed by glucose (20 mg). The solution was stirred under Ar until the $[Ru(NH_3)_5Cl]Cl_2$ dissolved, then an amount of urea, sufficient to yield a 2 M solution was added. The resulting solution was allowed to react at room temperature for 4 h. As the complex formed, the color of the solution changed from yellow to deep red (λ_{max} = 490 nm), a color typical of the $[Ru(NH_3)_5\text{isonicotinamide}]^{2+}$ ion[35]. The

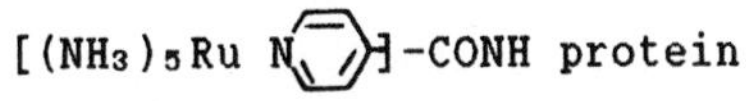

$[(NH_3)_5Ru$ N⟨py⟩]-CONH protein

modified glucose oxidase was then separated from the reaction mixture using a 15 cm long, 2 cm diameter column of Sephadex C-25 cation exchange resin, equilibrated with 0.1 M phosphate buffer (pH 7.0). This buffer was used also as eluent.

$Ru(NH_3)_5$ (pyridyl-azo-glucose oxidase)

The reaction sequence through which multiple pyridyl-azo groups, complexed with $Ru(NH_3)_5$, were bound to the enzyme is shown in Scheme 3. Glucose oxidase (100 mg) was dissolved in 2 mL carbonate buffer (0.5 M pH = 10.5) and chilled in an ice bath. 100 μL of 10 M sodium nitrite was

Scheme 3

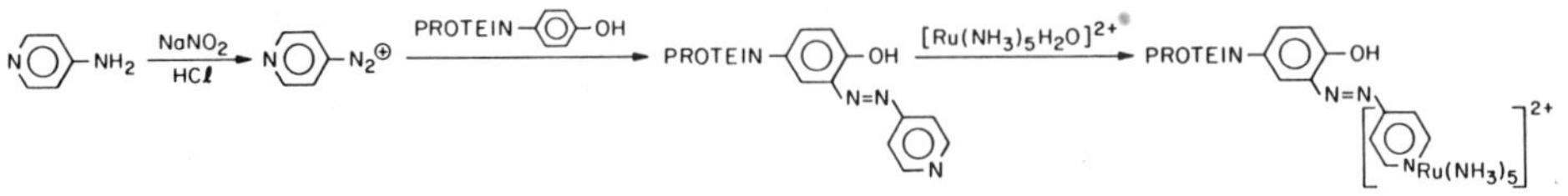

then added to the 4-aminopyridine solution. The solution turned orange-yellow promptly and gas evolution was observed. 100 μL of this solution was added dropwise to the vigorously stirred glucose-oxidase solution, which immediately turned red. The modified enzyme was separated from the reaction mixture by gel filtration chromatography (Sephadex G-15) and was subsequently coupled to $[Ru(NH_3)_5(H_2O)]^{2+}$, following the procedure described for the isonicotinoyl-modified enzyme. The blue-green enzyme solution shows absorption maxima at 460 nm and 610 nm.

Complexes of Ruthenium Pentaammine and Glucose Oxidase

The reaction sequence for this enzyme-modification is based on the procedure for binding of ruthenium pentammine to ferricytochrome C[36] and is outlined in Scheme 4. Freshly prepared amalgamated Zn (1 g) was

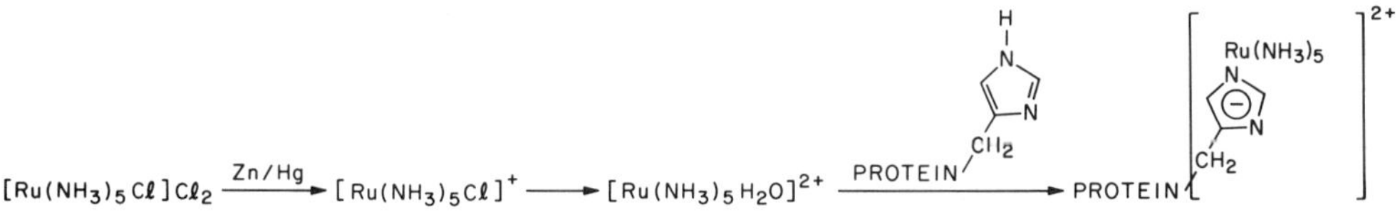

added to a vigorously stirred suspension of $[Ru(NH_3)_5Cl]Cl_2$ (200 mg) in 5 mL phosphate buffer (0.1 M, pH 7.0). The suspension was stirred under Ar for 15 min, then filtered (0.2 μm filter) to yield a deep yellow solution of $[Ru(NH_3)_5(H_2O)]^{2+}$. Glucose oxidase (100 mg) and urea (480 mg) were dissolved in the $[Ru(NH_3)_5(H_2O)]^{2+}$ solution and the resulting solution was kept under Ar in an ice bath for 48 h. Subsequently, O_2 was bubbled through the solution for 1 h. The ruthenium pentammine complex modified glucose oxidase was then separated from the reaction mixture by cation exchange chromatography (Sephadex C-25).

Assay of the Iron and Ruthenium Contents of the Chemically Modified Enzymes

To determine the average number of electron-transfer relays bound to each enzyme molecule, the iron or ruthenium content of the enzyme-containing solutions was assayed by atomic absorption spectroscopy. Determination of the number of relays per enzyme molecule was made possible by quantitative recovery of the enzymes in the gel-filtration chromatography or cation-exchange steps, where they were quantitatively separated from low molecular weight complexes of the metals.

Electrochemical Measurements

The electrochemical cell has two compartments. The inner, enzyme containing, compartment was a 1.0 cm ID pyrex tubing with a VF glass frit. The working electrode was an epoxy embedded gold, platinum or carbon disks of 1.5, 4.0, 3.0 mm diameter, respectively. The working electrode was polished sequentially with 1.0, 0.3, and 0.05 μm Al_2O_3, rinsed in deionized water, and dried in a nitrogen stream prior to each measurement. The reference electrode was NaCl-saturated calomel, and the counter electrode was a spectroscopic graphite rod. The electrolyte in the outer compartment was the buffer used for gel-filtratiion. The measurements were performed under a nitrogen atmosphere, using a PAR 173 potentiostat/galvanostat with a PAR 175 programmer, and an HP 7090 A recorder. The cyclic voltammograms of the native and modified enzymes were obtained at 25 units/mL (10 mg/mL) for both native and modified-glucose-oxidase. The solutions of D-amino-acid-oxidase were buffered with 0.1 M HEPES (pH 8.2) and those of glucose-oxidase, with 0.1 M phosphate (pH 7.0). All solutions contained 3000 to 6000 units/mL of catalase to decompose any hydrogen peroxide that might be formed in the presence of traces of oxygen.

TTF/TCNQ Electrodes

Discs of 6 mm diameter were made in graphite rods recessed in Teflon sleeves, following the reported procedure[23].

Measurement of the Diffusion Coefficient of the Native and Chemically Modified Enzymes.

The coefficients were measured by dynamic light scattering as described by Wiltzius, in his apparatus[37].

Measurement of the Stability of the Chemically-Modified Enzymes and of the Enzyme Electrode.

The enzymatic and electrochemical activity of the native or modified enzymes as a function of time was determined by incubating these in 30 mM glucose. Periodically retrieved samples were subjected to gel-filtration-chromatography, in order to eliminate all low-molecular-weight decomposition-products. Following such separation, the enzymes were assayed for their iron or ruthenium content. The enzymatic activity of the samples was determined by measuring the time required to bleach a blue (oxidized) indophenol solution. The rate of decrease of the electrochemical activity, i.e., the drop in the Faradaic current at 30 mM glucose concentration, was then measured by cyclic voltammetry.

RESULTS

Glucose oxidase and D-amino acid oxidase can be chemically modified both in the presence and the absence of 2 M urea. At 2 M concentration urea interacts with proteins, opening their structure. Modification in the presence of 2 M urea was preferred because of superior reproducibility of the modified-enzyme-properties and because of superior electrochemical characteristics of the modified enzymes. At higher (6 M) concentration, urea alters drastically, and often irreversibly, the protein structure[38].

The modified enzymes, irrespective of the chemical changes in their structure, continue to catalyse their normal reactions

$$\text{glucose} + O_2 \xrightarrow{\text{glucose oxidase}} \text{gluconolactone} + H_2O_2 \quad (2)$$

or

$$\text{D-alanine} + O_2 + H_2O \xrightarrow{\text{D-amino acid oxidase}} \text{ammonium pyruvate} + H_2O_2 \quad (3)$$

The changes in enzymatic activity upon chemical modification are summarized in Table 1.

Curve a of Figure 1, shows the cyclic voltammogram of a solution of glucose oxidase before modification. Identical curves were obtained with and without 5.0 mM glucose. The cyclic voltammograms observed under similar conditions with the ferrocenecarboxamide-modified enzyme are seen in curve b (no glucose), c (0.8 mM glucose) and d (5.0 mM glucose) of Figure 1. It is evident that while the unmodified enzyme does not respond to the addition of glucose, the modified enzyme does. The point at which half the glucose-concentration-dependent limiting current is reached is 0.5 V (SHE), near the redox potential of the ferrocenecarboxylate/ferrociniumcarboxylate couple[1].

Because the observed electrochemistry can originate from noncovalently attached ferrocene moieties (that would dissociate), these were removed in gel filtration steps. To this end, the solution of the ferrocene-

Table 1. Activity of the Chemically Modified Enzymes

RELAY/ENZYME	NUMBER OF RELAYS PER ENZYME MOLECULE	ACTIVITY OF MODIFIED ENZYME / ACTIVITY OF NATIVE ENZYME
Ferrocene–CO–NH–GLUCOSE OXIDASE	12 ± 1	0.60 ± 0.05
Ferrocene–CH_2CO–NH–GLUCOSE OXIDASE	13 ± 1	0.6 ± 0.1
$(NH_3)_5$ Ru N(pyridine)–CO–NH–GLUCOSE OXIDASE	6 ± 1	0.75 ± 0.05
$(NH_3)_5$ Ru N(pyridine)–N=N–GLUCOSE OXIDASE	2 ± 0.3	2.7 ± 0.3
$(NH_3)_5$ Ru–GLUCOSE OXIDASE	14 ± 1	0.70 ± 0.05
Ferrocene–CO–NH–D–AMINO–ACID OXIDASE	3 ± 1	0.27 ± 0.02

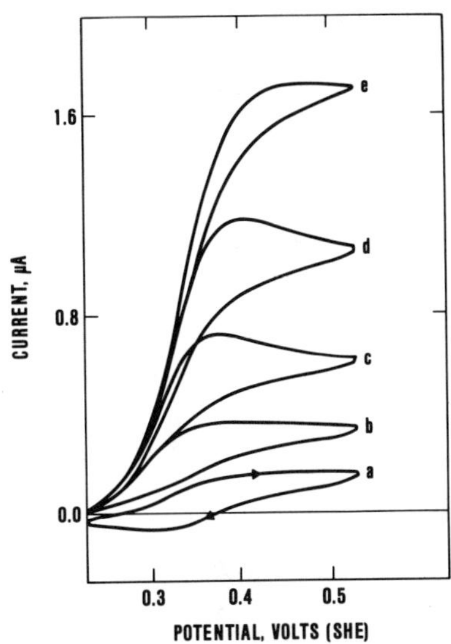

Figure 1. Cyclic voltammograms of glucose oxidase containing buffer solutions with and without glucose. Prior to modification, either without glucose or with 5.0 mM glucose **a**; after attaching the electron relays, with 0.8 mM glucose, **c**; and with 5 mM glucose, **d**. 1.5 mm diameter gold-disc electrodes, 1 mVs^{-1}.

carboxamide-modified glucose oxidase was incubated with 50 mM glucose at 25°C under nitrogen for 24 h and the gel filtration was repeated, until there was no colored band except that of the modified enzyme. The electrochemical behavior of the solution after the second gel filtration was already identical with that after the first, i.e., with that shown in curves b, c and d of Figure 1.

The cyclic voltammograms of the other modified enzymes (Figures 2-6), also show that enzymes that do not communicate prior to modification with metal or carbon electrodes, they do so after electron-transfer relays are chemically bound to their proteins. Direct electrical communication is established with gold, platinum, glassy-carbon or graphite electrodes after binding to the flavoenzymes any of a series of fast redox couples, having redox potentials 0.07 V to 0.55 V positive of the enzymes' redox potential (-0.05 V (SHE) at pH 7)[39]. Relays attached through amide links to protein amines (such as those of lysine), through azo-bonds to the proteins' activated aromatic rings (such as those of tyrosine or tryptophane), or through coordination to the proteins' heterocyclic rings (imidazole-rings of histidine) are all effective. Any of these is sufficient to alter the electrochemical characteristics of the enzyme electrodes that are, prior to enzyme-modification, insensitive to change in the concentration, or even to the very presence of substrate. In contrast, the relay-modified enzyme-electrodes sense the presence of their enzyme's substrate and show Faradaic currents that increase, across a defined concentration range, with substrate concentration. One thus observes, at potentials equaling or exceeding the redox potentials of the electron relays, glucose-concentration dependent currents in the case of modified glucose oxidase and response to D-alanine in the case of modified D-amino-acid oxidase.

The number of relays per enzyme molecule, the number of enzyme-daltons per relay, the observed redox-potentials of the enzyme-bound

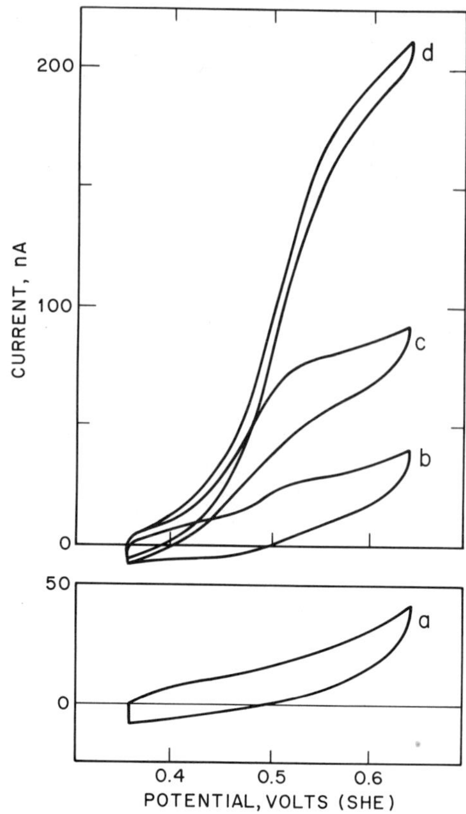

Figure 2. Cyclic voltammograms obtained with glucose oxidase modified with 13 ± 1 ferrocylacetamide functions at different glucose concentrations: a. 0 mM; b. 0.8 mM; c. 2 mM; d. 4 mM; e. 8 mM. 3 mm diameter glassy carbon electrodes; scan rate 2 mVs^{-1}.

relays, the potential difference between the flavoenzymes $FAD/FADH_2$ centers and those of the bound relays, and the period to 10% loss in substrate-dependent current at a potential 100 mV positive of the bound-relay's reversible redox potential, are summarized in Table 2. The glucose-concentation dependence of the current of a relay-modified glucose oxidase solution at 0.43 V (SHE) is shown in Figure 7. The enzyme used in this measurement was modified by forming amides between 13 ± 1 of its amines and ferrocenylacetic acid.

To disprove that relay-modified-enzyme molecules act simiarly to the low-molecular-weight diffusing redox couples, (i.e., as diffusing shuttles of electrons between enzyme molecules and electrodes) the glucose concentration dependence of the current-voltage characteristics of a cell with a solution containing equal amounts of two different enzymes was measured. One of these was the native, unmodified, active enzyme. The other was the chemically modified enzyme, with covalently attached ferrocene-carboxamide centers, that has been intentionally deactivated through removal of its FAD/FADH coenzymes. The current-voltage characteristics of the cell with the two enzymes, in contrast with that made with the chemically modified but active enzyme, did not change upon the addition of glucose. The characteristics of the cell with the modified but modified but deactivated enzyme did, however, show the normal reduction and oxidation waves of the ferrocene/ferrocinium couple (Figure 1, curve b). This

Table 2. Electron Transfer Relays in Flavoenzymes

RELAY	PROBABLE AMINO-ACIDS BOUND TO THE RELAY	RELAYS PER ENZYME MOLECULE	ENZYME DALTONS PER RELAY	E^0 OF ENZYME-BOUND RELAY, VOLTS (SHE)	$E^0_{RELAY}-E^0_{ENZYME}$ VOLTS AT pH 7	PERIOD TO 10% LOSS IN CURRENT
(ferrocenyl)$-CO-NH-$, $Fe^{2+/3+}$	LYSINES OF GLUCOSE OXIDASE	13 ± 1	11,500 ± 1,000	0.5	0.55	2 ± 0.5h
(ferrocenyl)$-CO-NH-$, $Fe^{2+/3+}$	LYSINES OF D-AMINO-ACID OXIDASE	3.5 ± 0.5	11,000 ± 2,000	0.5	0.55	2 ± 0.5h
(ferrocenyl)$-CH_2-CO-NH-$, $Fe^{2+/3+}$	LYSINES OF GLUCOSE OXIDASE	13 ± 1	11,500 ± 1,000	0.35	0.4	2 ± 0.5h
$(NH_3)_5\ Ru^{2+/3+}$ N(pyridyl)$-CO-NH-$	LYSINES OF GLUCOSE OXIDASE	6 ± 1	25,000 ± 5,000	0.3	0.35	10 ± 2h
$(NH_3)_5\ Ru^{2+/3+}$ N(pyridyl)$-N=N-$(HO-phenyl)$-$	TYROSINES OR TRYPTOPHANES OF GLUCOSE OXIDASE	2 ± 0.3	75,000	0.4	0.45	10 ± 2h
$(NH_3)_5\ Ru^{2+/3+}$ N(imidazolyl)N-H	HISTIDINES OF GLUCOSE OXIDASE	14 ± 1	11,000 ± 1,000	0.02	0.07	as Ru^{3+} >1 week as Ru^{2+} <5 min

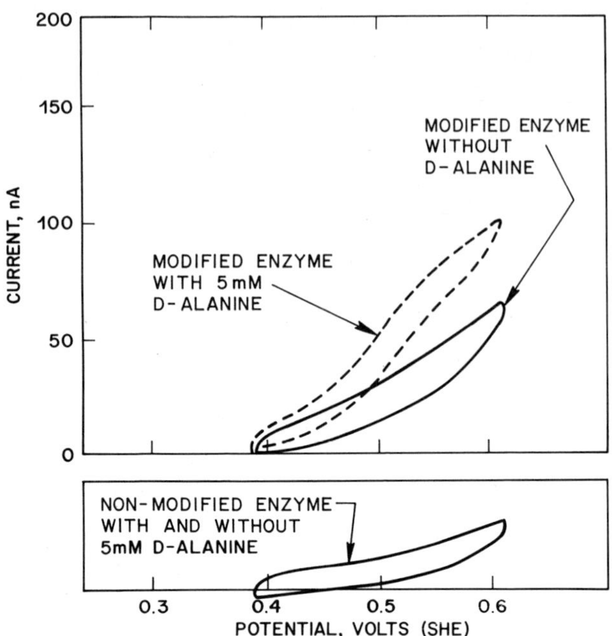

Figure 3. Cyclic voltammograms obtained with native D-amino-acid oxidase without D-alanine or with D-alanine; and with the same enzyme after attaching 3 ± 1 ferrocene-carboxamide functions, without D-alanine and at a 5 mM D-alanine concentration. 1.5 mm diameter gold-disc electrodes; scan rate 1 mVs^{-1}.

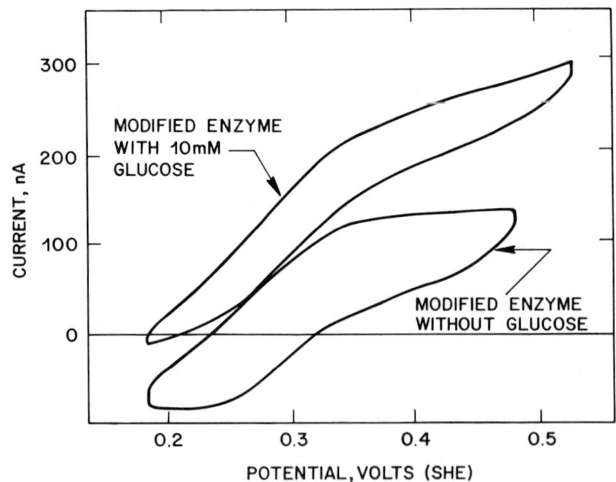

Figure 4. Cyclic voltammograms obtained with glucose oxidase modified with 6.5 ± 0.5 enzyme-bound ruthenium pentaammine isonicotinamide functions without glucose and at 20 mM glucose concentration. 3 mm diameter glassy carbon disk electrodes; scan rate 1 mVs^{-1}.

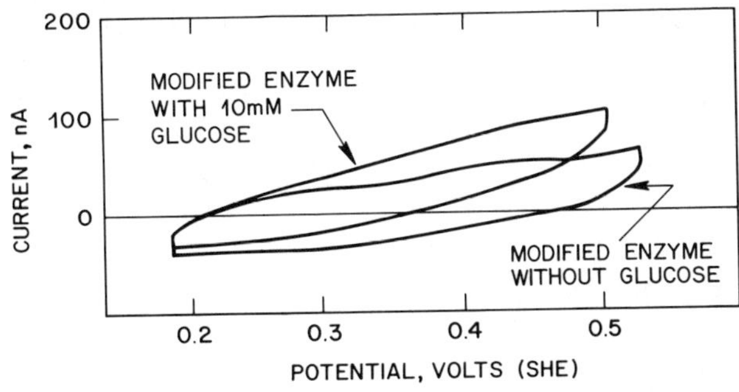

Figure 5. Cyclic voltammograms obtained with glucose oxidase modified with ~2 enzyme-bound ruthenium pentaammine pyridyl-azo functions without glucose and at 20 mM glucose concentration. 3 mm diameter glassy carbon disk electrodes, scan rate 1 mVs^{-1}.

proved that, although electrochemical reduction/oxidation was taking place and although the deactivated enzyme could freely diffuse to the active enzyme, electron exchange between the relays of one enzyme and FAD/$FADH_2$ centers of the other did not take place and the chemically modified enzyme did not act as a conventional diffusing mediator.

While glucose-reduced glucose oxidase is not oxidized on gold, platinum or carbon electrodes, it is oxidized on TTF/TCNQ electrodes[8,15,23]. To determine whether such oxidation is direct or mediated by a diffusing electron-shuttle originating in the TTF/TCNQ electrode, the following experiment was performed. The TTF/TCNQ electrode was operated at + 150 mV (SCE) in a cell containing 20 mL 10^{-4} M glucose oxidase and, initially, 5 mM glucose. It was periodically replaced with a 1.5 mm diameter gold disc electrode and the increment in current upon

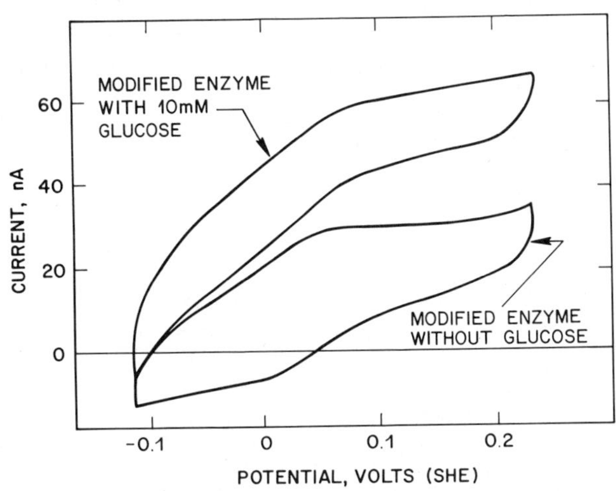

Figure 6. Cyclic voltammograms obtained with glucose oxidase modified with 14 ± 1 enzyme-bound ruthenium pentaammine functions without glucose and at 20 mM glucose concentation. 3 mm diameter glassy carbon disk electrodes; scan rate 1 mVs^{-1}.

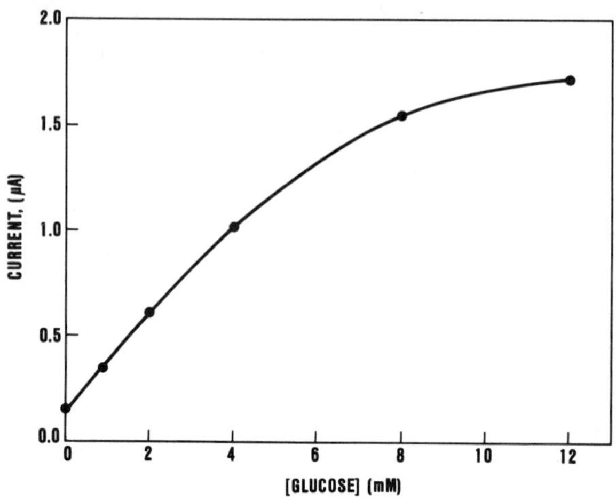

Figure 7. Glucose-concentration dependence of the current at 0.5 V (SHE) for the ferrocylacetamide modifed electrode.

increasing the glucose-concentration was determined. Although initially only the TTF/TCNQ electrode responded to added glucose, after 1 day of operation the gold electrode sensed the increment in glucose concentration equally well, showing that operation of the TTF/TCNQ electrode at 150 mV results in the release of an electron-shuttling mediator into the solution.

The relay-modified enzyme electrodes vary in their chemical and electrochemical stability. In the group listed in Table 2, the enzyme with 14 of its histidines bound to ruthenium-pentaammine was both the most and the least stable. When the enzyme-bound ruthenium was predominantly in its trivalent state, the modified enzyme showed stable electrochemistry for over a week. When the ruthenium was reduced, for example by adding glucose to the solution, some of the enzyme-bound ruthenium-pentaammine complex dissociated and the glucose-concentration dependent current dropped rapidly: The modified enzyme solution lost 10 % of its current in less than 5 min. Furthermore, assay of the number of bound ruthenium atoms after incubation of the modified enzyme (at 25°C and at 30 mM glucose concentration) for 20 min showed a drop from 14 ruthenium atoms/enzyme molecule to 7 ruthenium atoms/enzyme molecule.

Table 3 summarizes the changes observed in the diffusion coefficients of the enzymes upon chemical modification.

DISCUSSION

Organized arrays for intermolecular electron transfer abound in biological systems. For example, structurally and energetically characterized theoretically understood arrays are involved in bacterial photosynthesis[40-45]. In these arrays, long-range electron transfer is attributed to tunneling. The tunneling current depends on the distance between the electron accepting and transferring centers, on their chemistry (i.e., the phonon spectrum) on the medium through which the electrons tunnel, on the energetics of the electron transfer process, and on the electric field between the centers. Though nature has precisely

Table 3. Diffusion Coefficients of Modified Glucose Oxidase

MODIFICATION	$D \times 10^7$, $cm^2\ sec^{-1}$	COMMENTS
NONE	4.1 ± 0.4	NATIVE ENZYME
DEC	3.2 ± 0.3	TREATMENT OF THE ENZYME WITH THE COUPLING AGENT. NO RELAYS.
Fe(C5H4)(C5H5)–CO–NH–PROTEIN	2.9 ± 0.3	12 ± 1 RELAYS PER ENZYME MOLECULE.
Fe(C5H4)(C5H5)–CH_2–CO–NH–PROTEIN	2.9 ± 0.3	13 ± 1 RELAYS PER ENZYME MOLECULE.
$(NH_3)_5$ Ru N(C5H4)–C(=O)–NH–PROTEIN	4.9 ± 0.5	6 ± 1 RELAYS PER ENZYME MOLECULE.
$(NH_3)_5$ Ru N(C5H4)–N=N–PROTEIN	4.3 ± 0.4	2 RELAYS PER ENZYME MOLECULE.

microengineered numerous such arrays, and although there has been important work on insertion of electron donors and acceptors into biomolecules, chemical modification has done little to improve on functional charge transfer systems. Our results show, however, that such improvement is possible: Electrical communication between glucose oxidase or D-amino-acid oxidase and metal electrodes can be established by attaching electron relays to the protein part of the enzyme molecules.

Our observation that chemical modification of the enzymes causes only slight loss in activity leads us to propose that in the enzymes studied part of the protein or glycoprotein structure is not essential for enzymatic activity. Indeed it had been reported that periodate-oxidation of the polysaccharide of glucose oxidase decreases the thermal stability of the enzyme, but does not reduce its activity[46]. We now propose that beyond providing structural stability, part of the protein or glycoprotein has the function of electrically insulating the FAD/$FADH_2$ centers. Such insulation is essential for the survival of a living system: Were different redox enzymes able to transfer electrons to each other in an uncontrolled way, the electrons would cascade thermodynamically downhill. Because of the cascade, enzymes could not be maintained at their proper oxidation potential (i.e., proper oxidation state) to fulfill their role. As a result, the rich inventory of biochemicals in nature would be reduced and synthetic routes, as well as metabolic paths, would close. In addition, the cascase of electrons would drastically reduce the amount of chemically-stored free energy available to the organism. The biologically essential insulating shells prevent, however, electrical communication between $FADH_2$ centers of native enzymes and metal electrodes. Incorporation of electron-transfer relays in the shells opens channels for such electron-transfer. As is evident from the fact that the modified-enzymes retain most or all of their normal catalytic activity, the catalytic core of the enzyme is only mildly affected by the chemical changes in the insulating shell.

Following modification, a significant glucose-concentration dependent current flows at the redox potential of the relay. The limiting curent is usually controlled by the rate of either reaction 4 or 5. In the presence of excess glucose oxidase and at adequate glucose concentrations the rate of reaction 4 is rapid.

β-D-glucose + (glucose oxidase-FAD) $\rightarrow$ (4)

δ-D-gluconolactone + (glucose oxidase-$FADH_2$)

(glucose oxidase-$FADH_2$) $\rightarrow$ (glucose oxidase-FAD) + $2H^+$ + $2e^-$ (5)

Because glucose oxidase is a molecule of 43 Å radius[47] and because the FAD/$FADH_2$ centers are deep inside the enzyme, reaction 5 proceeds too slowly with the unmodified enzyme for an electrochemically detectable current to flow to electrodes: the distance between the FAD/$FADH_2$ centers and the electrodes is excessive (Figure 1(a)). Our results also show that this is the case also in TTF/TTNQ electrodes, where some or all of the glucose-concentration dependent current results not from direct electron transfer but from release of a diffusing, electron-shuttling mediator into the solution.

Key to electrical communication between glucose oxidase and the electrodes is the spacing, i.e., the density, of relays. The relays must be sufficiently close to both the FAD/$FADH_2$ centers and to the metal electrodes (and possibly also to each other) for the electron transfer to be rapid. The distance-dependence of electron transfer rates in biosystems has been the subject of intensive theoretical and experimental research in recent years[48-61]. It is now evident that for distances greater than 8 Å electron transfer rates (k) within and between molecules, as well as between electrodes and ions or molecules in their proximity, or between the ions themselves, decay exponentially with the distance (d) between the involved centers, i.e., $k = ce^{-\alpha d}$[48-50]. In proteins the electron transfer rates drop by ~10^3 when the distance between an electron donor and an acceptor is increased from 8 Å to 17 Å[48]. As shown schematically on the right side of Figure 8, the distances between the FAD/$FADH_2$ centers and the relays between the relays themselves, and between the relays and the electrodes, can all be shorter than those between the FAD/$FADH_2$ centers and the electrodes. When these distances are sufficiently short, electrical communication between the FAD/$FADH_2$ centers and the metal electrodes becomes possible. Thus, after incorporating electron transferring relays in the protein part of the enzyme, glucose reduces the FAD to $FADH_2$ and electron flow, via the relays, to the metal electrodes. Therefore, the addition of glucose (or of D-alanine in the case of modified D-

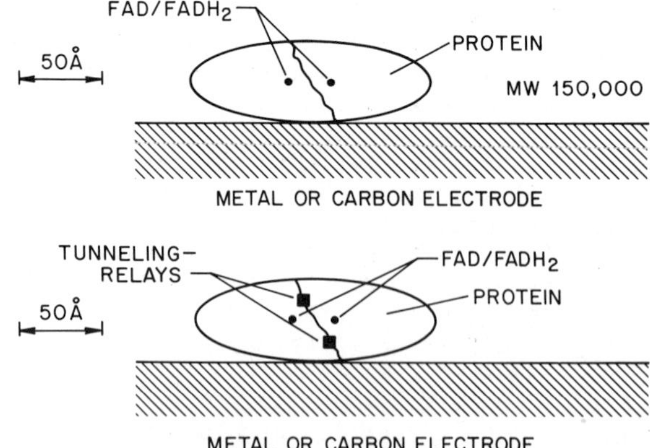

Figure 8. Schematic diagram showing a section along the long axis of glucose oxidase before (top) and after (bottom) chemical bonding of electron-transfer relays to the protein.

amino-acid oxidase) produces an anodic current at potentials positive of the redox potential of the protein-bound relay.

Electrons are relayed in the chemically-modified enzymes both by tunneling[44,45,57-70] and by protein-dynamics[71-74], i.e., motion in and of the protein chains. The tunneling rates increase when the relays are fast redox-couples, i.e., when there is little change in the structure of the relay and its solvent-environment upon oxidation or reduction.Furthermore, directional or vectorial tunneling, i.e., flow of current, requires that the potentials of the donor and the acceptor differ. Our design of modified enzymes, involving shortened tunneling distances, fast redox-couple relays and potential-gradients, is consistent with that required in electron-tunneling systems. Motion of the protein chains to which the relays are attached, i.e., protein-dynamics, contributes to electron-transfer through transiently reducing the tunneling distances between $FADH_2$ centers and the relays, and between the relays and the metal or carbon electrodes. Consistently with this model, relay-mediated electron-transfer from $FADH_2$ centers to electrodes is more efficient when the density of the relays is higher, i.e., when the tunneling distances are, on the average, shorter. We note in this context that the superior electron transfer seen in enzymes that were modified in the presence of 2 M urea, an agent known to reversibly open protein structures[38], is consistent with a larger number and better distribution of relays within the enzymes.

Direct electrical communication between electron-relay modifed enzymes and metal electrodes opens a route to safer *in vivo* electrochemical sensors, and to *in vivo* electrochemical synthesis of biochemicals. *In vivo* sensors based on relay-modified enzymes, will have advantages with respect to present oxygen or hydrogen-peroxide monitoring glucose-sensors, sensors based on diffusing electron-shuttles, and diffusing-shuttle releasing electrodes such as TTF/TCNQ. With respect to the first, their advantage is that the measurements are intrinsically independent of oxygen pressure in the presence of catalase and other H_2O_2 decomposition catalysts. With respect to the second and third, their advantage is in eliminating the risk associated with leakage of an electron-shuttling mediator into the living tissue. Although electron-shuttling mediator based glucose sensors are useful for *in vitro* assays, extended and safe in vivo monitoring of glucose requires, when diffusing mediators are used, biocompatible membranes that are not available. These membranes must prevent escape of the electron-shuttling mediator from the enzyme-containing electrode-compartment, yet allow in-diffusion of glucose and the out-diffusion of gluconolactone. Excape of mediator is unacceptable not only because its depletion from the membrane-contained compartment diminishes the response of the electrode but, more importantly, because uncontrolled release of an electron-transfer shuttle into a living organism can electrically "short" enzymes, i.e., cause uncontrolled electron transfer from one enzyme to another. As pointed out earlier, nature carefully avoids "shorting", that would narrow the range of possible biochemical reactions and would reduce the amount of free energy available to the organism. Indeed the release of a mediator can be both energetically and metabolically devastating: Diffusing electron-transfer mediators that are applied in herbicides, such as Diquat and Paraquat, are not only effective plant-killers[75], but are also highly toxic also to humans[76] and to animals[77]. Both are fast redox couples based on viologen. The risk associated with the release of an electron-transfer mediator into the living tissue is avoided when a relay-modified enzyme is used: Through tailoring the pores in membranes the enzyme can be contained in the electrode compartment with available membranes. Specifically, glucose oxidase, an enzyme with a molecular weight of about 160,000 and a hydrodynamic diameter of 86 Å[47], can be contained in biocompatible membranes of

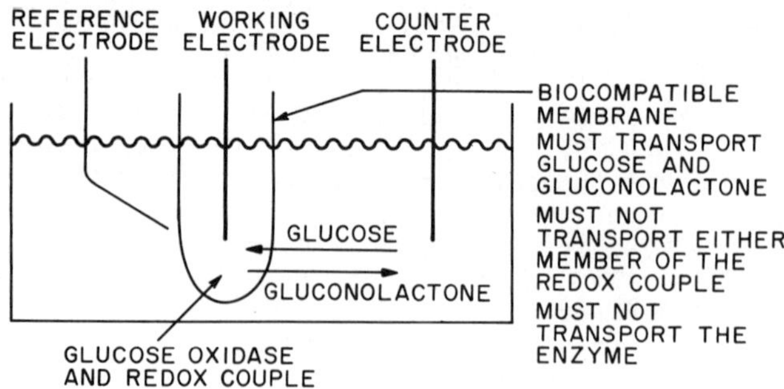

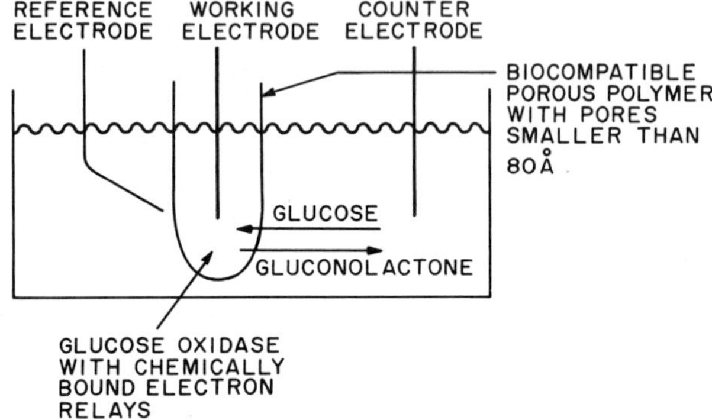

Figure 9. Membrane requirements in implanted, redox-enzyme-based electrodes. Top - electrodes made with diffusing redox couples; bottom - electrodes made with an enzyme modified by chemically bound electron-transfer relays.

pore-sizes of 80 Å or less. Such membranes are available[78]; they transport glucose and gluconolactone, yet prevent the release of the relay-modified enzyme from the electrode compartment (Figure 9).

A safe, implantable glucose-electrode would be a welcome development for diabetics who monitor their glucose-level and particularly for one-fifth of the diabetics who require insulin treatment. With its development it should also become possible to perfect an artificial pancreas built of three components: (a) an already existing, externally worn, or implanted, insulin pump[13,79], (b) a readily manufacturable microelectronic logic, memory and timing circuit, and (c) a safe, in vivo glucose sensor.

Implanted relay modified-enzyme electrodes might find application also in selective and controlled electrochemical oxidation or reduction of biochemicals in a specific organ (e.g., the oxidation of L-tyrosine to L-Dopa). This would open a route to electrochemotherapy.

CONCLUSIONS

From the results we conclude the following. First, that direct electrical communication can be established between redox enzymes (such as glucose oxidase or D-amino-acid oxidase) and metal electrodes, by chemically binding to the enzyme-proteins electron-transfer relays (Figures 1-6, Table 2). Second, that amperometric glucose and D-amino-acid sensors can be made with the chemically modified enzymes and conventional metal or carbon electrodes (Figures 1,2). The electron-transfer sequence, on which these sensors are based, involves (a) selective reduction of the FAD centers of the enzyme by the substrate to $FADH_2$; (b) oxidation of the $FADH_2$ centers by the protein-bound electron-transfer relays, which are reduced; and (c) electron transfer from the reduced relays to the electrode i.e., electrochemical reoxidation of the relays. In a sensor, the current associated with the latter process is measured. Over a range of glucose-concentrations, there is a linear relationship between the concentration of the substrate and the current (Figure 7). Third, that the relays must have redox-potentials that are oxidizing relative to the potential of the $FAD/FADH_2$ centers (Table 2). Otherwise step (b) cannot take place. Fourth, that the most effective relays are fast redox couples. Fifth, that electrodes made with enzymes having a greater density of relay, i.e., shorter tunneling distances, show greater current-increment when the substrate concentration is increased (Table 2, Figures 1-6). Sixth, that in relay-modified enzymes the structural and chemical changes affect the activity only slightly. The residual enzymatic activity after attachment of relays exceeds in glucose oxidase 60% of the activity of the native, unmodified enzyme and one modification triples the enzymatic activity.

ACKNOWLEDGMENTS

We thank Pierre Wiltzius for measuring the diffusion coefficients of the native and the chemically modified enzymes; Anthony M. Williams for assays of iron and ruthenium; Tetsuo Yamane for valuable discussions; and Robert Bittman of Queens College, CUNY for sharing with us his insight into conformational changes in proteins.

REFERENCES

1. Y. Degani and A. Heller, J. Phys. Chem., 91, 1285 (1987).

2. S.J. Updike and G.P. Hicks, Science (Washington, D.C.), 158, 270 (1967).

3. S.J. Updike and G.P. Hicks, Nature (London), 214, 986 (1967).

4. D.R. Thevenot, P.R. Coulet, R. Sternberg and D.C. Gautheron, Bioelectrochem. Bioenerg., 5, 548 (1978).

5. D.R. Thevenot, R. Sternberg, P.R. Coulet, J. Laurent and D.C. Gautheron, Anal. Chem., 51, 96 (1979).

6. K.S. Chua and I.K. Tan, Clin. Chem., 24, 150 (1978).

7. C. Bourdillon, J.P. Bourgeois and D. Thomas, J. Am. Chem. Soc., 102, 4231 (1980).

8. J.J. Kulys, A. Samalius and G.J.S. Svirmickas, FEBS Lett., 114, 7 (1980).

9. E. Lobel and J. Rishpon, Anal. Chem., 53, 51 (1981).

10. R.M. Ianniello and A.M. Yacynych, Anal. Chem., 53, 2090 (1981).

11. D.A. Gough, J.K. Leypoldt and J.C. Armour, Diabetes Care, 5, 190 (1982).

12. L.C. Clark, Jr and C.A. Duggan, Diabetes Care, 5, 174 (1982).

13. M. Shichiri, Y. Yamasaki, R. Kawamori and H. Abe, Lancet, 2, 1129 (1982).

14. N.K. Cenas, A.K. Pocius and J.J. Kulys, Bioelectrochem. Bioenerg., 11, 61 (1983); 12, 583 (1984).

15. J.J. Kulys and N.K. Cenas, Biochim. Biophys. Acta, 744, 57 (1983).

16. S.D. Varfolomeev and S.O. Bachurin, J. Mol. Catal., 27, 305 (1984).

17. T. Ikeda, I. Katasho, M. Kamei and M. Senda, Agric. Biol. Chem., 48, 1969 (1984).

18. A.E.G. Cass, G. David, G.D. Francis, H.A.O. Hill, W.J. Aston, J.I. Higgins, E.V. Plotkin, L.D.L. Scott and A.P.F. Turner, Anal. Chem., 56, 667 (1984).

19. D.A. Gough, J.Y. Lucisano and H.S. Tse, Anal. Chem., 57, 2351 (1985).

20. T. Ikeda, H. Hamada and M. Senda, Agric. Biol. Chem., 50, 883 (1986); 49, 541 (1985).

21. T. Ikeda, I. Katasho and M. Senda, Analytical Sciences (Japan), 1, 455 (1985).

22. W.J. Albery and D.N. Bartlett, J. Electroanal. Chem., 194, 211 (1985).

23. W.J. Albery, D.N. Bartlett and D.H. Craston, J. Electroanal. Chem., 194, 223 (1985).

24. A.P.F. Turner, World Biotech. Rep., 1, 181 (1985).

25. L.O. Gorton, F. Scheller and F. Johansson, Stud. Biophys., 109, 199 (1985).

26. K. Narashimhan and L.B. Wingard, Jr., Anal. Chem., 58, 2984 (1986).

27. D.J. Claremont, C. Penton and J.C. Pickup, J. Biomed. Eng., 8, 272 (1986).

28. M. Senda, T. Ikeda, K. Miki and H. Hiasa, Analytical Sciences (Japan), 2, 501 (1986).

29. A.L. Crumbliss, H.A.O. Hill and D.J. Page, J. Electroanal. Chem., 206, 327 (1986).

30. M. Senda, T. Ikeda, H. Hiasa and T. Katasho, Nippon Kagaku Kaishi, 358 (1987).

31. T.L. Shirey, Clin. Biochem., 16, 147 (1983).

32. H.G. Curme, et al., Clin. Chem., 24, 1335 (1978).

33. L.H. Vogt, J.L. Katz and S.E. Wiberley, Inorg. Chem., 4, 1157 (1965).

34. D. Lednicer, J.K. Lindsey and C.R. Hauser, J. Org. Chem., 23, 653 (1958).

35. P. Ford, De F.P. Rudd, R. Gaunder and H. Taube, J. Am. Chem. Soc., 90, 1187 (1968).

36. K.M. Yocom, J.B. Shelton, J.R. Shelton, W.A. Schroeder, G. Worosila, S.S. Isied, E. Bordignon and H.B. Gray, Proc. Natl. Acad. Sci. USA, 79, 7052 (1981).

37. P. Wiltzius, Phys. Rev. Lett., 58, 710 (1987).

38. C.N. Pace, "Determination and Analysis of Urea and Guanidine Hydrochloride Denaturation Curves", Chapter 14 in "Methods in Enzymology, Vol. 131. Enzyme Structure Part L", C.H.W. Hirs and S.N. Timasheff, Eds., Academic Press, Orlando, pp. 266-280 (1986).

39. F. Scheller, G. Strand, B. Neumann, M. Kuhn and W. Ostrowski, Bioelectrochem. Bioenerg., 6, 117 (1979).

40. "Photosynthesis: Energy Conversion by Plants and Bacteria", Govindjee, Ed., Academic Press, New York, 1982.

41. W. Zinth, E. Knapp, S.F. Fischer, W. Kaiser, J. Diesenhofer and H. Michel, Chem. Phys. Lett., 119, 1 (1985).

42. W. Zinth, M. Sander, J. Dobler, W. Kaiser and H. Michel, in "Antennas and Reaction Centers of Photosynthetic Bacteria", M.E. Michel-Beyerle, Ed., Springer-Verlag, Berlin, p. 97 (1985).

43. J.R. Miller in "Antennas and Reaction Centers of Photosynthetic Bacteria", M.E. Michel-Beyerle, Ed., Springer-Verlag, Berlin, p. 234 (1985).

44. M. Bixon and J. Jortner, J. Phys. Chem., 90, 3795 (1986); FEBS Lett., 200, 303 (1986).

45. J. Jortner, Biochim. Biophys. Acta, 594, 193 (1980); J. Chem. Phys., 64, 4860 (1976); J. Am. Chem. Soc., 102, 6676 (1980).

46. S. Nakamura, S. Hayashi and K. Koga, Biochim. Biophys. Acta, 445, 294 (1976).

47. H. Tsuge, O. Natsuaki and K. Ohashi, J. Biochem., 78, 835 (1975).

48. S.L. Mayo, W.R. Ellis, Jr., R.J. Crutchley and H.B. Gray, Science, 233, 948 (1986).

49. G. Mclendon, J.R. Miller, K. Simolo, K. Taylor, A.G. Grant and A.M. English, ACS Symp. Ser., 307, 150 (1986).

50. S.S. Isied, Prog. Inorg. Chem., 32, 443 (1984).

51. G. McLendon, T. Guarr, M. McGuire, K. Simolo, S. Strauch and K. Taylor, Coord. Chem. Rev., 64, 113 (1985).

52. S.R. Peterson-Kennedy, J.L. McGourty, P.S. Ho, C.J. Sutoris, N. Liang and H. Zemel; N.V. Blough, E. Margoliash, B.M. Hoffman, ibid, 125.

53. T. Takaka, K. Takenaka, H. Kawamura and Y. Beppu, J. Biochem. (Tokyo), 99, 844 (1986).

54. R.J. Crutchley, W.R. Ellis and H.B. Gray, J. Am. Chem. Soc., 107, 5002 (1985).

55. S. Larsson, J. Chem. Soc. Faraday Trans., 2, 79, 1375 (1983).

56. J.L. McGourty, N.V. Blough and B.M. Hoffman, J. Am. Chem. Soc., 105, 4470 (1983).

57. R.A. Marcus, Int. J. Chem. Kinetics, 13, 865 (1981).

58. R.M. Marcus and N. Sutin, Biochim. Biophys. Acta, 81, 265 (1985).

59. D.N. Beratan and J.J. Hopfield, J. Am. Chem. Soc., 106, 1584 (1984).

60. S.S. Isied, G. Worosila and S.J. Atherton, J. Am. Chem. Soc., 104, 7659 (1982).

61. J.J. Hopfield, Proc. Natl. Acad. Sci. USA, 71, 3640 (1974).

62. N. Sutin, Prog. Inorg. Chem., 30, 441 (1983).

63. N. Sutin and M.D. Newton, Ann. Rev. Phys. Chem., 35, 437 (1984).

64. A. Kuznetsov and J. Ulstrup, J. Chem. Phys., 75, 2047 (1981).

65. T.T. Li and M.J. Weaver, J. Am. Chem. Soc., 106, 6107 (1984).

66. M.J. Weaver and T.T. Li, J. Phys. Chem., 90, 3923 (1986).

67. L.I. Krishtalik, Elektrokhimiya, 11, 184 (1975).

68. H. Kuhn, Phys. Rev., A34, 3409 (1986).

69. B. Mann and H. Kuhn, J. Appl. Phys., 42, 4398 (1971).

70. R.R. Dogonadze, A.M. Kuznetsov and J. Ulstrup, Electrochim. Acta, 22, 967 (1977).

71. A.K. Churg and A. Warshel, "Modeling the Activation Energy and Dynamics of Electron Transfer Reactions in Proteins", in "Structure and Motion: Membranes, Nucleic Acids and Proteins", E. Clementi, G. Corongui, M.H. Sarma, R.H. Sarma, Eds., Adenine Press, p. 361-374 (1985).

72. Section II. "Structural Dynamics and Mobility in Proteins in "Methods in Enzymology. Volume 131. Part L. Enzyme Structure", C.H. Hirs and S.N. Timasheff, Eds., Academic Press, Orlando, 1986, p. 283-607.

73. R. Huber and W.B. Bennett, Jr., Biopolymers, 22, 262 (1983).

74. W.S. Bennett and R. Huber, CRC Crit. Rev. Biochem., 15, 291 (1984).

75. J. Doull, C.D. Klaassen and M.O. Amdur, "Casarett and Doull's Toxicology, The Basic Science of Poisons", Second Edition, Macmillan, New York, 1980, p. 390-392.

76. N.I. Sax, "Dangerous Properties of Industrial Materials", Van Nostrand Reinhold, New York, 1979, p. 628, 885-6.

77. Kirk-Othmer Encyclopedia of Chemical Technology, 3rd Edition, Wiley, New York, 1980, Vol. 12, p. 338.

78. N. Ueda, M. Sekiya, Y. Yamasaki, R. Kawamori and M. Shichiri, "Membrane Design of Glucose Senor for Long-Term Clinical Use: Application of Alginate-Polylysine-Alginate Membrane". Abstract, Internat. Conf. on Biosensors, June 23-26, 1987, Braunschweig-Stockheim, W. Germany.

79. C.S. Rosenberg, Mt. Sinai J. Med. (NY), 54, 217 (1987).

POTENTIAL CONTROLLED ENZYMATIC ACTIVITY OF CONDUCTING ENZYME MEMBRANE

Masuo Aizawa*, So-ichi Yabuki and Hiroaki Shinohara

Department of Bioengineering
Tokyo Institute of Technology
Ookayama, Meguro-ku
Tokyo 152 / Japan

INTRODUCTION

Bioelectrocatalysis is a unique combination of electrochemical and biochemical reactions, which rests not only on the ability to control at will the oxidizing and reducing ability of the electrode by changing its Fermi level, but on the specificity and selectivity of biochemical catalysis. Endeavors have been made to develop bioelectrocatalysis with enzymes that catalyze redox reactions.

The multiheme protein cytochrome c_3, an electron transfer protein from the sulfate-reducing bacteria *Desulfovibrio vulgaris* (strain *Miyazaki*), was the first example of a heme protein exhibiting a reversible electrode reaction at a mercury electrode[1]. Due to the sophisticated structure of proteins, there is great difficulty in most cases to achieve electron transfer of enzyme molecules via an electrode.

Modified electrodes for biocatalysis use either electron mediators or promoters immobilized on the electrode surface[2-7]. In both cases, redox enzyme molecules are in solution and in contact with the common electron mediators for redox enzymes such as cytochrome c and ferredoxin. An electron promoter is not a mediator since it does not take part in electron transfer in the potential region of interest. An electrode modified with promoter molecules has enables some redox enzymes to directly transfer electrons. It has been shown that 4,4'-bipyridyl, bis(4-pyridyl)sulfide, and bis(4-pyridyl)disulfide are excellent promoters of electron transfer of cytochrome c. Cytochrome c gives a reversible cyclic voltammogram at gold electrodes modified with these promoters.

EXPERIMENTAL

Chemicals

Glucose oxidase [EC 1.1.3.4] (grade II, 113 unit$\cdot$mg^{-1}) was purchased from Toyobo Co. Alcohol dehydrogenase [EC 1.1.1.1] (grade III, 240 unit$\cdot$mg^{-1}) was purchased from Boehringer Mannheim Co. Pyrrole was obtained from Tokyo Kasei Kogyo Co., and distilled prior to use.

Apparatus

A potentiostat (Hokuto Denko Co., HA-301) and function generator (Hokuto Denko Co., HB-105) were used for electrochemical measurements. Differential pulse voltammetry was performed with polarographic analyzer P-1100 (Yanako Co.).

Synthesis

A glucose oxidase (GOD) entrapped polypyrrole membrane (PP-GOD) and an ADH entrapped polypyrrole membrane (PP-ADH) were deposited on a platinum electrode by potential-controlled polymerization of pyrrole in the presence of GOD or ADH. The electrolyte contained 0.1 M pyrrole and 1 M KCl, and was deoxygenated by bubbling nitrogen gas. Either GOD ($30\ mg{\cdot}ml^{-1}$) or ADH ($20\ mg{\cdot}ml^{-1}$) was added to the electrolyte. A platinum wire diameter 0.5 mm, was used as working electrode. Then the potential of the Pt electrode was set at +0.7 V and +1.0 V vs Ag/AgCl for synthesis of an alcohol dehydrogenase-entrapped polypyrrole membrane and a glucose oxidase-entrapped polypyrrole membrane, respectively. Oxidative current was measured. Charge passed during electrochemical polymerization was controlled at $850\ mC{\cdot}cm^{-1}$.

Measurement of Entrapped Enzyme Activity

The enzyme-entrapped polypyrrole was rinsed, and measurement of entrapped enzyme activity was performed. Enzyme activity of a PP-GOD and PP-ADH membranes were spectrophotometrically determined. A PP-GOD membrane was soaked in a solution which contained 90 mM glucose, $7\ \mu g{\cdot}ml^{-1}$ peroxidase (about 7×10^{-3} unit ml^{-1}), 0.2 mM o-dianisidine[8]. The solution was adjusted to pH 5.5 by citrate buffer. Ten minutes later, the concentration of the oxidized form of dianisidine was measured.

A PP-ADH membrane was soaked in a solution containing 0.3 M ethanol and 8 mM NAD^+. After ten minutes the NADH concentration in the solution was spectrophotometrically measured at 340 nm.

RESULTS AND DISCUSSION

Enzymatic Activity of PP-GOD and PP-ADH Membranes

A PP-GOD membrane was prepared by electrochemical oxidation of pyrrole in the presence of 30 mg ml^{-1} GOD. The enzyme activity of the PP-GOD membrane was 30×10^{-3} unit cm^{-2}. A PP-ADH membrane was prepared at an enzyme concentration of 20 mg ml^{-1} ADH. The PP-ADH membrane had about 30×10^{-3} unit cm^{-2} of activity. All the PP-GOD and PP-ADH membranes were prepared at fixed concentrations of pyrrole and potassium chloride. Both membranes had electroconductivity of 10 S cm^{-1}.

Electron Transfer of Membrane-bound GOD

The electron transfer between the electrode and glucose oxidase was measured for the following PP-GOD membrane. At first, GOD was adsorbed on a platinum wire electrode at a potential of +0.5 V vs Ag/AgCl. After 30 min, the electrode was rinsed, and then polymerization started in the solution of 0.1 M pyrrole and 1 M KCl. The potential of this electrode was set at 0.7 V vs Ag/AgCl and charge passed during polymerization was

100 μC cm^{-2}. After rinsing, the electrode was dipped into a solution containing 1 M KCl and 0.1 M citrate buffer (pH = 5.5), in the absence of oxygen. Differential pulse voltammetry was performed with the PP-GOD membrane. Figure 1(a) shows the cathodic current caused by the reduction of the oxidized form of GOD. Figure 1(b) shows the anodic current due to the oxidation of the reduced form of GOD. These results indicate that electrons transfer between polypyrrole-bound GOD and the electrode.

Enzyme Cycling of GOD in the Absence of Oxygen

The enzyme reaction of GOD can be divided into two steps. In the first step, glucose is oxidized by GOD (while at the same time the cofactor FAD is reduced) to yield the reduced form of GOD. In the next step, in the presence of oxygen, reduced GOD is oxidized by oxygen so that hydrogen peroxide is generated.

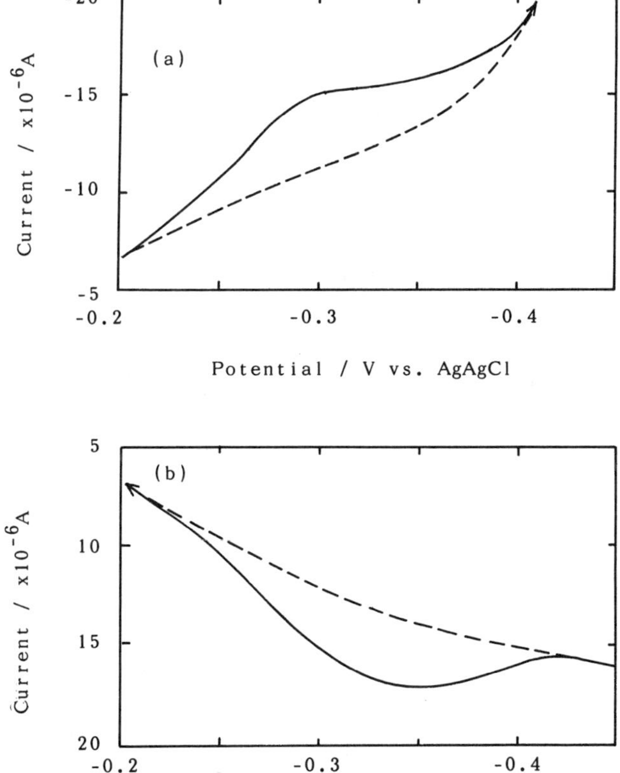

Figure 1. Differential pulse voltammograms of PP-GOD membrane. Solid lines show the differential pulse voltammograms (DPV) of PP-GOD membrane. Broken lines show the DPV of polypyrrole membrane in the absence of GOD. The electrolyte solution contained 0.1 M citrate buffer (pH = 5.5) and 1 M KCl. Scan rate was 2 mVs^{-1}. Pulse was 50 ms width, and 50 mV, its interval was 1.0 s. (a) The reductive direction of DPV. (b) Oxidative direction of DPV.

In the absence of oxygen, oxidized GOD cannot be regenerated from reduced GOD. PP-GOD membrane, which was synthesized at 850 mC cm^{-2}, was dipped into an oxygen-free solution containing 1 mM glucose. Electrolysis was done with the PP-GOD membrane at a potential of +0.4 V vs Ag/AgCl for 100 minutes. The reduced form of PP-GOD may be oxidized under this condition. The concentration of glucose decreased after electrolysis. It was found that 0.38×10^{-6} mol glucose per cm^2 PP-GOD membrane was consumed during electrolysis. This phenomenon should result from the regeneration of membrane-bound GOD from its reduced form. The turnover number of this cycling was estimated at 1.88×10^5. In contrast, the electrochemical regeneration of membrane-bound GOD did not proceed in the potential range below the redox potential of GOD. The enzyme activity of the PP-GOD membrane was potential-dependent in the absence of oxygen.

Electrochemical Control of Enzyme Activity of a PP-ADH Membrane

Membrane-bound ADH reacts with ethanol and NAD^+ to produce NADH and acetaldehyde. The diffusion of these substrates and products in a membrane might be influenced by the characteristics of the membrane matrix. Changes in the electrical charge of the polypyrrole membrane could influence the diffusion of the charged substrate in the polymer. Polypyrrole is positively charged when the polymer is in its doped state. If the polymer is undoped it should be neutral. Substrates which are charged, such as NAD^+, should exhibit diffusion velocity changes with changes in membrane potential.

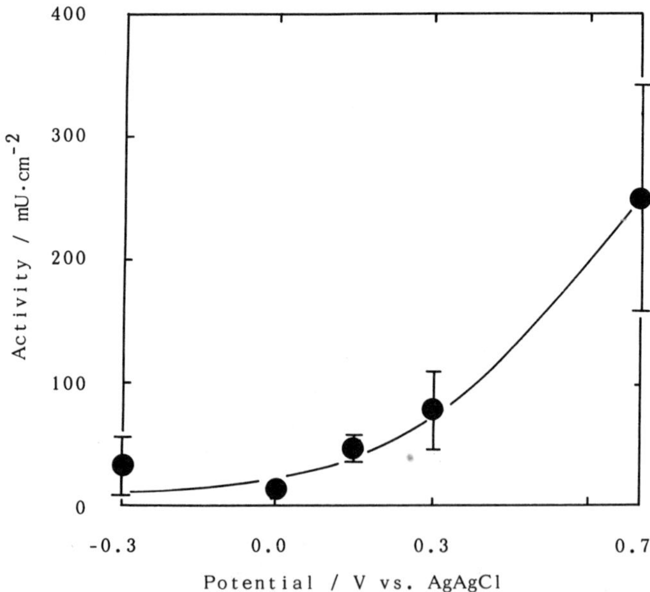

Figure 2. Enzyme activity of PP-ADH membrane at controlled potentials. The membrane was dipped into 0.67 M ethanol, 4 mM NAD^+ and 0.1 M phosphate buffer solution (pH = 7.5), and the potential was then applied. After 30 minutes, the activity was calculated by the absorption of NADH at 340 nm.

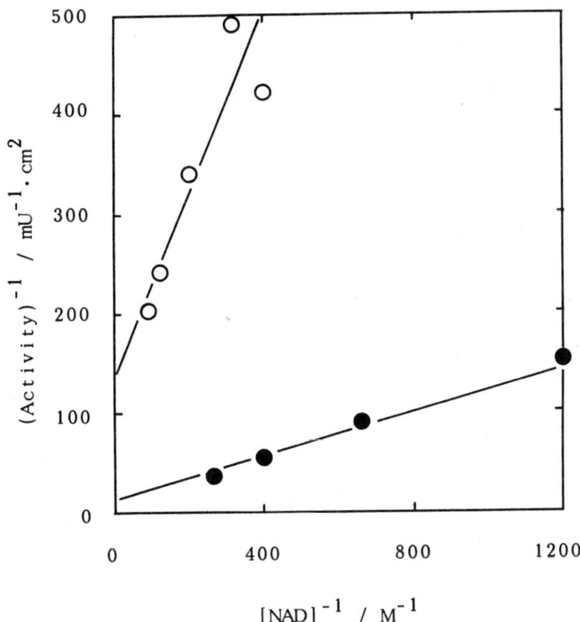

Figure 3. Lineweaver-Burk plots of PP-ADH membrane. The membrane was dipped into a solution containing 0.1 M phosphate buffer (pH = 7.5) and 0.67 M ethanol. Activity was measured at the membrane potential of +0.15 V (○) and +0.7 V (●) vs Ag/AgCl.

A PP-ADH membrane was immersed in a solution containing NAD^+ (4 mM) and ethanol (0.67 M). The potential of this membrane was set at various different potentials for 30 minutes. The potential dependence of enzyme activity is shown in Figure 2. The enzyme activity was remarkably dependent on the potential of the PP-ADH membrane. In particular, the enzyme activity of the PP-ADH membrane at 0.7 V was more than ten times that at 0.0 V. Lineweaver-Burk plots of this membrane was determined at 0.7 V and 0.15 V in a solution containing ethanol and NAD^+. Figure 3 shows Lineweaver-Burk plots at these two potentials. From these plots, the Michaelis constant and maximum velocity of this membrane was calculated and are shown in Table 1. The Michaelis constant was not strongly influenced by potential changes. However, the maximum velocity at 0.7 V is more than ten times greater than that at 0.15 V. It is concluded that changes in the enzyme activity with potential is caused by changes of maximum velocity.

Table 1. Kinetic Values of PP-ADH Membrane from Lineweaver-Burk Plots.

Potential (Volt vs Ag/AgCl)	K'_m (M)	V'_{max} (mU cm^{-2})
0.15	5.9×10^{-3}	7.0
0.70	8.0×10^{-3}	72

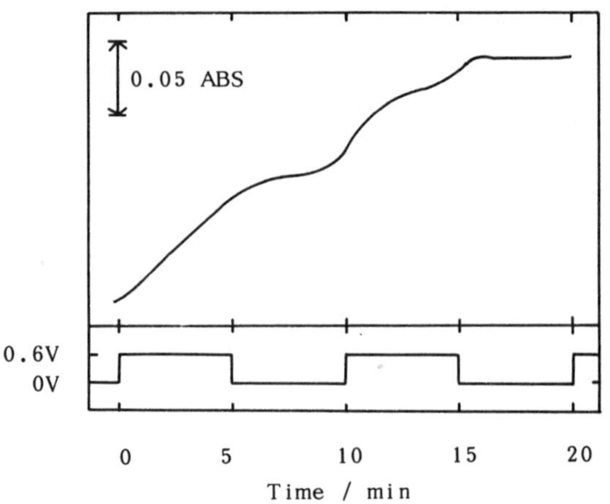

Figure 4. Electrically controlled enzyme activity of PP-ADH membrane. A 0.5 mm diameter PP-ADH membrane (7.5 mm length) was dipped into a phosphate buffer solution (0.1 M pH = 7.5) containing 4 mM NAD^+ and 0.67 M ethanol. The volume of the solution was 4.5 ml. 0.0 V to 0.6 V rectangular wave (lower figure) potential was applied to the membrane, and absorbance was measured at 340 nm.

Continuous electrochemical control of enzyme activity was performed with an enzyme reactor. The PP-ADH membrane was placed in a batch-type reactor. The potential of the membrane was controlled. The reaction contained 4 mM NAD^+ and 0.67 M ethanol. Absorbance at 340 nm was measured to determine the NADH formed. NADH increased at a controlled potential of 0.6 V, indicating that membrane-bound ADH remained active (Figure 4). An increase in NADH concentration, however, stopped when the potential was set at 0.0 V. Membrane-bound ADH lost its activity under this condition. Thus, it was proved that the activity of membrane-bound ADH was potential-controllable.

CONCLUSIONS

Electrochemically synthesized polypyrrole-enzyme membranes have been shown to be enzymatically active and electrically conductive. Membrane-bound GOD is electrochemically regenerated from the reduced form of GOD. The enzyme activity of membrane-bound ADH can be electrochemically regulated.

REFERENCES

1. K. Niki, T. Yagi, H. Inokuchi and K. Kimura, J. Electrochem. Soc., 124, 1889 (1977).

2. N.S. Lewis and M.S. Wrighton, Science, 211, 944 (1981).

3. H.L. Landrum, R.T. Salmon and F.M. Hawkridge, J. Am. Chem. Soc., 99, 3154 (1977).

4. E.F. Bowden, F.M. Hawkridge and H.N. Blount, Bioelectrochem. Bioenerg., 7, 477 (1980).

5. C.D. Crawley and F.M. Hawkridge, Biochem. Biophys. Res. Commun., 99, 516 (1981).

6. W.J. Albery, M.J. Eddowes, H.A.O. Hill and A.R. Hillman, J. Am. Chem. Soc., 103, 3904 (1981).

7. H.A.O. Hill, N.J. Walton and I.J. Higgins, FEBS Lett., 126, 282 (1981).

8. Bergmeyer, "Methods of Enzymatic Analysis", 2nd Ed., Vol. 1, Academic Press, New York (1974).

ELECTRO-ENZYMATIC STUDIES OF THE INTERFACIAL BEHAVIOUR OF PROTEINS

F.W. Scheller* and R. Hintsche

Central Institute of Molecular Biology
Academy of Sciences of the G.D.R.
DDR-115 Berlin-Buch
German Democratic Republic

V. Bogdanovskaja

Frumkin Institute of Electrochemistry
Academy of Sciences of the USSR

P. Ohlsson

Umea University / Sweden

INTRODUCTION

Biospecific electrodes for the determination of nine metabolites and for six enzyme activities have been brought into the clinical laboratory by our group[1]. These sensors contain a dialysis membrane and up to three sequentially acting enzymes. Worldwide commercial enzyme electrodes use the membrane technique which was adapted from the Clark-type oxygen electrode. More direct enzyme fixation is needed for affinity and immunosensors and enzyme FETs. In addition to the covalent binding at the electrode surface, adsorption of the protein also offers several advantages, i.e., simple enzyme loading and low diffusion resistance. In addition to analytical applications the adsorption of protein at interfaces is also of fundamental interest and plays an important role in other biotechnological processes such as chromatography or immobilization on solid carriers.

In this paper the interfacial behaviour of the enzymes glucose oxidase (GOD), peroxidase (POD) and laccase (Lacc) is characterized both at the dropping mercury electrode (dme) and at graphite electrodes. In the absence of the cosubstrate, peaks in the cyclic voltammograms indicate the direct electron exchange between the prosthetic group of the adsorbed enzymes and the electrode. The generation of catalytic currents is used in order to demonstrate the maintenance of the enzymatic activity inside the adsorbed layer. In these studies the hydroquinone (H_2Q)/benzoquinone (BQ) couple acts as an electron-mediating cosubstrate of these enzymes (Figure 1). The feasibility of these enzyme electrodes has been demonstrated in a flow device. In addition to mono-enzyme electrodes coadsorbed bi-enzyme systems of the sequence, anti-interference and amplification type have also been developed. Their analytical parameters are compared with those of the membrane sensors.

EXPERIMENTAL

Reagents

Catalase-free glucose oxidase (GOD) from Penicillium notatum (E.C. 1.1.3.4., 46 U/mg) was purchased from VEB Arzneimittelwerk Dresden. Laccase from Polyporus versicolor (E.C. 1.10.3.2., 180 000 U/ml) was obtained from the Institute of Biochemistry, Academy of Sciences, Armenian SSR. Peroxidase (POD) from milk (E.C. 1.11.1.7., 340 µmol/l) was a gift from the Institute of Physiological Chemistry, University of Umea (Sweden). Other reagents used were of analytical grade. The background solution was a Sörenson phosphate buffer of pH 5.0 - 7.0 or 0.1 M acetate buffer of pH 5.5.

Apparatus

The amperometric measurements were carried out with a computer aided Electrochemical Measuring System ECM 700 from the Academy of Sciences of the GDR. A potential of +400 mV vs SCE for H_2Q detection and of +100 mV for BQ reduction was applied at the enzyme electrode. Hexacyanoferrate(III) reduction was monitored at +150 mV. The electrodes were inserted into a measuring cell containing 2 ml of stirred buffer at room temperature.

Carbon rods (Ringsdorff-Werke GmbH, type RW 203, diam. 3.1 mm) were cut and polished on wet, fine emery paper just before the application of the enzyme. They were press-fitted into Teflon holders so that only the flat end surface contacted the solution.

Flow injection analysis (FIA)

The carrier electrolyte was pumped with a peristaltic pump through a single-channel FIA system consisting of an injector and the flow-through

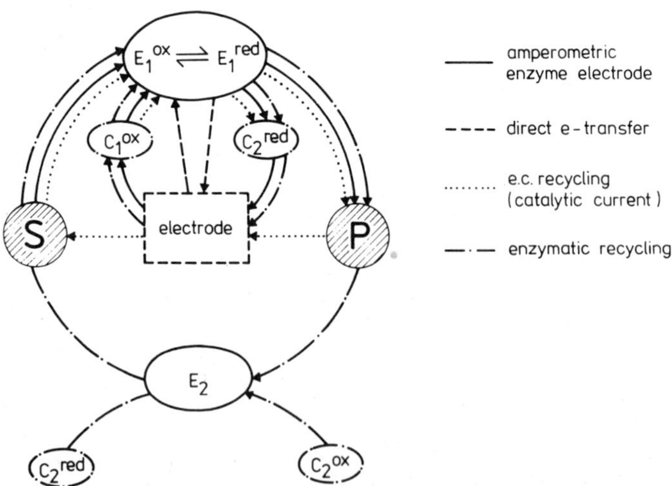

Figure 1. Schematic representation of electro-enzymatic principles.

electrode[2]. The samples (50 μl) were injected with a pneumatically operated valve (Cheminert, type SVA).

Enzyme adsorption

For loading, the electrodes were dipped for 5 min in a solution containing 20 U/ml GOD in acetate buffer or in a mixture of 10 U/ml GOD and 100 U/ml POD or 160 U/ml laccase[3]. In order to desorb weakly bound enzyme molecules each loading was followed by washing the electrode for 10 min in 50 ml intensively stirred buffer solution. The apparent activity of the enzymes absorbed at the electrode was determined by measuring the rate of product formation in the bulk solution by another electrode[4].

Membrane sensors

Laccase was immobilized alone or co-immobilized with GOD or cytochrome b_2 by entrapment in gelatin. The layer had a thickness of 20-30 μm. The enzyme membranes contained 120 U/cm^2 laccase, 56 U/cm^2 GOD and 30 U/cm^2 of cytochrome b_2, respectively[5]. For preparation of the sensor, the mono- or bi-enzyme layer was fixed on the dialysis membrane (17 μm thick) of the modified oxygen electrode, which comprised a platinum indicator electrode of 0.5 mm diameter covered by a dialysis membrane held by an O-ring.

RESULTS AND DISCUSSION

Adsorption at the mercury electrode

In the literature there are conflicting reports about the state of proteins adsorbed at the mercury electrode. Kuznetsov[6] concluded that all proteins are irreversibly denatured both at the dme and hdme to a layer corresponding to one peptide chain. On the other hand, Santhanam et al.[7] demonstrated the enzymatic activity of urease adsorbed at mercury covered thermistors.

In order to decide whether the enzyme is deactivated in the adsorption process we used the catalytic current approach. This method offers the opportunity to indicate the activity of the adsorbed enzyme even when it is also present in the bulk phase. This is based on the principle that the cosubstrate of the enzymatic reaction, e.g., BQ, is generated only at the electrode surface from the coproduct. Its regeneration in the enzyme catalyzed reaction leads to an increase of the original current. A well-known example of a catalytic current is the influence of catalase on oxygen reduction at the mercury electrode[8]. The H_2O_2 formed at the electrode is catalytically decomposed which is reflected in an increase of the oxygen reduction wave. This catalytic current is completely eliminated on addition of the catalase inhibitor azide[9]. A similar catalytic effect on the cathodic oxygen reduction by POD was found by Brown[10]. Furthermore, the addition of POD substrate resulted in a decrease of the catalytic current. Brown[10] explained this effect by the consumption of compound I in the POD catalyzed reaction. We studied this reaction using H_2Q as substrate. Contrary to the effect described in the literature, Strnad[11] found an increase of the catalytic current on addition of H_2Q. Obviously, the cathodic reduction of the BQ formed regenerates the substrate H_2Q. In this way, the H_2O_2 is catalytically reduced by the cyclic reaction. It was concluded that this reaction is mainly catalyzed by the absorbed POD because the current increase levels off at completion of the first adsorption layer. These results demonstrate that catalase and POD retain the enzymatic activity even at the expanding dme. On the other hand, we did not find catalytic currents with GOD or laccase using the BQ/H_2Q system. Also, no catalytic effect of laccase on the oxygen reduction was established at the dme. These results might be explained by dissociation of the protomeric enzymes into enzymatically inactive subunits during the adsorption process.

Direct electrochemical transformtions of enzymes adsorbed at carbon electrodes

Direct electron transfer between an electrode and the redox-active prosthetic group of enzymes is of interest both for the development of new types of biosensors or bio-fuel cells and for the fundamental understanding of biological processes. With the enzymes GOD, POD and laccase at carbon electrodes several redox processes have been found in cyclic voltammograms. For example, peroxidase, a heme-containing enzyme, exhibits redox processes in the potential ranges of 0.18 to 0.20 V and -0.33 to -0.25 V. The position of the more negative peak is close to the value of the redox potential of peroxidase. When the inhibitor potassium cyanide is added to the background solution, the latter peak in the current-voltage curve disappears. This result suggests that the cathodic maximum is related to the reduction of the active center of the enzyme. Absorbed laccase in the potential range from 0.28 to 0.15 V shows two small cathodic and corresponding anodic peaks. The shape and position of these peaks is independent of the potential sweep rate between 0.01 to 0.100 Vs^{-1}. The position of the peaks is far from the redox potential of laccase. With adsorbed GOD, redox processes occur near to the redox potential of the enzyme (-305 mV). The peak current increases proportionally to the bulk concentration of GOD during adsorption. When the GOD was denatured by heating the enzyme solution before absorption, a redox process at more positive potentials is observed in the current-voltage curve (Figure 2).

The nature of the electrochemical signals of oxidoreductases found in the absence of mediators and substrates is not yet clear. Probably the protein-bound prosthetic group is involved in the heterogeneous electron transfer since typically these signals differ from those of the free prosthetic group and also of the apoenzymes. Ikeda et al.[12] postulated that only a small fraction of the adsorbed molecules which is structurally changed and, therefore, enzymatically inactive is involved in the direct

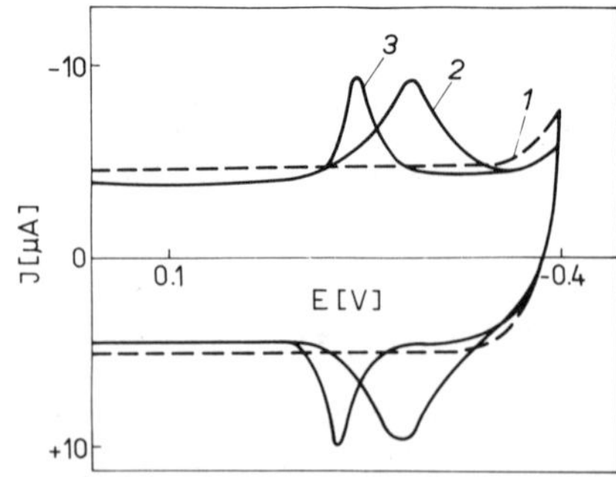

Figure 2. Cyclic voltammograms of glucose oxidase adsorbed at pyrolytic carbon. 0.1 mol/l acetate buffer pH 5.5, sweep rate 50 mv/s. 1-Bare electrode, 2-native GOD, 3-heat-denatured GOD.

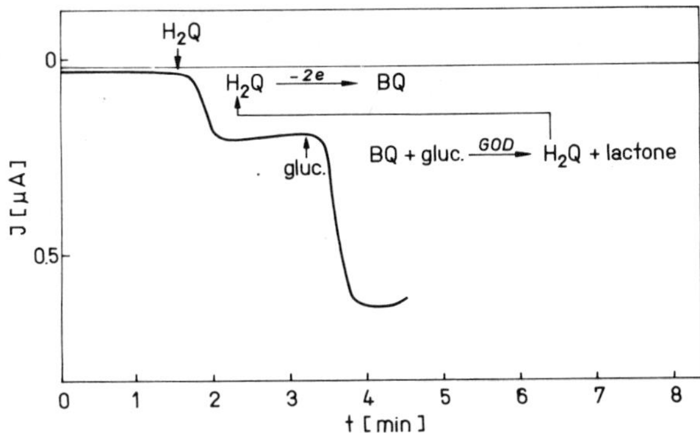

Figure 3. Catalytic current established by the enzymatic regeneration of hydroquinone on addition of glucose in presence of adsorbed POD. H_2Q: 20 µmol/l, glucose: 100 mmol/l, acetate buffer pH 5.5, E = +400 mV vs SCE.

electron transfer. Alternatively, we suppose that denaturation of the proteins is not necessarily involved in the direct electron transfer. Whilst the electron transfer requires a minimum distance to the electrode and an appropriate orientation of the protein all adsorbed molecules may be included in the substrate conversion.

Enzymatic properties of proteins adsorbed at carbon electrodes

Among current approaches to develop enzyme sensors, adsorption at carbon materials has attracted considerable interest[13]. By analogy to the dme experiments we demonstrated the enzymatic activity of adsorbed enzymes by the formation of catalytic currents. For this purpose only the co-product of the GOD-catalyzed reaction - H_2Q - is added to the bulk solution. It is converted at the electrode to BQ - the cosubstrate of the absorbed GOD. On addition of glucose the H_2Q is regenerated in the enzyme reaction. This is reflected by a threefold increase of the oxidation current of H_2Q (Figure 3). Similarly with adsorbed POD an eightfold current increase of BQ reduction is found on addition of hydrogen peroxide

In order to quantify the enzyme adsorption at the carbon surface we determined the apparent activity using a second BQ or H_2O_2 indicating electrode. From the rate of product accumulation in the bulk phase we calculated for GOD a value of 100 µU/cm^2. This value is in the region of the apparent activities reported for membrane covered electrodes[13,14]. The relatively high apparent activity is due to the absence of a protecting dialysis membrane which would restrict diffusional processes.

Another possibility to characterize the amount of the adsorbed activity is the determination of the "substrate-converting capacity". In these experiments the flux of substrate crossing the enzyme layer unreacted is indicated by the electrode. For adsorbed GOD the concentration dependence of the BQ reduction in the absence and in the presence of an excess of glucose is measured. The addition of glucose switches on the GOD; thus part of the BQ reaching the adsorption layer is converted to H_2Q. This is reflected by a decrease in the BQ reduction current by 25 %. This value indicates that the adsorption layer is not able to ensure the complete conversion of the cosubstrate.

When the product formation is directly indicated by the GOD-coated electrode, stepwise addition of the substrate gives the well-known stair curves. The enzymatic origin of this electrode signal was revealed by complete elimination of the current on addition of 5 mmol/l Cu^{2+} - an inhibitor of GOD.

The glucose sensor prepared by adsorption of GOD on a carbon electrode was mounted in a flow injection analysis system[2]. It was found that the best signal-to-background and signal-to-noise ratios were obtained at around +800 mV on the rising part of the voltammetric wave.

Freshly prepared enzyme electrodes, which had stabilized for 30 min, showed a constant response factor for hydrogen peroxide. The response was 15% of the sensitivity for H_2O_2, and it decreased during the first 15 min. After that the electrode showed an almost constant sensitivity for at least 180 min; the activity decreased at most 1% during the time. The sensitivity the next day was generally lower, i.e., about 75% of the initial value.

Injections of glucose samples using the enzyme electrode in the flow cell resulted in a linear response over at least two and a half concentration decades.

The low yield of glucose conversion in the adsorbed layer and the time dependent decrease of sensitivity indicate that the overall process is limited by the rate of the enzymatic reaction. Even when the activity of the adsorbed enzyme is relatively low the glucose conversion gives evidence for the maintenance of enzymatic activity of the adsorbed GOD molecules. Therefore, the dimeric structure of the GOD molecule is obviously not destroyed in the adsorption process.

The difference in the time-course of the response factors for H_2O_2 and glucose indicates a loss or inactivation of the enzyme. The initial fairly rapid decrease of enzyme activity at the time of glucose introduction can be explained by assuming that conformational changes in the enzyme during turn-over results in the loss of some loosely adsorbed glucose oxidase. The slow decrease of sensitivity later on may be due to desorption, possibly enhanced during turn-over, and to deactivation.

In order to realize coupled enzyme reactions we studied the co-adsorption of different enzymes, e.g., glucose oxidase and peroxidase by dipping the electrode in a solution containing both proteins. The individual apparent activities of 50 mU/cm² for GOD and 200 mU/cm² for POD were found. The competitive adsorption is reflected by a decrease of the GOD activity as compared with the simple GOD electrode. We demonstrated the sequential action of both enzymes by measuring the BQ formed after addition of glucose to an H_2Q-containing solution (Figure 4). In this sequence the GOD produces hydrogen peroxide on addition of glucose which in turn is converted in the POD-catalyzed reaction. To our knowledge this result is the first enzyme sequence electrode prepared by co-adsorption of two enzymes.

The blue copper enzyme laccase catalyzes the oxidation of para-diphenols or diamines but also of inorganic ions, e.g, ferrocyanide, at the expense of oxygen. In the H_2Q oxidation, depending on the electrode potential, either the H_2Q is regenerated at the electrode from the BQ or the remaining H_2Q permeating the laccase layer unconverted is indicated. Therefore, both the effect of substrate recycling or the breakthrough of the hydroquinone can be studied. Wasa et al.[15] demonstrated a signal amplification by electrochemical regeneration of hydroquinone in combination with immobilized laccase. They calculated a maximum amplification

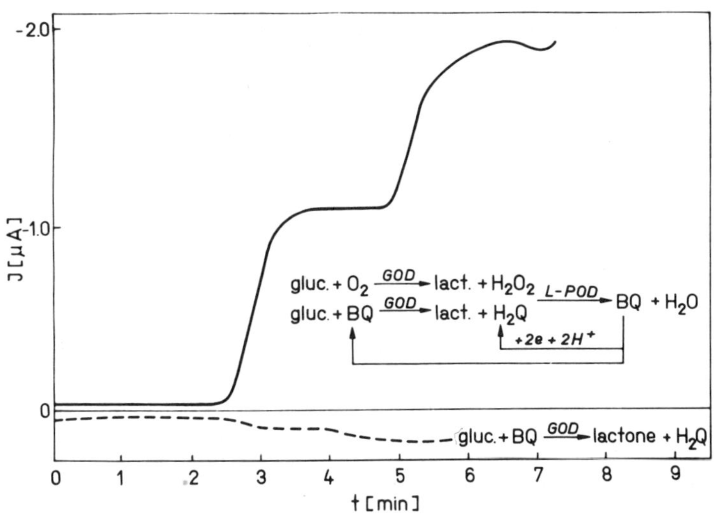

Figure 4. Demonstration of the sequential action of coadsorbed glucose oxidase and peroxidase. H_2Q: 20 μmol/l, glucose: 0.5 mmol/l, acetate buffer pH 5.5, E = +100 mV vs SCE.

factor of 20 by comparing the signals for a laccase-free electrode and one containing covalently attached laccase.

By analogy with the experiment described for adsorbed GOD we also studied the substrate-converting capacity of the laccase layer. For this reason the unreacted H_2Q reaching the electrode surface is indicated in the absence and in the presence of the reversible inhibitor azide. These experiments indicate that only 5 - 20 % of the H_2Q are enzymatically oxidized in the laccase layer. On the other hand, laccase adsorbed at carbon powder electrodes exhibits a pronounced catalytic effect on the cathodic oxygen reduction. Tarasevich[16] and Anson et al.[17] demonstrated that in neutral solution the stationary potential is close to the equilibrium potential of the oxygen electrode and that no intermediary hydrogen peroxide is formed. The polarization curve is drasticaly shifted in a positive direction indicating the acceleration of the electroreduction of oxygen within the potential range of 0.6 to 1.2 V. The enzymatic nature of the electrocatalytic effect was proven by the specific inhibition by fluoride and azide. This laccase electrode showed a stable performance for 50 h.

Enzyme electrodes based on membranes

We demonstrated the signal amplification by electrochemical regeneration of hydroquinone also in combination with gelatin-immoblized laccase. This enzyme layer has a thickness of 30 μm. In order to characterize the permeability of the laccase membrane we measured the concentration dependence of the anodic H_2Q oxidation (at +100 mV) in the presence of azide. On the other hand, in the absence of the inhibitor the enzymatically formed BQ is indicated at the electrode polarized at -100 mV. This reduction current increases linearly with the H_2Q concentration levelling off at a bulk concentration of 1.2 mM (Figure 5).

The higher sensitivity with active laccase is based on the electrochemical/enzymatic cycling of the H_2Q/BQ couple. However, this recycling

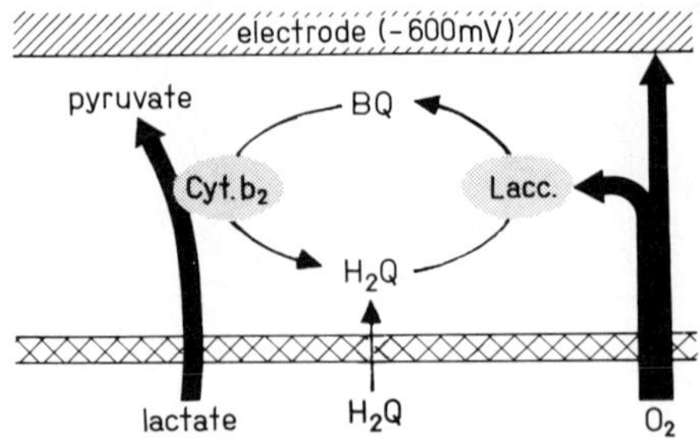

Figure 5. Comparison of the response for BQ and H_2Q of the membrane electrode containing active or inhibited laccase/phosphate buffer pH 5.0.

does not result in a real increase of sensitivity for H_2Q. The laccase membrane in front of the electrode decreases the current by a factor of about eight as compared with a bare electrode. Therefore, the overall result is a loss of sensitivity caused by covering of the electrode.

Analyte recycling

Recently several enzyme electrodes based on a recycling between two enzymes of the substrate to be measured have been investigated[1,18]. With horseradish peroxidase/glucose dehydrogenase a 40-fold amplification was obtained in NAD(H) measurement. Lactate recycling has also been accomplished using the lactate oxidase/lactate dehydrogenase couple in enzyme electrodes. In the former, an amplification factor as high as 4,100 was obtained.

In another recycling system we co-immobolized cytochrome b_2 and laccase in a gelatin layer in front of an oxygen electrode. In the presence of the cytochrome b_2 substrate, lactate, BQ is reduced to H_2Q. Therefore, this enzyme substitutes the cathode of the previous cycling system (Figure 6). However, the substrate recyling is not restricted to the phase boundary (as with electrochemical recyling) but it proceeds in the total volume of the enzyme layer. At the optimum pH of cytochrome b_2 (pH 6.5) and lactate saturation a maximum amplification of 500 has been obtained. In these experiments the loading of cytochrome b_2 was 30 U/cm² and that of

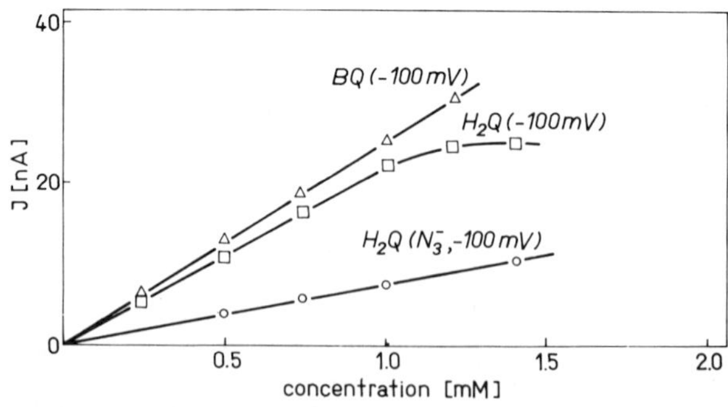

Figure 6. Scheme of the signal amplification for the BQ/H_2Q couple by enzymatic recycling.

laccase 120 U/cm². The linear concentration dependence of the amplified signal levels off at about 2.4 μM H_2Q. It reaches almost the saturation current found in the unamplified system in the absence of lactate. Therefore, the oxygen inside the bi-enzyme membrane is only partly consumed even under effective H_2Q recycling. The amplification depends markedly on the concentration of the cosubstrate of cytochrome b_2, lactate. This concentration dependence at a constant H_2Q concentration is curved. Principally this sensor represents a new lactate electrode.

The concentration dependence of the oxidation current at +100 mV shows a "threshold"-type. Thus, in the absence of lactate, addition of H_2Q up to 1 mM does not lead to a typical current increase. Obviously, the laccase converts its substrate completely to BQ which is invisible at +100 mV. The current increase above 1.2 mM H_2Q reflects the "breakthrough" of unreacted H_2Q (Figure 7). In the presence of lactate, part of the BQ formed in the laccase-catalyzed reaction is recycled to H_2Q. Therefore, a "short cut" is found at lower H_2Q concentrations.

Anti-interference layers

The capability of a diffusion-controlled enzyme membrane to completely convert the substrate permeating the membrane offers an interesting analytical application. In other words, interfering substances are converted into non-disturbing products by "filtering" the analyte flux to the indicator electrode.

We demonstrated that reducing substances disturbing the indicator reaction, e.g., ascorbic or uric acid, can be eliminated by a preceding laccase-catalyzed reaction[5]. The interfering substances are either directly oxidized by oxygen, as in the case of epinephrine, or a preceding oxidation reaction is carried out in the background solution by ferricyanide. The ferrocyanide formed in the chemical reaction is subsequently oxidized by laccase which is co-immobilized with the respective oxidase. In this

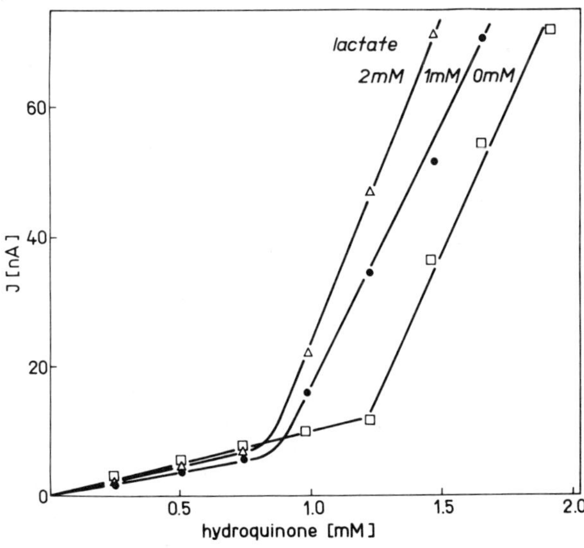

Figure 7. Influence of the lactate concentration on the H_2Q converting capacity of the laccase/cytochrome b_2 electrode/phosphate buffer pH 6.5, E = +400 mV vs SCE.

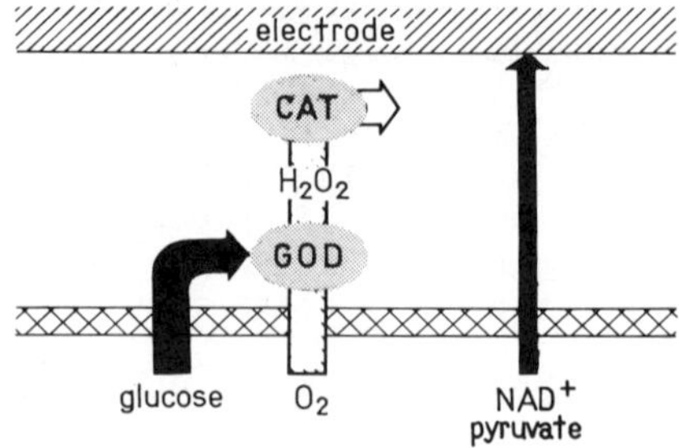

Figure 8. Scheme of GOD/catalase layer for the removal of oxygen in front of a flat electrode.

way the interference caused by ascorbic acid is completely eliminated up to 20 mmol/l. By co-immobilization of GOD and catalase in front of a flat membrane-covered mercury electrode the principle of an anaerobic sensor was developed. Oxygen is converted in the GOD-catalyzed reaction to H_2O_2 which in turn is destroyed by the catalase (Figure 8). In this way the O_2 concentration in the enzyme membrane is decreased to almost zero, which is reflected by a drastic background current decrease. Therefore, the sensitivity of the electrode can be considerably increased. Using the anaerobic electrode, concentration-proportional peaks were obtained for NAD^+, methyl viologen and pyruvate in the micromolar range. In this way, the determination of dehydrogenase or pyruvate kinase activity is possible by a simple polarographic device.

CONCLUSIONS

The electro-enzymatic studies presented in this paper indicate that all enzymes investigated retain their enzymatic activity after adsorption at carbon electrodes. The adsorption energy may lead to pronounced structural changes which influence the kinetic parameters K_M and V_{max}. Owing to the limited surface of the electrodes, the substrate-converting capacity and the functional stability are considerably lower than for optimized enzyme membranes. Furthermore, the enzymatic activities are not sufficient for the enzymatic elimination of interfacing substances or for an effective signal amplification by electro-enzymatic recycling of the analyte. Considerable progress may be achieved by using high-surface area materials in combination with chemically modified surfaces to facilitate the direct electron transfer between the enzyme and the electrode.

REFERENCES

1. F.W. Scheller, U. Wollenberger, F. Schubert, D. Pfeiffer and V. Bogdanovskaja, Proc. Work Shop on Biosensors, Braunschweig, May 1987, p. 39.

2. L. Gorton, F.W. Scheller and G. Johansson, Studia Biophys., 109, 199 (1985).

3. V. Bogdanovskaja, R. Hintsche, F.W. Scheller and M. Tarasevich, Bioelectrochem. Bioenerg., in press.

4. R. Hintsche and F.W. Scheller, Studia Biophys., 119, 178 (1987).

5. U. Wollenberger, F.W. Scheller, D. Pfeiffer, V. Bogdanovskaja, M. Tarasevich and C. Hanke, Analyt. Chim. Acta, 187, 39 (1986).

6. B. Kuznetsov, G. Shumakovich and N. Mestechkina, Bioelectrochem. Bioenerg., 4, 582 (1977).

7. K. Santhanam, N. Jespersen and A. Bard, J. Amer. Chem. Soc., 99, 274 (1977).

8. J. Koutecy and R. Brdicka, Coll. Czech. Chem. Commun., 18, 611 (1953).

9. F.W. Scheller, G. Strnad, R. Rennberg and D. Kirstein, in "Charge and Field Effects in Biosystems", M. Allen and P. Usherwood, Eds., Abaccus Press, Nottingham (1984), p. 483.

10. G. Brown, Arch. Biochem. Biophys., 49, 303 (1954).

11. G. Strnad, Electrochim. Acta, in press.

12. T. Ikeda, I. Katasho, M. Kamai and M. Senda, Agric. Biol. Chem., 48, 1969 (1984).

13. V. Razumas, J. Jasaitis and J. Kulys, Bioelectrochem. Bioenerg., 12, 297 (1984).

14. F.W. Scheller, R. Renneberg and F. Schubert, Methods Enzymol., 137, 39 (1988).

15. T. Wasa, K. Akimoto, T. Yao and S. Murao, J. Chem. Soc. Jpn., 9, 1398 1984).

16. M. Tarasevich, in "Comprehensive Treatise of Electrochemistry", Vol. 10, S. Srinivasan, Yu. Chizmadzhev, J. Bockris, B. Conway and E. Yeager, Eds., Plenum Press, New York and London (1985), p. 231.

17. C.-W. Lee, H. Gray, F. Anson and B. Malmström, J. Electroanal. Chem., 172, 289 (1984).

18. F. Schubert, D. Kirstein, K.L. Schröder and F.W. Scheller, Anal. Chim. Acta, 169, 391 (1985).

19. J. Kulys, V. Sorochinski and R. Vidziunaite, Biosensors, 2, 135 (1986).

CATALYTIC OXIDATION OF D-GLUCOSE AT AN ENZYME-MODIFIED ELECTRODE WITH ENTRAPPED MEDIATOR

Tokuji Ikeda, Hiroshi Hiasa and Mitsugi Senda*

Department of Agricultural Chemistry
Faculty of Agriculture
Kyoto University
Sakyo-ku, Kyoto 606 / Japan

INTRODUCTION

Biocatalyst electrodes, that is, modified electrodes carrying an immobilized enzyme in which the electrode behaves as a substitute for a chemical electron acceptor or donor in the enzyme reaction can be used in such novel applications as biosensors, bioreactors and biofuel cells[1,2]. The bioelectrocatalytic oxidation or reduction of substrates at biocatalyst electrodes can be accelerated by the presence of a small molecule which functions as an electron transfer mediator between the electrode and the prosthetic group of the immobilized enzyme[1-6]. Two types of the enzyme-modified electrode with entrapped mediator have been designed[7,8,9], a carbon paste electrode and a porous electrode, both modified with enzyme and a reservoir of mediator. In the former type of electrode[7,8], where glucose oxidase (GOD) was immobilized on the surface of a carbon paste electrode along with p-benzoquinone (BQ) by coating the enzyme-loaded surface with a semipermeable membrane, BQ was dissolved into the immobilized enzyme layer and retained there effectively to serve as an electron transfer mediator between the carbon paste electrode and the immobilized enzyme. It has been shown[8-11] that the kinetics of the bioelectrocatalytic oxidation of D-glucose (Glc) at the membrane-coated GOD-modified carbon paste electrode with mixed-in BQ can be explained by theoretical equations in which the diffusion enzyme reaction of the substrate and mediator in the immobilized-enzyme layer and the diffusion of the substrate (and mediators) in the coating membrane are taken into account[8-10]. In this study, a number of quinone derivatives and a few ferrocene derivatives were examined as electron transfer mediators in the GOD electrode, and the mediator activity of the compounds was evaluated on the basis of theoretical equations.

EXPERIMENTAL

Chemicals

GOD, EC 1.1.3.4 (Type VII, lot No. 39c-0255) was obtained from Sigma Chemical Co., BQ, 2-methyl-p-benzoquinone, tetrachloro-p-benzoquinone and α-naphthoquinone were purchased from Wako Pure Chemical Co., 2,5-dimethyl-p-benzoquinone, 2-isopropyl-5-methyl-p-benzoquinone, tetramethyl-p-benzoquinone, tetrabromo-p-benzoquinone, 2,5-dichloro-p-benzoquinone, 2,6-dichloro-p-benzoquinone, β-naphthoquinone, β-naphthoquinone-4-sulfonate and 1,1'-dimethyl-ferrocene from Tokyo Kasei Kogyo Co., tetrabromo-o-benzoquinone, ferrocene and ferrocenecarboxylic acid from Aldrich Chemical Co. These chemicals were used as received except for BQ and 2-methyl-o-benzoquinone which were purified by sublimation before use. The other chemicals used were of reagent grade.

Apparatus

A Yanagimoto P-1000 voltammetric analyzer was used in conjunction with a Nikko Keisoku NFG-3 function generator to apply the potential to a three-electrode system, and currents were measured with a Watanabe Keisoku WX4401 X-Y recorder.

Preparation of electrodes

The mediator-mixed carbon paste electrodes were prepared as described elsewhere[7,8]. The geometric surface area of the electrode was 9.0×10^{-2} cm^2. A known amount of GOD solution was syringed onto the surface of the carbon paste electrode with mixed-in mediator. After the solvent was allowed to evaporate the electrode surface was covered with a dialysis membrane (Seamless Cellulose Tubing, Union Carbide Co.) of 20 μm thickness. The whole electrode was covered by a nylon net to give the electrode physical strength. The construction of the GOD modified carbon paste electrode with mixed-in mediator coated with a dialysis membrane (a film-coated GOD CPE with mediator) is illustrated in Figure 4 in Reference 9 (see also Figures 1 in Reference 12).

Electrochemical measurements

All measurements were done at 25 °C in a deaerated 0.2 mol/dm^3 phosphate buffer solution of pH 7.0 containing 5% v/v ethanol (the basal solution) under stirring with a magnetic stirrer at 500 rpm, where concentration polarization in the solution appeared negligibly small. All potentials were measured against a saturated calomel electrode (SCE).

RESULTS AND DISCUSSION

Catalytic oxidation current of Glc with the film-coated glucose oxidase-immobilized carbon paste electrode (GOD CPE) with entrapped mediator

Previous experiments with film-coated GOD CPEs with BQ showed[7,8] that the BQ mixed in the carbon paste dissolved into the enzyme layer on the electrode surface to reach a steady-state concentration and that the concentration of BQ in the enzyme layer increased with increasing concentration of the mixed-in BQ in the carbon paste, m_{BQ}. Thus, the film-coated GOD CPE with BQ immersed in the basal solution produced cyclic voltammetric waves due to the electrochemical redox reaction of BQ in the enzyme layer, and the wave height of the voltammogram increased with increasing m_{BQ} (Figure 1 in Reference 8). When Glc was added to the buffer solution, the anodic current increased greatly (Figure 2 in Reference 8). The increase in the anodic current was due to the BQ-mediated electrocatalytic oxidation of Glc and the steady-state anodic current, I, was obtained when measured at a fixed electrode potential, E. Representative current-potential (I - E) curves are shown in Figures 1 and 7. The steady-

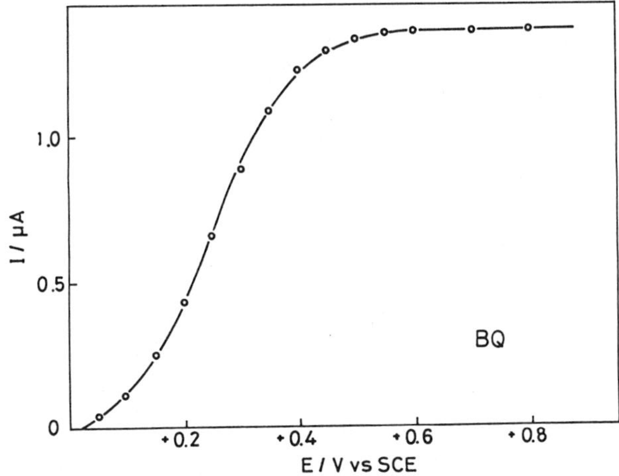

Figure 1. Dependence of the steady-state anodic current, I, on the electrode potential, E. I was measured with a film (20 μm)-coated GOD (0.5 μg) CPE with BQ (0.1%) in the basal solution containing 125 mmol/dm³ Glc.

state anodic current increased with increasing positive potential and reached a limiting current at potentials more positive than 0.5 V (Figure 1). The steady-state limiting current, I_l, obtained with the biocatalyst electrode with an entrapped mediator can be expressed by[8-11]

$$I_l/nFA = k'_{cat}(E)l/(1 + K'_{m1}/{}^{l}c_S + K'_{m2}/{}^{o}c_O) \quad (1)$$

where n, F and A are the number of electrons (here n = 2), the Faraday constant and the electrode surface area, respectively. (E) is the concentration of the enzyme in the enzyme layer on the electrode surface, l is the thickness of the layer, ${}^{l}c_S$ the concentration of the substrate (here Glc) in the enzyme layer at the boundary between the enzyme layer and the film (here dialysis membrane) coating the enzyme layer, ${}^{o}c_O$ the concentration of the mediator (here BQ) at the electrode surface of the enzyme layer; k'_{cat}, K'_{m1} and K'_{m2} are the apparent catalytic constant, and the apparent Michaelis constants for the substrate and mediator, respectively. These apparent constants include the effect of the diffusional resistance of the enzyme layer to the substrate and mediator and, when this effect is small, they are reduced to their true values, that is, the catalytic constant, k_{cat}, the Michaelis constants for the substrate K_{m1} and for the mediator, K_{m2}, respectively, defined by[4]

$$k_{cat} = k_2k_4/(k_2 + k_4) \quad (2)$$

$$K_{m1} = (k_4/(k_2+k_4))((k_{-1}+k_2)/k_1) \quad (3)$$

$$K_{m2} = (k_2/(k_2+k_4))((k_{-3}+k_4)/k_3) \quad (4)$$

for the enzyme reaction with the mediator,

$$S + E_{ox} \underset{k_{-1}}{\overset{k_1}{\rightleftharpoons}} ES \xrightarrow{k_2} E_{red} + P \quad (5)$$

$$M_{ox} + E_{red} \underset{k_{-3}}{\overset{k_3}{\rightleftharpoons}} EM \xrightarrow{k_4} E_{ox} + M_{red} \quad (6)$$

where S and P are the substrate and product, E_{ox} and E_{red} the oxidized and reduced forms of the enzyme, M_{ox} and M_{red} the oxidized and reduced forms of the mediator, respectively. In this reaction scheme the participation of the proton transfer associated with the redox reactions is not included for simplification.

Figure 2 shows the dependence of I_l on m_{BQ}, measured at 0.5 V with film (20 μm)-coated GOD (5 μg) BQ(m_{BQ})-mixed CPEs in the basal solution containing 0.3 mol/dm³ Glc. As seen in this figure and also in Figure 3 the reproducibility of the current response of the CPEs was at the level of 20% (coefficient of variation) when measured with different electrodes prepared by the same procedure with the same materials, whereas the current response measured with the same electrode (Figures 1, 4, and 7) was reproducible at the level of a 3% coefficient of variation. Figure 2 shows that I_l increases with increasing m_{BQ} to approach a saturation value, indicating that $K'_{m2}/{}^{o}c_o \ll 1$ in Equation 1 at $m_{BQ} > 5\%$ for this particular CPE, since ${}^{o}c_o$ increases with increasing m_{BQ}. Similar dependence of I_l on m_M, the concentration of the mediator mixed in the carbon paste, was observed with the film-coated GOD CPEs with 2-methyl-p-benzoquinone, 2-isopropyl-5-methyl-p-benzoquinone. Similarly, the saturation values were attained at $m_M > 2\%$ with 1,1'-dimethyl-ferrocene and ferrocene-carboxylic acid and at $m_M > 4\%$ with the other mediators examined. Figure 3 shows the dependence of I_l on (E)1A, obtained with film (20 μm)-coated GOD((E)1A) CPEs with BQ (6%) in the basal solution containing 0.3 mol/dm³ Glc. I_l increased first linearly, but deviated downward from linearity with increasing (E)1A. In the following experiments the amount of GOD immobilized on the CPE, that is, (E)1A, was kept at 0.5 μg unless otherwise stated. The current intensity is also controlled by the diffusion rate of the substrate through the coating membrane;

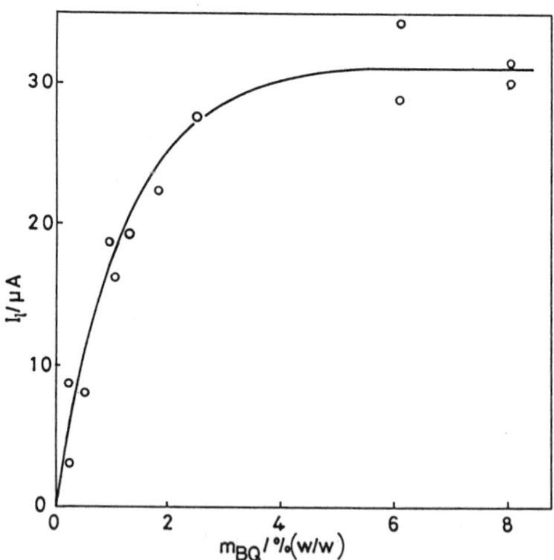

Figure 2. Dependence on the limiting anodic current, I_l, on the concentration of BQ in carbon paste, m_{BQ}. I_l was measured with film (20 μm)-coated GOD (5 μg) CPEs with BQ(m_{BQ}) in the basal solution containing 0.3 mol/dm³ Glc.

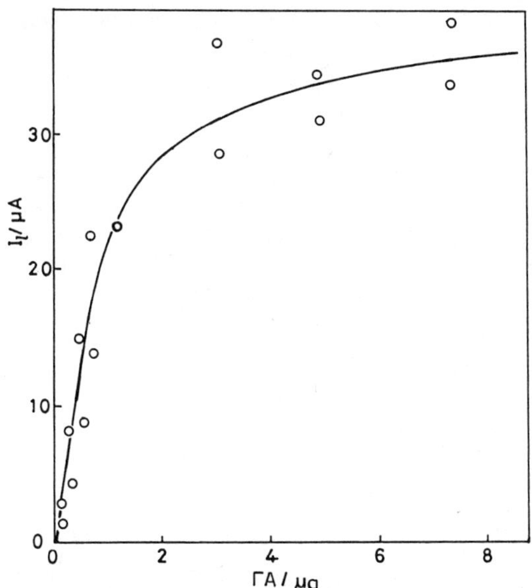

Figure 3. Dependence of the limiting anodic current, I_l, on the amount of GOD loaded on the carbon paste electrode surface, (E)1A = Γ A. I_l was measured with film (20 μm)-coated GOD (Γ A) CPEs with BQ (6%) in the basal solution containing 0.3 mol/dm^3 Glc.

$$I_l/nFA = P_{sm}(^*c_s - {}^1c_s/\beta_s) \tag{7}$$

where P_{sm} is the permeability of the dialysis membrane to Glc, $^*c_s$ the concentation of Glc in the solution immediately adjacent to the dialysis membrane, and β_s the distribution coefficient of Glc between the enzyme layer and the solution. Eliminating ^1c_s from Equation 1 but $K'_{m2}/^oc_o \ll 1$ with Equation 7, we obtain[8]

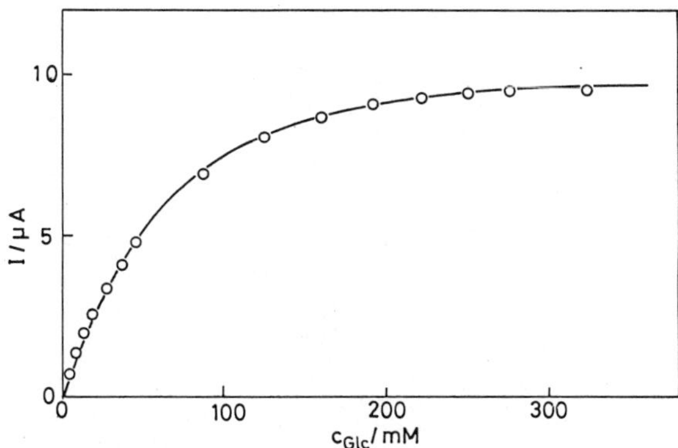

Figure 4. Dependence of the limiting anodic current, I_l, on the concentration of Glc, c_{Glc}, in the basal solution. I_l was measured with a film (20 μm)-coated GOD (0.5 μg) CPE with BQ (6%) in the basal solution containing Glc.

$$I_l/nFA = (k'_{cat}(E)l\beta_s{}^*c_s/K'_{ml})/(1 + (\beta_s{}^*c_s/K'_{ml}) + (\beta_s(k'_{cat}(E)l-I/nFA)/P_{sm}K'_{ml})) \qquad (8)$$

This Equation predicts[8] the dependence of I_l on $^*c_s$, the bulk concentra- ation of Glc since there is no concentration polarization in the solution outside the film (see Experimental), showing that I_l again approaches a saturation value at high values of $^*c_s$. This prediction was experimentally verified as shown in Figure 4, for example, for I_l obtained with a film-coated GOD CPE with BQ (6%), giving the saturation or the apparent maximum current, $I^{max\prime}$, which should be given by

$$I^{max\prime} = nFAk'_{cat}(E)l \qquad (9)$$

The k'_{cat} value calculated from the $I^{max\prime}$ value was 20 s^{-1} ($A(E)l$ = 0.5 μg and n = 2), which can be compared with k_{cat} = 360 s^{-1} obtained with the solubilized GOD[4]. This result is in line with the theoretical prediction on the ratio of k'_{cat}/k_{cat}[11], but suggests that the catalytic constant may be smaller in the immobilized state of GOD than in the solubilized state and/or that only a part of the immobilized GOD remains enzymatically active.

Dependence of the catalytic oxidation current of Glc on the electrochemical properties of mediators

Twelve quinone derivatives and three ferrocene derivatives were examined. Their cyclic voltammograms were measured with a film-coated GOD (0.5 μg) CPE without mixed-in mediator immersed in the basal solution containing 1 mmol/dm³ mediator. However, for some mediators, such as tetramethyl p-benzoquinone, 1,2-naphthoquinone, 1,4-naphthoquinone, ferrocene, and 1,1'-dimethyl-ferrocene a saturated solution (generally, of low solubility) was used. Figures 5 and 6 show the cyclic voltammograms of 2,6-dichoro-p-benzoquinone and BQ. The mid-point potential, E_{mid}, was obtained from the cathodic and anodic peak potentials, E_{pc} and E_{pa}, respectively, by $E_{mid} = (E_{pc}+E_{pa})/2$. All compounds examined were active as electron transfer mediators for bioelectrocatalysis with the GOD CPEs with mixed-in mediator (usually m_M = 5%), except tetramethyl-p-benzoquinone (that is,

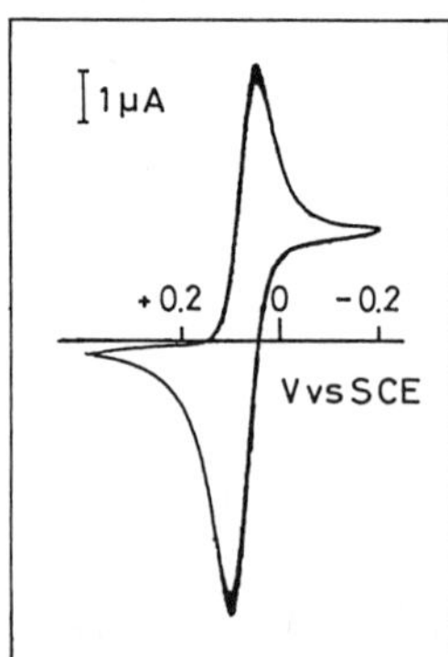

Figure 5. Cyclic voltammogram of 2,6-dichloro-p-benzoquinone recorded with a film-coated GOD (0.5 μg) CPE in the basal solution containing 1 mmol/dm³ 2,6-dichloro-p-benzoquinone. The potential scan rate: 10 mV/s.

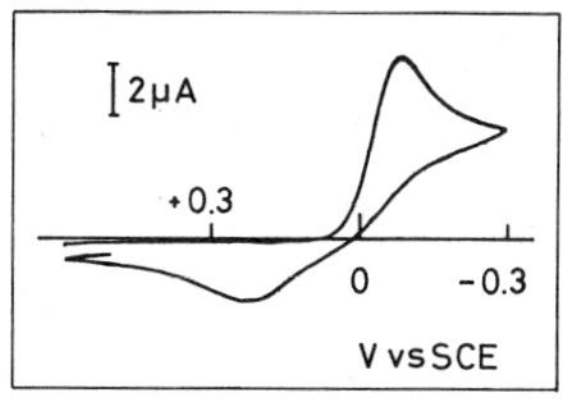

Figure 6. Cyclic voltammogram of BQ recorded with a film-coated GOD (0.5 μg) CPE in the basal solution containing 1 mmol/dm³ BQ. The potential scan rate: 10 mV/s.

showing no appreciable catalytic oxidation current) which had E_{mid} = -0.24 V vs SCE.

The current-potential curves of the catalytic oxidation current mediated by the mixed-in mediators depend on the electrode reaction properties of the mediator used, as theory has predicted[10,11,13]; those which gave reversible cyclic voltammograms, such as, 1,2-naphthoquinone and 2,6-dichloro-p-benzoquinone produced reversible catalytic current-potential curves, as shown in Figure 7, whereas those which gave quasi-reversible or irreversible cyclic voltammograms, such as BQ and 2-methyl-p-benzoquinone, gave quasi- or irreversible drawn-out waves, as seen in Figure 1. The half-wave potential of the catalytic current nearly coincided with the mid-potential for the former "reversible" compounds, whereas it was shifted to more positive potential than the mid-potential for the latter. In the same way, the enzyme kinetics should also influence the dependence of the catalytic current on the electrode potential. Further discussion will be given elsewhere.

The limiting catalytic current, I_l, of Glc at a film-coated GOD CPE with mixed-in mediator (m_M = 6%) was measured for various mediators in the basal solution containing Glc at various concentrations, $^*c_s$, and I_l was plotted against $^*c_s$ (see Figure 4) to get the apparent maximum current, $I^{max'}$, for each of the mediators examined. In Figure 8, $\log I^{max'}$ was plotted against $\Delta E = E_{mid} - E^{o'}_{GOD}$, $E^{o'}_{GOD}$ being the redox potential of GOD (0.41 V vs SCE at pH 7)[14]. The measured $I^{max'}$ may be discussed by

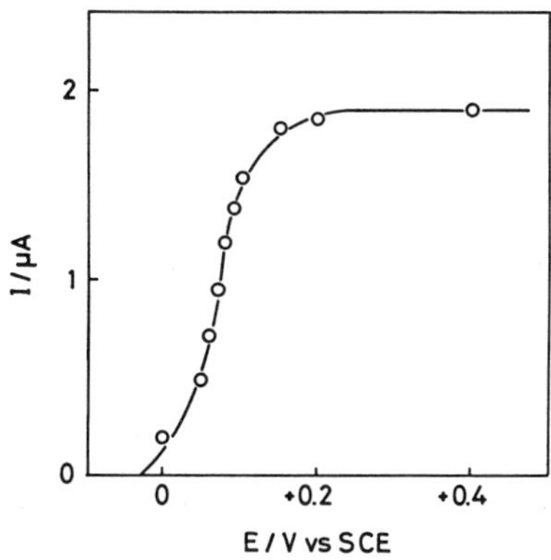

Figure 7. Current-potential curve obtained with a film (20 μm)-coated GOD (0.5 μg) CPE with 2,6-dichloro-p-benzoquinone (0.1%) in the basal solution containing 125 mmol/dm³ Glc.

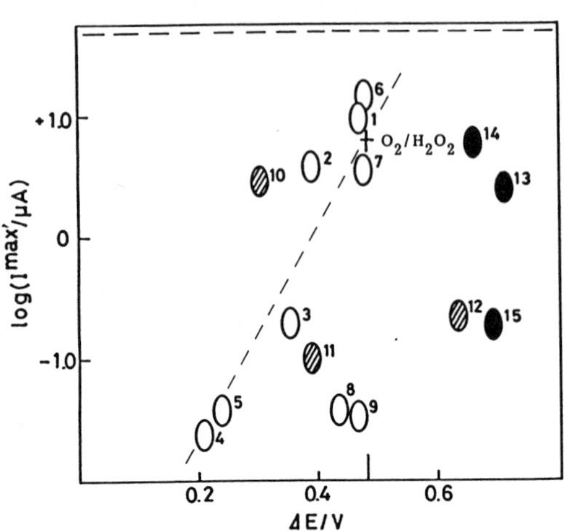

1. p-benzoquinone
2. 2-methyl-p-benzoquinone
3. 2,5-dimethyl-p-benzoquinone
4. 1,4-naphthoquinone
5. 2-isopropyl-5-methyl-p-benzoquinone
6. 2,6-dichloro-p-benzoquinone
7. 2,5-dichloro-p-benzoquinone
8. tetrabromo-p-benzoquinone
9. tetrachloro-p-benzoquinone
10. 1,2-naphthoquinone
11. 1,2-naphthoquinone-4-sulfonic acid
12. tetrabromo-o-benzoquinone
13. ferrocene
14. 1,1'-dimethylferrocene
15. ferrocenecarboxylic acid

Figure 8. Plot of $\log I^{max}$ against ΔE.

use of the approximation $I^{max'} \alpha\ k_{cat} = k_2 k_4/(k_2+k_4)$ (Equations 2 and 9), by assuming that $K'_{m2}/{}^{o}c_{o} \ll 1$ (see above). The last condition appeared to be satisfied for most of mediators examined. Figure 8 includes the $I^{max'}$ value for oxygen as the mediator. The steady-state catalytic oxidation current at 0.7 V was measured with a film (20 μm)-coated GOD (0.5 μg per 0.09 cm^2)-modified platinum disk electrode in air-saturated basal solution containing Glc. To obtain the I^{max} value, the I value was corrected for the oxygen concentration by use of K_{m2} = 0.48 mmol/dm^3 for oxygen[15]. Figure 8 indicates that in a series of p-quinone derivatives, I^{max} increases with increasing ΔE for compounds 1-7, but for compounds 8 and 9 it remains as small as those of compounds 4 and 5 and for compounds 12 as that of compound 3 in spite of their large ΔE values. The result indicates the structure factor that when all positions of benzo-quinone ring: 2, 3, 5 and 6 are occupied by a large substituent like chlorine or bromine, the mediator activity is greated reduced. Also, compounds 10 and 11 indicate that the mediator activity of negatively charged compounds is greatly reduced compared with the neutral homolog. The negative charge effect was observed also for ferrocene derivatives, compounds 13 and 15. In Figure 8 the broken line parallel to the abscissa shows the estimated upper limit of $I^{max'}$, that is, the $I^{max'}$ value corresponding to the case of $k_{cat} \approx k_2$ (when $k_4 \gg k_2$) by taking the k_2 value as ten times the k_4 value for oxygen[15]. After correction for the structure factor and the negative charge effect, the dependence of $\log I^{max'}$ on the redox potential difference ΔE appears true, indicating that k_4 (since $k_{cat} \approx k_4$ when $k_2 \gg k_4$) depends on ΔE, that is, the driving force of the redox reaction between the reduced enzyme and the oxidized mediator is the redox potential difference between the reactants. The inclined broken line in Figure 8 shows the slope of the dependence of $\log I^{max'}$ on ΔE/120 mV. Similar dependence of the rate of the half-reaction (Equation 6) in solubilized GOD reaction on the redox potential difference has been described by Kulys and Cenas[16].

CONCLUSION

The kinetics of the electrocatalytic oxidation of D-glucose at film-coated GOD CPEs with mixed-in mediator were investigated and explained by the theory of a biocatalyst electrode with an entrapped mediator. A number of quinone and ferrocene derivatives were examined as electron transfer mediators for the GOD reaction in the immobilized enzyme layer on the carbon paste electrode, and their mediator activity was evaluated.

ACKNOWLEDGEMENTS

This work was supported by a Grant-in-Aid for Scientific Research (No. 61560092) and Scientific Research on Priority Areas "Macromolecule-Complexes" (No. 6212003) from the Ministry of Education, Science and Culture.

REFERENCES

1. M. Senda and T. Ikeda, "Kobunshi Hyomen no Kiso to Ohyo (Polymer Surface, Fundamentals and Applications)", Vol. 2, Ed. Y. Ikada, Kagaku-Dojin, Kyoto, 1986, p. 201.

2. M.R. Tarasevich, "Comprehensive Treatise of Electrochemistry", Vol. 10, Eds. S. Srinivasan et al., Plenum Press, New York, 1985, p. 231.

3. J.J. Kulys, M.V. Presliakiene and A.S. Samalius, Bioelectrochem. Bioenerg., 8, 81 (1981).

4. T. Ikeda, I. Katasho, M. Kamei and M. Senda, Agric. Biol. Chem., 48, 1969 (1984).

5. C. Bourdillon, J.M. Laval and D. Thomas, J. Electrochem. Soc., 133, 706 (1986).

6. A.E.G. Cass, G. Davis, G.D. Francis, H.A.O. Hill, W.J. Aston, I.J. Higgins, E.V. Plotkin, L.D.L. Scott and A.P.F. Turner, Anal. Chem., 56, 667 (1984).

7. T. Ikeda, H. Hamada, K. Miki and M. Senda, Agric. Biol. Chem., 49, 541 (1985).

8. T. Ikeda, H. Hamada and M. Senda, Agric. Biol. Chem., 50, 883 (1986).

9. M. Senda, T. Ikeda, K. Miki and H. Hiasa, Anal. Sciences, 2, 501 (1986).

10. M. Senda, T. Ikeda, H. Hiasa and I. Katasho, Nipon Kagaku Kaishi, 357 (1987).

11. T. Ikeda, K. Miki and M. Senda, Anal. Sciences, 4, 91 (1988).

12. T. Ikeda, I. Katasho and M. Senda, Anal. Sciences, 1, 455 (1985).

13. T. Ikeda, K. Miki, F. Fushimi and M. Senda, Agric. Biol. Chem., 51, 747 (1987).

14. M.T. Stankovich, K.M. Schopfer and V. Massey, J. Biol. Chem., 253, 4971 (1978).

15. Q.H. Gibson, B.E.P. Swoboda and V. Massey, J. Biol. Chem., 239, 3927 (1964).

16. J.J. Kulys and N.K. Cenas, Biochim. Biophis. Acta, 744, 57 (1984).

ELECTROCHEMISTRY OF THE IRON-MOLYBDENUM COFACTOR OF NITROGENASE: EVIDENCE FOR MULTIPLE SPECIATION AND ELECTROCATALYTIC BEHAVIOR

Franklin A. Schultz*

Department of Chemistry
Indiana University-Purdue University at Indianapolis
Indianapolis, Indiana 46223

Stephen F. Gheller and William E. Newton

USDA/ARS Western Regional Research Center
Albany, California 94710
and
Department of Biochemistry and Biophysics
University of California
Davis, California 95616

INTRODUCTION

Iron-molybdenum cofactor (FeMoco)[1] is the Fe-Mo-S and O-containing cluster that serves as the active site for biological reduction of dinitrogen to ammonia. FeMoco may be extracted[2] from the iron-molybdenum protein of nitrogenase enzyme of nitrogen-fixing organisms (e.g. Azotobacter vinelandii) into N-methylformamide (NMF), where it exists as a low-molecular-weight entity of approximate composition $MoFe_{6-8}S_{8-10}O_{2-3}$. Study of this small chemical system is simpler than the large biological one; thus, considerable progress towards understanding the catalytic site at which nitrogen fixation occurs has been made by characterizing the composition, structure and spectroscopic properties of FeMoco[1]. These investigations have failed to address the principal role of the cofactor as a biological electron transfer catalyst. Therefore, we have undertaken studies of its electrochemical reactivity to understand better how the process of N_2 reduction is accomplished at this Mo-Fe-S-O center in biological systems.

In our initial publications on this topic[3-5] we demonstrated that FeMoco exchanges electrons with an electrode and that its electrochemical behavior is qualitatively consistent with the previously determined redox behavior of the Mo-Fe-S center in intact nitrogenase enzyme[6]. In this paper we present the results of a more detailed electrochemical examination of FeMoco and provide evidence for the existence of multiple forms of cofactor in its fully oxidized and semi-reduced oxidation states and for electrocatalytic activity associated with reduction to its fully reduced level. Both observations are significant in that they may lead to greater understanding of the structure and function of FeMoco.

EXPERIMENTAL

Molybdenum-iron protein from A. vinelandii was purified according to a previously published procedure[7]. FeMoco was isolated from this protein by an HCl/NaOH modification[8] of the original extraction procedure[2]. The FeMoco samples were concentrated by vacuum evaporation and stored anaerobically at -80°C until use. FeMoco samples were assayed spectrophotometrically by use of the iron-bathophenanthroline procedure described in reference 8b, and the concentration reported is based on the assumption that 6.5 Fe atoms are present in each FeMoco unit.

N-Methylformamide (Aldrich) was vacuum distilled immediately before use in electrochemical experiments. Solvent that was distilled without pretreatment is referred to as "acidic". Otherwise, the solvent was stirred over BaO for 24 hours before distillation; this material is referred to as "alkaline". The relative acidity of these materials was assessed by diluting 1 mL of NMF into 9 mL of water and measuring pH. "Acid" NMF samples gave pH readings of 6-8; "alkaline" NMF samples gave pH = 10-11. The supporting electrolyte for electrochemical experiments was tetra-n-butylammonium hexafluorophosphate ($TBAPF_6$) from Southwestern Analytical Chemicals.

All electrochemical experiments were carried out under an atmosphere of purified argon in a Vacuum Atmospheres glove box. Two small-volume cells were constructed for electrochemical experiments. The first cell consisted of a Kel-F collar threaded over an inverted glassy carbon (GC) working electrode from Bioanalytical Systems (0.071 cm^2 area). The auxiliary electrode was a Pt wire inserted through the Kel-F collar. A miniature Ag/AgCl reference electrode was immersed in the sample solution at the time of measurement. Cyclic (CV) and normal pulse (NPV) voltammetry experiments were conducted on 50-100 µL volumes of undiluted cofactor sample in this cell. The second cell was designed for controlled potential coulometry (CPC) as well as CV and NPV experiments of FeMoco at a large surface area reticulated vitreous carbon (RVC) working electrode. This cell was constructed by drilling a cylindrical cavity (14 mm diam x 10 mm deep) in a 1 1/4-inch diameter Kel-F rod. A horseshoe-shaped piece of 80 pores per inch RVC (Energy Research and Generation, Inc., Oakland, CA) was shaped to fit this cavity and attached to a Pt lead wire with graphite epoxy. A second, smaller cavity was drilled below the RVC electrode to provide room for a motor-driven stirring bar. The auxiliary electrode was a Pt coil immersed in 0.1 M $TBAPF_6$/NMF. The reference electrode was an aqueous Ag/AgCl (3M NaCl) half-cell. These electrodes were contained in short sections of 4 mm glass tubing isolated from the sample solution by Vycor frits and were located in the space provided by the open end of the RVC horseshoe. Experiments were conducted by filling the cell with ca. 1 mL of supporting electrolyte solution and adding aliquots of cofactor to provide ca. 0.05-0.5 mM solutions of FeMoco in 0.1 M $TBAPF_6$/NMF. Potentials measured against the Ag/AgCl reference electrodes were converted to the NHE scale using either the ferrocene/ferricenium[9] or $Mo[S_2C_2(CN)_2]_4^{3-/4-}$ [10] redox couples, as described previously[3].

A CV-27 potentiostat (Bioanalytical Systems) was used to conduct CV and CPC experiments; an EC-225 voltammetric analyzer (IBM Instruments) was used for NPV experiments. Samples for EPR spectroscopy were sealed in the glove box, removed and frozen in liquid nitrogen as quickly as possible and stored under liquid nitrogen until spectra were run. EPR spectra were obtained at 9°K on a Varian Associates Century Series E-Line spectrometer at the University of Michigan. We thank Dr. W.R. Dunham for assistance with EPR spectroscopy.

RESULTS AND DISCUSSION

General Electrochemical Behavior

Figure 1 shows a cyclic voltammogram at the RVC electrode of a 0.133 mM cofactor solution prepared by dilution of the oxidized form of FeMoco into acidic 0.1 M $TBAPF_6$/NMF and recorded at the relatively fast sweep rate of 0.4 V s^{-1}. Two quasi-reversible electrode reactions were observed at -0.30 and -0.98 V vs. NHE. These are assigned to sequential one-electron transfers involving conversion of FeMoco from its oxidized (ox) to semi-reduced (s-r) and semi-reduced to fully reduced (red) forms:

$$\underset{(S=0)}{\text{FeMoco(ox)}} \underset{(a)}{\overset{e^-}{\rightleftharpoons}} \underset{(S=3/2)}{\text{FeMoco(s-r)}} \underset{(b)}{\overset{e^-}{\rightleftharpoons}} \text{FeMoco(red)} \tag{1}$$

The electrode reactions in Equation (1) correspond to those observed previously for undiluted FeMoco at a glassy carbon electrode[3] and to redox state changes at the Mo-Fe-S centers of intact MoFe protein in nitrogenase[6].

In our previous investigations of FeMoco electrochemistry[3-5] we encountered two observations that could not be fully explained: (i) the height of the voltammetric wave corresponding to FeMoco(ox)-to-(s-r) reduction was significantly less than the value expected for a one-electron transfer in NMF, and (ii) the current following this voltammetric

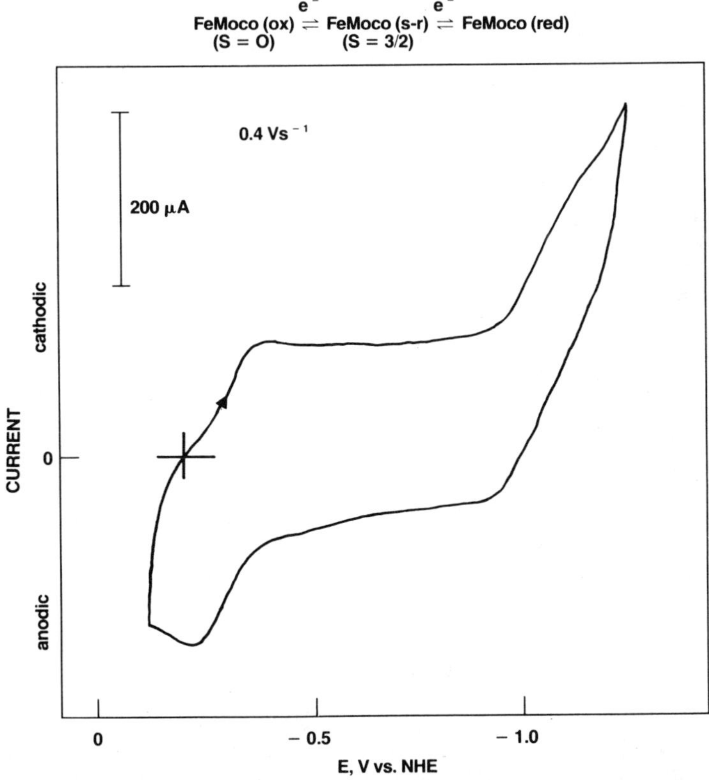

Figure 1. Cyclic voltammogram of 0.133 mM FeMoco(ox) in "acidic" 0.1 M $TBAPF_6$/NMF at RVC electrode. Sweep rate = 0.4 V s^{-1}.

Table 1. Controlled Potential Coulometry of FeMoco[a]

FeMoco, μ moles	Q_{red}, mC	n_{red}	Q_{ox}, mC	n_{ox}
0.116	8.9	0.80	9.0	0.80
0.116	11.0	0.98	10.1	0.90
0.137	18.3	1.38	19.4	1.47
0.084	6.1	0.75	5.1	0.63
0.288	30.7	1.10	28.0	1.01
0.143	14.0	1.01	12.3	0.89
0.152	13.7	0.93	10.8	0.74
		0.99±0.20		0.92±0.27

[a] Conducted by pipetting indicated quantity of FeMoco into 0.1 M $TBAPF_6$/NMF and electrolyzing at RVC working electrode at -0.6 to -0.7 V vs. NHE for FeMoco(ox) reduction or -0.1 to -0.2 V vs. NHE for FeMoco(s-r) oxidation.

reduction peak did not decay as rapidly as expected for a simple diffusion-controlled electrode reaction. In this paper we present the results of more detailed electrochemical investigations which provide partial answers to these questions and which reaffirm the one-electron character of FeMoco(ox) to (s-r) reduction.

Table 1 contains the results of controlled potential coulometry experiments in which samples of FeMoco(ox) were reduced at a potential

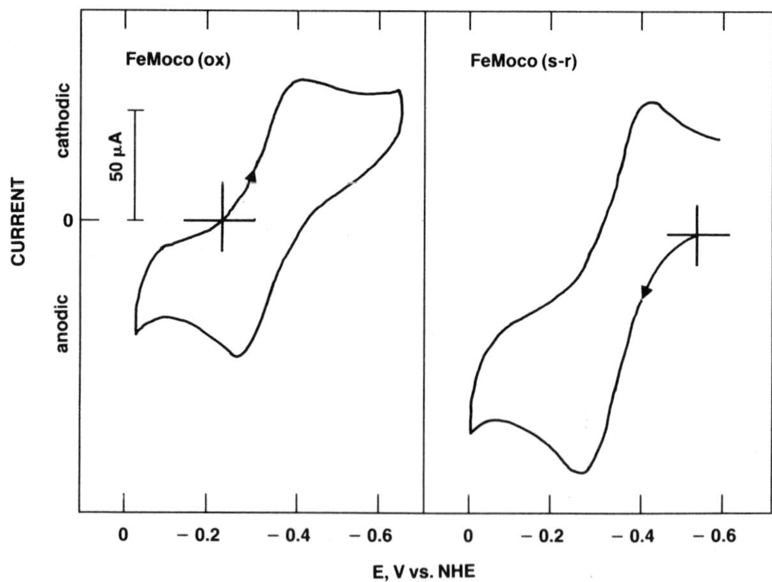

Figure 2. Cyclic voltammograms recorded at RVC electrode of FeMoco(ox) sample before and after coulometric reduction to FeMoco(s-r). 0.22 mM FeMoco(ox); "alkaline" 0.1 M $TBAPF_6$/NMF; sweep rate = 40 mV s^{-1}.

300-400 mV more negative than the potential of Equation 1a and then oxidized at a potential 100-200 mV more positive than this value. The result $n \approx 1$ for coulometric reduction and reoxidation confirms the one-electron character of the FeMoco(ox/s-r) couple. The relatively large uncertainties associated with the measurements are attributed to difficulties in precisely handling small quantities of FeMoco in the glove box and to relatively large steady-state currents at the end of electrolysis (typically 1-2 μA vs. 50-100 μA initial currents). We also observed that accumulation of charge is considerably more sluggish during (ox) to (s-r) reduction than the reverse (oxidation) process. This fact is consistent with other features of FeMoco electrochemistry which will be described below.

Evidence for Multiple Forms of Oxidized Cofactor

Figure 2 shows cyclic voltammograms of a FeMoco(ox) sample diluted into "alkaline" 0.1 M $TBAPF_6$/NMF before and after coulometric reduction to FeMoco(s-r). Despite the fact that equal quantities of charge are transferred in the reductive and oxidative coulometry experiments, the voltammetric peak current for FeMoco(ox) reduction is ca. two-thirds that for for FeMoco(s-r) oxidation. This is true before coulometry as well as for a FeMoco(ox) sample that has been carried through a complete cycle of coulometric reduction and reoxidation. The ratio of (ox) to (s-r) currents is between 0.50 and 0.75 in all FeMoco samples examined to date. Because the area of the RVC electrode in contact with the sample solution is uncertain, absolute peak currents cannot be interpreted in a meaningful way. However, their relative magnitudes suggest that only a fraction of the FeMoco(ox) present in solution is reduced in the cathodic wave at ca. -0.35 V vs. NHE, but that after conversion to the semi-reduced state all of the cofactor is oxidized in the anodic wave at -0.25 V. This result is confirmed by EPR spectroscopy. Double integration of the $S = 3/2$ signal observed after conversion of the sample in Figure 2 to the FeMoco(s-r) state gives 0.90 spins per cofactor center.

Figure 3a shows the cyclic voltammogram of an FeMoco(ox) sample in "acid" NMF. The experiment in question was performed in the GC electrode cell on an undiluted cofactor sample. We find that the solvent in such samples is rendered relatively acidic by cofactor isolation or storage procedures and exhibits behavior nearly identical to that of the "acid" NMF solvent obtained by distillation without treatment by BaO. Results for an undiluted FeMoco sample are shown here. Figure 3a shows that the principal cathodic peak at -0.35 V is still undersized and that several smaller peaks are observed at more negative potentials. The waves nearly coalesce into a single cathodic peak at slower sweep rates (Figures 3b). However, the cathodic peak current remains consistent with only about 2/3 of the total FeMoco(ox) in solution. Figures 3c shows the result of adding a tenfold molar excess of PhSH to the FeMoco(ox) sample in Figures 3a and 3b. This addition converts all cathodic peaks to a single wave at -0.32 V which now has a peak current appropriate to a one-electron transfer. Thiophenol is known to form a strong 1:1 complex with semi-reduced cofactor[8a]. We conclude that its addition to the FeMoco(ox) sample has converted all oxidized species to a single form reducible to thiolated FeMoco(s-r) by a diffusion-controlled one-electron transfer.

$$\text{PhSH} - \text{FeMoco(ox)} + e^- \rightleftharpoons \text{PhSH} - \text{FeMoco(s-r)} \qquad (2)$$

This result is substantiated by the data in Figure 4. The sample containing excess PhSH exhibits a plot of i_p vs. $\nu^{1/2}$ that is linear and passes through the origin. The slope of the line yields a CV current parameter of $i_p/\nu^{1/2}AC = 278\ \mu As^{1/2}V^{-1/2}cm^{-2}mM^{-1}$. A normal pulse

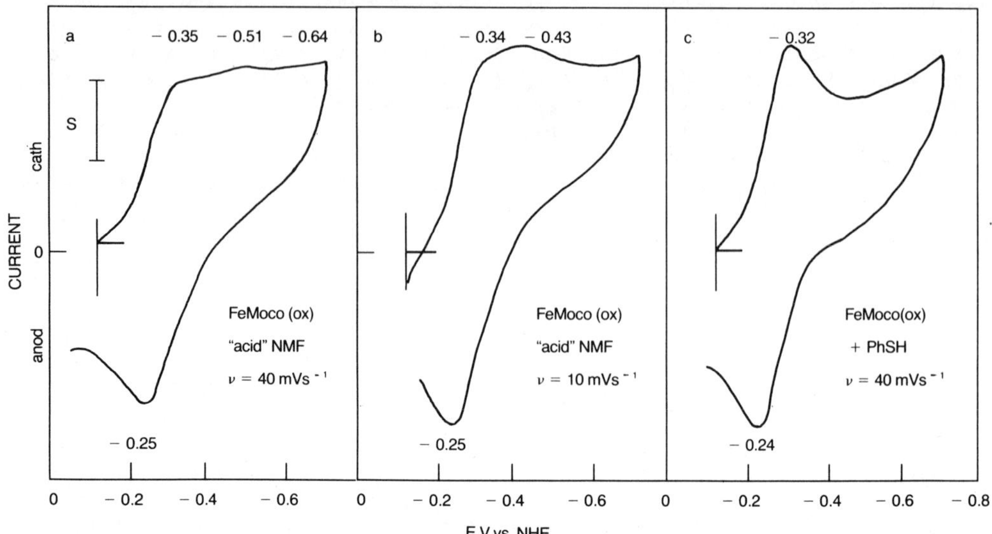

Figure 3. Cyclic voltammograms recorded at GC electrode on 2.3 mM sample of undiluted FeMoco(ox) (traces a, b). Trace c recorded in the presence of 0.021 M PhSH. S = 4,1 and 4 μA for traces a, b and and c, respectively.

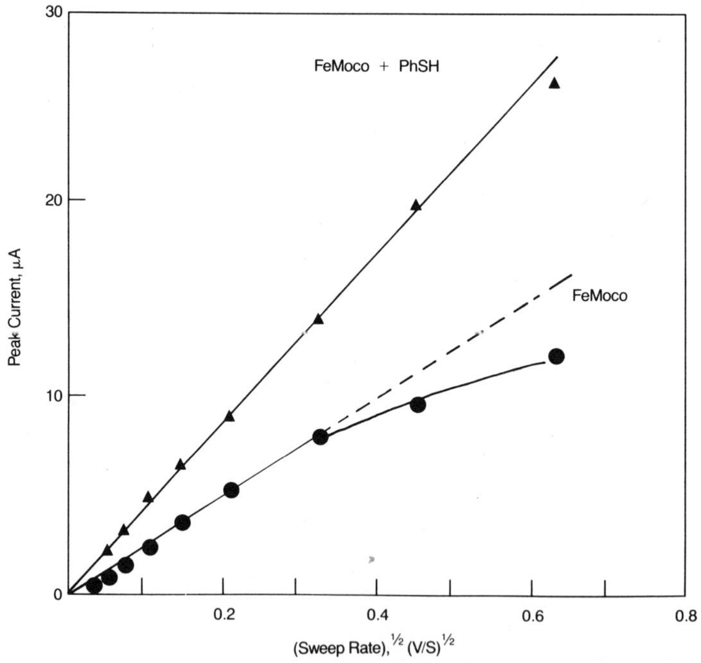

Figure 4. Plot of peak current vs. (sweep rate)$^{1/2}$ for the -0.35 V reduction wave of FeMoco(ox) at a glassy carbon electrode alone (filled circles) and in the presence of 0.021 M PhSH (filled triangles). FeMoco concentration = 2.3 mM. Dashed line represents extrapolation of linear $i_p - \nu^{1/2}$ behavior for the lower line to fast sweep rates.

voltammogram of this solution exhibits a single reduction wave with i_d/AC = 328 μA cm^{-2} mM^{-1}. These parameters are consistent with one-electron transfer in NMF[11]. For the sample containing no PhSH, the low-sweep-rate slope of the i_p vs. $\nu^{1/2}$ plot is 160 μA $s^{1/2}$ $V^{-1/2}$ cm^{-2} mM^{-1}. This result suggests that only about 60 % of FeMoco(ox) is active in the CV experiments without added PhSH. Furthermore, the missing FeMoco(ox) cannot be accounted for by the material found in the additional cathodic peaks in Figures 3a and 3b. At the faster sweep rates where these species appear, the current for the principal reduction wave falls below the value expected from the 160 μA $s^{1/2}$ $V^{-1/2}$ cm^{-2} mM^{-1} slope of the i_p-$\nu^{1/2}$ plot (Figure 4). At the slower sweep rates where the additional peaks coalesce with the principal reduction wave, the current response does not exceed this value. Consequently, acidification of NMF does not convert electroinactive FeMoco(ox) into additional electroactive material, but rather converts some of the electroactive fraction into other voltammetrically detectable species.

In summary, distribution of FeMoco(ox) into multiple forms appears to occur along three lines. The first produces fractions of materials, A_{ox} and C_{ox} (in "acid" NMF) in approximate 2:1 proportion, that are electroactive and -inactive, respectively, in cyclic voltammetry experiments. C_{ox} may be converted to an electroactive state by either reduction to FeMoco(s-r) or treatment with PhSH. Two other manifestations of FeMoco(ox) distribution are triggered by changes in solvent acidity. One results in the appearance of additional electroactive forms (A_{ox}', A_{ox}'' and A_{ox}''') of the principal oxidized species in "acid" NMF. The third observation is that the potential of the principal FeMoco(ox)-to-(s-r) reduction wave (determined as the average of cathodic and anodic peak potentials by CV) is sensitive to solvent acidity. In "acid" NMF (Figures 3a and 3b) we observe $E^{o\prime}$ = -0.31 ± .01 V vs. NHE, and in "alkaline" NMF (Figure 2), $E^{o\prime}$ = -0.36 ± .02 V. We assign these potentials to species A_{ox} and B_{ox}, respectively.

Evidence for Multiple Forms of Semi-Reduced Cofactor

Prior to our investigations there was no evidence for more than one species of cofactor in each of its oxidation states. However, having encountered multiple forms of FeMoco(ox) we examined EPR spectra of semi-reduced cofactor to determine if additional forms could be detected in this oxidation level. The upper half of Figure 5 shows the EPR spectrum of FeMoco(s-r) generated by one-electron coulometric reduction of FeMoco(ox) in "alkaline" NMF. A single S = 3/2 signal is observed with g-values of 4.6, 3.4 and 2.0, which are the parameters typically found for FeMoco(s-r) isolated in the presence of excess sodium dithionite[2]. We term this the R (for regular) form of FeMoco(s-r). Double integration of the R_{s-r} signal in Figure 5 yields an intensity of one S = 3/2 center per FeMoco unit. Thus, although not all of the FeMoco(ox) is visualized in a cyclic voltammetry experiment, all components of this oxidized solution are reduced to a single semi-reduced form, which has an EPR signal consistent with the entire FeMoco content of the solution.

The lower half of Figure 5 shows the results of one-electron coulometric reduction of FeMoco(ox) in "acid" NMF. Under these conditions two overlapping S = 3/2 EPR signals distinct from R_{s-r} are found. The g-value spacings of these signals lead us to identify them as the N (for narrow; g = 4.5, 3.6, 2.0) and W (for wide; g = 4.9, 3.1, 1.9) forms of FeMoco(s-r). Double integration of the spectrum yields an intensity of 0.7 spins per FeMoco center for the N_{s-r} signal and 0.3 spins per FeMoco center for W_{s-r}. This ratio is consistent with the proportion of FeMoco(ox) species observed to be voltammetrically active and inactive in "acid" NMF. Thus, we associate the N_{s-r} form of semi-reduced cofactor

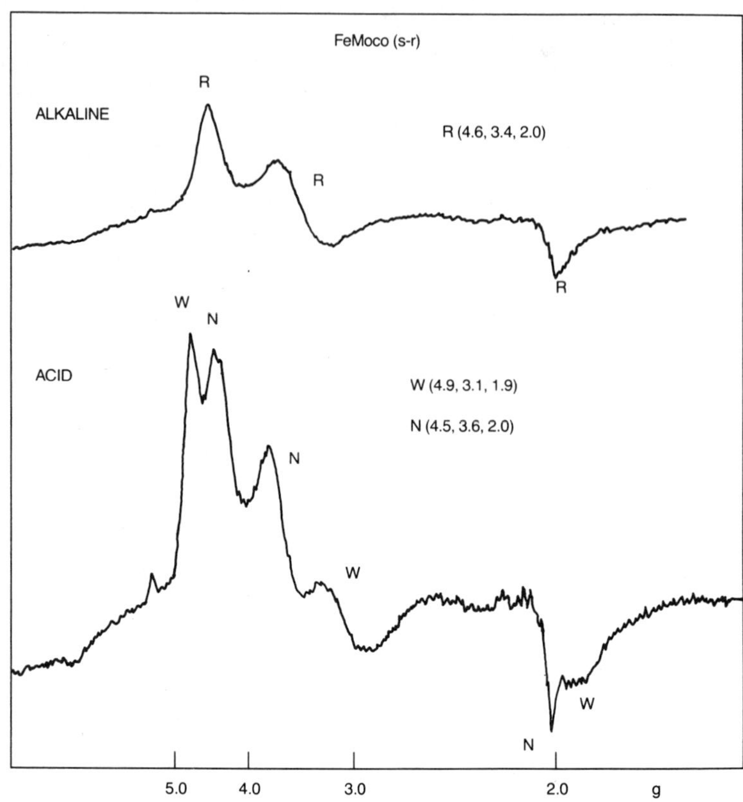

Figure 5. EPR spectra of FeMoco(s-r) produced by one-electron coulometric reduction of FeMoco(ox) in "alkaline" and "acidic" 0.1 M $TBAPF_6$/NMF.

with the electroactive oxidized form, A_{ox}, and the W_{s-r} form with electro-inactive C_{ox}. However, despite the observation of a mixture of semi-reduced species by EPR spectroscopy, voltammetric oxidation of FeMoco(s-r) solutions in "acid" NMF reveals only a single anodic wave (Figure 3).

Correlation of Electrochemical and EPR Results

The detection of multiple forms of oxidized and semi-reduced FeMoco has important implications with regard to the structure and function of cofactor. Derivation of these species from the iron-molybdenum center in the protein is illustrated in Scheme 1. This behavior is based on the electrochemical and EPR results presented in the previous sections. Whereas our studies have not progressed to a point from which we can provide detailed interpretation of all results, several important general observations are apparent and these are discussed below.

Cofactor appears to exist in two principal forms in approximate 2:1 ratio under all conditions except for alkaline solution where a single species (R_{s-r}) is observed in the semi-reduced state. We believe this distribution between major and minor forms represents two distinct and relatively inalterable structural states of cofactor produced upon its extraction from the protein. The oxidized species are identified as A_{ox} and C_{ox} in acid solution and B_{ox} and D_{ox} in alkaline solution. The semi-reduced species are identified as N_{s-r} and W_{s-r} in acid solution and R_{s-r} in alkaline solution. Only the major oxidized fraction (A_{ox} in acid solution, B_{ox} in alkaline solution) is detected in cyclic voltammetry experiments. However, the charge passed in coulometry experiments indicates full 1-electron stoichiometry. Therefore, species C_{ox} and D_{ox} either undergo electrochemical reduction at carbon electrodes very slowly or are converted to electroactive forms A_{ox} and B_{ox} by sluggish chemical equilibria. We believe the former explanation is more likely, and cite the broad tail of the voltammetic wave for FeMoco(ox) reduction (Figure 2) and show accumulation of charge in coulometry experiments as evidence of this fact. On the other hand, the entire fraction of semi-reduced cofactor is voltammetrically active, regardless of whether a single form (R_{s-r}, alkaline solution) or two forms (N_{s-r} and W_{s-r}, acid solution) are observed spectroscopically. This behavior suggests that a significant change in structure or composition occurs upon (ox) to (s-r) reduction which imparts more facile electrochemical behavior to the semi-reduced form. Structural or compositional change also is suggested by the transition from an S = 0 to S = 3/2 spin state upon one-electron reduction of FeMoco(ox) to FeMoco(s-r). Molecular features associated with these proposed changes remain to be identified.

SCHEME 1

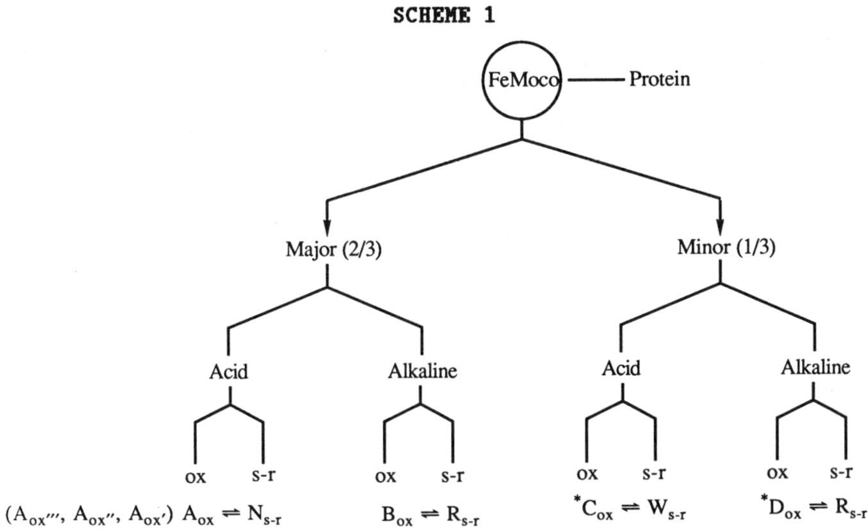

*Not detectable in cyclic voltammetry experiments

A second factor influencing the distribution of cofactor forms is solvent acidity. With one exception, the principal 2:1 distribution is uninfluenced by this variable. Thus, changes in solvent acidity appear to produce subsets of cofactor species with the following properties: (1) Different reduction potentials for the major fraction of oxidized cofactor in acid ($E°'_{A(ox)}$ = -0.31 V) and alkaline ($E°'_{B(ox)}$ = -0.36 V) solution; (2) changes in the number (R_{s-r} versus N_{s-r} and W_{s-r}) and EPR spectroscopic properties of semi-reduced cofactor species; (3) formation of additional forms (A_{ox}', A_{ox}'', A_{ox}''') of electroactive FeMoco(ox). These observations are summarized in Scheme 1.

Correlations among species in the (ox) and (s-r) oxidation states is made by considering electrochemical and EPR spectroscopic data. These relationships are summarized in Scheme 2.

The 2:1 ratios of A_{ox}:C_{ox} and of N_{s-r}:W_{s-r} suggest that A_{ox} is reduced to N_{s-r} and C_{ox} to W_{s-r} in acidic NMF. This fact is supported by chemical titrations of FeMoco(ox) with sodium dithionite under acidic conditions where both A_{ox} and C_{ox} species are present initially[3]. Forms N_{s-r} and W_{s-r} do not appear at the same time in the titration. Rather, the N_{s-r} signal appears first, and the increase in its intensity corresponds with a decrease in peak current for A_{ox}. The W_{s-r} signal appears later in the titration, where it is expected that the more sluggishly reducible C_{ox} would be consumed. Thus, the following correlations are established between oxidized and semi-reduced cofactor species in acid solution.

$$A_{ox} + e^- \rightleftharpoons N_{s-r} \tag{3}$$

$$C_{ox} + e^- \rightleftharpoons W_{s-r} \tag{4}$$

A similar correlation cannot be drawn for alkaline solution because only a single semi-reduced form (R_{s-r}) is present. Presumably, B_{ox} and D_{ox} are reduced concomitantly to this material.

$$B_{ox} + D_{ox} + e^- \rightleftharpoons R_{s-r} \tag{5}$$

SCHEME 2

Acid Solution

$$A_{ox'''} \rightleftharpoons A_{ox''} \rightleftharpoons A_{ox'} \rightleftharpoons A_{ox}$$

$$A_{ox} + e^- \rightleftharpoons N_{s-r}$$

$$C_{ox} + e^- \rightleftharpoons W_{s-r}$$

Alkaline Solution

$$B_{ox} + e^- \rightleftharpoons R_{s-r}$$

$$D_{ox} + e^- \rightleftharpoons R_{s-r}$$

Solvent acidity also causes the distribution of oxidized FeMoco into additional voltammetrically detectable forms (A_{ox}', A_{ox}'', A_{ox}'''). We assume all these species are reduced to N_{s-r}. Figure 3 shows that only a single reoxidation wave is observed when multiple oxidized forms are present, and that the multiple A_{ox} forms almost coalesce into a single reduction wave at sweep rates below 0.01 Vs^{-1}. Consequently, reduction of A_{ox} in acid media is described by a CE

$$A_{ox}''' \rightleftharpoons A_{ox}'' \rightleftharpoons A_{ox}' \rightleftharpoons A_{ox} + e^- \rightleftharpoons N_{s-r} \tag{6}$$

mechanism in which four oxidized forms are linked by chemical equilibria of intermediate mobility. At faster sweep rates, all A_{ox} forms are observed. At slower sweep rates, the chemical equilibria shift from left to right and voltammetric waves for only two species (A_{ox} and $A_{ox'}$) are observed. The newly detected forms of cofactor arise primarily from changes in solvent acidity. To date, cofactor has been studied primarily in the FeMoco(s-r) oxidation state[1]. To generate and maintain this state extraction procedures[2,8] call for treatment of samples with excess dithionite containing hydroxide ion to impart stability to $S_2O_4^{2-}$. Consequently, only the R_{s-r} form has been observed. We examined samples in both the ox and s-r oxidation states and at different acidity levels by allowing FeMoco to "self-oxidize"[3] prior to analysis and by employing different methods of solvent distillation. The new forms detected under these conditions presumably reflect changes in structure or composition of cofactor involving proton-dependent chemistry. This chemistry may involve direct protonation of the Mo-Fe-S center or changes in its ligation resulting from changes in acidity. N-deprotonated NMF has been proposed as a ligand[8b] of extracted cofactor and as a model[12] for the peptide environment at the Mo-Fe-S site in the protein. If deprotonated NMF is a ligand of extracted FeMoco, then simpler chemistry (as is observed) would be expected to prevail in alkaline media where a greater concentration of $HCONCH_3^-$ exists.

Several questions remain unanswered regarding other aspects of FeMoco speciation. One is the repeated observation that about 33 % of each FeMoco(ox) sample cannot be visualized by cyclic voltammetry, whereas all samples of semi-reduced cofactor display full electrochemical activity. We speculate that the inactive fraction must undergo a large structural change upon reduction to FeMoco(s-r) and therefore exhibits slow electron transfer kinetics. A second concern is the fact that only a single oxidation wave at -0.25 ± .02 V is observed for FeMoco(s-r) despite the fact that three forms of this oxidation state are detected by EPR spectroscopy. One explanation is that forms R_{s-r}, N_{s-r} and W_{s-r} are coupled by chemical equilibria which cause all species to react at the potential of the most easily oxidized one. However, E_{pa} is virtually independent of sweep rate and solution

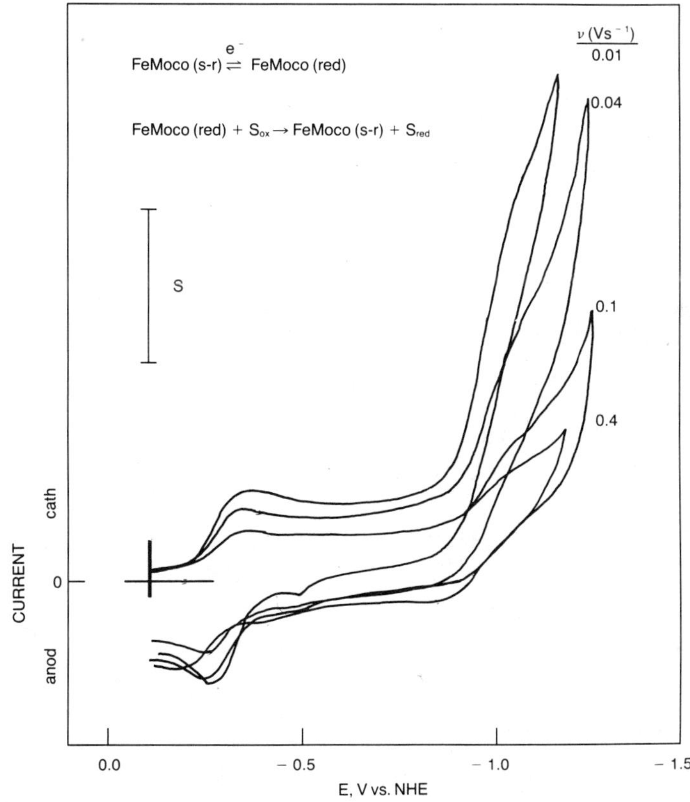

Figure 6. Cyclic voltammograms illustrating electrocatalytic behavior of of FeMoco(red). Sample: 0.13 mM FeMoco in "acidic" 0.1 M $TBAPF_6$/NMF. RVC electrode. S = 40, 80, 200 and 400 μA for 0.01, 0.04, 0.1 and 0.04 V s^{-1} traces, respectively.

conditions, which suggests that if these equilibria exist they are very rapidly established. More plausible is the explanation that the oxidation potentials of R_{s-r}, N^{s-r} and W_{s-r} are not significantly different from one another. This possibility is suggested by the observation that the reduction potentials of A_{ox} and B_{ox} differ by only 50 mV. Also, the structural and/or compositional changes that lead to changes in the EPR spectrum may occur in ways that do not affect the redox potential. EPR activity is an electronic spin state property, whereas electrochemical potentials reflect electronic charge density and orbital energies. These properties are not necessarily related. We have noted previous examples in molybdenum chemistry of a lack of correlation between EPR spectroscopic parameters and electrochemical redox potentials[13]. Similar behavior may exist in FeMoco chemistry.

Electrocatalytic Behavior

The cyclic voltammogram in Figure 1 shows the waves for FeMoco(ox)-to-(s-r) and FeMoco(s-r)-to-(red) reduction at a relatively fast sweep rate of 0.4 V s^{-1}. Under this condition, the currents for the two reductions are approximately equal when the smaller current for the first wave due to distribution into multiple species is taken into account. Figure 6 shows that if voltammograms are recorded at slower sweep rates, the current for the second reduction wave increases dramatically in proportion to the first. The observed behavior suggests that the FeMoco(red) product of electrode reaction 1b undergoes a catalytic reaction with a

$$\text{FeMoco(s-r)} + e^- \rightleftharpoons \text{FeMoco(red)} \qquad (7)$$

$$\text{FeMoco(red)} + S_{ox} \rightarrow \text{FeMoco(s-r)} + S_{red} \qquad (8)$$

component of the solution. Additional evidence for electrocatalytic behavior is observed. (1) The second reduction wave changes from diffusional peak-shaped response to catalytic plateau-shaped response as sweep rate is decreased. (2) The anodic peak observed on scan reversal after the second reduction wave is absent at slower sweep rates. (3) The half-peak or half-plateau potential of the second reduction wave shifts in the positive direction with decreasing sweep rates. (4) Current for the second reduction wave becomes almost independent of sweep rate at low values of ν. Observations 1-4 are diagnostic for an electrocatalytic reaction mechanism[14] of the type described by Equations (7) and (8). In addition, controlled potential coulometric reduction of FeMoco at a potential more negative than the second reduction wave results in almost continual accumulation of charge. In one experiment, a sample containing 285 nanomoles of FeMoco consumed 380 mC of charge after a 1-hour electrolysis at -1.2 V vs NHE. The 180 μA initial current decayed to a steady state value of 60 μA at this time. The total charge consumed corresponded to a transfer (i.e. turnover number) of 14 electrons per cofactor center.

Further data illustrating the electrocatalytic behavior are contained in Figure 7. Figure 7 contrasts the concentration dependence of the first and second wave reduction currents at different voltammetric sweep rates. For (ox) → (s-r) reduction, the plots of i_p vs FeMoco concentrations are linear and display the sweep rate dependence expected for a diffusion-controlled electrode reaction. Plots of i_p vs concentration for (s-r) → (red) reduction are likewise linear, but display almost no dependence on sweep rate. Such behavior corresponds to Equations (7) and (8) operating under full catalytic conditions with the sweep-rate independent current depending linearly on catalyst (FeMoco) concentration according to the relationship

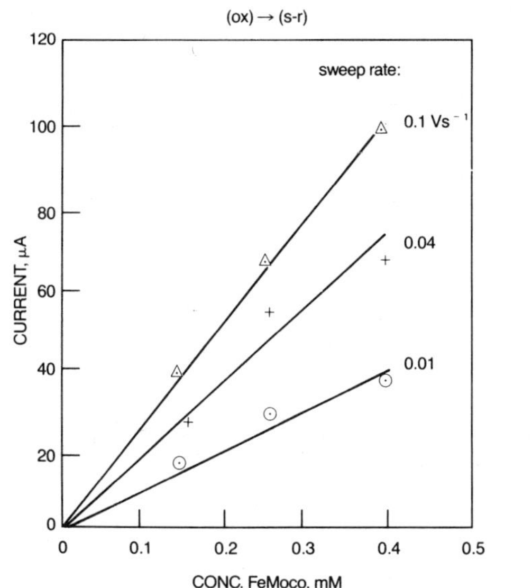

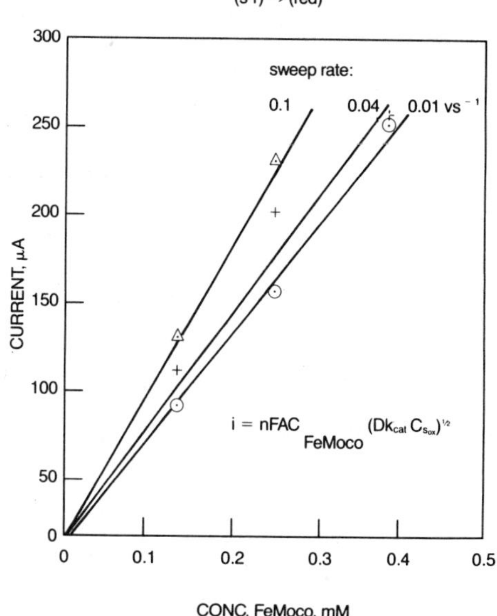

Figure 7. Voltammetric peak currents as a function of FeMoco concentration for (ox) → (s-r) and (s-r) → (reduction). RVC electrode, "acidic" 0.1 M $TBAPF_6$/NMF.

$$i = nFAC_{FeMoco}(Dk_{cat}C_{s_{ox}})^{1/2} \tag{9}$$

where k_{cat} is the rate constant of reaction 8 and $C_{s_{ox}}$ is the substrate concentration.

The identity of the substrate undergoing reduction in Equation (8) is unknown. No substrate was added to the solution deliberately. However, adventitious H^+ and H_2O, which are present in all FeMoco preparations, are alternative substrates of nitrogenase enzyme, and we suspect that these materials may act as substrates in the observed catalysis. Evidence for catalytic activity by FeMoco(red) is significant because this oxidation level is presumed to be responsible for substrate reduction in the intact enzyme. Furthermore, we have demonstrated that the process can be initiated electrochemically, which suggests that these techniques can be used to study the mechanism of dinitrogen and other substrate reductions by FeMoco.

ACKNOWLEDGMENTS

We are pleased to acknowledge support of this research by the National Science Foundation under Grant No. CHE 84-09594 (to F.A.S.) and the National Institutes of Health under Grant No. DK-37255 (to W.E.N.).

REFERENCES

1. (a) B.K. Burgess and W.E. Newton, in "Nitrogen Fixation: The Chemical-Biochemical-Genetic Interface", A Müller and W.E. Newton, Eds., Plenum Press, New York (1983), pp. 83-110. (b) E.I. Stiefel and S.P. Cramer, in "Molybdenum Enzymes", T.G. Spiro, Ed., Wiley, New York (1985), Chapter 2. (c) S.D. Conradson, B.K. Burgess, W.E. Newton,

K.O. Hodgson, J.W. McDonald, J.F. Rubinson, S.F. Gheller, L.E. Mortenson, M.W.W. Adams, P.K. Mascharak, W.A. Armstrong and R.H. Holm, J. Am. Chem. Soc., 107, 7935 (1985).

2. V.K. Shah and W.J. Brill, Proc. Natl. Acad. Sci. U.S.A., 74, 3249 (1977)

3. F.A. Schultz, S.F. Gheller, B.K. Burgess, S. Lough and W.E. Newton, J. Am. Chem. Soc., 107, 5364 (1985).

4. W.E. Newton, S. Gheller, F.A. Schultz, B.K. Burgess, S.D. Conradson, J.W. McDonald, B. Hedman and K.O. Hodgson, in "Nitrogen Fixation Research Progress", H.J. Evans, P.J. Bottomley and W.E. Newton, Eds., Martinus Nijhoff, Dordrecht (1985), pp. 605-610.

5. W.E. Newton, F.A. Schultz, S.F. Gheller, S. Lough, J.W. McDonald, S.D. Conradson, B. Hedman and K.O. Hodgson, Polyhedron, 5, 567 (1986).

6. (a) G.D. Watt, in "Molybdenum Chemistry of Biological Significance". W.E. Newton and S. Otsuka, Eds., Plenum Press, New York (1980), pp. 3-21. (b) G.D. Watt, in "Nitrogen Fixation Research Progress", P.J. Bottomley and W.E. Newton, Eds., Martinus Nijhoff, Dordrecht (1985), pp. 585-590.

7. B.K. Burgess, D.B. Jacobs and E.I. Stiefel, Biochim. Biophys. Acta, 614, 196 (1980).

8. (a) B.K. Burgess, E.I. Stiefel and W.E. Newton, J. Biol. Chem., 255, 353 (1980). (b) S.-S. Yang, W.-H. Pan, D.G. Friesen, B.K. Burgess, J.L. Corbin, E.I. Stiefel and W.E. Newton, J. Biol. Chem., 257, 8042 (1982).

9. R.R. Gagne, C.A. Koval and G.C. Lisensky, Inorg. Chem., 19, 2854 (1980).

10. D.A. Smith, J.W. McDonald, H.O. Finklea, V.R. Ott and F.A. Schultz, Inorg. Chem., 21, 3825 (1982).

11. For the reversible one-electron reduction of $Mo[S_2C_2(CN)_2]_4^{3-}$ [10] in NMF, values of $i_p/v^{1/2}AC$ = 360 μA $s^{1/2}$ $V^{-1/2}$ cm^{-2} mM^{-1} and i_d/AC = 430 μA cm^{-2} mM^{-1} are obtained by CV and NPV, respectively. It is expected that one-electron reduction of FeMoco would exhibit slightly smaller values than this because its greater molecular weight would result in a smaller diffusion coefficient.

12. M.A. Walters, S.K. Chapman and W.H. Orme-Johnson, Polyhedron, 5, 561 (1986).

13. (a) D.A. Smith and F.A. Schultz, Inorg. Chem., 21, 3035 (1982). (b) P.G. Perkins and F.A. Schultz, Inorg. Chem., 22, 1133 (1983).

14. R.S. Nicholson and I. Shain, Anal. Chem., 36, 706 (1964).

SERRS AS A PROBE OF ELECTRODE PROCESSES:

INTERACTION OF GLUCOSE OXIDASE AND FAD AT SILVER ELECTRODES

R.E. Holt and T.M. Cotton*

Department of Chemistry
University of Nebraska-Lincoln
Lincoln, Nebraska 68588-0304

INTRODUCTION

Direct electron transfer between redox proteins and metal electrodes has many advantages with respect to analytical applications. Hill and coworkers[1] have pointed out the similarities between heterogeneous electron transfer reactions of proteins at electrodes and catalysis. The sequence of events at the electrode include: 1) diffusion of reactant protein to the electrode surface; 2) adsorption of the protein in an orientation suitable for electron transfer; 3) electron transfer; 4) dissociation of the protein from the electrode surface; and 5) diffusion of the protein away from the surface. If all these requirements are not met, well behaved redox activity will not be observed. There have been various approaches to accomplishing reversible redox reactions in proteins based upon these requirements. Hill and coworkers have focused on the second step and have shown by their elegant promoter studies that the correct orientation of the protein at the electrode is crucial for rapid electron transfer. Others have utilized mediator-type electrodes or chemically modified proteins[2].

At least one report of direct electron transfer between glucose oxidase (GO) and an unmodified metal electrode (Hg) has been reported[3]. In other studies, however, efforts to observe reversible electrochemistry for this protein in the absence of mediators or chemical modification of the electrode surface have met with failure. The reasons for the slow electron transfer kinetics of GO are not understood. Our goal has been to apply surface enhanced resonance Raman scattering (SERRS) spectroscopy to the study of protein electrochemistry. The present study is concerned with GO, its interaction with the electrode and the location of the FAD group with respect to the electrode surface. These issues are related to steps two and three in Hill's mechanism.

Another important issue in the present study concerns the catalytic activity of adsorbed GO. SERRS has been employed in several flavoprotein studies[4,5,6]. Most striking in these studies was the observation of significant differences between the SERRS spectra at the electrode and the RR spectra of the flavoproteins in solution[7,8]. Such differences often characterize surface enhanced Raman scattering phenomena, but their significance has not yet been established. In the case of GO, one explanation is that the observed SERRS spectrum arises from free flavin produced by denaturation of the protein following adsorption to Ag. If this is the case, the protein should no longer be enzymatically active.

EXPERIMENTAL

Materials

Flavin adenine dinucleotide (FAD), and glucose oxidase (β-D-glucose: oxygen 1-oxidoreductase, EC 1.1.3.4.; Type II from *A. niger*) were purchased from Sigma Chemical Company. FAD was further purified by anion exchange chromatography (Whatman DE-23), using 0.1 M phosphate buffer, pH 7 as an eluant. GO was used as received or purified by molecular exclusion chromatography (Sephadex G-25) or dialysis (Spectrapore, MW cutoff 6000-8000). Flavodoxin from *M. elsdenii* was a gift from Professor Marion Stankovich of the University of Minnesota.

Instrumentation

The SERRS instrumention, electrochemical cell, and electrode (A = 0.2 cm^2) have been described previously[6]. Potential control during the SERRS experiment was accomplished using a potentiostat constructed in our laboratory. All potentials are reported with respect to the saturated standard calomel electrode (SSCE) reference or Ag/AgCl reference, as noted. Osteryoung square-wave and differential pulse voltammetries were performed using a BAS-100 electrochemical analyzer (Bioanalytical Systems Inc.). Square-wave parameters were as follows: sw amplitude = 25 mV, frequency = 15 Hz, step E = 4 mV. Differential pulse parameters were as follows: pulse amplitude = 50 mV, pulse width = 50 msec, pulse period = 1 sec, scan rate = 4 mVs^{-1}.

Procedures

The procedures used for electrochemical roughening of the Ag electrode and adsorption of free FAD and GO onto the electrode surface have also been described previously[6]. SERRS of FAD and GO were obtained using 488.0 or 457.9 nm excitation (Ar^+, Coherent INNOVA 90-5) and low laser powers (typically 10 mW). Spectra were acquired with an OMA II (Princeton Applied Research Corp.) system equipped with an intensified photodiode array (Princeton Applied Research Corp., model 1420). In most cases, the spectra were accumulated using 1 second integration time (60 delays) and signal averaging of 25 scans.

Osteryoung square-wave voltammetry of the Ag_2O/Ag electrodes was accomplished as follows: The polished electrode was placed into a cell containing the GO/glucose solution and stirred for 10 minutes. The electrode was then removed, rinsed briefly in electrolyte solution and placed into an electrochemical cell containing 10 mLs 1.0 M NaOH. The voltammetry was performed immediately following connection of the cell to the instrument.

Low temperature SERRS spectra were obtained by adsorbing the purified flavodoxin onto the electrochemically roughened Ag electrode and flash freezing the electrode in liquid N_2. The electrode was then transferred to a quartz Dewar flask containing liquid N_2. The Dewar flask was constructed with a transparent body to allow direct acquisition of the SERRS spectrum of the electrode in liquid N_2.

The procedure used for synthesis of the FAD-Ag complex and measurement of its resonance Raman (RR) spectrum was similar to that described previously[9].

RESULTS AND DISCUSSION

The first SERRS spectra reported for GO at Ag surfaces were similar to spectra reported for free FAD adsorbed at Ag colloids[4] and at Ag electrodes[10] (Figure 1). The spectral resemblance was initially ascribed to an interaction between the protein-bound FAD and the Ag surface which is similar to that of free flavin. The plausibility of this interpretation was supported by previously published results indicating that the ring III portion of the flavin isoalloxazine (shown below) nucleus is accessible to

Structure 1

solvent[11], and, therefore, sufficiently exposed to interact with the Ag surface.

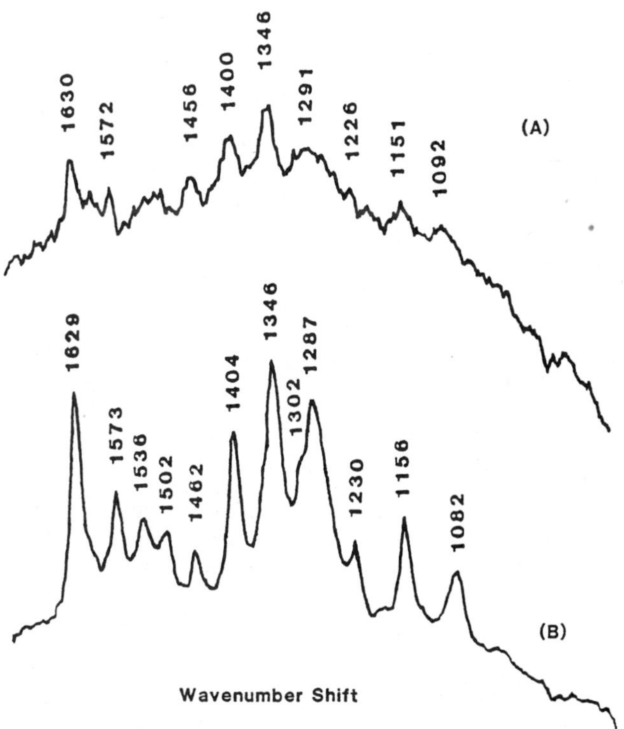

Figure 1. SERRS of GO (A) and FAD (B) adsorbed onto Ag.

Subsequent investigations have shown that free flavin is typically present in commercial preparations of GO and strongly interferes with attempts to observe SERRS spectra of protein-bound FAD at Ag electrodes[6]. Lee et al.[5] have demonstrated that the spectrum of free flavin is observed for a number of different flavoproteins adsorbed onto Ag colloids. There are three possible sources of the free flavin. First, flavins may be present as impurities in the protein preparation, as described above for GO. Second, in most flavoproteins, the FAD is noncovalently bound and partial dissociation of FAD from the protein is expected in accordance with the FAD-apoprotein binding constant. Although the binding constant is quite large for GO, even very low concentrations of free flavin can be detected[6]. A third possibility is that the protein denatures upon adsorption, allowing the flavin to come into direct contact with the surface. Although the source of free flavin was not determined in each of the previous flavoprotein studies, in the case of purified GO, release of FAD is not due to denaturation of the protein following its adsorption onto Ag electrodes[6].

On the basis of surface enhancement theories and investigations of other biological systems, we have devised a set of criteria for establishing that the SERRS spectra of GO are representative of the native enzyme:

1. The vibrational frequencies observed by SERRS are expected to be nearly identical to those observed by RR spectroscopy, if the FAD group is buried within the protein matrix. Large frequency shifts are indicative of a chemical enhancement mechanism, requiring direct contact of the prosthetic group with the Ag surface. Long range electromagnetic (EM) enhancement, on the other hand, is not expected to produce significant frequency shifts, but differences in relative band intensities may occur as a result of the orientation of the prosthetic group relative to the surface.

2. The SERRS excitation profile of protein-bound FAD is expected to follow closely the absorption spectrum of GO in solution, in accordance with SERRS investigations of dyes[12,13]. Significant differences in the excitation profile are an indication of charge transfer interactions between the adsorbate and Ag. Again, this would require direct contact of the FAD with the Ag surface.

3. The potential dependent behavior of the GO SERRS spectra is expected to be different from that observed for free FAD.

4. The SERRS spectrum of GO should be sensitive to catalytic activity of the enzyme in the presence of glucose.

In the results discussed below, the SERRS spectra of GO are evaluated in terms of these criteria.

Applying the first criterion to published SERRS spectra of GO indicates that these spectra are representative of free FAD. The SERRS spectrum of the enzyme is identical to that of free FAD and significantly different from CARS spectra reported for GO[7,8]. Large shifts in the SERRS spectra of FAD and protein-bound FAD can be ascribed to the specific interaction of flavin with the electrode surface. This requires close proximity of the flavin to the Ag surface[14,15]. This is not likely in the native enzyme, however. Although FAD is accessible to solvent, an investigation of the electron transfer rate from reduced GO to anionic electron acceptors suggests that the isoalloxyzine nucleus is ca. 13 Å from the surface of the apoprotein[16]. This would preclude an interaction with the Ag surface identical to that of free FAD. Thus, the GO SERRS frequencies should be closer to those observed by RR spectroscopy.

Evidence to support the expected similarity between SERRS and RR spectra of proteins containing buried chromophores may be found in recent studies of cytochrome c[17] and hemoglobin[18]. In the latter, identical values were observed in the SERRS and RR spectra for high frequency modes as well as the structure-sensitive $Fe-N_{His}$ stretching vibration when the adsorbed hemoglobin was demonstrated to be physiologically functional.

The second criterion for evaluating the SERRS spectrum of GO concerns the agreement between the SERRS excitation profiles for GO with the visible absorption spectrum of the protein. The excitation profiles for both FAD and lumiflavin are markedly different than their solution absorption spectra. This is further evidence for a chemical interaction between the flavin and silver. SERRS excitation profiles for GO have thus far been precluded by our inability to observe strong spectra from the native enzyme.

The third criterion is the agreement between the reduction potential of GO obtained by monitoring the potential dependent SERRS spectrum and the thermodynamic value reported from potentiometric measurements of the protein in solution. Potential dependent SERRS is a highly specific form of spectroelectrochemistry. As FAD is reduced on the Ag electrode, its absorption spectrum changes to that of the "leuco" form which is not in resonance with the laser wavelength used here. Thus, the SERRS intensity decreases dramatically. The potential at which GO reduction occurs in solution at pH 4.5-7.5 has been reported[19]. Reduction of GO by leucosafranin was found to occur at potentials shifted by +76 mV relative to that of FAD throughout this pH range. Although the reduction potentials for FAD in solution[20] are very close to those observed at Hg[3], Pt and carbon electrodes[21], a similar correspondence between the reduction potential of GO in solution and at electrodes may not exist. An overpotential may be required at the electrode to effect electron transfer through the protein matrix. This appears to be the case in a study of GO covalently attached to a graphite electrode in which the enzyme was reported to undergo direct electron transfer with the electrode at a potential that was shifted -60 mV with respect to adsorbed FAD[22]. In contrast, Scheller et al. have reported that GO reduction is shifted + 100 mV from that of FAD at an Hg electrode[3]. Thus, predictions of a potential shift based upon homogeneous electron transfer processes appear to be tenuous at best.

Figure 2 illustrates the bleaching of the 1345 cm^{-1} SERRS band, associated with reduction of adsorbed FAD and GO on Ag, as a function of applied potential. The midpoints for the bleaching of FAD and GO occur at approximately equal potentials (-285 ± 5 mV and -275 ± 10 mV vs Ag/AgCl respectively) and are coincident with reduction of FAD observed by differential pulse voltammetry. This similarity is consistent with other results that suggest that we are actually observing free FAD reduction in the adsorbed GO.

According to the fourth criterion, we would expect to observe bleaching of the SERRS spectrum of catalytically-active GO following the addition of glucose to the electrochemical cell under anaerobic conditions. However, experiments designed to test this hypothesis were negative. No change was observed in the SERRS spectrum of adsorbed GO when glucose was added to the cell in the absence of bulk enzyme. This result demonstrates that adsorbed GO is not catalytically active. There are two possible reasons for its inactivity. First, the active site of the enzyme may not be accessible to the substrate. Or, second, the protein may be denatured on the electrode surface.

As understood above, changes were not observed in the SERRS spectrum of GO when bulk enzyme was absent from the electrochemical cell. This was

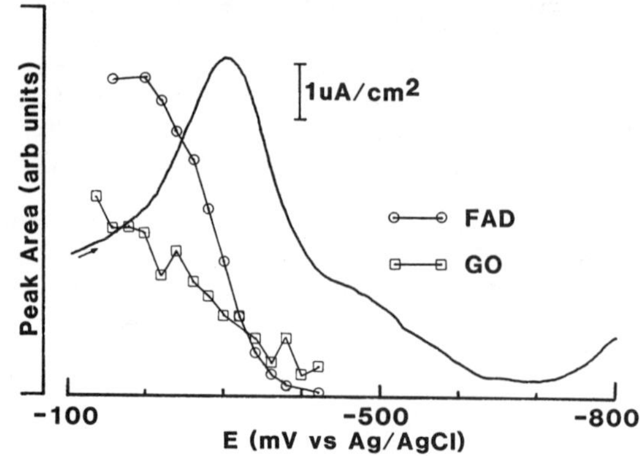

Figure 2. Intensity of the 1345 cm⁻¹ SERRS band of GO and FAD as a function of applied potential. Superimposed is the differential pulse voltammogram of FAD (smooth curve) recorded under similar conditions.

not the case, however, when bulk GO and O_2 were also present. Large changes were observed in the SERRS spectrum under these conditions as shown in Figure 3. The spectral changes were accompanied by a significant positive shift (ca. 100 mV) in the cell potential. The SERRS spectrum at the shifted potential was identified as that of a flavin-Ag complex by comparison with the resonance Raman spectrum of the chemically generated complex (Figure 3c). The shift in cell potential and the presence of the Ag complex suggest that oxidation of the electrode surface occurs during

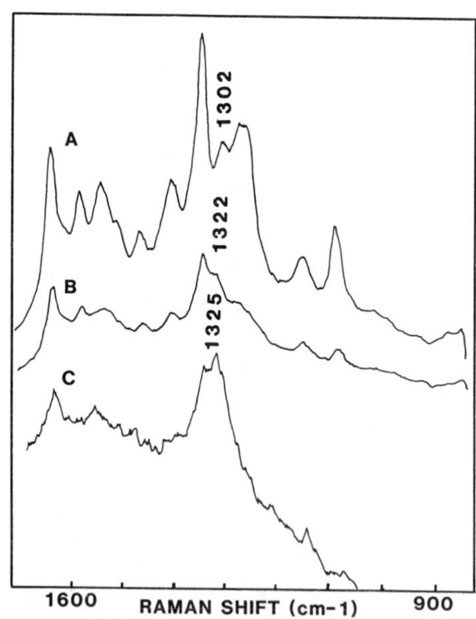

Figure 3. Change in the SERRS spectrum of GO (used without purification): before (A) and after (B) addition of glucose to the aerated system. (C)-RR spectrum of FAD-Ag complex.

expression of catalytic activity by bulk GO:

$$\text{glucose} + O_2 + H_2O \xrightarrow{GO} \text{gluconic acid} + H_2O_2$$

The H_2O_2 generated as a product of this reaction subsequently undergoes catalytic decomposition at the Ag surface, as reported in the early literature, leading to the formation of Ag_2O[23]:

$$H_2O_2 \longrightarrow HO_2^- + H^+$$

$$HO_2^- + 2Ag \longrightarrow Ag_2O + OH^-$$

Evidence for the formation of Ag_2O was obtained from Osteryoung square-wave voltammetry of the electrode following the enzymatic reaction[24]. Figure 4 verifies the presence of Ag_2O on the electrode following its exposure to the catalytic reaction. From a comparison with voltammetric results obtained for an electrode treated in H_2O_2 or an an electrode that was oxidized in basic media, we can attribute the peak at +20 mV (vs Ag/AgCl) to reduction of Ag_2O.

The formation of the FAD-Ag complex is facilitated by the hydrolysis of Ag_2O:

$$Ag_2O + H_2O \longrightarrow 2Ag^+ + 2OH^-$$

Although the dissociation constant for this reaction is small, the equilibrium is shifted to the right by the formation of the FAD-Ag complex. The overall equilibrium constant for the formation of the complex from Ag_2O is estimated to be $\approx 10^1$.

The above sequence of reactions is expected to interfere with attempts to observe SERRS from catalytically active GO at an Ag electrode. A similar problem may arise with other flavoprotein oxidases. Additionally, the Ag_2O layer formed at the electrode surface may inhibit electron transfer between these proteins and the Ag electrode.

In summary, an analysis of the above results provides strong evidence that the SERRS spectra of GO are due to free flavin. The spectra reported previously are probably not from FAD which is sufficiently exposed in the

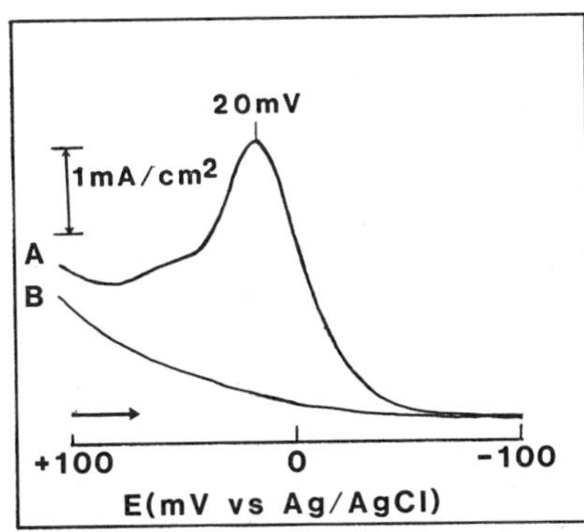

Figure 4. Osteryoung square-wave voltammogram of an Ag_2O/Ag electrode produced as a consequence of expression of catalytic activity by GO (A). (B) Polished Ag electrode.

native protein (e.g., at ring III) to interact with the electrode surface in the same manner as free FAD.

Careful purification of GO results in extremely weak and, under some circumstances, undetectable SERRS spectra. These results suggest two possibilities: either purified GO is not adsorbed onto the electrode or the flavin group is not sufficiently close to the surface to undergo appreciable EM enhancement. The former possibility was disproven by the following experiment[6]. The Ag electrode was anodized and dipped into GO solution for ca. 30 minutes. The electrode was then removed and placed into the electrochemical cell containing electrolyte only. A very weak SERRS spectrum was observed. Next, the electrode was reanodized, following which an intense SERRS spectrum identical with that of free flavin resulted. Thus, GO was initially present on the electrode surface and re-anodization resulted in disruption of the protein matrix, either releasing FAD or denaturing the protein structure sufficiently to allow direct contact between FAD and the electrode surface. Protein denaturation was also found to occur when the electrode potential was held at very negative values. Figure 5 illustrates the increase in the FAD SERRS spectrum when an Ag electrode with adsorbed GO is poised at a potential which is sufficiently negative to reduce the protein disulfide bonds. This treatment is not expected to produce a restructuring of the electrode surface, which could lead to a change in the surface enhancement. Rather, the increase in intensity of flavin SERRS bands verifies that GO is adsorbed on the electrode at more positive potentials, but the flavin is not observed until the protein is denatured at cathodic potentials.

Weak SERRS spectra of native GO might be expected based upon the exponential decrease in EM enhancement with increasing distance of the adsorbate from the electrode surface. In systems where only EM enhancement

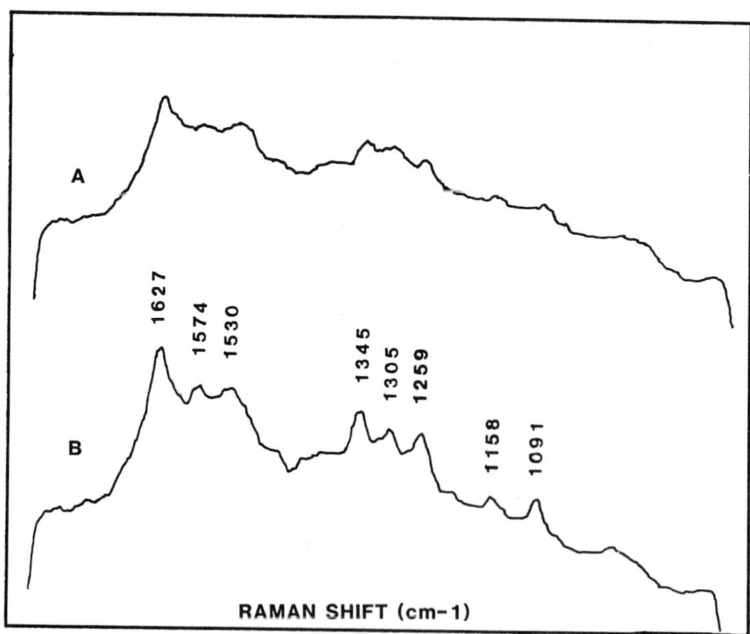

Figure 5. SERRS of GO adsorbed at an Ag electrode (E =-100 mV vs SSCE), before (A) and after (B) cathodization of the electrode at -900 mV vs SSCE for 90 s.

is present, the dependence of SERRS intensity upon distance has been reported for various silver surfaces using polymer spacers[25] and Langmuir-Blodgett multilayer assemblies[26]. The former results indicate that enhancement may decrease by as much as 10^3 with an increase in the adsorbate-metal separation of ca. 100 Å [25]. The decrease in EM enhancement with distance may preclude observations of SERRS from large proteins containing buried chromophores. Moreover, even in those cases where the chromophore is close to the surface of the protein, the surface coverage with respect to the chromophore may be too small to result in detectable SERRS spectra from large proteins. Glucose oxidase is reported to possess a hydrodynamic diameter of 86 Å [27]. Assuming square packing of the GO at the surface and an area of 2.8×10^{-18} m^2 for free FAD adsorbed flat on the electrode surface[28], the number of FAD molecules in the protein monolayer would be $\leq 10\%$ of that present in a monolayer of free FAD. Depending upon the orientation of the protein and the exact location of the two FAD molecules in GO, the distance between the chromophores and the surface may vary between ca. 10 and 80 Å (Figure 6). Thus, the low density of FAD molecules and their distance from the electrode surface in the adsorbed state contributes both to the poor electrochemical response of GO at metal electrodes and the absence of strong SERRS attributable to the intact protein.

The above results underscore the current limitation to the application of SERRS to large molecules when EM enhancement only is present. The enhancement factor may be too small to produce SERRS for those proteins containing low concentrations of chromophores that are positioned at considerable distances from the electrode surface. However, these systems are expected to be eventually amenable to SERRS as continued advances are made in detector sensitivity (e.g., charge-coupled devices).

SERRS has been used successfully to obtain spectra of chromophores in smaller proteins (cytochrome c[17]) or proteins that contain a higher chromophore/polypeptide ratio than GO (hemoglobin[18]). Among the flavoproteins, flavodoxin, with a MW ≈ 16,000 and one flavin mononucleotide (FMN) cofactor positioned near the periphery of the protein, is an ideal protein for SERRS studies. Initial results have shown that higher quality SERRS spectra of this protein may be obtained at low temperatures. No free flavin interference is observed because the protein is apparently stabilized under these conditions. Figure 7 showed the SERRS spectrum of flavodoxin at liquid N_2 temperature. It is obvious that the spectrum is quite different from that of FAD or from that reported previously for flavodoxin at room temperature[5]. The spectrum appears to arise from protein-bound flavin based upon a comparison of the observed peak

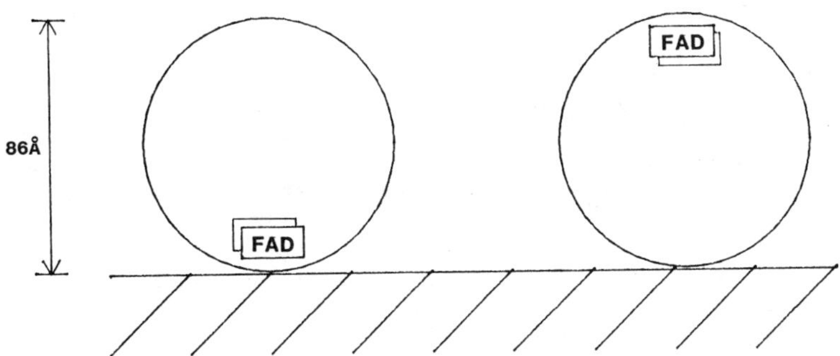

Figure 6. Adsorption of GO onto the AG surface in optimal and least optimal orientations for observation of SERRS from GO bound FAD.

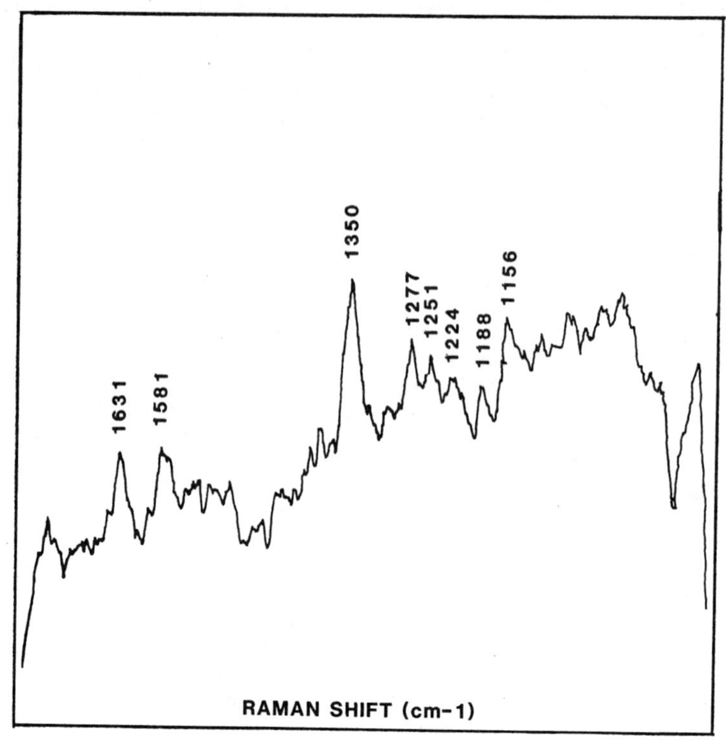

Figure 7. Low temperature (80 K) SERRS spectrum of flavodoxin (*M. elsdinii*).

positions with those previously obtained by coherent anti-Stokes Raman scattering (CARS) spectroscopy[7] of flavodoxin in solution. Thus, there does not appear to be a strong chemical interaction between the flavin and the Ag surface, which would imply that the protein is denatured.

CONCLUSIONS

SERRS investigations of GO are complicated by the presence of free flavin in the system, which produces an intense spectrum suggestive of a chemical enhancement mechanism (i.e., large shifts are observed for several bands in the SERRS spectrum as compared to the RR spectrum). The SERRS excitation profile for free FAD adsorbed onto Ag is also indicative of chemical enhancement.

Flavin is not released from GO as a result of its adsorption onto the electrode, but probably arises from dissociation of FAD from the enzyme prior to the adsorption step[6]. Very weak SERRS spectra are observed from highly purified GO. This is not due to lack of adsorption of the enzyme, however. Reanodization and cathodization experiments have shown that the GO is irreversibly adsorbed at the electrode surface.

SERRS spectroelectrochemistry of adsorbed GO and FAD have shown that their reduction potentials are identical. This suggests that the protein may be denatured on the surface. This possibility is supported by our inability to observe catalytic activity for the adsorbed enzyme. Reduction of the enzyme by glucose was not observed under anaerobic conditions.

However, the lack of catalytic activity may be due to the inaccessibility of the active site to glucose.

The potential of SERRS for obtaining mechanistic information about reaction at the electrode surface was demonstrated in the study of GO catalysis. When bulk GO is present in aerated glucose solutions, catalytically produced H_2O_2 reacts with the electrode surface to produce Ag_2O. This shifts the electrode potential. The Ag_2O subsequently interacts with free flavin present at the electrode surface to form an FAD-Ag complex. This surface reaction interferes with potentiometric measurements of glucose concentration. Similar reactions at other metal surfaces may contribute to irreproducibility in potentiometric methods for glucose analysis.

The SERRS results suggest that both irreversible adsorption and unfavorable orientation of GO with respect to the electrode surface (steps 2-4 in the Hill mechanism) contribute to the poor electron transfer response of native GO at Ag electrodes. The requirement for close proximity of the prosthetic group to the electrode surface is necessary for both the observation of strong SERRS spectra and rapid electron transfer kinetics.

ACKNOWLEDGEMENTS

We would like to thank the National Institutes of Health for support of this research in the form of grant GM 35108-03 and Professor Marion Stankovich for the gift of flavodoxin which was used in this work.

REFERENCES

1. F.A. Armstrong, H.A.O. Hill and N.J. Walton, Quart. Rev. Biophys., 18, 261 (1986).

2. Y. Degani and A. Heller, J. Phys. Chem., 91, 1285 (1987).

3. F. Scheller, G. Strnad, B. Neumann, M. Kühn and W. Ostrowski, J. Electroanal. Chem., 104, 117 (1979).

4. R.A. Copeland, S.P.A. Fodor and T.G. Spiro, J. Am. Chem. Soc., 106, 3872 (1984).

5. N.-S. Lee, Y.-Z. Hsieh, M.D. Morris and L.M. Schopfer, J. Am. Chem. Soc., 109, 1358 (1987).

6. R.E. Holt and T.M. Cotton, J. Am. Chem. Soc., 109, 1841 (1987).

7. A.J.W.G. Visser, J. Vervoort, D.J. O'Kane, J. Lee and L.A. Carriera, Eur. J. Biochem., 131, 639 (1983).

8. P.K. Dutta, J.R. Nestor and T.G. Spiro, Proc. Natl. Acad. Sci. USA, 74, 4146 (1977).

9. M. Benecky, T.-J. Yu, K.L. Watters and J.T. McFarland, Biochem. Biophys. Acta, 626, 197 (1980).

10. T.M. Cotton and R.E. Holt, in: "Spectroscopy of Biomolecules", A.J.P. Alix, L. Bernard and M. Manfait, Eds., John Wiley and Sons, N.Y. (1985), p. 38.

11. V. Massey and P. Hemmerich, in: "Flavins and Flavoproteins", V. Massey and C.H. Williams, Eds., Elsevier North Holland Inc., N.Y. (1982), p. 83.

12. P. Hildebrandt and M. Stockberger, J. Phys. Chem., 88, 5935 (1984).

13. O. Siiman, A. Lepp and M. Kerker, J. Phys. Chem., 87, 5319 (1983).

14. J. Xu, R.L. Birke and J.R. Lombardi, J. Am. Chem. Soc., 109, 5645 (1987).

15. N.-S. Lee, R.-S. Sheng, M.D. Morris and L.M. Schopfer, J. Am. Chem. Soc., 108, 6179 (1986).

16. J.J. Kulys and N.K. Cenas, Biochem. Biophys. Acta, 744, 57 (1983).

17. P. Hildebrandt and M. Stockberger, Proc. Second Eur. Conf. Spectrosc. Biol. Mols., (1987), in press.

18. J. deGroot, R.E. Hester and T. Kitagawa, in: "Second European Conference on Spectroscopy of Biological Molecules", Abstract P/24 (1987).

19. M.J. Walter, Biochem. Biophys. Acta, 128, 504 (1966).

20. H.J. Lowe and W.M. Clark, J. Biol. Chem., 221, 983 (1956).

21. L. Gorton and G. Johansson, J. Electroanal. Chem., 112, 151 (1983).

22. R.M. Ianello and A.M. Yacynych, Anal. Chem., 53, 2090 (1981).

23. J. Bagg, Aust. J. Chem., 15, 210 (1962).

24. R.E. Holt and T.M. Cotton, J. Am. Chem. Soc., submitted.

25. C.A. Murray, in: "Surface Enhanced Raman Scattering", R.K. Chang and T.E. Furtak, Eds., Plenum Press, N.Y. (1982), p. 203.

26. T.M. Cotton, R.A. Uphaus and D. Möbius, J. Phys. Chem., 90, 6071 (1986).

27. S. Nakamura, S. Hayashi and K. Koga, Biochem. Biophys. Acta, 445, 294 (1976).

28. Y. Asaki, J. Pharm. Soc. Jap., 76, 378 (1956).

SURFACE ELECTROCHEMISTRY OF AMINO ACIDS:

VOLTAMMETRY ASSISTED BY EELS, AUGER SPECTROSCOPY AND LEED

Arthur T. Hubbard*, Douglas G. Frank,
Michael J. Tarlov, Nikolas Batina, Nicholas Walton,
Edna Wellner and James W. McCargar

Department of Chemistry
University of Cincinnati
Cincinnati, Ohio 45221

INTRODUCTION

Recent studies by means of thin-layer electrochemistry of the chemisorption at polycrystalline Pt of hydroquinone, catechol and more than 50 related compounds have revealed that a layer of oriented molecules is formed in virtually all instances[1-4]. Variables affecting adsorbate orientation include: adsorbate molecular structure, adsorbate concentration, electrode potential, nature of the electrolyte anion, temperature, solvent and structure of the Pt surface. Adsorbate orientation strongly influences the course of electrocatalytic oxidation and reduction[3]. References 1 and 2 are recent reviews; Reference 4 includes subsequent work.

The present work brings some advances in technique to bear on such studies: well-defined Pt(111) and Pt(100) surfaces[5] were employed as substrates; surface molecular packing densities were measured by means of Auger spectroscopy[6] (rather than by thin-layer voltammetry), thus extending the range of experimentation to include non-electroactive compounds and concentrations outside the range of adsorption measurements with thin-layer electrodes; much useful insight into the nature of the adsorbed species was obtained from electron energy-loss spectra (EELS) of the adsorbed layers. Compounds studied in this work were: L-dopa (LD), L-tyrosine (TYR), L-cysteine (CYS), L-phenylalanine (PHE), L-alanine (ALA), dopamine (DA), catechol (CT) and 3,4-dihydroxykphenylacetic acid (DOPAC). These studies have revealed that evacuation does not remove chemisorbed aromatic molecules from the surface or alter their electrochemical behavior; accordingly, surface spectroscopy in ultra-high vacuum is directly applicable to characterization of chemisorbed species at electrode surfaces.

EXPERIMENTAL SECTION

The experimental procedures employed in this work have been described[5]. A specially-constructed instrument was employed to characterize the electrode surface and adsorbed layer by means of a combination of techniques: surface structure by LEED; elemental composition of Auger spectroscopy; vibrational bands by EELS; and electrochemical reactivity by cyclic voltammetry (argon atmosphere). Transfer between the UHV and ambient pressure phases of the experiment did not result in contamination of the surface as judged by Auger spectroscopy. The Pt(111) and Pt(100) surfaces employed for this work were oriented[7] and polished[8] such that all six faces were crystallographically equivalent. All six faces of the crystal were cleaned simultaneously by bombardment with Ar^+ ions (500 eV; 4×10^{-6} A/cm^2), and annealed by resistance heating in UHV. Cleaning of the surface was continued until Auger spectroscopy and LEED showed that the surface was free of detectable impurities and disorder. The clean, ordered Pt surface was isolated in an argon-filled antechamber for immersion into buffered aqueous electrolyte containing the adsorbate. Electrode potentials were measured and controlled by means of three-electrode electrochemical circuitry based upon operational amplifiers. The electrochemical cell was constructed of Pyrex glass and Teflon double-wall tubing. Purging between the inner and outer tubes with argon was necessary to minimize diffusion of air through the Teflon tubes transporting the solutions. The electrochemical reaction vessel, reference half-wall (Ag/AgCl, 1 M KCl) and Pt auxilary electrode were introduced into the apparatus by means of a UHV bellows assembly and gate valve.

Solutions employed for voltammetric and coulometric measurements contained 10 mM KF electrolyte, adjusted to pH = 4 by addition of HF. Prior to evacuation the surface was rinsed with a solution containing 0.1 M KF adjusted to pH = 4 with HF. Water used in the experiments was pyrolytically distilled in pure O_2 through a Pt gauze catalyst at 800°C, and once more distilled. An experiment was performed from time to time in which the sample was immersed into a blank solution consisting of pure electrolyte, rinsed, evacuated and analyzed. Purification and cleaning procedures were continued until all impurities detectable by Auger spectroscopy were consistently absent.

Supplies of the subject adsorbates were obtained as follows: L-DOPA (LD), cysteine (CYS), 3,4-dihydroxyphenylacetic acid (DOPAC) and catechol

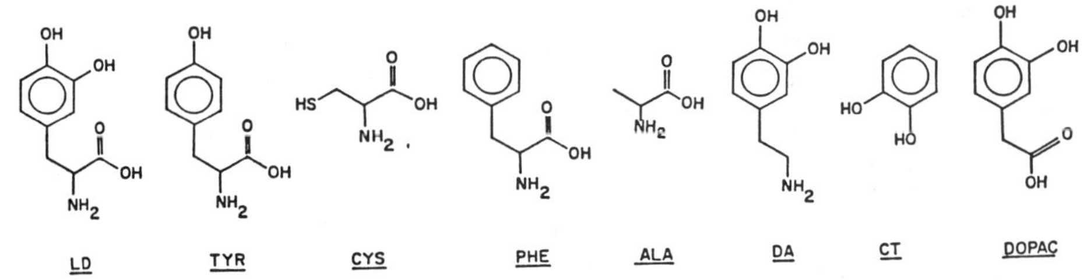

LD TYR CYS PHE ALA DA CT DOPAC

(CT) from Sigma Chemical Co., St. Louis, Missouri 63178, U.S.A. L-tyrosine (TYR), L-phenylalanine (PHE), alanine (ALA) and dopamine (DA) were obtained from Aldrich Chemical Co., Milwaukee, Wisconsin 53233, U.S.A.

Electron energy-loss spectra (EELS) were obtained by means of a Kesmodel EELS spectrometer (Bloomington, Indiana 47405, U.S.A.). Beam current at the sample was approximately 100 pA; beam energy was 4 eV. The spectrometer produced a resolution of about 10 meV F.W.H.M. (80 cm^{-1}) in these experiments.

Infrared spectra of solid compounds in KBr were obtained using a Perkin-Elmer model 1420 dispersive instrument operated at 4 cm^{-1} resolution.

RESULTS AND DISCUSSION

L-DOPA (LD)

Auger spectra obtained after immersion of Pt(100) into LD solution were as shown in Figure 1B; similar spectra were obtained with Pt(111). Auger data are collected in Table 1. Packing density, Γ (nmol/cm²) was calculated from Auger data as described in Reference 6. Elemental packing densities, such as Γ_C, Γ_N and Γ_O, were deduced from elemental Auger signals I_C, I_N and I_O by means of Equations 1-3:

$$\Gamma_C = (I_C/I^{o}_{Pt})/[B_C(5/9 + 4f_C/9)] \quad (1)$$

$$\Gamma_O = (I_O/I^{o}_{Pt})/B_O \quad (2)$$

$$\Gamma_N = (I_N/I^{o}_{Pt})/B_N \quad (3)$$

where the attenuation coefficients are $f_C = f_O = f_N = 0.70$, the proportionality constants are $B_C = 0.377$ cm²/nmol, $B_O = 0.476$ cm²/nmol, $B_N = 1.176$ cm²/nmol, and the Pt Auger signal for the clean surface was measured at 161 eV. Molecular packing density was found from the carbon packing density (for which the precision was generally the best):

$$\Gamma = \Gamma_C/9 \quad (4)$$

Molecular packing density was obtained independently from the ratio of Pt signals for the coated (I_{Pt}) and clean Pt surface (I^{o}_{Pt}):

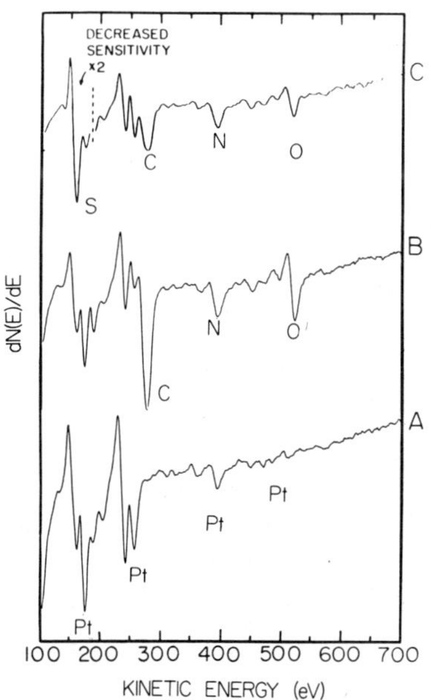

Figure 1. Auger Spectra. (A) Clean Pt(100); (B) Pt(100) treated with LD (3 mM); (C) Pt(100) treated with CYS (1 mM). Experimental conditions: incident beam, 10^{-7} A at 2000 eV, normal to the surface; modulation amplitude, 5 V peak-to-peak; electrolyte 10 mM KF adjusted to pH = 4 with HF; electrode potential, 0.20 V vs Ag/AgCl (1 M) reference.

Table 1. Auger and Electrochemical Data for Molecules Adsorbed at Pt(100).

-logC	Normalized Auger Intensity I_C/I°_{Pt}	I_O/I°_{Pt}	I_N/I°_{Pt}	I_S/I°_{Pt}	I_{Pt}/I°_{Pt}	Coulometric Charge Density $(Q_{ox} - Q'_b)/A$ microcoulombs/cm^2
L-DOPA (LD)						
6.00	0.204	0.235	0.108	-	0.957	-
5.52	0.314	0.226	0.128	-	0.847	-
5.00	0.498	0.314	0.152	-	0.671	-
4.52	0.492	0.298	0.175	-	0.662	-
4.00	0.609	0.355	0.192	-	0.624	-
3.52	0.736	0.402	0.242	-	0.600	-
3.00	0.666	0.286	0.213	-	0.568	389
2.52	0.713	0.316	0.337	-	0.608	-
2.10	0.719	0.485	0.258	-	0.566	-
L-TYROSINE (TYR)						
3.00	0.661	0.252	0.242	-	0.621	268
L-CYSTEINE (CYS)						
3.00	0.700	0.349	0.334	3.41	0.560	661
L-PHENYLALANINE (PHE)						
3.00	0.668	0.176	0.285	-	0.618	325
L-ALANINE (ALA)						
3.00	0.338	0.275	0.267	-	0.744	118
DOPAMINE (DA)						
3.00	0.612	0.267	0.295	-	0.639	594
CATECHOL (CT)						
3.00	0.649	0.422	-	-	0.677	452
DOPAC						
3.00	0.617	0.412	-	-	0.586	357

Experimental Conditions: Concentration of adsorbate is in units of mol/L; Auger beam was also 0.1 microampere at 2000 eV energy, incident normal to the surface; modulation amplitude at the cylindrical mirror was 5 V p-p; Auger peak heights for the coated surface were normalized by the peak height of the Pt peak at 161 eV for the clean Pt surface (except CYS for which the positive lobe of the 235 eV Pt peak was used); electrode potential was 0.2 V vs Ag/AgCl (1 M KCl) reference; coulometric charges (Q) were measured as described in the narrative.

$$I_{Pt}/I^{o}_{Pt} = (1-9K_C\Gamma)(1-5K_C\Gamma) \quad (5)$$

where K_C= 0.160 cm²/nmol. Packing density data for LD at Pt(100) is given in Table 2 and Figure 2. The packing density plateau based upon carbon Auger signal is $\Gamma = \Gamma_C/9$ = 0.23 nmol/cm², while that based upon Pt signal attenuation is 0.22 nmol/cm².

The ideal packing density of LD based upon the surface structure model shown in Figure 3A is 0.213 nmol/cm² (78.0 Å²/molecule). The similiarity between observed and calculated packing densities is evidence that LD is absorbed in an orientation of the type shown in Figure 3A. In contrast, other possible conformations led to calculated packing densities much smaller (ring horizontal, amino acid extended out over the Pt surface, 0.176 nmol/cm²) or much larger (all vertical ring orientations gave about 0.5 nmol/cm²) than observed (0.22 to 0.23 nmol/cm²).

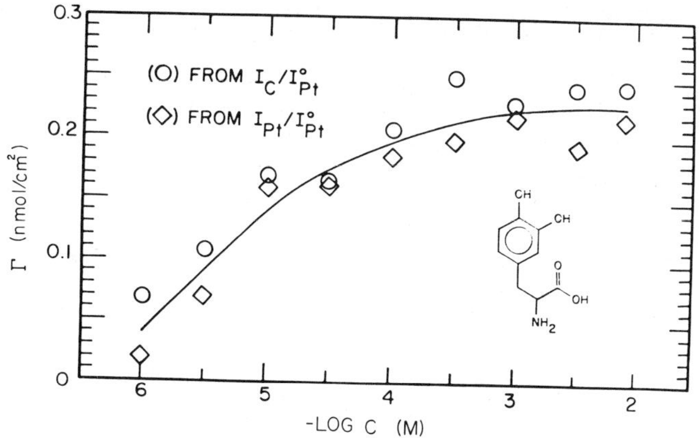

Figure 2. Packing density of LD absorbed at Pt(100). (O) - From Pt Auger signal attenuation; (◊) - From elemental Auger signals.

Vibrational spectra obtained by electron energy-loss spectroscopy (EELS) of LD adsorbed at Pt(111) were as shown in Figure 4A. Also shown in the Figure are the locations and relative intensities of bands in the mid-infrared spectrum of solid LD in KBr. The EELS spectra contain peaks corresponding to most of the IR bands of solid LD, although not all of the bands are resolved, and absence of the phenolic OH stretches is to be expected (see catechol, below), Table 3. Shown for comparison are the EELS spectra of the amino acids L-tyrosine (TYR), L-cysteine (CYS), phenylalanine (PHE) and alanine, and of the related compounds dopamine (DA) and

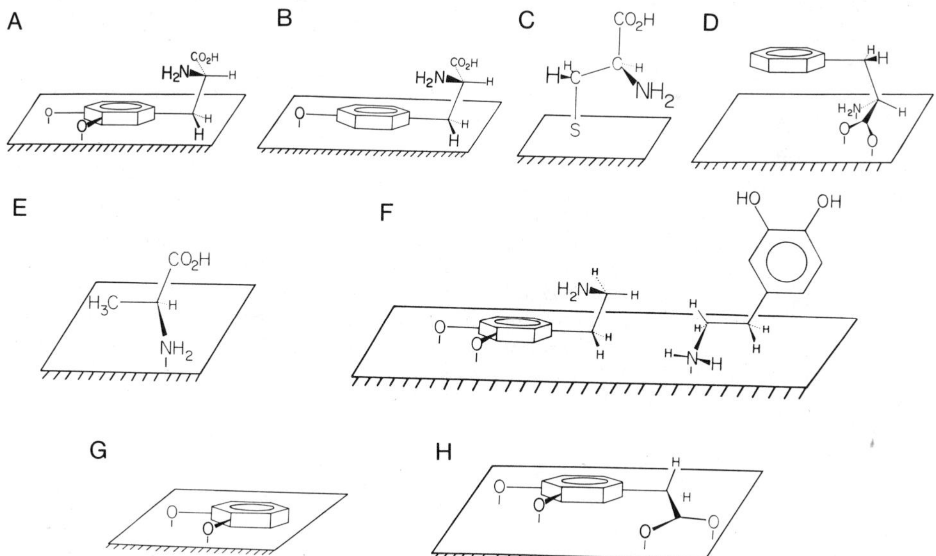

Figure 3. Structural models of adsorbed molecules at Pt(100) and Pt(111) surfaces. (A) L-dopa (LD); (B) L-Tyrosine (TYR); (C) L-cysteine (CYS); (D) L-phenylalanine (PHE); (E) L-alanine (ALA); (F) dopamine (DA); (G) catechol (CT); (H) 3,4-dihydroxyphenylacetic acid (DOPAC).

Table 2. Packing Densities and Oxidation Factors at Pt(100).

	Packing Densities from Elemental Auger Signals					from Pt Auger Signal Attenuation	
-logC	Γ_C, nmol/cm^2	Γ_O, nmol/cm^2	Γ_N, nmol/cm^2	Γ_S, nmol/cm^2	Γ, nmol/cm^2	Γ, nmol/cm^2	Oxidation Factor, $n_{ox} = (Q_{ox} - Q'_b)/FA\Gamma$
L-DOPA (LD)							
6.00	0.624	0.49	0.09	-	0.069	0.019	-
5.56	0.961	0.48	0.11	-	0.107	0.071	-
5.00	1.524	0.66	0.13	-	0.159	0.160	-
4.52	1.505	0.63	0.15	-	0.167	0.165	-
4.00	1.863	0.75	0.16	-	0.207	0.186	-
3.52	2.252	0.84	0.21	-	0.250	0.199	-
3.00	2.038	0.50	0.18	-	0.226	0.217	17.8
2.52	2.181	0.66	0.70	-	0.242	0.194	-
2.10	2.200	1.02	0.22	-	0.244	0.218	-
L-TYROSINE (TYR)							
3.00	2.022	0.53	0.21	-	0.225	0.202	12.3
L-CYSTEINE (CYS)							
3.00	1.032	0.275	0.143	0.368	0.344	0.460	19.9
L-PHENYLALANINE (PHE)							
3.00	1.93	0.41	0.30	-	0.214	0.241	15.7
L-ALANINE							
3.00	1.06	0.58	0.27	-	0.352	0.277	3.5
DOPAMINE (DA)							
3.00	1.755	0.56	0.25	-	0.219	0.217	28.1
CATECHOL (CT)							
3.00	1.72	0.89	-	-	0.287	0.252	16.3
DOPAC							
3.00	1.64	0.87	-	-	0.205	0.216	18.0

Experimental Conditions, molecular packing density, Γ, from C signal was employed to calculate n_{ox} (oxidation factor); other conditions as in Table 1.

catechol (CT), Figures 4-6. Interpretation of the EELS spectra in line with the accepted assignments of the IR bands of CH[9], PHE[10], ALA[11], CYS[12] and propionic acid[13] led to the EELS assignments given in Table 3. The EELS band of LD at 3558 cm^{-1}, Figure 4A, is due to O-H stretching, and those at 3338 and 3158 cm^{-1} are due to N-H stretching. The O-H band is evidence of a pendant amino acid moiety. The aromatic C-H stretching band of LD (3070 cm^{-1}) is distinguishable from the aliphatic C-H bands (2980 and 2923 cm^{-1}), evidence of retention of aromatic character is absorbed LD. A carboxylic C=O stretch is present in the EELS spectrum of LD (1758 cm^{-1}), as well as the spectra of TYR (1760 cm^{-1}), CYS (1740 cm^{-1}), PHE (1690 cm^{-1}) and ALA (1760 cm^{-1}).

L-Tyrosine (TYR)

Packing density of TYR was calculated from Auger data (Table 1) by means of Equations 1-4 and 6:

$$I_{Pt}/I^{o}_{Pt} = (1-8K_C\Gamma)\ (1-5K_C\Gamma) \qquad (6)$$

The molecular packing density from I_C/I^{o}_{Pt} data at 1 mM TYR concentration (on the adsorption plateau) was 0.22 nmol/cm^2, and that from

$I_{Pt}/\overset{o}{}_{Pt}$ was 0.20 nmol/cm². The ideal packing density for the model shown in Figure 3B is 0.213 nmol/cm².

The EELS spectrum of TYR in Figure 4B and the assignments summarized in Table 3 are evident for the model shown in Figure 3B.

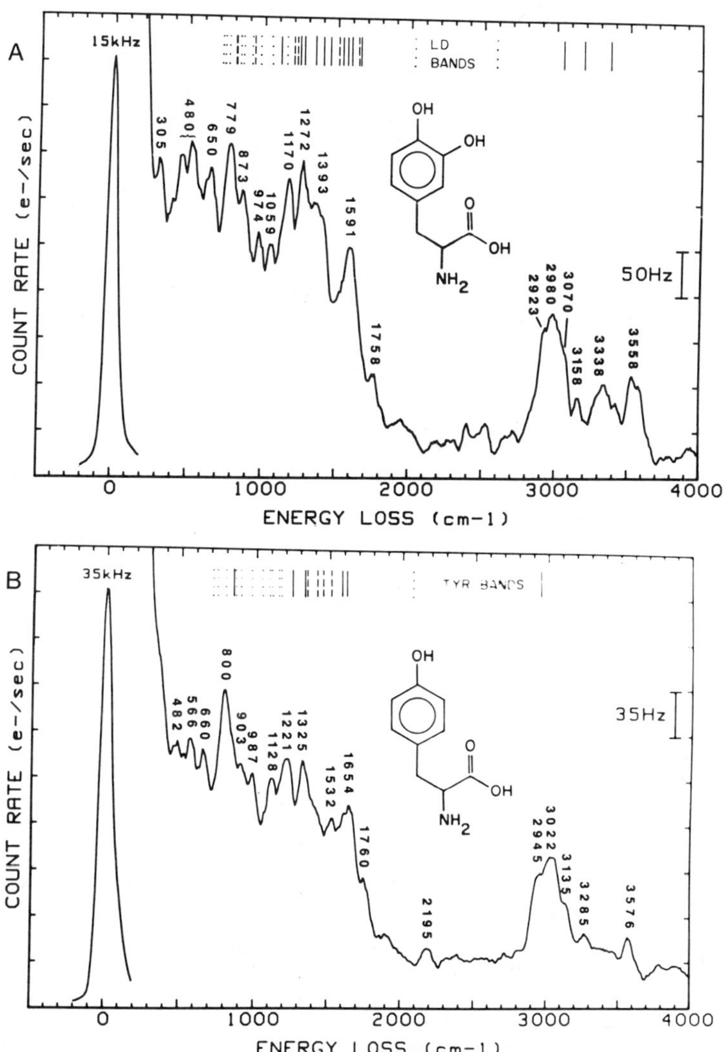

Figure 4. EELS spectra. (A) LD adsorbed at Pt(111); (B) TYR adsorbed at Pt(111); (C) CYS adsorbed at Pt(111).
Experimental conditions: electrolyte was 10 mM KF/HF (pH4); LD concentration, 5 mM; TYR and CYS concentrations were 1 mM; EELS incidence and detection angles 62° from surface normal; beam energy, 4 eV; beam current, 0.15 nA; 10 meV (80 cm^{-1}) FWHM resolution; adsorption potentials were +0.2, +0.1 and -0.1 V vs Ag/AgCl (1 M), respectively. Also shown for reference are the mid IR bands of the parent compound in KBr: | = strong bands, ¦ = medium strong, ⁝ = weak bands.

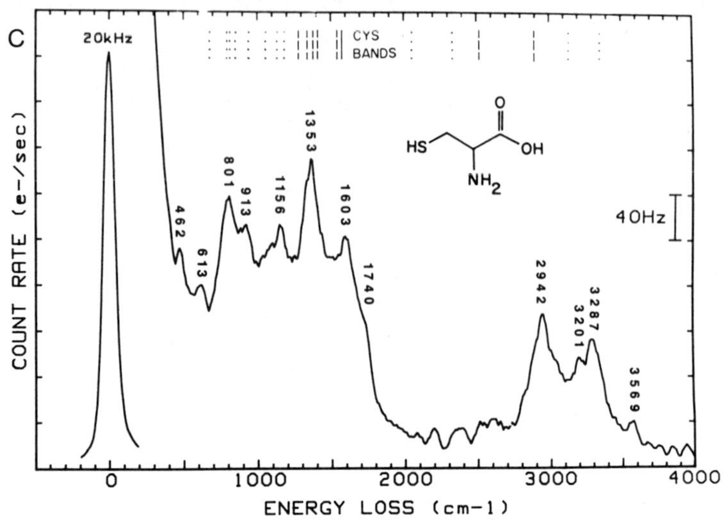

Figure 4. (Continued)

Phenylalanine (PHE)

Auger data for PHE adsorbed from 1 mM solution are given in Table 1. Packing density, Table 2, was obtained with the use of Equations 4 and 7-10:

$$\Gamma_C = (I_C/I^{o}_{Pt})/[B_C(7/9 + 2f_C/9)] \quad (7)$$

$$\Gamma_N = (I_N/I^{o}_{Pt})/[B_N(1/2 + f_N/2)] \quad (8)$$

$$\Gamma_O = (I_O/I^{o}_{Pt})/[B_O(3/4 + f_O/4)] \quad (9)$$

$$I_{Pt}/I^{o}_{Pt} = (1-5K_C\Gamma)(1-7K_C\Gamma) \quad (10)$$

The molecular packing density from I_C/I^{o}_{Pt} ws 0.21 nmol/cm², and that from I_{Pt}/I^{o}_{Pt} was 0.24 nmol/cm². The calculated packing density for the model structure shown in Figure 3D is 0.282 nmol/cm².

Further evidence for the model is found in the EELS spectrum of PHE adsorbed from 1 mM solution, Figure 5A, and the assignments given in Table 3. The N-H and carboxylate O-H bands (3100 to 3600 cm⁻¹) are very weak, continuing a trend followed also by LD and TYR. Evidently, the phenyl ring absorbs less readily than the phenol and catechol moieties[18], allowing adsorption of the amino acid moiety to predominate in attachment of PHE to the Pt surface. In keeping with this trend, the Ø-H bending mode increases in amplitude and frequency on going from LD (779 cm⁻¹) to TYR (800 cm⁻¹) to PHE (811 cm⁻¹), indicating decreasing ring-surface interactions in the same order. Similarly, the carboxylate C=O stretch is shifted to lower frequency on going from LD (1758 cm⁻¹) and TYR (1760 cm⁻¹) to PHE (1690 cm⁻¹).

L-Alanine (ALA)

Auger data for ALA absorbed from 1 mM solution are given in Table 1. Packing density, Table 2, was calculated with the use of Equations 2, 8 and 11-13:

Table 3. Assignments of EELS Bands of Adsorbed L-DOPA and Related Compounds.

Compound	Conc. mM	Peak Frequency cm^{-1}	Symmetry Species	Description
LD	5	3558	C_1:A	O-H stretch (CO_2H)
		3338		N-H stretch
		3158		N-H stretch
		3070		Ø-H stretch
		2980		Ø-H stretch
		2923		C-H stretch
		1758		C=O stretch (CO_2H)
		1581		CC stretch (Ø), NH_2 scissor
		1393		C-O stretch (CO_2H), CC stretch (Ø), C-H bend
		1272		Ø-H bend, C-O stretch
		1170		NH_2 rock, Ø-O stretch, CCN antisym. str.
		1059		Ø-H bend
		974		C-N stretch
		873		ring bend
		779		C-C stretch, CH_2 rock, CCN sym. str., CO_2H bend, O-H bend
		650		CO_2H rock
		480		CO_2H rock, NH_2 torsion, Pt-C, Pt-O
		305		skeletal def.
TYR	1	3576	C_1:A	O-H stretch (CO_2H)
		3285		N-H stretch
		3135		N-H stretch
		3022		Ø-H stretch
		2945		C-H stretch
		2195		(present in IR of TYR, but not assigned)
		1760		C=O stretch (CO_2H)
		1654		NH_2 scissor
		1532		CC stretch (Ø)
		1325		CC stretch (Ø), CO stretch (CO_2H), C-H bend
		1221		ØH bend, CO stretch
		1128		NH_2 rock, Ø stretch, CCN antisym. str.
		987		C-N stretch
		903		ring bend
		800		CC stretch, CH_2 rock, CCN sym. str., CO_2H bend, Ø bend
		660		CO_2H rock
		566		ring bend
		482		CO_2H rock, NH_2 torsion, Pt-C, Pt-O
PHE	1	3540(w)	C_1:A	OH stretch (weak)
		3290(w)		NH stretch (weak)
		3166(w)		N-H stretch (weak)
		3010		Ø-H stretch, C-H stretch
		1690		C=O stretch
		1595		CC stretch (O); NH_2 scissor
		1410		C-O stretch (CO_2H), CC stretch (O), C-H bend

(continued)

Table 3. (Continued)

Compound	Conc. mM	Peak Frequency cm^{-1}	Symmetry Species	Description
		1315		Ø-H bend, CO stretch
		1138		NH_2 rock, CCN antisym. str.
		969		C-N stretch
		905		ring bend
		811		CC stretch, CH_2 rock, CCN sym. str., CO_2H bend, ØH bend
		662		CO_2H rock
		531		CO_2H rock, NH_2 torsion, Pt-N
		320		skeletal def.
ALA	1	3489	C_1:A	O-H stretch
		3281		N-H stretch
		2931		C-H stretch
		2071		(not assigned)
		1760		C=O stretch
		1631		NH_2 def., CO_2H anti. str.
		1350		CH_3 sym. def.
		1120		NH_2 rock, CH_3 rock, CCN anti. str.
		969		CH_3 rock, C-N stretch
		790		CO_2H bend, CCN sym. str.
		508		CO_2H rock, NH_2 torsion, Pt-N
DA	1	3493	C_1:A	O-H stretch
		3302		N-H stretch
		3218		N-H stretch
		3050		Ø-H stretch
		2933		C-H stretch
		1617		NH_2 scissor, CC stretch (Ø)
		1402		C-H bend, CC stretch (Ø)
		1300		ØH bend, CH bend
		1225		NH_2 rock, Ø-O stretch, CCN antisym. str.
		909		ring bend
		778		C-C stretch, CH_2 rock, CCN sym. str., OH bend
		500		NH_2 torsion, Pt-N, Pt-C, Pt-O
		325		skeletal def.
CT		2981	C_{2v}:A	Ø-H stretch
		1618	A_1,B_1	CC stretch
		1378	A_1	CC stretch
		1254	B_1	Ø-O stretch, Ø-H bend
		1171	A_1,B_1	Ø-O stretch, Ø-H bend
		961	B_1,B_2	Ø-H bends
		857	B_1	ring in-plane def.
		704	A_1,B_2	ring in-plan def.; out-of-plane O-H bend
		508	B_2	ring bend
DOPAC		3000	C_s: A'	O-H stretch
		1620	A'	CC stretch (O)
		1211	A',A"	Ø-O stretch, Ø-H bend
		776	A"A'	CH_2 rock; ring def.
		469	A"	ring bend, Pt-C, Pt-O

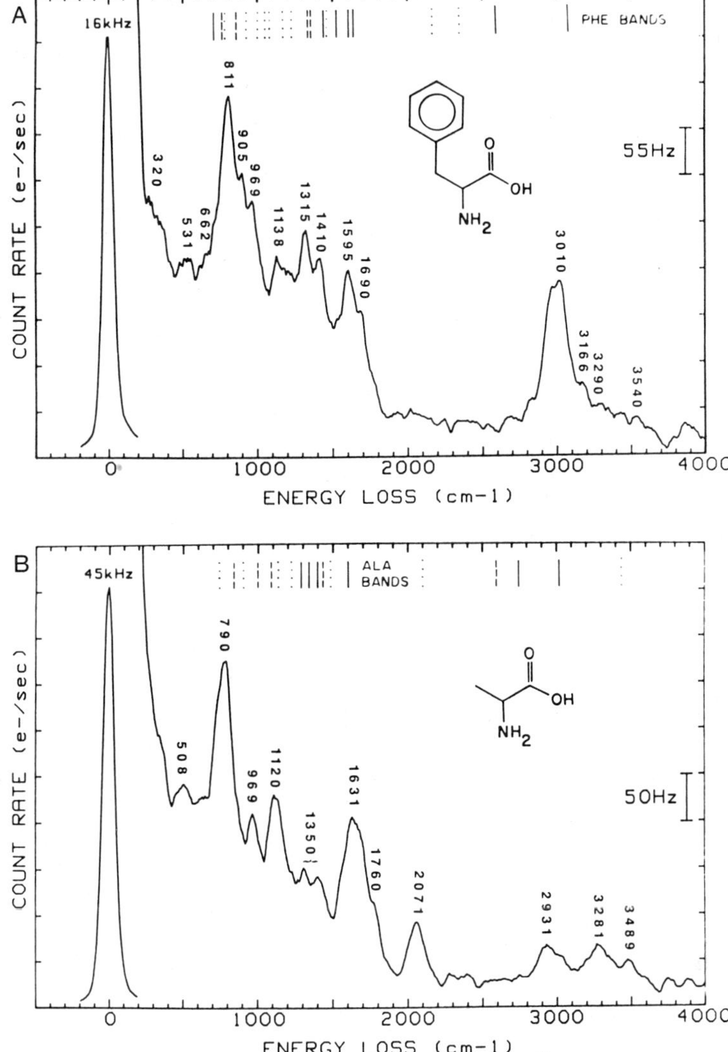

Figure 5. EELS spectra.
(A) PHE adsorbed at Pt(111); (B) ALA adsorbed at Pt(111).
Experimental conditions: PHE and ALA concentrations were 1 mM; adsorption potentials were +0.2 and +0.1 V vs Ag/AgCl (1 M), respectively; other conditions as in Figure 4.

$$\Gamma_c = (I_c/I^o_{Pt})/[B_c(1/2 + f_c/2)] \tag{11}$$

$$\Gamma = \Gamma_c/3 \tag{12}$$

$$I_{Pt}/I^o_{Pt} = (1-5K_c\Gamma)(1-K_c\Gamma) \tag{13}$$

The molecular packing density from I_c/I^o_{Pt} was 0.35 nmol/cm², and that from I_{Pt}/I^o_{Pt} was 0.28 nmol/cm². The calculated molecular packing design based upon the model structure shown in Figure 3E is 0.417 nmol/cm² (39.8 Å²/molecule). The packing density of ALA did not reach the limiting value calculated for the model structure when adsorbed from the millimolar solution employed in the present study.

An EELS spectrum of adsorbed ALA is shown in Figure 5B and assigned[9-13] in Table 3. The C-H stretching band of ALA (2931 cm^{-1}) is attenuated and shifted relative to LD, TYR and PHE as expected due to the absence of an aromatic ring. The O-H (3489 cm^{-1}) and C=O (1760 cm^{-1}) stretches are clearly evident. These are related characteristics indicative of a structure in which attachment to the surface occurs primarily through the NH_2 moiety, Figure 3E.

Dopamine (DA)

Packing density of DA, Table 2, was obtained from Auger data by means of Equations 2, 3 and 14-16:

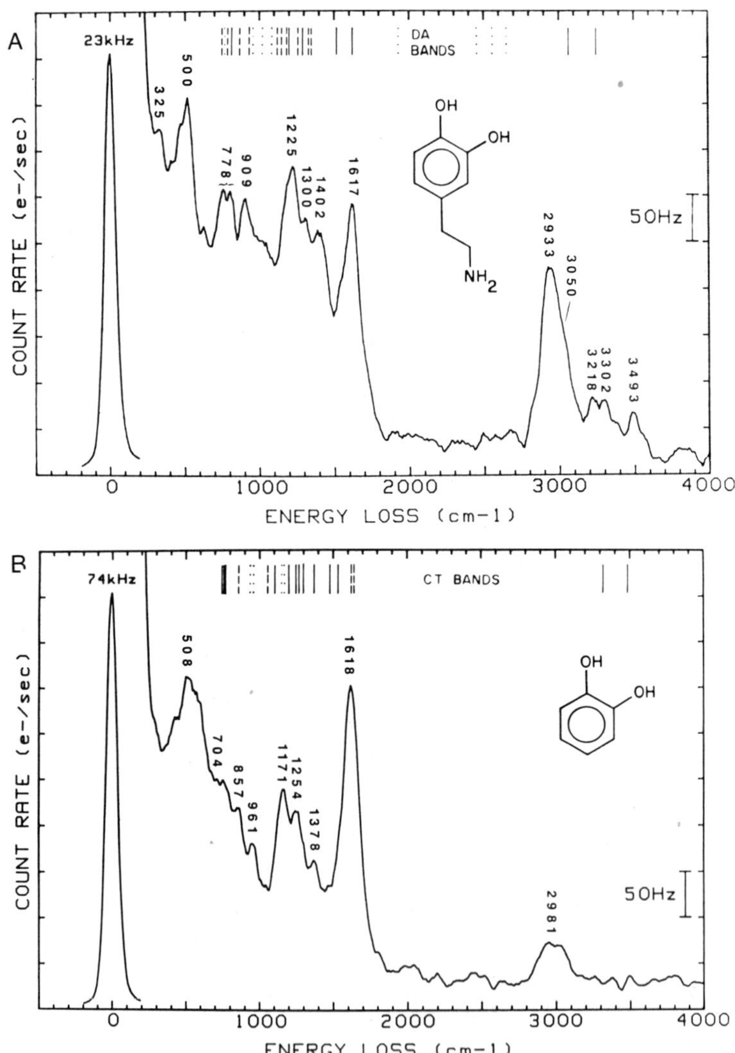

Figure 6. EELS spectra. (A) DA adsorbed at Pt(111); (B) CT adsorbed at Pt(111); (C) DOPAC adsorbed at Pt(111).
Experimental conditions: DA, CT and DOPAC concentrations were 1 mM; adsorption potentials were +0.2, +0.2 and +0.1 V vs Ag/AgCl (1 M), respectively; other conditions as in Figure 4.

$$\Gamma_C = (I_C/I^{o}_{Pt})/ B_C(3/4 + f_C/4) \quad (14)$$

$$\Gamma = \Gamma_C/8 \quad (15)$$

$$I_{Pt}/I^{o}_{Pt} = (1-9K_C\Gamma)(1-2K_C\Gamma) \quad (16)$$

Molecular packing density from I_C/I^{o}_{Pt} was 0.22 nmol/cm², and from I_{Pt}/I^{o}_{Pt} was also 0.22 nmol/cm². The calculated packing density based upon the model structure shown in Figure 3F (catechol ring parallel to the surface) is 0.213 nmol/cm² (78.0 Å²), while that of the N-bonded orientation is 0.438 nmol/cm² (37.9 Å²). In view of these findings the most probable surface condition is a mixture of the two orientations (ring-parallel, and N-bonded) (4m), Figure 3F.

Catechol (CT)

Auger data, Tables 1 and 2, and Equations 2 and 17-19 indicate absorption with the ring parallel to the Pt surface. Molecular packing density from

$$\Gamma_C = (I_C/I^{o}_{Pt})/B_C \quad (17)$$

$$\Gamma = \Gamma_C/6 \quad (18)$$

$$I_{Pt}/I^{o}_{Pt} = (1-8K_C\Gamma) \quad (19)$$

I_C/I^{o}_{Pt} was 0.29 nmol/cm², and based upon I_{Pt}/I^{o}_{Pt} a value of 0.25 nmol/cm² was obtained. The packing density calculated for the model structure in Figure 3G is 0.272 nmol/cm² (61.1 Å²).

The EELS spectrum of adsorbed CT, Figure 6B, accounts for many of the bands seen for LD, including 508 and 857 cm⁻¹ (ring bending), 1170 cm⁻¹ (Ø-O stretching), 1272 cm⁻¹ (Ø-H bending, C-O stretching), 1378 and 1618 cm⁻¹ (CC stretching) and 2980 cm⁻¹ (Ø-H stretching), Table 3. The

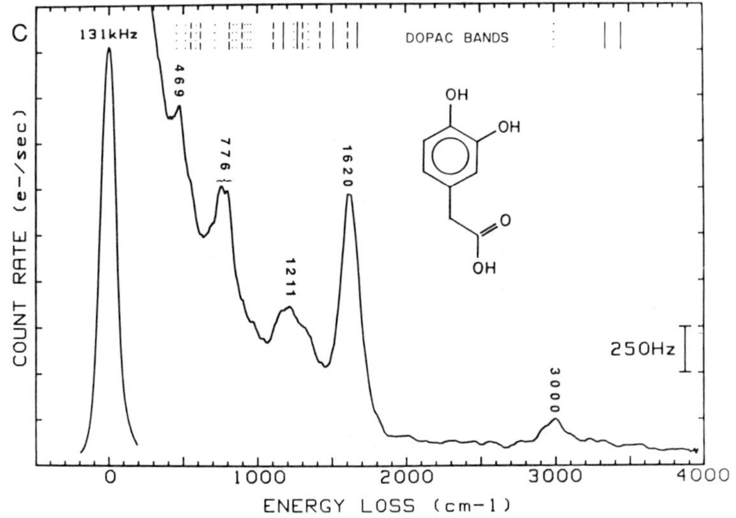

Figure 6. (Continued)

phenolic O-H stretches seen for solid CT in KBr (3485 and 3310 cm^{-1}) are absent from the EELS spectrum of adsorbed CT:

$$\text{C}_6\text{H}_4(\text{OH})_2 \longrightarrow \text{C}_6\text{H}_4(\text{O-SURFACE})_2 + 2H^+ + 2e^- \qquad (20)$$

3,4-Dihydroxyphenylacetic acid (DOPAC)

Auger data for absorbed DOPAC, Tables 1 and 2 and Equations 2, 15, 17 and 21, indicate orientation of DOPAC with the ring parallel to the Pt surface.

$$I_{Pt}/I^{o}_{Pt} = 1 - 12\ K_C\Gamma \qquad (21)$$

Molecular packing density based upon I_C/I^{o}_{Pt} was found to be 0.20 $nmol/cm^2$, and from I_{Pt}/I^{o}_{Pt} was 0.22 $nmol/cm^2$. The calculated packing density for the model structure in Figure 3H is 0.208 $nmol/cm^2$ (80.0 $Å^2$).

EELS spectra of adsorbed DOPAC, Figure 6C, do not display O-H stretching bands corresponding to those seen at 3340 and 3456 cm^{-1} in the IR spectrum of solid DOPAC[20]. That is, the carboxylic O-H stretch and the phenolic O-H stretches vanish or are substantially shifted as a result of adsorption. Furthermore, there is no carboxylic C=O stretch at 1700 cm^{-1} in the EELS spectrum of adsorbed DOPAC, additional evidence that the carboxylate moiety of DOPAC strongly interacts with the Pt surface. In keeping with these findings, the EELS spectrum of DOPAC is very similar to that of CT, Figure 6B, except for the aliphatic skeletal deformation band (776 cm^{-1}) of DOPAC.

L-Cysteine (CYS)

Auger data for CYS, Tables 1 and 2 and Equations 2, 8, 12 and 22-24,

$$\Gamma_C = (I_C/I^{o}_{Pt})/[B_C(1/3 + 2f_C/3)] \qquad (22)$$

$$\Gamma_S = (I_S/I^{o}_{Pt})/[B_S(f_S/2 + f^2/2)] \qquad (23)$$

$$I_{Pt}/I^{o}_{Pt} = (1-K_S\Gamma)(1-3K_C\Gamma)^2 \qquad (24)$$

where B_C = 0.848 $cm^2/nmol$ (positive lobe at 235 eV), B_O = 1.272 $cm^2/nmol$, B_N = 3.143 $cm^2/nmol$, B_S = 9.69 $cm^2/nmol$, F_S = 0.62 and K_C = 0.153 $cm^2/nmol$, gave a molecular packing density based upon I_C/I^{o}_{Pt} of 0.46 $nmol/cm^2$. The calculated packing density for the model structure shown in Figure 3C is 0.406 $nmol/cm^2$ (41.0 $Å^2$/molecule).

The EELS spectrum of CYS, Figure 4C, displays an OH stretch (3569 cm^{-1}) and an N-H stretching doublet (3201 and 3287 cm^{-1}, but no S-H stretch (2505 cm^{-1} in the IR spectrum of solid CYS), Table 3. The carboxylic acid C=O stretch is seen at 1740 cm^{-1} and the NH_2 deformation at 1603 cm^{-1}. Evidently, the CYS molecule adsorbs at Pt through the sulfur atom with loss of the sulfhydryl hydrogen[14], and having a pendent amino acid moiety, Figure 3C.

ACKNOWLEDGEMENT

This work was supported by the National Institutes of Health. Instrumentation was provided by the National Science Foundation, the Air Force Office of Scientific Research and the University of Cincinnati.

REFERENCES

1. A.T. Hubbard, J.L. Stickney, M.P. Soriaga, V.K.F. Chia, S.D. Rosasco, B.C. Schardt, T. Solomun, D. Song, J.H. White and A. Wieckowski, J. Electroanal. Chem., 168, 43 (1984).

2. A.T. Hubbard, V.K.F. Chia, D.G. Frank, J.Y. Katekaru, S.D. Rosasco, G.N. Salaita, B.C. Schardt, D. Song, M.P. Soriaga, D.A. Stern, J.L. Stickney, J.H. White, K.L. Vieira, A. Wieckowski and D.C. Zapien, in: "New Dimensions in Chemical Analysis", B.L. Shapiro, Ed., Texas A&M University Press, College Station, Texas, 1985, p. 135.

3. (a) M.P. Soriaga, J.L. Stickney and A.T. Hubbard, J. Molecular Catalysis, 21, 211 (1983).
   (b) M.P. Soriaga, J.L. Stickney and A.T. Hubbard, J. Electroanal. Chem., 144, 207 (1983);
   (c) M.P. Soriaga and A.T. Hubbard, J. Phys. Chem., 88, 1758 (1984);
   (d) M.P. Soriaga and A.T. Hubbard, J. Electroanal. Chem., 159, 101 (1983);
   (e) K.L. Vieira, D.C. Zapien, M.P. Soriaga, A.T. Hubbard, K.P. Low and S.E. Anderson, Anal. Chem., 58, 2964 (1986);
   (f) V.K.F. Chia, J.H. White, M.P. Soriaga and A.T. Hubbard, J. Electro-Chem., 217, 121 (1987).

4. (a) M.P. Soriaga, E. Binamira-Soriaga, A.T. Hubbard, J.B. Benziger and K.W.P. Pang, Inorg. Chem., 24, 65 (1985).
   (b) M.P. Soriaga, J.H. White, V.K.F. Chia, D. Song, P.O. Arrhenius and A.T. Hubbard, Inorg. Chem., 24, 73 (1985;
   (c) K.W.P. Pang, J.B. Benziger, M.P. Soriaga and A.T. Hubbard, J. Phys. Chem., 88, 4853 (1984).
   (d) J.H. White, M.P. Soriaga and A.T. Hubbard, J. Electroanal. Chem., 177, 89 (1984).
   (e) M.P. Soriaga, D. Song and A.T. Hubbard, J. Phys. Chem., 89, 285 (1985);
   (f) M.P. Soriaga, D. Song, D.C. Zapien and A.T. Hubbard, Langmuir, 1, 123 (1985);
   (g) J.H. White, M.P. Soriaga and A.T. Hubbard, J. Phys. Chem., 89, 3226 (1985);
   (h) J.H. White, M.P. Soriaga and A.T. Hubbard, J. Electroanal. Chem., 185, 331 (1985);
   (i) J.B. Benziger, F.A. Pascal, S.L. Bernasek, M.P. Soriaga and A.T. Hubbard, J. Electroanal. Chem., 198, 65 (1985);
   (j) D. Song, M.P. Soriaga, K.L. Vieira, D.C. Zapien and A.T. Hubbard, J. Phys. Chem., 89, 3999 (1985);
   (k) D. Song, M.P. Soriaga and A.T. Hubbard, J. Electroanal. Chem., 201, 153 (1986).
   (l) D. Song, M.P. Soriaga and A.T. Hubbard, Langmuir, 2, 20 (1986);
   (m) D. Song, M.P. Soriaga and A.T. Hubbard, J. Electroanal. Chem., 193, 255 (1985);
   (n) V.K.F. Chia, M.P. Soriaga and A.T. Hubbard, J. Phys. Chem., 91, 78 (1987);
   (o) D. Song, M.P. Soriaga and A.T. Hubbard, J. Electrochem. Soc., 134, 874 (1987).

5. (a) A.T. Hubbard, J.L. Stickney, S.D. Rosasco, M.P. Soriaga and D. Song, J. Electroanal. Chem., 150, 165 (1983);
(b) J.L. Stickney, S.D. Rosasco, D. Song, M.P. Soriaga and A.T. Hubbard, Surf. Sci., 130, 326 (1983);
(c) J.L. Stickney and A.T. Hubbard, J. Electrochem. Soc., 131, 260 (1984);
(d) J.L. Stickney, S.D. Rosasco, B.C. Schardt and A.T. Hubbard, J. Phys. Chem., 88, 251 (1984);
(e) A. Wieckowski, S.D. Rosasco, B.C. Schardt, J.L. Stickney and A.T. Hubbard, Inorg. Chem., 23, 565 (1984);
(f) A. Wieckowski, B.C. Schardt, S.D. Rosasco, J.L. Stickney and A.T. Hubbard, Surf. Sci., 146, 115 (1984);
(g) T. Solomun, B.C. Schardt, S.D. Rosasco, A. Wieckowski, J.L. Stickney and A.T. Hubbard, J. Electroanal. Chem., 176, 309 (1984);
(h) J.L. Stickney, S.D. Rosasco, G.N. Salaita and A.T. Hubbard, Langmuir, 1, 66 (1985);
(i) S.D. Rosasco, J.L. Stickney, G.N. Salaita, D.G. Frank, J.Y. Katekaru, B.C. Schardt, M.P. Soriaga, D.A. Stern and and A.T. Hubbard, J. Electroanal. Chem., 188, 95 (1985);
(j) D.G. Frank, J.Y. Katekaru, S.D. Rosasco, G.N. Salaita, B.C. Schardt, M.P. Soriaga, D.A. Stern, J.L. Stickney and A.T. Hubbard, Langmuir, 1, 587 (1985);
(k) B.C. Schardt, J.L. Stickney, D.A. Stern, D.G. Frank, J.Y. Katekaru, S.D. Rosasco, G.N. Salaita, M.P. Soriaga and A.T. Hubbard, Inorg. Chem., 24, 1419 (1985);
(l) G.N. Salaita, D.A. Stern, F. Lu, H. Baltruschat, B.C. Schardt, J.L. Stickney, M.P. Soriaga, D.G. Frank and A.T. Hubbard, Langmuir, 2, 20;
(m) D.A. Stern, H. Baltruschat, M. Martinez, J.L. Stickney, D. Song, S.K. Lewis, D.G. Frank and A.T. Hubbard, J. Electroanal. Chem., 207, 101 (1987);
(n) J.L. Stickney, D.A. Stern, B.C. Schardt, D.C. Zapien, A. Wieckowski and A.T. Hubbard, J. Electroanal. Chem., 213, 293 (1986);
(o) J.L. Stickney, B.C. Schardt, D.A. Stern, W. Wieckowski and A.T. Hubbard, J. Electrochem. Soc., 133, 648 (1986);
(p) B.C. Schardt, J.L. Stickney, D.A. Stern, A. Wieckowski, D.C. Zapien and A.T. Hubbard, Langmuir, 3, 239 (1987);
(q) B.C. Schardt, J.L. Stickney, D.A. Stern, A. Wieckowski, D.C. Zapien and A.T. Hubbard, Surf. Sci., 175, 520 (1986);
(r) H. Baltruschat, M. Martinez, S.K. Lewis, F. Lu, D. Song, D. Stern, A. Datta and A.T. Hubbard, J. Electroanal. Chem., 217, 111 (1987);
(s) F. Lu, G.N. Salaita, H. Baltruschat and A.T. Hubbard, J. Electroanal. Chem., 222, 305 (1987).
(t) H. Baltruschat, F. Lu, D. Song, S.K. Lewis, D.C. Zapien, D.G. Frank, G.N. Salaita and A.T. Hubbard, J. Electroanal. Chem., 234, 229 (1987).
(u) G.N. Salaita, F. Lu, L. Laguren-Davidson and A.T. Hubbard, J. Electroanal. Chem., 229, 1 (1987).
(v) L. Laguren-Davidson, F. Lu, G.N. Salaita and A.T. Hubbard, Langmuir, 4, 224 (1988).

6. D.G. Frank, L. Laguren-Davidson, F. Lu, G.N. Salaita and A.T. Hubbard, Anal. Chem., in press.

7. E.A. Wood, "Crystal Orientation Manual", Columbia University Press, New York, 1963.

8. L.E. Samuels, "Metallographic Polishing by Mechanical Methods", Pitman, London, 1967.

9. H.W. Wilson, Spectrochim. Acta, 30, 2141 (1974).

10. A.W. Herlinger, S.L. Wenhold and T.V. Long II, J. Amer. Chem. Soc., 92, 6474 (1970).

11. S. Suzuki, T. Ohshima, N. Tamiya, K. Fukushima, T. Shimanouchi and S.I. Mizushima, Spectrochim. Acta, 11, 969 (1959).

12. Y.K. Sze, A.T. Davis and G.A. Neville, Inorg. Chem., 14, 1969 (1975).

13. P.J. Corish and D. Chapman, J. Chem. Soc., 1746 (1957).

14. D.A. Stern, E. Welner, G.N. Salaita, L. Laguren-Davidson, F. Lu, N. Batina, D.G. Frank, D.C. Zapien, N. Walton and A.T. Hubbard, J. Amer. Chem. Soc., in press.

ELECTROCHEMISTRY AND SPECTROELECTROCHEMISTRY OF ADENINE AND ADENOSINE COMPLEXES WITH 3d TRANSITION METALS

Tadeusz Malinski* and Judith R. Fish

Oakland University
Rochester, Michigan 48063

Chester N. Mikulski and Nicholas M. Karayannis

Beaver College, Glenside, Pennsylvania 19033 and
Amoco Chemicals Corp., Naperville, Illinois 60566

INTRODUCTION

Nucleic acids are involved in a variety of interactions with metals in biological systems. Binding studies of metal ions to nucleic acid constituents are of interest due to the crucial role played by metal ions at some state of gene expression or their ultimate combination with the gene product to produce an active metalloenzyme or other metal protein complex[1]. In addition, several metal complexes of nucleic acid constituents, especially platinum, show antitumor activity[1]. In order to thoroughly understand the mechanism of the interactions of metals with nucleic acids requires a knowledge of redox properties of the system in the range of potential available through metabolic processes.

In this work, well-defined complexes of biologically important 3d transition metals (Cu(II), Fe(III), Fe(II), and Ni(II)) with either neutral or monodeprotonated anionic adenine or adenosine, synthesized[2,3] and characterized[4,5] as described previously, have been used as a model system to study the effects of the interaction of transition metals with purine and purine nucleoside components of nucleic acids on redox properties of the system. The structures of the complexes is simpler than that of nucleic acids and facilitates evaluation of the electrochemical results. The non-phosphorylated monomeric units are suitable model ligands as the use of nucleotides offers complicating factors associated with phosphate due to self-association and self-complexation[6] and preference for the PO_4 moiety as the site for complexation[7].

As shown in Figure 1a, adenine has five potential binding sites, all of which have been reported as having been used in coordination to transition metal ions in various adenine metal complexes[4]. The anionic adenine in the complexes used in this study is deprotonated at the N-9 site (Figure 1a). Adenosine has the four potential binding sites shown in Figure 1b.

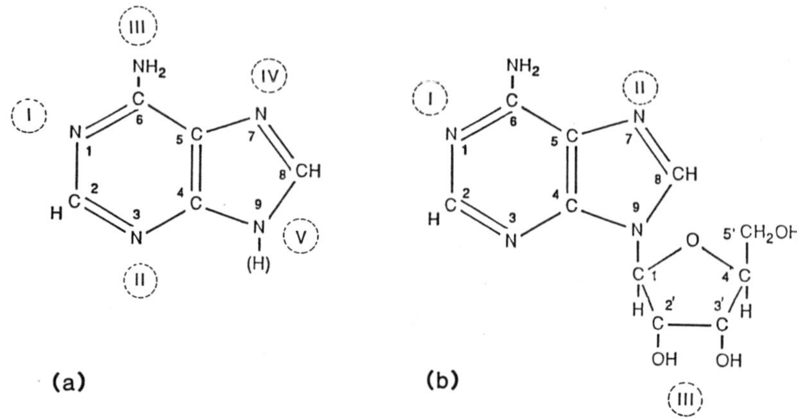

Figure 1. Structures of a) adenine and b) adenosine used as biological ligands showing potential binding sites for complexation (numbers in circles).

Figures 2 and 3 show simplified structural formulas of adenine and adenosine complexes, respectively. The structures of the complexes reflect the pronounced tendency of purines to act as bridging ligands. Binding sites of adenine differ with metal ion variation[4]. Evidence from IR, electronic spectra, and magnetic susceptibilities reported previously suggest structures for the complexes vary both with the central metal and the nature of the ligand.

The Cu(II) adenine complex (2a) has been formulated to be a binuclear dimer, with quadruple bridges of bidentate, N(3), N(9) bonded neutral adenine with coordinated and ionic perchlorate and lattice type ethanol.

Both iron adenine complexes are considered to be binuclear dimers with a double N_y, N(9)-bonded bridge (y sugested to be (1)) of deprotonated adenine. The Fe(III) adenine complex (2c) has two terminal N(9)-bonded neutral adenine ligands and both coordinated and ionic perchlorate. The Fe(II) complex (2d) contains only the anionic bridging adenine and coordinated perchlorate, ethanol and water.

The Ni(II) complex (2b) is a linear polymeric, single-bridged structure containing N(6)-N(z) bonded anionic adenine (z=1, 3, or 9 with 9 likelier than 1 or 3). This complex contains coordinated perchlorate, ethanol and water.

The adenosine complexes are most probably linear polymeric species, involving single bridges of N(1), N(7)-bonded adenosine between adjacent metal ions with additional coordination sites occupied by perchlorate, ethanol, water, or adenosine and additional ionic perchlorate as needed for electric neutrality.

The most likely structural type for the Fe(III) adenosine complex is shown in Figure 3c. This is the only complex studied with 3:2 adenosine to metal stoichiometry. The terminal adenosine ligand would be expected to bind through N(7). This complex contains coordinated water and perchlorate.

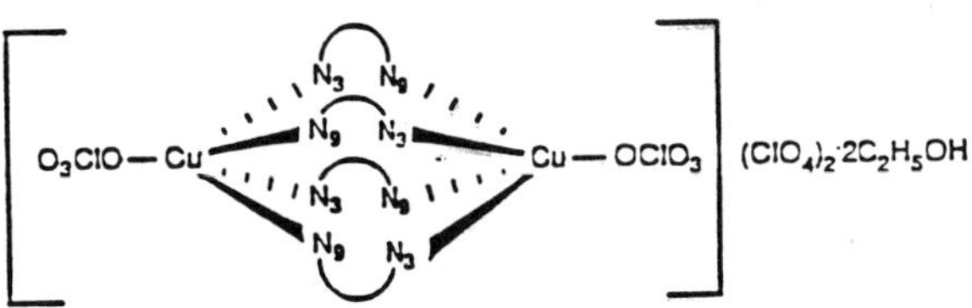

(a) $[(O_3ClO)\,Cu(adH)_4\,Cu(OClO_3)]\,(ClO_4)_2 \cdot 2\,C_2H_5OH$

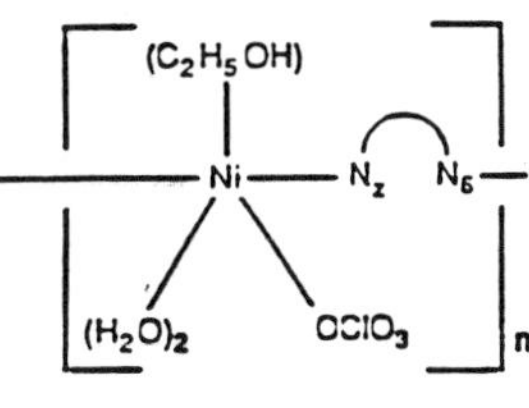

(b) $Ni\,(ad)\,(ClO_4) \cdot C_2H_5OH \cdot 2\,H_2O$

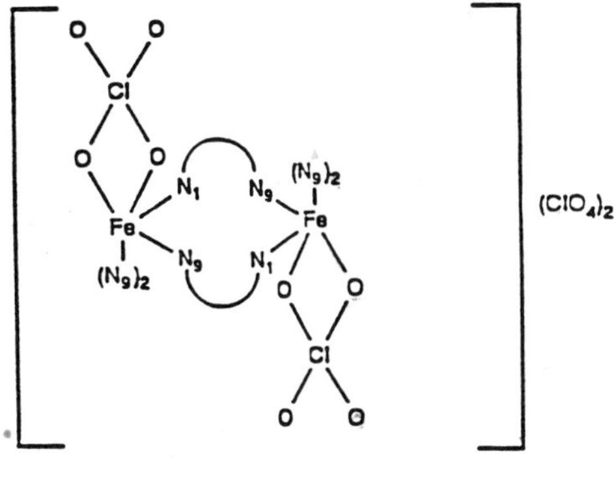

(c) $Fe\,(ad\,H)_2\,(ad)\,(ClO_4)_2$

(d) $Fe\,(ad)\,(ClO_4) \cdot C_2H_5OH \cdot 2\,H_2O$

Figure 2. Structural formulas of adenine complexes: (a) Cu(II), (b) Ni(II), (c) Fe(III), (d) Fe(II). Simplified designations of ligands with only binding sites shown: N_mN_n ~ bridging adenine; N_9 ~ terminal, presumably N(9), bonded adenine; (adH) ~ neutral adenine; (ad) ~ anionic adenine.

The similar linear polymeric structural type for the Ni(II) complex which contains coordinated perchlorate as well as water and an analogous structure without coordinated perchlorate for the Cu(II) complex are shown in 3b and 3a, respectively. The remaining adenosine complex, Fe(II), (3d) is anhydrous with coordination number 4 achieved through tridentate bridging of adenosine. Structure 3d shows coordination of adenosine through ribose hydroxyl oxygens as indicated by IR data and consistent with other experimental evidence.

Although crystal structures have been elucidated for a variety of complexes of adenine with metal ions[4], there is relatively little information regarding complexes of 3d metal ions with adenosine, especially synthesized solid complexes[5]. The other studies which have been reported tend to deal with syntheses[2,3], potentiometric[8,9] or electrophoretic[10] investigations and characterizations[4,5]; there is a dearth of electrochemical and spectroelectrochemical studies.

However, the chemistry of transition metal complexes with macrocycles[11,12] and protein mimetic[13,14,15] ligands is extensive due to the close relationship to molecules of biological interest and their extensive role as metalloenzymes[16]. While these studies are not directly applicable to this work, they serve to identify some crucial issues in the electrochemistry of transition metal complexes, e.g., the mechanism of electron transfer, the nature of the electron transfer, whether to and from the central metal or the ligand, and the reversibility of the electron transfer processes. Other considerations which are pertinent to this study are the electrochemical generation of intermediate species, stability of complexes and intermediates, disproportionation (in binuclear complexes), and ease of demetallation vs stability of formally zerovalent macrocyclic complexes[16].

According to Elving and coworkers[17], adenine is rather strongly adsorbed at the mercury electrode at potentials between -0.200 and -0.700 V reaching a maximum at -0.40 to -0.45 V. This is not in the range of potential at which reduction has been observed to occur[18-24] so presumably is not an issue in the electrochemistry of the free molecule. However, it does overlap the potential range in which electron transfer to and from the central metal would be expected to occur in transition metal complexes.

While studies in aqueous media are precluded due to the narrow range of potential available, the solid complexes are poorly soluble in organic solvents. However, DMSO provided sufficient solubility and was used as the medium for both spectrophotometric and electrochemical investigations. The aim of this investigation was to identify the redox processes, determine the nature of the species present in each oxidation state, and, when possible, elucidate the mechanism of the processes observed via a systematic electrochemical and spectroelectrochemical study. Redox processes have been observed for both the central metal and the biological ligand(s). Several of the complexes have been found to represent chemically reversible redox systems whose significance in biological processes has not yet been considered. The mechanism of the reduction processes observed varies depending upon whether or not aqueous ligands are structural components of the complex.

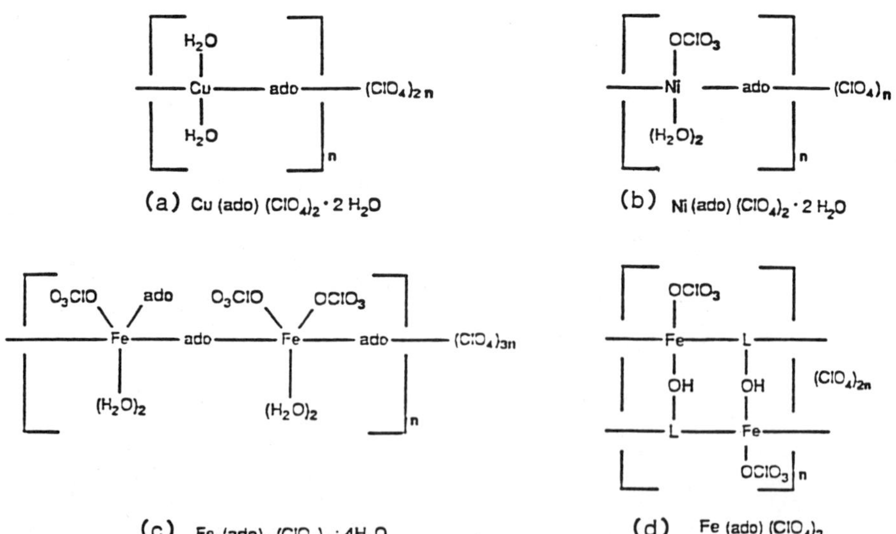

Figure 3. Structural formulas of adenosine complexes: (a) Cu(II), (b) Ni(II), (c) Fe(III), (d) Fe(II). Simplified designations of ligands: ado ~ N(1)-N(7) bonded adenosine.

EXPERIMENTAL SECTION

Chemicals

Complexes of 3d transition metals with adenine (Ad) and/or the monodeprotonated anion (Ad-) or adenosine (Ado) were prepared by refluxing a mixture of Ad or Ado with the appropriate hydrate metal perchlorate dissolved in triethylorthoformate and ethanol. Both the syntheses[2,3] and characterization[4,5], have been described previously. Hydrated metal perchlorates, adenine, and adenosine used in electrochemical experiments were used as received from Aldrich. All electrochemical experiments were performed in dimethyl sulfoxide (Me_2SO) (Aldrich) containing 0.1 M tetrabutylammonium perchlorate (TBAP) (Eastman Kodak). Me_2SO was purified by fractional crystallization using a method similar to that described for purification of pyridine[25]. TBAP was recrystallized and dried in vacuo prior to use.

Solution concentrations for cyclic and differential pulse voltammetry and macroscale electrolysis were about 5.0 x 10^{-1} mM. Similar concentrations were used in spectroelectrochemical experiments. Products of macroscale electrolysis were separated by thin layer chromatography, using silica gel, and identified by using FTIR and NMR spectroscopy.

Instrumentation

A platinum button and a platinum wire served as the working and auxilary electrodes respectively for cyclic and differential pulse voltammetric measurements made using a conventional three-electrode configuration and an IBM Model EC225 voltammetric analyzer. A saturated calomel electrode (SCE), separated from the bulk of the solution by a fritted glass disk, was used as the reference electrode. A Princeton Applied Research Model 173 potentiostat with Model 179 digital coulometer was used for controlled-potential electrolysis (CPE) and coulometry. All potentials are reported versus aqueous SCE.

UV-visible spectroelectrochemistry was performed either in a thin-layer[26] or bulk[27] cell which followed the design of Lin et al. or Fajer et al., respectively[26,27]. The SCE was separated from the rest of the solution by a fritted bridge containing supporting electrolyte and solvent. UV-visible spectra were obtained with a Tracor Northern optical multichannel analysis system composed of a Tracor Northern 6050 spectrometer containing a crossed Czerny-Turner spectrograph in conjunction with the Tracor Northern 1710 multichannel analyzer. Each spectra resulted from signal averaging of 100 single, 190-490 nm, 2.5-ms spectral acquisitions simultaneously recorded by a double-array detector with a resolution of 0.6 nm/channel. Values of λ are accurate to +/- 0.5 nm. While ε is good to +/- 5% of the absolute value presented, some increases in ε are due to increasing solubilities of monomers produced during electrochemical reaction.

Proton NMR and FTIR spectra were collected on Varian T60 and IBM Instruments, Inc. IR-98 spectrometers, respectively.

RESULTS AND DISCUSSION

Table 1 summarizes the half wave potentials for the reduction processes occurring on platinum in the potential range +0.4 to -1.8 V obtained by differential pulse and cyclic voltammetry for the 3d metal perchlorates, the adenine complexes, and the adenosine complexes.

Table 1. Potentials for reduction of adenine, adenosine, 3d metal perchlorates, adenine complexes and adenosine complexes, in DMSO, 0.1 M TBAP solution.

Compound	Peak I	Peak potential, V vs SCE Peak I'	Peak II	Peak II'	Peak III
adenine					-1.58
adenosine					-1.60
$Cu(ClO_2)_2 \cdot 6H_2O$	0.18[a]		-0.32		
$Cu(ad)_2$-$(ClO_4)_2 \cdot C_2H_5OH$	0.15[a]		-0.24		-1.68
$Cu(ado)_2$-$(ClO_4)_2 \cdot 2H_2O$	0.17[a]		-0.24		-1.45
$Fe(ClO_4)_3 \cdot 6H_2O$	0.19[a]		-0.88		
$Fe(ad)_2(ad^-)$-$(ClO_4)_2$	0.16[a]		-0.92	-1.06	-1.54
$Fe_2(ado)_3$-$(ClO_4)_6 \cdot 4H_2O$	0.05[a]		-1.00		-1.62
$Fe(ad^-)$-$(ClO_4) \cdot 2H_2O$	-		-1.04		-1.54
$Fe(ado)$-$(ClO_4)_2$	0.18[a]		-0.78	-1.06	-1.68
$Ni(ClO_4)_2 \cdot 6H_2O$	-1.12	-1.32			
$Ni(ad^-)$-$(ClO_4) \cdot 2H_2O$	-1.15	-1.25			-1.60
$Ni(ado)$-$(ClO_4)_2 \cdot 2H_2O$	-1.04	-1.16			

a-$E_{1/2}$

A typical series of cyclic voltammograms for the Cu(II), Fe(III) and Fe(II), and Ni(II) compounds are shown in Figures 4, 5, and 6, respectively.

$Cu(ClO_4)_2 \cdot 6H_2O$ is reduced in Me_2SO at $E_{1/2}$ = 0.18 V in a one-electron step to form $CuClO_4$ (peak Ic) (Figure 4a). A second one-electron reduction to Cu(0) is observed at potential -0.32 V (peak IIc). In the return potential scan, a stripping peak, IIa, corresponding to the two-electron oxidation of Cu(0) to Cu(II) is observed at E_{pa} = -0.24 V.

A similar pattern in the cyclic voltammetry of $Fe(ClO_4)_3$ and $Fe(ClO_4)_2$ shows the reversible reduction of Fe(III) to Fe(II) (peak I) at $E_{1/2}$ = 0.19 V and further two-electron reduction to Fe(0) (peak II) at E_{pc} = -0.88 V.

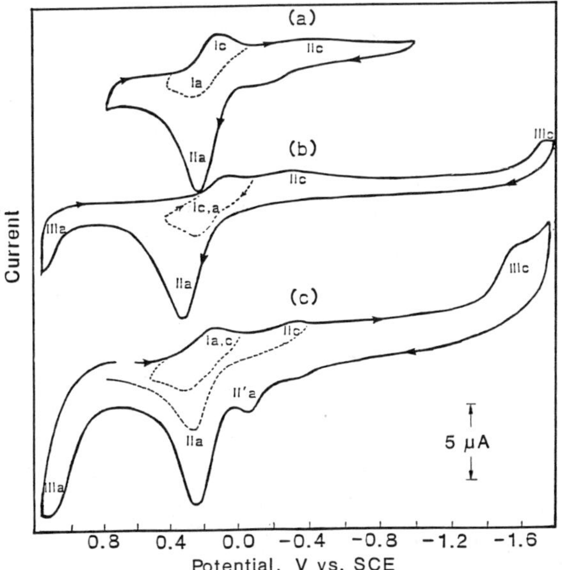

Figure 4. Cyclic voltammograms of Cu(II) complexes in Me_2SO, 0.1 M TBAP solution; scan rate 0.1 V s^{-1}: (a) $Cu(ClO_4)_2 \cdot 6H_2O$, (b) $Cu(ad)_2(ClO_4)_2 \cdot C_2H_5OH$, (c) $Cu(ado)_2(ClO_4)_2 \cdot 2H_2O$.

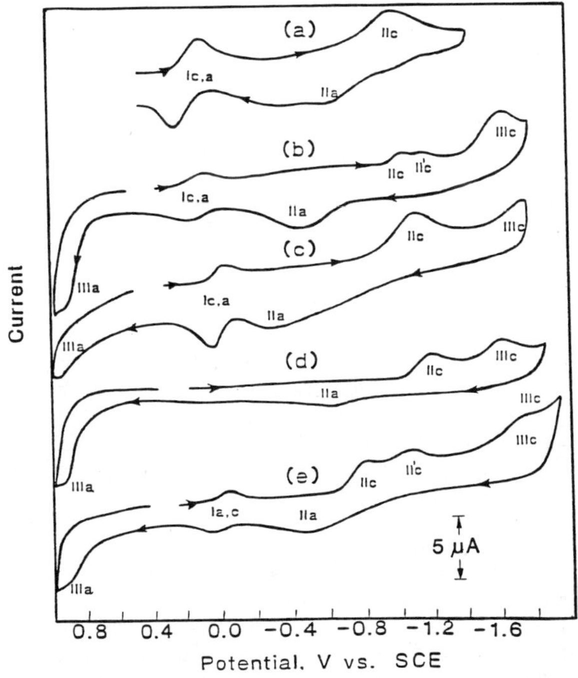

Figure 5. Cyclic voltammograms of Fe(III and II) complexes in Me_2SO, 0.1 M TBAP solution; scan rate 0.1 V s^{-1}: (a) $Fe(ClO_4)_3 \cdot 6H_2O$, (b) $Fe(ad^-)_2(ad)(ClO_4)_2$, (c) $Fe_2(ado)_3(ClO_4)_6 \cdot 4H_2O$, (d) $Fe(ad^-)(ClO_4) \cdot 2H_2O$, (e) $Fe(ado)(ClO_4)_2$.

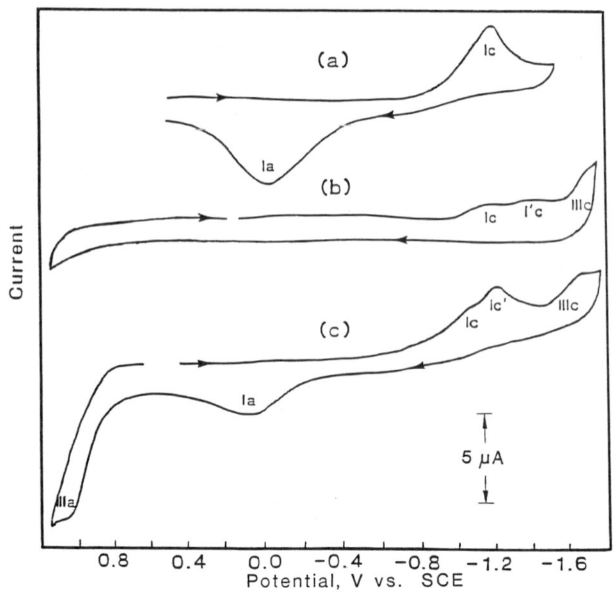

Figure 6. Cyclic voltammograms of Ni(II) complexes in Me_2SO, 0.1 M TBAP; scan rate 0.1 V s^{-1}: (a) $Ni(ClO_4)_2 \cdot 6H_2O$, (b) $Ni(ad^-)(ClO_4) \cdot 2H_2O$, (c) $Ni(ado)(ClO_4)_2 \cdot 2H_2O$.

The irreversible two-electron reduction peak of $Ni(ClO_4)_2 \cdot 6H_2O$ is split into peaks I and I' due to the presence of Ni(II) coordinated to Me_2SO and ClO_4^-, respectively. Me_2SO with Gutman donor number 39 is considered to be a moderately strong ligand.

The complexes of Cu(II) and Fe(III) with adenine and adenosine and Fe(II) with adenosine show a quasirreversible couple corresponding to the one-electron reduction of Cu(II) to Cu(I) or Fe(III) to Fe(II). All other redox processes (peaks II, II' and III) are irreversible due to following chemical reactions, adsorption, and film formation. The low currents and poorly defined peaks in cyclic voltammetry for these complexes reflect steric effects imposed by the bulky biological ligands and dimeric and polymeric structures.

Comparison of the redox potentials of the complexes with those of the metal perchlorate suggest the reduction occurring at the most negative potential can be attributed to the ligand. While studies in aqueous systems have revealed adenine to be reducible only when protonated at the N(1) nitrogen[18-23], Santhanam and Elving[24] examined the redox behavior of purine and 6-substituted purines including adenine (6-aminopurine) in non-aqueous media and concluded that neutral purines are reducible at potentials beyond those available in aqueous media of suitable pH, due to prior reduction of hydrogen ion, background electrolyte cation, or water itself, undergoing an initial one-electron reduction to form the corresponding anionic free radicals, which dimerize as indicated by lack of an anodic peak observed for radical oxidations on potential scan reversal, with the dimers oxidizable at considerably more positive potentials to regenerate the original purines[24]. Radical formation was attributed to the prewave occurring at −1.58 V for purine in DMF. Also postulated was the abstraction of a proton forming the protonated radical which undergoes dimerization.

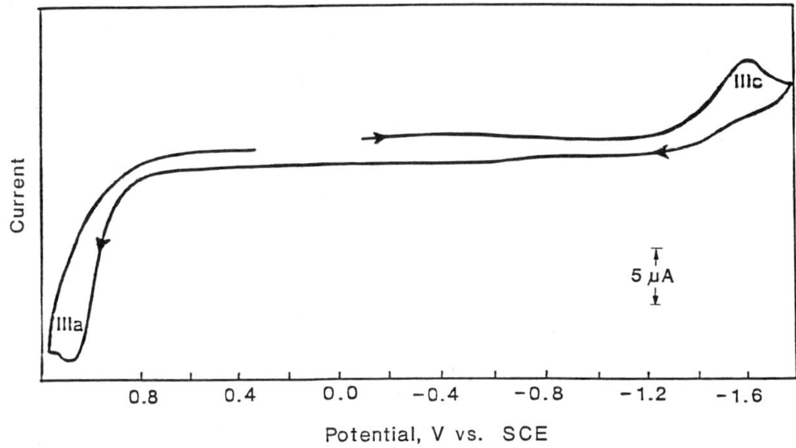

Figure 7. Cyclic voltammogram of adenine in Me_2SO, 0.1 M TBAP; scan rate 0.1 V s^{-1}.

In this work, the differential pulse voltammogram and cyclic voltammogram obtained for neutral adenine show an irreversible peak IIIc at E_{pc} = -1.58 V and, in the reverse scan, peak Ia at E_{pa} = 1.08 V (Figure 7). Spectra obtained during controlled potential electrolysis of adenine at -1.75 V (Figure 8a) show a single peak at λ_{max} = 266 nm decreased and a new peak appears at λ_{max} = 276 nm. The spectrum obtained after electrolysis at 1.15 V is the same as the original.

Differential pulse and cyclic voltammograms of adenosine are similar to those observed for adenine, with peak potentials E_{pc} = -1.60 and E_{pa} = 1.08 V. Spectral changes obtained during controlled potential electrolysis of adenosine at -1.75 V are shown in Figure 8b. The initial spectrum of adenosine shows two peaks at λ_{max} = 271 nm and 279 nm. Upon reduction, the

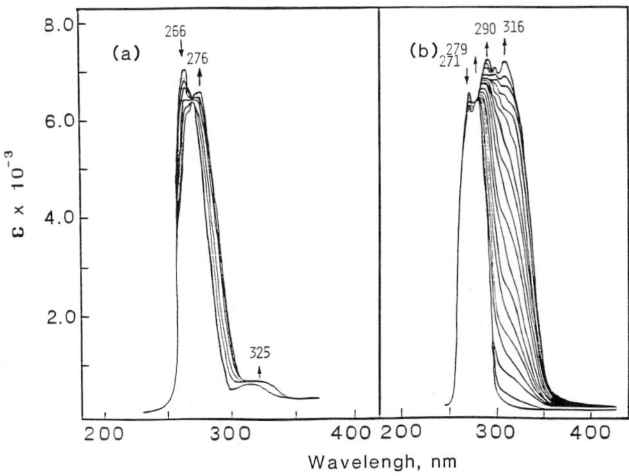

Figure 8. Thin-layer spectra of (a) adenine and (b) adenosine in Me_2SO containing 0.1 M TBAP during controlled potential electro-reduction at -1.75 V.

peak at 271 nm decreased and a broad band appeared in the region 290 nm to 316 nm. As with adenine, the spectrum obtained after electrolysis at +1.15 V is identical to the original.

These observations support a mechanism of radical generation in a one-electron transfer followed by protonation and dimerization with the dimer being reoxidized at high overpotentials. For adenine, this reaction would be as follows

$$\mathrm{Ad} \xrightarrow{+e^-} [\mathrm{Ad}]^- \xrightarrow{+H^+} \mathrm{AdH}\cdot \xrightarrow{+\ \mathrm{AdH}\cdot} (\mathrm{AdH})_2 \qquad (1)$$

and similar processes would be observed for adenosine.

UV-visible spectra of the complexes have several general characteristics which allow the redox processes involved to be monitored during controlled potential electrolysis. All complexes have a strong absorption band which can be assigned to the ligand adenine or adenosine, respectively. The 266 nm peak for adenine is split by complexation and appears at λ_{max} = 268 and 276 nm. Upon demetallation of the complexes, also evidenced by the appearance of metal plated on the platinum mesh electrode, the peak at 276 nm decreases and disappears. When reduction is then carried out at more negative potentials, -1.75 V, the single peak at 268 nm decreases and the peak at 276 nm increases. Complexes also show band broadening (monitored at 305 nm) and charge transfer bands around 343 nm which decrease upon demetallation. These parameters were used as an indication of continued complexation in those complexes which do not demetallate.

Table 2 summarizes the spectral changes for the reduced species obtained during contolled potential electrolysis of the adenine complexes.

$Cu(adH)_2(ClO_4)_2 \cdot C_2H_5OH$

The five-coordinate complex of Cu(II) and adenine is geometrically constrained by the quadruply bridged adenine. One would predict instability of the Cu(I) state of the complex since an enforced rigid square planar structure destabilizes Cu(I) with respect to Cu(II) because of the difficulty in achieving the change in configuration necessary for a Cu(I) coordination state.

Cyclic voltammograms of wave $I_{a,c}$ for varying scan rates at $E_{1/2}$ = 0.15 V allowed calculation of a value of 1.7×10^{-3} cms^{-1} for K, the heterogeneous electron transfer rate constant, for the Cu(II) to Cu(I) process in the adenine complex, employing the method of Shain and Nicholson[28].

Figure 9a-c shows the spectral changes obtained during controlled potential electrolysis at peaks Ic and IIc, (a); IIa, (b); and IIIa, (c); at 0.0, -0.9; 0.9 and +1.15, respectively. Initially, the adenine peak is split by complexation and appears at λ_{max} = 268 and 276 nm. During controlled potential electrolysis at 0.0 V, the Cu(I) intermediate is electrochemically generated by a one-electron reduction of the starting complex. A new peak appears at λ_{max} = 280 nm as well as an isosbestic point at 305 nm which reflects an equilibrium between the mono and divalent states of the complex. Upon reoxidation at 0.4 V, the original spectra is regenerated. Coulometry verified a transfer of one faraday of charge per mole of copper for both the oxidation and the reverse process. Peaks I_a and I_c can be assigned to the oxidation-reduction process of the Cu(II) adenine complex according to the reaction

Table 2. Absorption maxima (nm) and molar absorptivities of reduced adenine, adenosine, and adenine complexes in Me_2SO, 0.1 M TBAP solution.

	λ_{max} nm (10^{-3} ε)			
COMPOUND	Neutral	Red(I)	Red(II)	Red(III)
Adenine	266 (6.9)			266 (6.1)
	271 (6.5)			271 (6.3)
	276 (5.6)			276 (6.4)
Adenosine	271 (6.3)			271 (6.0)
	279 (6.2)			
				290 (7.1)
				316 (7.1)
$Cu(ad)_2(ClO)_4O_2\cdot C_25$	268 (7.4)	268 (7.0)	268 (7.0)	a
	276 (7.4)	276 (7.7)	276 (6.2)	
	280 (7.1)	280 (7.4)	280 (4.8)	
	305 (2.4)	305 (3.3)	305 (0.3)	
$Fe(ad)_2(ad^-)(ClO_4)_2$	267 (8.8)	267 (8.4)	267 (8.4)	a
	276 (9.0)	276 (8.6)	276 (7.8)	
	280 (8.6)	280 (8.6)	280 (5.9)	
	343 (1.5)	343 (0.6)	343 (0.4)	
$Fe(ad^-)(ClO_4)\cdot 2H_2O$	268 (7.8)		268 (7.3)	268 (7.3)
	276 (8.0)		276 (7.1)	276 (7.5)
	280 (7.6)		280 (5.9)	280 (7.3)
	343 (0.6)		343 (0.2)	343 (0.4)
$Ni(ad^-)(ClO_4)\cdot 2H_2O$	268 (5.0)	268 (5.3)		268 (6.7)
	276 (5.1)	276 (5.4)		276 (7.1)
	280 (5.0)	280 (5.3)		280 (7.3)
	284 (4.7)	284 (5.1)		284 (7.6)
	305 (2.1)	305 (3.2)		305 (6.8)
	343 (1.4)	343 (1.3)		343 (3.8)
	400 (2.0)	400 (3.3)		400 (8.2)

a = same as adenine

$$Cu(II)(Ad)_4Cu(II) + 2e^- \longleftrightarrow Cu(I)(Ad)_4Cu(I) \qquad (2)$$

The process at II_c corresponds to the reduction of complexed Cu(I) to uncomplexed Cu(0) and free neutral adenine. While it is not clearly apparent from either differential pulse or cyclic voltametry, spectroelectrochemistry provides evidence that the copper centers in the dimeric complex may be reduced separately and the broad peak at -0.2 V may consist of overlapping peaks. Upon reduction at -0.4 V, band broadening monitored at 305 nm and charge transfer bands around 343 nm decrease, current reaches a minimum, and there are no further spectral changes. Copper is present on the electrode but the adenine peak remains split. Cyclic voltammograms of the reduced solution show a clearly defined peak at E_{pc} = -0.6 V as well as peak III_c, a peak at E_{pc} = -0.48 V, peak I_a and I_c and a poorly defined peak III_a. All are indications that some copper remains complexed. Upon further reduction at -0.9 V, absorption at 305 nm and 343 nm continues to decrease and the peak at 276 nm decreases and dis-

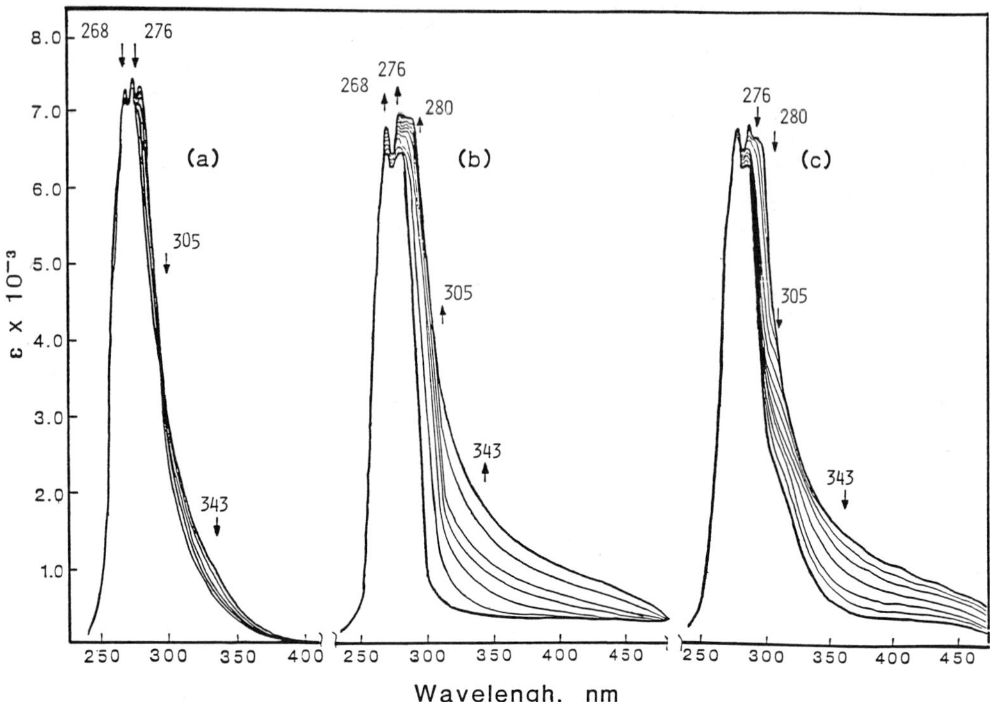

Figure 9. Thin-layer spectra of $Cu(ad)_2(ClO_4)_2 \cdot C_2H_5OH$ in Me_2SO containing 0.1 M TBAP during controlled potential electroreduction at (a) -0.9 V and electroreoxidation at (b) 0.9 V and (c) 1.15 V.

appears indicating total demetallation. The final spectra is markedly similar to that of neutral adenine. Cyclic voltammograms of the reduced solution reveal no peaks other than peak III_c and III_a, again well and poorly defined, respectively. The demetallation process occurring at peak II_c may be described as

$$Cu(I)(Ad)_4Cu(I) + 1e^- \longrightarrow Cu(I)(Ad)_x + Cu(0) \qquad (3)$$

$$Cu(I)(Ad)_x + Cu(0) + 1e^- \rightleftharpoons 2\ Cu(0) + 4\ Ad \qquad (4)$$

Upon reduction at more negative potential, -1.75 V, peak III_c, changes are identical to those observed in the reduction of neutral adenine (Figure 5a). Cyclic voltammograms obtained using the reduced solution show no peaks in the negative potential region, 0 to -1.8 V, or on a positive potential scan prior to the potential previously observed for the reoxidation of the protonated radical.

Figure 9b shows the spectra produced upon oxidation at potentials more positive than peak II_a. Metallic copper is stripped from the electrode and apparently complexes with the still reduced ligand, i.e., the protonated dimer. A new peak appears at λ_{max} = 280 nm along with increases at 305 nm and 345 nm characteristic of complexation.

$$Cu(0) - 2e^- \longrightarrow Cu(II) \xrightarrow{+(AdH)_2} Cu(II)(AdH)_2 \qquad (5)$$

Further oxidation at 1.15 V, peak III_a, results in the spectral changes depicted in Figure 9c. While the final spectrum is similar to the original, it is not identical, particularly in the relative absorptivities of the peaks at 268 and 276 nm (Table 2). The cyclic voltammogram obtained following reoxidation is also different from that of the original complex. Presumably, the final product is a complex of metal with reoxidized adenine, for which the stoichiometry and binding sites have yet to be elucidated.

$$Cu(II)(AdH)_2 + 2e^- \longrightarrow Cu(II)(Ad)_n \qquad (6)$$

$Fe(adH)_2(ad)(ClO_4)_2$

The complex of Fe(III) and adenine rather easily changes to the divalent state of Fe as evidenced by a clearly defined - for these complexes - reversible couple at $E_{1/2} = 0.16$ V. A value of 3.9×10^{-3} cm s^{-1} was calculated for κ, the heterogeneous electron transfer rate constant, for the reduction of Fe(III) to Fe(II) in this complex. Both the Fe(III) and Fe(II) complexes have octahedral coordination. Even though the complex is geometrically constrained by the doubly bridged adenine, there is no change in configuration necessary to achieve the Fe(II) state. The Fe(II) intermediate was easily electrochemically generated by a 1-electron reduction of the starting complex by controlled potential electrolysis at 0.0 V. Figure 10a shows the spectral changes obtained during this process. The peaks at Λ_{max} = 268 and 276 nm and the charge transfer band at 343 nm decrease and a shoulder at Λ_{max} = 280 nm appears. Upon reoxidation at 0.4 V, the original spectra is regenerated.

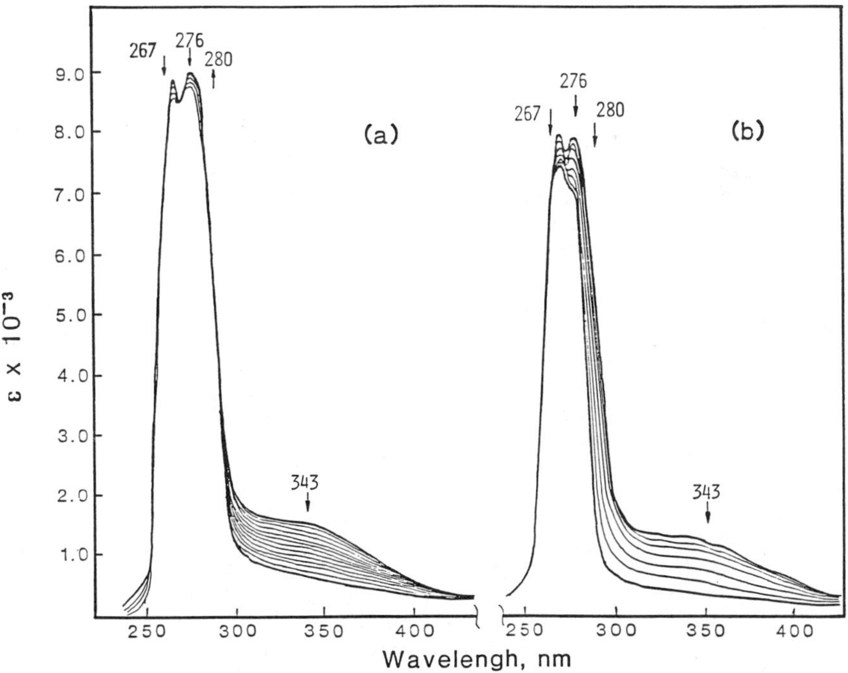

Figure 10. Thin-layer spectra of $Fe(ad^-)_2(ad)(ClO_4)_2$ in Me_2SO containing 0.1 M TBAP during controlled potential electroreduction at (a) 0.0 V and (b) -1.3 V.

Figure 10b shows the spectral changes obtained during controlled potential electrolysis at peak II_c, -1.3 V. The process at I_c corresponds to the reduction of complexed Fe(II) to uncomplexed Fe(0) and free neutral adenine. Splitting of the peak into II and II' is presumably due to the presence of a form of the Fe(II) complex in which Me_2SO has replaced perchlorate. Spectral parameters show the same pattern of response as the Cu(II) adenine complex which is reduced at the more positive potential of the two. Metallic iron is present on the electrode and the final spectra is markedly similar to that of neutral adenine. Cyclic voltammograms of the reduced solution reveal no peaks other than peak III_c and III_a which are due to the reduction of neutral adenine and dimer oxidation, respectively.

$$Fe(III)Ad + e^- \rightleftharpoons Fe(II)Ad \tag{7}$$

$$Fe(II)Ad + 2e^- \longrightarrow Fe(0) + Ad \tag{8}$$

Spectra obtained during oxidation are not shown as the data obtained and processes observed parallel to those obtained for Cu(II) adenine complex. At potentials more positive than peak II_a, metallic iron is stripped from the electrode and complexes with the protonated dimer. A new peak appears λ_{max} = 280 nm along with increases at 305 nm and 345 nm, characteristic of complexation. Further oxidation at 1.15 V, peak III_a, results in a final spectrum similar but not identical to the original Table 2. As with copper, the structure and binding sites of the complex formed following reoxidation have not yet been determined, but the product again is assumed to be a complex of metal with neutral adennine.

$$Fe(III)(AdH)_2 + xe^- \longrightarrow Fe(III)(Ad)_x \tag{9}$$

$Fe(ad)(ClO_4)\cdot 2H_2O$

The Fe(II) adenine complex is structurally similar to that of Fe(III) in that both are binuclear dimers bound together by double anionic adenine bridges. However, the Fe(III) complex discussed above is water-free, containing coordinated neutral adenine while the Fe(II) complex contains additional proton donating ligands, water and ethanol. The electrochemical and spectroelectrochemical data exhibit several disinguishing features. The reversible couple which would correspond to the Fe(III)/Fe(II) process in the uncomplexed metal perchlorate is absent in freshly prepared solutions of the complex (Figure 5d; Table 1). Cyclic voltammograms in which the potential scan is reversed immediately following peak II_c (E_{pc} = -1.04 V) reveal an accompanying anodic wave at E_{pa} = -0.98 V, even at relatively slow scan rates -- 100 mV s^{-1}, and a peak at E_{pa} = 0.2 V, the expected potential of peak I_a. The original spectra is similar to that obtained with the Fe(III) adenine complex. During controlled potential electrolysis of peak II_c, -1.3 V, the potential at which the Fe(III) complex is reduced to uncomplexed Fe(0) and free adenine, reduction to uncomplexed Fe(0) is not observed. Metallic iron is not present on the electrode and the final spectra retains the split peak at λ_{max} = 276 nm characteristic of complexed adenine. Coulometric data indicate charge transfer to one faraday per mole of iron. The presence of an Fe(I) oxidation state has been observed with porphyrins[29]. Another possibility is that the situation is similar to that observed in the binuclear copper complex and only one of the metal centers is reduced at this potential leaving the other complexed. This complex is poorly soluble and the amount of metal deposited might not be observable on the electrode. Cyclic voltammograms of the reduced solution show only peak III_c and a high current at -1.75 V in a negative potential scan and an anodic peak at E_{pa} = -0.62

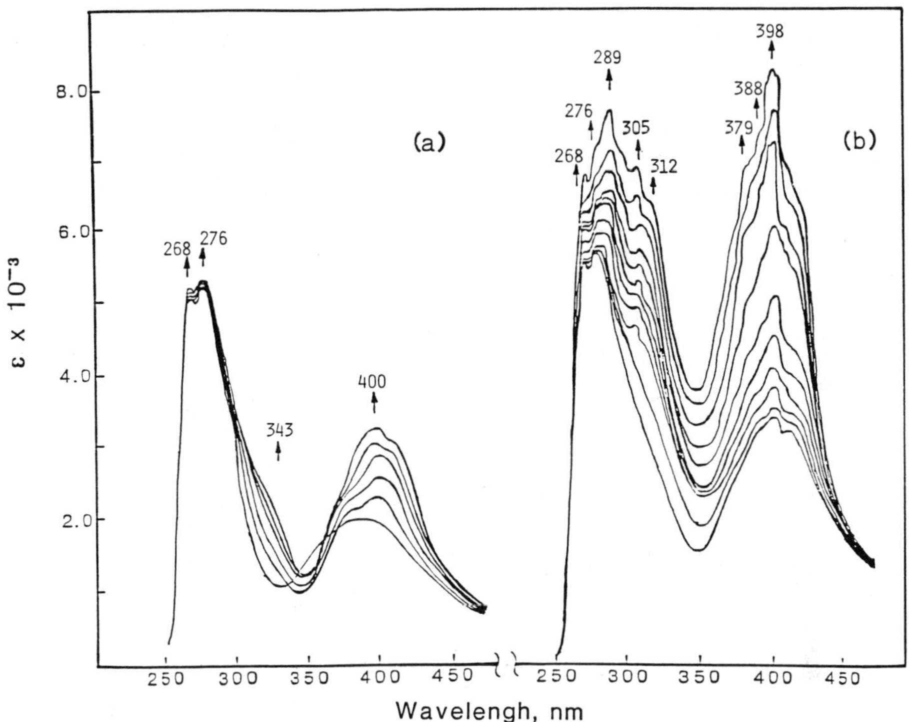

Figure 11. Thin-layer spectra of Ni(ad$^-$)($ClO_4 \cdot 2H_2O$ in Me_2SO containing 0.1 M TBAP during controlled potential electroreduction at (a) -1.25 V and (b) -1.75 V.

which could be attributed to the reoxidation of still complexed Fe. During the reduction at -1.75 V, peak III_c, the single peak at 268 nm decreases and the peak at 276 nm increases as does the shoulder at 280 nm and the peak width monitored at 305 nm. Cyclic voltammograms obtained using the reduced solution show no peaks in the negative potential region, 0 to -1.8 V and a new clearly-defined oxidation peak at E_{pa} = 0.8 V.

The apparent lack of demetallation, a coulometric n value of 2, spectral evidence of continued complexation and the newly observed high current at -1.75 V indicate the reduction process is markedly different for this complex than those discussed previously. In as much as the major difference between the Fe(III) and Fe(II) adenine complexes is the inclusion of water and ethanol in the structure of the latter and the Cu(II) adenine complex is water-free as well, the Ni(II) adenine complex, which contains coordinated H_2O, was investigated.

Ni(ad)(ClO_4)·$2H_2O$

The linear polymeric complex of nickel(II) and adenine (Figure 2b) contains N(6)-N(z) bonded anionic adenine (z = 1, 3, or 9 with 9 likelier than 1 or 3) and coordinated perchlorate, ethanol and water.

Cyclic voltammograms of the Ni(II) adenine complex show peaks at $E_{1/2}$ = -1.15 and -1.25 V (Table 1) and a ligand reduction peak, III_c, at E_{pc} = -1.60 V (Figure 6b).

Figure 11 shows the spectral changes obtained during controlled potential electrolysis at peaks a) I_c, -1,35 V and b) III_c, -1.75 V. The complex is very poorly soluble and seems to form a colloid rather than a true solution as an intact complex. Therefore, the ordinate of Figure 11 is an "apparent ε" calculated based on the mass of complex added to solvent and supporting electrolyte. Before reduction at -1.75, not all of the complex is dissolved. Changes which occur during reduction of the biological ligand result in greatly increased solubility, most likely due to depolymerization.

The peak for the ligand, deprotonated adenine, similar to that observed for neutral adenine is split by complexation and appears at λ_{max} = 268 and 276 nm. There is the familiar peak broadening and charge transfer band as well as an additional peak at λ_{max} = 400 nm. Upon reduction at -1.35 V, the peaks at 305 nm, 343 nm, and 400 nm all increase. The current reaches a minimum, and there are no further spectral changes. No metal is observed on the electrode. Cyclic voltammograms of the reduced solution show no peaks prior to peak III_c in the negative potential region. A scan of the positive potential range shows a poorly defined peak III_a and high currents at potentials greater than 0.8 V. No faradaic currents or spectral changes are observed during attempts at reoxidation at potentials as positive as 0.8 V. Coulometric data obtained give a value of one electron per formula unit for this reduction indicating the formation of Ni(I).

Further reduction at -1.75 V, produces the rich spectra shown in Figure 8b. Adsorption increases at 268, 276, and 400 nm; however, new peaks also appear at 284, 305, 312, 379, 388, and 398 nm. During this process, the solution becomes bright yellow and the complex totally dissolves. Cyclic voltammograms of the reduced solution show no peaks in the negative potential region, 0 to -1.8 V. A scan of positive potential reveals a ramp-like current from 0.1 to 0.7 V followed by peak III_a.

An n value of 4 electrons per formula unit obtained for the reduction indicates the ligand reduction involved cannot be limited to radical formation. A similar enhancement in limiting current and increase in coulometric n value for the reduction of purine had been observed by Elving and co-workers[24] during addition of H_2O to non-aqueous solvents. The enhancement was attributed to the abstraction of a proton by the free radical to form the protonated radical which having a higher electron affinity than the parent base was further reduced, resulting in an ECE (electron transfer-chemical reaction-electron transfer) process. Similar results were obtained in studies involving protonation by benzoic acid. Maximum enhancement in limiting current was four times that observed in the absence of a proton donor and the number of electrons involved in the reduction estimated to be four for all the purines. These data suggest that under conditions of proton donor availability, the reduction procedes via the mechanism observed in aqueous solvents: an initial $2e^-$, $2H^+$ reduction of the N-1=C-6 double bond forming 1-6 dihydro adenine followed by further reduction of the C-2=N-3 double bond at the same potential forming 1,2,3,6-tetrahydro-6-amino purine.

Figure 12 shows a proposed reaction pathway for the electrochemical reduction of the Ni(I) adenine complex assuming octahedral coordination is retained following the reduction of Ni(II). During controlled potential electrolysis at -1.75 V, the adenine ligand undergoes intracomplex proton transfer and a 4 electron reduction (per formula unit) forming 1,2,3,6-tetrahydroadenine with Ni(I) as shown.

The increased solubility can be attributed to depolymerization. The Ni(II) complex, unlike those of Cu(II), Fe(III), and Fe(II) discussed

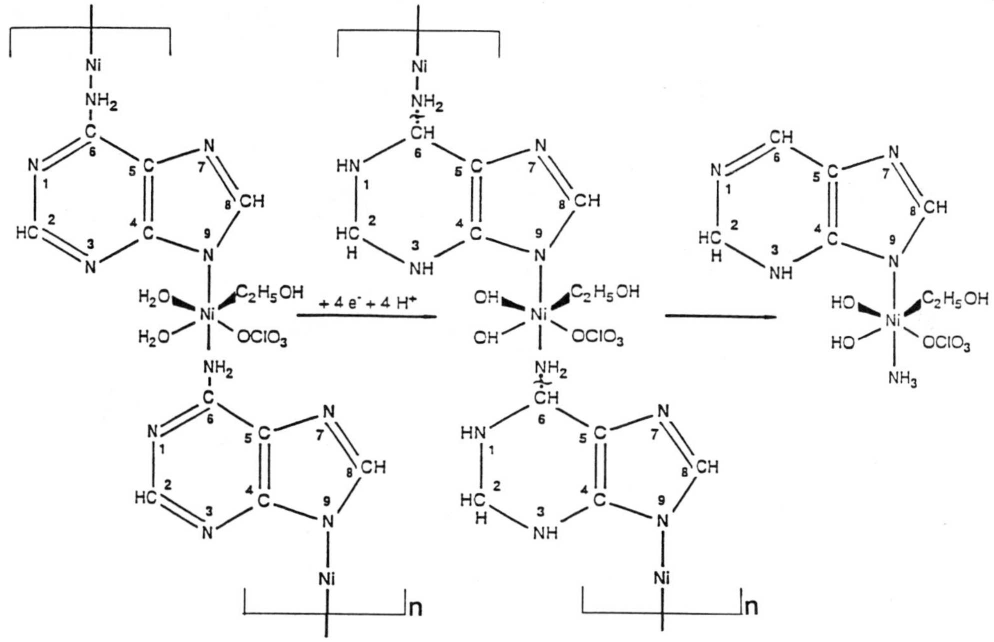

Figure 12. Reaction pathway for electrochemical reduction of $Ni(ad^-)(ClO_4) \cdot 2H_2O$.

previously is coordinated through the amino nitrogen (N-6). Deamination results in depolymerization forming monomeric units of a complex of Ni(I) with 2,3-dihydropurine, ammonia, and presumably OH^- and C_2H_5OH although, again, the exact structure of the units has not been elucidated. However, complexation between ammonia and Ni would account for the intense band at λ_{max} = 400 nm (Figure 11b).

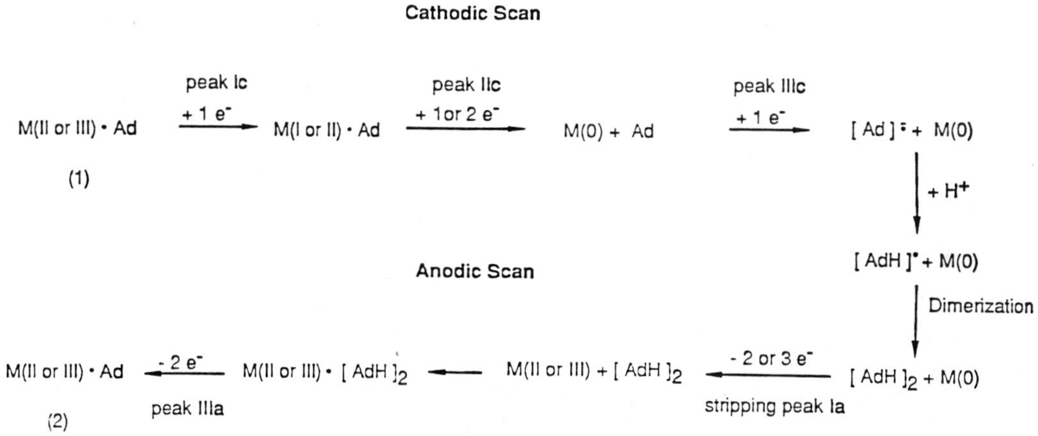

Figure 13. General reaction pathway for electrochemical reduction and reoxidation of metal adenine complexes (MAd) not containing aqueous ligand(s).

Figure 13 summarizes the pathway for the reduction of the water-free adenine complexes. In this scheme, reduction of the metal results in de-metallation which is followed by reduction of the ligand forming a radical followed by protonation and dimerization analogous to the reduction of adenine in non-aqueous solvents.

By analogy with the Ni(II) adenine complex which contains aqueous ligands, an "aqueous solvent" mechanism can be proposed for the reduction of the Fe(II) adenine complexes. In as much as the N(1) site is involved in complexation, during controlled potential electrolysis at -1.75 V (peak III), the adenine ligand might undergo intracomplex proton transfer and 2 electron reduction (per formula unit) of the C-2=C-3 double bond forming 2,3-dihydroadenine complexed with Fe(I). This mechanism would be consistent with the bathochromic shift in spectral wavelengths and the single, well-defined oxidation peak at E_{pc} = 0.8 V following controlled potential electrolysis at -1.75 V. The latter could be the reoxidation of the C-2=C-3 bond.

In the case of the complexes under consideration, presumably in those complexes containing ligands which are proton donors (H_2O and, possibly, enthanol), the close proximity of the proton donor is the controlling factor in determining which mechanism prevails: a "non-aqueous solvent" mechanism or an "aqueous solvent" mechanism.

Adenosine complexes

The half wave and peak potentials summarized in Table 1 and the cyclic voltammograms shown in Figures 4c, 5c and f, and 6c indicate the general pattern observed for redox processes for the adenosine complexes parallels that for the adenine complexes. Reversible one-electron transfer occurs for Cu(II) and Cu(I) at $E_{1/2}$ = 0.17 V and at $E_{1/2}$ = 0.05 and 0.18 V for Fe(III) to Fe(II) in the tri and divalent Fe adenosine complexes, respectively. There are subsequent electron transfers to the metal followed by reduction of the biological ligand at the most negative potential. In addition, the Cu(II) adenosine complex shows a reoxidation peak for Cu(I) to Cu(II) at E_{pa} = -0.1 V as well as the stripping peak for Cu(O) to Cu(II) at E_{pa} = 0.24 V.

All the adenosine complexes studied in this work have linear polymeric structures and N(1)-N(7) bonded adenosine and all but the Fe(II) complex contain coordinated water. Spectra obtained during reduction of the Cu-adenosine complex show some copper demetallation with some Cu(I) remaining complexed. A complication is that the mechanism of reduction may be dependent upon the way potential is applied, i.e., products of a single step controlled potential electrolysis at -1.75 V may be different than those obtained following step-by-step reductions prior to peak III. Studies are underway to elaborate the structures of intermediates and path-ways of electron transfer and to characterize the structures and binding sites of products obtained upon reoxidation.

This investigation has shown that complexes undergo both reversible and irreversible electron transfer processes to and from the central metal. The mechanism of biological ligand reduction seems to be dependent upon the presence of aqueous ligands in the structure as well as the binding site. Water-free complexes tend to demetallate and undergo reduc-tion of the biological ligand in a mechanism expected for non-aqueous solvents. Complexes with aqueous ligands undergo more complicated intra-molecular proton transfers and reduction of the pyrimidine ring, the nature and extent of which depends on the binding site of the complex. The stability of the associations between the transition metals studied and the biological ligand is evidenced by the reformation of complexes upon reoxidation of metals both with reduced and reoxidized biological ligands.

REFERENCES

1. R.W. Gellert and R. Bau, in: "Metal Ions in Biological Systems, Vol. 8", H. Sigel, Ed., Marcel Dekker, New York (1979), p. 1.

2. A.N. Speca, C.M. Mikulski, F.J. Iaconianni, L.L. Pytlewski and N.M. Karayannis, Inorg. Chim. Acta, 37, L551 (1979).

3. C.M. Mikulski, R. Minutella, N. DeFranco and N.M. Karayannis, Inorg. Chim. Acta, 106, L33 (1985).

4. A.N. Speca, C.M. Mikulski, F.J. Iaconianni, L.L. Pytlewski and N.M. Karayannis, J. Inorg. Nucl. Chem., 43, 2271 (1981); and references cited therein.

5. C.M. Mikulski, R. Minutella, N. DeFranco, G. Borges, Jr. and N.A. Karayannis, Inorg. Chim. Acta, 123, 105 (1986).

6. T.A. Kaden, K.H. Scheller and H. Sigel, Inorg. Chem., 25, 1313 (1986).

7. R.M. Izatt, J.J. Christensen and J.H. Rytting, Chemical Reviews, 71, 441 (1971).

8. M.M. Taqui Khan and C.R. Krishnamoorthy, J. Inorg. Nucl. Chem., 33, 1417 (1971).

9. M.M. Taqui Khan and M.S. Jyoti, Indian J. Chem., 15A, 1002 (1977).

10. S.K. Mishra, A.P. Mishra, V.K. Mishra and K.L. Yadaua, Nat. Acad. Sci. Letters, 8, 351 (1985).

11. J.P. Gisselbrecht and M. Gross, in: "Electrochemical and Spectrochemical Studies of Biological Redox Components", K.M. Kadish, Ed., American Chemical Society (1982), p. 111.

12. A.M. Bond and M.A. Khalifa, Inorg. Chem., 26, 413 (1987).

13. K.D. Karlin and J. Zubeita, "Biological and Inorganic Copper Chemistry", Adenine Press, New York (1985).

14. R.R. Gagne, R.P. Kreh and J. Dodge, in: "Electrochemical and Spectrochemical Studies of Biological Redox Components", K.M. Kadish, Ed., American Chemical Society (1982), p. 139.

15. A. ElJammal, E. Graf and M. Gross, J. Electroanal. Chem., 214, 507 (1986).

16. E. Freiden, J. Chem. Ed., 62, 917 (1985).

17. M.A. Jensen, T.E. Cummings and P.J. Elving, Bioelectrochem. Bioenerg., 4, 447 (1977).

18. G. Dryhurst and P.J. Elving, Talanta, 16, 825 (1969).

19. G. Dryhurst, "Electrochemistry of Biological Molecules", Academic Press, New York (1977), pp. 88-92.

20. D.L. Smith and P.J. Elving, J. Am. Chem. Soc., 84, 1412 (1962).

21. P. Valenta, H.W. Nurnberg and D. Krznaric, Bioelectrochem. Bioenerg., 3, 418 (1976).

22. D. Krznaric, P. Valenta, H.W. Nurnberg and M. Branica, J. Electroanal. Chem., 93, 41 (1978).

23. B. Janik and P.J. Elving, Chem. Rev., 68, 295 (1968).

24. K.S.V. Santhanam and P.J. Elving, J. Amer. Chem. Soc., 96, 1653 (1974).

25. D.A. Hall and P.J. Elving, Anal. Chim. Acta, 39, 141 (1967).

26. X.Q. Lin and K.M. Kadish, Anal. Chem., 57, 1498 (1985).

27. J. Fajer, D.C. Borg, A. Forman, D. Dolphin and R.H. Felton, J. Am. Chem. Soc., 92, 3451 (1970).

28. R.S. Nicholson and I. Shain, Anal. Chem., 36, 706 (1964).

29. C.A. Reed, in "Electrochemical and Spectrochemical Studies in Biological Redox Components", K.M. Kadish, Ed., American Chemical Society (1982), p. 333.

ELECTROCHEMICAL STUDY OF THE MECHANISM AND KINETICS OF OXYGEN REDUCTION MEDIATED BY ANTHRACYCLINE ANTIBIOTICS ADSORBED ON ELECTRODE SURFACE

Kenji Kano*, Tomonori Konse, Bunji Uno
and Tanekazu Kubota

Gifu Pharmaceutical University
5-6-1 Mitahora-higashi
Gifu 502 / Japan

INTRODUCTION

It is well known that anthracycline antibiotics having an anthraquinone group in the molecule are an important class of widely used antitumor antibiotics[1]. Their clinical efficacy is limited by severe dose-related cardiotoxicity[1,2]. It has been considered that these undesirable side effects as well as the clinical benefits result from their ability to shuttle electrons from NADPH-cytochrome P-450 reductase to molecular oxygen that is, in turn, transformed into activated oxygen species such as superoxide anion radical, hydroxyl radical, and hydrogen peroxide[1-3]. The activated oxygen species have been proposed to play a fatal role in DNA scission[4] and peroxidation of mitochondrial membranes[5]. However, the detailed mechanism is not yet fully understood.

Electrochemical approaches are useful in the study of redox chemistry of biomolecules. Considerable recent attention has been focused on electrocatalytic reactions using electrode surfaces modified with electron-transfer catalysts[6]. A catalyzed electrode reaction of molecular oxygen is a most interesting subject[7]. The catalysts, in most cases, have been metal porphyrins and phthalocyanines, and redox polymer films[8]. Quinoid compound-modified electrodes have also been used though to a lesser extent[9]. Some kinetic models have been presented and discussed[8,9]. In these investigations, rotating-disk voltammetry (RDV) has frequently been used in order to quantitatively characterize the kinetic behavior[8-10]. Cyclic voltammetry (CV) has also been applied, although the analysis of catalytic voltammograms are necessarily qualitative owing to the lack of analytical equations of current-potential curves.

In previous papers[11] we have shown that anthracycline adriamycin (ADM) and its derivatives are strongly adsorbed on electrode surfaces and that the two-electron redox reaction of the anthraquinone moiety occurs through its semiquinone radical intermediate even in aqueous media. In addition, we have found characteristic reduction waves of molecular oxygen catalyzed by ADM and 7-deoxyadriamycinone (7-DADMN) adsorbed on basal-plane pyrolytic graphite electrode (BPGE) surfaces[12]. In this report, we will analyze RDV and CV curves of quinoid compound-mediated reduction of oxygen by means of a curve-fitting technique, and then discuss the mechanism in detail.

EXPERIMENTAL

Chemicals and electrodes

Adriamycin hydrochloride (ADM) was kindly supplied by Kyowa Hakko Co. (Japan) and used as received. 7-DADMN was derived from ADM[12]. The other quinoid compounds used are listed in Table 1. All of them except **9** were commercially available and purified by sublimation and recrystallization[13]. Compound **9** was synthesized by alkylation of **1** with dimethyl sulfate. After being extracted with ether, it was recrystallized from methanol. The elemental analysis data for all the samples agreed well with the calculated values. Superoxide dismutase (SOD) (from bovine erythrocytes) was purchased from Sigma and used as received. Other chemicals were of reagent grade quality.

Naked BPGE's were prepared and pretreated electrochemically as described in the previous paper[12]. The electrode area (A) of the BPGE was 0.126 cm^2. The naked BPGE's were immersed in methanol solutions containing ca. 2 x 10^{-5} mol dm^{-3} of the quinoid compounds for a few seconds and air-dried at room temperature. These electrodes were used as modified electrodes. The adsorption on the electrode surface under these conditions was confirmed to be a monomolecular layer on the basis of the measurement of the total surface concentration (Γ_t) (vide infra).

Apparatus and electrochemical measurements

All electrochemical measurements were performed with a Yanagimoto P-1000 voltammetric analyzer and/or a Hokuto Denko HA-501 potentiostat equipped with an NF Circuit FG-121B function generator and a Hokuto Denko HB-104 function generator. For an RDV, a Yanagimoto SM6S2 synchronous motor was employed. All electrode potentials (E) were referred to a saturated calomel electrode (SCE), the temperature being 25 ± 0.2 °C. The electrolysis solutions used (pH 2.2 - 11.7) were the same as described in the previous paper[11b], the ionic strength being adjusted to 0.5 with KNO_3. The buffer solutions were saturated with air in the case of oxygen reduction current measurements, the bulk concentration of oxygen (C) being ca. 2.5 x 10^{-4} mol dm^{-3} [9c]. However, in other cases, they were deaerated by bubbling N_2 gas. All other electrochemical measurements have been described in previous papers[11-12].

Current-potential calculation

Current readings of RDV and CV curves were made every 20 mV by correcting for background currents. Values of Γ_t of adsorbed quinones were estimated by measuring charge, which was determined from areas under the quinone/hydroquinone surface-redox waves in CV. The reading of the current and area of voltammograms were carried out with an NEC PC-9801 personal computer equipped with a Graphtec KD4030 digitizer. The normal explicit finite difference digital simulation technique[14] was employed in the calculation of current-potential curves[15]. Values of the diffusion coefficient (D = 1.7 x 10^{-5} cm^2 s^{-1}), considered to be common for oxygen and superoxide, and the kinematic viscosity of the solution (ν = 0.01 cm s^{-1}) were taken from the literature[9c]. In curve-fitting analysis of voltammograms a non-linear least-squares analysis program[11c-f] was linked to a digital simulation program. The calculations were carried out with an NEC PC-9801 personal computer.

RESULTS AND DISCUSSION

Electrochemical behavior of 7-DADMN-modified BPGE

7-DADMN adsorbed on a BPGE exhibits a symmetrical surface-redox wave in CV in the absence of oxygen in the pH range examined. Examples at pH 7.00 and 7.95 are depicted in Figures 1 and 2 as dotted lines. The reversible surface wave should be ascribed to a surface-redox reaction of the anthraquinone moiety of the adsorbed 7-DADMN[11,12]. The behavior of these surface-redox waves can be interpreted in terms of a two-step one-electron surface-redox reaction (surface EE mechanism)[11,16].

$$O_{ad} + n_s e \rightleftharpoons S_{ad}, \quad S_{ad} + n_s e \rightleftharpoons R_{ad} \tag{1}$$

the notations O_{ad}, S_{ad}, and R_{ad} stand for the quinone, semiquinone, and fully reduced hydroquinone species, respectively, and n_s is the number of electrons (n_s = 1 in this case). Following the theory regarding the surface EE mechanism, the half-peak width $\Delta E_{p/2}$ can be converted to a semiquinone formation constant (K)[11,16]. Here K is defined as:

$$K = [S_{ad}]^2/[O_{ad}][R_{ad}] = \exp[(n_s F/RT)(E_1^{o'} - E_2^{o'})] \tag{2}$$

$E_1^{o'}$ and $E_2^{o'}$ are the surface-redox potential of the couples O_{ad}/S_{ad} and S_{ad}/R_{ad}, respectively. Other symbols have their common meanings. The values of K in Figures 1 and 2 are 0.490 and 0.565,

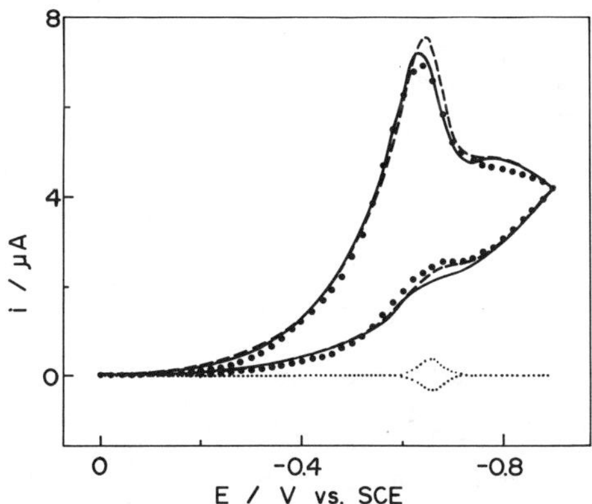

Figure 1. Background-corrected cyclic voltammograms with 7-DADMN-modified BPGE in the absence of oxygen (dotted line) and in air-saturated buffer (full circles) at pH 7.00 (phosphate buffer), v = 20 mV s^{-1}, and 25 °C. The solid line represents the regression curve based on the mechanism involving the disproportionation reaction of superoxide (see text) using the following refined data: $k_c = 1.14 \times 10^9$ cm^3 mol^{-1} s^{-1} and $k_{disp} = 1.77 \times 10^5$ dm^3 mol^{-1} s^{-1}. The broken line represents the regression curve based on the simiplified mechanism (see text) using $k_c = 3.67 \times 10^8$ cm^3 mol^{-1} s^{-1}. The other common data used in the calculation were: $E^{o'} = -0.661$ V, K = 0.490, $\Gamma_t = 4.18 \times 10^{-11}$ mol cm^{-2}, $k_d^o = 2.35 \times 10^{-3}$ cm s^{-1}, $\alpha = 0.115$, $C = 2.73 \times 10^{-4}$ mol dm^{-3}.

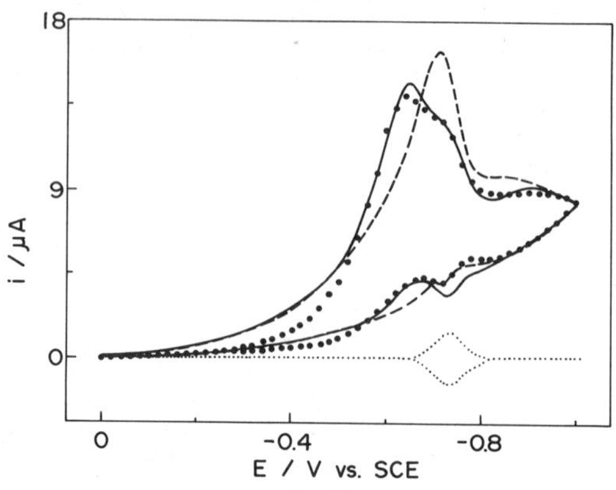

Figure 2. Background-corrected cyclic voltammograms with 7-DADMN-modified BPGE in the absence of oxygen (dotted line) and in air-saturated buffer (full circles) at pH 7.95 (phosphate buffer), v = 100 mV s^{-1}, and 25 °C. The solid line represents the regression curved based on the mechanism involving the disproportionation reaction of superoxide (see text) using the following refined data: $k_c = 2.43 \times 10^{10}$ cm^3 mol^{-1} s^{-1} and $k_{disp} = 4.12 \times 10^4$ dm^3 mol^{-1} s^{-1}. The broken line represents the regression curve based on the simplified mechanism (see text) using $k_c = 7.23 \times 10^8$ cm^3 mol^{-1} s^{-1}. The other common data used in the calculation were: $E^{o'} = -0.728$ V, K = 0.565, $\Gamma_t = 4.07 \times 10^{-11}$ mol cm^{-2}, $k_d{}^o = 5.46 \times 10^{-3}$ cm s^{-1}, $\alpha = 0.096$, C = 2.65×10^{-4} mol dm^{-3}.

respectively; the K value increases with an increase in pH[11b]. Combination of the K value with the peak potential, which should be equal to the standard redox potential ($E^{o'} = (E_1^{o'} + E_2^{o'})/2$), leads to $E_1^{o'}$ and $E_2^{o'}$. Inflection points of these potentials vs pH plots mean the acid dissociation constants (K_a) of the adsorbed species[11b]. In our experimental conditions, pK_a's of O_{ad}, S_{ad}, and R_{ad} ($pK_{a,o}$, $pK_{a,s}$, and $pK_{a,r}$) have been obtained as ca. 11, 8.8, and 8.2, respectively. These values are somewhat larger than those evaluated for ADM adsorbed on a mercury electrode[11b].

Electrocatalytic reduction of oxygen with 7-DADMN-BPGE

RDV curves in an air-saturated buffer of pH 7.0 measured with a 7-DADMN-modified BPGE and a naked BPGE are depicted in Figure 3 as full and open circles, respectively. The limiting current at the naked BPGE ($i_{d,l}$) corresponds to the diffusion-controlled two-electron reduction process of oxygen on the basis of the Levich equation[17]; $i_{d,l} = n_d FA(D/\delta)C$ where $\delta = 1.61 D^{1/3} \omega^{-1/2} \nu^{-1/6}$, n_d: the number of electrons per oxygen reduced (that is, $n_d = 2$), and ω: the angular velocity in RDV. At the 7-DADMN-modified BPGE, a well-defined catalytic reduction current is observed. However, the currents due to the catalytic and direct reduction of oxygen partially overlap each other. The catalytic peak in the RDV is symmetrical about the vertical line at $E = E^{o'}$, making allowance for the partial overlap of the catalytic and direct reduction current. The catalytic peak height at $E^{o'}$ (i_p) increases and approaches $i_{d,l}$ with increasing Γ_t. The second wave, due to the direct reduction of oxygen, is independent of Γ_t, and coincides with the wave observed at the naked BPGE. This means that the direct reduction of oxygen is independent of the presence of adsorbed 7-DADMN. In higher pH medium, 7-DADMN is liable to be desorbed from the BPGE during the electrode rotation.

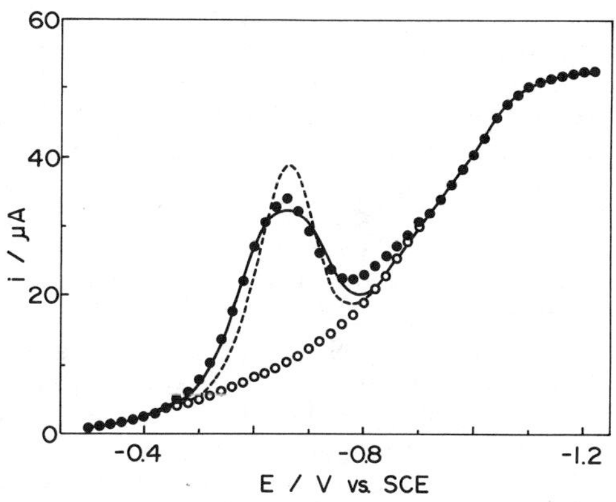

Figure 3. Background-corrected rotating-disk voltammograms with 7-DADMN-modified BPGE (full circles) and naked BPGE (open circles) in air-saturated phosphate buffer at pH 7.00, ω = 1200 rpm, v = 20 mV s^{-1}, and 25 °C. The solid line represents the regression curve based on the mechanism involving the disproportionation reaction of superoxide (see text) using the following refined data: $k_c = 5.32 \times 10^9$ cm^3 mol^{-1} s^{-1} and $k_{disp} = 7.75 \times 10^4$ dm^3 mol^{-1} s^{-1}. The broken line represents the regression curve based on the simplified mechanism (see text) using $k_c = 6.58 \times 10^8$ cm^3 mol^{-1} s^{-1}. The other common data used in the calculation were: $E^{o'} = -0.661$ V, K = 0.511, $\Gamma_t = 1.26 \times 10^{-10}$ mol cm^{-2}, C = 2.20×10^{-4} mol dm^{-3}. See Equation (12) for k_d values.

Full circles in Figures 1 and 2 show CV curves in an air-saturated buffer with 7-DADMN-modifed BPGE's. A pronounced catalytic reduction current is observed again. The catalytic reduction current appears around the $E^{o'}$ of the adsorbed 7-DADMN at pH 7.00 (Figure 1), while the peak position at pH 7.95 shifts to a value less negative than the $E^{o'}$ (Figure 2). At pH less than 4, the catalytic wave is no longer observed. Analogous results have been reported for the catalytic reduction of oxygen with quinoid compound-modified electrodes[9c,d]. At the stationary electrode, the adsorption of 7-DADMN is relatively stable, though it is also apt to be desorbed at pH higher than 10.5.

Here it is worth noting that the cathodic catalytic wave does not exhibit the usual broad tailing resulting from a simple diffusion-limited process, but shows the sharp adsorption-like peak. This behavior corresponds to the symmetrical catalytic wave in the RDV (Figure 3). Here it should be emphasized that the distribution of the surface concentration of the semiquinone species (Γ_s) is symmetrical about the line of $E = E^{o'}$ as predicted by Equation (3) on the basis of the theory of the surface EE mechanism[11,16].

$$\Gamma_s = \Gamma_t \eta \sqrt{K} / (\eta^2 + \eta \sqrt{K} + 1) \qquad (3)$$

where η is given by $\eta = \exp[(n_s F/RT)(E - E^{o'})]$. Therefore, a clear explanation of the phenomena mentioned above is that the one-electron-reduced semiquinone intermediate of 7-DADMN catalyzes oxygen reduction.

Effect of SOD on the catalytic current

It is predicted that superoxide anion radical $O_2^{\bar{\cdot}}$ and/or its conjugate acid, perhydroxyl radical $HO_2^{\cdot}$ (its $pK_{a,so}$ being 4.5-4.9[18]), are generated by one-electron transfer from the semiquinone radical to molecular oxygen. In a very weakly acidic medium the superoxide radical is very quickly converted into O_2 and hydrogen peroxide H_2O_2 by a spontaneous disproportionation reaction. However, with increasing pH, the reaction rate decreases greatly and superoxide becomes somewhat stable[18]. Therefore, in alkaline medium, if the generated superoxide is not reduced directly or catalytically and then exists in the vicinity of the modified electrode surface, the addition of SOD will cause an increase in the catalytic current. This is because SOD brings about an increase in the O_2 concentration at the expense of $O_2^{\cdot-}$ near the electrode surface. This prediction has been justified by the following experimental finding; at pH 8.5, the catalytic peak current in CV increases by 14.3% after the addition of 0.6 mg cm^{-3} (2100 U cm^{-3}) of SOD in spite of the 10.8 % decrease in Γ_t of 7-DADMN[19]. The decrease in Γ_t is attributable to a competitive adsorption between SOD and 7-DADMN. This observation confirms that molecular oxygen undergoes a one-electron catalytic reduction, and the disproportionation reaction of the generated superoxide may contribute to the intensity of the catalytic current. A direct reduction and/or a successive catalytic reduction of superoxide, if any, might be relatively very slow.

Electrode reaction kinetics

A non-linear least-squares curve-fitting of the voltammograms was attempted in order to verify the proposed mechanism and to evaluate kinetic parameters. Based on the above experimental results, the following assumptions were adapted; a) the electrode process of the 7-DADMN monolayer on a BPGE obeys a reversible surface EE mechanism; b) the semiquinone radical intermediate causes the reduction of molecular oxygen to generate superoxide radical ($O_2^{\cdot-}$ or $HO_2^{\cdot}$) as well as the quinone species of 7-DADMN; c) superoxide radical undergoes disproportionation into O_2 and H_2O_2, and d) the direct two-electron reduction of oxygen is independent of the presence of adsorbed quinone.

Under these conditions, the problem is formulated by ordinary kinetic diffusion equations:

$$\partial[O]/\partial t = D\partial^2[O]/\partial x^2 + k_{disp}[R]^2 \qquad (4)$$

$$\partial[R]/\partial t = D\partial^2[R]/\partial x^2 - 2k_{disp}[R]^2 \qquad (5)$$

In these equations, [O] and [R] are the concentrations of oxygen and superoxide, respectively, t is the time, x the distance from the electrode surface, and k_{disp} the disproportionation reaction rate constant. The boundary conditiions at x = 0 in this mechanism are:

$$i_c = n_c FAn_c k_c \Gamma_s [O]_{x=0} = n_c FAJ_c \qquad (6)$$

$$i_d = n_d FAk_d [O]_{x=0} = n_d FAJ_d \qquad (7)$$

$$(\partial[O]/\partial t)_{x=0} = (J_c + J_d)/D \qquad (8)$$

$$(\partial[R]/\partial t)_{x=0} = -J_c/D \qquad (9)$$

where the subscripts c and d are for the catalytic and direct reduction, respectively, and i, n, and J are the current, the number of electrons, and the flux of oxygen at the electrode surface, respectively ($n_c = 1$ in this case). The terms k_d and k_c are the rate constants for the direct reduction and the electron cross-exchange reaction between semiquinone radical and molecular oxygen, respectively. The boundary conditions at

$x = \infty$ and the initial conditions will be expressed by the usual equations. In order to obtain i_c and i_d, Equations (4) and (5) have been solved by a digital simulation technique[14,15].

The surface EE mechanism shows that the current brings about the reversible surface-redox reaction of the quinone mediator (i_s) and is written as follows[11,16]:

$$i_s = (n_s{}^2F^2Av\Gamma_t/RT)\eta(\eta^2\sqrt{K}+4\eta+\sqrt{K})/(\eta^2+\eta\sqrt{K}+1)^2 \tag{10}$$

Here v is the potential scan rate and in our case is positive for the cathodic scan. The total current (i) is given by $i = i_c + i_d + i_s$, although $|i_c + i_d| \gg |i_s|$ in RDV.

In the curve-fitting analysis of voltammograms, $E^{o'}$, K, and Γ_t were estimated from analysis of CV-curves recorded under anaerobic conditions. Γ_s in Equation (6) is a function of E as expressed by Equation (3). In CV-curve calculation, k_d in Equation (7) is given by Equation (11):

$$k_d = k_d{}^o\exp[(-\alpha F/RT)(E - E^o)] \tag{11}$$

where $k_d{}^o$ is the rate constant at the standard redox potential of oxygen (E^o), and α is the product of the transfer coefficient and the number of electrons pertinent to the irreversible process. For convenience, in actual calculation $E^o = E^{o'}$. The curve fitting of a direct reduction current recording using a naked BPGE gives k_d, α, and C. In RDV-curve calculations, however, k_d is estimated by:

$$k_d = (D/\delta)i_d/(i_{d,l} - i_d) \tag{12}$$

The i_d value is of course the direct reduction current for a bare BPGE. The value of C is estimated from $i_{d,l}$.

In line with the above theoretical treatment, the curve-fitting of the catalytic waves in CV and RDV has been carried out using two parameters: k_c and k_{disp}. Examples of the final regression curves are drawn in Figures 1-3 with solid lines. The characteristic waves seem to be reproduced very well. The relative standard deviations of the refined k_c and k_{disp} are in most cases less than 10 % and less than 30 %, respectively. The k_{disp} values thus obtained are comparable to the value available in the literature[18a]. These results confirm our proposed mechanism for the electrocatalytic reduction of molecular oxygen.

A simplified mechanism for this catalytic reaction will now be considered with the assumption that the catalytically generated superoxide undergoes a one-electron reduction immediately at the electrode surface, that is, $n_c = 2$ in Equation (6), and the term on the left hand side of Equation (9) as well as $[R]_{x=0}$ is zero[22]. Based on this simplified mechanism, curve-fitting of the voltammograms was tried again. The resulting regression curves are given in Figures 1-3 as broken lines. The regression curve at pH 7.00 (Figure 1) also reproduces the experimental curve. In the analysis of RDV curves at pH 7.00, the fittings is again good, although a gradual deviation occurs at large Γ_t (Figure 3). At higher pH, however, the discrepancy between the experimental curve and the regression curve, based on the simplified mechanism, turns out to be large (Figure 2). So, the mechanism involving the disproportionation reaction is considered to be more general. This model was used in the following analysis. The simplified model seems to be useful if k_{disp} is sufficiently large, that is, in weakly acidic or neutral media.

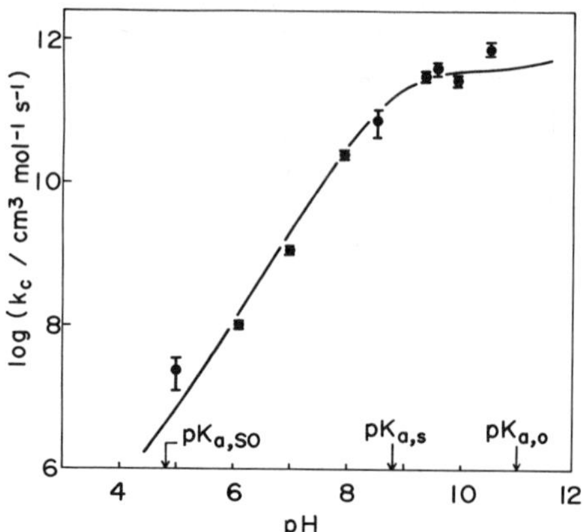

Figure 4. Dependence of log k_c on pH of 7-DADMN-modified BPGE. The solid line represents the calculated curve according to Equations (14)-(15) using the following data: log k_c^o = 1.71, β = 0.271, $pK_{a,s}$ = 8.80, $pK_{a,o}$ = 11.0, and $pK_{a,so}$ = 4.8. The error bars represent the standard deviation of log k_c estimated by the curve fitting of CV-curves.

pH Dependence of k_c

The catalytic wave depends to a great extent upon pH as mentioned above. We have thus investigated the pH dependence of k_c, which has been estimated by the curve fitting of CV curves[21]. The results are illustrated in Figure 4. Log k_c and pH are in a positive linear relation with the slope ca. 1.3 up to pH $\approx$ 8.5, but it converges to some constant value at pH > 9.5. Note here that the inflection point of the log k_c vs pH plot almost coincides with $pK_{a,s}$. This means that the semiquinone anion radical, i.e., the conjugate base of the adsorbed semiquinone intermediate, serves as a true mediator in this catalytic reaction. Furthermore, it is necessary to take into account the difference between the redox potentials of the quinone/semiquinone couple and the oxygen/superoxide couple on the ground of linear free-energy relationships[22]. Accordingly,

$$k_c = k_c^{int}(\Gamma_{s,b}/\Gamma_s)\exp[(\beta F/RT)(E^o - E_1^{o'})] \tag{13}$$

where k_c^{int}, $\Gamma_{s,b}$, and β are, respectively, the intrinsic catalytic activity of the semiquinone anion, the surface concentration of the conjugate base of the semiquinone, and the transfer coefficient. After $\Gamma_{s,b}$, E^o, and $E^{o'}$ are rewritten as a function of the proton concentration ([H])[23], Equations (14) and (15) can be derived,

$$k_c = \frac{k_c^o}{([H] + K_{a,s})}\left[\frac{([H] + K_{a,so})([H] + K_{a,o})}{[H]([H] + K_{a,s})}\right]^{\beta} \tag{14}$$

$$k_c^o = k_c^{int}K_{a,s}\exp[(\beta F/RT)(\bar{E}^o - \bar{E}_1^{o'})] \tag{15}$$

where k_c^o is the catalytic rate constant at pH = 0, $\bar{E}^o$ and $\bar{E}_1^{o'}$ being E^o and $E_1^{o'}$ at pH = 0. Equations (14) and (15) predict the slopes of a log k_c vs pH plot to be (1 + β) at $pK_{a,so}$ < pH < $pK_{a,s}$ and zero at $pK_{a,s}$ < pH < $pK_{a,o}$. The solid line in

Figure 4 is the calculated curve with β = 0.271 and log $k_c{}^o$= 1.71[24,25]. The agreement between the calculated and experimental values is good.

Our proposed mechanism is in fundamental accord with the mechanisms presented by many biochemists and pharmacologists[1-5]. Nagaoka et al.[9d] have reported a similar catalytic oxygen reduction wave with a 5,8-dihydroxy-1,4-naphthoquinone-modified glassy carbon electrode, and they have also suggested the semiquinone anion radical as a catalytically active form based on semiquantitative analysis. However, the possibility of participation of the hydroquinone species in the catalytic reaction cannot be completely excluded. The hydroquinone species would react with oxygen slowly, if at all, compared with the semiquinone radical at least under the voltammetric conditions used. Some authors have studied the electrocatalytic reduction of oxygen with electrodes modified with naphthoquinone derivatives[9a], naphthoquinone polymers[9b], and anthraquinone polymer[9c]. They proposed that the fully-reduced hydroquinone species is the active center for the catalytic reaction[9a-c]. Although the quasi-reversible behavior of the mediators and/or the overlapping of the catalytic and direct reductions of oxygen seem to be ambiguous points in their discussions, the hydroquinone species might also serve as the mediator under their experimental conditions.

Electrocatalysis with other quinone-modified BPGE

An analogous catalytic oxygen reduction wave is observed with an ADM-modified BPGE as shown in Figure 5. This catalytic wave seems to be explained fundamentally by the above reaction mechanism. Strictly speaking, however, ADM undergoes reductive deglycosidation and is converted to 7-DADMN[11f,26]. In Figure 5, the small peak at -0.75 V on the cathodic wave may be considered as the typical current due to the deglycosidation[11f]. The oxygen reduction will be thus mediated by both the semiquinones of ADM

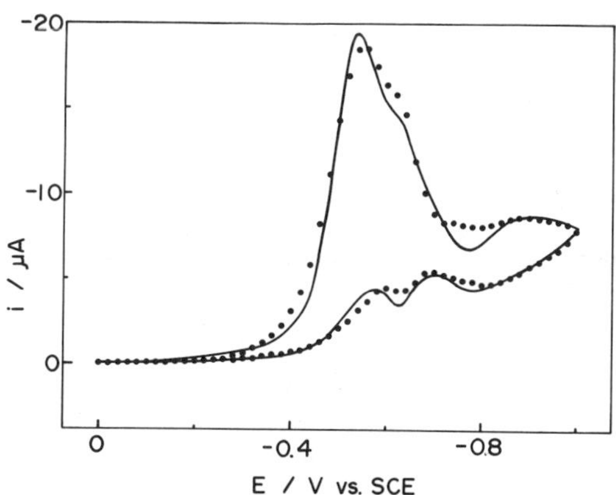

Figure 5. Background-corrected cyclic voltammograms with ADM-modified BPGE in air-saturated phosphate buffer (full circles) at pH 7.00, v = 100 mV s^{-1} and 25 °C. The solid line represents the regression curve based on the mechanism involving the disproportionation reaction of superoxide (see text) using the following refined data: k_c = 3.99 x 10^{10} cm^3 mol^{-1} s^{-1} and k_{disp} = 8.71 x 10^4 dm^3 mol^{-1} s^{-1}. The other data used in the calculation were: $E^{o'}$ = -0.630 V, K = 0.481, Γ_t = 4.78 x 10^{-11} mol cm^{-2}, $k_d{}^o$ = 1.63 x 10^{-3} cm s^{-1}, α = 0.102, C = 2.74 x 10^{-4} mol dm^{-3}.

Table 1. Surface Redox Potential ($E^{o'}$) and Catalytic Rate Constant (k_c) of Quinone-Modified BPGE at pH 7.0

Abbr. or No.	Compound	$E^{o'}$ V vs SCE	k_c [a)] 10^9 cm^3 mol^{-1} s^{-1}
7-DADMN	7-Deoxyadriamycinone	-0.661	1.14
ADM	Adriamycin	-0.630	39.9[b)]
1	1,4-Dihydroxyanthraquinone	-0.623	5.16
2	1,5-Dihydroxyanthraquinone	-0.600	4.36
3	1,8-Dihydroxyanthraquinone	-0.591	1.83
4	5,12-Naphthacenequinone	-0.664	-[c)]
5	2-Aminoanthraquinone	-0.608	-[c)]
6	1-Hydroxyanthraquinone	-0.577	-[c)]
7	2-Methylanthraquinone	-0.558	-[c)]
8	Anthraquinone	-0.551	-[c)]
9	1,4-Dimethoxyanthraquinone	-0.383[d)]	-[c)]
10	9,10-Phenanthraquinone	-0.230	-[c)]
11	1,4-Naphthoquinone	-0.200	-[c)]

a) The k_c values are estimated by the curve fitting of the CV curve at Γ_t = 5.7 (± 1.2) x 10^{-11} mol cm^2 (see Reference 21).
b) Effect of the reductive deglycosidation of ADM is not taken into consideration. See also Reference 27.
c) The catalytic current is not observed.
d) The surface wave is quasi-reversible; that is, the peak separation is 40 mV even at v = 10 mV s^{-1}.

and the partially generated 7-DADMN. These complexities make it difficult to quantitatively analyze the waves. Nevertheless, tentative curve-fitting analysis has been done in a similar manner as described above (Figure 5 and Table 1). The refined k_c value is very large compared with that of 7-DADMN at pH 7.0. This may be mainly attributed to small $_pK_{a,s}$ value of ADM[27].

The catalytic activity of various quinone-modified BPGE's have been also examined. In Table 1, the $E^{o'}$ and the catalytic activity of all the quinone-modified BPGE's are collected. Compounds **1-3** give similar catalytic waves, which are interpreted as in the case of 7-DADMN. The other quinone-modified electrodes here studied exhibit no definite catalytic wave. It is interesting that all of the catalytically active quinones (**1-3**) as well as ADM and 7-DADMN have hydroxyl groups in the peri-positions and their E^o's are more negative than E^o of $O_2/O_2^{\overline{\cdot}}$ (-0.57 V at pH > $pK_{a,so}$[18a,c]), that is, $(E^{o} - E_1^{o'}) > 0$ in Equation (13)[28]. These structural features might be essential for catalytic activity although other factors, such as an interaction between neighboring molecules, might also play important roles.

CONCLUSIONS

Based on the mechanistic and kinetic investigation described above, it may be concluded that the semiquinone radical species, particularly its conjugate base, of anthracyclines serves as a catalyst for the reduction of oxygen in aqueous media. In addition, the disproportionation reaction of the generated superoxide radical plays an important role in this reaction mechanism, though a true reaction mechanism of superoxide radical might be more complicated than that proposed here.

From the point of view of chemotherapeutics, these observations are important. ADM strongly intercalates into or binds to DNA[1,29] and the affinity of 7-DADMN as well as ADM for the phospholipid membrane is high[30]. In such a more or less hydrophobic environment, the semiquinone radical is stable[11e]. Our conclusions based on an electrochemical approach predict that an increase in semiquinone concentration will bring about an increase in the production of superoxide radical. Nevertheless, our results suggests the possibility of reduction of catalytic activity by drug design. On this point it is important to note $pK_{a,s}$, $E^{o'}$ (or $E_1^{o'}$), and the presence of a peri-hydroxyl group. If the relation between the structure and the catalytic activity can be clarified by the electrochemical approach, a new drug design for the anthracyclines would be possible.

ACKNOWLEDGMENT

We are grateful to Kyowa Hakko Kogyo Co. for kindly supplying adriamycin.

REFERENCES

1. F. Arcamone, "Doxorubicin Anticancer Antibiotics", "Medicinal Chemistry", Vol. 17, Academic Press, New York (1981); F. Arcamone, Med. Res. Rev., 4, 153 (1984); R.C. Young, R.F. Ozols and C.E. Myers, N. Engl. J. Med., 305, 139 (1981).

2. M.D. Green, J.L. Speyer and F. Muggia, Eur. J. Cancer Clin. Oncol., 20, 293 (1984); and the references sited therein.

3. H. Nohl, W. Jordan and R.J. Youngman, Adv. Free Radical Biology & Medicine, 2, 211 (1986); J.H. Peters, G.R. Gordon, D. Kashiwase, J.W. Lown, S.-F. Yen and J.A. Plambeck, Biochem. Pharmacol., 35, 1309 (1986); T. Komiyama, T. Kikuchi and Y. Sugiura, Biochem. Pharmacol., 31, 3651 (1982); S. Tero-Kubota, Y. Ikegami, K. Sugioka and M. Nakano, Chem. Lett., 1583 (1984).

4. J.W. Lown, S.-K. Sim, K.C. Majundar and R.Y. Chang, Biochem. Biophys. Res. Commun., 76, 705 (1977); V. Berlin and W.A. Haseltine, J. Biol. Chem., 256, 4747 (1981); and the references cited therein.

5. E.G. Mimnaugh, K.A. Kennedy, M.A. Trush and B.K. Sinha, Cancer Res., 45, 3296 (1985); E.G. Mimnaugh, M.A. Trush, M. Bhatnagar and T.E. Gram, Biochem. Pharmacol., 34, 847 1985); and the references cited therein.

6. R.W. Murray, in: "Electroanalytical Chemistry", A.J. Bard, Ed., Vol. 13, Marcel Dekker, New York (1984), pp. 191-368; R.W. Murray, A.G. Ewing and R.A. Durst, Anal. Chem., 59, 379A (1987); L.R. Faulkner, Chem. Eng. News, 62, 28 (1984).

7. E. Yeager, J. Mol. Catal., 38, 5 (1986).

8. P.A. Forshey, T. Kuwana, N. Kobayashi and T. Osa, Am. Chem. Soc. Adv. Chem., 201, 601 (1982); R. Durand and F.C. Anson, J. Electroanal. Chem., 134, 273 (1982); R. Durand, C.S. Bencosme, J.P. Collman and F.C. Anson, J. Am. Chem. Soc., 105, 2710 (1983); P. Martigny and F.C. Anson, J. Electroanal. Chem., 139, 383 (1982); N. Oyama, N. Oki, H. Ohno, Y. Ohnuki, H. Matsuda and E. Tsuchida, J. Phys. Chem., 87, 3642 (1983); K. Itaya, N. Shoji and I. Uchida, J. Am. Chem. Soc., 106, 3423 (1984); and the references cited therein.

9. a) G.S. Calabrese, R.M. Buchanan and M.S. Wrighton, J. Am. Chem. Soc., 104, 5786 (1982); idem., ibid., 105, 5594 (1983); b) M.-C. Pham and J.-E. Dubois, J. Electroanal. Chem., 199, 153 (1986); c) C. Degrand, J. Electroanal. Chem., 169, 259 (1984); d) T. Nagaoka, T. Sakai, K. Ogura and T. Yoshino, Anal. Chem., 58, 1953 (1986).

10. C.P. Andrieux, J.M. Dumas-Bouchiat and J.A. Saveant, J. Electroanal. Chem., 131, 1 (1982); idem., ibid., 169, 9 (1984).

11. a) K. Kano, T. Konse, N. Nishimura and T. Kubota, Bull. Chem. Soc. Jpn., 57, 2382 (1984); b) K. Kano, T. Konse and T. Kubota, ibid., 58, 424 (1985); c) idem. ibid., 58, 1879 (1985); d) T. Konse, K. Kano and T. Kubota, ibid., 59, 265 (1986); e) T. Konse, K. Kano, R. Kano and T. Kubota, ibid., 59, 3299 (1986); f) K. Kano, T. Konse, K. Hasegawa, B. Uno and T. Kubota, J. Electroanal. Chem., 225, 187 (1987).

12. T. Konse, K. Kano and T. Kubota, J. Electroanal. Chem., in press; Paper presented at the 32nd Annual Meeting on Polarography and Electroanalytical Chemistry in Tokyo, Nov. 12-14, 1986: Abstract, Rev. Polarogr. (Kyoto), 32, 48 (1986).

13. B. Uno, K. Kano, T. Konse, T. Kubota, S. Matsuzaki and M. Kuboyama, Chem. Pharm. Bull., 33, 5155 (1985).

14. S.W. Feldberg, in: "Electroanalytical Chemistry", A.J. Bard, Ed., Vol. 3, Marcel Dekker, New York (1969), pp. 199-296.

15. In our case, a dimensionless diffusion coefficient ($DNORM = D*\Delta TT/(\Delta RR)^2$) was taken as 0.45 ($\Delta TT$ and ΔRR are respectively the time unit and the distance unit). In CV-curve calculations, the potential interval (ΔPOT) was 1 mV and then $\Delta TT = |\Delta POT/v|$. In RDV-curve calculations, a dimensionless parameter VZERO ($= v_o*\Delta TT*\Delta RR$) was taken as 0.001 ($v_o = 0.51(\omega^3/\nu)^{1/2}$) and then $\Delta TT = (VZERO/v_o)^{2/3}(DNORM/D)^{1/3}$. Other operations were done in line with the literature[14].

16. T. Kakutani and M. Senda, Bull. Chem. Soc. Jpn., 53, 1942 (1980); E. Laviron, in: "Electroanalytical Chemistry", A.J. Bard, Ed., Vol. 12, Marcel Dekker, New York (1982), pp. 53-157.

17. A.J. Bard and L.R. Faulkner, "Electrochemical Methods - Fundamentals and Applications", Wiley, New York (1980), pp. 283-298.

18. a) D.T. Sawyer and J.S. Valentine, Acc. Chem. Res., 14, 393 (1981); b) J.M. McCord, J.D. Crapo and I. Fridovich, in: "Superoxide and Superoxide Dismutases", A.M. Michelson, J.M. McCord and I. Fridovich, Ed., Academic Press, London, New York (1977), pp. 11-17; c) J.A. Fee and L.S. Valentine, in Reference 18b, pp. 19-60.

19. The curve-fitting analysis (see the latter section of text) of the CV curves reveals that SOD accelerates the disproportionation reaction rate; $k_c = 4.65 \times 10^{10}$ cm^3 mol^{-1} s^{-1}, $k_{disp} = 7.04 \times 10^4$ dm^3 mol^{-1} s^{-1} at $\Gamma_t = 7.66 \times 10^{-11}$ mol cm^{-2} in the absence of SOD, but $k_c = 3.67 \times 10^{10}$ cm^3 mol^{-1} s^{-1}, $k_{disp} = 5.03 \times 10^5$ dm^3 mol^{-1} s^{-1} at $\Gamma_t = 6.84 \times 10^{-11}$ mol cm^{-2} after the addition of SOD.

20. In simplified model, the computation time of the CV-curve is greatly reduced and the analytical equation of the RDV curve can be obtained[12]. Furthermore, a Koutecky-Levich-type linear relation between i_p^{-1} and $\omega^{1/2}$ is expected, which has been justified experimentally[12].

21. The k_c value decreases with Γ_t owing to the attractive interaction between adsorbed molecules[12]. In order to avoid the complexity, the pH dependence k_c was examined at a constant Γ_t as possible ($5.7 \pm 1.2 \times 10^{-11}$ mol cm^2).

22. T. Ikeda, C.R. Leidner and R.W. Murray, J. Am. Chem. Soc., 105, 7422 (1981); idem., J. Electroanal. Chem., 138, 343 (1982); R.J. Klingler and J.K. Kochi, J. Am. Chem. Soc., 102, 4790 (1980).

23. $\Gamma_{s,b}$, $E^{o'}$ and $E_1^{o'}$ are given by: $\Gamma_{s,b} = \Gamma_s K_{a,s}/(K_{a,s} + [H])$, $E^o = \bar{E}^o + (RT/F)\ln([H] + K_{a,so})$, and $E_1^{o'} = \bar{E}_{1}^{o'} + (RT/F)\ln([H]^2 + [H]K_{a,s})/([H] + K_{a,o})$. The $\bar{E}_1^{o'}$ is reported as -0.29 V[18a] and $\bar{E}_1^{o'}$ is estimated as -0.234 V under our experimental conditions.

24. The value of β is somewhat smaller than that predicted from the simplified Marcus relationship[25] (0.5). From the log k_c^o thus obtained, k_c^{int} can be estimated to be 5.8×10^{10} cm^3 mol^{-1} s^{-1} based on Equation (15).

25. R.A. Marcus, Annu. Revs. Phys. Chem., 15, 155 (1964).

26. D.L. Kleyer and T.H. Koch, J. Am. Chem. Soc., 106, 2380 (1984); J. Fischer, B.R.J. Abdella and K.E. McLane, Biochemistry, 24, 3562 (1985); A. Anne and J. Moiroux, Nouv. J. Chim., 9, 83 (1985).

27. The values of $pK_{a,s}$, $pK_{a,o}$, and $E_1^{o'}$ of ADM adsorbed on a mercury electrode have been reported as 6.93, 8.53 and -0.200 V, respectively. Putting these values in Equations (14)-(15) gives the k_c value of ADM at pH 7 as 3.86×10^{10} cm^3 mol^{-1} s^{-1}, by using the values of k_c^{int} and β estimated for 7-DADMN (see Reference 24 and text). This calculated value coincides with the found value.

28. Theoretically, however, non-spontaneous catalytic reactions with $(E^o - E_1^{o'}) < 0$ is anticipated if both rate constants of the electron self-exchange reactions of the catalyst and the substrate are sufficiently large[22,25].

29. S. Neidle, in: "Topics in Antibiotic Chemistry", P.G. Sammes, Ed., Vol. 2, Wiley, New York (1978), pp. 240-278.

30. T.G. Burke and T.R. Tritton, Biochemistry, 24, 1768 (1985); D. Cheneval, M. Muller, R. Toni, S. Ruetz and E. Carafoli, J. Biol. Chem., 260, 13003 (1985).

STUDIES ON REDOX PHOTOTRANSFORMATIONS OF REDUCED FORMS OF COENZYME $NADP^+$ AND RELATED COMPOUNDS

Barbara Czochralska, Elzbieta Bojarska and Krzysztof Pawlicki

University of Warsaw
Institute of Experimental Physics
Department of Biophysics
93 Zwirki & Wirgury
02-089 Warszawa / Poland

INTRODUCTION

The increasing interest in electrochemical techniques to study the molecular redox mechanisms of compounds associated with energy conversion in living systems is based on the analogy between *in vivo* and polarographic conditions, which enable one to investigate biological compounds under conditions approximating those *in vivo.*

The foregoing analogy postulates that a study of the steps involved in the electrochemical mechanism of biologically active compounds may be extrapolated to some conclusions regarding reduction or oxidation mechanisms *in vivo.* This is supported by a series of investigations on the electrochemical and enzymatic oxidation of several purines[1,2].

It is known that the couple $NAD(P)^+/NAD(P)H$ plays a dominant role as a cofactor of many enzymatic redox reactions related to energy transfer and substrate metabolism in biological systems. Although $NAD(P)^+/NAD(P)H$ is considered to operate enzymatically through hydride transfer, the possibility that primary reactions involve one-electron transfer is supported by electrochemical[3] and pulse radiolysis[4] studies. High intensity excitation of NADH was found to produce two-photon ionization of NADH[5] to radical cations.

It was previously shown that two reduced forms of NAD^+, viz. NADH and the dimer $(NAD)_2$, may revert photochemically to the parent, enzymatically active NAD^+. Similar photochemical ability has been found for electrochemically generated dimers of pyrimidine and purine derivatives[6]. In the presence of oxygen this photochemical oxidation is accompanied by formation of H_2O_2, and the superoxide radical $O_2^{\overline{\cdot}}$ is an intermediate in the reaction[7].

Interest in the physiological role of H_2O_2 has been considerably enhanced by observations on its generation by a variety of cells and cellular organisms, e.g., during the course of oxidation of NADH by the hepatic plasma membranes[8].

The foregoing prompted us to extend our previous observations to the system NADP/NADPH, which also plays a key role in photosynthesis[9].

MATERIALS AND METHODS

$NADP^+$, NADPH, NAD^+, MNM^+, N_1-Methylnicotinamide (iodide salt), NBT and horseradish peroxidase (HRP) were purchased from Sigma Chemical Co. (St. Louis, MO / USA). Catalase and superoxide dismutase (SOD) were obtained from Boehringer (Mannheim, FRG). Glucose-6-phosphate dehydrogenase and glutathione reductase were products of Serva (Heidelberg, FRG).

The dimers $(NADP)_2$, $(NAD)_2$, $(NMN)_2$ and $(N_1$-Methylnicotinamide$)_2$ were prepared by electrolysis of 5×10^{-4} M solutions of $NADP^+$, NAD^+, NMN^+ and N_1-Methylnicotinamide, respectively, in carbonate buffer (pH 9.5), as previously described[7,10].

Electrochemical reduction was conducted in a three compartment cell, at constant potentials corresponding to the $E_{1/2}$ of the first one-electron wave for each compound. The cathode was a mercury pool (area 12 cm^2) stirred magnetically with a teflon-coated bar, with a saturated calomel electrode as reference electrode. The cell was continuously flushed with argon during electrolysis. Model OH-404 C integrator (Radelkis, Hungary) was used to determine the number of electrons involved in the reduction process.

D.C. polarography was carried out with a Radiometer Polariter PO-4 instrument (Copenhagen, Denmark), the characteristics of the dropping mercury electrode being m = 2.54 mg/s, t = 3.7 s at 60 cm Hg. Curves were recorded at room temperature, and all potentials are relative to the saturated calomel electrode. Solutions were rendered approximately oxygen-free by flushing with argon.

Irradiation of solutions, in 10-mm pathlength quartz cuvettes, at 254 nm, was conducted with a Philips TUV 6 W lamp; radiation below 240 nm was eliminated with the aid of a 5-mm filter of 35% acetic acid. An Osram HQV 125-W Wood's lamp with major emission at 365 nm was employed for irradiation at $\lambda \geq 320$ nm ($I_o = 6 \times 10^{-6}$ E x min^{-2} x cm^{-2} at $\lambda = 365$ nm).

Ultraviolet adsorption spectra were run on a Specord UV-VIS, using 1- and 10-mm pathlength quartz cuvettes.

Quantum yields for photooxidation of NADPH and $(NADP)_2$ at 365 nm were determined with the aid of the ferrioxalate actinometer, according to Parker[11]. Quantum yields at 254 nm were based on the use of the photohydration reaction of uridine, as previously described[12].

Enzymatic assay for H_2O_2

An irradiated solution of NADPH or $(NADP)_2$ was divided into two equal portions, to one of which was added catalase. Both solutions were subjected to polarography, and the amount of H_2O_2 formed was estimated from the difference in heights between the polarographic waves for the two solutions.

Assay for $O_2^{\cdot-}$ with superoxide dismutase

The amount of $O_2^{\cdot-}$ formed in irradiated solutions of NADPH and $(NADP)_2$ was determined by spectral was estimated at 560 nm of the amount of the added NBT which underwent reduction. Since reduction of NBT occurs to a small extent even in the absence of the superoxide anion, a control reaction was conducted simultaneously in the presence of an excess of enzyme to eliminate the $O_2^{\cdot-}$ rapidly enough so that it reacted only negligibly with the NBT.

Enzymatic assey for NADPH and $(NADP)_2$ with horseradish peroxidase (HRP)

The assay was performed in a 3-ml reaction mixture buffered with 50 mM phosphate (pH 7.0) or acetate (pH 5.5), containing 0.15 mM NADPH or $(NADP)_2$ and 0.3 mM H_2O_2. The reactions were initiated by the addition of HRP (final concentratiion 5 µm). Oxidation to $NADP^+$ was followed by the decrease in absorbance at 340 nm[13].

Enzymatic assay for $NADP^+$

This was carried out with the glucose-6-phosphate dehydrogenase system by following the appearance of the NADPH λ 340 nm adsorption band[14].

Enzymatic assay for NADPH in irradiated $(NADP)_2$

20 µl 0.1 M GSSG were added to 1 ml of an irradiated solution of $(NADP)_2$, A_{340}^{1cm} = 0.9. The reaction was initiated by addition of glutathione reductase[15]. The resulting decrease in adsorbance at 340 nm, relative to a non-irradiated control, gave the amount of NADPH generated during irradiation.

RESULTS

Photooxidation reactions

Scheme 1

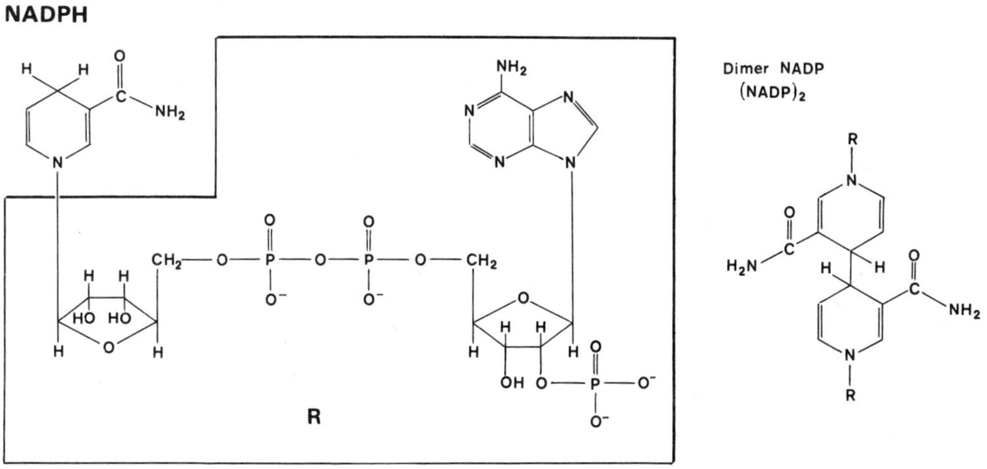

NADPH

Dimer NADP $(NADP)_2$

R

Irradiation of NADPH or $(NADP)_2$ (Scheme 1) at concentrations of 1-2 x 10^{-4} in 0.1 M carbonate buffer pH 9.5 at 254 nm or at 365 nm (Wood's lamp), led in both instances to appearance of $NADP^+$, monitored by the decrease in height of the 340 nm band and the smaller increase in absorption at 260 nm (Figure 1 and 2), and by the concomitant appearance of the two polarographic waves (with $E_{1/2}$ = -0.96 V and -1.7 V) characteristic for $NADP^+$ at pH 9.5[16].

The extent of photochemical conversion to $NADP^+$ was also measured enzymatically with the glucose-6-phosphate dehydrogenase system, as well as polarographically. Under anaerobic conditions the calculated quantum yield for phototransformation of NADPH and $(NADP)_2$ at 365 nm was $\emptyset$ = 5 x 10^{-4} and $\emptyset$ = 4 x 10^{-3}, respectively. In the presence of oxygen the values for $(NADP)_2$ were unaltered whereas the values for NADPH were increased to 3 x 10^{-3}.

From Tables 1 and 2 it will be further noted that conversion of $(NADP)_2$ to $NADP^+$, either in the presence or absence of O_2, was 80-90%

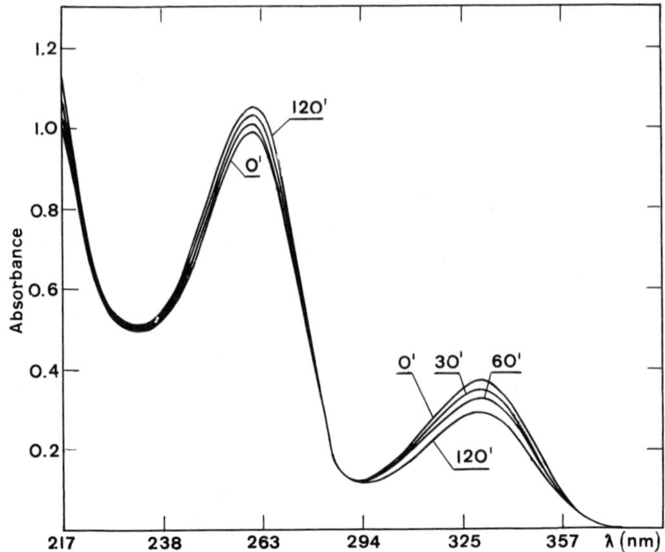

Figure 1. Changes in absorption spectrum of NADPH (5.8×10^{-5} M) at pH 9.5 (carbonate buffer) with time of irradiation at λ = 365 nm in absence of oxygen (argon atmosphere).

following irradiation at 365 nm depending on the extent of the reaction. On irradiation at 254 nm, conversion of the dimer to $NADP^+$ was somewhat lower (about 10%).

By contrast, the radiation-induced conversion of NADPH to $NADP^+$ was significantly O_2-dependent, amounting to ~ 50% under anaerobic conditions. The oxygen effect for NADPH and $(NADP)_2$ photooxidation is shown in Figure 3.

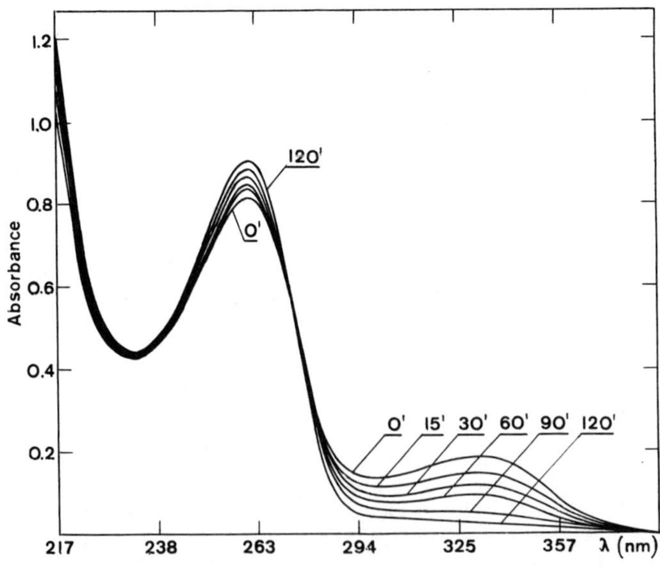

Figure 2. Changes in absorption spectrum of $(NADP)_2$ (3.1×10^{-5} M) at pH 9.5 (carbonate buffer) with time of irradiation at λ = 365 nm in absence of oxygen (argon atmosphere).

Table 1. Irradiation of deareated solutions of $(NADP)_2$ and NADPH in carbonate buffer, pH 9.5 (initial concentration for both compounds 1.5×10^{-4} M).

Compound	Time of irradiation (min)	Concentration of photochemically transformed $(NADP)_2$ or NADPH	Concentration of regenerated $NADP^+$	% regenerated $NADP^+$	Concentration of H_2O_2 formed during irradiation	$\frac{C_{H_2O_2}}{C_{NADP^+}} \times 100$
			λ = 365 nm (argon)			
$(NADP)_2$	60'	8.4×10^{-5} M	1.5×10^{-4} M	89	5.0×10^{-5} M	33%
$(NADP)_2$	90'	1.1×10^{-4} M	1.8×10^{-4} M	81	6.5×10^{-5} M	36%
$(NADP)_2$	120'	1.3×10^{-4} M	2.1×10^{-4} M	81	8.0×10^{-5} M	38%
NADPH	60'	1.1×10^{-5} M	0.7×10^{-5} M	60	$\sim 3 \times 10^{-6}$ M	45%
NADPH	120'	2.1×10^{-5} M	1.1×10^{-5} M	52	$\sim 5 \times 10^{-6}$ M	46%
			λ = 365 nm (oxygen)			
$(NADP)_2$	90'	1.2×10^{-4} M	1.9×10^{-4} M	79	7.0×10^{-5} M	37%
$(NADP)_2$	120'	1.4×10^{-4} M	2.3×10^{-4} M	82	9.0×10^{-5} M	30%
NADPH	60'	7.1×10^{-5} M	5.8×10^{-5} M	82	4.5×10^{-5} M	78%
NADPH	120'	1.2×10^{-4} M	9.6×10^{-5} M	80	7.0×10^{-5} M	73%

For both reactions the failure to observe quantitative conversion to $NADP^+$ is due to side reactions involving acid catalyzed hydration of the C_5-C_6 double bond of both reduced forms of coenzyme NAD(P) which occurs even in neutral or alkaline solutions[10,16-18]. For hydration of the pyridine ion in the excited state, addition of water at C_2 has been proposed[19].

Table 2. Irradiation of deareated solutions of $(NADP)_2$ and NADPH in carbonate buffer (pH 9.5) at λ = 365 nm).

Initial concentration of $(NADP)_2$ or NADPH	Time of irradiation (min)	Concentration of photochemically transformed $(NADP)_2$ or NADPH	Concentration of generated $NADP^+$	% of regenerated $NADP^+$	Concentration of H_2O_2 formed during irradiation
			$(NADP)_2$ argon		
1.2×10^{-4} M	60'	8.5×10^{-5} M	1.5×10^{-4} M	88	5.0×10^{-5} M
1.7×10^{-4} M	60'	1.1×10^{-4} M	1.8×10^{-4} M	82	5.5×10^{-5} M
1.5×10^{-4} M	75'	9.5×10^{-5} M	1.7×10^{-4} M	89	5.5×10^{-5} M
2.0×10^{-4} M	90'	1.4×10^{-4} M	2.2×10^{-4} M	79	7.0×10^{-5} M
1.7×10^{-4} M	120'	1.6×10^{-4} M	2.6×10^{-4} M	81	6.0×10^{-5} M
			NADPH argon		
1.2×10^{-4} M	60'	1.1×10^{-5} M	0.7×10^{-5} M	60	5×10^{-6} M
1.5×10^{-4} M	90'	1.5×10^{-5} M	0.8×10^{-5} M	53	5×10^{-6} M
2.0×10^{-4} M	90'	4.1×10^{-5} M	2.1×10^{-5} M	51	1×10^{-5} M
1.7×10^{-4} M	120'	3.8×10^{-5} M	1.8×10^{-5} M	48	8×10^{-6} M

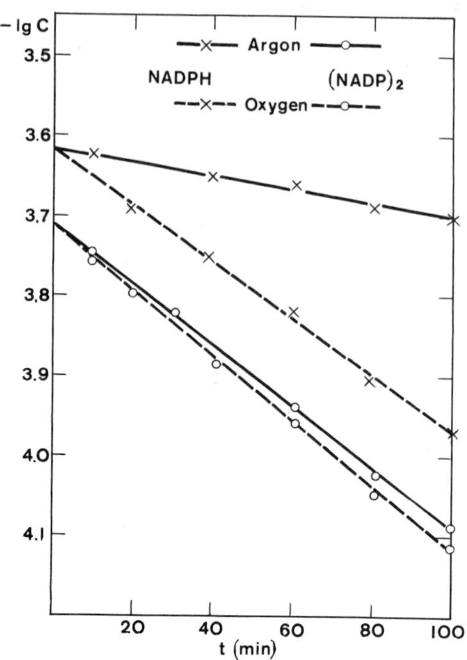

Figure 3. Time-dependence of log C of the NADPH and $(NADP)_2$ in presence and absence of oxygen (irradiation at λ = 365 nm at pH 9.5).

Formation of H_2O_2 during irradiation

The similar rates of photooxidation of $(NADP)_2$ in the presence or absence of oxygen pointed to the presence of an electron acceptor other than the latter, suggesting the presence of H_2O_2 as a possible intermediate. And, in fact, as previously noted for photooxidation of NADH and $(NAD)_2$[7], irradiation of NADPH or $(NADP)_2$ in the presence of O_2 was accompanied by appearance of H_2O_2, the level of which was proportional to the amount of $NADP^+$ product. In line with the mechanism proposed for photooxidation of NADH and $(NAD)_2$[7], the level of generated H_2O_2 would be expected to equal the amount of $NADP^+$ formed from irradiated NADPH, and 50% of that formed from irradiated $(NADP)_2$. The observed level of H_2O_2

Table 3. Irradiation at λ = 365 nm of deareated solutions of the NADP dimers in various buffers at pH 9.5 (time of irradiation 90').

Buffer	Concentration of photochemically transformed dimer	Concentration of regenerated $NADP^+$	% of regenerated $NADP^+$	Concentration of H_2O_2 formed during irradiation	$\frac{C_{H_2O_2}}{C_{NADP^+}} \times 100$
Carbonate	7.45×10^{-5} M	1.25×10^{-4} M	84	4.1×10^{-5} M	34%
Tris	7.62×10^{-5} M	1.23×10^{-4} M	81	5.3×10^{-5} M	43%
Phosphate	6.35×10^{-5} M	1.02×10^{-4} M	81	4.5×10^{-5} M	44%
Ammoniate	6.45×10^{-5} M	1.09×10^{-4} M	84	5.0×10^{-5} M	46%

was, however, 20% lower than the anticipated values (Tables 1 and 2), presumably due to radiation decomposition of H_2O_2 at pH 9.5[20], which is found to proceed twofold more rapidly at 254 nm than at longer wavelengths. This would account for the lower yield of H_2O_2 on irradiation at 254 nm.

Under anaerobic conditions, formation of peroxide for NADPH was drastically reduced. Irradiation of NADPH led to an H_2O_2 level of ~ 5 x 10^{-6} M, comparable to that expected from residual traces of oxygen in the solution. This was 8- to 10-fold higher with irradiated $(NADP)_2$, and comparable to the level of H_2O_2 observed during irradiation of $(NADP)_2$ in the presence of oxygen (Tables 1 and 2).

Influence of buffer media and pH

In the pH range 9.0 - 9.5 the rate of photooxidation of $(NADP)_2$ was unaffected by the nature of the buffer medium (ammoniate, phosphate, tris, carbonate), but the amount of H_2O_2 formed in carbonate buffer was slightly (10%) lower, probably due to reaction of H_2O_2 with HCO_3^- to give the carbonate peroxide[21]. Reduction in pH to 7.5 also reduced H_2O_2 formation by 10%; and at pH 4 only traces (~ 1 x 10^{-5} M) of H_2O_2 were observed, due to predominance of the hydration reaction of the dimer, dependent on the H^+ concentration.

Possible involvement of $O_2^{\bar{\cdot}}$ radical

SOD were employed to determine whether oxygen was the electron acceptor in the photooxidation of $(NADP)_2$ and NADPH. The procedure was the same as that previously used with $(NAD)_2$ and NADH[7]. It was found that the $O_2^{\bar{\cdot}}$ radical appeared only during photooxidation of both compounds in the presence of oxygen.

Peroxidase-catalysed oxidation

Oxidation of NADPH by HRP in the presence of H_2O_2 at pH 7, followed by the decrease in absorption at 340 nm, proceeded at a twofold higher rate under anaerobic conditions relative to that in the presence of O_2.

Oxidation of $(NADP)_2$ by HRP in O_2-saturated medium occurred in the absence of H_2O_2, but was appreciably accelerated in the presence of the latter (Figure 4). Under anaerobic conditions the reaction proceeded only

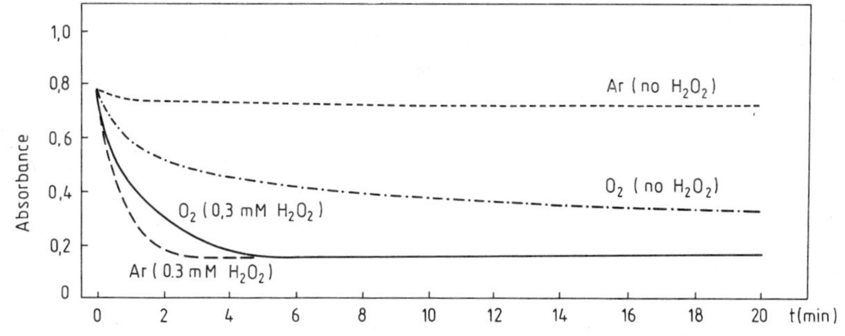

Figure 4. Oxidation and peroxidation of $(NADP)_2$ by HRP at pH 7.0 (phosphate buffer).

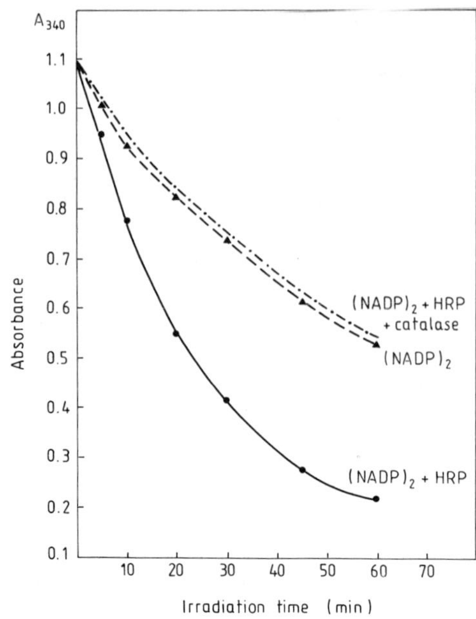

Figure 5. Effect of horseradish peroxidase with and without catalase on the oxidation of $(NADP)_2$ during irradiation of $\lambda = 365$ nm (pH 7.0, phosphate buffer).

in the presence of H_2O_2, and, as in the case of NADPH, was twice as fast as in the presence of O_2.

At pH 5.5 the rate of HRP-catalyzed oxidation of NADPH and $(NADP)_2$ was also twofold higher under anaerobic conditions. These findings point to the role of free radicals in these reactions, as previously noted for NADH and $(NAD)_2$[7].

Irradiation of $(NADP)_2$ at 365 nm and pH 7 under anaerobic conditions in the presence of HRP led to a twofold higher rate of oxidation relative to a control without HRP (Figure 5), clearly due to the combined effects of enzymatic and photo-oxidation (activity of HRP was unaffected by irradiation).

Since the oxidation activity of HRP is expressed only in the presence of H_2O_2 under anaerobic conditions, this constitutes evidence for generation of H_2O_2 during irradiation of the dimer under anaerobic conditions. And, in fact, on irradiation of the dimer in the presence of both peroxidase and catalase, the rate of oxidation was the same as in a control without the enzymes, since the former H_2O_2 generated is rapidly removed by catalase (Figure 5). The behavior of NADPH under the foregoing conditions were similar to that of $(NADP)_2$.

The foregoing results are consistent with those reported by Bodaness and Chan[22] for irradiation of NADPH at pH 7 in an air-saturated medium in the presence of HRP.

Detection of NADPH during irradiation of $(NADP)_2$

With the aid of the glutathione reductase system, it was found that irradiation of $(NADP)_2$ under anaerobic conditions led to appearance of NADPH at a level of about 10 % of the photochemically transformed dimer.

Table 4. Irradiation of deareated solutions of the dimers: $(NADP)_2$, $(NAD)_2$, $(NMN)_2$, $(N\text{-Methylnicotinamide})_2$ in carbonate buffer (pH 9.5) at λ = 365 nm (initial concentration was 1.5×10^{-4} for all dimers).

Dimer	Time of irradiation (min)	Concentration of photochemically transformed dimers	Concentration of monomers	Concentration of H_2O_2 formed during irradiation
$(NADP)_2$	60'	1.0×10^{-4} M	1.2×10^{-4} M	6×10^{-5} M
$(NAD)_2$	60'	1.1×10^{-4} M	1.3×10^{-4} M	5×10^{-5} M
$(NMN)_2$	60'	1.2×10^{-4} M	1.5×10^{-4} M	$\sim 1 \times 10^{-5}$ M
(N-Methylnicotinamide)$_2$	60'	1.4×10^{-4} M	2.2×10^{-4} M	–

Other nicotinamide dimers

The level of generated H_2O_2 during photooxidation under anaerobic conditions is in the order $(NADP)_2 > (NAD)_2 \gg (NMN)_2$ (traces) (Table 4). The absence of H_2O_2 during photooxidation of N(1)-methylnicotinamide dimer is likely due to its rapid rate of acid-catalyzed hydration under anaerobic conditions[23] and the fact that N(1)-alkylnicotinamide and partly $(NMN)_2$ undergo 1,6 dimerization as compared to 1,4 for NAD^+[23,24]. That it is not due solely to the absence of the sugar component was shown by the fact that all the foregoing reactions were unaltered in the presence of added free ribose or ribose phosphate.

DISCUSSION

Both reduced forms of $NADP^+$, i.e., NADPH and $(NADP)_2$, are photochemically converted to $NADP^+$, both in the presence and absence of oxygen, with evolvement of H_2O_2, as in the case of NADH and $(NAD)_2$[7,25], but the two processes differ significantly in that for NADPH but not $(NADP)_2$ the rate of photoconversion to $NADP^+$ and the level of generated H_2O_2 are both strongly dependent on the presence of oxygen.

The generation of $O_2^{\overline{\cdot}}$ under aerobic conditions points to participation of molecular oxygen as the electron acceptor in both reactions, as for the photochemical conversion of NADH to NAD^+[7]. However, as shown by Cunningham[26], generation of $O_2^{\overline{\cdot}}$ during photooxidation of NADPH is minimal, the quantum yield varying from 2×10^{-9} at 405 nm to 2×10^{-7} at 290 nm.

Avigliano et al.[25] proposed that photooxidation of $(NAD)_2$ in the presence of oxygen proceeds via dissociation to radicals, which are then oxidized to NAD^+ with generation of $O_2^{\overline{\cdot}}$.

The foregoing does not explain why the rate of photoconversion of the dimer $(NADP)_2$ is unaltered under anaerobic conditions, where the only remaining source of oxygen is water, so that the source of H_2O_2 would be photooxidation of water by excited NADP· via generation of OH radicals, as proposed for oxidation of water in chloroplasts[27].

The calculated redox potential of the NAD· radical (λ_{max} 400 nm) in the excited state[28], as follows:

$$(NAD\cdot) + e + H^+ \rightarrow NADH \quad (1)$$

is 3.05 V, hence adequate for the one-electron decomposition of water according to the reaction

$$H_2O \rightarrow OH + H^+ + e^- \quad (2.31\ V)^{27} \quad (2)$$

or the calculated redox potential E° (pH 7) is 2.18 V[29] and oxidation of water to H_2O_2 is 2.68 V[27].

Generation of H_2O_2 by this pathway would account for the lack of a difference in rates of photooxidation of the dimer in the presence or absence of oxygen. Generation of only traces of H_2O_2 by NADH irradiated under anaerobic conditions confirms the role of the NAD· radical in generation of H_2O_2 under such conditions, since photoionization of NADH leads to production of about 10% NAD[5].

Because irradiation of the dimer $(NADP)_2$ also leads to the appearance of NADPH (~ 10% of the initial dimer concentration), this may lead to involvement in the photoreduction reaction of the NAD· radical (Scheme 2, below, Equation (1A, 1C) or disproportionation of the dimer in the excited state (Equation 1B) or both. The formation of radicals during photodissociation of $(NAD)_2$ at pH 9.5 by pulsed N_2 laser excitation at 337 nm has been confirmed by Reverse Pulse Polarography[30].

Catalysis of photooxidation of $(NADP)_2$ by peroxidase in the absence of oxygen and H_2O_2, and lack of inhibition of this process by catalase, indicate that H_2O_2 is generated during photooxidation.

From the foregoing it appears to us that the mechanism of photooxidation of the dimers under aerobic and anaerobic conditions is as shown in Scheme 2 and that in the presence of oxygen there is the additional reaction

$$O_2 + NAD\cdot \rightarrow O_2^- + NAD^+$$

Formation of the maximum level of H_2O_2 following 75% dimer photodissociation and subsequent decrease of the level of H_2O_2 suggests partial conversion of H_2O_2 to OH· radicals[31], which may participate in oxidation of the remainder of the dimer or NADPH to $NADP^+$ and consistent with the observation that the dimer may be oxidized in the ground state by OH· radicals generated by radiolysis of water[32].

The foregoing does not, however, indicate what is the alternate electron acceptor leading to appearance of $NADP^+$, along with H_2O_2, under anaerobic conditions. Possible candidates are the products of acid decomposition (hydration) of the reduced forms NADPH and $(NAD)_2$ protonated in the excited state. Such products give polarographically a reduction wave with $E_{1/2}$ = -1.10 V close to the potential for reduction of $NADP^+$, so that they are good electron acceptors[33]. It is also known that the products of decomposition of $(NMN)_2$ in aprotic solvents undergo disproportionation with partial formation of the oxidized form[34].

The mechanism formulated in the present investigation for the photooxidation of the dimer $(NADP)_2$ may be related to the model of photosynthesis in which generation of O_2 proceeds via OH· radicals, H_2O_2 and $O_2^{\overline{\cdot}}$ [27].

Scheme 2

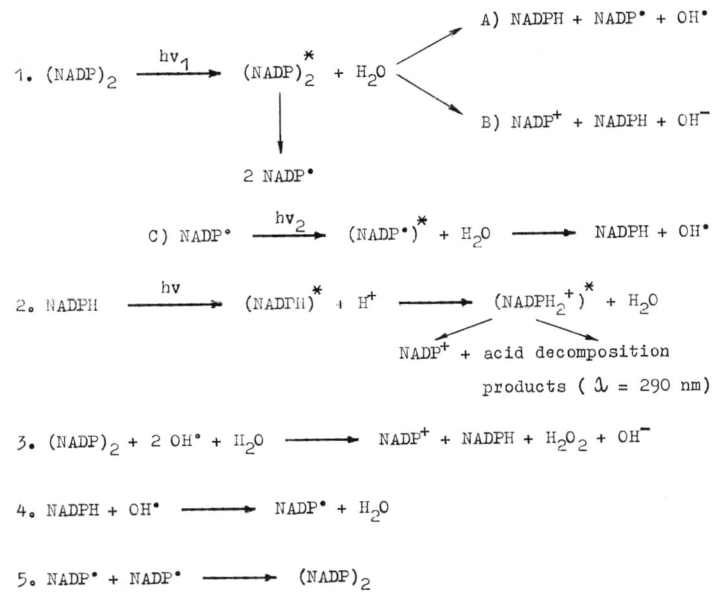

1\. $(NADP)_2 \xrightarrow{hv_1} (NADP)_2^* + H_2O$ → A) $NADPH + NADP^\bullet + OH^\bullet$

→ B) $NADP^+ + NADPH + OH^-$

$(NADP)_2^* \rightarrow 2\ NADP^\bullet$

C) $NADP^\bullet \xrightarrow{hv_2} (NADP^\bullet)^* + H_2O \longrightarrow NADPH + OH^\bullet$

2\. $NADPH \xrightarrow{hv} (NADPH)^* + H^+ \longrightarrow (NADPH_2^+)^* + H_2O$

→ $NADP^+$ + acid decomposition products (λ = 290 nm)

3\. $(NADP)_2 + 2\ OH^\bullet + H_2O \longrightarrow NADP^+ + NADPH + H_2O_2 + OH^-$

4\. $NADPH + OH^\bullet \longrightarrow NADP^\bullet + H_2O$

5\. $NADP^\bullet + NADP^\bullet \longrightarrow (NADP)_2$

The possibility of generation of the radical and dimer of coenzymes such as NAD^+, $NADP^+$ and NMN^+ in living organisms by either ionizing radiations[4,35,36] or UV[37,38] points to the biological significance of the present observations. And, finally, the generation of H_2O_2 and other oxygen species under both aerobic and anaerobic conditions may conceivably lead to some cytotoxic effects[39].

ACKNOWLEDGMENTS

We are indebted to Professor D. Shugar for discussions and his interest in this work. This investigation was supported by the Polish Ministry of Science and Higher Education under Project CPBP 01.06.

REFERENCES

1. M.Z. Wrona and G. Dryhurst, Biochim. Biophys. Acta., 570, 371 (1979).

2. M.Z. Wrona and R.N. Goyal and G. Dryhurst, Bioelectrochem. Bioenerg., 7, 433 (1980).

3. P.J. Elving, W.T. Bresnahan, J. Moiroux and Z. Samec, Bioelectrochem. Bioenerg., 9, 65 (1982).

4. E.J. Land and A.J. Swallow, Biochim. Biophys. Acta, 234, 34 (1981).

5. B. Czochralska and L. Lindqvist, Chem. Phys. Letter, 101, 297 (1983).

6. B. Czochralska, M. Wrona and D. Shugar, Topics in Current Chemistry, 30, 133 (1986).

7. B. Czochralska, W. Kawczynski, G. Bartosz and D. Shugar, Biochim. Biophys. Acta, 801, 403 (1984).

8. T. Ramasarma, A. Swaroop, W. MacKellar and F.L. Crane, J. Bioenerg. Biomembrane, 13, 241 (1981).

9. J.D. Rawn, "Biochemistry", Harper and Row Publishers, New York (1983).

10. M. Miller, B. Czochralska and D. Shugar, Bioelectrochem. Bioenerg., 9, 287 (1982).

11. C.A. Parker, "Photoluminescence of Solutions", Elsevier, London (1968).

12. B. Czochralska and D. Shugar, Biochim. Biophys. Acta, 281, 1 (1972).

13. K. Yokota and I. Yamazaki, Biochim. Biophys. Acta, 105, 301 (1965).

14. B.L. Horecker and A. Kornberg, in "Methods in Enzymology", S. Colowick and N.O. Kaplan, Eds., Academic Press, Inc., New York (1955), Vol. III, p. 879.

15. E. Racker, ibid., Vol. I, p. 879.

16. C.O. Schmakel, K.S.V. Santhanam and P.J. Elving, J. Am. Chem. Soc., 97, 5083 (1975).

17. E.M. Kosower, A. Teurstein, H.D. Burrows and A.J. Swallow, J. Am. Chem. Soc., 100, 5185 (1978).

18. J.T. Wu, L.H. Wu and J.A. Knight, Clinical Chemistry, 32, 314 (1986).

19. K. Takagi and Y. Ogata, J. Chem. Soc. Perkin II, 10, 1402 (1979).

20. A. Nadezhdin and H.B. Dunford, J. Phys. Chem., 83, 1957 (1979).

21. H. Metzner, in: "Biochemistry I", G. Milazzo and M. Blank, Eds., Plenum Press, London-New York (1983), p. 51.

22. R.S. Bodaness and P.C. Chan, J. Biol. Chem., 252, 8554 (1977).

23. C.O. Schmakel, K.S.V. Santhanam and P.J. Elving, J. Electrochem. Soc., 121, 1033 (1974).

24. H. Hanschmann, Stud. Biophys., 45, 183 (1974).

25. L. Avigliano, V. Carelli, A. Casini, A. Finazzi-Agrò and F. Liberatore, Biochim. Biophys. Acta, 723, 372 (1983).

26. M.L. Cunningham, J.S. Johnson, S.M. Giovanazzi and M.J. Peak, Photochem. Photobiol., 42, 125 (1985).

27. M.K. Raval and U.C. Biswal, Bioelectrochem. Bioenerg., 12, 57 (1984).

28. J. Joussot-Dubien, R.A.C. Albrecht, H. Gerischer, R.S. Knox, R.A. Marcus, M. Schott, A. Weller and F. Willing, in: "Light - Induced Charge Separation in Biology and Chemistry", H. Gerischer and J.J. Katz, Eds., Berlin: Dahlem Konferenzen (1979), p. 129.

29. W.H. Koppenol and J. Butler, Adv. Free Radical Biology and Medicine, 1, 91 (1985).

30. J. Hermolin, E. Kirowa-Eisner and E.M. Kosower, J. Am. Chem. Soc., 103, 1591 (1981).

31. B. Halliwell, Biochem. J., 167, 317 (1977).

32. B. Czochralska, E. Bojarska, K. Pawlicki and D. Shugar, in preparation.

33. A. Webber and J. Osteryoung, Anal. Chim. Acta, 157, 17 (1984).

34. W.J. Blaedel and R.G. Haas, Anal. Chem., 42, 918 (1970).

35. B.H.J. Bielski and P.C. Chan, J. Am. Chem. Soc., 102, 1713 (1981).

36. S.S. Chan, I.M. Nordlund, H. Frauenfelder, J.E. Harrison and I.C. Gunsalus, J. Biol. Chem., 250, 716 (1975).

37. H. Hanschmann and H. Berg, Bioelectrochem. Bioenerg., 8, 71 (1981).

38. P. Kovacic, J.R. Ames, P. Lumme, H. Elo, O. Cox, H. Jackson, L.A. Rivera, L. Ramirez and M.D. Ryan, Anti-Cancer Drug Design, 1, 197 (1986).

ANTICANCER QUINONES AND QUINOLINES:

MODE OF ACTION VIA ELECTRON TRANSFER AND OXIDATIVE STRESS

Peter Kovacic*, James R. Ames and James W. Grogan

Department of Chemistry
University of Wisconsin-Milwaukee
Milwaukee, Wisconsin 53211

Banasri Hazra

Organic Chemistry Section
Jadavpur University
Calcutta 700 032 / India

Michael D. Ryan

Department of Chemistry
Marquette University
Milwaukee, Wisconsin 53233

INTRODUCTION

General background material may be found in our companion article in these Proceedings dealing with other anticancer agents[1].

The aim of this study was to gain insight into the feasibility of electron transfer (ET) *in vivo* by examining the electrochemical characteristics of two categories of anticancer agents: quinones (diospyrin and 2,3-bis(bromomethyl)1,4-naphthoquinone) and conjugated quinolinium (iminium) ions from Dup 785 and camptothecin. Mechanistic ramifications are discussed.

EXPERIMENTAL

The bis(naphthoquinones) **1a, b** were obtained by published methods[2,3]. Bis(2,3-bromomethyl)1,4-naphthoquinone **2** was donated by Professor W.K. Musker, University of California, Davis. Antitumor agent **6** (Dup 785) was procured from E.I. DuPont Pharmaceuticals. Samples **7a, b** and 2-phenylquinoline (Aldrich Chemical Company) were used without further purification. Amide **7c** was prepared from the acid by standard methods, m.p. 197-198°C, lit.[4] m.p. 199°C. Camptothecin **8** was acquired from Dr. Monroe Wall, Research Triangle Institute.

Cyclic voltammetric measurements were performed as previously reported[5] in the absence of artificial light; a platinum electrode (Sargent) was also used as a working electrode for compounds **1** and **2**. Sodium hydroxide (0.125 M) was added to the indicated solutions. Reported potentials are referenced to the normal hydrogen electrode (NHE).

RESULTS AND DISCUSSION

1. Quinones

Quinones are widely distributed in nature and participate in various reactions of biological significance. Quinones can be readily generated *in vivo* from phenolic precursors[6].

A. Diospyrin

The bis(naphthoquinone), diospyrin **1a**, and its dimethyl ether derivative **1b** produce significant regression in the growth of Ehrlich ascites carcinoma in Swiss albino mice[2,3]. To gain information on the redox properties of **1**, we performed cyclic voltammetric measurements in aprotic solvent. The quinone **1a** gave several reversible waves with Pt or HMDE as the working electrode. Discussion will generally be limited to the first

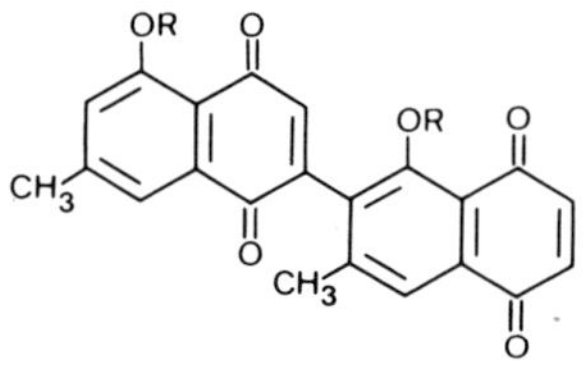

1

a) R = H
b) R = CH_3

reduction process. With Pt, $E_{1/2}$ was -0.12 V (Table 1). The E_{pc} and current function (CF) were constant at all scan rates employed (10 to 200 mV/s). Comparison of the CF with that of benzil, a compound known to undergo a reversible one-electron reduction[7], gave a value of 0.85, consistent with formation of a semiquinone. Radical anion generation was also indicated by the ΔE_p value of 60 mV, very close to theory (59 mV) for such a process[8]. The one-electron reduction species was 95% (i_{pa}/i_{pc} x 100%) reoxidizable, and the second wave gave $E_{1/2}$ of -0.34 V. Upon switching to mercury, only minor changes were observed; both the ΔE_p and percent reversibility increased slightly. One electron was transferred, c.f., CF ratio (Table 1). Diospyrin dimethyl ether **1b** reacted at -0.40 V (Table 1). Evidence from the difference in cathodic and anodic potentials (65 mV), current ratio (1.0), and CF ratio (0.8) indicated semiquinone formation, similar to **1a**.

The difference (280 mV) in $E_{1/2}$ for the two compounds can be attributed to intramolecular H-bonding of the hydroxyl group in **1a**. This interaction stabilizes the resultant radical anion. In **1b** the resonance effect of the methoxyl group hinders reduction, and H-bonding is eliminated. Data for several related naphthoquinones assist in elucidating important chemical effects. Diospyrin reduces at a value close to that (-0.15 V)[9] reported for the anticancer[10] 5-hydroxy 1,4-naphthoquinone

Table 1. Cyclic Voltammetry of 1,4-Naphthoquinone Antitumor Agents[a]

Quinone	Electrode	$-E_{1/2}$[b]	ΔE_p[b]	i_{pa}/i_{pc}[b]	CF_{ratio}[b,c]	$-E_{1/2}$[d]
1a	Pt	0.12	60	0.95	0.85	0.34
	HMDE	0.12	70	1.1	0.70	0.35
1b	Pt	0.40	70	1.0	0.82	0.56
	HMDE	0.39	65	1.0	0.75	0.56
2	Pt	0.0[e]	70[f]	-	1.83	-

[a] 0.1 V/s, tetraethylammonium perchlorate (0.1 M), quinone (0.5mM) vs NHE, reversible.
[b] First wave
[c] $CF_{ratio} = CF_{quinone}/CF_{benzil}$; CF_{benzil} $A/(V/s)^{1/2}M = 16.27$; (0.5 mM, Pt), 0.185 (0.5 mM, HMDE) DMF.
[d] Second wave.
[e] Irreversible, E_p value.
[f] $E_{pp/2}$ value.

(juglone). The 5-methoxyl analog gave a figure of -0.44 V[9]. The difference (290 mV) in $E_{1/2}$ values, similar to our case, has also been attributed to H-bonding and resonance effects. An $E_{1/2}$ of -0.34 V is reported for the parent 1,4-naphthoquinone[9]. The additional conjugation from the second aromatic ring attached to the quinone moiety of diospyrin might be expected to favorably influence reaction. However, the $E_{1/2}$ for naphthoquinone and its 2-phenyl derivative differ by only 5 mV[11]. This is another illustration of the pitfalls one may encounter in making electrochemical predictions.

Despite the differences in reduction potential, replacement of hydroxyl by methoxyl does not result in appreciable change in antitumor activity[2,3]. The potency of the ether may be due to permeability and/or adsorption effects, or **1b** may be converted to the hydroxyl analogue *in vivo*. Metabolic dealkylation of phenolic methyl ethers is a known transformation[12]. The compounds most likely exert their toxic action via redox cycling; a threshold value of about -0.5 V for biological activity with related antitumor compounds has been suggested[9].

Their are several prior studies which provide support for our proposed ET mechanism. Diospyrin produces morphological changes and also lyses cancer cells[2], similar to the action of another anticancer quinone, β-lapchone, on trypansoma parasites[13]. These ultrastructural alterations might be accounted for by the action of superoxide and other oxygen derived species. Alternatively, these agents may function as electron sinks, thus disrupting normal electron flow. Respiration is inhibited by diospyrin[2]. Several antimalarial naphthoquinones have been proposed to act in a similar, disruptive manner[14].

In addition to antitumor activity, diospyrin exhibits antileishmanial properties[15], perhaps by ET[16]. Based on common functionalities (quinone and phenol), diospyrin is structurally related to several other compounds that are physiologically active. The antiviral agent, sakyomicin A, which contains a hexose residue and the juglone skeleton, generates H_2O_2 after being reduced by NADH[10]. The antibiotic nanaomycin A[17], antitumor plumbagin[2], and mutagenic anthrarufin[18] also show structural resemblance. The much studied anthracyclines, e.g., adriamycin, $E_{1/2}$ -0.43 V[9], are also akin to diospyrin in functional groups. In relation to the mechanism[19], generally these quinone antitumor antibiotics cause strand breaks in tumor DNA as a result of superoxide formation[20]. The toxicity presumably results from redox cycling and is oxygen dependent[21]. The process, designated "site specific" free radical generation[22], includes metabolic reduction to the semiquinone[19]. Other potential quinone mechanisms include sulfhydryl arylation[6], or topoisomerase involvement[1].

An analogous situation pertains to the electronically related heterocyclic di-N-oxides derived from phenazines, e.g., iodinin, and quinoxalines, e.g., dioxidine, which function as bactericides[23,24]. Evidence indicates a correlation involving hydrogen bonding, reduction potential, and activity.

Certain metal ions are required as cofactors for some antitumor agents[19]. Often oxy radicals are the result. An example of a drug that falls in this category is streptonigrin. Superior antineoplastic characteristics appear to be associated with the daunorubicin$_3$-Fe(II) complex. Adriamycin is capable of chelating metal ions[19,21]. The iron complexes[25] apparently generate their toxic effects via reduced oxygen species. By analogy, it is reasonable to visualize similar involvement in complex formation in the case of diospyrin; antimicrobial metal complexes of juglone are known[26].

B. 2,3-Bis(bromomethyl)1,4-naphthoquinone

Another naphthoquinone antitumor agent is 2,3-bis(bromomethyl)1,4-naphthoquinone, **2**[27]. The presence of halogen makes for interesting mechanistic ramifications.

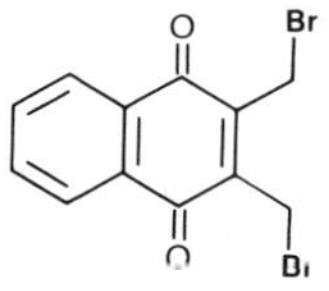

2

Agent **2** reduced at 0.0 V in DMF (Table 1), in a diffusion limited, irreversible manner. The $E_{pp/2}$ (E_p-$E_{p/2}$) value at 100 mV/s was 70 mV. A figure of -0.03 V has been reported in pH 7 buffered solution[28]. The inductive effect of bromine has a positive influence on the reduction potential, c.f., 2,3-dimethyl 1,4-naphthoquinone,-0.61 V (CH_3CN)[29]. Similar correlations have been observed for mesoionic compounds[30], substituted iodobenzenes[31], and benzaldehydes[31].

In relation to our theoretical approach, the chloro analog of compound **2** produced an initial burst of oxygen uptake[32,33], apparently leading to superoxide, followed by inhibition of respiration. Activity under hypoxic conditions may be due to a detour of normal electron flow (*vide supra*). A free radical mechanism of action is postulated to be responsible for the selective toxicity of the chloro analogue of **2**, as well as mitomycin C[34].

Radical anions of alkyl halides rapidly eliminate halide ion with formation of organic radicals[35]. α-Halonitro aliphatic antibacterial agents are likely to reduce in an analogous fashion[36]. However, 6-halomethyl 1,4-naphthoquinones do not lose halide before hydroquinone formation[34].

Compound **2** and its analogues are thought to act as bioreductive alkylating agents due to loss of bromide from hydroquinone with formation of a reactive quinone methide which presumably crosslinks biological molecules[37]. Our results (Table 1) indicate involvement of two electrons in the reduction process, c.f., the current function ratio. Subsequent chemical reaction was also indicated by the irreversible nature of the voltammogram. Alternatively, **2** may undergo SN2 displacement reactions with tissue sulfhydryl groups, thus disrupting sulfhydryl enzymes and/or influencing electron transport pathways[38]. With all of the possible mechanisms, it is unlikely that **2** would simply serve as a catalytic ET agent per se. It seems capable of exerting its action by several different routes.

2. Quinoline Derivatives (Iminiums)

The main features of the iminium anticancer theory as applied to this category involve protonation of **3** *in vivo* forming the cation **4**, usually conjugated. Reduction of **4** by an electron donor, such as protein, DNA, or an enzyme generates the radical **5** which can convey an electron to oxygen

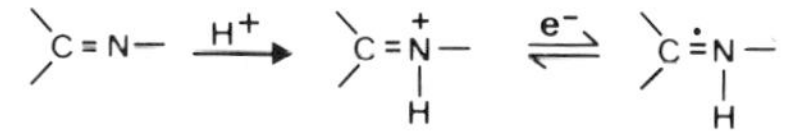

$$\gt C{=}N{-} \xrightarrow{H^+} \gt C{=}\overset{+}{N}(H){-} \underset{}{\overset{e^-}{\rightleftharpoons}} \gt C{\dot{=}}N(H){-}$$

3 4 5

completing the redox cycle. Superoxide produced by this process is converted to more reactive oxy radicals whose role in anticancer activity has received considerable attention[39-43]. Alternatively, the iminium ion may act by diverting electrons from vital electron transport systems.

In the current study cyclic voltammetry was performed on two quinoline-type anticancer drugs, Dup 785 **6** and camptothecin **9** to ascertain whether they possess the requisite characteristics of ET agents. Emphasis is placed on reduction potentials obtained for the quinolinium (iminium) forms. Several model compounds were examined in order to determine the effect of substituents. Reduction potentials were obtained in both protic and aprotic solvents to simulate environments which may pertain at the active site. In addition, $E_{pp/2}$ and ΔE_p values were calculated (Table 2). The relationship of electrochemical characteristics to drug activity is addressed.

A. Dup 785 and Model Compounds

The best results were generally obtained in aqueous ethanol as shown in Table 2. Reduction potentials for the free bases were too negative to be of significance for *in vivo* ET, ranging from -0.90 V for **7b** to

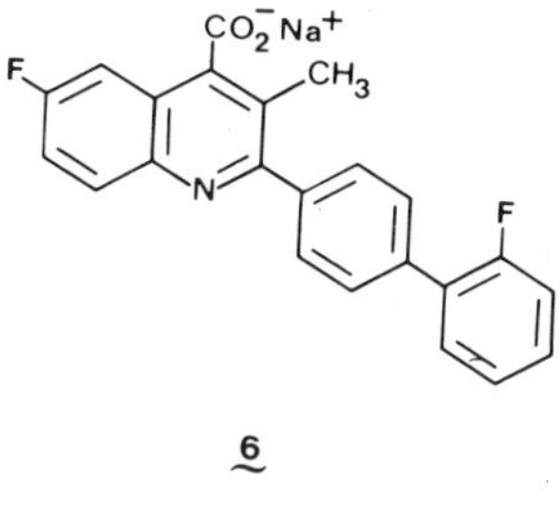

$CO_2^- Na^+$, CH_3, F, N, F

6

Table 2. Cyclic Voltammetry of Anticancer Quinoline Derivatives and Related Compounds[a]

Compound	$[OH^-]$mM	$[HClO_4]$mM	[HOAc]mM	$-E_p$ (V) DMSO	$-E_p$ (V) $EtOH_{aq}$	$-E_{p\,p/2}$ (mV) DMSO	$-E_{p\,p/2}$ (mV) $EtOH_{aq}$
6	-	-	-	1.84[b,c]	1.41	60[b]	70
	-	1.14	-	0.58[b,c]	0.73[d]	130[b]	80
	-	-	2.52	-	0.72[e]	-	60
7a	-	-	-	1.15	0.54[b]	90	70[b]
	0.50	-	-	>1.70[f]	1.23	60[b]	70
	0.50	1.14	-	0.50[b]	0.50[b]	70[b]	60[b]
	0.50	-	2.52	-	0.58[d]	-	100[f]
7b	-	-	-	1.22[b]	0.90	60[b]	50
	-	1.14	-	0.50[b]	0.45[b]	120[b]	40[b]
	-	-	2.52	1.10	0.55[b]	60	40[b]
7c	-	-	-	1.53	1.02	80	60
	-	1.14	-	0.60[c]	0.56	80	40
	-	-	2.52	1.40[c]	0.68	60	60
2-Phenyl-quinoline	-	-	-	1.86	1.34	80	120
	-	1.14	-	0.75	0.56	70	40
	-	-	2.52	1.71[c,g]	0.63	90	40
Quinoline	-	-	-	-	0.86[h]	-	-
9	-	-	-	1.22	-	60	-
	-	1.14	-	0.52	-	60	-

[a]0.5 mM substrate, 0.1 M tetraethylammonium perchlorate, 100 mV/s scan rate, HMDE vs NHE.
[b]Reversible, $E_{1/2}$ or ΔE_p value.
[c]Single trial.
[d]Extensive precipitation, adsorption onto electrode.
[e]Trace of precipitate, adsorption onto electrode during one trial.
[f]Extensive adsorption for anodic peak.
[g]Broad shoulder preceding peak, 20% height of peak.
[h]Aqueous pH 4 buffer, Reference 50.

-1,41 V for **6**. Protonation by either perchloric or acetic acid produced shifts to much more favorable values. With perchloric acid an irreversible reduction wave was obtained for **6** with E_p of -0.73 V. The weaker acetic acid, which more closely emulates *in vivo* acids, gave a similar result (E_p of -0.72 V).

An $E_{1/2}$ of -0.50 V was obtained for the reversible reduction of **7a** after protonation with perchloric acid. Acetic acid produced a quasi-reversible wave with $E_{1/2}$ equal to -0.58 V. Since **7a** exists in neutral solutions as zwitterion **8**[44] (E_p = -0.54 V), an equivalent of NaOH was first introduced to obtain the free base.

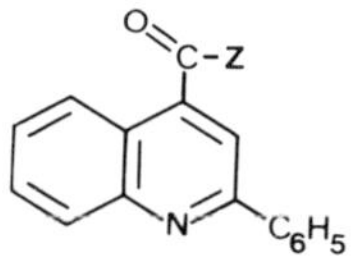

7 Z = a) OH b) OCH_3 c) NH_2

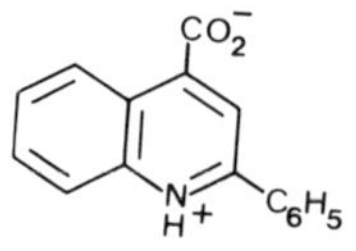

8

Addition of perchloric acid to the ester **7b** gave a reversible reduction wave with an $E_{1/2}$ of -0.45 V. The amide **7c** reduced irreversibly with perchloric acid (E_p = -0.56 V). An E_p of -0.68 V was obtained for the irreversible reduction with acetic acid. 2-Phenylquinoline also reduced irreversibly in perchloric and acetic acids resulting in E_p values of -0.56 V and -0.63 V, respectively.

Analogous results were obtained in DMSO with reduction potentials generally more negative than those found in aqueous ethanol. The free bases gave values ranging from -1.22 to -1.86 V, while addition of perchloric acid increased the potentials to a range between -0.50 to -0.75 V. However, **6** reduced more positively in this medium (-0.58 V) than in aqueous ethanol (-0.73 V).

In comparison of **6** with zwitterion **8**, also biologically active[45,46], both underwent substantial positive shifts in reduction potential on conversion from free base to protonated form. As quinolinium ions, **8** gave somewhat more positive values than **6**. Various substituent effects interact to determine the energetics of reduction for **6**. The 3-methyl group has an adverse inductive effect, whereas halides favorably affect reduction[30,31]. The 2-phenyl entity in **8** is known to be coplanar and conjugated with the quinoline nucleus[44], which stabilizes the resulting radical. Although the 3-methyl entity might be expected to inhibit coplanarity involving the biphenyl substituent, hindrance to rotation in the related N-methyl-3-methyl-2-(2'-halophenyl)quinolines is not prohibitive[47]. Resonance stabilization by the 2-aryl moiety has previously been proposed as a favorable feature in the mechanism of benzodiazepine action[5].

The effect of carboxyl derivatives on reduction can be seen by comparing **7a**, **7b** and **7c** with 2-phenylquinoline (Table 2). The 4-substituted compounds generally reduced at more positive potentials than the parent. Electron withdrawing substituents would be expected to favor reduction by enhancing the electron deficient character of the nucleus, an effect previously demonstrated with 4-substituted derivatives of 2-phenylquinoline[48]. The differences in reduction potentials for the derivatives compared to the parent were much greater for the free base than for the quinolinium forms. In aqueous ethanol, free base values ranged from -0.90 V for **7b** to -1.34 V for 2-phenylquinoline. In the presence of perchloric acid the spread decreased to 0.11 V for the two compounds. This narrow range stands in contrast to results from a related study involving substrates with iminium and carboxyl functionalities directly joined in an otherwise unconjugated system[49]. For the free bases, the negative charge on the anion of **7a** would be expected to have an adverse effect on reduction as observed, compared to **7b** or **7c** which are both neutral species. In the absence of acid, compound **7b** reduced more positively than **7c**, which can be rationalized by the greater electronegativity of ester oxygen and by the resonance effect of amide nitrogen. Similar results were observed for the protonated forms of **7b** and **7c**. However, **8** reduced more positively relative to **7b** and **7c** than expected on theoretical grounds; the zwitterion present under mildly acidic conditions is a neutral species. Positively charged **7b** and **7c** would be expected to have a much greater affinity for electrons. A rationale may lie in the differing degrees of protonation. A pKa value of 4.8 is reported for **8** (anion-zwitterion equilibrium) compared to 2.8 for **7b** (cation-neutral species equilibrium)[44].

Quinoline has previously been reported to reduce at -0.86 V (corrected to NHE) in an aqueous pH 4 buffer[50]. The more positive value for 2-phenylquinoline reflects the effect of conjugation.

The reduction potentials obtained for **6** under acidic conditions support the contention that electron transfer may be involved in its anticancer activity via the iminium form (*vide supra*). In a related study similar results were obtained for other quinolinium compounds with anticancer activity[51]. Protonation of imine nitrogen is a key aspect of our hypothetical approach. The favorble *in vivo* results obtained in acetic acid for **6** and its model compounds, and the existence of zwitterion **8**, indicate that even weak proton donors are capable of generating the requisite iminium species. Protonation *in vivo* would be facilitated if the drug is situated at the active site in proximity to a hydrogen ion donor. Generation of the salt form in a biological system has been proposed for the antimalarial quinine, also in the quinoline class[52]. Iminium formation would further be favored by the lower pH prevailing in tumor cells[53].

A number of biologically active compounds share basic structural features with **6**, and may also function by electron transfer via the iminium moiety. These similarities include the quinoline nucleus conjugated with an unsaturated, often aryl entity, at the 2-position, and a carboxyl group (or derivative) at the 4-position. Cinchophen **7a** and derivatives have demonstrated CNS[45,46,54], virucidal[55], and bactericidal[56,57] activity. Other agents bearing unsaturated 2-substituents fall in the fungistatic[58], tuberculostatic[58], schistosomicidal[59], CNS[60], bactericidal[60] protozoacidal[60], or antitumor class[61,62]. Quinoline derivatives with no 4-carboxy group, but with conjugation through the 2-position, such as camptothecin and pyrvinium cation[64], possess anticancer and anthelmintic activity, respectively.

As an alternative mode of action, the anticancer activity of **6** may result from inhibition of a key enzyme in *de novo* pyrimidine biosynthesis, depleting nucleotide and presumably deoxynucleotide pools[65]. Tumor death would result from inability to synthesize RNA and/or DNA. It is noteworthy that at least two of the three general classes of compounds known to inhibit this enzyme function by disruption of electron transport. Dup 785 reportedly is being studied to determine possible interference with electron transport[65].

B. Camptothecin

Camptothecin **9**, insoluble in aqueous ethanol, reduced irreversibly in DMSO with an E_p of -1.22 V (Table 2). Addition of perchloric acid gave an irreversible wave (E_p = -0.52 V). A complicated set of waves was obtained with acetic acid. Reduction of the quinolinium ion would be favored by conjugation extending through two additional fused rings into the pyridone carbonyl group.

Electon transfer has been associated with the antineoplastic activity of camptothecin in several studies[39,66-68]. Thus, photoexcitation of the drug generated activated oxygen species linked to DNA

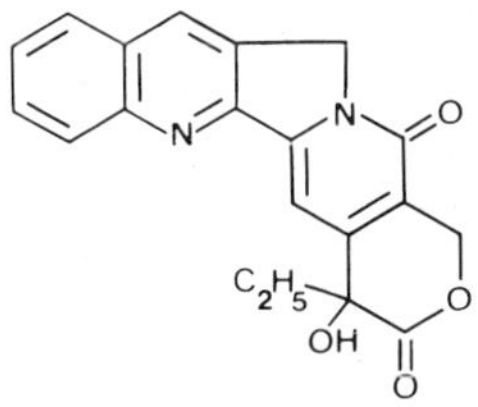

9

scission[67,69]. It is our contention that the iminium moiety is a conduit for ET. Various portions of the molecule are designated as primarily involved in electron uptake depending upon the pH. The quinoline nucleus was proposed in a mechanism involving irreversible reduction with formation of bioalkylating agent[66]. A later study, performed with sodium salt generated by hydrolysis of the α-hydroxylactone moiety, reported a reduction potential of -0.71 V at 37°C in neutral solution. The authors designated the pyridone ring as the electroactive functionality[68]. Under the mildly acidic conditions employed in the current study, the reduction potential of -0.52 V is presumably a result of reaction at the quinolinium site.

Three regions of camptothecin have been associated with activity, namely, the conjugated planar fused ring system, the α-hydroxylactone group and substituents on the homocyclic quinoline ring[63]. The fused ring portion has been linked with inhibition of RNA syntheses[70], and possibly weak intercalation into DNA[67]. The α-hydroxylactone moiety appears to be involved in oxygen dependent DNA strand scission, following irradiation, via a diradical species that generates oxy radicals under aerobic conditions and directly degrades DNA anaerobicaly[67]. Another postulate (*vide supra*) employs formation of two reactive sites for alkylation of nucleic acid bases[66]. Alkylation of guanosine nucleotides presumably generates oxy radicals through ET by the iminium salt[71]. A preference for the cytosine-guanine regions of DNA has been noted for camptothecin[67]. Substituents on the homocyclic quinoline ring can drastically affect activity depending on type, location, and number, apparently as a result of steric or electronic factors[63]. A recent proposal involves inhibition of a topoisomerase by complexation and production of DNA breaks[72]. Unlike the action of other topoisomerase inhibitors, not all DNA breaks were protein concealed, leading the authors to suggest that more than one mechanism may pertain.

Although Dup 785 and camptothecin reduced irreversibly under several sets of conditions in the current study, behavior *in vivo* may be different for the drug immobilized at the active site, and the energetics may be more favorable. These aspects are treated in more detail elsewhere[24,30,73].

CONCLUSIONS

Further insights into the electron transfer mechanism of action proposed for various types of antitumor agents is gained by examining the redox characteristics of several quinones (substituted 1,4-naphthoquinones) and quinolinium (iminium) ions from Dup 785 and camptothecin. Reduction potentials ranged from 0.0 to -0.73 V. Electrochemical data are related to structure and physiological activity. For example, in the quinone category the influence of hydrogen bonding and ability to lose halide are discussed. Both Dup 785 and camptothecin exhibited appreciable increases in reduction potentials at higher acidities. The values for the iminium species are of the same magnitude as for known ET agents. Model compounds show that certain substituents in the quinoline ring favor reduction. Many substances similar to Dup 785 possess biological activity. Electron transfer and oxy radical generation have been associated with the action of camptothecin. The results for the quinones and the quinoline drugs are in accord with our theoretical approach involving ET.

ACKNOWLEDGMENT

We are grateful to Drs. W.K. Musker, M. Wall, and J.G. Whitney for generously providing several of the antineoplastic agents. Stephen Bejvan is also acknowledged for helpful assistance.

REFERENCES

1. P. Kovacic, J.R. Ames and M.D. Ryan, these Proceedings (and references therein).
2. B. Hazra, P. Sur, D.K. Roy, B. Sur and A. Banerjee, Planta Med., 51, 295 (1984) (and references therein).
3. B. Hazra, A. Banerjee and D.K. Roy, IRCS Med. Sci., 14, 35 (1986).
4. I.M. Roushdi and A.I. El-Sebai, J. Pharm. Sci. U. Arab. Rep., 2, 109 (1961); Chem. Abstr., 58, 13912 (1963).
5. P.W. Crawford, P. Kovacic, N.W. Gilman and M.D. Ryan, Bioelectrochem. Bioenerg., 16, 407 (1986).
6. R.D. Irons and T. Sawahata, in: "Bioactivation of Foreign Compounds", M.W. Anders, Ed., Academic Press, New York (1985), Chapter 9.
7. M.D. Ryan and D.H. Evans, J. Electroanal. Chem., 67, 333 (1976).
8. A.J. Bard and L.R. Faulkner, in: "Electrochemical Methods, Fundamentals and Applications", Wiley, New York (1980), Chapter 6.
9. A. Ashnagar, J.M. Bruce, P.L. Dutton and R.C. Prince, Biochim. Biophys. Acta, 801, 351 (1984).
10. Y. Take, M. Sawada, H. Kunai, Y. Inouye and S. Nakamura, J. Antibiotics, 39, 557 (1986).
11. G. Klopman and N. Doddapaneni, J. Phys. Chem., 78, 1820 (1974).
12. T.R. Sweeney and R.E. Strube, in: "Burger's Medicinal Chemistry", Part 2, 4th ed., M.E. Wolff, Ed., Wiley, New York (1979), p. 383.
13. R. Docampo and S.N.J. Moreno, in: "Free Radicals in Biology", Vol. VI, W.A. Pryor, Ed., Academic Press, Orlando, FL (1984), pp. 257-263.
14. Y.C. Martin, T.M. Bustard and K.R. Lynn, J. Med. Chem., 16, 1089 (1973).
15. B. Hazra, A. Banerjee and D.K. Roy, IRCS Med. Sci., 14, 593 (1986); Chem. Abstr., 105, 75768 (1986).
16. J.R. Ames, M.D. Ryan and P. Kovacic, Life Sci., 41, 1895 (1987).

17. O.H.W. Decker and H.W. Moore, J. Org. Chem., 52, 1174 (1987).

18. T. Matsushima, M. Muramatsu, O. Yagame, A. Araki, L. Tikkanen and S. Natori, Prog. Clin. Biol. Res., 209B, 133 (1986); Chem. Abstr., 105, 147907 (1986).

19. J.W. Lown, Adv. Free Rad. Biol. Med., 1, 225 (1985).

20. G.R. Buettner and L.W. Oberley, in: "Oxygen and Oxy-Radicals in Chemistry and Biology", M.A.J. Rodgers and E.L. Powers, Eds., Academic Press, New York (1981), p. 607.

21. B. Halliwell and J.M.C. Gutteridge, in: Free Radicals in Biology and Medicine, Clarendon Press, Oxford (1985), pp. 306-313.

22. N.R. Bachur, M.V. Gee and R.D. Friedman, Cancer Res., 42, 1078 (1982).

23. M.D. Ryan, R.G. Scamehorn and P. Kovacic, J. Pharm. Sci., 74, 492 (1985).

24. P.W. Crawford, R.G. Scamehorn, U. Hollstein, M.D. Ryan and P. Kovacic, Chem.-Biol. Interactions, 60, 67 (1986).

25. L. Gianni, J.L. Zweier, A. Levy and C.E. Meyers, J. Biol. Chem., 260, 6820 (1985).

26. B.A. Kulkarni, V.D. Kelkar and P.L. Kulkarni, Indian J. Pharm. Sci., 45, 21 (1983); Chem. Abstr., 99, 50185 (1983).

27. M.S. Berger, R.E. Talcott, M.L. Rosenblum, M. Silva, F. Ali Osman and M.T. Smith, J. Toxicol. Environ. Health, 16, 713 (1985).

28. A.J. Lin and A.C. Sartorelli, Biochem. Pharmacol., 25, 206 (1976).

29. R.D. Rieke, W.E. Rich and T.H. Ridgway, J. Am. Chem. Soc., 93, 1962 (1971).

30. J.R. Ames, K.T. Potts, M.D. Ryan and P. Kovacic, Life Sci., 39, 1085 (1986).

31. P. Zuman, "Substituent Effects in Organic Polarography", Plenum Press, New York (1967), pp. 76-77.

32. L.A. Cosby, R.S. Pardini, R.E. Biagini, T.L. Lambert, A.J. Lin, Y.-M. Huang, K. M. Hwang and A.C. Sartorelli, Cancer Res., 36, 4023 (1976).

33. R.E. Biagini, R.S. Pardini, A.J. Lin and A.C. Sartorelli, Cancer Biochem. Biophys., 3, 129 (1979); Chem. Abstr., 91, 83335 (1979).

34. I. Wilson, P. Wardman, T.-S. Lin and A.C. Sartorelli, J. Med. Chem., 29, 1381 (1986) (and references therein).

35. E. Canadell, P. Karafiloglou and L. Salem, J. Am. Chem. Soc., 102, 855 (1980).

36. J.R. Ames, M.D. Ryan and P. Kovacic, J. Free Rad. Biol. Med., 2, 377 (1986).

37. T.-S. Lin, I. Antonini, L.A. Cosby and A.C. Sartorelli, J. Med. Chem., 27, 813 (1984).

38. R.E. Talcott, A. Ketterman and D.G. Giannini, Biochem. Pharmacol., 33, 2663 (1984).

39. L.W. Oberley, in: "Superoxide Dismutase", Vol. II, L.W. Oberley, Ed., CRC Press, Boca Raton, FL, (1982), Chapter 6.

40. P. Kovacic, J.R. Ames, P. Lumme, H. Elo, O. Cox, H. Jackson, L.A. Rivera, L. Ramirez and M.D. Ryan, Anti-Cancer Drug Design, 1, 197 (1986).

41. L.W. Oberley, S.W.C. Leuthauser, R.F. Pasternack, T.D. Oberley, L. Schutt and J.R.J. Sorenson, Agents and Actions, 15, 535 (1984).

42. L.W. Oberley and G.R. Buettner, Cancer Res., 39, 1141 (1979).

43. P. Kovacic, Age, 6, 144 (1983).

44. B. Zalis, A.C. Capomacchia, D. Jackman and S.G. Schulman, Talanta, 20, 33 (1973).

45. S.G. Vainstein and V.T. Tyurikov, Deposited Doc. VINITI 1839 (1976); Chem. Abstr., 88, 187500 (1978).

46. M. Boffill Cardenas and R. Marrero Perez, Rev. Cubana Farm., 17, 1 (1983); Chem. Abstr., 100, 694 (1984).

47. C.E. Kaslow and H. Moe, J. Org. Chem., 25, 1512 (1960).

48. R. Andruzzi, A. Trazza, L. Greci and L. Marchetti, J. Electroanal. Chem., 108, 49 (1980).

49. P. Kovacic, J. Jawdosiuk, J.R. Ames and M.D. Ryan, Bioorg. Chem., 15, 423 (1987).

50. R.E. Cover and J.T. Folliard, J. Electroanal. Chem., 30, 143 (1971).

51. P.W. Crawford, W.O. Foye, M.D. Ryan and P. Kovacic, J. Pharm. Sci., 76, 481 (1987).

52. J.R. Ames, M.D. Ryan, D.L. Klayman and P. Kovacic, J. Free Rad. Biol. Med., 1, 353 (1985) (and references therein).

53. W.C.J. Ross, Biochem. Pharmacol., 8, 235 (1961).

54. M.C. Dubroeucq, G. Le Fur and C. Renault, Eur. Pat. Appl. EP 112, 776 (1984); Chem. Abstr., 101, 191716 (1984).

55. R.S. Belen'kaya, E.I. Boreko, M.N. Zemtsova, M.I. Kalinina, M.M. Timofeeva, P.L. Trakhtenberg, V.M. Chelnov, A.E. Lipkin and V.I. Votyakov, Khim.-Farm. Zh., 15, 29 (1981); Chem. Abstr., 95, 80680 (1981).

56. J.M. Maheshvari and K.A. Thaker, J. Inst. Chem. (India), 55, 189 (1983); Chem. Abstr., 100, 156473 (1984).

57. S.M. El-Khawass and N.S. Habib, Sci. Pharm., 46, 49 (1978); Chem. Abstr., 89, 59827 (1978).

58. N.M. Sukhova, I. Sprunka, M. Lidaka and A. Zidermane, Khim.-Farm. Zh., 16, 169 (1982); Chem. Abstr., 96, 199484 (1982).

59. B. Chen, L. Zhang and M. Xu, Yiyao Gongye, 16, 66 and 91 (1985); Chem. Abstr.,103, 87754 (1985).

60. F. Savelli, F. Sparatore and G. Cordella, Chim. Ind. (Milan), 59, 300 (1977); Chem. Abstr., 87, 152101 (1977).

61. N.M. Sukhova, M. Lidaks, A. Zidermane, V.A. Voronova, V. Kauss and I. Katlaps, Vses. Nauchn. Konf. Khim. Tekhnol. Furanovykh Soedin., [Tezisy Dokl.], 3rd, Y.P. Stradyn, Ed., Riga, USSR (1978), pp. 97-98; Chem. Abstr., 92, 157718 (1980).

62. A. Zidermane, I. Sablina, S. Hillers, M. Lidaks, N.M. Sukhova and A.V. Eremeev, Tezisy Dokl.-Vses. Konf. Kimioter. Zlokach. Opukholei, 2nd, V.I. Astrakhan, Ed., Akad. Med. Nauk. SSSR, Moscow, USSR (1974), pp. 73-74; Chem. Abstr., 86, 183114 (1977).

63. M.C. Wani, A.W. Nicholas and M.E. Wall, J. Med. Chem., 29, 2358 (1986).

64. P. Dickie, A.R. Morgan, D.G. Scraba and R.C. von Borstel, Mol. Pharmacol., 29, 427 (1986).

65. S.-F. Chen, R.L. Ruben and D.L. Dexter, Cancer Res., 46, 5014 (1986).

66. H.W. Moore, Science, 197, 527 (1977) (see ref. 68).

67. J.W. Lown and H.-H. Chen, Biochem. Pharmacol., 29, 905 (1980).

68. J.W. Lown, H.-H. Chen and J.A. Plambeck, Chem.-Biol. Interactions, 35, 55 (1981).

69. J. Kuwahara, T. Suzuki, K. Funakoshi and Y. Sugiura, Biochemistry, 25, 1216 (1986).

70. M.E. Wall and M.C. Wani, Natural Products and Drug Development, Alfred Benzon Symposium 20, P. Krogsgaard-Larson, S.B. Christensen and H. Kofod, Eds., Munksgaard, Copenhagen (1984), p. 253.

71. P. Kovacic, P.W. Crawford, M.D. Ryan and V.C. Nelson, Bioelectrochem. Bioenerg., 15, 305 (1986).

72. M.R. Mattern, S.-M. Mong, H.F. Bartus, C.K. Mirabelli, S.T. Crooke and R.K. Johnson, Cancer Res., 47, 1793 (1987).

73. J.R. Ames, S. Brandänge, B. Rodriguez, N. Castagnoli, Jr., M.D. Ryan and P. Kovacic, Bioorg. Chem., 14, 228 (1986) (and references therein).

OXIDATION PATHWAYS OF PURINE DRUGS: ELECTROCHEMICAL, LIQUID CHROMATOGRAPHIC AND MASS SPECTROMETRIC INSIGHTS

K.J. Volk, T. Peterson, K. McKenna
P.J. Kraske and A. Brajter-Toth*

Department of Chemistry
University of Florida
Gainesville, Florida 32611

INTRODUCTION

Purine drugs, especially thiopurines, are potent antineoplastic agents[1]. 6-Thiopurine is also an effective immunosuppressant[2]. Purine drugs are used extensively in treatment of various leukemias. Their activity and toxicity are affected by metabolic redox processes which, despite extensive biochemical studies, are still only partially understood.

Electrochemical methods are uniquely suited to provide insight into redox and related chemical reactivity of compounds of biological interest. In electrochemical studies of enzymatic redox reactions, an electrode poised at a suitable potential is used to simulate the redox enzyme. Although the similarity is superficial, the value of such studies has been verified by the unique insights into redox reactivity of compounds of biological interest[3,4]. These include purines[3,4] and, recently, the drugs mitomycin[5] and acetaminophen[6].

Biological (enzymatic) reactivity may involve simple outer-sphere electron transfer which, as has been demonstrated, can be simulated electrochemically. Evidence for this has been based on the identical intermediates and products in the electrochemical and peroxidase-catalyzed oxidation of uric acid[3]. This is in agreement with reports of the outer-sphere nature of the peroxidase-catalyzed reaction[7]. In cases where simple outer-sphere processes are not involved, electrochemical studies provide information which can also form the basis for better understanding of enzymatic redox reactivity.

Redox reactivity of purine drugs, although of considerable biological importance, has not been systematically studied. Purine drugs are obtained as simple derivatives of naturally-occurring purines. Our initial objective was to determine pathways of redox reactions of purine drugs. Effects of structural changes on redox reactivity were then evaluated by comparing reactions of drugs with those of naturally occurring purines, which have been studied extensively[4]. Insights into enzymatic redox activity were also obtained.

The information power of electrochemistry can be expanded by coupling it with methods which can determine the chemical identity of intermediates and products of electrode reactions. *In-situ* information can be obtained by on-line spectroscopic methods such as ultraviolet/visible (UV/vis) thin-layer spectroelectrochemistry[8]. Recently, effective on-line coupling of an electrochemical cell with a mass spectrometer has been demonstrated[9] and its application to the study of biological redox reactions has been described[10]. Electrochemistry/mass spectrometry as well as off-line methods were used in determining redox and related chemical reactivity of purine drugs.

EXPERIMENTAL

Chemicals

2,6-Diaminopurine, 6-thioguanine, 6-thioxanthine and tubercidin-5'-monophosphate were obtained from Sigma Chemical Co. and were used without purification. 2,6-Diamino-8-purinol was obtained from Aldrich. For electrochemical measurements, unless otherwise noted, solutions were prepared in 0.5 M phosphate buffers which were prepared from reagent grade chemicals. In electrochemistry thermospray mass spectrometry the mobile phase was 0.1 M ammonium acetate (pH ~ 7.0) which was prepared from reagent grade chemicals.

Apparatus

Electrochemical measurements were carried out using a Princeton Applied Research (PAR) Model 173 potentiostat, a PAR Model 175 function generator and a PAR Model 179 digital coulometer. A Tracor Northern diode array spectrophotometer was used in UV/vis spectrometric studies. The absorbance was measured in a 1.0 cm quartz cell.

A three-electrode system was used in all electrochemical measurements. The electrochemical cell contained the saturated calomel electrode (SCE), a platinum counter electrode and a graphite working electrode. In all electrochemical measurements, except coulometry, the working electrodes were either rough pyrolytic graphite (RPG) (Pfizer, Easton, PA) of ca. $0.050 cm^2$ or glassy carbon (GC) (High Performance) of ca. $0.84 cm^2$. The RP and GC electrodes were prepared by sealing a piece of carbon into glass or Teflon, respectively, with epoxy. The RPGE was resurfaced before each experiment on 600-grit silicon carbide (SiC) paper (Buehler) using a metallographic polishing wheel, then washed with a spray of doubly distilled water and gently dried. The GCE was polished with Gamal gamma alumina (Fisher Scientific) and copiously washed before use.

For coulometric measurements a three-compartment electrochemical cell was used in which the compartments were separated by cation-exchange membranes (RAI Research, Hauppauge, NY). Working electrodes were large plates (6.3 x 1.8 cm) of pyrolytic graphite (Pfizer, Minerals Division, Easton, PA) which were cleaned by polishing on 600-grit SiC paper followed by copious washing. All solutions were deaerated with N_2 for ca. 10 min before experiments. The solutions were deaerated with N_2 throughout electrolysis and the solutions were continuously stirred. All measurements were run at 25°C.

In thin-layer spectroelectrochemical cells reticulated vitreous carbon with 60 pores cm^{-2} was used as the working electrode. The cell design has been described[11].

Liquid Chromatography Separations

The products were isolated and separated from buffer components, which interfere with analysis, by gel-permeation liquid chromatography on a G-10 (MW cut-off 700) Sephadex (Pharmacia) columns (40 x 3.0 cm). Water was used as the eluting solvent at a flow rate of ca. 0.10 mL min^{-1}. The separations were monitored by UV spectroscopy at 210 nm. The separated products were collected with a Buehler LC 200 fraction collector; 3.5 mL fractions were collected. The separated samples were freeze-dried using a Labconco-8 Freeze Drier. For high performance liquid chromatography (HPLC) analysis, an Altex 110A pump and solvent programmer were coupled to a Kratos spectroflow UV detector. An HPLC C-18 4.6 mm x 25 cm column (Resolvex, Fisher Scientific) was used, and the sample volume was 20 µL. The separations were isocratic with 0.02 M phosphate buffer (pH adjusted with NaOH or HCl) as the mobile phase. The mobile phase was filtered through a 0.45 µm filter (Rainin Co.) before use. The flow rate was 1.0 mL min^{-1}, and the separations were monitored at 220, 240, 265 and 305 nm.

Off-Line Mass Spectrometric Product Identification Procedures

The intermediates and products were identified by gas chromatography/mass spectrometry (GC/MS) using a Finnigan Model 4021 System. A DB5 (0.32 mm i.d. x 15 m) capillary column (J & W) was used for the separations. The dry samples were derivatized before the GC/MS analysis. Typically 70 µL of silylating reagent, N,O-bis(trimethylsilyl)trifluoroacetamide (BSTFA) and 70 µL of silylation grade acetonitrile (Supelco) were added to approximately 0.1 mg of product in a 3 mL reaction vial. The silylation was carried out at ca. 120°C for 25-30 min. Five µL samples were used for GC/MS analysis. After sample injection the column temperature was maintained at 100°C for 12 min and then the temperature was increased at a linear rate of 6°C min^{-1} to 300°C. Mass spectra of the separated samples were not recorded until after 12 min. In EI, a 70 eV beam was used. For CI, methane was used as the reagent gas. The molar masses of the underivatized compounds were verified using the program SILYL[12].

Fast atom bombardment mass spectra (FAB MS) were obtained with a Kratos MS-50 mass spectrometer equipped with an RF magnet and a DS-55 data system. Fast xenon atoms of 8 keV were used and the spectra were obtained in a thioglycerol matrix.

On-Line Electrochemistry Thermospray Tandem Mass Spectrometry (EC/TSP/MS/MS).

The results were obtained with an ESA coulometric cell which was coupled via a Vestec thermospray interface to a Finnigan MAT TSQ 4500 triple quadrupole mass spectrometer with an INCOS data system. The coulometric cell had a volume of 5 µL, a 12 cm^2 reticulated vitreous carbon working electrode, and palladium counter and reference electrodes. The thermospray probe tip and source block temperatures were 250 and 290°C, respectively. In EC/TSP/MS/MS analysis, the mobile phase was 0.1 M ammonium acetate (pH ~ 7), at a flow rate of 2.0 mL min^{-1}. Twenty five µL of a 300 µM solution were injected.

The conditions for TSP/MS were: scan range m/z 125 to 300, electron multiplier voltage 1000 V, and preamplifier gain 10^8 VA^{-1}. For MS/MS, the scan range was m/z 15 to 300, electron multiplier voltage 1600 V, preamplifier gain 10^8 VA^{-1}, collision energy 25 eV, and N_2 collision gas pressure 1.8 mTorr.

Other Methods

The photochemical experiments were carried out in a 1 cm quartz cell using a deuterium lamp. The spectral changes were monitored with a Tracor Northern diode array spectrophotometer.

FTIR spectra were obtained with a Nicolet 5-DX spectrometer. The samples were prepared in Nujol. ^1H NMR spectra were obtained with a Nicolet NT-300 instrument. Typically 0.5 to 1 mg samples were dissolved in 100 µL of d_6-DMSO.

RESULTS AND DISCUSSION

Oxidation Pathways of 2,6-Diaminopurine and 2,6-Diamino-8-Purinol

In vivo and *in vitro* studies have shown that oxidations play a major role in purine drug metabolism[1]. Electrochemical oxidation pathways of purine drugs were, therefore, evaluated.

Oxidation pathways of the purine antimetabolite 2,6-diaminopurine and its metabolite 2,6-diamino-8-purinol have recently been reported[13,14]. Electrochemistry was used to determine the number of electrons and protons involved in the oxidation as well as the sequence, rate of formation, and stability of the reaction intermediates and products. Stability and rate of formation of intermediates and products were confirmed by UV/vis thin-layer spectroelectrochemistry. The products and stable intermediates were isolated at different stages in the oxidation and were identified by gas chromatography/mass spectrometry and by FAB MS.

The overall electrochemical oxidation pathway of 2,6-diaminopurine (I) and 2,6-diamino-8-purinol (II) which was proposed based on the results of these studies is shown in Figure 1. 2,6-Diamino-8-purinol was identified as an intermediate in the oxidation of 2,6-diaminopurine. 2,6-Diamino-8-purinol also forms in the xanthine oxidase catalyzed reaction of the drug[7,13]. There is clearly a chemical similarity between the electrochemical and xanthine oxidase catalyzed reaction.

In the electrochemical oxidation, 2,6-diamino-8-purinol (II) is further oxidized in a 2e, $2H^+$ reaction to an unstable product which was proposed to be a diimine (III). This product has not been identified directly, but its formation is strongly supported by the identification[14] of a diol (VI, Fig. 1) which must form through hydrolysis of the diimine intermediate (III → VI, Fig. 1). The diol decomposes to the final product which was identified as the bicyclic carboxylic acid (IX, Fig. 1B). In the identification, separations were used to isolate intermediates and products, and their chemical identity was confirmed by gas chromatography/mass spectrometry. Volatile derivatives were formed before GC/MS analysis. The off-line measurements have drawbacks associated with necessary manipulation of samples[13,14].

As shown in Figure 1, intermediates and products in the oxidation of 2,6-diaminopurine and 2,6-diamino-8-purinol and the intermediates and products identified in the oxidation of naturally occurring purines such as uric acid[3] are similar. This is in agreement with the results of *in vitro* and *in vivo* studies which show that 2,6-diaminopurine can enter metabolic pathways of naturally occurring purines[1,15,16]. However, the products and their distribution are not exactly the same, which may reflect differences in reactivity.

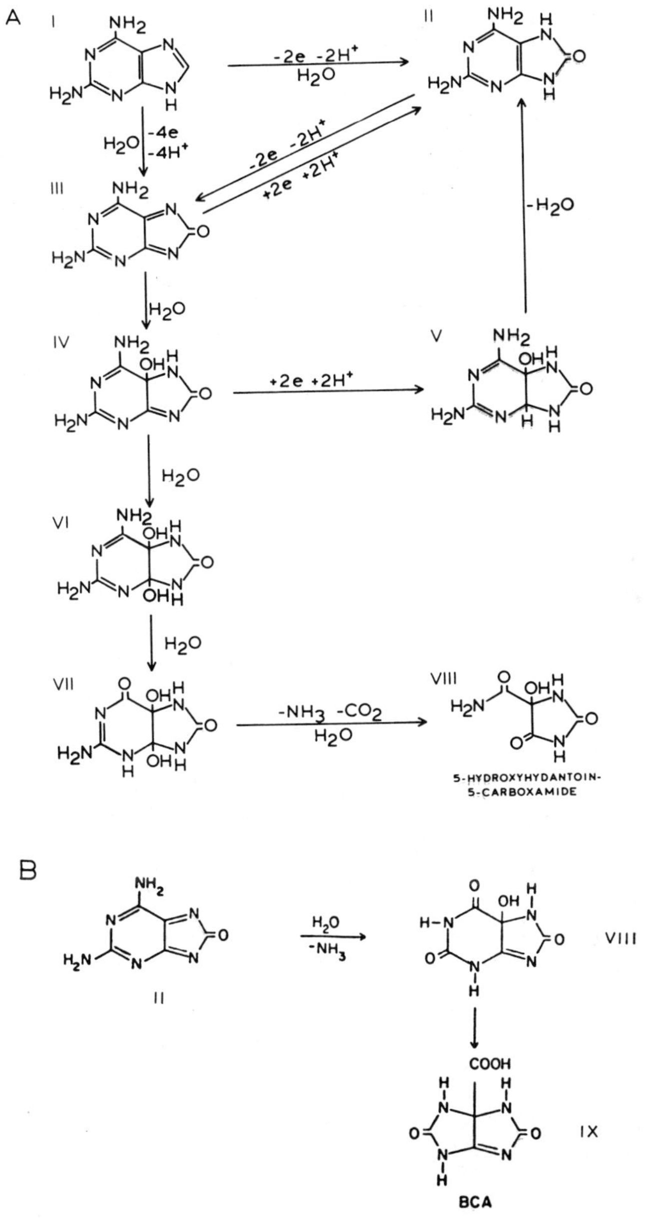

Figure 1. Pathway proposed for the electrochemical oxidation of 2,6-diaminopurine and 2,6-diamino-8-purinol in 0.5 M phosphate buffer pH 7.0 at a pyrolytic graphite electrode. Reactions I → V account for the cyclic voltametric behavior. Reactions IV → VIII (A) and II → IX (B) explain formation of the final products which were identified following exhaustive oxidation (by permission, D. Ashwood, C.N. D'Amico, T. Lippincott and A. Brajter-Toth, J. Electroanal. Chem., 198, 283 (1986)).

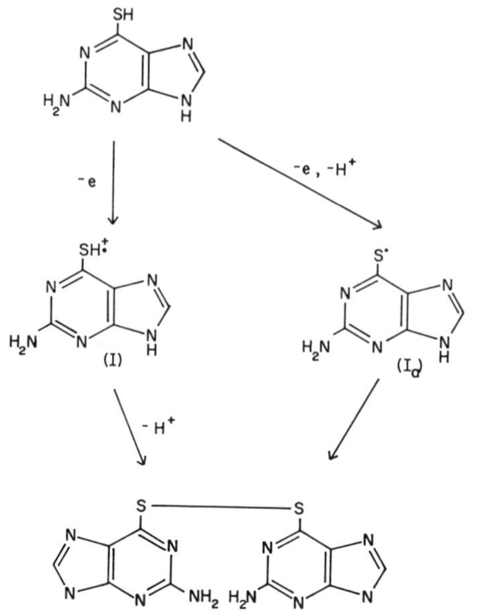

Figure 2. Proposed 6-thioguanine oxidation pathway at moderate potentials in 0.5 M phosphate buffer, pH 7.0, at pyrolytic graphite electrode (by permission, P.J. Kraske and A. Brajter-Toth, J. Electroanal. Chem., 207, 101 (1986).

Electrochemical Oxidation of 6-Thioguanine and 6-Thioxanthine

Overall electrochemical oxidation pathways of 6-thioguanine, an active drug[1], and 6-thioxanthine, an inactive metabolite[1], were evaluated by combination of electrochemistry, mass spectrometry and UV/vis spectroscopy[17,18]. The thio group of both compounds can be oxidized at moderate potentials. From cyclic voltammetry, coulometry, UV/vis spectroscopy recorded during and after electrochemical oxidation, and mass spectrometry of the isolated products, it was determined that the immediate product of 6-thioguanine oxidation must be a radical which forms a disulfide in a rapid following dimerization reaction. The proposed oxidation pathway[17] is shown in Figure 2.

It was only possible to obtain indirect electrochemical evidence for this pathway in the oxidation of 6-thioxanthine because of the apparent instability of the postulated dimeric oxidation product. The pathway which was proposed, which is also consistent with the known chemistry of 6-thioxanthine, is shown in Figure 3[18]. At potentials where one electron oxida-

$$2\,RS^- \xrightarrow{-2e} RS-SR$$

$$2\,RS-SR + 4\,OH^- \longrightarrow RSO_2^- + 3\,RS^- + 2\,H_2O$$

R = xanthine

Figure 3. Proposed oxidation pathway of 6-thioxanthine in 0.5 M phosphate buffer, pH 7.0, at rough pyrolytic graphite electrode. The first step in the pathway accounts for the cyclic voltammetric results. The overall pathway explains results of the exhaustive oxidation (by permission, K. McKenna and A. Brajter-Toth, J. Electroanal. Chem., 233, 49 (1987).

tion of the thio group was postulated, xanthine was the only product identified at the end of exhaustive oxidation[18].

Coulometry, UV/vis spectroscopy and product analysis show that during exhaustive oxidation at more positive potentials further oxidation of the thio group of 6-thioguanine and 6-thioxanthine must occur. There is also evidence for the oxidation of the purine ring. Hydrolysis reactions must follow electrochemical oxidation to account for the formation of the final products which were identified by mass spectrometry. In the oxidation of 6-thioguanine, 5-guanidinohydantoin (VII, Fig. 4) was the final product[17], the same product which has been identified in the electrochemical oxidation of guanine[19]. In the oxidation of 6-thioxanthine, allantoin (VII, Fig. 4), the final product identified in the electrochemical oxidation of xanthine[4], was identified as the final product[18]. The proposed pathways are shown in Figure 4. In the oxidation of 6-thio-xanthine, formation of the proposed sulfonic acid intermediate (II) was supported by the identification of sulfite[18].

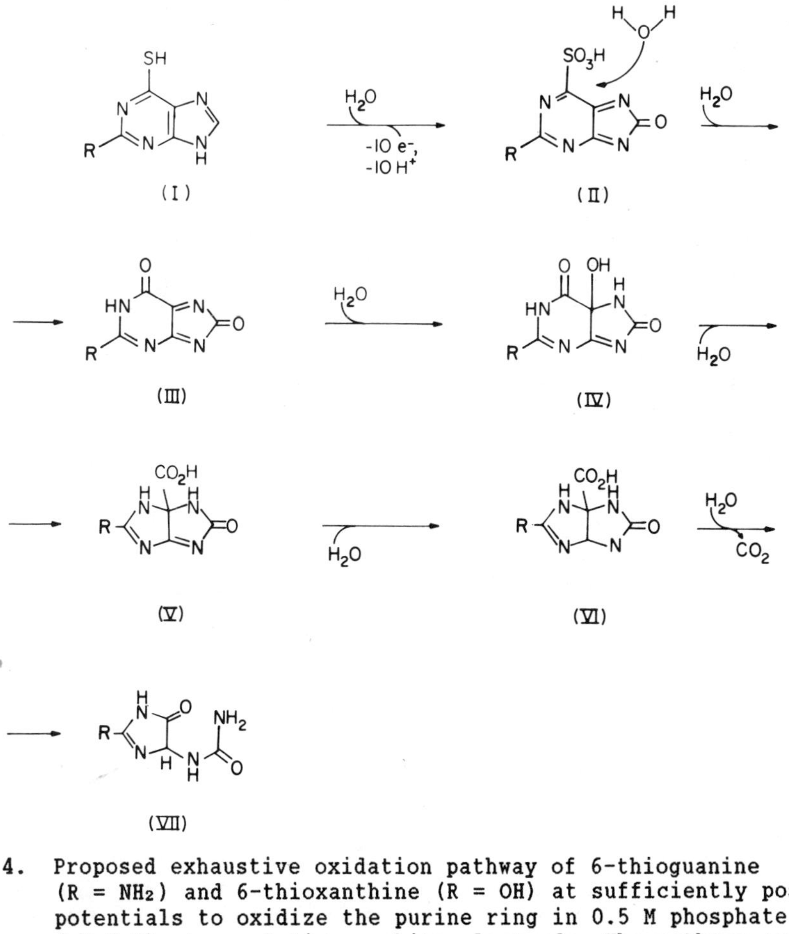

Figure 4. Proposed exhaustive oxidation pathway of 6-thioguanine ($R = NH_2$) and 6-thioxanthine ($R = OH$) at sufficiently positive potentials to oxidize the purine ring in 0.5 M phosphate buffer (pH 7.0) at pyrolytic graphite electrode. The pathway explains formation of the identified product (VII): 5-guanidino-hydantoin ($R = NH_2$); allantoin ($R = OH$).

It is of interest that the electrochemical study shows different reactivity of the apparent, easily formed, thio group oxidation products of compounds of different biological reactivity. High reactivity of the thio group towards one electron oxidation and the apparent ease of formation of disulfides are also in agreement with recent reports of biological reactivity of thiopurines[20].

Tubercidin-5'-Monophosphate

The electrochemical oxidation pathway of tubercidin-5'-monophosphate (TMP) is of interest because TMP, a structural analog of adenosine, is not degraded by enzymes of purine metabolism[21]. It was, therefore, anticipated that the redox rectivity of this compound should reflect its deazapurine structure and should provide considerable insight into structure-reactivity relations of purine derivatives.

Electrochemical oxidation of TMP is not facile and occurs[22] at pH 7.0 at ca. 0.8 V at a rough pyrolytic graphite electrode at similar potential to that of adenine[4]. The electrochemical oxidation is irreversible. The insight into the chemistry of the oxidation was, therefore, obtained during exhaustive oxidation and through product analysis. The apparent electrochemical reactivity of the oxidation products of TMP, which form after chemical reactions following electron transfer, leads to a complex oxidation pathway. This was first indicated by the extreme length of electrolysis and was confirmed by the UV/vis spectral changes which occurred long after TMP was exhaustively oxidized. The complicated nature of the oxidation was verified through HPLC assay[23]. The assay was devised to determine complete oxidation of TMP, since the interference by electroactive and UV/vis absorbing products made cyclic voltammetry and UV/vis spectroscopy unsuitable for this purpose[23]. The sequence of formation of the separated products was also followed with HPLC. Five products were clearly separated, and three of them were identified. Following separation by gel-permeation liquid chromatography, the products were identified by FTIR, NMR and mass spectrometry. The pathway in Figure 5 shows the

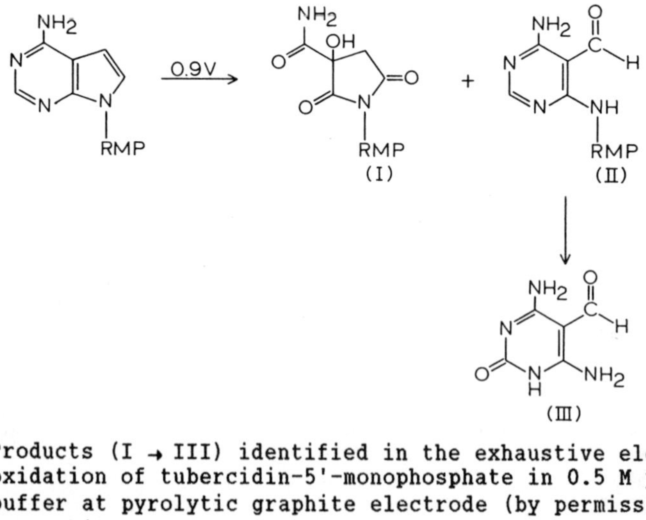

Figure 5. Products (I → III) identified in the exhaustive electrochemical oxidation of tubercidin-5'-monophosphate in 0.5 M phosphate buffer at pyrolytic graphite electrode (by permission, T.E. Childers-Peterson and A. Brajter-Toth, J. Electroanal. Chem., 239, 161 (1988).

proposed structures of the isolated products. Oxidation products of the deazapurine TMP (Fig. 5) are considerably different from those formed in the electrochemical oxidation of structurally similar purines[4]. Photochemical evidence indicates that the oxidation pathways may also be different[22]. This is in agreement with the known biological differences between the metabolism of TMP and that of related purines[21].

On-Line Electrochemistry Mass Spectrometry in the Study of Redox Reactivity of Purine Drugs

It is clear from the results discussed in previous sections that in the case of complex electrochemical reactions the key chemical changes caused by electron transfer must be followed by other methods. Electrochemical measurements provide chemical information indirectly from numbers of electrons exchanged per mole of substance, the effect of pH on current-potential (i-E) response and, in some cases, from theoretical analysis of i-E curves[24].

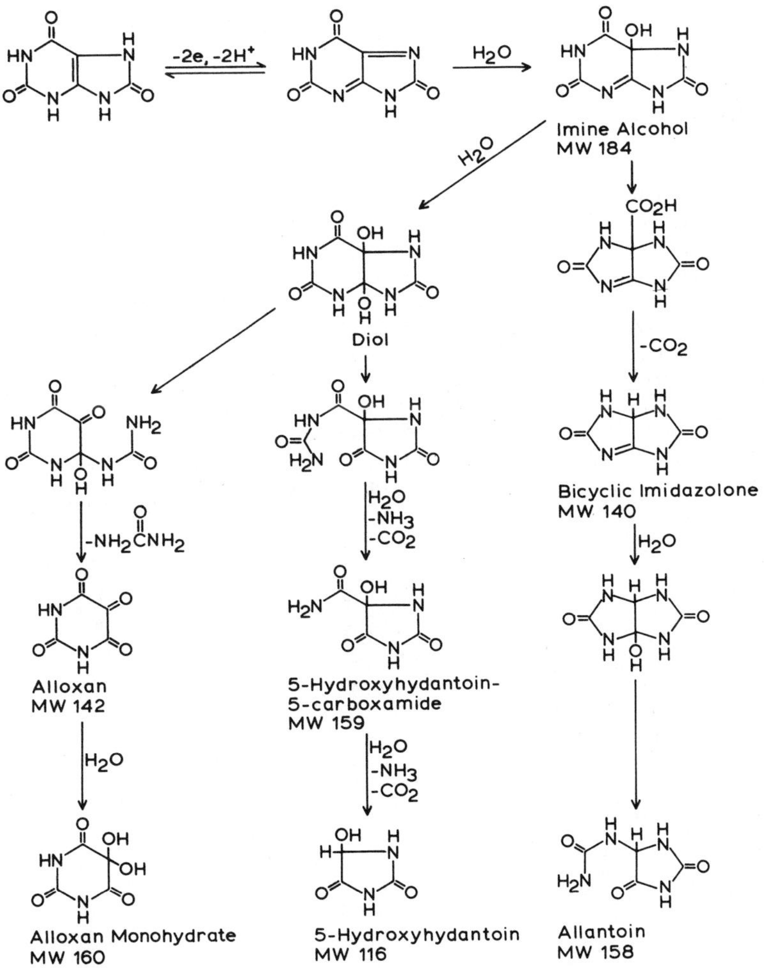

Figure 6. Electrochemical oxidation of uric acid. Molecular weights are shown for intermediates and products observed by on-line EC/TSP/MS/MS.

On-line thin-layer spectroelectrochemistry can provide chemical information but by itself it is not sufficient to confirm chemical identity. Typically the supporting evidence has been obtained by off-line methods such as mass spectrometry, FTIR and NMR.

The feasibility of on-line electrochemistry mass spectrometry in the study of electrode processes has recently been demonstrated by Heitbaum et al.[9]. We have tested the potential of on-line mass spectrometry in the study of redox reactivity of biological compounds with uric acid[10] as a probe. Electrochemical oxidation of uric acid has been studied extensively[3]. The scheme in Figure 6 shows the electrochemical oxidation pathway of uric acid and indicates intermediates and products which were identified by on-line electrochemistry thermospray mass spectrometry (EC/TSP/MS/MS)[10]. In our studies, tandem mass spectrometry (MS/MS) was used to obtain structurally informative fragmentation patterns (daughter spectra) of standards for comparison to the mass spectra of intermediates and products obtained by EC/TSP/MS/MS. This, for example, allowed identification of allantoin through its characteristic daughter spectrum. It also allowed confirmation of the structural features of the intermediate, bicyclic carboxylic acid, which apparently forms from the imine alcohol in the oxidation of uric acid. The intermediates and products which were identified in this way are indicated in the scheme, and mass spectral results are summarized in Table 1.

EC/TSP/MS/MS in the Study of Thiopurines

Results which were obtained by electrochemistry thermospray mass spectrometry in the oxidation of 6-thioxanthine are summarized in Table 1[10]. The EC/TSP/MS/MS shows formation of xanthine at potentials where only the thio group can be oxidized. Xanthine has been identified[18] by derivatization and GC/MS as the product of exhaustive oxidation at these potentials. In addition to xanthine, 2-hydroxypurine was identified as a product. Simultaneously with these products, the parent compound, 6-thioxanthine, was detected. The results support the pathway (Fig. 3) which has been proposed on the basis of the off-line methods[18]. The two products, xanthine and 2-hydroxypurine, may form as a result of nucleophilic or electrophilic attack, respectively, on the initial oxidation product, sulfenic acid (Fig. 7). Control experiments confirmed that the two products form only during oxidation at these potentials.

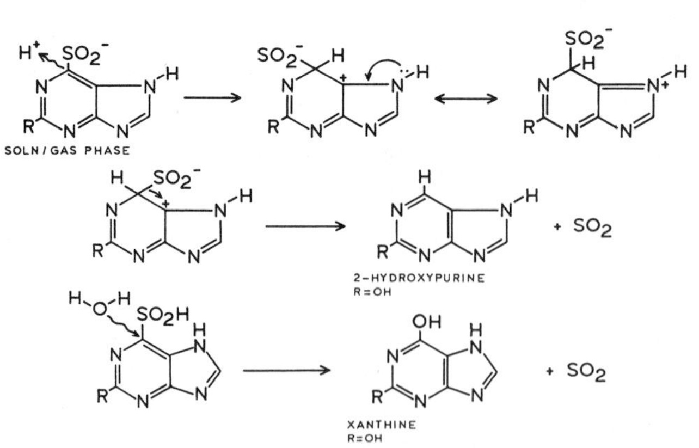

Figure 7. Pathway proposed for the formation of products identified in the oxidation of 6-thiopurine (R = H) and 6-thioxanthine (R = OH) in EC/TSP/MS/MS at moderate potentials.

Table 1. Products of the Electrochemical Oxidation of Purines Identified by On-Line EC/TSP/MS/MS

Purine	Electrode Potential	Oxidation Product	MW	Negative Ions: MS (%RA)[a]	Negative Ions: MS/MS Daughter Spectrum (%RA)[b]	Positive Ions: [b]MS(%RA)[a]	Positive Ions: MS/MS Daughter Spectrum (%RA)
Uric Acid	>+0.20V	Imine Alcohol	184	183(100)[M-H]$^-$	183(80),140(100),139(20),97(10)	d	e
		Allantoin[c]	158	157(61)[M-H]$^-$	157(100),156(30),140(35),114(22),97(22),59(30)	159(5)[M+H]$^+$ 176(12)[M+NH_4]$^+$	f
		Alloxan monohydrate[c]	160	175(5) [116+CH_3COO]$^-$	115(100),59(25)	d	e
		5-Hydroxyhydantoin-5-carboxamide	159	158(47)[M-H]$^-$	158(40),115(100),97(10)	d	e
		Bicyclic-imidazolone	140	139(28)[M-H]$^-$	e	141(16)[M+H]$^+$	e
		5-Hydroxhydantoin	116	g	e	134(100)[M+NH_4]$^+$	e
		?	143	142(80)]M-H]$^-$	142(100),141(10),125(18),99(35),98(10)	d	e
		Alloxan	142	g	e	143(40)[M+H]$^+$	e
6-Thioxanthine	>+0.40 and <+0.60V	6-Thioxanthine[c]	168	167(100)[M-H]$^-$	167(100),133(20)	169(100)[M+H]$^+$	169(100),152(50)
		Xanthine[c]	152	151(7)[M-H]$^-$	151(100)	153(20)[M+H]$^+$	153(100),136(50),110(30)
		2-Hydroxypurine	136	135(7)[M-H]$^-$	e	137(50)[M+H]$^+$	e
6-Thioxanthine	>+0.60V	6-Thioxanthine[c]	168	167(20)[M-H]$^-$	167(100),133(20)	d	e
		Imine Alcohol	184	183(10)[M-H]$^-$	183(40),140(100),139(20)	d	e
		?	143	142(12)[M-H]$^-$	142(50),141(20),125(20),112(100),99(50)	d	e

a All ions with relative abundance ≥ 5% in the normal mass spectrum tabulated
b All ions with relative abundance ≥ 10% in the daughter spectrum tabulated
c Assignments confirmed by comparison of MS spectra and MS/MS daughter spectra with those of standards
d No ions with ≥ 1% abundance relative to m/z 131 ($NH_4CH_3COO \cdot NH_4 \cdot 2H_2O$)$^+$ background peak
e No daughter spectra obtained
f No daughter spectra obtained. Allantoin standard yielded daughter spectra of [M+H]$^+$ 159(100), 116(75), 99(94), 73(22), and of [M+NH_4]$^+$ 159(43), 142(18), 116(100)
g No ions with ≥ 1% abundance relative to m/z 155 ($CH_3COOH \cdot CH_3COO \cdot 2H_2O$)$^-$ background peak

In the EC/TSP/MS/MS of 6-mercaptopurine at potentials at which the thio group can be oxidized, purine, hypoxanthine and the parent compound, 6-mercaptopurine, were identified. This indicates that the pathway in Figure 7 may be more general.

We have previously inferred from the identified final products that the oxidation of 6-thioxanthine at potentials more positive than 0.7 V must lead to the oxidation of the purine ring (Fig. 4)[18]. Using EC/TSP/MS/MS this was confirmed by identification of the imine alcohol intermediate (IV, Fig. 4), the same intermediate which forms in the oxidation of uric acid (Fig. 6)[10].

CONCLUSIONS

Oxidation pathways of purine drugs show, as expected, chemical similarities to the oxidation pathways of naturally occurring purines. There are also significant differences, especially in product distribution, which may reflect different biological reactivity. Interesting correlations have been observed between electrochemically determined redox reactivity and the known biological reactivity. The capabilities of on-line electrochemistry thermospray mass spectrometry in the study of biological reactions are clearly demonstrated.

ACKNOWLEDGEMENTS

This research was partially supported by the National Institutes of Health through a grant GM35451-01 A2, by the Division of Sponsored Research at the University of Florida and by the Research Corporation. Partial support by the U.S. Army Research, Development and Engineering Center (No. DAAA15-85-C0034) is also gratefully acknowledged. K. McKenna acknowledges the support of Merck-Dohme through a graduate fellowship and the support of Dow Chemical through a summer fellowship administered by the Division of Analytical Chemistry of the American Chemical Society.

We wish to thank Dr. J.P. Toth for assistance with mass spectral analysis.

REFERENCES

1. P. Langen, "Antimetabolites of Nucleic Acid Metabolism", Gordon and Breach, New York (1975).

2. A. Winkelstein, J. Immunopharmacol., 1, 429 (1979).

3. G. Dryhurst, K.M. Kadish, F. Scheller and R. Renneberg, Eds., "Biological Electrochemistry", Vol. 1, Academic Press, New York (1982).

4. G. Dryhurst, "Electrochemistry of Biological Molecules", Academic Press, New York (1977).

5. P.A. Andrews, S. Pan and N.R. Bachur, J. Am. Chem. Soc., 108, 4158 (1986).

6. D.J. Miner, J.R. Rice, R.M. Riggin and P.T. Kissinger, Anal. Chem., 53, 2258 (1981).

7. T.G. Traylor, W.A. Lee and D.V. Stynes, Tetrahedron, 40, 553 (1984).

8. F.M. Hawkridge, in: "Laboratory Techniques in Electroanalytical Chemistry", P.T. Kissinger and W.R. Heineman, Eds., Marcel Dekker, New York (1984).

9. G. Hambitzer and J. Heitbaum, Anal. Chem., 58, 1067 (1986).

10. K.J. Volk, M.S. Lee, R.A. Yost and A. Brajter-Toth, Anal. Chem., 60, 720 (1988).

11. V.E. Norvell and G. Mamantov, Anal. Chem., 49, 1470 (1977).

12. R.J. Anderegg, A. Brajter-Toth and J.P. Toth, Anal. Chem., 56, 1351 (1984).

13. D. Astwood, T. Lippincott, M. Deysher, C. D'Amico, E. Szurley and A. Brajter-Toth, J. Electroanal. Chem., 159, 295 (1983).

14. D. Astwood, C. D'Amico, T. Lippincott and A. Brajter-Toth, J. Electroanal. Chem., 198, 283 (1986).

15. J.B. Wyngaarden, J. Biol. Chem., 224, 453 (1957).

16. A. Bendich, S.S. Furst and G.B. Brown, J. Biol. Chem., 185, 423 (1950).

17. P.J. Kraske and A. Brajter-Toth, J. Electroanal. Chem., 207, 101 (1986).

18. K. McKenna and A. Brajter-Toth, J. Electroanal. Chem., 233, 49 (1987).

19. R.N. Goyal and G. Dryhurst, J. Electroanal. Chem., 135, 75 (1982).

20. R.M. Hyslop and I. Jardine, J. Pharmacol. Exp. Ther., 218, 621 (1981).

21. C.A. Nicholas, "Handbuch der experimentellen Pharmakologie", Vol. 38, Springer-Verlag, Berlin (1975).

22. T.E. Childers-Peterson and A. Brajter-Toth, J. Electroanal. Chem., 239, 161 (1988).

23. T.E. Childers-Peterson and A. Brajter-Toth, Anal. Chim. Acta, 202, 167 (1987).

24. A.J. Bard and L.R. Faulkner, "Electrochemical Methods", Wiley, New York (1980).

ELECTRON TRANSFER MECHANISM FOR COCAINE ACTION

Peter Kovacic*, James R. Ames and Mikolaj Jawdosiuk

Department of Chemistry
University of Wisconsin-Milwaukee
Milwaukee, Wisconsin 53211

Michael D. Ryan

Department of Chemistry
Marquette University
Milwaukee, Wisconsin 53233

INTRODUCTION

Although much progress has been made, there are many unanswered questions concerning the chemistry of the brain, including mechanistic features of central nervous system (CNS) drugs. It is reasonable to invoke the involvement of electrochemical phenomena in some cases, as was done in the pioneering investigations of Szent-Györgyi and co-workers[1]. The principal categories of electron transfer (ET) agents consist of quinones, metal complexes, $ArNO_2$, and iminium ions[2]. In our prior work in the CNS area, iminium participation via ET was suggested for the action of benzodiazepines[3], phencyclidine (PCP)[4], nicotine[4], mesoionics[5], and 1-methyl-4-phenyl-1,2,3,6-tetrahydropyridine (MPTP)[6].

Cocaine 1 is mainly metabolized by two distinct pathways in humans[7]. The major biotransformation consists of various hydrolytic reactions. Since these are apparently not important in the toxic manifestations, there will be no additional discussion. The minor route is an oxidative one involving the amine moiety of the tropane ring. It is this pathway that appears to be responsible for some of the toxic responses of the opiate.

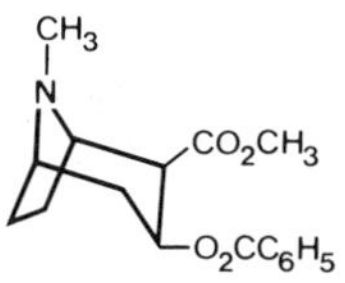

1

Possible entities for participation in the CNS effects of **1** are the drug itself, norcocaine **2**, norcocaine nitroxide **3**, and an iminium ion generated from **1**. To test the potential involvement of ET, we investigated the electrochemical characteristics (reduction and/or oxidation) of cocaine, norcocaine, and nitroxide 3. Relevent prior literature is also discussed.

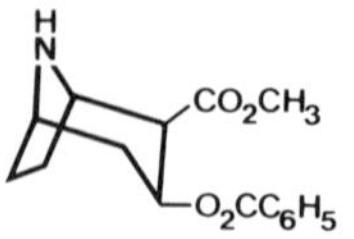

EXPERIMENTAL

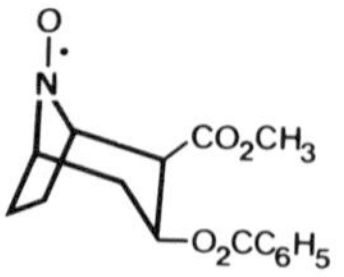

Cocaine **1** hydrochloride (Mallinckrodt Chemical Works) was converted to the free base in the electrochemical cell by addition of NaOH (0.125 M). Norcocaine **2** (used as oil) was prepared by the method of Lazer et al.[8]; hydrochloride, mp 114-115°C, lit. mp 115-117°C; m/z(EI) 289 (**2**-HCl). Norcocaine nitroxide **3** was obtained as an oil from **2** using m-chloroperbenzoic acid[9]; m/z(EI) 304(3); the sample also contained some m-chlorobenzoic acid. Di- and tri-n-butylamines (Eastman and K and K Laboratories) were distilled from caustic pellets. For electrochemical analysis the solvents acetonitrile (Chempure or Aldrich) and dimethylformamide (DMF) (Aldrich) were used without further purification. The electrolyte was tetraethylammonium perchlorate (TEAP) (G.F. Smith Chemical Co.).

Electrochemical measurements were performed as previously described[10]. The working electrode was either a platinum flag (Sargent) (oxidation) or hanging mercury drop (HMDE) (reduction). The reference was a saturated calomel electrode (SCE). All potentials are reported vs. NHE by addition of 0.24 V to the SCE reduction values, while the internal ferrocene/ferrocenium couple (E° = 0.40 V)[11] was used as the standard in oxidations.

RESULTS AND DISCUSSION

1. Norcocaine Nitroxide

Norcocaine **2**, the demethylated metabolite of cocaine **1**, is detected soon after administration of the parent drug[12]. This transformation[12] may be the initial step in a metabolic sequence leading to the observed toxicity. Microsomes in the liver and brain further oxidize **2** to the nitroxide 3[7,8,13] which may be intimately involved in a variety of free radical reactions[8,13]. There are two potential mechanisms for ET by nitroxide **4**, namely, reduction to the hydroxylamine anion **5** (discussed first), or oxidation to the electrophilic nitrosonium ion **6**.

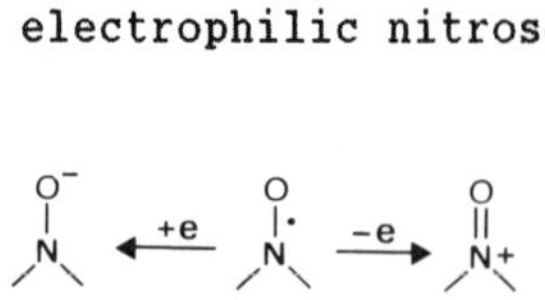

Norcocaine nitroxide **3** reduced at about -0.5 V in aprotic media (Table 1) in an irreversible fashion with $E_{p-p/2}$ = 70-80 mV (HMDE), but no reaction was observed with Pt as the working electrode. The magnitude of the reduction potential may be related to delocalization of the radical on both nitrogen and oxygen[14] as in **7a-c**, as well as on the carbonyl group of the methyl ester (transannular interaction). Similarly, various

Table 1. Electrochemical Characteristics of Norcocaine Nitroxide[a]

Process	E_p (V)		$E_{pp/2}$ (mV)[b]	
	CH_3CN	DMF	CH_3CN	DMF
Reduction	-0.55[c]	-0.48[c]	80	70
Oxidation	0.80[d]	0.75[d]	100	85

[a]100 mV/s, 0.1 M TEAP, 0.5 mM 3, versus NHE.

[b]For irreversible, $E_{pp/2} = E_p - E_{p/2}$; reversible, $\Delta E_p (E_{pc}-E_{pa})$ values.

[c]Irreversible, HMDE.

[d]Reversible, $E_{1/2}$, Pt.

resonance forms can be drawn for the indolinine nitroxide (reduction potential of about -0.45 V)[15]. Other five- and six-membered nitroxides possessing diverse functionalities reduce in the range of -0.1 to -0.4 V[16]. An adverse influence on electron uptake may be the energetically unfavorable interaction of hydroxylamine anion with the nucleophilic oxygen of methyl ester carbonyl. The magnitude of the reduction potential (about -0.5 V) is similar to that of other biologically active agents known to

$$\rangle N\dot{-}O \longleftrightarrow \rangle \ddot{N}-O\cdot \longleftrightarrow \rangle \overset{\delta+}{\cdot N}-\overset{\delta-}{\ddot{O}}:$$

7a 7b 7c

effect ET, including quinones[17,18], nitroheterocycles[19], triarylmethane dyes[19] (iminium), and metal complexes[2,19].

The hydroxylamine derivative is metabolically formed from norcocaine[7]. Since reduction of nitroxides in protic solvents yields this derivative in a reversible manner[15], the transformation evidently does not pertain to our irreversible case, although it may be applicable in vivo. A futile redox cycle involving 3 and the hydroxylamine form has been proposed to account for the hepatotoxicity of 1[7]. Irreversibility may be due to 1,5-intramolecular nucleophilic attack by the oxygen anion 5 on the methyl ester carbonyl. Related nitrone species are known to undergo similar intramolecular interaction with unsaturation[20]. The absence of reoxidation of the anion observed in solution might not pertain in vivo due to the active site binding.

Our proposed mechanism is consistent with the observations of lipid peroxidation during incubation of N-hydroxynorcocaine which is converted to 3 by liver microsomes[7]. Superoxide is generated during the reverse process. It is significant that recent evidence points to metabolism of cocaine to 3 in brain microsomes, albeit the enzyme activity is only about 10% of that in the liver[13]. Nitroxide 3 can be reversibly converted to the N-hydroxy compound. Production of superoxide was observed during interconversion of 3 and the N-hydroxy form[13].

Alternatively a redox cycle may pertain in which **3** and nitrosonium ion **6** participate. Although supportive evidence exists, this pathway is not well established in vivo[21]. Our results (Table 1) are in general agreement with a prior study ($E_{1/2}$ = 0.78 V, DMF)[21]. However, in CH_3CN our value was somewhat more negative than reported[21]. All of the oxidations involved one electron and were reversible. The methyl ester conceivably exerts an adverse influence on the oxidative process. For example, introduction of a carbonyl group in the 4-position of a piperidine nitroxide increases the oxidation potential by 0.17 V versus the parent compound[22]. The interaction was rationalized by unfavorable charge interaction of carbonyl with nitroxide.

2. Cocaine and Norcocaine

Another possibility for catalytic ET involves one electron oxidation to yield a radical cation[23]. To ascertain the ease of formation of these reactive intermediates for **1** and **2**, electrooxidation was used. Cocaine **1** reacted at about 1.3 V in both CH_3CN and DMF (Table 2) in an irreversible manner. The irreversibility may be due to participation of the adjacent methyl group to form an iminium species (vide infra). The secondary amine **2** oxidized at about 1.4 V (Table 2). Based on Bredt's rule, iminium formation from **2** is unlikely due to the presence of only bridgehead hydrogens in the α positions. Irreversibility of the process may be due to reaction of the radical cation with the aprotic solvent, as was proposed for simple tert-amines[23] and nortropane[24].

The potentials (E_p) observed for **1** and **2** are more positive than for the tropane (1.07 V) and nortropane (1.35 V) models[24]. The difference is likely due to the methyl ester group. Thus, tropacocaine, which is identical to **1** except for the absence of that substituent, oxidizes at 1.05 V[24]. Previous studies[25] of bicyclic amines containing a γ-carbonyl have shown an adverse effect on oxidation. For example, with 9-azabicyclo [3.3.1]nonan-3-one compounds (cyclic ketones with carbonyl γ to amine) oxidation was 0.2 V more difficult than for the corresponding methylene

Table 2. Oxidation of Cocaine and Norcocaine[a]

Amine	E_p (V)	
	CH_3CN	DMF
1·HCl[b]	1.33	1.29
2	1.39	1.46
Tri-n-butylamine	0.91	-
Di-n-butylamine	1.10	-

[a]100 mV/s, TEAP, (0.1 M), substrate (0.5 mM), irreversible, Pt flag electrode, vs NHE; internal standard, ferrocene/ferrocenium; E^o = 0.40 V.

[b]NaOH (0.5 mM) added.

analogues. The effect is attributed to energetically unfavorable transannular interactions. A lesser influence (0.02 V) has been observed for simple acyclic amines containing γ carbonyl[25], e.g., **8**, supporting the importance of transannular effects. In our case the carbonyl of the methyl

Structur

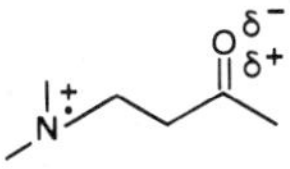

8

ester is fixed by the bicyclic ring system. Simple sec- and tert-amines have also been investigated[26]. The results for **1** and **2** are in line with those for tri-(E_p 0.91 V) and di-(E_p 1.11 V)-n-butylamines (Table 2). tert-Amines are known to oxidize more readily than secondary amines[24,26]. The more positive potentials in our case may be due in part to the bicyclic structure. The oxidation of simple amines apparently results in geometrical reorganization. The requisite movement in a bicyclic system could be restricted. The crystal structure of 9,9'-bis-9-azabicyclo-[3.3.1]nonane showed that the C-N-C bond angle increased about 6 degrees upon electron removal[27]. Our more constrained system is expected to hinder energetically favorable angular alterations.

The very positive nature of the oxidations for **1** and **2** make them unlikely participants in ET. However, other physiologically active compounds possess fairly positive oxidation potentials. For example, the tert-amine of amorphine **9** reacts at about 1.1 V[28]. It is reasonable to expect stabilization of the cation radical by transannular interaction with the benzenoid nucleus. An ET mechanism for the CNS effects has been proposed[29,30]. Related opium alkaloids in the codeine (a methyl ether of **9**) class react in the range of 1.0 to 1.1 V[28]. The psychotherapeutic agent chlorpromazine is similarly transformed at about 0.9 V in a reaction which is believed to occur *in vivo*[31]. This radical cation may be responsible for the psychotropic activity of chlorpromazine[32]. The redox catalysts SOD (Cu) and bleomycin (Fe) show redox properties at 0.42[33] and 0.13 V[34], respectively.

Str

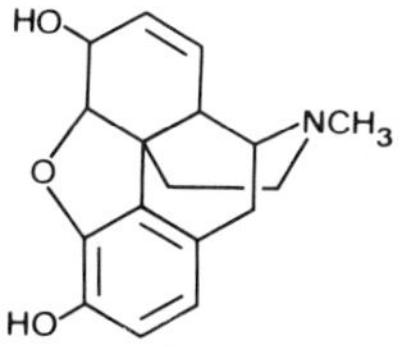

9

3. Cocaine Iminium

Finally, another possibility for ET involves iminium **10**. This general type of oxidative transformation for tert-amines is observed *in vivo*[35]. The morphine radical cation may undergo further reaction to form an iminium ion at the active site[29].

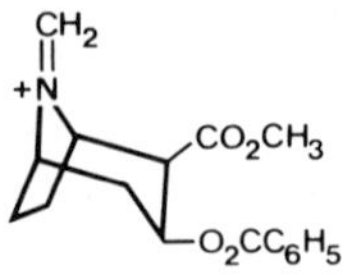

$$\underset{\underset{\sim}{10}}{\text{[structure: N-methylene iminium cation of cocaine, }CH_2{=}N^{+}\text{, }CO_2CH_3\text{, }O_2CC_6H_5\text{]}}$$

Apparently **10** has not been reported as a metabolic product of 1 although it is reasonable to designate the cation as a plausible intermediate during demethylation on the basis of formaldehyde formation[36]. An estimate of the reduction potential may be made by examination of data for simple model compounds[4] and the iminium salt generated from N-methyl-9-azabicyclo-[3.3.1]nonane, $E_p = -0.71$ V (CH_3CN)[37]. The value may be even more positive for **10** due to transannular interaction of the intermediate radical with the ester moiety. Such transannular facilitation by carbonyl has been observed in other bicyclic systems, e.g., the camphor type[38]. Carboxy carbonyl adjacent to iminium also enhances electron uptake[10,39]. If **10** is acting in ET at the active site, then there must be absence of nucleophiles, e.g., water, which could destroy it.

4. Other Mechanisms of Action

Cocaine has been shown to inhibit the reuptake of norepinephrine in the sympathomimetic nerve terminal[7]. Local anesthesia probably results from a membrane effect. Small doses cause an increase in respiratory rate whereas large amounts lead to death from respiratory depression or cardiac arrest.

5. Other Aspects of ET

Although some of the reactions observed in this study were irreversible, behavior in vivo may be different for the drug immobilized at the active site. Also, the energetics may be more favorable. These aspects are treated in greater detail elsewhere[4,10]. The ET theory has been broadly applied in our laboratory to other areas including various drugs and carcinogens[40,41].

CONCLUSION

This report deals with application of the ET concept to the mode of action of cocaine and its metabolites. Possible agents for participation in electrochemical transformations are cocaine, norcocaine, norcocaine nitroxide, and cocaine iminium. Electroreduction of cocaine nitroxide occurred in the energetically favorable range of -0.48 to -0.55 V. Electro-oxidation was observed at the indicated values: cocaine (1.29 V), norcocaine (1.39 V), and norcocaine nitroxide (0.75 V). Relevant prior literature is discussed.

ACKNOWLEDGMENTS

We thank Philip Crawford, Dr. Vera M. Kolb and William K. Hoffman, M.D. for generous assistance.

REFERENCES

1. G. Karreman, I. Isenberg and A. Szent-Györgyi, Science, 130, 1191 (1959).

2. J.R. Ames, U. Hollstein, A.R. Gagneux and P. Kovacic, Free Rad. Biol. Med., 3, 85 (1987) (and references therein).

3. P.W. Crawford, P. Kovacic, N.W. Gilman and M.D. Ryan, Bioelectrochem. Bioenerg., 16, 407 (1986).

4. J.R. Ames, S. Brandänge, B. Rodriguez, N. Castagnoli, Jr., M.D. Ryan and P. Kovacic, Bioorg. Chem., 14, 228 (1986).

5. J.R. Ames, K.T. Potts, M.D. Ryan and P. Kovacic, Life Sci., 39, 1085 (1986).

6. J.R. Ames, N. Castagnoli, Jr., M.D. Ryan and P. Kovacic, Free Rad. Res. Comms., 2, 107 (1986).

7. M.W. Kloss, G.M. Rosen and E.J. Rauckman, Biochem. Pharmacol., 33, 169 (1984).

8. E.S. Lazer, N.D. Aggarwal, G.J. Hite, K.A. Nieforth, R.T. Kelleher, R.D. Spealman, C.R. Schuster and W. Wolverton, J. Pharm. Sci., 67, 1656 (1978).

9. E.J. Rauckman, G.M. Rosen and J. Cavagnaro, Mol. Pharmacol., 21, 458 (1982).

10. P.W. Crawford, R.G. Scamehorn, U. Hollstein, M.D. Ryan and P. Kovacic, Chem.-Biol. Interactions, 60, 67 (1986).

11. R.R. Gagne, C.A. Koval and G.C. Lisensky, Inorg. Chem., 19, 2854 (1980).

12. A.L. Misra, P.K. Nayak, M.N. Patel, N.L. Vadlamani and S.J. Mule, Experientia, 30, 1312 (1974).

13. M.W. Kloss, G.M. Rosen and E.J. Rauckman, Psychopharmacol., 84, 221 (1984).

14. Y. Deguchi, Bull. Chem. Soc. Jap., 35, 260 (1962).

15. R. Andruzzi, A. Trazza, L. Greci and L. Marchetti, J. Electroanal. Chem., 107, 365 (1980).

16. V.L. Varand, I.A. Grigor'ev, L.I. Vasil'eva and L.B. Volodarsky, Izv. Sibirsk. Otdel. Akad. Nauk SSR, 6, 99 (1982).

17. A. Ashnagar, J.M. Bruce, P.L. Dutton and R.C. Prince, Biochem. Biophys. Acta, 801, 351 (1984).

18. P. Kovacic, J.R. Ames, P. Lumme, H. Elo, O.Cox, H. Jackson, L.A. Rivera, L. Ramirez and M.D. Ryan, Anti-Cancer Drug Design, 1, 197 (1986).

19. J.R. Ames, M.D. Ryan and P. Kovacic, J. Free Rad. Biol. Med., 2, 377 (1986).

20. J.J. Tufariello, G.B. Mullen, J.J. Tegeler, E.J. Trybulski, S.C. Wong and S.A. Ali, J. Am. Chem. Soc., 101, 2435 (1979).

21. J.C. Charkoudian and L. Shuster, Biochem. Biophys. Res. Communs., 130, 1044 (1985).

22. W. Sümmermann and U. Deffner, Tetrahedron, 31, 593 (1975).

23. R. Lines in: Organic Electrochemistry, M.M. Baizer and H. Lund (Eds.), Marcel Dekker, New York (1983), Chap. 15.

24. B.L. Laube, M.R. Asirvatham and C.K. Mann, J. Org. Chem., 42, 670 (1977).

25. S.F. Nelsen, C.R. Kessel, L.A. Grezzo and D.J. Steffek, J. Am. Chem. Soc., 102, 5482 (1980).

26. C.K. Mann, Anal. Chem., 36, 2424 (1964).

27. S.F. Nelsen, W.C. Hollinsed. C.R. Kessel and J.C. Calabrese, J. Am. Chem. Soc., 100, 7876 (1978).

28. P. Surmann, Arch. Pharm. (Weinheim), 312, 734 (1979).

29. V.M. Kolb, J. Pharm. Sci., 73, 715 (1984).

30. B. Belleau and P. Morgan, J. Med. Chem., 17, 908 (1974).

31. G. Dryhurst, in: Comprehensive Treatise of Electrochemistry, Vol 10, S. Srinivasan, Yu. A. Chizmadzhev, J. O'M. Bockris, B.E. Conway and E. Yeager (Eds.), Plenum Press, New York (1985), Chap. 2.

32. G. Vincow, in: Radical Ions, E.T. Kaiser and L. Kevan, Eds., Interscience, New York (1968), p. 201.

33. B.G. Malmström, L.-E. Andreasson and B. Reinhammar, in: The Enzymes, Vol. XII, P.D. Boyer (Ed.), Academic Press, New York (1975), pp. 551-552

34. D.L. Melnyk, S.B. Horwitz and J. Peisach, Biochemistry, 20, 5327 (1981).

35. M. Overton, J.A. Hickman, M.D. Threadgill, K. Vaughan and A. Gescher, Biochem. Pharmacol., 34, 2055 (1985).

36. A.R. Dahl and W.M. Hadley, Toxicol. Appl. Pharmacol., 67, 200 (1983).

37. S.F. Nelsen, C.R. Kessel and D.J. Brien, J. Am. Chem. Soc., 102, 702 (1980).

38. P. Zuman, Substituent Effects in Organic Polarography, Plenum Press, New York (1967), pp. 311-312.

39. P. Kovacic, M. Jawdosiuk, J.R. Ames and M.D. Ryan, Bioorg. Chem., 15, 423 (1987).

40. P. Kovacic, J.R. Ames and M.D. Ryan, in Proceedings of the Third International Symposium of Redox Mechanisms and Interfacial Behaviors of Molecules of Biological Importance (G. Dryhurst and K. Niki, Eds.), Plenum Press, New York (and references therein).

41. P. Kovacic, J.R. Ames, J.W. Grogan, B. Hazra and M.D. Ryan, in Proceedings of the Third International Symposium of Redox Mechanisms and Interfacial Behaviors of Molecules of Biological Importance (G. Dryhurst and K. Niki, Eds.), Plenum Press, New York (and references therein).

STUDIES ON ACID-BASE, HYDRATION-DEHYDRATION AND KETO-ENOL EQUILIBRIA IN AQUEOUS SOLUTIONS OF α-KETOACIDS BY POLAROGRAPHY, LINEAR SWEEP VOLTAMMETRY AND SPECTROSCOPY

J. Kozlowski and P. Zuman*

Department of Chemistry
Clarkson University
Potsdam, New York 13676

INTRODUCTION

Even when α-ketoacids play an important role as a link between the metabolism of carbohydrates and proteins and their solution chemistry should be of general interest, quantitative information regarding positions of acid-base, keto-enol, and hydration-dehydration equilibria established in their dilute aqueous solutions is rather limited. Only properties of pyruvic acid were studied extensively[1], but those of α-ketoglutaric acid received only limited attention[1c,2]. For the most complex system represented by oxalacetic acid attention was paid predominantly to keto-enol equilibria[3] and only recently the role of hydration yielding a geminal diol has been considered[1c,1f,4].

In these studies potentiometric, polarographic, spectral (electronic, proton-NMR, Raman, even IR), calorimetric, relaxation and stop-flow methods were applied. In most cases in each study only a single technique was used. Only recently NMR studies were combined with spectrophotometric measurements[3j,4b,4c]. This approach is complicated by substantial differences in concentrations of the studied solutions: In proton NMR studies 0.1 - 1.7 M solutions were used, compared to 10^{-4} M solutions studied spectrophotometrically. For structurally related 1,2-diketones[5] the values of equilibrium constants for the dehydration obtained by NMR in more concentrated solutions are always higher than those obtained in millimolar and more dilute solutions. Thus di- or polymerization or other types of solute-solute interactions in more concentrated solutions cannot be excluded[6].

In this study α-ketoglutaric and oxalacetic acids were investigated by UV-spectrophotometry, polarography and linear sweep voltammetry in dilute solutions at a controlled ionic strength. To facilitate the interpretation of the data, some of the esters of these acids were included. Results obtained in these dilute solutions show some differences when compared to data obtained with NMR for more concentrated solutions. Presented data were obtained under conditions closer to those existing in biological systems and should have an impact on interpretation of kinetic investigations[4c,4d,7].

EXPERIMENTAL

Compounds and Solutions

Oxalacetic acid (072397 Aldrich Chem. Co.), pyruvic acid sodium salt (040687 Aldrich Chem. Co.), α-ketoglutaric acid (102487 Aldrich Chem. Co.), diethyl oxalacetate sodium salt (118 ICN Nutritional Biochemicals Co.), and thallium (1) sulfate (Alfa Inorganic) were all more than 95% pure and used as received. Oxalacetic acid (012397 Aldrich Chem. Co.) contained some pyruvic acid as proved by TLC, diethyl oxalacetate sodium salt (1991581179 Fluka Chem. Co.) and monoethyl oxalacetate (13311 United States Biochemicals) contained impurities yielding additional UV absorption bands and polarographic waves, proved also by TLC. Monoethyl oxalacetate sodium salt (2199 Research Organics) was found to be only 47.5 ± 0.3% pure using potentiometric titration. The presence of large amounts of impurities was confirmed by TLC. Since the impurities did not absorb at wavelengths greater than 220 nm, this compound was used with correction for the actual concentration of the monoethyl ester. Diethyl α-ketoglutarate was prepared by refluxing α-ketoglutaric acid in absolute ethanol with 0.1 ml of concentrated sulfuric acid added, for 24 h. Diethyl α-ketoglutarate extracted into diethyl ether from the reaction mixture was then neutralized with saturated sodium bicarbonate. The ether was removed and the diethyl ester distilled at b.p. 111.2°C at 2.5 mm Hg. The ester was identified by NMR and IR spectra. The product gave a single TLC spot at R_f = 0.86, indicating absence of α-ketoglutaric acid.

Sulfuric and hydrochloric acid solutions were prepared by dilution of the reagent grade acid. Carbonate-free solutions of sodium hydroxide were prepared from a saturated solution of ACS reagent pellets of sodium hydroxide dissolved in CO_2-free water. Those of lithium hydroxide (G. Frederick Smith Chem., reagent grade) were prepared similarly in carbonate-free water and standardized. Acetone-d_6 (Aldrich Chem. Co.) was 99.5% pure; Chloroform-d_3 (Norell Chemical Co.) was 99.8% pure; 1,4-dioxane-d_8 (Aldrich Chem. Co.) was 98.5% pure; all were used as received.

Simple buffer solutions for the pH range from 1.8 to 12, unless otherwise stated in the text, were adjusted to an ionic strength of 0.20 M by an addition of sodium chloride. The buffer solutions used had a total concentration of the buffer components of 0.04 M and were prepared from J.T. Baker reagent grade chemicals (unless otherwise stated). The following buffers were used: pH 1.9 to 3.0, phosphate; pH 3.5 to 5.5 acetate; pH 6.0 to 8.0, phosphate; pH 8.0 to 10.0, ammonia; pH 8.0 to 10.0, borate and pH 10.5 to 11.8, phosphate. Bicine buffers pH 7.2 to 9.2 were prepared from N,N-bis(2-hydroxyethyl)glycine; CHES buffers pH 8.3 to 10.3 from cyclohexylamino-ethanesulfonic acid; CAPS buffers pH 9.5 to 10.5 from cyclohexylaminopropanesulfuric acid; all from United States Biochemicals Corp. Solutions with pH values smaller than 1.9 were obtained using solutions of sulfuric acid in varying concentrations. The acidity of these solutions was described using the amide acidity function, H_A[8]. Solutions of pH greater than 11.8 were obtained by using sodium hydroxide and their acidity (at pH ≥ 13) was described using the J_acidity scale[9].

Apparatus

For d.c. polarography three instruments were used: A Princeton Applied Research Model 174 Polarographic Analyzer with an MFE model 815 M Plotamatic X-Y Recorder, an IBM Mark 225 Voltammetric Analyzer with an IBM X-Y-T Recorder, and a Sargent model XVI polarograph. For the recording of differential pulse polarograms, the Princeton Applied Research model 174 with mechanical drop control (one second drop time and a pulse amplitude of 125 mV) was used with the MFE recorder mentioned above. For LSV experiments a custom-built voltage ramp generator allowing scan rates from 2 to 2000 V min^{-1} was used and the current-voltage curves were recorded on a Tektronic Type 564 B Storage Oscilloscope, equipped with type 2A63 and type 3A9 differential amplifiers.

D.c. and differential pulse polarographic electrolysis were carried out in a Kalousek cell with a two electrode system. In solutions of pH 0 to 14 a saturated calomel electrode was used as the reference electrode. In strongly acidic solutions (pH < 0) a Hg/Hg_2SO_4 reference electrode, prepared with concentrated sulfuric acid, was employed. In strongly basic media (J_- > 14) a carbon rod reference electrode in 5 M NaOH was used with phthiocol as an internal potential standard. Several dropping mercury electrodes with slightly differing characteristics were used in the course of this research. All currents were converted to those obtained with a capillary with the following characteristics: m = 3.32 mg s^{-1} and t_1 = 2.3 s in 1.0 M KCl (0.0 V vs SCE) at a height of 64 cm. Unless otherwise stated, polarographic curves were recorded at 25°C.

Linear sweep voltammetry (LSV) was carried out using a two-electrode Kalousek cell in conjunction with a slowly dropping mercury electrode with m = 0.67 mg and natural drop-time t_1 = 11.4 s (in 1.0 M KCl at 0.0 V) at h = 25 cm as the working electrode and a saturated calomel electrode, separated by liquid junction, as the reference electrode. In sulfuric acid solutions of pH less than 0 an $Hg/HgSO_4$ reference electrode was used, as was previously described. The voltage scans were started 6 s after the previous drop was dislodged.

Potentiometric titrations were performed with a Radiometer PHM84 research pH meter equipped with a Radiometer G 202 B (low alkaline error) glass electrode and a Radiometer K 401 calomel reference electrode. Carbon dioxide was rigorously excluded during all titrations by a stream of nitrogen. Both the standard solution and the titrated sample were adjusted to a constant ionic strength, such as 0.1 M, with sodium chloride (J.T. Baker Chemical Co., reagent grade). All titrations were done at 25°C unless otherwise stated.

Ultraviolet and visible absorption spectra were recorded using a Perkin-Elmer Model 559 Spectrophotometer using a 1 cm quartz cell. The temperature of the cell compartment was kept constant, with a Precision Scientific Model 66600 circulating temperature bath, at 25°C. Proton NMR spectra were obtained using an Hitachi Perkin-Elmer R-24B NMR spectrometer.

Procedures

To prepare stock solutions of the α-keto acids and the monoester studied the required amount of the compound was dissolved in distilled water, and then diluted to the mark. The stock solutions of diethyl esters were prepared in dimethyl sulfoxide or freshly distilled acetonitrile. Stock solutions of all compounds were stored at 5°C and kept in an ice bath during use. Oxalacetic acid stock solutions were only stable for a day, the diethyl ester stock solutions, in dimethyl sulfoxide or acetonitrile, were stable for four days. Stock solutions of all other α-keto acids were found to be stable for at least one week and were prepared freshly every week. Concentrated stock solutions of α-ketoglutaric acid (greater than 0.5 M) were found to undergo measurable changes in about four hours.

In a typical electrochemical experiment 10.0 ml of the supporting electrolyte was pipetted into the electrochemical cell and oxygen removed by passing a stream of nitrogen through the solution for 5 minutes. A sample of a stock solution (50 to 200 μl) was added, purged with nitrogen for an additional 45 seconds and the current-voltage curve recorded. Several current-voltage curves were recorded and if changes with time was observed, the limiting current was measured at the various times and extrapolated to t = 0. All measured currents were corrected for changes in the viscosity of the solution by multiplying the measured current by $(\nu/\nu_0)^{1/2}$.

In order to record UV-visible spectra, 50 to 200 μl of a stock solution was added to 10 ml of the buffer, sulfuric acid, or sodium hydroxide solution. Several spectra were recorded in order to show changes with time, if necessary measured quantities were extrapolated back to t = 0. Values of pK and pK' (or pK") were reproducible to about ± 0.1, unless otherwise stated.

T.L.C. was carried out on silica gel coated plastic sheets (Eastman Chromagram 13181 silica gel) using a mixture of diethyl ether (Malinckrodt), formic acid (J.T. Baker): acetone (J.T. Baker): distilled water in the ratio 12:2:2:1 (by volume).

RESULTS AND DISCUSSION

UV spectra

The interpretation of UV spectra, attribution of individual bands and molar absorptivities of individual species are described elsewhere[10].

Acid-Base Equilibria

Equilibrium constants for various acid-base reactions were determined based on inflection points of plots of absorbances at selected wavelengths on pH or acidity function, by potentiometry and from changes in shapes of $E_{1/2}$-pH plots in polarography or E_p-pH plots in differential pulse polarography (Table 1).

Attribution of chemical reactions to experimentally accessible dissociation constants is indicated by Equations (1)-(7) (shown here for oxalacetic acid, the same assignments are used for α-ketoglutaric acid):

$$\mathrm{HOOC\overbrace{COHCH_2}^{+}COOH} \xrightleftharpoons{K_1^{Ke}} \mathrm{HOOCCOCH_2COOH} + \mathrm{H^+} \quad (1)$$

$$\mathrm{HOOC\overbrace{COHCH_2}^{+}COOH} \xrightleftharpoons{K_1^{En}} \mathrm{HOOC\underset{\displaystyle OH}{\underset{|}{C}}{=}CHCOOH} + \mathrm{H^+} \quad (2)$$

$$\mathrm{HOOC\underset{HO\ \ OH_2^+}{\underset{/\ \backslash}{C}}CH_2COOH} \xrightleftharpoons{K_{OH_2^+}} \mathrm{HOOC\underset{HO\ \ OH}{\underset{/\ \backslash}{C}}CH_2COOH} + \mathrm{H^+} \quad (3)$$

$$\text{Acid} \xrightleftharpoons{K_2^{meas}} \text{Monoanion} + \mathrm{H^+} \quad (4)$$

$$\text{Monoanion} \xrightleftharpoons{K_3^{meas}} \text{Dianion} + \mathrm{H^+} \quad (5)$$

$$\text{Dianion} \underset{}{\overset{K_4^{meas}}{\rightleftharpoons}} \text{Carbanion-enolate} + H^+ \quad (6)$$

```
     (-)                  KOH          (-)
-OOCCOCHCOO-   +  OH-   <====>   -OOCCCHCOO-          (7)
                                     /\
                                   HO  O-
```

Symbols Acid, Monoanion and Carbanion-enolate in Equations (4) - (6) refer to the total concentration of the given species, including keto, enol and hydrated forms.

In addition to the above mentioned reactions spectra of oxalacetic acid and α-ketoglutaric acids in a very strongly acidic media (at $H_A < -5$) indicate the presence of a second protonation step, probably involving protonation of one of the carboxylic groups[11]. Observation of such processes for the esters was prevented by the rapidly occurring acid catalyzed hydrolysis.

Taking into account that pK_a-values obtained from polarographic measurements ($E_{1/2}$, E_p = f(pH) are at best accurate to ± 0.2 pK-units, the practical values of pK_i (Table 1) obtained by various methods at the same ionic strength show acceptable agreement. The values of most pK_i increase with decreasing ionic strength.

The experimental values of pK_i^{meas} reported in Table 1 are suitable for general comparisons, as shown above, but for more detailed discussion of structural effects attention must be paid to the roles of hydration and keto-enol equilibria as discussed in Reference 10.

The rate constants of recombination (k_r), i.e., for the protonation of the conjugate base preceding the electron transfer, correspond to reverse forms of reactions (4) through (6) and can be obtained from polarographic data. When the reaction is sufficiently fast and occurs like a volume reaction (as shown by the potential independent limiting current), it is possible[14] to calculate from the difference between the pH in the inflection point of the dependence of the limiting current on pH (polarographic pK') and the equilibrium value of pK (as obtained, e.g. by spectrophotometry or potentiometry) the values of the recombination rate constants k_r using Equation (8):

$$\log k_r = 2pK_i' - pK_i - 2 \log 0.886 - \log t_1 \quad (8)$$

where t_1 is the drop-time and other symbols have been described above. Rate constants calculated in this way from polarographic data (Table 2) indicate that both recombination (k_r) and dissociation (k_d) rate constants for oxalacetic and α-ketoglutaric are similar, both for the first and second dissociation step. Introduction of a carbethoxy group in 4-ethyl oxalacetate results in a considerable increase in both k_r and k_d.

Keto-Enol Equilibria

Keto-enol equilibrium constants K_{En} = [enol form]/[keto form] were obtained from spectral data[10] and from ratios of two polarographic waves of oxalacetic acid and its esters in acidic media (Table 3).

Table 1. Experimental Values of pK_a for Acid-Base Reactions of Some α-Keto Acids and Their Esters at 25°C.

Compound	Reaction No.	pK_i	Method[a]	pK_ameas[b]	μ,M[c]
Oxalacetic acid	(1)	pK_1^{Ke}	$E_{1/2}^{1a}$	0.5	0.2
	(2)	pK_1^{En}	A^{264}	-3.3	-
			$E_{1/2}^{1b}$	-3.7	-
	(4)	pK_2^{meas}	A^{264}	2.1	0.2
			E^o	2.4	0.1
			$E_{1/2}^{1a}$	1.9	0.2
	(5)	pK_3^{meas}	A^{264}	4.1	0.2
			E^o	4.2	0.1
			$E_{1/2}^{2}$	3.9	0.2
	(6)	pK_4^{meas}	A^{264}	13.3	0.2
	(7)	pK_{OH}	A^{264}	15.7	0.2
			A^{283}	15.6	-
Monoethyl oxalacetate	(1)	pK_1^{Ke}	$E_{1/2}^{1a}$	0.4	0.2
	(2)	pK_1^{En}	A^{268}	-3.0	-
			$E_{1/2}^{1b}$	-2.5	-
	(4)	pK_2^{meas}	A^{268}	2.2	0.2
			$E_{1/2}^{1a}$	1.9	0.2
	(6)	pK_4^{meas}	A^{268}	9.3	0.2
			$E_{1/2}^{2}$	9.3	0.2
			E^o	9.7	0.1
Diethyl oxalacetate	(6)	pK_4^{meas}	A^{270}	7.4	0.2
			E^o	7.6	0.1
			$E_{1/2}^{2}$	7.0	0.2
α-Ketoglutaric acid	(1)	pK_1^{Ke}	$E_{1/2}^{1}$	0.0	0.2
	(4)	pK_2^{meas}	A^{310}	1.9_5	1.0
			E^o	2.03	0.1
			$E_{1/2}^{1}$	1.9	0.2
			E_p^{1}	1.8	0.2
	(5)	pK_3^{meas}	A^{310}	4.4	1.0
			E^o	4.81[d]	0.1
			$E_{1/2}^{2}$	4.1	0.2
	(6)	pK_4^{meas}	A^{320}	15.5	-
			A^{263}	15.8	-
Diethyl α-keto-glutarate	(6)	pK_4^{meas}	A^{315}	7.4	0.2

[a] Methods: A^x - UV absorbance at x nm; E^o - potentiometry with a glass electrode; $E_{1/2}^{n}$ - d.c. polarographic half-wave potentials of wave n; E_p^{1} - differential pulse polarographic peak potential of wave 1;

[b] values of pK_a smaller than 0.0 were obtained using the acidity function H_A;

[c] ionic strength;

[d] most reliable value obtained by analysis of titration curves using a curve-fitting program (Reference 15).

Table 2. Recombination and Dissociation Rate Constants for α-Keto Acids and Esters Determined Polarographically

Conjugate Base	k_r, $\ell mol^{-1}s^{-1}$	k_d, s^{-1}
$^-OOCCOCH_2COOH$	7.0×10^5	8.8×10^3
$^-OOCCOCH_2COO^-$	1.8×10^7	2.3×10^3
$^-OOCCOCH_2COOR$	1.1×10^{10}	1.4×10^8
$^-OOCCOC^{(-)}HCOOR$	pK' = pK slow	
$ROOCCOC^{(-)}HCOOR$	3.5×10^{11}	3.5×10^4
$^-OOCCOCH_2CH_2COOH$	2.8×10^5	3.5×10^3
$^-OOCCOCH_2COO^-$	1.1×10^7	8.7×10^2
$CH_3COCOOH$	4.6×10^9 [a]	1.87×10^7 [a]

[a] Reference 1e

Hydration-Dehydration Equilibria

Even though the mechanism of reversible hydration of carbonyl compounds is rather complex[12], the overall process can be depicted as a reversible addition to water to the carbonyl group (Equation 9):

$$HOOCC(=O)CH_2COOH + H_2O \rightleftharpoons HOOCC(OH)_2CH_2COOH \tag{9}$$

The equilibrium constant for the reaction may be expressed as $K_d = [CO]/[C(OH)_2]$ using CO as a symbol for the keto and $C(OH)_2$ for the hydrated form.

Values of K_d can be obtained by electrochemical or spectroscopic techniques. In polarography, the equilibrium between the hydrated and carbonyl form is perturbed and limiting currents governed both by the position of equilibrium (Equation 9) and by the rate of dehydration[12,13]. D.c. polarographic limiting currents thus offer only information about the upper limit of the value of K_d. In linear sweep voltammetry (LSV) the measurement of the current can be carried out over such a short period of time that the equilibrium cannot be reestablished[5,14]. The current is then proportional to the concentration of the carbonyl form in the bulk of the solutions and can be used for determination of values of K_d, as discussed in detail elsewhere[5].

Absorbance at 315-320 nm (A_{320}) corresponding to the $n \rightarrow \pi^*$ transition involving the carbonyl group can also be used for determining K_d.

Values of K_d obtained by electrochemical and spectroscopic methods are summarized in Table 4. For comparison, data found by proton NMR are

Table 3. Values of the Enol Equilibrium Constant K_{En} Obtained by Various Techniques (in 2×10^{-4} M solutions, $\mu = 0.2$, at 25°C).

Species	UV Spectra	Polarography	NMR
$HOOCC\overset{(+)}{O}HCH_2COOH$	3.55	-	-
$HOOCCOCH_2COOH$	0.32	0.12	0.47[a]
$^{-}OOCCOCH_2COOH$	0.13	-	-
$^{-}OOCCOCH_2COO^{-}$	0.13	-	0.085[a]
$HOOCC\overset{(+)}{O}HCH_2COOR$	1.39	-	-
$HOOCCOCH_2COOR$	0.38	0.47	0.32[b]
$^{-}OOCCOCH_2COOR$	0.34	-	-
$ROOCCOCH_2COOR$	0.30	0.42	-

[a]Reference 4b C = 1.14 M, at 38°C;
[b]Reference 3i, pH, μ, and C uncertain

included, which were obtained at concentrations of the acid two to three orders of magnitude higher, and often in unbuffered systems. Decrease in values of K_d with increasing concentration of α-ketoglutaric and pyruvic acid can, therefore, be due to a decrease in pH.

Voltammetric, polarographic and spectrophotometric data show general agreement (Table 4). This indicates that the assumptions made about the similarity of diffusion coefficients of α-ketoglutaric and other keto acids and esters in polarography, about diffusion control of reduction of α-ketoglutaric acid in strongly acidic media in linear sweep voltammetry, and about the estimates of the molar absorptivities of the free carbonyl forms in spectrophotometry are plausible. Generally higher values of K_d obtained by NMR than those by other techniques are attributed to the high concentrations used, favoring dimerizations and other higher order competitive reactions (such as Claisen condensation). Such an increase is also demonstrated for values obtained by spectrophotometry for α-ketoglutaric acid. The estimates of the upper limits from polarographic data are useful in the choice of the most reliable value of k_d with exception of the two diethyl esters. It cannot be excluded that the five times higher concentration used in LSV than in polarography is responsible for the higher values obtained by LSV than were expected from d.c. polarography.

Electrochemical Reduction

Polarographic reduction studied both by d.c. and differential pulse polarography and linear sweep voltammetry are discussed in detail elsewhere[15], in particular with respect to pH-dependence of limiting currents and half-wave and peak potentials. The results can be summarized as follows.

Table 4. Values of K_d for Some α-Keto Acids and Esters Obtained by Various Techniques (at 25°C).

Compound	LSV[a] C/mM	LSV[a] K_d	Polar[b] C/mM	Polar[b] k_d	NMR[c] C/mM	NMR[c] K_d	Spec[d] C/mM	Spec[d] K_d
α-Ketoglutaric Acid[e]								
$HOOCCOCH_2CH_2COOH$	0.99	0.56	0.20	< 0.79	1000	0.83[f]	10	0.38[f]
					2000	0.63[f]	25	0.52[f]
					3000	0.45[f]	50	0.68[f]
							100	0.70[f]
$HOOCCOCH_2CH_2COO-$							4.8	3.8
Diethyl α-Ketoglutarate								
$ROOCCOCH_2CH_2COOR$	0.99	0.41	0.20	< 0.36	-	-	-	-
Oxalacetic Acid								
$HOOCCOCH_2COOH$	0.99	0.14	0.20	< 0.20	1140	0.15[g]		
$HOOCCOCH_2COO^-$					1140	2.4[g]		
$^-OOCCOCH_2COO^-$					1140	16.81[g]		
					1700	15.87[g]		
4-Monoethyl Oxalacetate								
$ROOCCOH^+CHCOOH$	0.47	0.70	0.093	<0.96				
$ROOCCOCH_2COOH$	0.47	0.13	0.093	<0.16				
$ROOCCOCH_2COO^-$	0.47	0.32	0.093	<0.39				
Diethyl Oxalacetate								
$ROOCCOCH_2COOR$	0.99	0.07	0.20	<0.05				
Pyruvic Acid[e]								
$CH_2COCOOH$	0.99	0.34	0.20	0.70	1300	0.64[h]	20	0.42[j]
					2000	0.54[l]		
CH_3COO^-					1300	17.5[h]		
					2000	18.5[l]		

[a]Linear sweep voltammetry; [b]d.c. polarography; [c]proton NMR; [d]spectrophotometry at λ = 310-320 nm; [e]P-Jump at c = 10-100 mM, K_d = 0.74 (pressure jump method); [f]Reference 2; [g]Reference 4b at 38°C; [h]Reference 1g; [j]Reference 1b; [k]Reference 1e; [l]Reference 1h.

Oxalacetic acid is reduced in four two-electron waves: in acidic media the protonated keto form and the protonated enol form, in medium pH-range the monoanion pre-protonated on the carbonyl group and in slightly alkaline solutions the unprotonated dianion. The carbanion-enolate present at pH > 14 is not reducible.

4-Monoethyloxalacelate is reduced in acidic media in protonated keto and protonated enol form, but in the medium pH-range is not preprotonated before the uptake of the first of the two electrons. The carbanion-enolate present at pH $>$ 9 is not reducible.

Diethyl oxalacetate is reduced in acidic media in protonated keto and protonated enol form, at pH $>$ 6 in the unprotonated form. The carbanion-enolate present at pH $>$ 9 is not reducible.

α-Ketoglutaric acid is reduced in the protonated keto form in the acidic medium, pre-protonated monoanion form in the medium pH-range and as unprotonated dianion in alkaline solution. Effect of carbanion-enolate formation has not been observed up to pH 15.

Diethyl-α-ketoglutarate is reduced up to pH 7 in the pre-protonated form. Hydrolysis prevented studies at pH $<$ 2 and pH $>$ 7.

Detailed discussion of the role of the individual acid-base, hydration-dehydration and keto-enol equilibria on the dependence of limiting currents and half-wave potentials is found elsewhere[15] together with discussion of structural effects.

CONCLUSIONS

The similarities of the behavior of the studied keto acids and their esters are far more pronounced than their differences. This applies in particular to the acid-base equilibria. The dissociation of the first carboxylic group of all keto acids compared is predominantly affected by the adjoining carbonyl group, as reflected by the same range in which the values of pK_2^{meas} are observed (Table 1). Replacement of CH_3 by $HOOCCH_2CH_2$ results in a decrease in the value of pK_2^{Ke} by 0.36 that by $HOOCCH_2$ by 0.66 pK-units, indicating a decrease of the inductive effect of the carboxylic group with increasing separation of the carboxylic group from the carbonyl group.

The dissociation of the second carboxylic group is similarly affected by mutual interaction of the two carboxylate groups through the chain. This interaction increases with decreasing number of the methylene groups, as indicated by the decrease in the value of pK_3^{meas} by 0.7 pK-units, when $COCH_2CH_2COO^-$ is replaced by $COCH_2COO^-$. Similarity of behavior of studied α-keto acids is indicated also by the dramatic decrease in values of pK_4^{meas} or pK_4^{Ke} resulting from esterification of both carboxylate groups (Table 1). Substitution of COO^- by COOR causes a decrease in pK_4 by about 6 pK-units for oxalacetate and about 18 pK-units for α-ketoglutarate. The higher pK-values of the dicarboxylate when compared with dicarbalkoxy groupings are due at least partly to the accumulation of the negative charge in the resulting trianion derived from the acids when compared with the monoanion formed from dialkyl esters.

Similarity is also marked for the covalent hydration of the carbonyl group of studied α-ketoacids. In all experimentally accessible cases hydration of the protonated form, predominating in strongly acidic media, is considerably smaller than that of the uncharged acid molecule (Table 4). This is similar to the trend observed for other carbonyl compounds[5]. The free acids and corresponding uncharged forms of esters are the most strongly hydrated among the species studied. The hydration is considerably weaker in monoanions and often negligibly small in dianions. This effect of the carboxylate group is opposite to the effect of halogens and its interpretation must be postponed until structural effects on rate constants of both hydration and dehydration are available. Replacement of the first carboxyl group results in a small decrease in the value of K_d in the uncharged form (Table 4), but in a considerable increase in the hydration of the monoanion. Replacement of two COOH groups for two COOR groups results in a decrease in concentration of the keto form, from 12% to 6% for oxalacetic and from 35% to 29% for α-ketoglutaric acid.

The most striking difference between α-ketoglutaric and oxalacetic acid is the presence of significant concentration of enol forms in solutions of the latter. This reflects the fact that oxalacetic acid is not only an α-ketoacid, but also the β-ketoacid. In the enol form of oxalacetic acid resonance stabilization can occur involving both carboxylic groups. Such stabilization cannot occur if the second carboxylic group is separated from the double bond by another sp^3 carbon. In the uncharged form the position of the keto-enol equilibrium is only slightly affected by esterification of one or both carboxy groups (Table 3).

Differences in enolization between oxalacetic and ketoglutaric acid result also in some differences in spectral and polarographic behavior. The responses, in both of these techniques, of the keto group are similar for both keto acids. On the other hand, absorbance in the 260-270 nm range and polarographic reduction waves at more negative potentials are characteristic for the enol form and are absent in solutions of α-ketoglutaric acid.

REFERENCES

1. a) K.J. Pedersen, Acta Chem. Scand., 6, 243 (1952).
   b) M. Becker and H. Strehlow, Z. Elektrochem., 64, 813, 818 (1960).
   c) S. Ono, M. Takagi and T. Wasa, Collect. Czechoslov. Chem. Commun., 26, 141 (1961).
   d) M. Eigen, K. Kustin and H. Strehlow, Z. Physikal. Chem., 31, 140 (1962).
   e) H. Strehlow, Z. Elektrochem., 66, 392 (1962).
   f) M. Takagi, S. Ono and T. Wasa, Review of Polarography (Japan), 11, 210 (1963).
   g) M. Becker, Ber. Bunsenges. Phys. Chem., 68, 669 (1964).
   h) V. Gold, G. Socrates and M.R. Crampton, J. Chem. Soc., 5888 (1964).
   i) V.S. Griffith and G. Socrates, Trans. Faraday Soc., 63, 673 (1967).
   j) G. Öjelund and I. Wadsö, Acta Chem. Scand., 21, 1408 (1967).
   k) Y. Pocker, J.E. Meany, B.J. Nist and C. Zadorojny, J. Phys. Chem., 73, 2879(1969).
   l) Y. Pocker and J.E. Meany, J. Phys. Chem., 74, 1486 (1970).
   m) N. Hellstrom and S.O. Almquist, J. Chem. Soc. B, 1396 (1970).
   n) G.A. Gachko, L.N. Kivach, S.A. Maskevich, Yu.M. Ostrovskii, S.G. Podtynchenko, Dokl. Akad. Nauk BSSR, 27, 946 (1983).
   o) W. Knoche, M.A. Lopez-Quintela and J. Weiffen, Ber. Bunsenges. Phys. Chem., 89, 1047 (1985).

2. J. Jen and W. Knoche, Ber. Bunsenges. Phys. Chem., 72, 539 (1969).

3. a) K.H. Meyer, Chem. Ber., 45, 2843 (1912).
   b) A. Hantzsch, Chem. Ber., 48, 1407 (1915).
   c) E. Gelles and R.W. Hay, J. Chem. Soc., 3673 (1958).
   d) B.E.C. Banks, J. Chem. Soc., 5043 (1961).
   e) W.D. Kumler, E. Kun and J.N. Shoolery, J. Org. Chem., 27, 1165 (1962).
   f) G.W. Kosicki, Canad. J. Chem., 40, 1280 (1962).
   g) G.W. Kosicki and S.N. Lilpovac, Canad. J. Chem., 42, 403 (1964).
   h) S.S. Tate, A.K. Grzybowski and S.P. Datta, J. Chem. Soc., 1372 (1964).

i) C.S. Tsai, Y.T. Lin and E.E. Sharkawi, J. Org. Chem., 37, 85 (1972).
j) H.L. Hess and R.E. Reed, Arch. Biochem. Biophys., 153, 226 (1972).
k) D.W. Schiering and J.E. Keaton, J. Molec. Struct., 144, 71 (1986).

4. a) C.I. Pogson and R.G. Wolfe, Biochem. Biophys. Res. Commun., 46, 1048 (1972).
b) F.C. Kokesh, J. Org. Chem., 41, 3593 (1973).
c) M. Emly and D. Leussing, J. Am. Chem. Soc., 103, 628 (1981).
d) P.Y. Bruice, J. Am. Chem. Soc., 105, 4982 (1983).

5. J.P. Segretario, N. Sleszynski, R.E. Partch, P. Zuman and V. Horak, J. Org. Chem., 54, 1986.

6. For dianioin of oxalacetic acid it has been claimed (Reference 4b) that the relative content of all forms is independent of concentration, but data in Table 3 (Reference 4b) indicate for the content of the hydrated form a significant difference, even when concentration of the acid was changed only from 1.14 M to 1.7 M.

7. a) P.Y. Bruice and T.C. Bruice, J. Am. Chem. Soc., 100, 4793 (1978).
b) P.Y. Bruice and T.C. Bruice, J. Am. Chem. Soc., 100, 4802 (1978).

8. R. Boyd, "Acidity Functions" in "Solvent-Solute Interactions", Vol. 1 (J.F. Coetzee and C.D. Ritchie, Eds.), M. Dekker, New York, 1969, p. 97-228.

9. a) W.J. Bover and P. Zuman, J. Am. Chem. Soc., 95, 2531 (1973).
b) T.J.M. Pouw, W.J. Bover and P. Zuman, Advances in Chemistry, 155, 343 (1976).

10. J. Kozlowski and P. Zuman, J. Am. Chem. Soc., submitted.

11. E.M. Arnett, Prog. Phys. Org. Chem. (S.G. Cohen, A. Streitwieser, Jr. and R.W. Taft, Eds.), 1, 223-402 (1963).

12. a) D. Barnes and P. Zuman, J. Electroanal. Chem., 46, 323 (1973).
b) P. Zuman, ibid., 75, 523 (1977).
c) J. Rusling and P. Zuman, ibid., 143, 283 (1983).

13. a) K. Vesely and R. Brdicka, Collect. Czechoslov. Chem. Commun., 12, 313 (1947).
b) R. Bieber and G. Trümpler, Helv. Chim. Acta, 30, 706, 971, 1109, 1286, 1534, 2000 (1947).

14. a) P. Valenta, Collect. Czechoslov. Chem. Commun., 25,853 (1960).
b) J. Volke and P. Valenta, ibid., 25, 580 (1960).
c) E. Laviron, Bull. Soc. Chim. Fr., 2325 (1961).
d) J.P. Segretario, J.F. Rusling and P. Zuman, J. Electroanal. Chem., 143, 291 (1983).

15. J. Kozlowski and P. Zuman, J. Electroanal. Chem., 226, 69 (1987).

ELECTRON TRANSFER-OXY RADICAL MECHANISM FOR ANTICANCER AGENTS: ETOPOSIDE, Cu DIPS, AND BIS(9-AMINOACRIDINES)

Peter Kovacic* and James R. Ames

Department of Chemistry
University of Wisconsin-Milwaukee
Milwaukee, Wisconsin 53211

Michael D. Ryan

Department of Chemistry
Marquette University
Milwaukee, Wisconsin 53233

INTRODUCTION

A mechanism of action for many drugs which is gaining acceptance involves redox cycling accompanied by catalytic electron transfer (ET) and oxy radical formation. Several reviews on this subject have appeared[1-3]. The initial oxy radical hypothesis[4-5] for anticancer agents has recently received increasing support[6-9]. There appear to be several categories of ET agents including quinones, aromatic nitro compounds, metal complexes, and iminium moieties. A theory has been advanced that the iminium ion 1, which may form endogenously[10,11], plays an important role in biological functions[7]. The involvement of 1 in ET includes generation of oxidative stress or interference with normal electron transfer. Application of our general theme has been made to anticancer drugs[8,12-14], carcinogens[15] [O-alkylguanine salt in DNA], antibacterial agents[16-18], CNS agents [benzodiazepine[19]; 1-methyl-4-phenyl-1,2,3,6-tetrahydropyridine (MPTP)[20]; phencyclidine[21]; nicotine[21]], spermine[21], antimalarial agents[22], mesoionic compounds[23], and antiamebic agents[24]. The natural phagocytic response to foreign bodies involves attack by activated oxygen[25].

In relation to antineoplastic action, the mechanism in some cases presumably involves reduction of 1, usually conjugated, by a substrate, such as, protein or DNA, entailing electron abstraction. Oxidation of the radical intermediate 2 by electron donation to an acceptor, for example, oxygen, resulting in superoxide formation, completes the ET process

$$-\overset{+}{N}=C- \underset{-e}{\overset{+e}{\rightleftharpoons}} -N\overset{\bullet}{\text{---}}C- \qquad (1)$$

$\underset{\sim}{1}$ $\qquad$ $\underset{\sim}{2}$

(Equation 1). Derived oxy species, such as the hydroxyl radical, then exert a lethal effect on the tumor cell by attacking vital cellular constituents. The well-known paradox of oncology that generally the substances which are antineoplastic may also induce cancer gives support to this mechanism of action, since oxy radicals have been implicated in carcinogenesis[8,15]. The ability to combat this condition may then be linked to the same property. Also many tumor cells are more susceptible than normal cells to elevated concentrations of oxy radicals. A more complete treatment of this subject is presented elsewhere[8].

The objective of the present study was to determine the electrochemical characteristics of several main categories of antineoplastic agents: quinones (etoposide model), metal complexes [Cu(II)(diisopropylsalicylate)$_2$ (Cu Dips)], and iminium ions from bis(9-aminoacridines). The results are relevant to the feasibility of ET in vivo. In addition, the relationship of electrochemical behavior to structure and physiological activity is addressed.

MATERIALS AND METHODS

Reagents for the syntheses were obtained from Aldrich Chemical Co. (organic) and MCB ($CuCl_2 \cdot 2H_2O$). The six bis(9-aminoacridine) compounds were supplied by the Drug Synthesis and Chemistry Branch, Division of Cancer Treatment, National Cancer Institute. 3-Methoxy-o-benzoquinone was prepared by a published method[26], m.p. 114-116°C, reported 107-110°C[27]. The infrared spectrum was identical to that reported. Cu Dips was synthesized by the procedure of Sorenson[28] with slight alteration (dried at 110°C for 4 h); m.p.136-138°C (dec.), reported 142-144°C (dec.). Elemental analysis for $C_{26}H_{34}O_6Cu$ was, calc. %C, 61.7.; %H, 6.7; found %C, 60.3; %H, 6.6. Some Cu Dips solutions were prepared by adding $CuCl_2 \cdot 2H_2O$ to a solution of 3,5-diisopropylsalicylic acid (1.0 mM) containing sodium hydroxide (1.0 mM), so that the final concentration of Cu was 0.5 mM, in accord with prior procedures (see references in the metal section).

Cyclic voltammetry was performed as previously described[17,19]. The electrodes consisted of either a platinum flag (Sargent-Welch) or a hanging mercury drop (HMDE) working electrode, with a platinum wire as the counter electrode. The Pt electrode was soaked for 10 minutes in 50/50 (v/v) concentrated HNO_3/H_2O, washed well with water, and dried prior to each measuremnt in the alcohol solutions. For each scan with HMDE, a new mercury drop was used. Observed potentials (our work and literature values) were converted to the NHE by addition of 0.24 V to the SCE values.

RESULTS AND DISCUSSION

Quinones

Podophyllotoxin 3 is a naturally occurring antitumor antibiotic[29]. Attempts to reduce the toxicity have resulted in the synthesis of glucosidic phenolic derivatives, VP-16 (etoposide, a semisynthetic analog) 4a and VM-26 (teniposide) 4b. Their mode of action is thought to be associated with DNA breakage[30,31]. Haim et al.[32] proposed a mechanistic pathway entailing oxidation with subsequent O-demethylation to form the o-quinone metabolite 5a (Equation 2). In vivo conversion of methoxyl to hydroxyl has been previously observed[33]. Electrochemical oxidation of VP-16[34] and other methoxyphenols[35] also leads to quinone products.

To test the viability of 5a involvement in redox processes, we performed cyclic voltammetry on a model, namely, 3-methoxy-o-benzoquinone,

5b. This compound was studied in nonaqueous media because of its participation in side reactions in aqueous solution[27]. The quinone gave two reduction waves with Pt and HMDE, with slight differences between the two cases. With Pt the first, $E_{1/2}$ = -0.16 V (Table 1) was quasireversible and diffusion controlled. ΔE_p doubled in magnitude (80 to 160 mV) upon change in the sweep rate from 10 to 200 mV/s. Comparison of the current function (CF) with that of benzil, a compound known to undergo a reversible one-electron reduction[36], gave a value of 1.08 (Table 1) indicating the transfer of one electron. The percent reoxidizability (i_{pa}/i_{pc} x 100%) increased from 79 to 100% as the sweep rate increased. The second wave (E_p = -0.91 V) was broadened, reduced in height, and showed no sign of an anodic peak. Similar results were previously observed with 3,5-di-tert-butyl-o-benzoquinone[37]. The results were attributed to disproportionation of the semiquinone and basicity of the dianion. The same process may be operating with 5b since the pKa of the reduced catechol form is 9.32[38].

Upon switching to HMDE, a change was observed in ΔE_p of the first couple, namely, a decrease from 70 mV at 10 mV/s sweep rate to 60 mV for 20-200 mV/s with the average being 62 ± 4 mV, very close to the theoretical value of 59 mV for a one-electron process[39]. The CF ratio was still about one (Table 1). The second peak, however, was shifted anodically to E_p = -0.75 V, and there was indication of a broad reoxidation peak centered at -0.51 V.

Table 1. Cyclic Voltammetry of 3-methoxy o-benzoquinone (5b) and Cu Dips[a].

Substrate	Electrode	Reduction Potential (V)	CF ratio[b]	i_{pa}/i_{pc}	$E_{pp/2}$(mV)
5b	Pt	-0.16[c], -0.91[d]	1.08	0.98	-
	HMDE	-0.16[c], -0.75[d]	0.96	0.99	-
Cu Dips	Pt	+0.27[d]	-	-	203
$CuCl_2\cdot 2H_2O$-Dips anion[e]	Pt	+0.28[d]	-	-	190
	Pt	+0.21[d]	-	-	260
$CuCl_2\cdot 2H_2O$	Pt	+0.31[d]	-	-	-

[a] 100 mV/s, tetraethylammonium perchlorate (0.1 M, DMF; 0.05 M, 90% aqueous ethanol), substrate (0.5 mM) vs NHE.

[b] $$CF_{ratio} = \frac{CF_{5b}}{CF_{benzil}};\quad CF_{benzil} \frac{A}{(V/s)^{1/2}M} = 16.27 \text{ (0.5mM, Pt, DMF)},$$ 0.185 (0.5 mM, HMDE, DMF).

[c] Quasireversible, Pt; reversible, HMDE, DMF.

[d] Irreversible, DMF.

[e] 1:2 solution (see Materials and Methods).

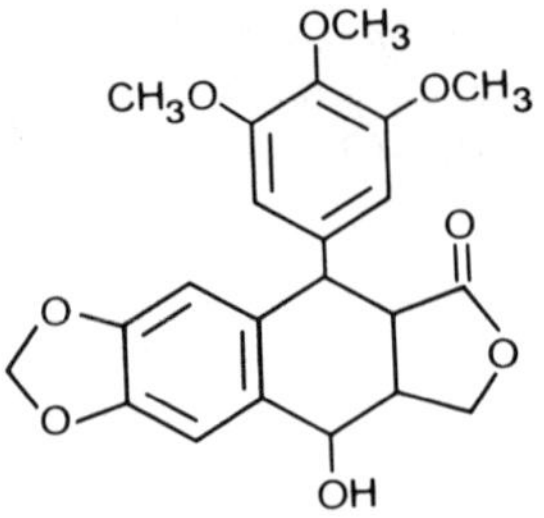

3

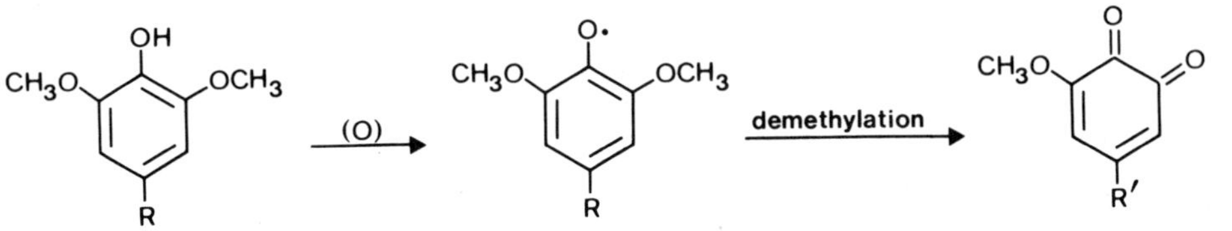

4

5

a) R′ = R in 4
b) R′ = H

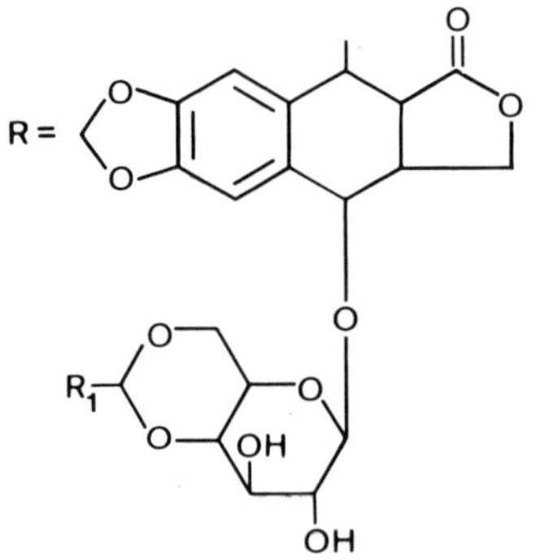

a) R_1 = CH_3 (VP - 16, etoposide)

b) R_1 = 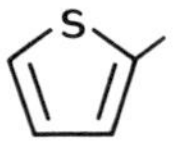 (VM - 26, teniposide)

The reduction potential for 5b has recently been measured at pH 7.1. The reported value[38] ($E°_{H_2}$ = +0.340 V) was for a two-electron, two-proton process. A study of the reversible reduction of 5a ($E_{1/2}$ of about +0.44 V at pH 7) was also carried out involving reduction to the catechol[34]. The oxidation/O-demethylation of 4a to form 5a has been studied in depth. It is difficult to compare these data to our results because of the differences in the electron uptake and in the conditions, for example, solvent. A similar situation has previously been observed with the quinone antibiotics and their models[40]. Compound 5a should reduce at a slightly more negative potential than 5b due to substituent effects. Alkyl groups are known to make the substance more resistant to reduction[23,41a]. The methoxy moiety has a similar effect[42] versus the parent[37].

A possible model of action for 5a would entail redox cycling with formation of superoxide. Subsequently, H_2O_2 and oxy radicals can arise. The antitumor antibiotic quinones and their model analogues display favorable reduction potentials[40,43-45]. Evidence suggests that quinone 5a is the species responsible for the bioactivity. Thus, the phenolic hydroxyl group is crucial, and metabolic activation is necessary for induction of strand breaks[29,46]. In another study, free radical scavengers such as DMSO, thiourea, and disulfiram inhibited the cleavage of DNA by 4a[46]. More recently, etoposide was found to be more toxic under a normal or oxygen-enriched environment than under hypoxic conditions[47], similar to the antitumor antibiotic quinones (vide infra). Also, 4a produced DNA strand breaks[30]. In relation to the mechanism proposed, generally quinone antitumor antibiotics have been shown to cause strand breaks in DNA as a result of superoxide formation[48]. The toxicity presumably results from redox cycling and is oxygen-dependent[3]. The process includes metabolic reduction to the semiquinone[3,49].

o-Quinones may operate via the derived metal chelates which are known to exist[50,51]. Several other mechanisms have been proposed for this class[31,32,52,53].

A number of other quinone drugs possess antitumor activity, for example, β-lapachone[54] and those derived from levodopa and dopamine[55]. An oxy radical mechanism[56] for the anticancer action was invoked[54,55].

Metal Complexes

Metal species elicit a variety of chemical responses in vivo that are relevant to our study, such as oxygen radical formation[57], strand cleavage and formation of complexes with DNA[58]. There has been prior suggestion that electron transfer may play a mechanistic role[59].

1. Copper Complexes

In relation to our approach involving ET, we examined the electrochemical behavior of Cu Dips. The solid complex exhibited reduction at +0.27 V in 90% aqueous ethanol (Table 1). The reduction was characterized as irreversible and diffusion controlled on the basis of the $E_{pp/2}$ value of approximately 203 mV and the constant value of the CF at the scan rates employed (data not shown). Complexes prepared in solution [1:2 Cu(II) to Dips] gave variable irreversible, diffusion-controlled reductions at +0.28 V (six determinations) or +0.21 V (two determinations) (Table 1) in the potential range of +0.84 to 0.0 V. The average $E_{pp/2}$ values varied from about 190 to 260 mV for the two sets of determinations. Copper(II) chloride dihydrate gave a value of +0.31 V under the same conditions.

The variability observed with Cu Dips is in keeping with results from electrochemical studies on the related copper complexes of salicylic acid. Habashy[60] found complex irreversible polarographic waves for solutions at various pH values above 5.4, which was attributed to indefinite equilibrium between several complexes. It is conceivable that a number of chelate species are also being formed under our conditions, but some entities may not be present in high enough concentration to be observed. Also with unbuffered aqueous solutions, the pH and the composition at the electrode surface may determine the complex which participates. Other investigators have also examined the electrochemistry of coordination compounds derived from Cu(II) and salicylic acid. Reported reduction potentials range from +0.27 to -0.33 V[61-63]. The values for Cu Dips are expected to be slightly more negative than those for the salicylate complexes, due to the effect of alkyl substituents[23,41a]. We were unable to observe reduction waves in organic solution (DMF, DMSO) using the solid complex of Cu Dips. Our approach involving oxy radical generation via electron transfer is supported by the recent report of H_2O_2 production by Cu Dips, which may be associated with its SOD-like activity[64].

The related antitumor copper salicylaldoxime complexes gave irreversible reduction with potentials of -0.47 to -0.64 V[8]. Another class of Cu(II) coordination compounds, the thiosemicarbazones, is known to possess anticancer activity[65] and favorable $E_{1/2}$ values[22]. The well-known chelating agent, dimethylglyoxime, exhibits anticancer activity when complexed with copper[41b], which may be due to electron transfer; $E_{1/2}$ is about -0.08 V[66]. α,α'-Dipyridyl is effective against neoplasms[67]. When chelated to copper, the complex generates superoxide in the presence of O_2 and a reducing agent[68]; $E_{1/2}$ is +0.12 V[41c].

2. Others

Recently Clarke et al.[69] have advanced the proposal that antitumor Ru(II)ammine complexes may excert their action by redox cycling in the presence of oxygen and a reducing agent with formation of active oxygen species which cleave DNA. The most prominent member of the metal group is cis-DDP. Evidence for electron transfer is provided by electrochemical behavior[8], radiosensitization[70,71] and nephrotoxicity[72-74]. Anticancer activity is displayed by a number of other metals and metalloids in compound form[75-78]. The possibility exists that the ultimate active forms of the various metals may be the DNA complexes, for example, with copper[76] and ruthenium[69]. ET and oxy radical formation are known to occur with enzymes that contain copper[79].

Diacridines

Compound 6a exhibited an irreversible, diffusion-controlled process with E_p of -1.50 V. Further indication of irreversibility is based on the $E_{pp/2}$ value of 120 mV (Table 2). Addition of acetic acid produced reduction at -0.77 V, $E_{pp/2}$ = 40 mV, usually accompanied by adsorption. Diacridine 6b gave similar results, except in the presence of acid. Calculations from the voltammogram indicate potential reversibility[39]. Sample 6c reduced at -0.88 V, accompanied by adsorption. The difference between the values for 6c and 6a is likely due to the presence of methoxyl groups which are known to adversely affect reduction potential[42]. Reversibility was indicated by an $E_{pp/2}$ value of about 50 mV, and the absence of a change in E_p upon a tenfold increase in sweep rate. Upon addition of 1.0 mM hydroxide ion no reduction was observed until -1.52 V. The potential changed about 30 mV when sweep rate was increased tenfold. The initial data were then reproduced on addition of acetic acid. The octamethylene compound 6d, which gave results similar to 6c, was not studied further in the presence of base.

On introduction of the quinacrine nucleus, as in 6e and 6f, and alteration of the chain linkage, the potentials generally became more positive. For example, 6e gave E_p of -0.71 V for the dihydrochloride salt. Again computations point to reversibility (Table 2). The free base gave E_p at -1.36 V which increased to -0.83 V (Table 2) on addition of acid. Compound 6f provided results analogous to 6e (Table 2); however, the initial potential was -0.62 V, possibly due to the enhanced positive charge.

It is important to note the favorable influence of protonation (iminium formation) on the ease of reduction. The salt form is expected to be present in vivo, similar to other aminoacridines[80]. Although the potentials are fairly negative under our conditions, we believe that they may be appreciably more positive in vivo due to stereochemical effects. The adverse influence of 7 on reduction can be demonstrated with simple models. Thus, at pH 5.5, the 9-amino derivative of acridine gave $E_{1/2}$ of -0.83 V while the parent took up charge at -0.16 V[81]. The diacridines, such as 6a-d and 6f, are known to be very strong binders to DNA by bis-intercalation[82] which would decrease the contribution by 7. As the side chain is removed from coplanarity with the nucleus, the resonance con-

Table 2. Cyclic Voltammetry of Diacridines[a]

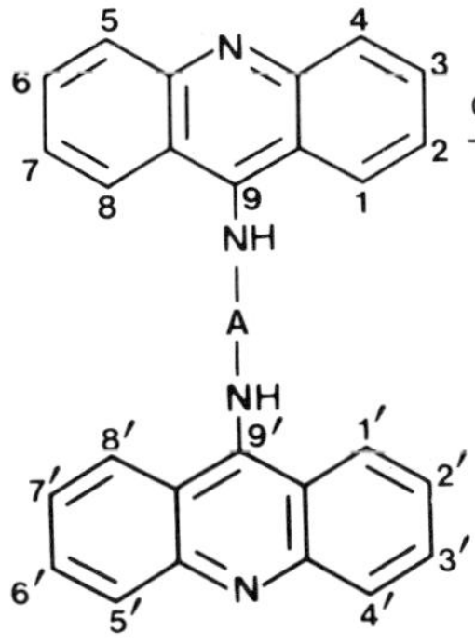

Compound	NCS number	Substituent: Ring	Substituent: A
a	219733	H	$(CH_2)_6$
b	219734	H	$(CH_2)_8$
c	260610	3,3′- OCH_3	$(CH_2)_6$
d	263434	3,3′- OCH_3	$(CH_2)_8$
e	294370	3,3′- Cl 7,7′- OCH_3	$(CH_2)_4 \cdot 2HCl$
f	194906	3,3′- Cl 7,7′- OCH_3	$(CH_2)_2NH(CH_2)_2NH(CH_2)_2 \cdot 4HCl$

Structure 6: acridine (positions 1–9, N) –NH–A–NH– acridine (positions 1′–9′, N)

Substrate	$[OH^-]$mM	[HOAc]mM	$-E_p$	$E_{p\,p/2}$
6a	-	-	1.50	120
	-	24	0.77[b]	40
	-	46	0.77	40
6b	-	-	1.50	100
	-	24	0.82	-
	-	46	0.78	55
6c	-	-	0.88[b]	-
	-	24	0.88	50
	1.0	-	1.52	80
	1.0	46	0.86	50
6d	-	-	0.88	-
6e	-	-	0.71	50
	1.0	-	1.36	70
	1.0	24	0.83	60
6f	-	-	0.62	60
	2.0	-	1.41	100
	2.0	24	0.80	70

[a]100 mV/s, tetraethylammonium perchlorate (0.1 M), substrate (0.5 mM) vs NHE, DMF, HMDE.

[b]With adsorption.

tribution from 7 lessens. A similar situation exists for the related 9-anilinoacridines[12,14]. Interaction with DNA places the side chain in the minor groove.

These agents are known to cleave DNA, perhaps partly from oxidative stress[12,14]. Other acridine moieties and related structures produce oxygen

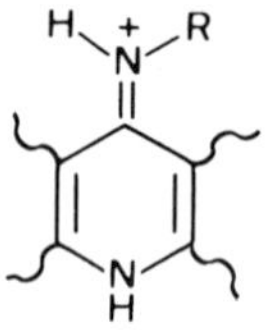

7

radicals under various conditions[14,22]. Nucleic acid cleavage has not been observed for the octamethylene linked diacridine[83].

Prior authors have proposed that the diacridines exert their effects by interference with nucleic acid synthesis[84]. However, there is no correlation between drug activity and growth parameters[82,85]. Alternatively, cell membranes may be involved[82,85,86]. These agents might also act by stimulation of a DNA-topoisomerase II complex[87], or by combination with protein thiol[87], similar to the 9-anilino counterparts.

Other Considerations

Although several of the reactions in this study were irreversible, behavior in vivo might be different for the drug immobilized at the active site. Also the energetics may be more favorable. Specific aspects regarding in vivo versus in vitro electrochemistry are discussed elsewhere[21,88].

The feasibility of electrochemical characteristics translating into physiological activity has been reported for antitumor iminobenzoquinones[89], aromatic and heterocyclic nitro compounds[90], mitomycins[91] and aminoacridines[92].

The present theory is clearly oversimplified. Other important factors include inhibition of DNA synthesis, DNA defect repair, antimetabolite action, and immunological responses[8]. One drug that is not reasily accommodated by our approach is 5-fluorouracil[93]. It is conceivable that several mechanisms operate in concert for certain drugs.

Based on the evidence from our work and prior contributions, the various agents appear to fit the working hypothesis: binding (intercalation) to DNA, involvement in ET, and production of activated oxygen. However, in certain cases there are gaps that require filling.

Companion articles on application of the ET theory to other anticancer drugs and to CNS agents are included in these Proceedings.

CONCLUSIONS

An electron transfer mechanism is advanced for various classes of anticancer agents. The neoplasm is presumably destroyed by induction of oxidative stress. Cyclic voltammetry was performed on some of the main categories: quinones (3-methoxy-o-benzoquinone, etoposide model), metal complexes [Cu(II)(3,5-diisopropylsalicylate)$_2$] (Cu Dips) and iminium from bis(9-aminoacridines). Reduction potentials ranged from +0.28 to -0.88 V. Electrochemical behavior is related to structure and physiological activity.

ACKNOWLEDGEMENTS

We thank Dr. Ven L. Narayanan (Drug Synthesis and Chemistry Branch, Division of Cancer Treatment, National Cancer Institute) for his generosity in supplying the diacridine compounds.

REFERENCES

1. H. Kappus, Biochem. Pharmacol., 35, 1 (1986) (and references sited therein).

2. H. Sies, Ed., "Oxidative Stress", Academic Press, New York (1985).

3. B. Halliwell and J.M.C. Gutteridge, "Free Radicals in Biology and Medicine", Clarendon Press, Oxford (1985), pp. 279-315.

4. R.A. Holman, Lancet, ii, 515 (1956).

5. P. Kovacic, Ohio J. Sci., 59, 318 (1959) (and references sited therein).

6. L.W. Oberley, in: "Superoxide Dismutase", Vol. II, L.W. Oberley, Ed., CRC Press, Boca Raton, FL., Chap. 6 (1982), pp. 127-165.

7. P. Kovacic, Kem. Ind., 33, 473 (1984) (and references therein).

8. P. Kovacic, J.R. Ames, P. Lumme, H. Elo, O. Cox, H. Jackson, L.A. Rivera and M.D. Ryan, Anti-Cancer Drug Design, 1, 197 (1986) (and references therein).

9. J.W. Lown, Adv. Free Rads. Biol. Med., 1, 225 (1985).

10. M. Overton, J.A. Hickman, M.D. Threadgill, K. Vaughan and A. Gescher, Biochem. Pharm., 34,2055 (1985).

11. J. Knabe, in: "Iminium Salts in Organic Chemistry", Part 2, H. Böhme and H.G. Viehe, Ed., in: "Advances in Organic Chemistry", E.C. Taylor, Ed., Wiley-Interscience, New York (1979), p. 733-758.

12. P.W. Crwford, P. Lumme, H. Elo, M.D. Ryan and P. Kovacic, Free Rad. Res. Commun., 3, 347 (1987) (and references therein).

13. P.W. Crawford, W.O. Foye, M.D. Ryan and P. Kovacic, J. Pharm. Sci., 76, 481 (1987).

14. P. Kovacic, J.R. Ames and M.D. Ryan, Anti-Cancer Drug Design, 2, 37 (1987) (and references therein).

15. P. Kovacic, P.W. Crawford, M.D. Ryan and V.C. Nelson, Bioelectrochem. Bioenerg., 15, 305 (1986) (and references therein).

16. J.R. Ames, M.D. Ryan and P. Kovacic, J. Free Rad. Biol. Med., 2, 377 (1986) (and references therein).

17. P.W. Crawford, R.G. Scramehorn, U. Hollstein, M.D. Ryan and P. Kovacic, Chem.-Biol. Interactions, 60, 67 (1986).

18. M.D. Ryan, R.G. Scamehorn and P. Kovacic, J. Pharm. Sci., 74, 492 (1985).

19. P.W. Crawford, P. Kovacic, N.W. Gilman and M.D. Ryan, Bioelectrochem. Bioenerg., 16, 407 (1986).

20. J.R. Ames, N. Castagnoli, Jr., M.D. Ryan and P. Kovacic, Free Rad. Res. Commun., 2, 107 (1986).

21. J.R. Ames, S. Brandänge, B. Rodriguez, N. Castagnoli, Jr., M.D. Ryan and P. Kovacic, Bioorg. Chem., 14, 228 (1986) (and references therein).

22. J.R. Ames, M.D. Ryan, D.L. Klayman and P. Kovacic, J. Free Rad. Biol. Med., 1, 353 (1985) (and references therein).

23. J.R. Ames, K.T. Potts, M.D. Ryan and P. Kovacic, Life Sci., 39, 1085 (1986) (and references therein).

24. J.R. Ames, U. Hollstein, A.R. Gagneux, M.D. Ryan and P. Kovacic, J. Free Rad. Biol. Med., 3, 85 (1987).

25. R.L. Baehner, L.A. Boxer and L.M. Ingraham, in: "Free Radicals in Biology", Vol. V, W. Pryor, Ed., Academic press, New York (1982), pp. 91-93.

26. E.M. Acton and G.L. Tong, J. Hetero. Chem., 18, 1141 (1981).

27. E. Adler, R. Magnusson, B. Berggren and H. Thomelius, Acta Chem. Scand., 14, 515 (1960) (and references therein).

28. J.R.J. Sorenson, J. Med. Chem., 19, 135 (1976).

29. B.H. Long, S.T. Musial and M.G. Brattain, Biochemistry, 23, 1183 (1984).

30. A.J. Wozniak and W.E. Ross, Cancer Res., 43, 120 (1983).

31. S.D. Horwitz and J.D. Loike, Lloydia., 40, 82 (1977).

32. N. Haim, J. Roman, J. Nemec and B.K. Sinha, Biochem. Biophys. Res. Commun., 135, 215 (1986) (and references therein).

33. T.R. Sweeney and R.E. Strube, in: "Burger's Medicinal Chemistry", Part II, 4th ed., M.E. Wolff, Ed., Wiley, New York (1979), p. 383.

34. J.J.M. Holthuis, W.J. Van Oort, F.M.G.M. Römkens, J. Renema and P. Zuman, J. Electroanal. Chem., 184, 317 (1985).

35. M. Petek, S. Bruckenstein, B. Feinberg and R.N. Adams, J. Electroanal. Chem., 42, 397 (1973).

36. M.D. Ryan and D.H. Evans, J. Electroanal. Chem., 67, 333 (1976).

37. M.D. Stallings, M.M. Morrison and D.T. Sawyer, Inorg. Chem., 20, 2655 (1981).

38. B.W. Carlson and L.L. Miller, J. Am. Chem. Soc., 107, 479 (1985).

39. A.J. Bard and L.R. Faulkner, in: "Electrochemical Methods. Fundamentals and Applications", Wiley, New York (1980), pp. 213-248.

40. A. Ashnagar, J.M. Bruce, P.L. Dutton and R.C. Prince, Biochim. Biophys. Acta, 801, 351 (1984).

41. A. Albert, in: "Selective Toxicity", 7th ed., Chapman and Hall, New York (1985); a) p. 51; b) p. 485; c) p. 463.

42. P. Zuman, in: "Substituent Effects in Organic Polarography", Plenum, New York (1967), p. 148.

43. G.M. Rao, J.W. Lown and J.A. Plambeck, J. Electrochem. Soc., 124, 195 (1977).

44. G.M. Rao, A. Begleiter, J.W. Lown and J.A. Plambeck, J. Electrochem. Soc., 124, 199 (1977).

45. G.M. Rao, J.W. Lown and J.A. Plambeck, J. Electrochem. Soc., 125, 534 (1978).

46. A.J. Wozniak, B.S. Glisson, K.R. Hande and W.E. Ross, Cancer Res., 44, 626 (1984) (and references therein).

47. B.A. Teicher, S.A. Holden and C.M. Rose, J. Natl. Cancer Inst., 75, 1129 (1985).

48. G.R. Buettner and L.W. Oberley, in: "Oxygen and Oxy Radicals in Chemistry and Biology", M.A.J. Rodgers and E.L. Powers, Eds., Academic Press, New York (1981), p. 607.

49. J.W. Lown, Molec. Cell. Biochem., 55, 17 (1983).

50. R.D. Irons and T. Sawahata, in: "Bioactivation of Foreign Compounds", M.W. Anders, Ed., Academic Press, New York (1985), pp. 259-281.

51. C.G. Pierpont and R.M. Buchanan, Coord. Chem. Rev., 38, 45 (1981).

52. M. Gosalvez, J. Perez-Garcia and M. Lopez, Eur. J. Cancer, 8, 471 (1972).

53. W. Ross, T. Rowe, B. Glisson, J. Yalowich and L. Liu, Cancer Res., 44, 5857 (1984) (and references therein).

54. R. Docampo and S.N.J. Moreno, in: "Free Radicals in Biology", Vol. VI, W.A. Pryor, Ed., Academic Press, Orlando (1984), pp. 243-288.

55. M.M. Wick, Cancer Res., 40, 1414 (1980).

56. G. Cohen and R.E. Heikkila, J. Biol. Chem., 249, 2447 (1974).

57. A. Stern, in: "Oxidative Stress", H. Sies, Ed., Academic Press, New York (1985), pp. 340-341.

58. A. Furst and S.B. Radding, J. Environ. Sci. Health, C2, 103 (1984).

59. P.O. Lumme and H.O. Elo, Inorg. Chim. Acta, 107,L15 (1985).

60. G.M. Habashy, J. Electroanal. Chem., 21, 357 (1969).

61. V. Srivastava and H.L. Nigam, Bioelectrochem. Bioenerg., 9, 627 (1982).

62. C.-L. O'Young and S.J. Lippard, J. Am. Chem. Soc., 102, 4920 (1980) (and references therein).

63. L.R. de Alvare, K. Goda and T. Kimura, Biochem. Biophys. Res. Commun., 69, 687 (1976).

64. L.W. Oberley, S.W.C. Leuthauser, R.F. Pasternack, T.D. Oberley, L. Schutt and J.R.L. Sorenson, Agents and Actions, 15, 535 (1984).

65. D.H. Petering, in: "Metal Ions in Biological Systems", H. Sigel, Ed., Vol. 11, Marcel Dekker, New York (1980), p. 197-229.

66. G.V. Prokhorova, L.K. Shpigun and E.N. Vinogradova, J. Anal. Chem. USSR, 27, 685 (1972).

67. R.M. Hellman, J. Glass and M.T. Nunez, in: "Structure and Function of Iron Storage, Transport Proteins", Proceedings of the 6th International Conference, I. Urushizaki, P. Aisen and I. Listowsky, Eds., Elsevier, Amsterdam (1983),p. 479-480; Chem. Abstr., 101, 48280 (1984).

68. D.R. Graham, L.E. Marshall, K.A. Reich and D.S. Sigman, J. Am. Chem. Soc., 102, 5419 (1980).

69. M.J. Clarke, B. Jansen, K.A. Marx and R. Kruger, Inorg. Chim. Acta, 124, 13 (1986).

70. G.P.Jacobs, Inter. J. Radiat. Biol., 49, 887 (1986).

71. V.L. Narayanan and W.W. Lee, Adv. Pharmacol. Chemother., 19, 155 (1982).

72. Y. Katsukura, N. Abe, N. Watabe, W. Oowada and H. Ochi, Yakuri to Chiryo, 14, 1285 (1986); Chem. Abstr., 105, 72244 (1986).

73. K. Sugihara and M. Gemba, Japn. J. Pharmacol., 40, 353 (1986); Chem. Abstr., 104, 81676 (1986).

74. M.E. De Broe and R.P. Wedeen, Eur. J. Cancer Clin. Oncol., 22, 1029 (1986).

75. P.J. Sadler, Chem. Brit., 18, 182 (1982).

76. B. Halliwell and J.M.C. Gutteridge, Molec. Aspects Med., 8, 89 (1985).

77. J.G. Goddard, D. Basford and G.D. Sweeney, Biochem. Pharmacol., 35, 2381 (1986).

78. P. Lumme, H. Elo and J. Jänne, Inorg. Chim. Acta, 92, 241 (1984).

79. B.G. Malmström, L.-E. Andreasson and B. Reinhammar, in: "The Enzymes", Vol. XII, P.D. Boyer, Ed., Academic Press, New York (1975), pp. 507-579.

80. D.C. Warhurst and S.C. Thomas, Biochem. Pharmacol., 24, 2047 (1975).

81. B. Breyer, G.S. Buchanan and H. Duewell, J. Chem. Soc., 360 (1944).

82. T.K. Chen, R. Fico and E.S. Canellakis, J. Med. Chem., 21, 868 (1978).

83. E.S. Canellakis, R.M. Fico, A.H. Sarris and Y.H. Shaw, Biochem. Pharmacol., 25, 231 (1976).

84. W.A. Denny, G.J. Atwell, B.C. Baguley and L.P.G. Wakelin, J. Med. Chem., 28, 1568 (1985) (and references therein).

85. R.E. Elliott, N.S. Karadsheh, J. Kole and E.S. Canellakis, Biochem. Pharmacol., 34, 2123 (1985).

86. A. Adams, B. Jarrott, B.C. Elmes, W.A. Denny and L.P.G. Wakelin, Molec. Pharmacol., 27, 480 (1985).

87. A. Wong, C.-H. Huang, S.-M. Hwang, A. Prestayko and S.T. Crooke, Biochem. Pharmacol., 35, 1655 (1986).

88. J.K. Barton, C.V. Kumar and N.J. Turro, J. Am. Chem. Soc., 108, 6391 (1986).

89. E.M. Hodnett, G. Prakash and J. Amirmoazzami, J. Med. Chem., 21, 11 (1978).

90. C.F. Chignell, Environ. Health Perspec., 61, 133 (1985).

91. S. Kinoshita, K. Uzu, K. Nakano, M. Shimiuzu, T. Takahashi and M. Matsui, J. Med. Chem., 14, 103 (1971).

92. V.A. Shapovalov, A.N. Gaidukevich and V.D. Bezuglyi, J. Gen. Chem. USSR, 50, 2264 (1980).

93. C.L. Will and B.J. Dolnick, Molec. Pharmacol., 29, 643 (1986).

SURFACE ENHANCED RAMAN SCATTERING (SERS) SPECTROSCOPY OF GUANINE DERIVATIVES

E. Koglin* and J.-M. Sequaris

Institute of Applied Physical Chemistry
Nuclear Research Center (KFA) Jülich
P.O. Box 1913
D-5170 Jülich / F.R.G.

INTRODUCTION

The discovery that Raman scattering from molecules adsorbed on metal surfaces (electrodes, colloids and island films) were enhanced by 10^5-10^6 has caused extraordinary interest and excitement in recent years[1,2]. What is the advantage of this Surface Enhanced Raman Scattering (SERS)?

1. The high enhancement factor of the Raman scattering intensity creates a new technique for obtaining high resolution vibrational spectra of molecules (organic, inorganic, biomolecules) from very dilute solutions down to 10^{-9} M.

2. SERS investigations have shown that this surface spectroscopy is an extemely powerful in-situ vibrational method of studying the interfacial behavior of biomacromolecules in their natural environment: water[2].

3. Combining SERS spectroscopy and high-performance thin-layer chromatography (HPTLC) has permitted in situ analysis of HPTLC spots down to nanogram amounts of substances[3].

4. It was also demonstrated that ring breathing SERS bands of nucleic acids bases are conformation sensitive for B $\rightarrow$ Z transitions in DNA[4].

The knowledge of structural characteristics of the nucleotides, nucleosides, and bases in aqueous solution provides fundamental information for the understanding of the conformation of nucleic acids. Modified bases in DNA cause mutagenic, carcinogenic and cytotoxic effects on the cell. A large group of mutagens and carcinogens are alkylating agents of which the electrophilic groups react in particular with guanine sites of DNA.

For the interpretation of vibrational spectra of methylated nucleic acids it is therefore necessary to measure precise frequencies of the guanine base and the methylated guanine derivatives. Recording normal Raman spectra of guanine and its derivatives under physiological salt conditions is impossible because of their very low solubility in water

($\leq 5 \times 10^{-4}$ g in 100 g H_2O). However, vibration spectra of these molecules have been obtained by means of colloid-surface-enhanced Raman spectroscopy (Colloid-SERS)[5,6].

The purpose of this contribution is to show the recent advances in the application of SERS from guanine derivatives adsorbed on silver electrode surfaces and HPTLC activated plates.

In addition the laser Raman microprobe has been used in combination with HPTLC/SERS spectroscopy for the investigation of HPTLC spots down to 1 μm in size or other forms of microsamples approaching the femtogram level in mass.

MATERIALS AND METHODS

Chemicals

Guanine (Gua), 1-methylguanine (1-MeGua), 3-methylguanine (3-MeGua), 7-methylguanine (7-MeGua), and 9-methylguanine (9-MeGua) were obtained from Fluka A.G. and 2-amino-6-chloro-purine or 6-chloroguanine (6-ClGua) from Sigma Chemical Company. O^6-Methylguanine (O^6-MeGua) was prepared from 6-chloroguanine according to Balsinger and Montgomery[11]. All other chemical reagents were of analytical quality and were purchased from E. Merck. HPTLC plates of silica gel 60 (10 cm x 10 cm) without fluorescent indicator are produced by E. Merck.

The spectroelectrochemical cell consisted of a quartz glass cylinder with a diameter of about 1.5 cm and a capacity of about 2 ml of solution. The working electrode was a polycrystalline silver disk, approximately 4 mm in diameter, enclosed in a Teflon holder. This silver electrode was prepared before each experiment by mechanical polishing, ultrasonic treatment and an electrochemical cleaning by H_2 evolution. The ex-situ electrode roughening procedure was carried out in the absence of the adsorbate, i.e., in a 0.1 M KCL electrolyte solution.

The silver colloids were prepared by the use of Creighton's procedure[7] by the reduction of $AgNO_3$ with $NaBH_4$. In a typical experiment, one volume part of 10^{-3} M $AgNO_3$ is added dropwise to three volume parts of 2×10^{-3} M $NaBH_4$, cooled in an ice bath, and mixed vigorously. The yellow-brownish silver colloidal solution was stored at 5°C for weeks without any change in color.

Applications of volumes of solution down to 100 nl has been undertaken with a 1 μl Hamilton syringe in conjunction with a micrometer. After drying, HPTLC plates are sprayed to wetness with silver colloidal solution by a spray atomizer. Colored spots arising on the HPTLC plate are used to locate the separated molecules. The spot color varies from pale yellow through orange to violet depending on the nature and concentration of the compounds.

Electrode surfaces, colloidal sols and HPTLC plates are analyzed at room temperature by a computer controlled double beam spectrometer: Spex double monochromator 14018 (0.85 m, f/7.8), Datamate DM 1, cold photomultiplier (RCA 31034 A) operated in the photon counting mode. Monochromator slits were selected so as to provide better than 8 cm^{-1} bandpass. The excitation wavelength was the 514.5 nm line of an argon ion laser (Spectra Physics, Model 2020-03) with 10 m W of power. HPTLC-SERS spectra were measured in a typical 90°C scattering arrangement. Spot diameters are about 2-4 mm and the laser beam focused on HPTLC plates is approximately 1.5 mm in length and 0.1 mm in width. Spectra were obtained with a scanning speed (accumulation time of 2 cm^{-1}/1.5 s for ratios of substance quantity/spot from 10 to 60 ng.

Micro-Raman HPTLC/SERS spectra were obtained with an Instruments S.A. MOLE-S 3000 spectrometer. The MOLE-S 3000 is a new, fully computerized, triple spectrometer system with multichannel (E-IRY 1024) acquisition of data.

RESULTS AND DISCUSSION

To illustrate the high sensitivity of SERS spectroscopy we have selected the guanine molecule, because this molecule has a very low solubility in H_2O (5 x 10^{-4} g in 100 g H_2O). Therefore, under these conditions it is impossible to obtain a normal Raman spectrum of this important molecule in biochemical research. But by means of electrode SERS-spectroscopy[2] such a very insoluble molecule shows a very strong vibration spectrum (cf. Figure 1). Thus, SERS spectroscopy opened up a new field of vibration spectroscopy for very low solubility molecules in water.

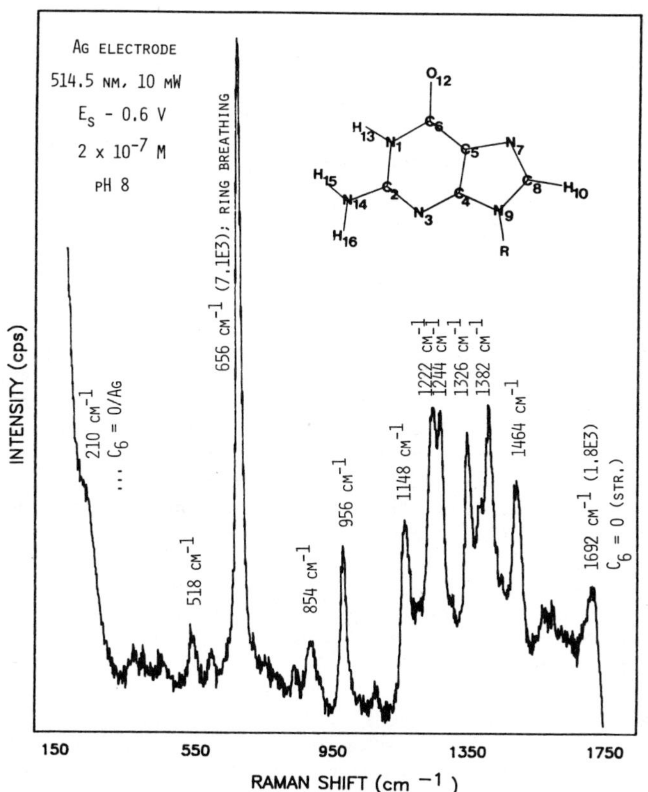

Figure 1. SERS spectrum of guanine. Guanine concentration 2 x 10^{-7} M; 0.1 M KCL, 2 x 10^{-3} M Tris buffer (pH 8.0), laser excitation line 514.5 nm, laser power at the electrode spot 10 mW, prior activation of the Ag-electrode (in the absence of guanine): 15 mCb between -0.1 V and +0.2 V). Adsorption potential E_s = -0.6 V vs (Ag/AgCl) reference electrode.

The strongest line in the guanine SERS spectrum at 656 cm^{-1} can be assigned to the inphase ring-breathing vibration. This SERS band is thus of great interest in detecting modified nucleic acid bases by the SERS method.

For example, the ring-breathing mode of guanine at 656 cm^{-1} is shifted to shorter frequencies by substitution of the original carbonyl group $C_6 = O$ with C_6-OCH_3 to 623 cm^{-1} and with C_6-Cl to 585 cm^{-1}.

a) Electrode-SERS spectra of methylated guanine

Alkylation of nucleic acid components and nucleic acids has been of considerable interest in recent years. SERS spectroscopy on Ag electrode is a useful technique for the identification of methyl derivatives of nucleic acids. The methylation of guanine leads to a specific change in the SERS spectrum.

For the interpretation of vibrational spectra of methylated nucleic acids it is therefore necessary to measure precise frequencies of methylated guanine derivatives adsorbed on electrode surfaces. The observed frequencies and mode assignments of the electrode SERS bands of guanine and its methylated guanine derivatives 1-Me-Gua, 3-Me-Gua, 7-Me-Gua, and 9-Me-Gua are given in Table 1. The substantially higher sensitivity of these electrode SERS method by a factor 10^5-10^6 in comparison with normal Raman scattering (NRS) permits rapid spectra recordings with a conventional Raman spectrometer of guanine derivatives at low bulk concentrations down to 10^{-7} M. Most of the electrode SERS bands are very similar to their known NRS counterparts. The largest frequency shift is about 30 cm^{-1}. However, it must be remarked that the electrode SERS spectra of methylated guanine bases are obtained in their neutral forms while the already published frequencies by NRS spectroscopy[8,9] only concern the ionized forms in solution. Indeed, in the case of normal Raman spectroscopy, the very low solubility of guanine and its methylated derivates in H_2O requires extreme pH values to establish the then necessary bulk concentration range from 10^{-2} to 10^{-1} M.

The most intense bands in the electrode SERS spectra of guanine and its methylated derivates are due to breathing modes in the 600 cm^{-1} region and to coupled ring and double bond stretching vibrations in the region 1500 to 1700 cm^{-1}. These bands are thus of interest in detecting modified nucleic acid bases by the electrode SERS methode. For example, the strong enhanced ring modes at 1382 cm^{-1} and 1464 cm^{-1}, characteristic for guanine, disappear in 7-methylguanine, the guanine ring breathing mode at 656 cm^{-1} decreases and new enhanced bands appear at 692 cm^{-1} and 1352 cm^{-1} (cf. Table 1).

The fixation of the electrophilic CH_3^+-group at N positions of the purine cycle also considerably modifies the SERS spectrum in the spectral range of 500 cm^{-1} (cf. Figure 2).

In the 9-methylguanine SERS spectrum, for instance, we observe new bands at 506, 536 and 556 cm^{-1}. The assignment of these relatively low frequency bands to N-CH_3 deformation modes is not yet very well understood.

Another striking feature of the electrode SERS spectra of N-methylated guanines is the dependence of the carbonyl stretching vibration on the site of methylation. In Table 1 it is observed that the SERS spectra of 9-Me-Gua and 3-Me-Gua depart substantially from the spectra of 1-Me-Gua and 7-Me-Gua in the 1700 cm^{-1} region (the C_6=0 stretching vibration is missing in the 3-Me-Gua and 9-Me-Gua spectra).

A first suggestion would assume a tautomeric form of guanine at neutral pH. Indeed the SERS spectra may be interpreted by a displacement of the tautomeric equilibrium from the predominant keto form in solution $-N_1H-C_6=O$ to the enol form $-N_1=C_6-OH$ in the adsorbed state.

The absence of the $C_6=O$ band in the SERS spectra of 9-Me-Gua and 3-Me-Gua would thus suggest their enolic forms. On the other hand, guanine and 7-Me-Gua preserve their keto form in the adsorbed state just as 1-Me-Gua.

When the hydrophobic[10] and electrical properties do not favor enolization and ionization of $C_6=O$ then an orientation effect should be considered. Indeed, another possible interpretation is based on the short-range effects[1,2] of the SERS enhancement factors and on specific orientations of the guanine derivates at the silver surface. As the electromagnetic model predicts[1], a very rapid decrease of SERS effects with distance on the Å scale would limit the enhancement factors to bond vibrations in the immediate vicinity of the silver surface. The disparition of the carbonyl stretching vibrations in SERS spectra of 3-Me-Gua and 9-Me-Gua would thus suggest that the $C_6=O$ bond lies far away from the surface. The hypothesis of an "inactive" SERS carbonyl vibration would consider orientations for both methylted guanines in the adsorbed state different from the other derivatives. In 1-Me-Gua and 7-Me-Gua the $C_6=O$

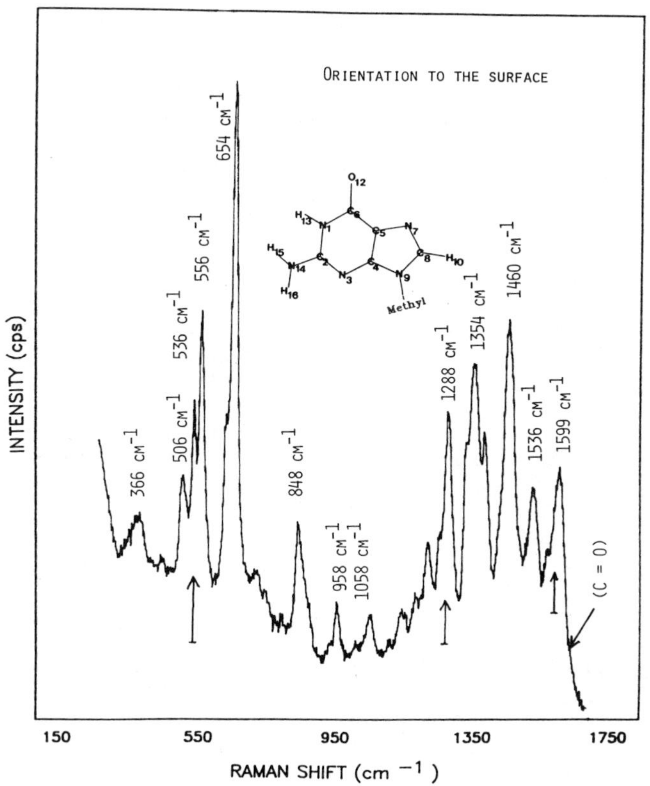

Figure 2. Ag-electrode SERS spectrum of 9-Me-Gua. Concentration 1×10^{-4} M; 0.1 M KCl, pH 8.0, laser excitation line 514.5 nm; laser power at the electrode: 10 mW. Adsorption potential E_s = −0.6 V vs (Ag/AgCl) reference electrode.

Table 1. Raman Band Frequencies (cm^{-1}) and Tentative Vibrational Mode Assignment of Guanine and Its Methylated Derivatives Adsorbed On A Silver Electrode Surface (E_s = -0.6 V).

Note: w, m, s, vs indicate weak, medium, strong and very strong intensities; sh = shoulder; str = bond stretching; d = angle deformation; b = bending; r = rocking

Gua	1-MeGua	3-MeGua	7-MeGua	9-MeGua	Mode Assignment
226sh	224sh		240sh	216sh	Ag/O; N
332w	342m				
378w		372m		366w	Co(b)+C_2H_{14}(b)
	394m		404m	416w	C_2N_{14}(b)+N_9R(b)
458w	474w	468w			
510w	516w	508w	502m	506m	CC_6N(d)-CN_3C(d)+CC_5C(d)
				536m	N-CH_3(b)
		558w		556s	N-CH_3(b)
656vs	644vs	632s	652vs	654vs	Ring Breathing
		654vs	692w		
	710w	690w			N_9R(str)+C_2N_{14}(b)+C_5N_7C(d)
		730w	726w		
	806m				
852m	862m	825m		848m	C_5N_7(str)-C_2N_{14}(str)+NC_2N(d)
	928m				
958m		950m		958m	N_1C_2(str)+C_2N_3(str)-N_9C_4(str)
	970s				
	1022m				
1048w		1050w	1042w	1058w	
	1090w	1094w	1080w		
1148m	1142w	1152w		1146w	C_6N_1(str)+C_5N_7-NH_2(r)
			1080s	1182w	
1222s	1212s	1226m	1230w	1224m	N_9R(str)+N_3C_4(str)+C_4C_5(str)
	1238s				
1272m	1272m	1274m	1282s	1280s	C_8H(b)+C_8N_9(str)
1328s		1330m		1332s	
1352w	1354w	1350w	1352vs	1354s	C_8N_9(str)+N_7C_8(str)
1382vs	1378m	1392vs		1386s	N_1H(b)+C_2N_{14}(str)
1454sh	1458m	1444vs	1422m		C_5N_7(str)+C_5C_6(str) +C_8N_9(str)-N_9C_4(str)
1464vs			1468s	1460vs	N_7C_8(str)+C_4C_5(str)+C_8H(b)
1532m	1528vs	1518m	1492sh		N_9C_4(str)+N_7C_8(str) +N_1C_2(str)+C_8H(b)
	1548sh		1530m	1536m	
1598m		1584m		1596m	N_3C_4(str)+N_9C_4(str)+N_9C_4C(d)
1692m	1702s		1704s		C_6=O(str)-C_5C_6(str)+C_4C_5(str)

group is oriented direct to the surface, in 3-Me-Gua and 9-Me-Gua the C_6=O bond lies away from the surface.

The detailed analysis of these investigations can be summarized as follows:

1. Ring-breathing mode vibrations are very sensitive to CH_3 substituents; 656 cm^{-1} in guanine; splitting in 3-Me-Gua (632 cm^{-1}, 654 cm^{-1}) and 7-Me-Gua (652 cm^{-1}, 692 cm^{-1}).

2. The absence of a carbonyl Raman band in the SERS spectra of 3-methylguanine and 9-methylguanine (cf. Figure 2) suggests specific chemical species or orientations at the electrode surface.

3. The sterical interpretation of these investigations indicated that the strong enhancement is caused by the short range mechanism for distances smaller than 5 Å from the surface.

b) Comparison of the Electrode and Colloid SERS spectra

The colloid SERS spectroscopy has a simple experimental pretreatment procedure and the Raman measurements can be carried out with conventional liquid cells or capillaries[5,6].

The advantage of the electrode SERS technique is that the surface charge can be determined by the applied potential. Therefore, we only have to change this electrode potential finding out the best correspondence in the electrode and colloid spectra. In the 7-Me-Guanine (cf. Figure 3) we see a very good correspondence at an electrode potential at 0.0 V. That is a very high positively charged silver surface and therefore this investigation shows that the adsorption process of 7-methyl-guanine on the Ag colloid takes place on a very positive charged surface.

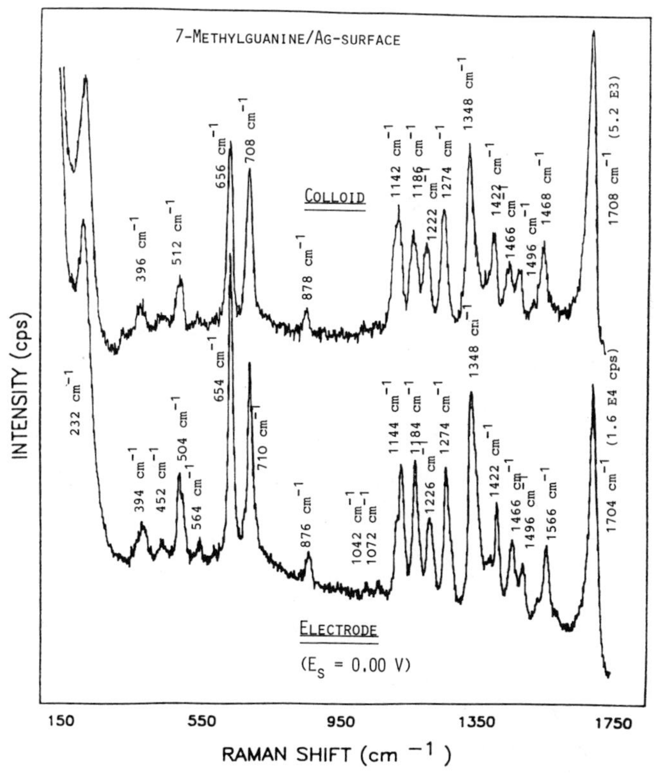

Figure 3. SERS spectrum of 7-Me-Gua adsorbed on Ag colloids (above): λ_{ex} = 514.5 nm, power 20 mW, concentration 3 x 10^{-6} M. SERS spectrum of 7-Me-Gua adsorbed on an Ag electrode (below): λ_{ex} = 514.5 nm, power 10 mW, E = -0.00 V, pH 8, concentration 9 x 10^{-6} M.

c) Combining SERS and Thin-Layer Chromatography (HPTLC)

Direct identification methods for vibrational spectra of high-performance thin-layer chromatography (HPTLC) spots such as infrared and Raman spectrometries are limited by the required high quantities of substances. However, recent findings of an enhanced Raman scattering for various compounds adsorbed at metal surfaces give new potential for analytical applications of Raman spectroscopy[2].

Combining SERS spectroscopy and HPTLC technique has permitted in situ analysis of HPTLC spots down to nanogram amounts of substances[3]. Moreover, SERS/HPTLC-spectroscopy creates a new technique for obtaining high resolution vibrational spectra of molecules with a solubility lower than about 5×10^{-4} g in 100 g H_2O. Thus, SERS/HPTLC-spectroscopy opened up a field of vibration spectroscopy for very low soluble molecules in water.

Finally, the combination of molecular resonance and surface enhancement (SERRS-spectroscopy) results in an extremely high Raman scattering enhancement of 10^8 to 10^{10} together with a strong quenching of the fluorescence[12]. Thus, SERRS/HPTLC-spectroscopy is of great interest for strong fluorescent molecule studies, because it offers a general means of obtaining high-quality Raman spectra from strong fluorescent dye molecules.

Surface enhanced Raman microspectroscopy has been developed as a technique for characterizing processes on nano-TLC plates in the pico- and femtogram region. In the reflected-light mode, the Raman microscope allows detailed examination of the HPTLC spot morphology and precise placement of

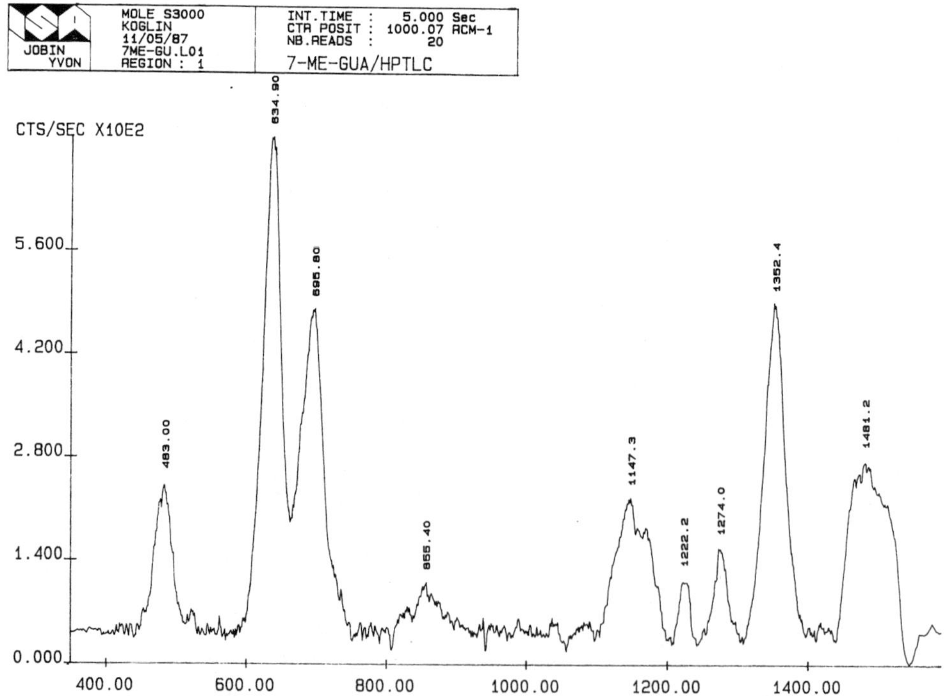

Figure 4. SERS/HPTLC microprobe analysis of 7-methyl-guanine (Instruments S.A. MOLE-S 300 Micro-Raman equipment). Laser spot focus 1 µm of the 10 ng substance spot of 7-methyl-guanine, femtogram in mass, integration time: 5 s, number of reads 8.

the laser focus with 1 μm spatial resolution. Figure 4 demonstrates a result with this Raman microprobe equipment in conjunction with Ag colloidal SERS activated HPTLC plates.

This HPTLC/SERS microprobe analysis has shown that the methylated DNA building stone guanine exhibits strong SERS bands when adsorbed on Ag colloidal activated nano-TLC plates. Prominent and characteristic bands are the SERS ring-breathing modes in the spectral range of 600 to 700 cm^{-1}. For example, the corresponding bands of 7-Me-Gua in the electrode spectrum are at 652 cm^{-1}, 692 cm^{-1}, in the colloid SERS spectrum at 651 cm^{-1}, 703 cm^{-1} and in the HPTLC/SERS microspectrum at 634 cm^{-1}, 695 cm^{-1}.

CONCLUSIONS

With the extremely high enhancement of the Raman scattering signals in the immediate vicinity of the charged metal-surface (electrode, Ag colloidal particles, activated nano-TLC plates) it is possible to identify the adsorbed parts of the methylated guanine derivatives and to study their configuration at very low concentrations. These results demonstrate new possibilities of Raman-spectroscopy to obtain high resolution vibrational spectra of adsorbed biomolecules.

REFERENCES

1. A. Otto, "Surface Enhanced Raman Scattering, Classical and Chemical Origins", in: Light Scattering in Solids, Vol. 4, M. Cardona, G. Güntherhods, Eds., Springer 1983.

2. E. Koglin and J.-M. Sequaris, "Surface Enhanced Raman Scattering of Biomolecules, in: Topics in Current Chemistry, Vol. 134, pp. 1-57, Springer-Verlag, Berlin Heidelberg 1986.

3. J.-M. Sequaris and E. Koglin, Anal. Chem., 59, 525 (1987).

4. A.J.P. Alix, L. Bernard and M. Manfait, Eds., "Spectroscopy of Biological Molecules", Wiley-Interscience Publications, 1985.
   a) E. Koglin and J.-M. Sequaris, p. 221.
   b) J.-M. Sequaris, H.H. Lewinsky and E. Koglin, p. 237.

5. E. Koglin, J.-M. Sequaris, J.C. Fritz and P. Valenta, J. Mol. Struct., 114, 219 (1984).

6. J.-M. Sequaris, J.C. Fritz, H.H. Lewinsky and E. Koglin, J. Coll. Interf. Sci., 105, 417 (1985).

7. J.A. Creighton, C.G. Blatchford and M.G. Albrecht, J. Chem. Soc., Faraday Trans., 2, 1979, 75, 790-798.

8. R.C. Lord and G.J. Thomas, Spectrochim. Acta, A23, 2551 (1967).

9. J.-M. Delabar, J. Raman Spectrosc. 7, 261 (1978).

10. G. Valette, J. Electroanal. Chem., 139, 285 (1982).

11. R.N. Balsinger and J.A. Montgomery, J. Org. Chem., 25, 1573 (1960).

12. E. Koglin and J.-M. Sequaris, J. Mol. Struct., 141, 405 (1986).

ELECTROCHEMISTRY IN CANCER: APPLICATIONS TO PHARMACOLOGICAL STUDIES OF QUINONE-CONTAINING ANTITUMOR AGENTS

Peter L. Gutierrez* and Binh Nguyen

University of Maryland Cancer Center
Division of Developmental Therapeutics
University of Maryland School of Medicine
655 West Baltimore Street
Baltimore, Maryland 21201

INTRODUCTION

Electrochemistry has been used in cancer pharmacology in a variety of ways. Practical applications include pharmacokinetics where drug concentrations in plasma are measured as a function of time. More fundamental studies deal with drug synthesis and mechanism(s) of drug action. Perhaps the great impetus to apply electrochemistry to cancer pharmacology stems from the fact that a family of relatively good anticarcinogens contain the quinone moiety. Adriamycin (doxorubicin) and daunorubicin are antibiotics which have been widely studied by electrochemical methods because of their clinical activity against a wide variety of malignancies and because they are quinones.

Akpofure et al.[1] used HPLC and electrochemical detection at an oxidative potential of 0.65 V to quantify daunorubicin and its metabolites in serum and plasma. Using a different method, Chaney and Baldwin[2] determined Adriamycin concentrations in urine. The determination was performed by adsorption of the compound at a carbon paste electrode and then applying pulse voltammetric analysis at the resulting surface.

In other applications, Anne et al.[3] used electrochemical reduction to synthesize new daunorubicin analogs. Kano et al.[4] studied the electrochemical properties of Adriamycin adsorbed on a mercury electrode, while Konse et al.[5] studied the electrochemical properties of Adriamycin adsorbed on pyrolyte graphite electrodes modified by phospholipid membranes. This latter work has relevance to the interaction of Adriamycin with cell membranes.

The use of electrochemistry to study mechanism(s) of action include determination of oxidation-reduction potentials and the detection of free radicals by electron spin resonance (ESR) spectroscopy.

In a series of naphthoquinone and benzoquinone derivatives containing an alkylating side chain (bioreductive alkylators), Lin and Sartorelli[6] found that, with a few exceptions, the compound with more negative half-wave potential generally possessed the most potent antitumor properties.

Semiquinones and oxygen-free radicals have been reported to arise during the metabolism of quinone and hydroquinone xenobiotics. Mason[7] has reviewed the subject well using a variety of xenobiotics as examples. Quinone-containing antitumor agents belong to a class of xenobiotics whose activity has been associated with the production of free radicals that include the semiquinone of the drug and oxygen radicals. Oxygen-free radicals such as superoxide anion and hydroxyl radicals can be generated during the quinone-semiquinone redox cycle, or during the oxidation by molecular oxygen of the fully reduced quinol form. The value of electrochemistry in studying these redox systems is that it is a relatively clean chemical system, it is relatively easy to control and can be studied in aprotic and aqueous solutions which allows one to evaluate the behavior of free radicals generated in biological systems.

Land et al.[8] and Svingen and Powis[9] used pulse radiolysis to determine one-electron reduction potentials for Adriamycin and other quinone antitumor drugs. The latter group found that the half-life of the semiquinone from Adriamycin in the presence of oxygen was relatively short (50 µs) when compared to the half-life of the diaziquone semiquinone (200 µs) (AZQ, Figure 1). Using polarography, Berg et al.[10] studied reduction and reoxidation of several anthracycline antibiotics including Adriamycin. The products of these reactions were identified and were useful in proposing reductive and oxidative pathways. An important result obtained was that only iremycin and roseorubicin A which contain sugar residues in the C-10 position are reversibly reduced by 2-electrons and produce H_2O_2 during the reoxidation of the hydroquinones. All other anthracyclines including Adriamycin are irreversibly reduced because the aglycone is formed during this process. Anne and Moiroux[11] studied the one-electron reduction of substituted anthraquinones including Adriamycin. Among their findings was the fact that electrochemically produced semiquinones react with oxygen to produce $O_2^{\cdot-}$ which was detected by amperometry. The amount of $O_2^{\cdot-}$ detected was at least 85 % of the value expected from stoichiometric analysis. In polarographic and coulometric studies on 15 analogs of Adriamycin, Berg et al.,[12] found that the reduction of the quinone mediates the splitting of the glycosidic bond to produce the aglycone and that the half-wave potentials which were distributed over a range of 0.3 V had no direct relationships to structures. Nakazawa et al.[13] studied the electrochemical reduction of the non-quinone containing anticarcinogen actinomycin D and two phenoxazone analogs. Formation of the free radical anion of these three compounds was confirmed by ESR. Their ESR spectra indicated that most of the unpaired electron density resides in the benzenoid ring of the phenoxazone system. Therefore, if free radicals are involved in the drug's binding to DNA, then the benzenoid portion of the phenoxazone system will be important in the binding mechanism.

DNA-drug interactions have also been studied by electrochemistry. Plambeck and Lown[14] devised a method to measure association constants between an electrochemically reducible antitumor agent and DNA in aqueous solutions. Similarly, Berg et al.[12] studied complexes of DNA with several anthracycline analogs and found that the therapeutic activity of such complexes depends on a strong binding constant and slow dissociation rate. Other factors that modify the therapeutic activity were found by Berg et al.[12] to be: i) stability of the glycosidic bond against hydrolytic or reductive splitting; ii) a low generation of H_2O_2 and iii) membrane permeability.

This short review is intended to give an idea of the extent that electrochemistry is used in the pharmacology and mechanistic studies of antitumor agents. This paper deals with a recent application of electrochemistry to study the mechanism of action of diaziquone (AZQ) a quinone-containing alkylating agent. Electrochemistry was used in conjection with ESR to detect free radicals and solve their ESR spectra, to investigate the effect of water in the reduction-oxidation reactions and to study AZQ-DNA interactions.

EXPERIMENTAL

Diaziquone (AZQ, Figure 1) was supplied by the Drug Synthesis and Chemistry Branch, Division of Cancer Treatment, National Cancer Institute, Bethesda, MD. Analogs RQ2 and RQ14 were a gift of Dr. J. Driscoll, Drug Development Branch, National Cancer Institute.

Cyclic voltammetry (CV) was performed with a CV-27 voltammograph (Bioanalytical Systems, Lafayette, IN) connected to an IBM electrochemical cell at room temperature. A glassy carbon electrode was used as the working electrode. The IBM cell was equipped with a platinum wire counter electrode and an Ag/AgCl reference electrode. Solutions of AZQ at a concentration of 5×10^{-4} M were prepared in spectroscopic grade Me_2SO containing 0.1 M tetraethylammonium perchlorate (TEAP, Kodak Co.,

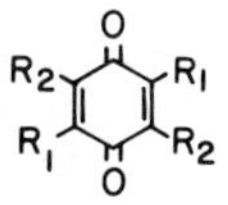

COMPOUND	R_1	R_2
AZQ	N◁ (aziridinyl)	$NHCOOCH_2CH_3$
RQ 1	N◁	H
RQ 2	N◁	Cl
RQ 3	N◁	F
RQ 4	N◁	OC_2H_5
RQ 5	N◁	SC_2H_5
RQ 6	N◁	$NHCH_2CH_3$
RQ 7	N◁	$N(CH_3)_2$
RQ 10	N◁	N⟨ ⟩N-CH_2CH_2OH
RQ 11	N◁	N⟨ ⟩
RQ 12	N◁-CH_3	$NHCOOCH_2CH_3$
RQ 13	N◁(CH_3)(CH_3)	$NHCOOCH_2CH_3$
RQ 14	Cl	$NHCOOCH_2CH_3$

Figure 1. Chemical structures of AZQ and analogs.

Rochester, N.Y.) as supporting electrolyte. Aqueous samples were run in phosphate buffered saline pH 7.5 (PBS), Hank's balanced salt solution (HBSS) pH 7.5 or unbuffered water adjusted to pH 7.5. The electrolyte was 0.05 M KCl.

One-electron anaerobic reductions were performed in an ESR electrolytic cell as follows: AZQ and analogs were prepared freshly at 5×10^{-4} M in Me_2SO containing 0.1 M TEAP. Mercury at the bottom of the ESR flat cell provided the cathode contact through a platinum wire electrode. A second platinum wire electrode inserted through the side port of the cell was used as a counter electrode. The reference electrode was an Ag/AgCl electrode. The voltage across the solution was measured with a Fluke 800 A digital multimeter and set with an AMEL-550 potentiostat (ECO Instruments, Newton, MA). One-electron reductions were carried out at voltages between the first and second electron waves which were determined by cyclic voltammetry. Another way to obtain ESR spectra of electrochemically produced free radicals is to use a flow-through coulometric reactor first designed by Miner and Kissinger[15]. This flow cell uses a graphite working electrode, a Pt counter electrode and an Ag/AgCl reference electrode. The compound to be reduced is dissolved in Me_2SO with 0.1 M TEAP. The reduction voltage (previously determined by CV was set slightly more negative than the first reduction wave. Using a peristaltic pump, the solution is pumped through the flow cell and into an ESR flat cell where it is directly detected and the ESR spectrum of the free radical recorded. ESR spectra were obtained with an X-band (9.3 GHz) Varian E-109 Century Series spectrometer. A dual rectangular cavity was used which contained strong pitch (g = 2.0028) in one section, and the samples in the other.

Differential pulse voltammetry (DPV) was used to investigate the interaction between AZQ and polynucleotides. DPV was performed on an IBM 225 voltammetric analyzer with pulse amplitude of 100 mV and a sweep rate of 5 Vs^{-1}. Current sampling was done with the drop-time control of the IBM analyzer set at 2. Current-voltage and current-time data were recorded and integrated with a Bascom Turner model 3120T integrator. The working electrode was a glassy carbon electrode which was resurfaced between runs by polishing with silica gel (0.05 microns). The electrochemical cell was equipped with a Pt wire counter electrode and an Ag/AgCl reference electrode. Voltammetric results were obtained at room temperature in 0.2 M phosphate buffer pH 7.4. Nucleic acid concentrations ranged between 333 µg/ml to 500 µg/ml. AZQ was added to 10 ml of buffer described above to achieve the appropriate final concentrations. After the graphite electrode was inserted into the test solution, it was allowed to stand 10 s without applied potential. Then, a pre-sweep potential (0.2 V) was applied for 128 s before the beginning of the sweep (the 128 s pre-sweep time gave the best results).

Cytoxic Evaluation

To ascertain drug activity, the concentration that inhibits cell growth by 50 % (IC_{50}) was evaluated. The evaluation was performed on L1210 or P388 murine leukemia cell lines which were maintained in serial cultures as described by Gutierrez et al.[16]. Cell cultures were incubated with and without drug in slanted test tube racks at 37°C in an atmosphere of 95 % air, 5 % CO_2 and 95 % humidity. After 72 h incubation time, the cells were counted and IC_{50} values obtained from plots of cell survival versus drug concentration.

RESULTS AND DISCUSSION

Electron Spin Resonance of Electrochemically Reduced Diaziquone

Diaziquone (AZQ, Figure 1) is a quinone antitumor agent which contains two alkylating group (aziridine) and two lipophilic groups (carboethoxyamino). AZQ was designed so that it could cross the blood-brain barrier. The carboethoxyamino groups confer to AZQ this property and the aziridines provide alkylating power. The quinone moiety places this drug among the xenobiotics discussed earlier whose activity is believed to be associated with a bioreductive process that involves free radicals. The aziridines and the quinone provide diaziquone with several possibilities for chemical reactions which may depend on such conditions as pH and buffer. These possibilities can lead to drug mechanism(s) that involve alkylation, drug semiquinone intermediates and/or reductive alkylation where the semiquinone facilitates alkylation. As shown by Gutierrez et al.[17,18] diaziquone has the property that it can be very easily reduced to its free radical anion by rat liver microsomes, NADPH cytochrome c reductase and live cells. Under these conditions, the ESR spectrum of this free radical was characterized by a quintet of broad lines at g = 2.0046.

The first task in understanding the reduction of diaziquone in biological systems and the free radicals resulting from this reduction was to study the effect of water in the cyclic voltammetry of AZQ.

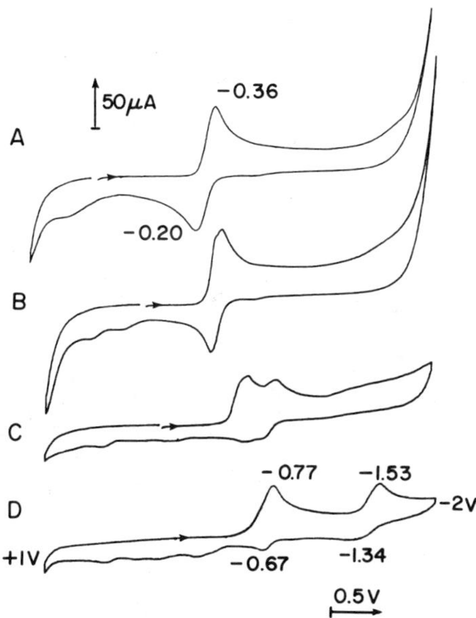

Figure 2. (A) Cyclic voltammograms of 5 x 10^{-4} M AZQ in aqueous medium (pH 7.5). (B) Same as (A) except with 20 % Me_2SO. (C) Same as (A) except 75 % Me_2SO. (D) Same as (A) except 100 % Me_2SO. The aqueous solution was in phosphate buffered saline (pH 7.5) using 0.05 M KCl as an electrolyte (A and D). For (C) and (D) above, 0.1 M tetraethylammonium perchorate was used as an electrolyte. Scan rate was 50 mVs^{-1} with glassy carbon as the working electrode. (Taken from Mossoba et al.[19] and reproduced with permission of the copyright owner, the American Pharmaceutical Association).

The cyclic voltammogram of AZQ (1-0.5 mM) in aqueous medium (phosphate buffered saline pH 7.5 or water pH 7.5) exhibits one reduction peak at a cathodic potential of E_p^c = -0.36 V (Figure 2A) vs Ag/AgCl electrode. This one peak, gradually resolves into two peaks at the ratio of Me_2SO to water increases (Figures 2B and C). The voltammogram in 100 % Me_2SO consists of two one-electron quasi-reversible peaks with cathodic potentials at E_p^c = -0.77 V and -1.53 V; and anodic potentials E_p^c = -0.67 and -1.34 V vs Ag/AgCl electrodes (Figure 2D). The coalescing of the two reduction waves in aprotic media to one wave in aqueous media implies a 2-electron reduction wave. Controlled potential electrolysis in aqueous solutions performed by Mossoba et al.[19] indicated that approximately 2 coulomb equivalents are required to reduce one mole of AZQ.

Reducing AZQ at a potential slightly more negative than -0.36 V yielded a colorless compound shown by Mossoba et al.[19] to be the fully reduced AZQ ($AZQH_2$). Under anaerobic conditions, this quinol product is diamagnetic and therefore has no ESR spectrum. Exposure to air produces the 5-line aqueous ESR spectrum indicating the autoxidation of $AZQH_2$ to the semiquinone (Figure 3).

The effect of water on the ESR spectrum of AZQ was studied by reducing AZQ in Me_2SO at a static potential of 0.9 V which produced the 11-line free radical spectrum shown on Figure 4A. Adding 20 % H_2O by volume to the ESR cell produced a broadening in the lines which with time resulted in the same ESR spectrum as obtained in the presence of live cells (compare Figure 4D with Figure 6A) and in the oxidation of $AZQH_2$ (Figure 3). This result indicated that the electrochemically produced and the biologically produced free radical were the same and the electrochemical studies can be used to draw conclusions on biologically obtained free radicals.

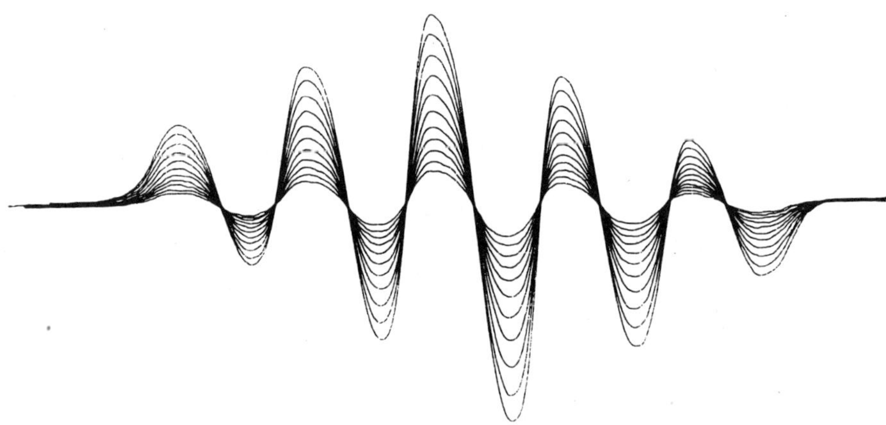

Figure 3. Electron spin resonance spectra of the oxidation of $AZQH_2$ produced electrochemically in aqueous solutions (HBSS pH 7.5). The sample was quickly transferred to the ESR flat cell and repetitive scans were taken. The ESR conditions at room temperature were: 9.3 GHz microwave frequency, 10 mW incident microwave power and 1 gauss modulation amplitude. A small free radical signal is observed at the initial tracing due to slight oxygen contamination. (Taken from Mossaba et al.[19] and reproduced with permission of the copyright owner, the American Pharmaceutical Association.)

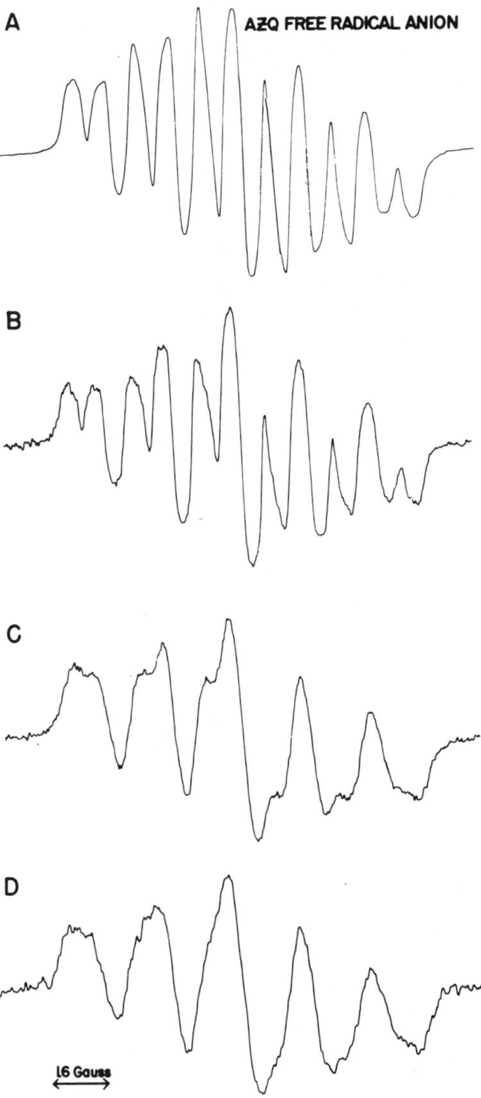

Figure 4. Electron spin resonance spectra of the one-electron electrochemical reduction of AZQ (A). Subsequent to the reduction in DMSO, water was added (20 % by volume) and slow mixing of the solvents is observed in B and C. Final mixing is represented in D. The ESR conditions at room temperature were as in Figure 3. (Taken from Gutierrez et al.[23] and reproduced with permission of the copyright owner, American Society for Pharmacology and Experimental Therapeutics.)

The two-electron reduction peak observed in the CV of AZQ in aqueous medium at -0.36 V (Figure 2) was found in the pH range of 4 to 9. At pH 3.5-4.0 this peak begins to shift towards more positive potentials with a cathodic peak at E_p^c = -0.12 V and anodic peak at E_p^a = +0.4 V (vs Ag/AgCl, Figure 5). This behavior is consistent with protonation as shown by Sawyer and Roberts[20] and Mossoba et al.[19]. Two additional reduction peaks (in addition to the E_p^c = -0.12 V peak) appear at this pH. They have reduction potentials at E_p^c = -0.41 V and E_p^c = -0.54 V, and represent the reduction of the aziridine rings. Wagner and Berg[21] obtained similar peaks for structurally similar quinone compounds. At pH 2.5, the cyclic voltammogram of AZQ, still exhibits the -0.12 V quinone/hydroquinone peak but does not exhibit the peaks attributed to the reduction of the aziridine rings. They show a rather broad irreversible peak at -0.7 V (Figure 5). This peak was shown by Mossoba et al.[19] to be due to opened aziridine rings. At pH 3.5 the peaks at -0.41 V and -0.54 V represent the reduction of closed aziridine rings because of the presence of their oxidizing waves at -0.39 V and -0.15 V. At pH 2.5 the -0.7 V peak does not have a corresponding oxidizing peak, indicating the irreversibility of the opening of the aziridine ring. Wagner and Berg[21] determined that the reduction of the aziridine ring with its subsequent opening requires 2 electrons. In basic medium (pH 11.5) the CV of AZQ showed that the -0.36 V

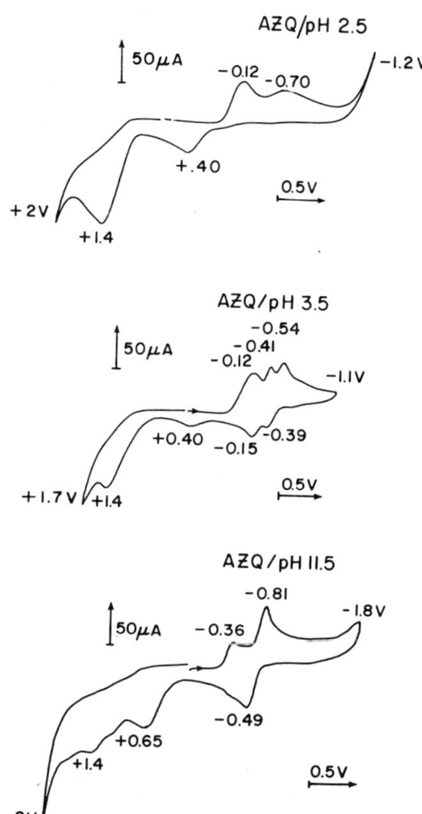

Figure 5. Cyclic voltammograms of 5 x 10^{-4} M AZQ as a function of pH. The two-electron voltammetric wave observed for AZQ in aqueous media (Figure 2) at pH 7.5 was also found in the pH range of 4-9. In the pH range of 3.5-4.0 this wave began to shift towards more positive potentials (E_p^c=-0.12 V). At pH 3.5, two additional waves appeared at -0.41 V and -0.54 V. At pH 2.5, these two additional waves were replaced by a broad wave at -0.70 V, but the -0.12 V wave remained. In basic medium, a new sharp wave at -0.81 V appeared while the -0.36 V wave found at pH 7.5 remained unchanged. Cyclic voltammetry conditions were as in Figure 2, except for using PBS as the solvent and 0.05 M KCl as the electrolyte. (Taken from Mossoba et al.[19] and reproduced with permission of the copyright owner, the American Pharmaceutical Association.)

medium (pH 11.5) the CV of AZQ showed that the -0.36 V quinone/semiquinone peak seen at pH 7 is still present, but a new as yet unidentified reduction peak at -0.81 V appeared. These results suggest that in acidic media, the aziridine rings are protonated and can be attacked by nucleophiles leading to the open ring group - $NHCH_2CH_2X$. In basic media, NMR data collected by Mossoba et al.[19] suggests a breakdown of AZQ.

To further understand the nature of the AZQ free radical in terms of ESR parameters, two analogs of AZQ were used. These two analogs contain chlorine atoms that substitute either the carboethoxyamino groups (RQ2) or the aziridine groups (RQ14). Gutierrez and Bachur[23] showed that these two analogs can be reduced to their free radicals by liver microsomes and NADPH-cytochrome c reductase. Figure 6B and 6C show the ESR spectra of the biologically reduced RQ2 and RQ14 respectively. Cyclic voltammograms obtained by Gutierrez et al.[22] provided the appropriate potential at which to reduce aprotic solutions of RQ2 and RQ14 to obtain stable anerobic concentrations of free radicals. Figure 6B' and 6C' show the ESR spectra of the electrochemically reduced RQ2 and RQ14 analogs. The spectra showed well defined lines which can be readily interpreted and used to solve the more complicated 11-line ESR spectrum of $AZQ^{\dot{-}}$ (Figure 6A'). Using computer simulation the hyperfine couplings were determined to be those given in Table 1. These ESR parameters indicate that for the most

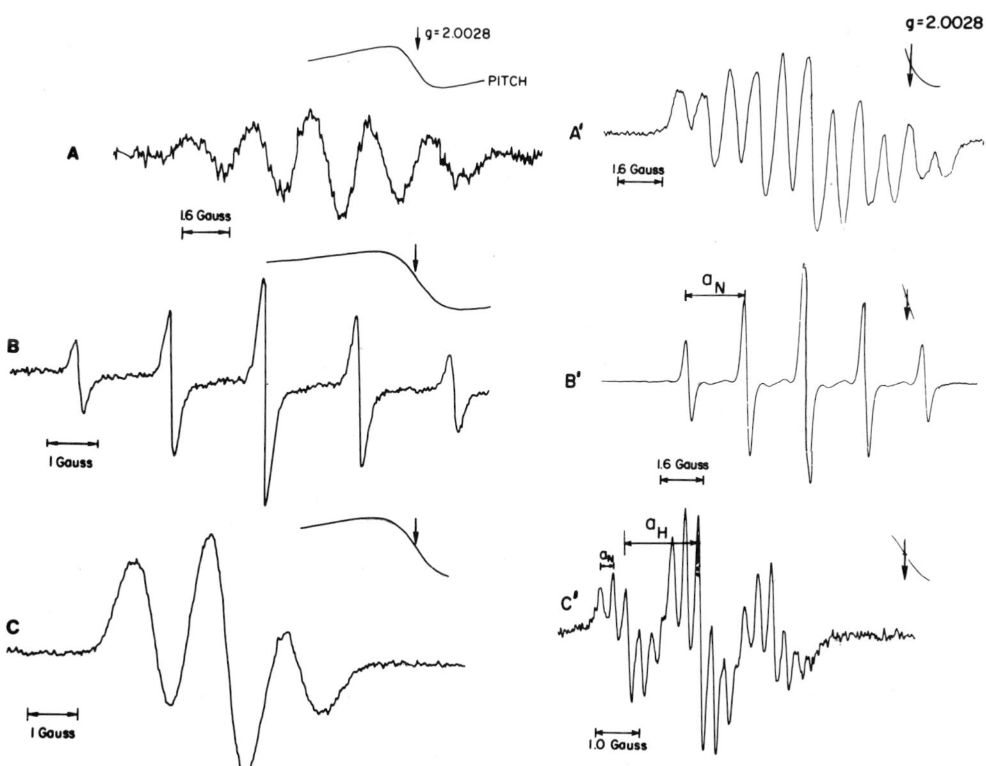

Figure 6. Electron spin resonance spectra of (A) AZQ free radical; (B) RQ2 free radical and (C) RQ14 free radical obtained by reduction with NADPH cytochrome C reductase. (A'), (B'), and (C') are the free radicals of AZQ, RQ2 and RQ14 obtained electrochemically as described in Methods. (Reproduced with permission from Gutierrez and Bachur[22], and Gutierrez et al.[23].)

Table 1. Electron Spin Resonance Parameters for the Electrochemically* Generated Free Radicals of Diaziquone and Its Analogs RQ2 and RQ14

FREE RADICAL	g-VALUE		HYPERFINE COUPLING (GAUSS)			
			AZIRIDINE		CARBOETHOXYAMINO	
	ELECTRO CHEM.	ENZYM.	ELECTRO CHEM.	ENZYM.	ELECTRO CHEM.	ENZYM
O, ROCHN, $O^{-1/2}$, N, O, NHCOR, N, $O^{-1/2}$	2.0048	2.0046	1.88 (N) 1.78 (N)	1.8 (2N)	0.87 (H) 0.73 (H) 0.15 (2N)	—
$O^{-1/2}$, Cl, N, N, Cl, $O^{-1/2}$	2.0049	2.0046	2.10 (2N)	1.9 (2N)		
$O^{-1/2}$, O, Cl, NHCOR, O, ROCHN, Cl, $O^{-1/2}$	2.0055	2.0050			1.64 (2H) 0.30 (2N)	1.6 (2H)

*The hyperfine coupling constants of the electro chemically generated radicals were obtained by computer simulation.

part, most of the spin density is found in the quinone carbons attached to the aziridine rings because the nitrogen hyperfine couplings from the azridines are the largest.

Once the ESR spectrum of AZQ was solved and hyperfine couplings obtained, it was relatively easy to study a series of 12 analogs

Table 2. Comparison of IC_{50}, Aziridine Nitrogen Couplings (a_N^{Az}) and Cathodic Peak Potentials (E_p^c) for AZQ and Analogs

RQ	IC50 (uM)[a]	a_N^{Az}, Gauss	E_p^c, volt[b]
AZQ	1.00	1.88 (N), 1.78 (N)	-0.77
RQ1	0.01	1.75 (2N)	-0.66
RQ2	0.20	2.10 (2N)	-0.42
RQ3	0.20	1.44 (2N)	-0.41
RQ4	1.00	1.20 (2N)	-0.71
RQ5	1.10	1.97 (2N)	-0.64
RQ6	NA[c]	1.28 (2N)	-0.94
RQ7	4.5	1.36 (2N)	-0.91
RQ10	0.25	1.43 (2N)	-0.87
RQ11	8.8	1.40 (2N)	-0.90
RQ12	0.48	2.11 (2N)	-0.42
RQ13	0.07	2.26 (N), 2.17 (N)	-0.48
RQ14	0.38	—	-0.15

a) The dose require to inhibit growth to 50% of control growth by 72 hrs after drug addition (low IC50 means high drug activity). The cells used were L1210. Points represent the results of triplicate measurements. S.E. was less or equal to 20%.

b) Measured versus Ag/AgCl.

c) Not available.

(Figure 1) in terms of their reduction potential peaks (E_p^c, obtained from their cyclic voltammograms) their ESR parameters (obtained from the ESR spectra) and their antitumor activity (measured by IC_{50} as defined in Methods). Table 2 shows the relationship between IC_{50} values and E_p^c and A_N^{Az}. In general, there is no correlation between drug activity and the two physicochemical parameters obtained. This is in contrast with previous findings by Lin and Sartorelli[6] for a different set of bioreductive alkylators.

It is of interest to compare the activity of RQ14 and RQ2 because in these two compounds, the two groups in AZQ can be studied separately. The IC_{50} values for RQ2 and RQ14 are relatively close at 0.20 μM and 0.38 μM, respectively (Table 2). An IC_{50} of 0.38 μM for RQ14 (Table 2) is a relatively low value (high activity) for a compound with no aziridine moieties. This is of particular interest because the carboethoxyamino groups are not alkylating groups by themselves. It is possible that RQ14 acts directly as an electrophile, a statement which is in harmony with RQ14's low reduction potential (-0.15 V) and its fast reduction kinetics by NADPH cytochrome c reductase as shown by Gutierrez and Bachur[22].

These findings suggest that AZQ has several modes of action and that these modes may be influenced by a particular chemical or biochemical environment. For instance, AZQ may act directly as an electrophile, or through the aziridines as an alkylating agent whose activity may be accentuated by its reduction to a semiquinone.

Interaction of AZQ with DNA.

Differential pulse voltammograms of single stranded and double stranded DNA show two well defined peaks at +0.9 V and +1.19 V (Figure 7). As Brabec[24] has previously observed, the peak at +0.9 V corresponds to the electrooxidation of guanine residues (peak G) and the peak at +1.19 V corresponds to the electrooxidation of adenine residues (peak A). A differential pulse voltammogram of poly(G) is also shown in Figure 7 for comparison. A marked increase in sensitivity of the voltammetric oxidation current of DNA can be obtained by plotting current versus data-acquisition time and recording the points at intervals of 0.5 s. Increased sensitivity was also obtained by adsorbing polynucleotides or DNA at the GCE at +0.2 V for 180 s before beginning to sweep. The peak areas of the DPV obtained were integrated as described in EXPERIMENTAL, and plotted as a function of AZQ concentration. Adding AZQ to poly(G) and poly(A) in 0.2 M phosphate buffer pH 7 led to a decrease in both voltammetric peak areas G and A (Figure 8A).

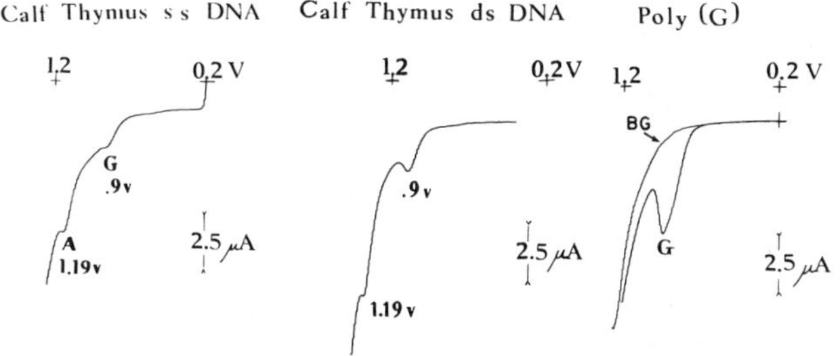

Figure 7. Differential pulse voltammograms at the GCE of 333 μg/ml single stranded DNA; 500 μg/ml double stranded DNA and 500 μg/ml poly G. Solutions were in phosphate buffer at pH 7.4.

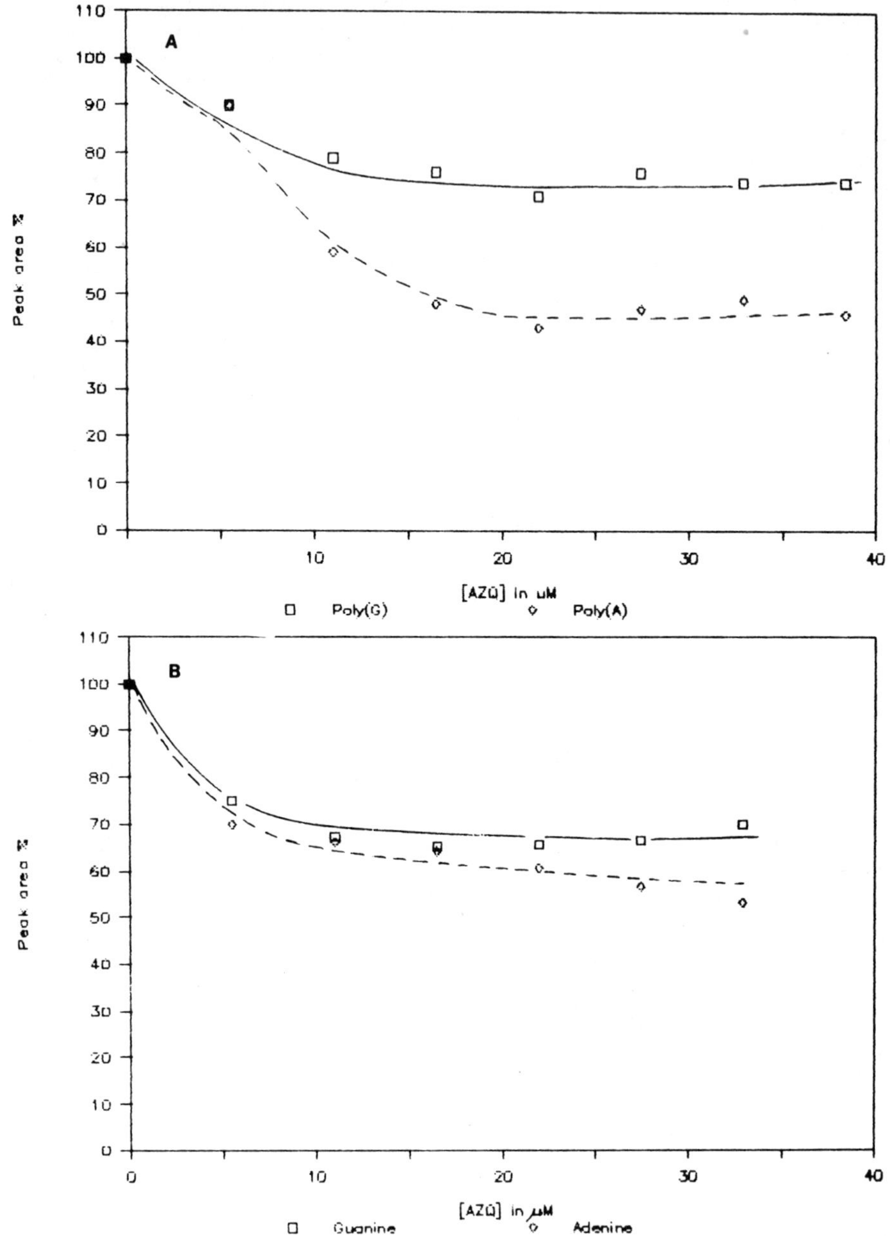

Figure 8. Variation of the differential pulse voltammetric peak area with AZQ concentration (% from no drug exposure) at the GCE in pH 7.4 phosphate buffer. (A) Poly G and poly A; (B) single-stranded DNA.

The peak area due to the electrooxidation of A decreased more than the peak area due to the electrooxidation of G reaching a plateau for both peak areas near 15 µM AZQ. When AZQ was added to single stranded DNA under the same conditions as above, the results were similar to those obtained for poly(G) and poly(A), except that at 15 µM AZQ, the peak from

A did not decrease as much as with poly(A) (50% from poly(A) and 60% for ssDNA). The decrease in area of peaks A and G can be explained in terms of the reaction of AZQ with purine residues. This reaction depletes the electroactive species available for the oxidation process at the GCE. The more pronounced decrease in peak area A implies that AZQ binds preferentially to adenine. The actual chemical structure of the AZQ-adenine is beyond the scope of this paper. However, a strong possibility is that there is a reaction between the N-7 of adenine and the aziridine group of AZQ. The stability of this complex is increased by hydrogen bonding of the carbonyl group of AZQ to the amino group of adenine.

CONCLUSIONS

The data presented here is an example of the variety of ways that electrochemistry can be used to study the pharmacology of AZQ. This paper emphasizes the use of electrochemistry in conjunction with ESR to study the free radicals generated during the one-electron reduction of AZQ, or the oxidation of the fully reduced quinone. Both of these reactions are important in the activity of AZQ as shown by several groups including Gutierrez et al.[25], Szmigiero and Kohn[26], and Alhonen-Hongisto et al.[27].

ACKNOWLEDGEMENTS

This work was supported by PHS grant CA 33681-04 awarded by the National Cancer Institute, DHHS, and by USPHS training grant HL-07618-01.

REFERENCES

1. C. Akpofure, C.A. Riley, J.A. Sinkule and W.E. Evans, J. Chromatography, 232, 377 (1982).

2. E.N. Chaney, Jr. and R. Baldwin, Anal. Chem., 54, 2556 (1982).

3. A. Anne, F. Bennani, J. Florent, J. Moiroux and C. Monneret, Tetrahedron Lett., 26, 2641 (1985).

4. K. Kano, T. Konse, N. Nishimura and T. Kubota, Bull. Chem. Soc. Jpn., 57, 2383 (1984).

5. T. Konse, K. Kano, R. Kano and T. Kubota, Bull. Chem. Soc. Jpn., 59, 3299 (1986).

6. A.J. Lin and A.C. Sartorelli, Biochem. Pharmacol., 25, 206 (1976).

7. R.P. Mason, in: "Free Radicals in Biology", W.A. Pryor, ed., Acad. Press, N.Y. (1982), Vol. V, p. 161.

8. E.J. Land, T. Mukherjee, A.J. Swallow and J.M. Bruce, Arch. Biochem. Biophys., 225, 116 (1983).

9. B.A. Svingen and G. Powis, Arch. Biochem. Biophys., 209, 119 (1981).

10. H. Berg, G. Horn and W. Ihn, J. Antibiot., 35, 800 (1982).

11. A. Anne and J. Moiroux, Nouveau Journal de Chimie, 8, 259 (1984).

12. H. Berg, G. Horn, U. Luthardt and W. Ihn, J. Electroanal. Chem., 128, 537 (1981).

13. H. Nakazawa, N.R. Bachur, F.E. Chou, M.M. Mossoba and P.L. Gutierrez, Biophy. Chem., 21, 137 (1985).

14. J.A. Plambeck and J.W. Lown, J. Electrochem. Soc., 131, 2556 (1984).

15. D.J. Miner and P.T. Kissinger, Biochem. Pharmacol., 28, 3285 (1979).

16. P.L. Gutierrez, B.M. Fox, M.M. Mossoba, M.J. Egorin, H. Nakazawa and N.R. Bachur, Biophys. Chem., 22, 115 (1985).

17. P.L. Gutierrez, R.D. Friedman and N.R. Bachur, Cancer Treat. Rept., 66, 339 (1982).

18. P.L. Gutierrez, M.J. Ergorin, B.M. Fox, R. Friedman and N.R. Bachur, Biochem. Pharmacol., 34, 1449 (1985).

19. M. Mossoba, M. Alizadeh and P.L. Gutierrez, J. Pharmaceut. Sci., 74, 1249 (1985).

20. D. Sawyer and J. Roberts Jr., in: "Experimental Electrochemistry for Chemists", Wiley, N.Y. (1974), p. 190.

21. H. Wagner and H. Berg, J. Electroanal. Chem., 2, 452 (1961).

22. P.L. Gutierrez and N.R. Bachur, Biochim. Biophys. Acta, 758, 37 (1983).

23. P.L. Gutierrez, B.M. Fox, M.M. Mossoba and N.R. Bachur, Molec. Pharm., 26, 582 (1984).

24. V. Brabec, Bioelectrochem. Bioenerg., 7, 69 (1980).

25. P.L. Gutierrez, S. Biswal, R. Nardino and N. Biswal, Cancer Res., 46, 5779 (1986).

26. L. Szmigiero and K.W. Kohn, Cancer Res., 44, 4553 (1984).

27. L. Alhonen-Hongisto, D. Deen and L. Marton, Cancer Res., 44, 39 (1984).

ON THE ELECTROCATALYTIC OXIDATION OF D-GLUCOSE AND RELATED POLYOLS ON LARGE Pt AND Au ELECTRODES

O. Enea

Laboratoire de Chimie
U.A. au CNRS no. 350
Universite de Poitiers
40 avenue du Recteur Pineau
86022 Poitiers Cedex / France

INTRODUCTION

Initially, the electrocatalytic oxidation of glucose on gold and platinum was studied in connection with the development of electrochemical sensors for determining the concentration of sugars in blood[1] and the realisation of implantable glucose fuel cells as power sources for cardiac pacemakers[2]. Unfortunately, glucose sensors failed to respond to variations in glucose concentrations because platinum electrodes lose their activity with time. Also, glucose anodes have had a low efficiency and stability of operation. These facts gave impetus to a large number of studies, mainly based on potentiodynamic voltammetric curves and the spectroscopic analysis of the reaction products, devoted to the anodic oxidation of glucose and of related compounds on small beads of noble metal electrodes[3-18].

The remarkable ability of glucose to be anodically oxidized in the so-called hydrogen adatom region on platinum was first explained by Ernst, Heitbaum and Hamann[3,4]. By comparing the anodic oxidation of glucose on platinum with that of other organic compounds like saccharose, lactose and acetaldehyde, they found evidence for the hemiacetalic group of glucose as the reactive center. The dehydrogenation of the C_1-atom adjacent to the hemiacetal OH^- group was assumed by them to be the rate determining step. A glucono-lactone was identified as the primary oxidation product.

De Mele, Videla and Arvia have analysed, through changes in potential/current (U/I) profiles, the influence of various experimental factors (temperature, pH, electrolyte composition, etc.) on the electro-oxdiation of glucose and related compounds on smooth platinum and gold electrodes[5-7]. From the comparative analysis of the electrochemical behavior of glucose with respect to that of inositol, fructose, xylose, arabinose and glucose-1-phosphate, it appears that the hemiacetalic group of glucose plays an important role in this process[5].

The glucose electroadsorption process (measured through potentiostatic current transients[6]) and the interaction between the electro-adsorbed species and H-adatoms, suggest a quite complex reaction pattern involving the formation of a large variety of intermediate species. Under potentiodynamic conditions, the contribution of each stage to the overall reaction pattern depends on the potential sweep rate and switching potential. Moreover, the main reactant may change with the solution pH and the adsorption of anions (and intermediates) greatly influences the reaction pathway. The influence of the adsorption of anions (phosphate, sulphate), of increasing temperature (from 30 to 64°C) and of potential perturbation conditions, were carefully examined[7] on the basis of changes in U/I profiles. The potentiodynamic results show that the overall reaction pathway should consist of different stages involving electro-adsorption, chemical and electrochemical processes. The main reaction product, detected by using spectrometric analysis, is glucono-δ-lactone, which yields gluconic acid through a hydrolysis reaction. Both the lactone and the acid (and CO_2) were found to show a strong inhibition for the electrooxidation of glucose: this itself explains the progressive decrease of the oxidation currents with time.

A qualitatively similar inactivation occurs during the electro-oxidatin of glucose on bright polycrystalline gold electrodes[8] and is presumably due, as in the case of platinum, to the increasing interaction of one of the reaction intermediates with the electroadsorbed layer containing oxygen.

Vassilyev, Khazova and Nikolaeva[9] have examined the process of adsorption and electrooxidation of glucose on smooth platinum over a wide range of pH values and have compared these results with those found with various other electrodes like Rh, Ir, Au, Cu and glassy carbon[10]. On platinum, the adsorption of glucose molecules occurs in addition to the formation of strongly chemisorbed particles and is accompanied by the dehydrogenation of the adsorbed molecules. Under steady state conditions, the rate of glucose electrooxidation is determined by the interaction of the chemisorbed carbon-containing residues with OH_{ads}, via a mechanism analogous to those known for the elementary fuels like formaldehyde or formic acid. Chloride ions (which are among the major anionic components of all the physiological liquids) adsorb on platinum and strongly inhibit the electrosorption of glucose on this metal. The effect is strongest in acid solution, whereas in alkaline media it plays a minor role.

In neutral and alkaline solution, the rates of glucose electro-oxidation on gold considerably exceeds those on platinum[10]. Moreover, physiological concentrations of chloride ion have practically no effect on glucose oxidation currents in alkaline solution. On gold, the electro-oxidation of glucose is inhibited by gluconic acid, only if its concentration is comparable to that of glucose, but the introduction of amino acids leads, as in the case of platinum, to an almost complete inhibition in neutral and in alkaline media.

Kokkinidis and Xonoglou[11] have examined the electrooxidation of glucose and of other monosaccharides like galactose and fructose on Pt surfaces modified by submonolayers of heavy metals (Tl, Bi and Pb) deposited at underpotentials. The catalytic action due to the adatom submonolayers has been interpreted in terms of the decrease of the electrode poisoning, resulting from a stable glucono-lactone type adsorbate, according to the model of the "third body effect"[11].

The steric and structural characteristics of the investigated mono-saccharides[11,12] have a marked influence on their oxidation on Pt electrodes modified by underpotential submonolayers. The hemiacetalic group was again thought to play the most important role in the electrooxidation process of monosaccharides[12].

The electrocatalytic oxidation of glucose in alkaline media on various noble metals modified by upd ad-layers decrease in the following order:

Pt > Pd > Rh > Ir.

Without upd ad-layers, Pd is however a better electrocatalyst than Pt towards glucose oxidation because the poisoning effects, resulting from strongly bound intermediates, are less important. The upd adatoms hinder the formation and adsorption of the poisoning intermediate, thus improving the electrocatalytic activity of Pt electrodes[13].

The effect of adatoms greatly depends on the arrangement of the catalytic sites on the surface: this was recently evidenced[14] by using monocrystalline Pt electrodes. No poisoning was observed during the electrooxidation of glucose on Pt 111 surfaces and therefore no catalytic effect was produced by submonolayers of upd-adatoms (Pb, Bi, Tl). In contrast, the activity of Pt 100 was enhanced several times by Tl and Bi adatoms, becoming thus as high as that of Pt 111 electrodes[14].

More recently, the influence of various factors like temperature, concentration, anions, adatoms (Bi, Cd, Tl, Pb, Re) was systematically reexamined in acidic and alkaline media[15]. The complexity of the electrooxidation of glucose on platinum, mainly due to the hemiacetal group, was again evidenced.

Clearly, the transformation into electricity of the high chemical energy of glucose, needed for glucose sensors and for glucose fuel cells, requires new anodic materials in order to avoid the poisoning due to the hemiacetal group and to allow a more complete oxidation of C-OH groups.

A longchain polyol like sorbital, which has six C-OH groups but no hemiacetal group, does not poison gold or platinum surfaces and displays relatively high oxidation currents[16], noticeably higher than those measured for glucose in the same conditions[17]. Moreover, the presence of several adjacent C-OH groups in the hydrocarboneous chain of C_2-C_6 polyols was shown to play an important role in the adsorption and in the electrocatalytic oxidation of these molecules.

One of the purposes of this study was, therefore, to compare the behavior of glucose, which has five C-OH groups, with that of related polyols, xylitol and sorbitol, having five and six C-OH groups, respectively.

Most of the previous studies have used small electrodes (tenths of cm^2) in large volumes of solution containing relatively high amounts of reactants: the modification of the initial concentrations is thus insignificant and the formation of the intermediate species cannot be followed. In contrast, large area electrodes (tens of cm^2) allow the detection of organic species in the bulk of solution by chromatographic methods (HPLC, GC) and also that of the quantity of OH^- (or H^+) ions involved in the whole oxidation process.

Coupled voltammetric and pH measurements, performed as previously described[19] during the prolonged electrolysis, are necessary to obtain a complete balance between the quantity of electricity measured and that of the compounds consumed or produced. Such information, concerning both C- and O-containing species, are particularly useful when the electrochemical process is accompanied or followed by significant chemical transformations like hydrolysis, aldolisation, Cannizzaro reaction, enolisation, etc., involving reactants and the intermediate or final products. At room temperature, the chemical transformation of "blank" solutions of 0.1 M glucose is even faster and more complex than under prolonged electrolysis[20]. Such chemical transformations do not occur in the bulk of solution if the initial concentration of OH^- ions is too low, but can arise near the surface of the electrode as soon as the electrostatic field produces an accumulation of OH^- ions. The purpose of this study was to show, by coupled voltammetric and pH measurements on large gold and platinum electrodes, the importance of chemical transformations accompanying the electrocatalytic oxidation of polyols and glucose.

EXPERIMENTAL

The electrolytic solutions were prepared from "suprapur" Merck compounds (NaOH, K_2SO_4, H_2SO_4) and p.a D-glucose, D-sorbital (EGA-Chemie) or D-xylitol (EGA Chemie) dissolved in deionized, Millipore quality water.

The supporting electrolyte (0.0325 M K_2SO_4) generally contained only 0.01 M NaOH (or 0.01 N H_2SO_4) or less in order to avoid fast transformations of the organic compounds in the bulk of solution and to allow accurate measurements of the pH changes (at least 0.2 units) during the electrooxidation process.

Coupled voltammetric and pH measurements were carried out at 25°C or 50°C in several three-electrode cells having two compartments separated by large (3 cm in diameter) sheets of Nafion membrane. A smooth polycrystalline gold foil (23.4 cm^2), a thin gold film (6.0 cm^2) deposited on glass, or a platinum grid (36.4 cm^2) were alternatively placed in the same position, in the large compartment, close to a saturated calomel electrode (SCE) used as the reference electrode. Several graphite sheets (2.5 x 2.5 cm), interconnected by a gold wire, were used as counter electrode introduced into the smaller compartment. Both compartments were filled with the same electrolyte solution and carefully deoxygenated by bubbling ultrapure nitrogen throughout the whole experiment.

A Tacussel combination electrode (glass and Ag/AgCl/Cl^-) which was always placed in the same position (in the middle of the larger compartment) was used to measure pH values before (pH^i) and after (pH^f). The prolonged electrolysis was performed under mechanical stirring (at 500 rpm) and nitrogen bubbling conditions. The values of ΔpH were obtained as differences between pH^i and pH^f values and were always measured when the working electrode was disconnected in order to avoid the influence of the electrostatic field fluctuations on the indications of glass electrode[19].

Various potential sequences were used and the ratios between the number of OH^- ions consumed, Δn_{OH^-} (or the number of H^+ ions produced, Δn_{H^+}), and the number of electrons, n_{e^-}, involved in the electrocatalytic oxidation, $\Delta n_{OH^-}/n_{e^-}$ (or $\Delta n_{H^+}/n_{e^-}$), were obtained for different experimental conditions (potential limits, rates of sweep, temperatures, concentrations, etc.).

The evolution of current-potential profiles with the number of scans was followed at various sweep rates (from 2 mVs^{-1} to 500 mVs^{-1}) and appropriate lower and upper potential limits, E_c and E_a.

RESULTS AND DISCUSSION

The voltammograms of xylitiol, sorbitol and glucose on the large (gold or platinum) electrodes shown in Figures 1 and 2 display several oxidation peaks, named A (positive sweep) or B (negative sweep).

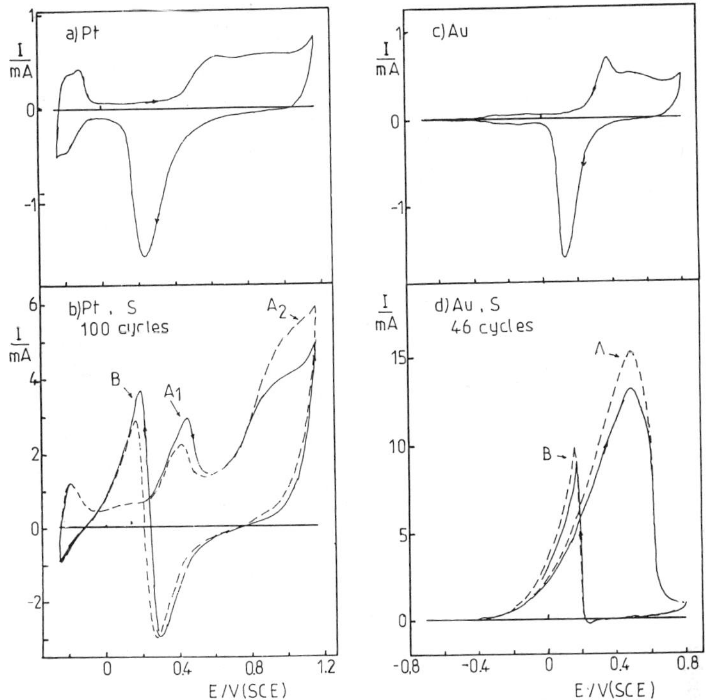

Figure 1. Current-potential profiles for a Pt electrode (36.4 cm^2) in 0.0325 M K_2SO_4 and 0.001 N H_2SO_4: a) supporting electrolyte alone or b) with 0.01 M sorbitol and for an Au film deposited on glass (6.0 cm^2) in 0.0325 M K_2SO_4 and 0.01 M NaOH; c) supporting electrolyte alone or d) with 0.01 M sorbitol at the initial (dotted lines) or the final (full lines) stages of the electrolysis (50 mVs^{-1}, 25°C).

Their shapes are similar to the respective voltammograms recorded with small electrodes[16-18] in the same experimental conditions, except higher current intensities, I_A and I_B, were observed.

On platinum electrodes, sorbitol is more efficiently oxidized in alkaline media (Figure 2c) than in acidic media (Figure 1b) and so is the case for xylitol or glucose. None of these compounds is oxidized on gold electrodes in acidic media, but in alkaline media (Figure 1d) their oxidation is more efficient than for platinum electrodes.

The voltammograms recorded in the supporting electrolyte alone (Figure 1c and 2a) do not change with the number of potential cycles, and the value of initial pH is constant as long as the potential limits do not allow the evolution of H_2 or O_2. In contrast, if an oxidizable compound is introduced into the cell, the voltammograms progressively change with the number of cycles (Figures 1b, 1d, 2b, 2c and 2d) and the value of pH decreases.

When gold electrodes are used, I_A (and to some extent I_B) values decrease with the number of cycles because of the consumption of OH^- ions and polyol (or glucose) molecules. Conversely, when platinum electrodes are used the repetitive cycling generally leads to higher currents (except for the oxidation peak occurring in the hydrogen adatom region),

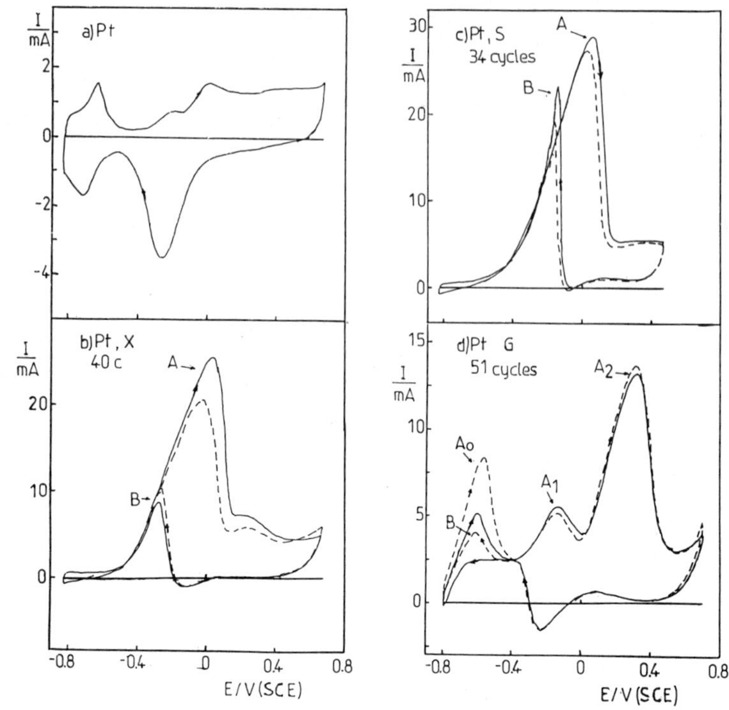

Figure 2. Current-potential profiles for a Pt electrode (36.4 cm^2) in 0.01 M NaOH, 0.0325 M K_2SO_4: a) supporting electrolyte alone; b) with 0.01 M D-Xylitol; c) with 0.01 M D-Sorbitol; d) with 0.01 M D-Glucose; at the initial (dotted lines) or final (full lines) stages of the electrolysis (50 mVs^{-1}, 25°C).

an effect presumably due to the activation of a surface covered at open circuit by poisoning species[21].

The values of peak currents greatly depend on the area of the working electrode and on the geometry of the cell, but those of $\Delta n_{OH^-}/n_{e^-}$ ratios are practically not affected. Besides, the oxidation currents

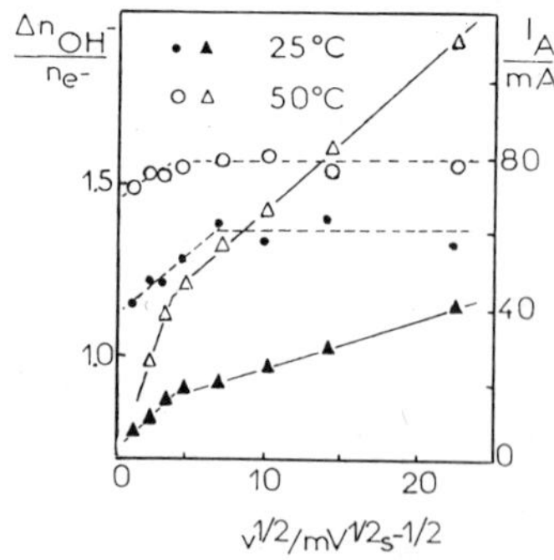

Figure 3. $\Delta n_{OH^-}/n_{e^-}$ vs $v^{1/2}$ (dotted lines, left scale) and I_A vs. $v^{1/2}$ (full lines, right scale) plots for the oxidation of 0.01 M D-Sorbitol on Pt grid (36.4 cm^2) in 0.01 M NaOH at 25°C and 50°C.

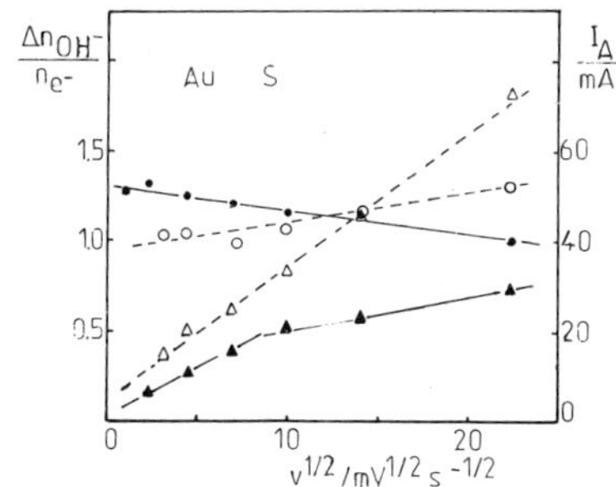

Figure 4. $\Delta n_{OH^-}/n_{e^-}$ vs $v^{1/2}$ (circles, left scale) and I_A vs $v^{1/2}$ (triangles, right scale) plots for the oxidation of 0.01 M D-Sorbitol on Au foil (23.4 cm^2, open circles and triangles, dotted line) and on Au film deposited on glass (6.0 cm^2, full circles and triangles, full lines), 25°C.

significantly increase with the sweep rate of potential, while only minor changes are observed for Δ_{OH^-}/n_{e^-} ratios (Figures 3 and 4).

In the same range of potentials, the values of $\Delta n_{OH^-}/n_{e^-}$ ratios for experiments under repetitive cycling at sweep rates from 20 to 200 mVs^{-1} are very close to those obtained when the working electrode was submitted to potential sequences in which a relatively long pause (up to 10 minutes) at one of the oxidation potentials was followed by a short incursion (at a sweep rate of 100 mVs^{-1}) at the upper and lower limits, E_a and E_c. In addition, the change of the working temperature from 25°C to 50°C produces an increase of only 0.2 units for the values of $\Delta n_{OH^-}/n_{e^-}$ ratios, while the oxidation currents are 2 or 3 times higher (Figure 3).

The values of $\Delta n_{H^+}/n_{e^-}$ for the oxidation of 0.01 M sorbitol on platinum in acidic media (Figure 5) are always lower than those of $\Delta n_{OH^-}/n_{e^-}$ determined for alkaline media (Figure 7), and this is also true for the corresponding oxidation currents.

In acidic media the potential range in which OH_{ads} species are adsorbed on platinum is narrower than for alkaline media where OH^- ions

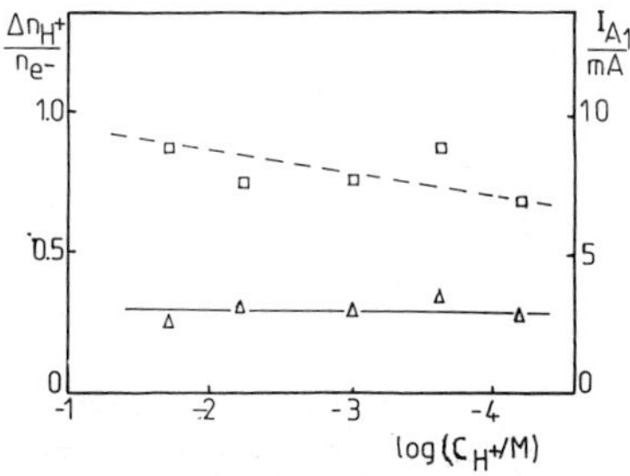

Figure 5. $\Delta n_{H^+}/n_{e^-}$ vs log C_{H^+} (dotted line, left) and I_{A1} vs log C_{H^+} (full line, right) plots for the oxidation of 0.01 M D-Sorbitol on Pt grid (36.4 cm^2) in acidic media 0.0325 M K_2SO_4 at 50 mVs^{-1} and 25°C.

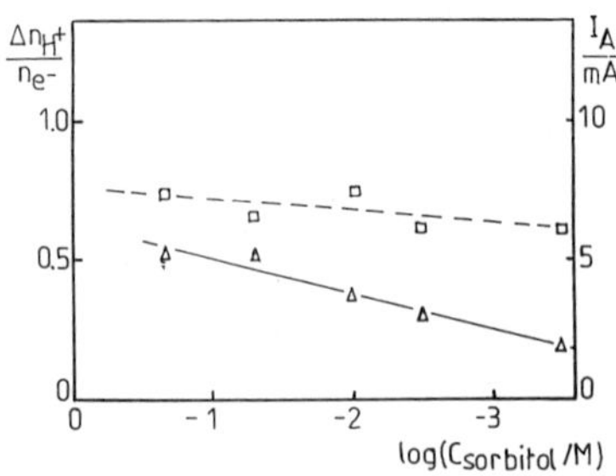

Figure 6. $\Delta n_{H^+}/n_{e^-}$ vs log C_{sor} (dotted line, left scale) and I_{A1} vs log C_{sor} (full line, right scale) plots for the oxidation of D-Sorbitol on Pt grid (36.4 cm^2) in acidic media (0.0325 M K_2SO_4, 0.001 M H_2SO_4) at 50 mVs^{-1} and 25°C.

accumulate near the charged surface. Consequently, $\Delta n_{H^+}/n_{e^-}$ values are lower than one (around 0.75) and remain practically constant when the initial concentrations of H^+ or of sorbitol are changed (Figures 5 and 6). Besides, the oxidation currents do not change with C_{H^+} but only with the concentration of sorbitol.

In contrast, the variation of $\Delta n_{OH^-}/n_{e^-}$ values and of peak oxidation currents significantly depends on both the initial concentrations of OH^- ions or sorbitol (Figures 7 and 8). Peak oxidation currents, I_A, reach a maximum for $C_{OH^-} \approx 3.10^{-3}$ M and suddenly decrease at $C_{OH^-} \leq 10^{-3}$ M with the concentration of OH^- ions in the bulk of solution (Figure 7). In contrast, $\Delta n_{OH^-}/n_{e^-}$ vs log C_{OH^-} plots display a minimum value close to 1.2 for $C_{OH^-} \approx 5.10^{-3}$ but increase sharply for solutions containing more than 0.01 M NaOH. At relatively high C_{OH^-} concentrations (more than 0.01 M) in the bulk, the accumulation of OH^- ions near the charged surface favors various chemical transformations of polyols or of the intermediate species issued from the electrooxidation process. The consumption of OH^- ions is thus increased with respect to the quantity of electricity measured.

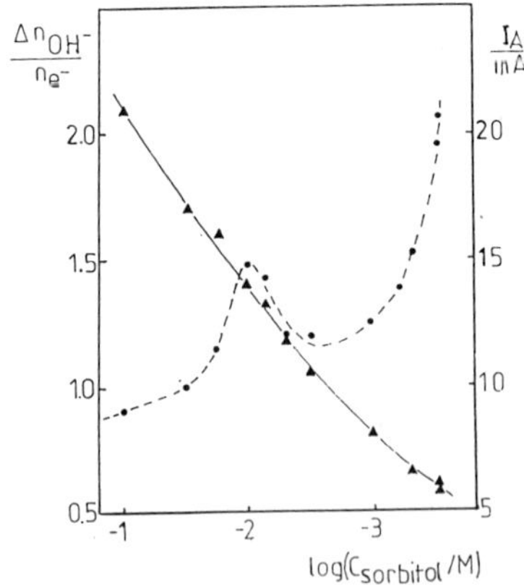

Figure 7. $\Delta n_{OH^-}/n_{e^-}$ vs log C_{OH^-} (dotted line, left scale) and I_A vs log C_{OH^-} (full lines, right scale) for the oxidation of 0.01 M D-Sorbitol on Pt grid (36.4 cm^2) in alkaline media 0.0325 M K_2SO_4 at 50 mVs^{-1} and 25°C.

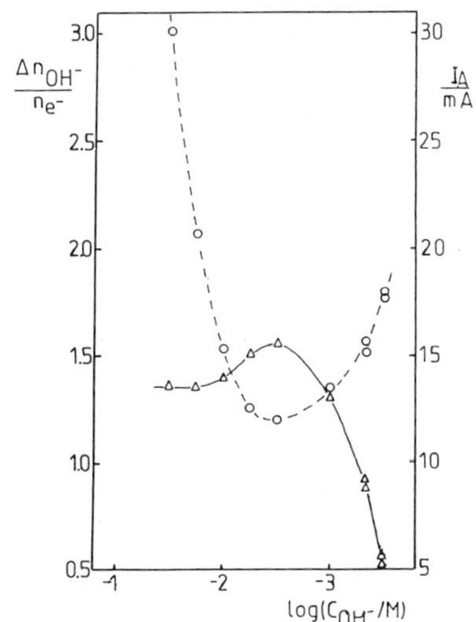

Figure 8. $\Delta n_{OH^-}/n_{e^-}$ vs log C_{sor} (dotted line, left scale) and I_A vs log C_{sor} (full line, right scale) for the oxidation of D-Sorbitol on Pt grid (36.4 cm^2) in alkaline media (0.0325 M K_2SO_4, 0.01 M NaOH) at 50 mVs^{-1}, 25°C.

In solutions containing 0.01 M NaOH and various quantities of sorbitol, the oxidation currents (I_A) progressively increase with the quantity of sorbitol (Figure 8). A reverse trend is observed for $\Delta n_{OH^-}/n_{e^-}$ vs log C_{sor} plots which display a particular shape in the range where $C_{OH^-} \approx C_{sor}$ (Figure 8). A higher bulk concentration of sorbitol leads to the accumulation of the polar molecules near the charged surface and to the decrease of the concentration of OH^- ions near

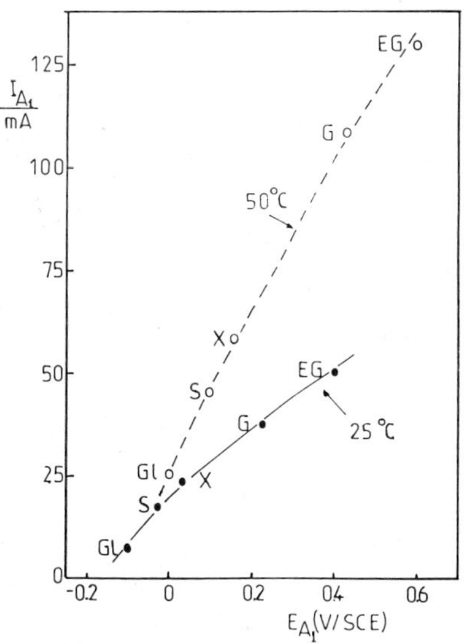

Figure 9. Dependences of peak currents, I_{A1} on peak potentials, E_{A1} for the oxidation of D-Glucose and of several polyols on Pt grid (36.4 cm^2) in 0.01 M NaOH at 50 mVs^{-1} and 25°C or 50°C.

the electrode. Consequently, chemical transformations occur to a lesser extent and thus $\Delta n_{OH^-}/n_{e^-}$ values approach those of $\Delta n_{H^+}/n_{e^-}$ values.

A comparative examination of the plots shown in Figures 7 and 8 clearly shows that the maximum oxidation currents can be reached for an optimal value of the ratio C_{OH^-}/C_{sor} between the initial concentrations of reactants.

The general trends described above for the influence of various factors on the electrooxidation of sorbitol are more or less valid for the other polyols and, to some extent, even for glucose. So, when the peak oxidation currents, I_{A1}, are plotted against the peak potentials, E_{A1}, for the compounds having a different number of C-OH groups (Figure 9), the behavior of glucose at 25°C and 50°C appears to be closer to that of higher polyols like sorbitol than to that of lower polyols like ethylene glycol.

The dependences of $\Delta n_{H^+}/n_{e^-}$, $\Delta n_{OH^-}/n_{e^-}$, I_A or I_B values with the number of C-OH groups, n_{OH}, are shown for various polyols in Figures 10 and 11.

As in the case of small Pt electrodes[18], I_A values for the electrocatalytic oxidation of polyols on large Pt electrodes in 0.1 M NaOH generally decrease with the number of C-OH groups in the molecule (Figure 10). The tendency of the CH and CH_2 groups to adsorb on the Pt surface at open circuit appears to be reduced by the opposite tendency of C-OH groups to keep the molecule in the bulk of the solution (unless the surface is positively charged and thus attracts them). Therefore, long chain polyols like sorbitol are less oxidized at room temperature than shorter ones like ethylene glycol, and this is also true at 50°C (Figure 10). The case of glucose is somewhat different because the hemiacetal group reinforces the adsorption of the molecule on platinum, the surface of which becomes progressively poisoned. Consequently, peak oxidation currents are small (I_A = 3 mA at 50°C) and their progressive decrease within a relatively small number of scans does not allow accurate measurements for $\Delta n_{H^-}/n_{e^-}$.

A similar dependency of peak currents with n_{OH} is observed for the electrooxidation of polyols on Pt in alkaline solutions. In all cases, I_A values corresponding to the negative sweep peaks are significantly higher than I_B values of the respective positive sweep peaks (Figure 11). The number of molecules oxidized and the quantity of electricity produced on B peaks is limited by the oxygen up-take of the surface and by the efficiency of OH_{ads} species formed from oxide species in a narrow range of potentials (Figures 1 and 2) to react with the organic adsorbates. Conversely, in the case of A peaks, the progressive coverage of the surface with OH_{ads} species in a large range of potentials (especially in alkaline media) allow several oxidation sequences to occur and leads to a higher turn-over per catalytic site, i.e., to a higher quantity of electricity.

The dependency of $\Delta n_{H^+}/n_{e^-}$ with n_{OH} (Figure 10) has a different shape compared to that of $\Delta n_{OH^-}/n_{e^-}$ vs n_{OH} (Figure 11). In both cases, an increase of temperature (known to enhance chemical transformations) produces higher $\Delta n_{H^+}/n_{e^-}$ and $\Delta n_{OH^-}/n_{e^-}$ values. Besides, for xylitol and sorbitol, $\Delta n_{H^+}/n_{e^-}$ and $\Delta n_{OH^-}/n_{e^-}$ ratios are somewhat higher than for ethylene glycol (Figures 10 and 11) the electrooxidation of which is accompanied by fewer chemical transformation than in the case of polyols possessing many C-OH groups. Peak currents corresponding to the oxidation of glucose on Pt in 0.01 M NaOH at 25°C (I_{A1} = 6 mA, I_{A2}= 11 mA) and 50°C (I_{A1} = 25 mA and I_{A2} = 24 mA) are generally lower than those of the polyols. Conversely, $\Delta n_{OH^-}/n_{e^-}$

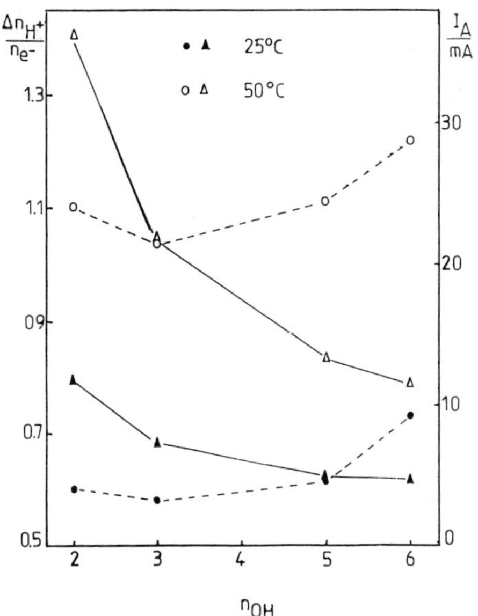

Figure 10. Dependences of $\Delta n_{H^+}/n_{e^-}$ (dotted lines, left scale) and I_A (full line, right scale) with the number of C-OH groups, n_{OH}, for the oxidation of 0.01 M polyol on Pt grid (36.4 cm^2) in acidic media (0.0325 M K_2SO_4, 0.003 M H_2SO_4) at 50 mVs^{-1} and 25°C or 50°C.

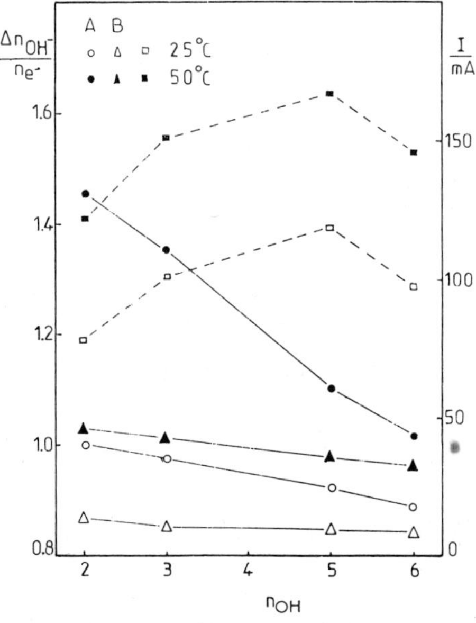

Figure 11. Dependences of $\Delta n_{OH^-}/n_{e^-}$ (dotted line, left scale) and I_A (circles) or I_B (triangles, full line, right scale) with the number of C-OH groups, n_{OH}, for the oxidation of 0.01 M polyol on Pt grid (36.4 cm^2) in alkaline media (0.0325 M K_2SO_4, 0.01 M NaOH) at 50 mVs^{-1} and 25°C or 50°C.

values measured at 25°C (1.67) and 50°C (1.74) are higher. Glucose is certainly more sensitive to the chemical transformations than the related polyols.

CONCLUSIONS

The main conclusions resulting from this comparative study of the electrocatalytic oxidation of glucose and related polyols on large Pt and Au electrodes can be summed up as follows:

a) The electrooxidation of polyols and sugars is accompanied, especially in alkaline media, by the chemical transformation, enhanced by an increase of the working temperature of the reactants and intermediate products due to the accumulation of a relatively high concentration of hydroxyl ions near the charged surface of the electrode. Therefore, the analytical results obtained by HPLC or other techniques for the intermediate species should be used with very great care in explaining the path of the electrooxidation process. The coupled voltammetric and pH measurements allow the possibility to estimate, from the values of $\Delta n_{OH^-}/n_{e^-}$ and $\Delta n_{H^+}/n_{e^-}$ ratios, the extent of the chemical transformations in the range of concentrations used for reactants.

b) On gold surfaces, the electrooxidation process mainly involves C-OH groups rather than species absorbed at more negative potentials. Therefore, the behavior of glucose is rather similar to that of xylitiol and sorbitol. On platinum, the activity of the C-OH groups of glucose is masked by that of the hemiacetal group which strongly adsorbs and poisons the surface. The electrooxidation of glucose can, therefore, be enhanced by the modification of the surface (adatoms, alloys, mixed deposits, etc.) in order to obtain a lower poisoning. Some of these mixed surfaces having a high-surface area are actually examined in our laboratory as glucose anodes for fuel cells applications.

ACKNOWLEDGEMENT

The financial support by CNRS and AFME (PIRSEM grant No. 8680) is acknowledged.

REFERENCES

1. U. Gebhardt, G. Luft, K. Mund, N. Preidel and G.J. Richter, Siemens Forsch. und Entwickl.-Ber., Bd. 12, 91 (1983).

2. I.R. Rao, G.J. Richter, F. van Sturm and E. Weidlich, Ber. Bunsenges. Phys. Chem., 77, 787 (1973).

3. S. Ernst, J. Heitbaum and C. Hamann, J. Electroanal. Chem., 100, 50 (1980).

4. S. Ernst, J. Heitbaum and C. Hamann, Ber. Bunsenges. Phys. Chem., 84, 50 (1980).

5. M.F.L. de Mele, H.A. Videla and A.J. Arvia, Bioelectrochem. Bioenergetics, 9, 469 (1982).

6. M.F.L. de Mele, H.A. Videla and A.J. Arvia, J. Electrochem. Soc., 129, 2207 (1982).

7. M.F.L. de Mele, H.A. Videla and A.J. Arvia, Bioelectrochem. Bioenergetics, 10, 239 (1983).

8. M.F.L. de Mele, H.A. Videla and A.J. Arvia, Bioelectrochem. Bioenergetics, 10, 213 (1986).

9. Yu. B. Vassilyev, J.A. Khazova and N.N. Nikolaeva, J. Electroanal. Chem., 196, 105 (1985).

10. Yu. B. Vassilyev, J.A. Khazova and N.N. Nikolaeva, J. Electroanal. Chem., 196, 127 (1985).

11. N. Xonoglou and G. Kokkinidis, Bioelectrochem. Bioenergetics, 12, 485 (1984).

12. G. Kokkinidis and N. Xonoglou, Bioelectrochem. Bioenergetics, 14, 375 (1985).

13. N. Xonoglou, Z. Moumtzis and G. Kokkinidis, J. Electroanal. Chem., in press.

14. G. Kokkinidis, private communication.

15. L.-H. Essis-Yei, Doctor Thesis, University of Poitiers, June 1987.

16. J.P. Ango, O. Enea and C. Lamy, Bioelectrochem. Bioenergetics, submitted.

17. J.P. Ango, B. Beden, O. Enea, H. Essis-Yei, C. Lamy and J.M. Leger, Proceedings of the 4th European Conference on Biomass, Orleans 1985.

18. J.P. Ango, O. Enea and C. Lamy, ISE Extended Abstracts 38 (1987).

19. O. Enea, Proceedings of the 170th Meeting of the Electrochemical Soc., 2, 230 (1986).

20. B. Kokoh Kouakou, DEA Report, University of Poitiers, 1987.

21. M.F.L. de Mele and H.A. Videla, Acta Cientifica Venezuela, 36 (1985).

ELECTROCHEMISTRY OF SOME SYNTHETIC OXYGEN CARRIER MOLECULES IN WATER

J.P. Ciccone, L.A. Deardurff, E.S. DeCastro,
J.B. Kerr* and B.D. Zenner

Aquanautics Corporation
4560 Horten Street, Q-111
Emeryville, California 94608

INTRODUCTION

In recent years the chemistry of oxygen binding transition metal complexes has attracted considerable attention[1,2]. Examples of these complexes range from naturally occurring compounds such as hemoglobin and complex synthetic models like iron porphyrins to very simple molecules such as cobalt pentaamine complexes[3]. In addition to their importance in fundamental chemistry these transition metal complexes are of great interest for oxygen separation processes and have been used in membrane systems[4], pressure swing systems[5] and electrochemical separation systems[6-8].

For this last example Aquanautics Corporation has developed a practical working system for the separation of oxygen from fluids like air or sea water. The device is illustrated in Figure 1 and it consists of: 1) an aqueous solution of oxygen-binding carrier complex which is circulated; 2) a hollow-fiber membrane where the complex picks up oxygen from a fluid on the other side of the membrane; 3) an electrochemical cell where the oxidation state of the complex is altered at the anode to

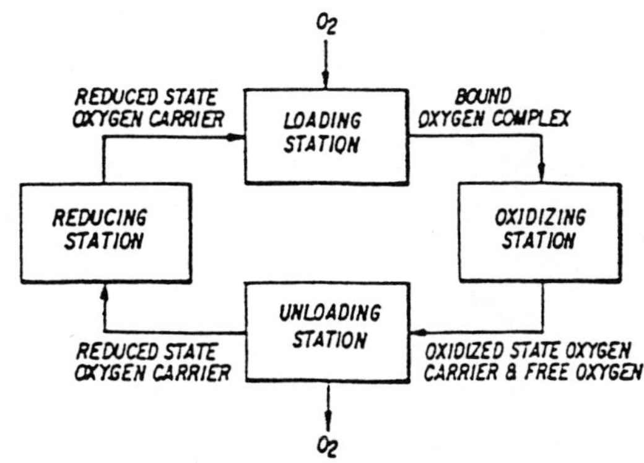

Figure 1. Schematic of oxygen extractor.

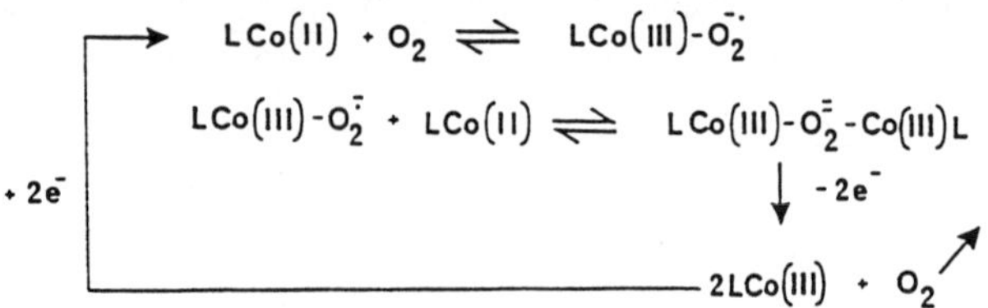

Figure 2. Schematic of chemistry involved in the oxygen extractor.

release the oxygen. After separating the oxygen the complex is then returned to its original state at the cathode so that it can recirculate to the oxygen loading membrane to repeat the cycle. The goal of such a system is to produce high purity oxygen with minimum power and stable performance over an extended period of time. Water is used as the solvent in order to avoid high voltage drops associated with the solution resistance of non-aqueous solvents. Since the system is open to moisture through the loading membrane costly gas-stream drying procedures are avoided. Solvent degradation commonly found with electrochemistry in non-

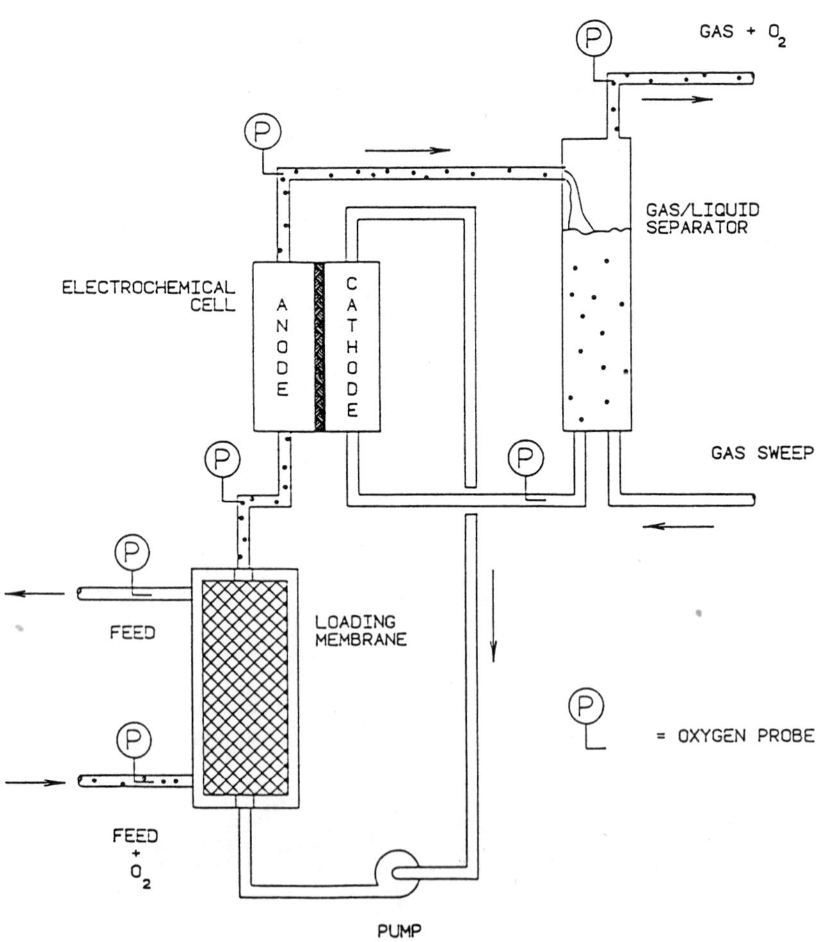

Figure 3. Experimental apparatus. Oxygen extractor.

aqueous solvents is also avoided. Since the electrolyte must circulate many times without degradation this is an important property.

In most cases cobalt complexes form a μ-peroxo dimer with oxygen in water. Figure 2 shows this reaction together with the electrochemical reactions required to complete the cycle in the extraction system. It is clear that the anode reaction involves a two-electron cleavage reaction while the cathodic reaction involves a "simple" electron transfer. Since the power consumed by the oxygen extraction system depends on the potential difference between the two reactions the factors which affect them are of practical as well as theoretical interest.

The binding of oxygen by cobalt complexes is often represented to involve at least partial electron transfer from the metal to the oxygen. It is thus an obvious step to attempt to correlate the properties of this reaction with the redox potentials for the complexes. Such correlations have been attempted using both water[1,9,10] and pyridine[1,11,12].

The correlations are relatively successful for complexes where the charge is delocalized within the molecule and where there are no steric constraints. Even more encouraging are the relatively good correlations observed for cobalt complexes in water where more ligand substitution chemistry occurs[9,10]. However, these measurements were made at the dropping mercury electrode. The purpose of this paper is to examine further the effects of solution pH and electrode material upon the electrochemistry of these complexes.

EXPERIMENTAL

Cobalt chloride (anhydrous) was obtained from Aesar, L-histidine hydrochloride from Sigma, N-benzyl-L-histidine from Vega, tetraethylene pentamine (tetren) was obtained from Aldrich. Water was deionized via a Barnstead Nanopure II ion exchange unit. Cells and electrodes for voltammetry were standard Bioanalytical Systems or Pine Instrument controlled by BAS 100A, BAS-CV27 or Pine Instrument RDE4 potentiostats and instrumentation. Spectroelectrochemistry was accomplished using a standard

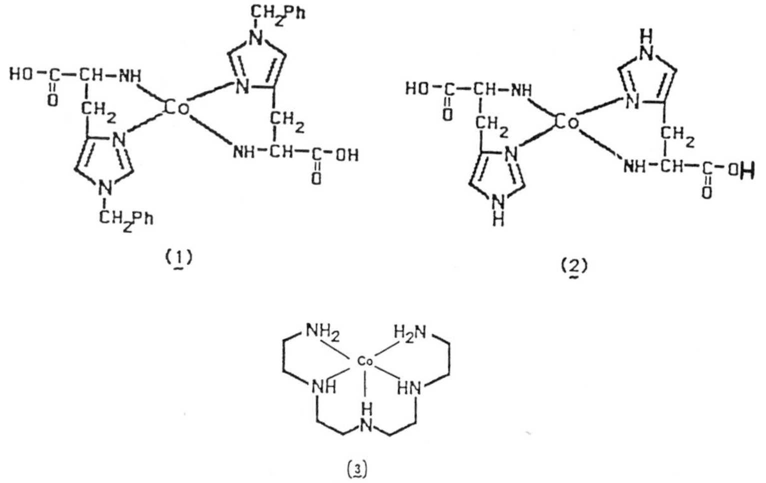

(1) (2)

(3)

Figure 4. Cobalt complexes used in this study.

gold minigrid optically transparent thin layer cell (vol. = 60 μl[13]) positioned in the beam of a HP 8451A diode array spectrophotometer.

The oxygen extractor system (Figure 3) is comprised of an array of hollow fiber membranes, an electrochemical cell and a gas/liquid separator which used about 0.5 L of carrier solution. The membrane array consists of spirally wound oxygen permeable tubules. The packing configuration results in 5 m^2 of effective surface area in 1.8 L including manifolds and packaging or 2800 m^2/m^3. The electrochemical cell has 25 cm^2 separator surface area and both anode and cathode are comprised of high surface area electrodes. The gas/liquid separator is simply a carrier reservoir sparged with a nitrogen sweep gas. However, pure oxygen has been generated with this separator. The oxygen output is measured by means of Clark oxygen probes and is limited by the power supply to 15 ml O_2/min. The system employs a Zenith PC and a Tecmar Labmaster to control and record potential, monitor current and monitor oxygen output from the various solution and gas stream Clark oxygen probes.

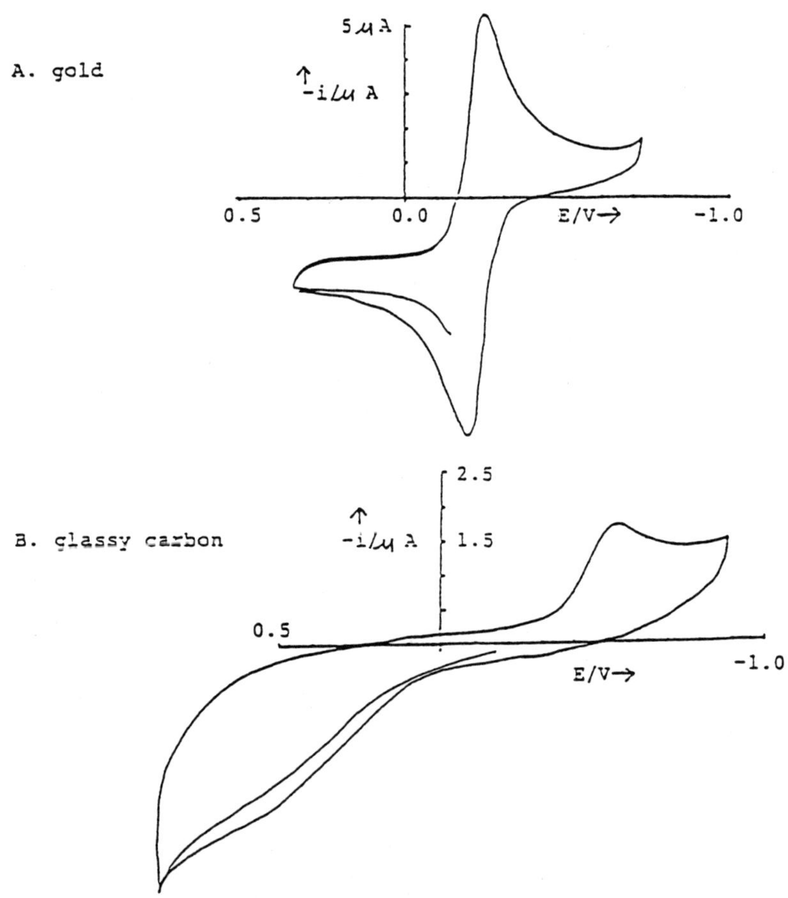

Figure 5. Cyclic voltammetry of cobalt bishistidine (1 mM) in aqueous saline solution containing 0.1 M sodium borate (pH 9). Sweep rate 50 mVs^{-1}. Reference electrode SCE. Solution degassed with N_2.

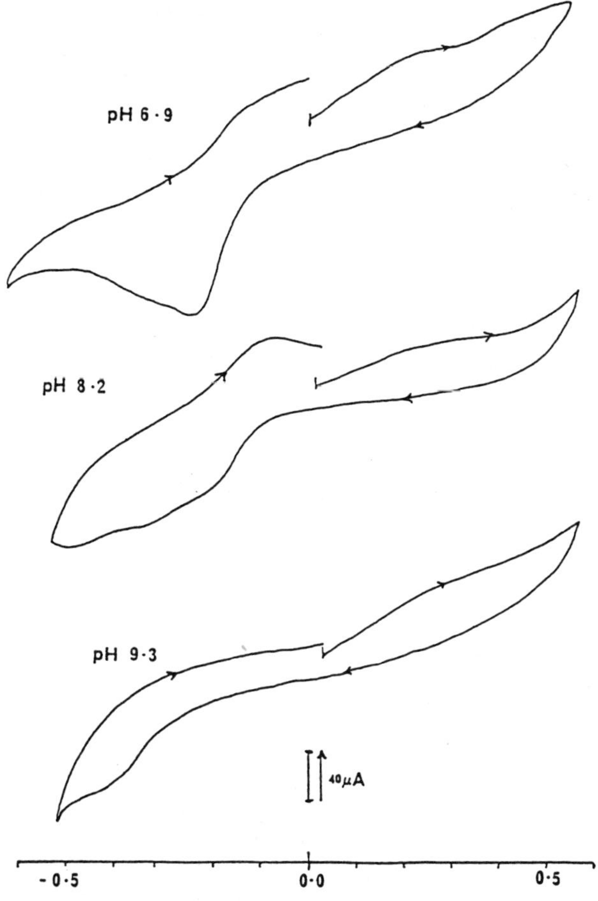

Figure 6. Cyclic voltammetry of cobalt tetrendichloride (3) (5 mM) at a a gold disc electrode in 0.5 M KCl-H_2O. Reference is SCE and sweep rate is 0.05 Vs^{-1}. Solution degassed with argon.

RESULTS AND DISCUSSION

The structures of the ligand-cobalt complexes used are shown in Figure 4. The voltammetry of these complexes show considerable sensitivity to the electrode material. Figure 5 illustrates this point for cobalt bis-histidine. Under certain conditions adsorption processes can be clearly seen. In addition the pH of the solution clearly has a major effect on the composition of the complex in solution as shown by the voltammetry illustrated in Figure 6. Behavior like this has been observed before for Vitamin B_{12}. Even the choice of counterion can significantly alter the shapes of the voltammetric responses and the associated rate constants. For this reason it is very difficult to ascertain E° values for the Co(III)/Co(II) couple by voltammetric methods and consequently these determinations were accomplished by spectroelectrochemistry. Figure 7 illustrates the effect of pH on the E° values measured by this method for cobalt bisbenzylhistidine which shows a clear discontinuity indicating some radical change in the structure occurs at around pH 6. Clearly some protonation of carboxylic acid groups has occurred upon lowering the pH which causes a major change in the complex. Less drastic changes are observed with the non-acid containing complexes which are similar to those pH variations described for Vitamin B_{12} [14].

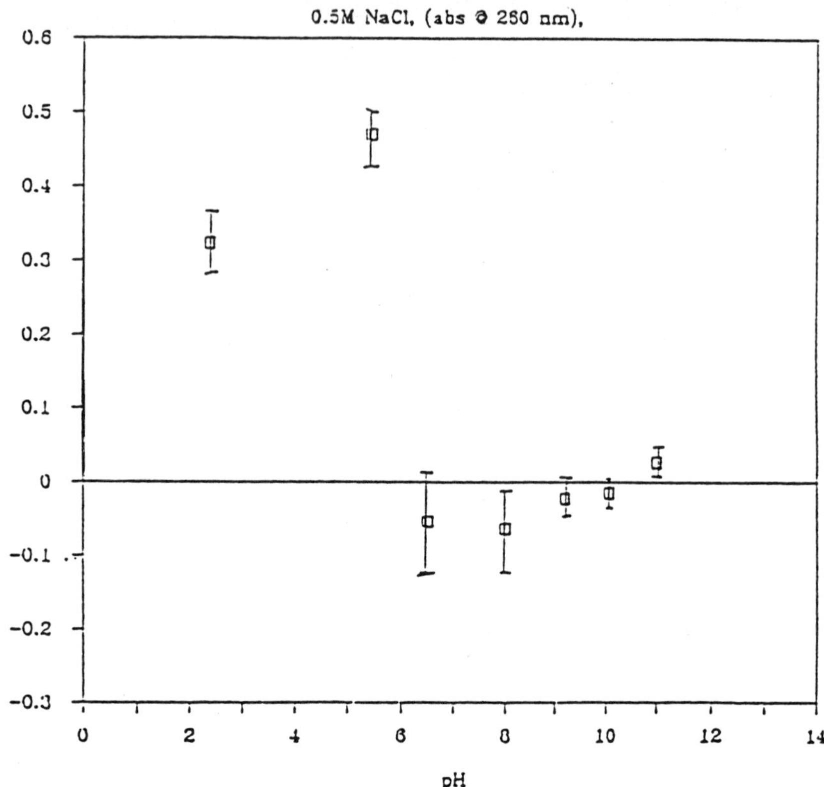

Figure 7. Effect on pH on the E° value for the Co(II)/Co(III) couple of cobalt bisbenzylhistidine. E° determinations carried out by spectroelectrochemistry.

The introduction of oxygen into the solution causes significant changes in the shapes of the voltammograms as can be seen for the case of cobalt bisbenzylhistidine (Figure 8). The currents increase by an order of magnitude. Figure 9 shows the voltammetry of Co tetren Cl_2 at pH 6. The reduction wave at -0.35 V is due to the reduction of the μ-peroxodimer while the reversible couple at +0.6 V is due to the formation of the superoxodimer. Spectroelectrochemistry was performed on this couple (Figure 10). The superoxodimer is less stable at higher pH as seen by the loss of reversibility of the couple and in a preparative experiment oxygen evolution is detected by the use of a Clark oxygen probe. The complex which is left in solution is the Co(III) tetren which can be reduced back to Co(II) tetren for further reaction with oxygen and the cycle may be repeated.

Cobalt tetren and cobalt bisbenzylhistidine have been used as carriers in the working oxygen extractor described in Figure 3. In particular cobalt tetren was used to extract 1 ml/min of oxygen from air continuously for 28 hours at a power cost of 300 $WminL^{-1}$ and it continued to produce oxygen at higher power levels (400-500 $WminL^{-1}$) for 12 days. These experiments demonstrate the principle of the oxygen extractor technology which has since been improved to operate in the region of 35-100 $WminL^{-1}$. The details of this latest work will be the subject of further publications.

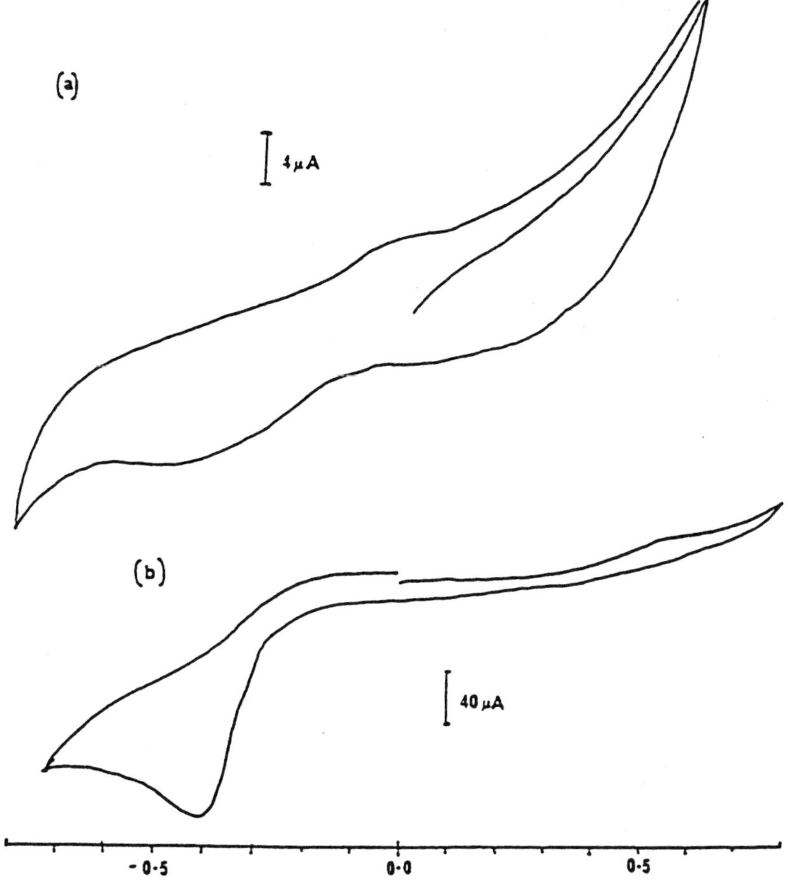

Figure 8. Cyclic voltammetry of cobalt bisbenzyl-L-histidine dichloride (1) (5 mM) at a gold disc electrode in 0.5 M KCl-H_2O (pH 7). Reference is SCE and sweep rate is 0.05 Vs^{-1}. a) solution degassed with argon; b) saturated with oxygen.

CONCLUSIONS

The first and most important conclusion of this work is that the principle of the oxygen extraction process has been demonstrated in a practical manner. However, to take the process from a demonstration to a practical technology involves several major hurdles. It is quite clear from the examples given here that the electochemistry of the cobalt carrier complexes is not straightforward and involves a number of factors which are hard to control. Indeed the electrochemical reaction which operates best is precisely that which is undesirable, namely, reduction of the oxygenated complex.

The most serious difficulty lies in the low current density observed for the cobalt complexes. The reasons for this are not clear. It is not a mass transfer problem since the oxygenated complexes give reasonable current densities and therefore the reason must lie in the fundamental kinetics of the electron transfer itself. Fortunately much work has been carried out by Weaver[15] in the area of the "simple" electron transfer reaction and by Saveant[16] in the area of the associated electron transfer

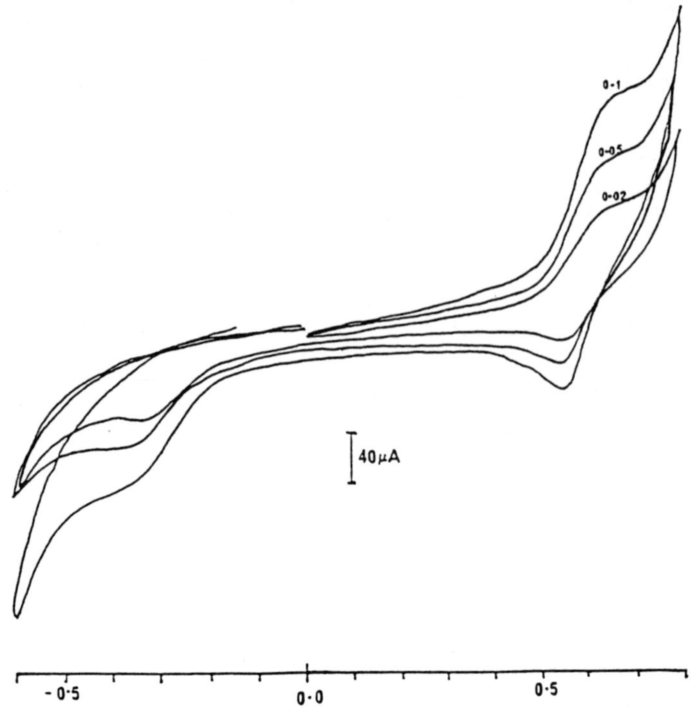

Figure 9. Cyclic voltammetry of cobalt tetren dichloride (3) (5 mM) at a gold disc vs SCE in O_2 - saturated 0.5 M KCl-H_2O (pH 6). Sweep rates (Vs^{-1}) are shown on curves.

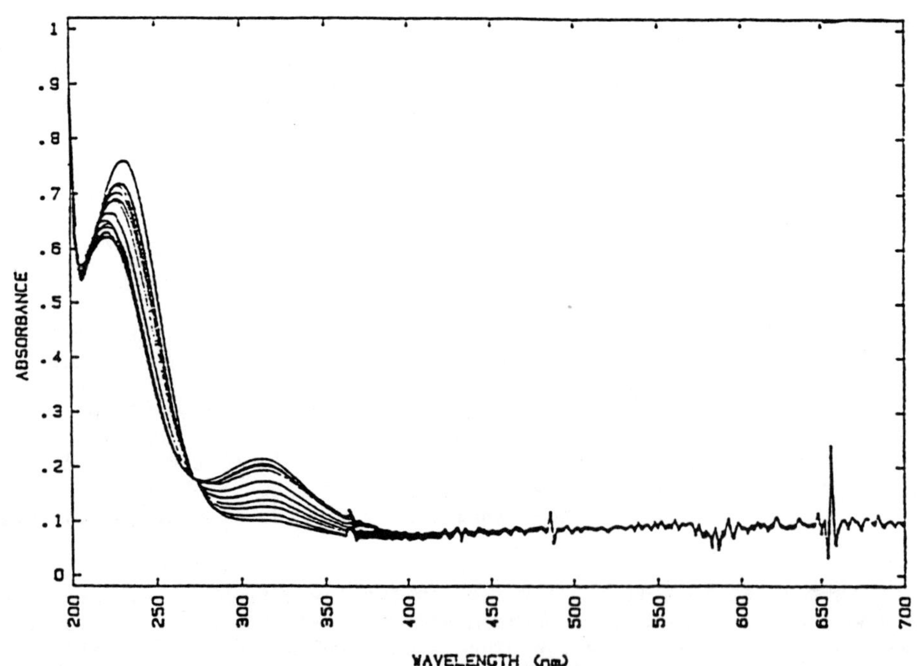

Figure 10. Spectroelectrochemistry of cobalt tetren dichloride (3) at a gold minigrid electrode in 0.5 M KCl-L_2O (pH 2.4). Potential range 0.65 - 0.72 V.

induced bond cleavage. Clearly the role of electrocatalysis will be vital to the development of the extractor technology but the structure-function relationships of the carrier complexes will be an additional dimension for experimentation. Both of these areas are currently being explored.

With respect to correlations of electrochemical potentials with properties such as binding constants or spectroscopic data it is clear that there is a considerable kinetic component in the electrochemical response of techniques like voltammetry. Although a general trend may indeed fall out of the free energy relationships there has been more than one complaint in the literature about the variability of electrochemical data for similar types of complexes[17]. It is obvious from the results outlined here that reversible conditions are not the usual state of affairs and that some caution should be applied when using the data. However, the kinetic character of the electrochemical response does allow much insight to be gained into the stability and structure of the complexes and that this area will be a fruitful one for those prepared to cope with the experimental difficulties.

ACKNOWLEDGEMENTS

This work has been funded by the Defense Advanced Research Projects Agency and the Office of Naval Research. We gratefully acknowledge their support.

REFERENCES

1. E.C. Niederhoffer, J.H. Timmons and A.E. Martell, Chemical Reviews, 84, 137 (1984).

2. S. Fallab and P.R. Mitchell in "Advances in Inorganic and Bioorganic Mechanisms", Academic Press (1984), p. 311.

3. R.D. Jones, D.A. Summerville and F. Basolo, Chemical Reviews, 79, 139 (1979).

4. B.M. Johnson, R.W. Baker, S.L. Matson, K.L. Smith, I.C. Roman, M.E. Tuttle and H.K. Lonsdale, J. Membrane Science, 31, 31 (1987).

5. A.J. Adduci, Chemtech, 575 (1976).

6. U.S. Patent 4,475,994.

7. U.S. Patent 4,605,475.

8. U.S. Patent 4,602,987; 4,629,544.

9. S.R. Pickens and A.E. Martell, Inorg. Chem., 19, 15 (1980).

10. W.R. Harris, A.L. McLendon, A.E. Martell, R.C. Bess and M. Mason, Inorg. Chem., 19, 21 (1980).

11. M.J. Carter, D.P. Rillema and F. Basolo, J. Am. Chem. Soc.,96, 392 (1974).

12. A. Puxeddu and G. Costa, J. Chem. Soc., Dalton Trans., 1115 (1981).

13. T.P. DeAngelis and W.R. Heineman, J. Chem. Ed., 94, 535 (1976).

14. D. Lexa and J.M. Saveant, Acc. Chem. Res., 16, 235 (1983).

15. M.J. Weaver, Ber. Bunsenges. Phys. Chem., 91, 450 (1987).

16. C.. Andrieux, I. Gallardo, J.M. Saveant and K.B. Su, J. Am. Chem. Soc., 108, 638 (1986).

17. C.L. Wong, J.A. Switzer, K.P. Balakrishnan and J.F. Endicott, J. Am. Chem. Soc., 102, 5511 (1980).

ELECTROCHEMICAL STUDIES ON THE GENERATION OF ACTIVE OXYGEN SPECIES IN BIOLOGICAL SYSTEMS WITH THE USE OF MEDIATORS

Sianette Kwee

Institute of Medical Biochemistry
University of Aarhus
DK-8000 Aarhus C / Denmark

INTRODUCTION

The process of the reduction of oxygen to water in the mitochondria has been widely studied in the course of time. It is generally accepted that oxygen is reduced by cytochrome oxidase, the terminal enzyme in the electron transport chain, in a single 4-electron 4-proton step. However, the discussion about the true mechanism of the process is still not closed. The electrochemical reduction of oxygen has also been studied extensively by direct methods[1] as well as indirectly in the presence of mediators[2-4]. At the electrode the reduction of oxygen proceeds in two 2-electron steps. This process would imply hydrogen peroxide as the first reduction product, which is then reduced to water in the second step. In the presence of one specific porphyrin mediator, practically no hydrogen peroxide was found, which in this case could indicate that a direct reduction of oxygen to water has taken place[5].

In mitochrondial oxygen reduction no free hydrogen peroxide has ever been detected. However, current evidence indicates that an intermediate peroxide-metal complex is involved, in which the peroxide is bridged between an iron and a copper atom from the cytochrome oxidase enzyme[6].

Oxygen can also be reduced to the superoxide anion radical by a 1-electron transfer from 1-electron donors as well as by other radicals. The best known example is the radical from the herbicide paraquat[7]. It is widely accepted that oxygen radicals cause various tissue damages. Peroxide, superoxide, peroxyl and hydroxyl radicals belong to this group of active oxygen species. Hydroxyl radicals, the most reactive species, are generated from the reaction between peroxide and superoxide[8].

In a previous work it was shown that tetrahydropterins could transfer electrons to oxygen in 1-electron steps, resulting in both superoxide and hydrogen peroxide[9]. At the same time they could serve as mediators in the electrochemical reduction of oxygen.

Since it was shown that biopterin can be a constituent of rat liver mitochondria, the next step was to investigate if tetrahydropterins could induce the generation of the above mentioned active oxygen species in whole mitochondria.

Electrochemical measurements on mitochondria have been performed in the presence of mediators[10] or by modified electrodes[11] with the purpose to determine formal redox potentials. In this study interest was mainly centered on changes under electrolysis conditions.

EXPERIMENTAL

Chemicals

Bovine superoxide dismutase and catalase were purchased from Sigma Chemical Co., glutathione peroxidase from Boehringer Mannheim and quinoxaline from Fluka AG. The synthesis of 6,7-dimethylperin and its 7,8-dihydro derivative has been described before[12]. 6,7-Dimethyl-5,6,7,8-tetrahydropterin was prepared by controlled potential electrolysis[12] in a 0.1 M phosphate buffer pH 7.0 at a glassy carbon or a mercury pool cathode. After complete electrolysis the tetrahydropterin was used *in situ*. 1,2,3,4-Tetrahydroquinoxaline was prepared by reducing quinoxaline in dry ether with lithium aluminium hydride[13] or in ethanol with solid sodium[14]. M.P. 97°C (litt. 96.5 - 97.5[14]).

Beef heart mitochondria was prepared according to Smith[15] and sub-mitochondrial particles according to Gregg[16].

Apparatus

The cells and apparatus used for cyclic voltammetry and large-scale electrolyses have been described before[12]. For micro-scale electrolyses a H-type cell was used, consisting of 2 cylindrical compartments (12 mm diameter), serving respectively as cathodic and anodic compartments, with separate side arms for reference electrode and gas-inlet. The cells were connected by a short glass tube with a sintered glass frit in the center. As a working electrode a glassy carbon disc (7 mm diameter) sealed into a glass tube was used. The counter electrode consisted of a platinum foil and the reference electrode (Ag/Ag^+) was separated by a thirsty glass frit from the catholyte. The working volumes were approximately 1.5 ml.

For spectroscopic measurements an Aminco DW-2a TM UV/VIS and Beckmann Acta MVI were used.

Procedure

The mitochondria were suspended in 0.25 M sucrose to give approximately 25-30 mg protein per milliliter of mitochondrial suspension. Protein was determined according to Lowry[17]. The mitochondrial suspension was stored at -20°C and thawed only immediately prior to use. Difference spectra were recorded according to the following standard procedure[18]: to both the reference (R) and sample (S) cuvettes were added 0.5 ml of mitochondrial suspension, 0.2 ml 0.1 M sodium phosphate buffer (pH 7.4) and 0.2 ml 10% sodium deoxycholate. After thorough mixing of the contents, 0.1 ml 0.05 M potassium ferricyanide and 0.1 ml of water were added to cuvette R. To cuvette S were added 0.1 ml 0.05 M sodium ascorbate and 0.1 ml of water. After mixing both cuvettes by inversion, a few grains of sodium hydrosulphite were added to cuvette S. After mixing by inversion, the difference spectrum was recorded from 500 to 630 nm. In subsequent experiments difference spectra were recorded after addition to cuvette S, after solubilization, either ascorbate only, or 6,7-dimethyltetrahydropterin in varying concentrations, or solid 1,2,3,4-tetrahydroquinoxaline. In a similar way, spectra were recorded in the presence of the enzymes superoxide dismutase or peroxidase. For micro-scale electrolyses the same solutions as described for spectroscopy were used: 12-15 mg per milliliter mitochondrial protein solubilized with 10 % sodium deoxycholate and 0.01 mM 6,7-dimethyl-5,6,7,8-tetrahydropterin were reduced at -1.6 V vs Ag/Ag^{+}. The reduction was also followed spectroscopically. Reduction of the mitochondria was usually completed within 5 min. Recycling was performed by flushing the catholyte with oxygen and the reoxidized mitochondria could be reduced in the same way.

RESULTS AND DISCUSSION

Spectroscopy

Figure 1A shows the difference spectrum of the completely reduced beef heart mitochondria. As was shown previously[9], oxygen had a competing effect on the reduction of ferricytochrome c (Fe^{3+}cyto c) by tetrahydropterin (PtH_4), since:

$$PtH_4 + Fe^{3+}\text{cyto c} \longrightarrow PtH_3^{\bullet} + Fe^{2+}\text{cyto c} + H^{+} \quad (1)$$

$$\text{and} \quad PtH_4 + O_2 \longrightarrow PtH_3^{\bullet} + O_2^{\bar{\cdot}} + H^{+} \quad (2)$$

This is also seen in Figure 1B, in which the tetrahydropterin is present in nearly stoichiometric amounts. Therefore, only an approximate 40% reduction of the cytochrome c, present in the mitochondria, is achieved. As a comparison, around 49% was reduced in a solution, containing only cytochrome c under the same conditions[9].

However, it is more striking to observe that no reduction of cytochrome b had taken place. Even in the presence of large amounts of tetrahydropterin hardly any cytochrome b was reduced. On the other hand, cytochrome a/a_3 is easily reduced up to 75-80 %, even in the presence of

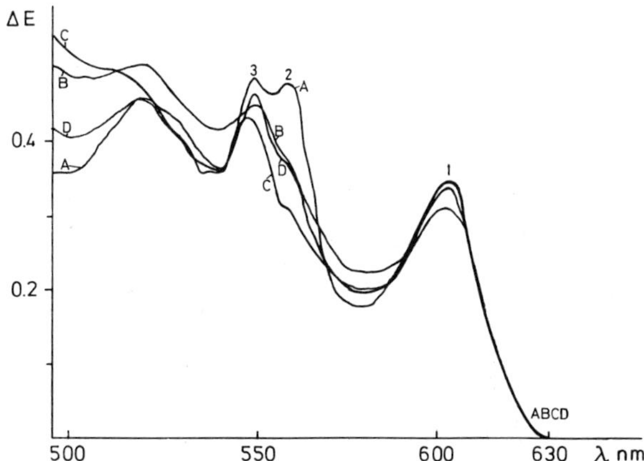

Figure 1. Difference spectrum (reduced vs oxidized) of beef heart mitochondria. Peak 1 represents cytochrome oxidase, peak 2 cytochrome b, peak 3 cytochromes c and c_1. Average concentrations calculated according to Williams[18]. Cytochrome oxidase 26.8 μM, cytochrome b 20.3 μM, cytochrome c_1 3.1 μM and cytochrome c 5.1 μM. Protein 13.4 $mg \cdot ml^{-1}$. Reductants: A. Ascorbate + Na dithionite, B. 6,7-Dimethyl-5,6,7,8-tetrahydropterin, C. 1,2,3,4-tetrahydroquinoxaline, D. Ascorbate.

low amounts of tetrahydropterin. This might be explained by assuming that tetrahydropterin is transported more slowly into the mitochondria or that cytochrome b is less accessible, due to the latter's position in the membrane. Therefore, another more apolar mediator was tried, i.e., 1,2,3,4-tetrahydroquinoxaline, which should penetrate the membrane more easily. As can be seen in Figure 1C, no reduction of cytochrome b was obtained either, even in the presence of large amounts of tetrahydroquinoxaline.

The experiments were repeated, using electron transport particles instead of whole mitochondria, because in this preparation the mitochondrial membrane is disrupted, which should make the cytochrome b more accessible. Also in this case, no reduction of the cytochrome b by the tetrahydropterin was observed.

So there must be another explanation, why cytochrome b is not reduced. Figure 2 shows the values for the midpoint potentials of the components of the electron transport chain. From this it can clearly be seen that the difference between the potentials of both the dihydro/tetrahydropterin and the dihydro/tetrahydroquinoxaline redox couples and that of cytochrome b, is too small for reduction to take place. On the other hand, the potential differences between the mediator couples and both the midpoint potentials of cytochrome c and cytochrome a/a_3 are large enough to drive the reduction. For the same reason, ascorbate alone cannot reduce cytochrome b (Figure 1D). It had been shown earlier[19] that the dihydro/-tetrahydropterin mediator system can be used for the oxidation of NADH, but naturally not for the reduction of NAD^+.

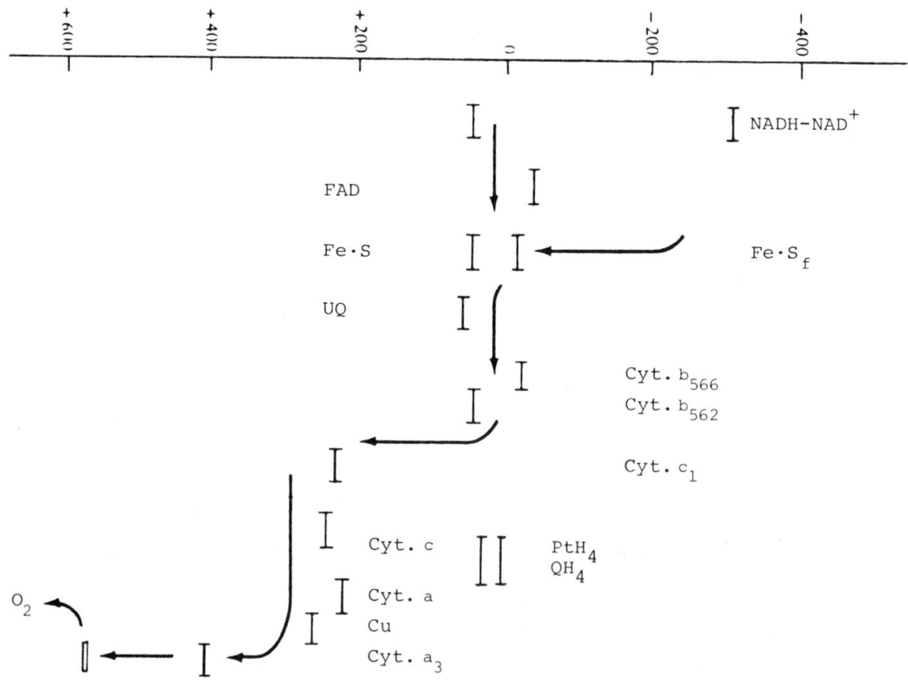

Figure 2. Midpoint potential values of the components of the electron transfer chain and pterin and quinoxaline mediator systems.

As mentioned before, in a system containing tetrahydropterin, oxygen and cytochrome c, superoxide anion radical can be generated by tetrahydropterin. The superoxide anion radical can in turn also reduce ferricychrome c or be reduced further to peroxide:

$$Fe^{3+}\text{cyto c} + O_2^{\bar{\cdot}} \longrightarrow Fe^{2+}\text{cyto c} + O_2 \qquad (3)$$

and

$$PtH_4^{\cdot+} + O_2^{\bar{\cdot}} \longrightarrow PtH_3OOH \qquad (4)$$

$$PtH_3OOH \longrightarrow PtH_2 + H_2O_2 \qquad (5)$$

In analogy, the effect of the enzyme superoxide dismutase was also studied in a system containing mitochondria. Figure 3 shows that also in the mitochondria the enzyme superoxide dismutase decreases the rate of reduction. This can be explained in a similar way, i.e., the rate of reaction (3) is inhibited, since the superoxide is consumed in the enzymatic reaction. This is most clearly seen at low tetrahydropterin concentrations, since the contribution of reaction (2) became greater than (1) to the overall reaction rate.

To detect if any hydrogen peroxide was formed in the mitochondrial system, peroxidase enzymes were included. This did not result in any change in the rate of reduction and neither could any peroxide be detected.

Cyclic voltammetry

Figure 4A shows the cyclic voltammogram of oxygen at the HMDE. In the presence of the mediator tetrahydropterin (Figure 4B), the first reduction peak of oxygen is increased, showing the catalytic effect of the pterin. No effect is seen on the second reduction peak. This is quite obvious since the redox potential of the tetrahydropterin is much more negative than that of hydrogen peroxide. Moreover, the tetrahydropterin is also oxidized by hydrogen peroxide:

$$PtH_4 + H_2O_2 \longrightarrow PtH_2 + 2H_2O \qquad (6)$$

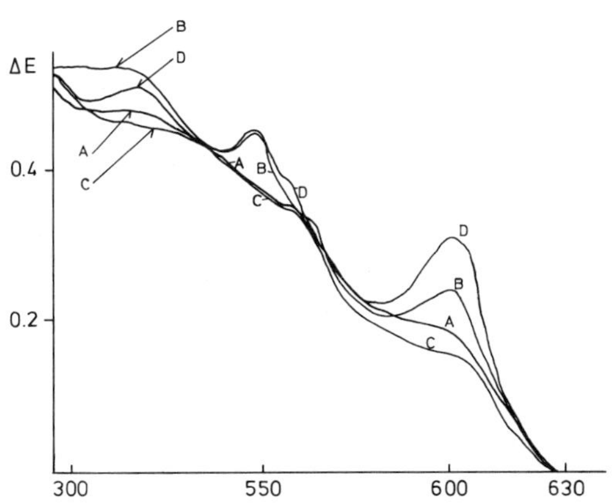

Figure 3. Difference spectrum of beef heart mitochondria. In the presence of: A. Superoxide dismutase + 6,7-dimethyltetrahydropterin (stoichiometric amount); B. Superoxide dismutase + 6,7-dimethyltetrahydropterin (excess); C. Reoxidized mitochondria; D. 6,7-dimethyltetrahydropterin.

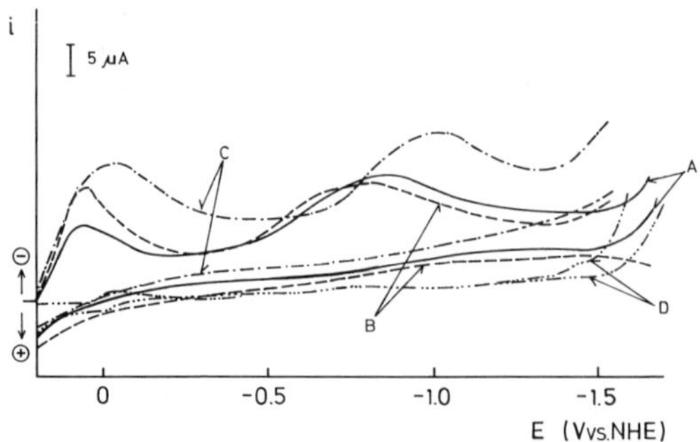

Figure 4. Cyclic voltammetric curent-potential curves at the hanging mercury drop electrode (HMDE) vs Ag/Ag^+. A. Oxygen saturated buffer solution pH 7.4. B. 0.25 mM PtH_4 oxygen saturated. C. 0.25 mM PtH_4 + 0.05 ml mitochondrial suspension (29.5 $mg \cdot ml^{-1}$ protein) oxygen saturated. D. 0.25 mM PtH_4 + 0.05 ml mitochondrial suspension, N_2-saturated. Scan rate 200 mVs^{-1}.

So the solution is continuously depleted of H_2O_2 near the electrode. In the presence of mitochondria an increase in both peaks is seen (Figure 4C) as well as a shift in potential of the second peak to more negative values. At first sight this could be ascribed to an increase in oxygen concentration due to the endogenous oxygen present in the mitochondria. However, a cyclic voltammogram under the same conditions, but in the absence of oxygen, i.e., in a deaerated solution (Figure 4D), does not show any oxygen reduction peaks from the mitochondria. So the effect of the mitochondria could be an increase in the rate of reduction of oxygen at the electrode. The shift in potential could be due to differing absorption conditions.

Figure 5 shows that absorption effects do play a role. Here the same measurements were done at a different electrode, i.e., a rotating Pt

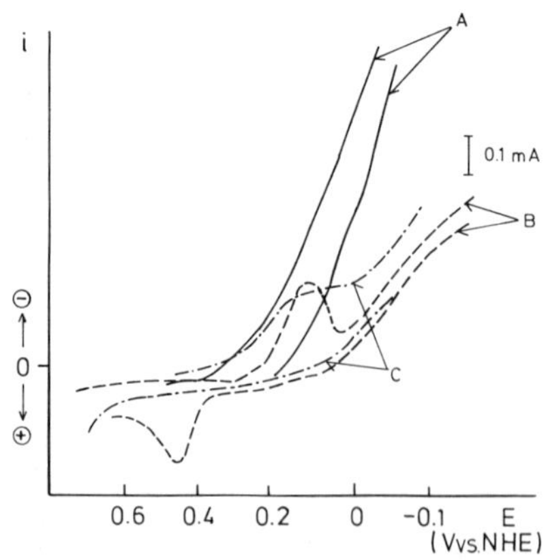

Figure 5. Cyclic voltammetric current-potential curves at the rotating Pt electrode vs Ag/Ag^+. Scan rate 200 mVs^{-1}. A. Oxygen saturated buffer pH 7.4. B. 0.25 mM PtH_4 oxygen saturated. C. 0.25mM PtH_4 + 0.05 ml mitochondrial suspension.

electrode. The effect of the mitochondria (Figure 5C) is almost similar to lowering the oxygen concentration[20]. At the same time, the peak for the tetrahydropterin has disappeared. All measurements were repeated using electron transport particles instead of mitochondria. No large differences were seen compared to the previous voltammograms.

Electrolyses

Solubilized mitochondria were also electrolyzed in the presence of the tetrahydropterin mediator.

Figure 6 shows the difference spectra of the catholyte before and after electrolysis. Complete reduction was achieved after a few minutes. A high mediator concentration was necessary to achieve a measurable current density. Since endogenous oxygen is always present in the mitochondria, a coulometric measurement of the amount of electricity, consumed solely for the reduction of the mitochondria was not possible. A simultaneous reduction of the oxygen took place, partly to the superoxide anion, which in turn could reduce the cytochromes in the mitochondria. This chemical reduction is slower[21] and so all of the electricity consumed could be accounted for as being used in the reduction of the cytochromes.

CONCLUSION

Previous studies had shown that tetrahydrobiopterin could be a cofactor in mitochondrial electron transfer[22]. The tetrahydropterin enters the chain at the level of cytochrome c and is able to catalyze oxygen reduction, but without the generation of ATP. The pool of tetrahydropterins can be continuously regenerated through the action of an NADPH dependent dihydropterin reductase, which catalyzes the reduction of the quinoid dihydropterin. It is not known if this enzyme is also present in the mitochondria. However, it has been shown that tetrahydropterin can easily cross the mitochondrial membrane. The physiological function of this system, which apparently couples NADPH/NADH pools directly through tetrahydropterin, cytochrome c, cytochrome oxidase to oxygen, but without simultaneous oxidative phosphorylation, is still not known.

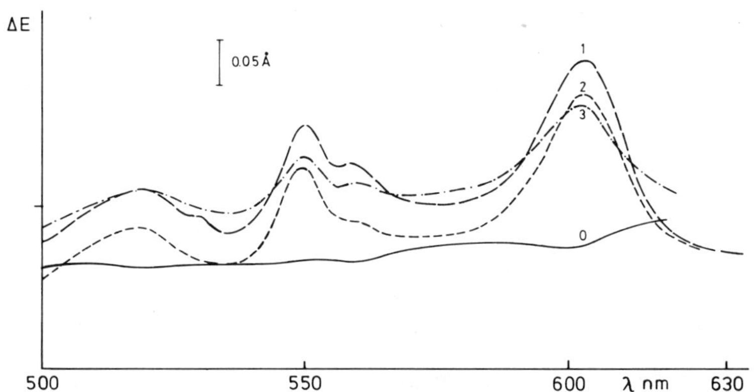

Figure 6. Difference spectrum of catholyte containing mitochondria (12-15 mg·ml^{-1} protein) mediator (0.01 mM). 0: before electrolysis. 1: after electrolysis mediator with tetrahydropterin. 2: after electrolysis with mediator tetrahydroquinoxaline. 3: after electrolysis of electron transport particles (18 mg·ml^{-1} protein) mediator PtH_4.

It has been shown here that in the presence of tetrahydropterin as a 1-electron donor, oxygen can be reduced in the mitochondria to superoxide anion radical. A further reduction to hydrogen peroxide did not seem to take place in this system, since no effect of added peroxidase was seen. This could, of course, also be due to the fact that an enzyme system is present in the mitochondria that effectively catalyzes the breakdown of the hydrogen peroxide. This study also showed the conditions under which hydroxyl radicals are generated according to the Fenton reaction:

$$O_2^{\cdot -} + H_2O_2 \xrightarrow{Fe} OH^{\cdot} + OH^- + O_2 \qquad (7)$$

are not present in the mitochondria. Although superoxide could be formed, it would certainly be oxidized by the cytochromes under in vivo conditions.

REFERENCES

1. E. Yeager in "Mechanisms of Electrochemical Reactions on Nonmetallic Surfaces", NBS special publication 455., (1976), p. 203.

2. T. Kuwana, M. Fujuhira, K. Sunakawa and T. Osa, J. Electroanal. Chem., 88, 299 (1978).

3. A. Bettelheim, R.J.H. Chan and R. Kuwana, J. Electroanal. Chem., 99, 391 (1979).

4. A. Bettelheim and T. Kuwana, Anal. Chem., 51, 2257 (1979).

5. J.P. Collman, M. Marocco, P. Denisevich, C. Koval and F.C. Anson, J. Electroanal. Chem., 101, 117 (1979).

6. D.F. Wilson, Bioelectrochem. Bioenerg., 18, 51 (1987).

7. J.A. Farrington, M. Ebert, E.J. Land and K. Fletcher, Biochem. Biophys. Acta, 314, 372 (1973).

8. D.A. Rowley and B. Halliwell, FEBS Letters, 142, 39 (1982).

9. S. Kwee, Bioelectrochem. Bioenerg., 18, 79 (1987).

10. R. Szentirmay and T. Kuwana, Anal. Chem., 50, 1879 (1978).

11. J.O.D. Coleman, H.A.O. Hill, N.J. Walton and F.R. Whatley, FEBS Letters, 154, 319 (1983).

12. S. Kwee and H. Lund, Biochem. Biophys. Acta, 297, 285 (1973).

13. F. Bohlman, B., 85, 390 (1952).

14. V. Merz and C. Ris, B., 20, 1190 (1987).

15. A.L. Smith in "Methods in Enzymology", Vol. X, R.W. Estabrook and M.E. Pullman, Eds., Acad. Press, New York, (1967), p 81.

16. C.T. Gregg in "Methods in Enzymology", Vol.X, R.W. Estabrook and M.E. Pullman, Eds., Acad. Press, New York, (1967), p. 181.

17. O.H. Lowry, N.J. Rosebrough, N.J. Farr, A.L. and A.J. Randall, J. Biol. Chem., 193, 265 (1951).

18. J.N. Williams Jr., Arch. Biochem. Biophys., 107, 537 (1964).

19. S. Kwee in "Chemistry and Biology of Pteridines", W. Pfleiderer, Ed., W. de Gruyter, Berlin, (1976), p. 671.

20. R.C. Long, F.M. Hawkridge and C.R. Hartzell, J. Electroanal. Chem., 198, 89 (1986).

21. S. Kwee, Bioelectrochem. Bioenerg., 16, 99 (1986).

22. H. Rembold and K. Buff, Eur. J. Biochem., 28, 586 (1972).

INTRACELLULAR VOLTAMMETRY WITH ULTRASMALL CARBON RING ELECTRODES

Jennifer B. Chien, Reginaldo A. Saraceno and
Andrew G. Ewing*

Department of Chemistry
Penn State University
University Park, Pennsylvania 16802

INTRODUCTION

Direct measurements of neurotransmitter concentrations inside single nerve cells can provide unique information concerning neurotransmitter storage and dynamics. Intracellular measurement of cell membrane potential[1-3] and inorganic ion concentration[4,5] are possible with ultrasmall glass pipettes filled with a saline solution or a liquid ion exchanger, respectively. We[6] and others[7,8] have developed voltammetric electrodes small enough to place inside single cells while maintaining cell viability. The aim of this paper is to present data characterizing ultrasmall ring-shaped carbon electrodes and their application to monitoring the concentration of the neurotransmitter dopamine inside single nerve cells.

There has been intense interest recently in the development of ultrasmall electrodes for voltammetry[9-15]. However, very few of these electrodes have small structural supports in addition to small electroactive area. We have used carbon deposition by hydrocarbon pyrolysis to fabricate ultrasmall ring-shaped carbon electrodes[6]. Although hydrocarbon pyrolysis has been used to construct carbon electrodes[16,17], the novel aspect of our work is the fabrication of structurally ultrasmall electrodes.

Structurally ultrasmall electrodes can be used in samples contained in very small environments. These include the cytoplasm of single cells where the electrode must be small enough for the cell membrane to seal around it following implantation. The relatively large and easily located dopamine neuron located on the left pedal ganglion of the pond small Planorbis corneus[18-20] provides an ideal model system for investigating intracellular dopamine neurochemistry. Dopamine in this cell is apparently stored in vesicles, metabolized and synthesized in much the same manner as in mammalian cells[21]. Ultrasmall carbon ring electrodes have been used to probe the dynamics of intracellular dopamine levels following application of extracellular, exogenous dopamine. Additionally, endogenous levels of dopamine have been monitored inside the dopamine neuron of Planorbis corneus.

EXPERIMENTAL

Chemicals

In vitro work was performed in pH 7.4 snail ringer solution (39 mM NaCl, 1.3 mM KCl, 4.5 mM $CaCl_2$, 1.5 mM $MgCl_2$, 6.9 mM $NaHCO_3$) or in pH 7.4 citrate-phosphate buffer. Solutions of catechols were thoroughly deoxygenated (N_2). Dopamine-HCl (DA), 4-methylcatechol (4-MC), and dihydroxyphenylacetic acid (DOPAC) (Sigma Chemical Co.) were used as received.

Electrodes and Apparatus

Carbon ring electrodes with structurally ultrasmall tip diameters were fabricated as described elsewhere[6]. Electrodes with 3-5 μm tip diameters were used for intracellular voltammetry. Electrodes used for characterization ranged from 3 to 15 μm in tip diameter.

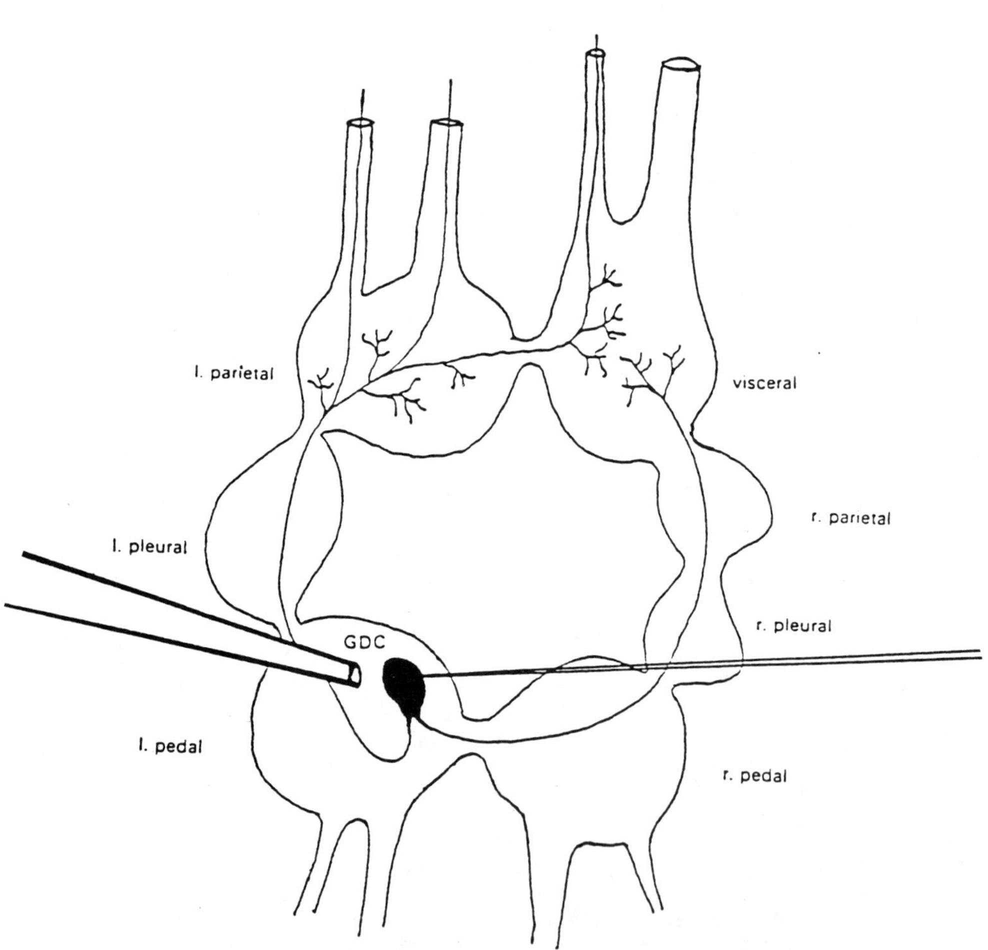

Figure 1. Schematic of the brain of Planorbis corneus showing electrode placement and microinjector.

Cyclic voltammetry was carried out with a locally constructed low-current potentiostat[22] and a waveform generator. All electrochemical measurements were carried out in a copper mesh Faraday cage. In vitro experiments were performed in a 30-mL glass bottle with three holes drilled in a plastic cap to accommodate the three electrode system. A sodium saturated calomel electrode (SSCE) served as the reference electrode and a platinum wire as the auxiliary electrode.

In Vivo Preparation and Measurements

Planorbis corneus were anesthetized by external application of chloral hydrate (2% wt/wt), pinned and dissected to reveal the left and right pedal ganglia (Figure 1). The preparation was immersed under snail ringer solution in a wax-filled petri dish. Microdissection was performed under a stereomicroscope. A micromanipulator (de Fonbrune) was used to place a carbon ring electrode into the identified dopamine neuron. The electrode potential was monitored (voltage follower and oscilloscope) vs a platinum wire during implantation. A negative shift in potential was indicative of cell penetration. Following implantation, the carbon ring working electrode was connected to the potentiostat. The SSCE reference was placed via a salt bridge into the ringer solution with a platinum wire serving as the auxiliary electrode. Application of exogenous dopamine or a 50% ethanol solution to the outside of the cell was accomplished with a micropipette manipulated (WPI or Prior) to a distance of approximately 50 μm from the identified cell.

Staircase voltammograms (-0.2 to 0.8 V, 50 steps, 20 mV/step and 100 ms/step, 2-s delay between scans) were collected using a locally constructed potentiostat and an IBM PC. The experimental protocol involved repeated voltammograms in series of 100. Voltammograms obtained during a stimulus represented the compound(s) changing concentration during stimulus. Plots of oxidation current at one chosen potential step for each voltammogram provided data concerning changes in concentration vs time.

RESULTS AND DISCUSSION

Electrode Fabrication and Characterization

Electrodes with tip diameters as small as 1 μm have been fabricated using hydrocarbon pyrolysis inside quartz microcapillaries. A schematic

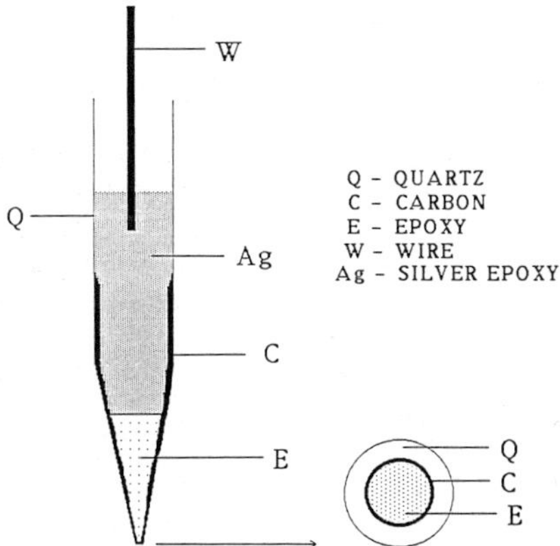

Figure 2. Schematic of a structurally small carbon ring electrode.

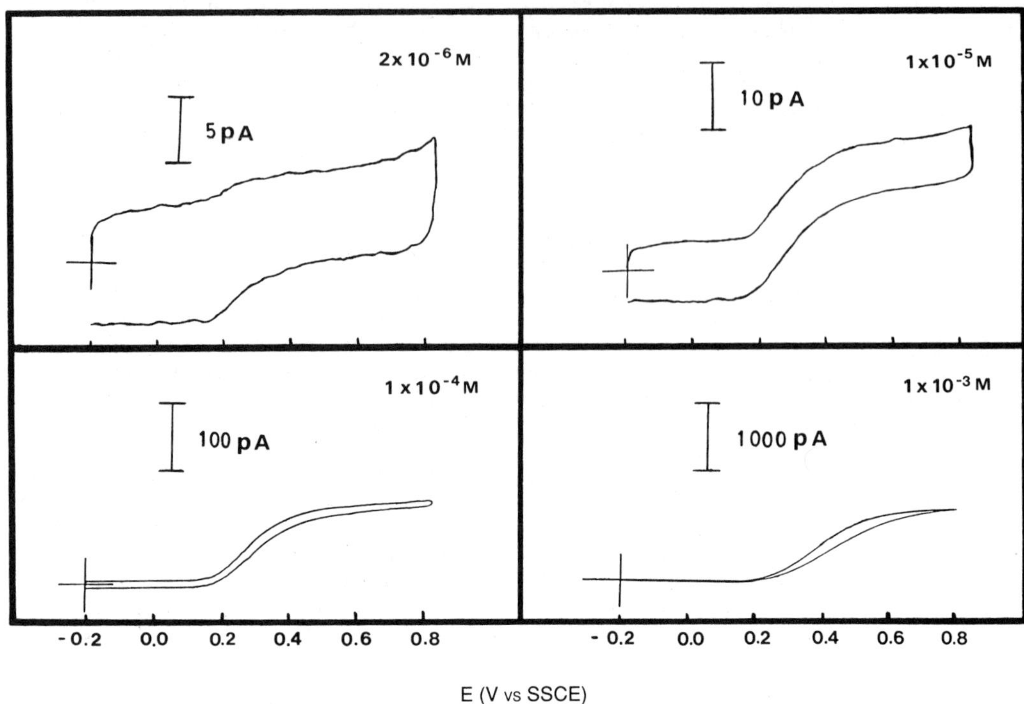

Figure 3. Cyclic voltammograms of DA at a carbon ring electrode in pH 7.4 citrate-phosphate buffer. Scan rate: 100 mVs^{-1}.

diagram of a carbon ring electrode is shown in Figure 2. Methane is pyrolyzed at reasonably high temperatures inside quartz microcapillaries leaving a conductive film of pyrolytic carbon inside the capillary tip. Cleaving the epoxy-filled tip with a scalpel exposes a ring of carbon.

Voltammetry of dopamine at a 4 μm tip diameter carbon ring electrode is shown in Figure 3. A key feature of this electrode is the useful dynamic range for dopamine concentration measurement (micromolar to millimolar). Additionally, sigmoidal shaped voltammograms are observed at moderate scan rates as expected for ultrasmall electrodes[10].

Anodic Pretreatment and Voltammetric Selectivity

Voltammograms obtained at carbon ring electrodes vary from one electrode to another. This necessitates careful calibration procedures prior to and following each in vivo experiment. Electrodes that appear to have poor charge transfer characteristics (large wave slope) when calibrated can be improved by use of an oxidative pretreatment. Linearly changing (200 mV/s) the electrode potential from -0.2 V to 1.8 V and back to -0.2 V in a solution containing only citrate-phosphate buffer results in improved charge transfer characteristics. Voltammograms in solutions of DA, 4-MC and DOPAC are shown both before and after pretreatment in Figure 4. Interestingly, the wave slope for voltammograms in DA solutions improves (becomes more Nerstian) following pretreatment. Little apparent difference is noted for voltammograms in solutions of 4-MC. The wave slope increases 88% (n = 8) of the time when comparing voltammetry in DOPAC

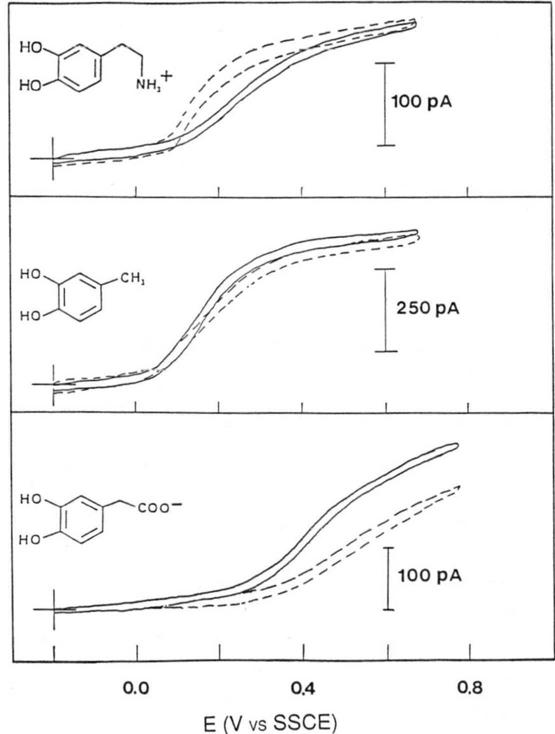

Figure 4. Cyclic voltammograms of 10^{-4} M DA, 4-MC and DOPAC at carbon ring electrodes before (solid lines) and after (dashed lines) electrochemical pretreatment. Scan rate: 100 mVs^{-1}.

solutions before and after pretreatment. At pH 7.4 these species are cationic, nonionic and anionic (DA, 4-MC, DOPAC, respectively), indicating a possible electrostatic effect between solute and electrode surface during charge transfer. DOPAC is the primary metabolite of DA and electrochemically treated electrodes offer a means to more selectively monitor DA over its interfering metabolite.

Intracellular Monitoring of Exogenous Dopamine

Following electrode implantation into the cell body of the dopamine neuron of Planorbis corneus, the cell was bathed with a DA solution. DA entering the cell was monitored intracellularly with repeated voltammograms. Figure 5 shows oxidation current monitored at the 0.76 V step of repeated voltammograms. Additionally, Figure 5 shows the response to repeated applications of exogenous DA. The increase in oxidation current monitored intracellularly following application of extracellular DA appears to represent active transport of DA across the cell membrane. The rapid decay in monitored oxidation current following DA increase appears to represent DA metabolism and/or vesicularization. Intracellular voltammetry can be used to monitor the magnitude and rates of DA uptake, metabolism and storage in this cell.

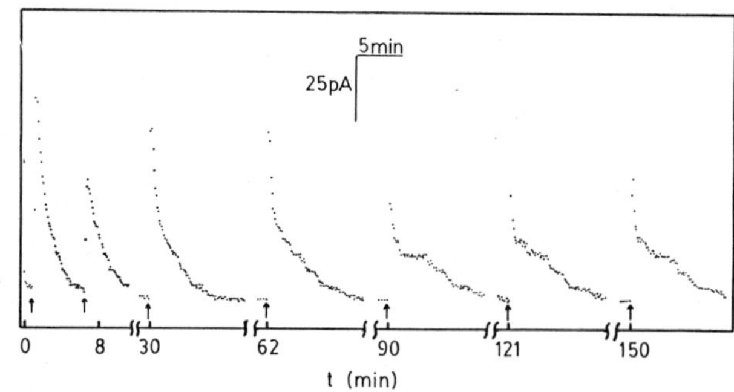

Figure 5. Oxidation current observed at 0.76 V vs SSCE at a carbon ring electrode implanted in the DA neuron of Planorbis corneus. Current observed following extracellular DA application (30 µL, 10^{-4} M at each arrow). Data obtained from staircase voltammograms (see text).

Intracellular Monitoring of Endogenous Dopamine

Voltammograms obtained following electrode implantatins in the dopamine cell of Planorbis corneus, but prior to manipulation of the chemical environment of the cell, showed no oxidation wave for DA. Application of 50 µL of a 50% ethanol solution results in a large increase in intracellularly monitored oxidation current (Figure 6). Voltammograms obtained intracellularly following ethanol application match closely in half-wave potential and voltammetric shape to those obtained in vitro in solutions of DA following the intracellular experiment (Figure 7).

The average maximum increase in intracellular DA observed following ethanol exposure was 6.5×10^{-5} M (standard error of the mean; S.E.M. = 2.0×10^{-5}; n = 6). Although the exact nature of the response caused by ethanol on the cell is unclear, it appears that bound DA (bound in vesicles or elsewhere and inaccessible to the electrode in the cytoplasm) is released into the cytoplasm by this stimulation. Hence, this procedure permits a direct measurement of DA stores in this single cell. Since no intracellular DA is observed initially, it must be assumed that free (cytoplasmic) DA is only present at levels below the detection limit

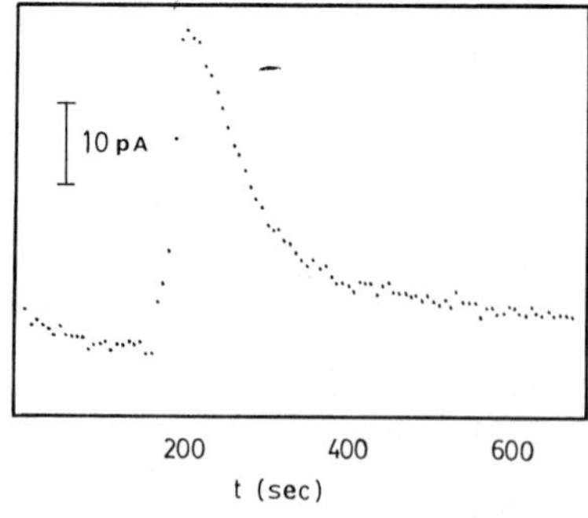

Figure 6. Oxidation current observed at 0.78 V vs SSCE at a carbon ring electrode implanted in the DA neuron of Planorbis corneus. Current observed following extracellular application of 40 µL of a 50% ethanol solution.

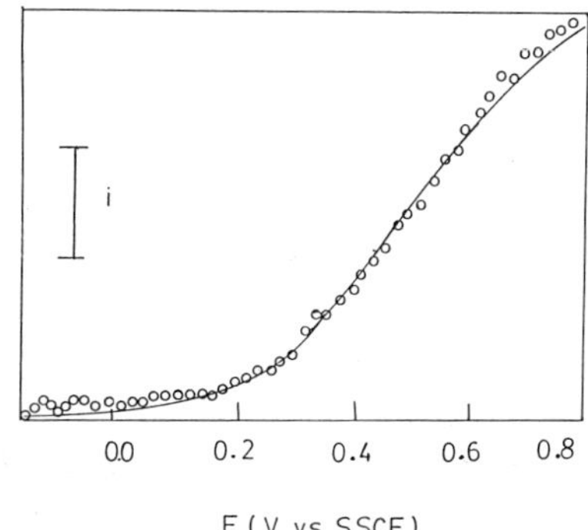

Figure 7. Staircase voltammogram (background corrected) obtained at the peak of the response shown in Figure 6 (circles) vs staircase voltammogram obtained in 20 μM DA in vitro (solid line) at the same carbon ring electrode following the intracellular experiment. i = 10 pA, intracellular; 5 pA, in vitro.

(1.9×10^{-6} M; S.E.M. = 0.30×10^{-6}; S/N = 3; n= 6). Therefore, the amount of DA that is bound and inaccessible to the cytoplasm must be at least 97%. This is in direct contrast to results obtained by Meulemans et al.[8] for serotonin in the metacerebral neuron of Aplysia suggesting dramatic differences in the functioning of these two neurons.

CONCLUSIONS

Carbon ring electrodes can be fabricated with structurally small (1 μm) tip diameter. These electrodes are useful for voltammetric monitoring of catecholamine neurotransmitters in extremely small environments. The selectivity for monitoring DA in the presence of its metabolite DOPAC is increased following electrochemical oxidative pretreatment in pH 7.4 citrate-phosphate buffer. Structurally small carbon ring electrodes can be used to monitor the concentration of DA in the cytoplasm of single cells. Transport (uptake) of exogenous DA into the DA neuron of Planorbis corneus has been monitored intracellularly. Intracellular endogenous DA levels following extracellular application of ethanol have also been monitored. These experiments permit estimation of stored DA levels in the cell body of this neuron.

ACKNOWLEDGEMENT

Support of this work by the National Science Foundation (BNS-8705777) is gratefully acknowledged. A.G.E. is the recipient of a Presidential Young Investigator Award (CHE-8657193).

REFERENCES

1. R.C. Thomas and V.J. Wilson, Science, 151, 1538 (1966).

2. D.W. Parsons, A.T. Maat and H.M. Pinsker, Science, 221, 1203 (1983).

3. A.A. Grace and B.S. Bunney, Science, 210, 654 (1980).

4. J.L. Walker, Anal. Chem., 43, 89A (1971).

5. R.C. Thomas and W.J. Moody, Trends in Biol. Sci., 5, 86 (1980).

6. Y.-T. Kim, D.M. Scarnulis and A.G. Ewing, Anal. Chem., 58, 1782 (1986).

7. A. Meulemans, B. Poulain, G. Baux, L. Tauc and D. Henzel, Anal. Chem., 58, 2088 (1986).

8. A. Meulemans, B. Poulain and G. Baux, Brain Res., 414, 158 (1987).

9. M.A. Dayton, J.C. Brown, K.J. Stutts and R.M. Wightman, Anal. Chem., 52, 946 (1980).

10. R.M. Wightman, Anal. Chem., 53, 1125A (1981).

11. K. Aoki and J. Osteryoung, J. Electroanal. Chem., 125, 315 (1981).

12. A.M. Bond, M. Fleischmann and J. Robinson, J. Electroanal. Chem., 168, 299 (1984).

13. M.R. Cushmann, B.G. Bennet and C.W. Anderson, Anal. Chim. Acta., 130, 323 (1981).

14. J.-L. Ponchon, R. Cespuglio, F. Gonon, M. Jouvet and J.F. Pujol, Anal. Chem., 51, 1483 (1979).

15. J.S. Symanski, S. Bruckenstein, Extended Abstract, 165th Meeting of the Electrochemical Society; Electrochemical Society: Pennington, NJ, May 1984, p. 627.

16. A.L. Beilby, W. Brooks, Jr. and G.L. Lawrence, Anal. Chem., 36, 22 (1964).

17. W.J. Blaedel and G.A. Mabbott, Anal. Chem., 50, 933 (1978).

18. C. Marsden, G.A. Kerkut, Comp. Gen. Pharmacol., 1, 101 (1970).

19. M.S. Berry, G.A. Cottrell, J. Physiol., 244, 589 (1975).

20. B. Powell and G.A. Cottrell, J. Neurochem., 22, 605 (1974).

21. N.N. Osborne, E. Priggemeier and V. Neuhoff, Brain Res., 90, 261 (1975).

22. A.G. Ewing, M.A. Dayton and R.M. Wightman, Anal. Chem., 53, 1842 (1981).

OXIDATION CHEMISTRY OF CNS INDOLES

Monika Z. Wrona, Keith Humphries and
Glenn Dryhurst*

Department of Chemistry
University of Oklahoma
Norman, Oklahoma 73019

INTRODUCTION

The aromatic amino acid L-tryptophan (TPP) undergoes a variety of metabolic transformations in mammals which result in the formation of a large number of compounds in which the indole nucleus is retained[1]. A most important series of metabolic transformations occur in the central nervous system (CNS) where dietary TPP, which can cross the blood-brain barrier, is first converted to 5-hydroxytryptophan (5-HTPP) by the enzyme tryptophan hydroxylase which in turn is converted, by 5-hydroxytryptophan decarboxylase, to 5-hydroxytryptamine (5-HT). Following the initial detection of 5-HT in brain[2,3] and because of its structural resemblence to certain psychoactive drugs[4] the indoleamine began to be suspected of being a chemical neurotransmitter which, of course, it is now known to be. Among the presumed roles of 5-HT in the CNS are the regulation of body temperaature, sleep and certain emotional states[5]. The normal metabolic pathways for 5-HT are outlined in Figure 1[6-12], which clearly shows that many 5-hydroxylated indoles are formed. However, 5-HIAA is the major metabolite formed by this pathway. At about the time that 5-HT was first discovered in brain, suggestions began to appear in the literature that a defect in the normal CNS metabolism of 5-HT might in some way be related to certain types of pyschotic behavior[13,14]. More recent reports have implicated such faulty metabolic pathways with mental illnesses such as schizophrenia and depression[15-18]. A recurring suggestion is that a defect in the metabolism of 5-HT leads to the formation of more highly hydroxylated compounds which, for some reason, are more reactive in the CNS and hence lead to neutronal damage[10,19,20]. While such compounds have never been detected, or apparently sought, in humans, such a suggestion has considerable merit because of the more recent discovery that 5,6-, 5,7-[21-27] and 4,5-DHT[25,28] are neurotoxins. Furthermore, 5-HT and other 5-hydroxylated indoles are known to be oxidized by human serum and ceruloplasmin concentrates to give colored solutions of unidentified products[29-35]. In view of the fact that ceruloplasmin might play a role in the regulation of biogenic amine levels in the CNS, it is surprising that no attempts have been made to identify such products and to investigate their biological properties. A hydroxyl group substituted in the benzene ring is a necessary condition for tryptamine derivatives to be oxidized by ceruloplasmin[33,36,37]. It has been suggested[38] that some melanins found in the CNS might be derived from 5-hydroxyindoles and *in vitro* experiments with 5-HT, 5-HTOl and 5-HIAA (see Figure 1 for structures) support this suggestion[39]. However, the

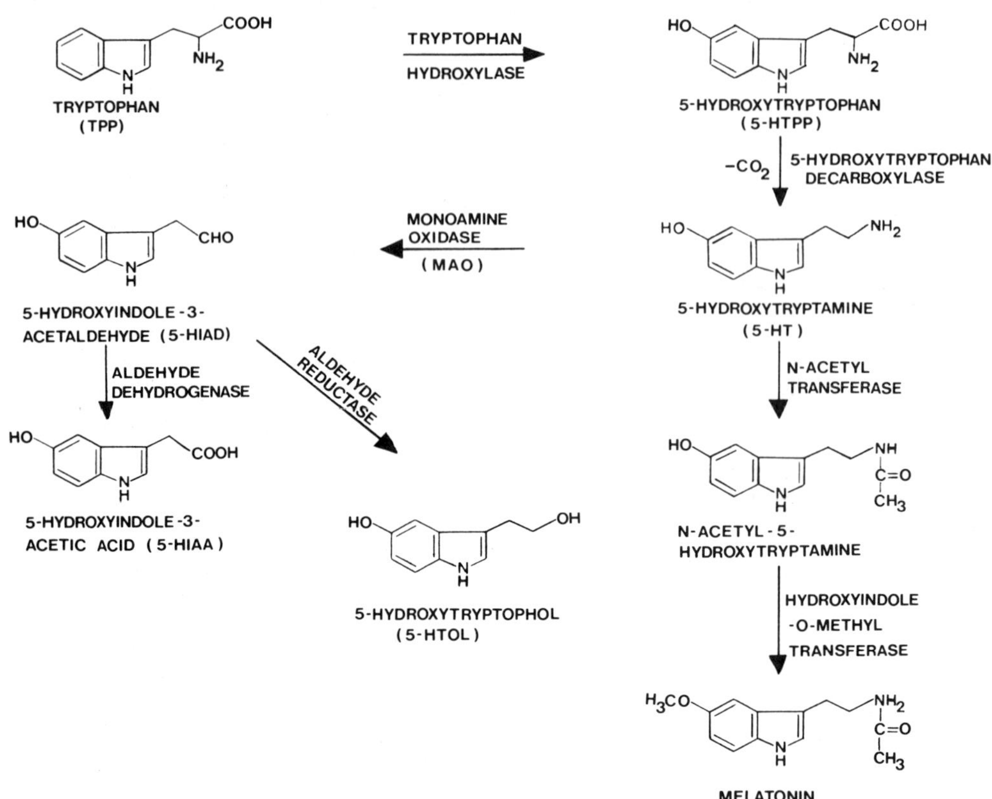

Figure 1. Biosynthesis of 5-hydroxytrypatamine in the CNS and subsequent metabolic pathways[6-12].

structures of these melanin pigments or mechanisms for their formation are not known although it has been suggested[39] that microsomal melanogenesis from 5-hydrooxyindoles proceeds through a quinonc imine intermediate. Hemolysates of erythrocytes also oxidize 5-HT and other 5-hydroxyindoles giving colored products of unknown identity[40]. It has been speculated that in such processes the 5-hydroxyindoles are first hydroxylated and then further oxidized[40] or that the primary oxidation results in formation of a quinone imine[36]. Molecular oxygen and Ag^+ also oxidize 5-HT to colored products, one being a melanin-like pigment; another has been speculated to be a relatively long-lived but unknown dimeric intermediate[20]. Oxidations of 5-HT by ferricytochrome c[41], alkaline permanganate[42], autoxidation in basic solution[43], ceruloplasmin[44] and during metabolism[45,46] appear to generate radical species. The reaction chemistry of these putative radicals and the ultimate products formed in the latter oxidation reactions are not known.

Thus, while the main metabolic fate of 5-HT is known (Figure 1), this indolic neurotransmitter can clearly undergo a variety of other biochemical and chemical oxidation reactions but, in fact, little is known about the products or mechanisms of these oxidations. Suggestions for oxidative intermediates include radicals, quinone imines and dihydroxytryptamines which lead to unknown ultimate products speculated to include dimers, oligomers and polymeric melanin-like pigments. Furthermore, anomalous oxidation reactions of 5-HT in the CNS have been speculated to play a role in the etiology of mental illnesses. Such speculations have focused on 5-HT because it is widely believed to be an important chemical neurotransmitter. However, it seems that its precursor in the CNS (i.e., 5-HTPP, Figure 1) and its many 5-hydroxylated indolic metabolites might also be potential candidates for anomalous oxidation reactions leading to neurotoxic species.

In an attempt to help clarify the apparently complex oxidation chemistry of indolic compounds found in the CNS we have begun a systematic study of this chemistry using electrochemical and other analytical approaches. Modern electroanalytical and related techniques are particularly well suited to studies of the oxidation (and reduction) of biologically significant molecules[47,48]. In this report some aspects of the electrochemical oxidation of 5-HT and 5-HTPP will be described.

EXPERIMENTAL

The suppliers of chemicals and a description of the equipment used for electrochemical and spectroscopic studies have been described elsewhere[49-51]. All linear sweep and cyclic voltammograms were obtained using pyrolytic graphite microelectrodes (PGE) having approximate surface areas of 2-4 mm^2. Similarly, all controlled potential electrolyses were performed at plates of pyrolytic graphite. Voltammetric measurements and controlled potential electrolyses were always performed using solutions which had been thoroughly deoxygenated with nitrogen. All potentials are referred to the saturated calomel reference electrode (SCE) at 25±3°C.

RESULTS AND DISCUSSION

In our initial studies of the electrochemical oxidation of 5-HT and 5-HTPP dilute HCl was generally employed as the supporting electrolyte. There were several reasons that this electrolyte was chosen. First, unlike conventional buffer systems, HCl is readily removed from product mixtures by freeze-drying. Second, preliminary experiments revealed that cyclic voltammograms, coulometric n-values and ultimated products formed upon oxidiation of 5-HT or 5-HTPP were essentially identical using dilute HCl or, for example, a phosphate buffer as the supporting electrolyte. Finally, the chromatographic separations of electrochemical oxidation products of 5-HT and 5-HTPP were much easier to accomplish in the absence of large excesses of inorganic buffer constituents. Work currently underway suggests that the oxidation chemistry of 5-HT and 5-HTPP is not greatly different in a buffered solution as physiological pH.

1. 5-HYDROXYTRYPTAMINE (5-HT)

Analyses in various regions of rat brain, for example, indicate that concentrations of 5-HT in wet tissue range from about 0.7 μM in the cerebellum to 42 μM in the pineal gland[52,53]. Thus, in initial studies the electrochemical oxidation of 5-HT at micromolar concentration levels (≤ 30 μM) was investigated[50]. However, it should be noted that in both central and peripheral serotonergic neurons 5-HT is stored in synaptic vesicles at much higher, perhaps molar, concentrations[54].

A typical cyclic voltammogram of 20 μM 5-HT in 0.01 M HCl is shown in Figure 2. At this very low concentration two oxidation peaks (I_a and

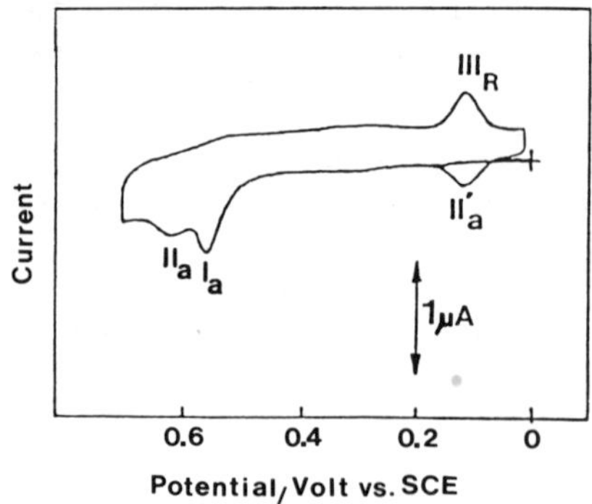

Figure 2. Cyclic voltammogram at the PGE of 20 μm 5-hydroxytryptamine in 0.01 M HCl. Sweep rate: 20 mVs^{-1}. (Reprinted by permission of the American Chemical Society from M.Z. Wrona and G. Dryhurst, J. Org. Chem., 52, 2817 (1987).

II_a) appear. On the reverse sweep reduction peak III_R and, on the second anodic sweep, oxidation peak II'_a appear as a reversible couple. Cyclic voltammograms of 20-30 μM 5-HT in 0.01 M HCl or in pH 2.0 phosphate buffer (μ = 0.5) at sweep rates up to 5 Vs^{-1} (the fastest sweep rate which could be employed where I_a could be observed above the charging current) did not show a reduction peak coupled to oxidation peak I_a. This implied that the initial peak I_a product is unstable on the latter time scale and hence is unavailable for reduction. Sweep rate and concentration studies indicate that the peak current for peak I_a is strongly controlled by the adsorption of 5-HT at the surface of the PGE. Thus, estimations of voltammetric n-values and other mechanistic parameters from the peak I_a current and shape were not possible. However, the peak potential (E_p) for peak I_a is dependent upon pH (Equation 1). Furthermore,

$$E_{p\,(pH\ 2.0-8.1)} = [0.615-0.047\ pH]V^* \qquad (1)$$

with increasing concentrations of 5-HT peak II_a grows relative to peak I_a so that at concentrations ≥ 1 mM peak I_a appears only as a poorly defined inflection on the rising portion of peak II_a. These results suggest that the peak I_a reaction generates at least one major product as a result of a second order reaction of a primary intermediate which is then oxidized in the peak II_a process.

Controlled potential electrooxidations at potentials corresponding to the half-peak potential ($E_p/2$) for oxidation peak I_a caused the characteristic UV spectrum of 5-HT (λ_{max} in 0.01 M HCl = 285 (sh), 272, 215 nm) to disappear and the solution to become purple. The spectrum of this solution is quite complex (λ_{max} =530, 350, 300, 272, 220 nm). The measured coulometric n-values at $E_{p/2}$ were 3.0 ± 0.3. Cyclic voltammograms recorded after such an electrolysis showed that the reversible peaks III_R/II'_a are present along with a small peak II_a and a new oxidation peak III_a (Figure 3).

* Data for this equation were obtained with 20 μM 5-HT in phosphate buffers (μ = 0.5) at a sweep rate of 20 mVs^{-1}. The intercept in this equation has been extrapolated to pH = 0, so to calculate E_p only the desired pH value must be inserted.

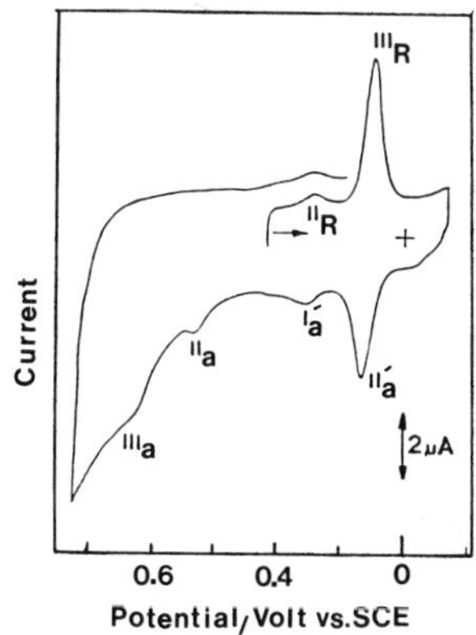

Figure 3. Cyclic voltammogram at the PGE after exhaustive electrochemical oxidation of 30 μM 5-hydroxytryptamine in 0.01 M HCl. Controlled potential electrolysis was performed at 0.52 V ($E_{p/2}$ for peak I_a). Sweep rate: 200 mVs^{-1}. (Reprinted by permission of the American Chemical Society from M.Z. Wrona and G. Dryhurst, J. Org. Chem., 52, 2817 (1987).

Figure 4B shows a high performance liquid chromatogram of the product solution formed by electrochemical oxidation of 25 μM 5-HT at 0.52 V ($E_{p/2}$ for peak I_a) in 0.01 M HCl. Clearly, three major products are formed responsible for chromatographic peaks A, B and C along with minor products responsible for peaks E, F and D. Controlled potential electrolyses at slightly more positive potentials (E_p for peak I_a, 0.56 V) resulted in similar spectra and cyclic voltammograms to those observed at $E_{p/2}$ but the coulometric n-value increased to ca. 3.3. However, HPLC analysis showed that peak B grew relative to all of the other chromatographic peaks (Figure 4C). Electrooxidation at 0.68 V (i.e., 50 mV more positive than E_p for peak II_a) gave an n-value of 3.9 ± 0.2. The product solution was bright purple with a relatively simple UV-visible spectrum (Figure 5A) (λ_{max} = 535, 350 (sh), 232 nm). HPLC analysis of the purple solution showed only peak B (Figure 4D). A cyclic voltammogram of the pure compound B in 0.01 M HCl (Figure 5B) showed that this compound is reversibly reduced in the peak III_R process and, apparently, irreversibly oxidized in the peak III_a process. Controlled potential electrochemical reduction of the freshly-formed purple product solution caused the purple color to disappear. The spectrum of the reduced solution showed λ_{max}in 0.01 M HCl at 293, 257, 208 nm. Coulometry showed that the n-value for this reduction was 2.0 ± 0.3. The reduced product could be quantitatively oxidized back to the purple product responsible for chromatographic peak B either by air or electrochemically (0.16 V).

Cyclic voltammetric and spectral studies showed that solutions (≤ 30 μM) of the purple compound responsible for chromatographic peak B were stable in 0.01 M HCl for about 1 h at room temperature. At longer times HPLC analysis showed the appearance of a new product responsible for chromatographic peak E (Figure 4E). When the solution contained only the product responsible for the latter peak it was red-brown in color (λ_{max} in 0.01 M HCl = 463, 345, 290, 230 nm, Figure 6A). A cyclic voltammogram of this solution showed two reversible couples at 0.10 V and -0.16 V (Figure 6B).

Based upon spectral studies (UV-visible spectrophotometry, fast atom bombardment mass spectrometry, liquid chromatography-mass spectrometry, gas chromatography-mass spectrometry and ^{1}H-NMR spectrometry) and electro-

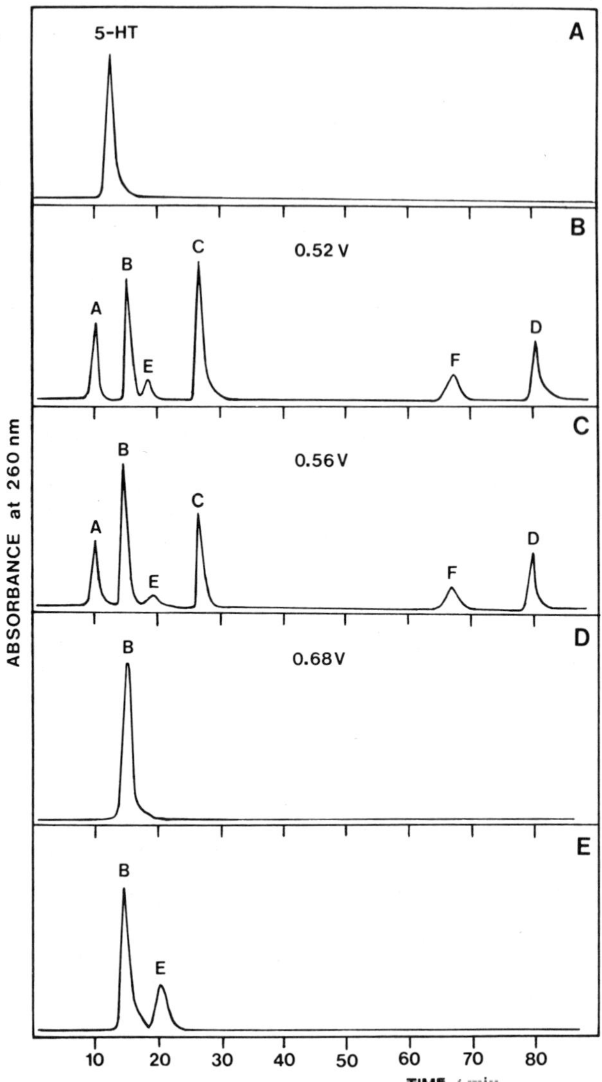

Figure 4. High performance liquid chromatograms of the product solutions formed upon electrochemical oxidation of 20 μm 5-hydroxytryptamine in 0.01 M HCl after exhaustive electrolysis at (B) 0.52 V, (C) 0.56 V, (D) 0.68 V. (A) Chromatogram obtained before the electrolysis. (E) Chromatogram obtained after standing the solution for (D) for 18 h at room temperature. A Brownlee Laboratories RP-18 (25 x 0.7 cm) column was employed. Between 0-17 min 100% solvent A (H_2O/MeCN; 100:7 v/v, containing 1.5 μM Et_3N and HCOOH to pH 3.1) at a flow rate of 2.8 ml min^{-1}. Between 17-65 min 100% solvent B (H_2O/MeCN; 100:40 v/v, containing 0.94 μM Et_3N and HCOOH to pH 3.1) at a flow rate of 2.5 ml min^{-1}.

chemical studies[50] the compounds responsible for each of the chromatographic peaks shown in Figure 4 have been identified and are shown in Figure 7.

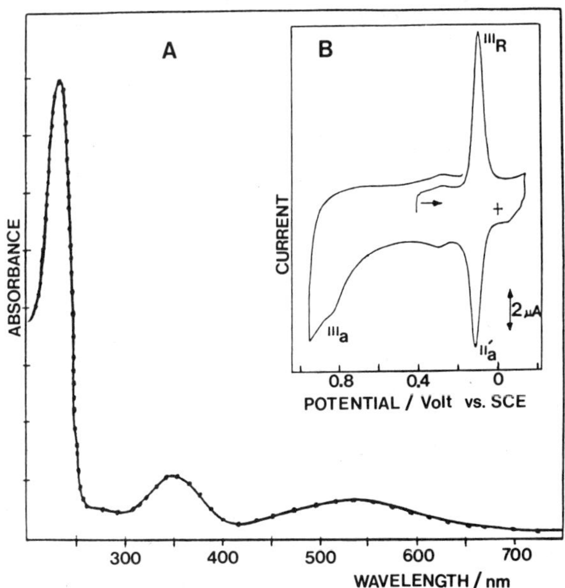

Figure 5. (A) Spectra and (B) cyclic voltammogram of the purple product solution obtained following controlled potential electro-oxidation of 26 μM 5-HT in 0.01 M HCl at 0.68 V. In (B) the sweep rate was 200 mVs^{-1}. (Adapted by permission of the American Chemical Society from M.Z. Wrona and G. Dryhurst, J. Org. Chem., 52, 2817 (1987)).

The results outlined briefly above indicate that the electrochemical oxidation of 5-HT is a pH-dependent process. HPLC analysis reveals that at low potentials (e.g., $E_{p/2}$ for peak I_a) one major dimer (4,4'-bi-5-hydroxy-tryptamine) and one minor dimer (3,4'-bi-5-hydroxytryptamine) are products. With increasingly positive applied potentials, however, the yields of these dimers decrease and, at sufficiently positive potentials, only tryptamine-4,5-dione is formed as a product (Figure 4D). These findings indicate that the initial oxidation of 5-HT must give a radical intermediate. This is likely to be a phenoxyl radical[50-55] which exists in at least two important resonance forms and is represented as the resonance hydrid structure **1** in Figure 8. On the basis of the dimeric products formed when micromolar concentrations of 5-HT are oxidized one resonance form of **1** must be **2** in which the unpaired electron is located at the C(4)-position. Dimerization of radical **2** then gives **3** which upon the expected rearrangement yields 4,4'-bi-5-hydroxytryptamine (**A**, Figure 8).

Another resonance form of **1** must be radical **4** in which the unpaired electron is located at the C(3)-position. Radical **4** then couples with radical 3 to give dimer **5** and hence 3,4'-bi-5-hydroxytryptamine (**F**). At potentials at which 5-HT is oxidized, dimer **F**, is also oxidized to dimer D[55] (Figure 8). Indeed, the very small reversible peaks (II_R/I_a') observed in cyclic voltammograms of product solutions formed upon oxidation of 5-HT at low potentials (e.g., Figure 3) are due to the **D/F** couple[55].

Even at very low applied potentials oxidations of 5-HT leads to formation of some tryptamine-4,5-dione (**B**, Figure 8). The way in which this can be formed is by electrochemical oxidation ($1e^-$, $1H^+$) of primary radical **1** to quinone imine **6** (Figure 8). This electron-deficient molecule cannot be detected by fast-sweep cyclic voltammetry at low pH and so must be rapidly attacked by water to give **4,5-DHT** or by HCl to give 4-chloro-5-hydroxy-

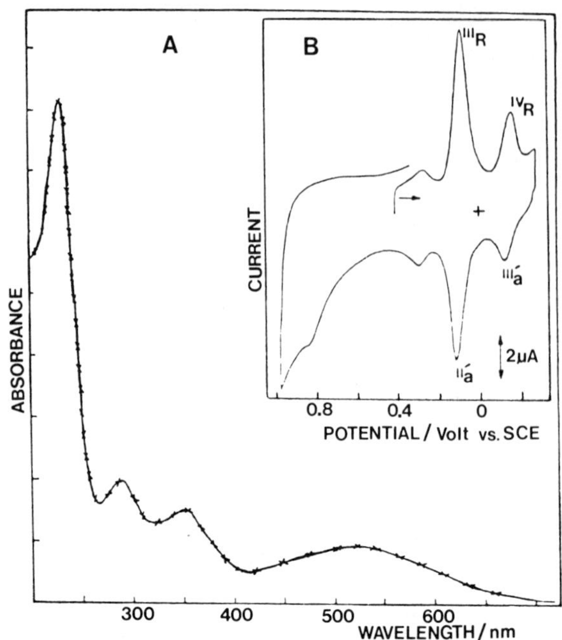

Figure 6. (A) Spectrum and (B) cyclic voltammogram at the PGE (200 mVs^{-1}) of the red-brown product responsible for chromatogrpahic peak E (see Figure 4). (Adapted by permission of the American Chemical Society from M.Z. Wrona and G. Dryhurst, J. Org. Chem., 52, 2817 (1987).

tryptamine **C** (Figure 8). At peak I_a potentials 4,5-DHT is rapidly and reversibly oxidized to tryptamine-4,5-dione (**B**), the compound responsible for the purple color of the electrolyzed solution of 5-HT. The reversible couple characterized by peaks III_R/II'_a in cyclic voltammograms of 5-HT are due to the B/4,5-DHT couple.

When 5-HT is oxidized at potentials equal to or more positive than E_p for peak II_a only a single product, tryptamine-4,5-dione (**B**), is initially formed. This indicates that at such potentials the primary oxidation product, radical **1**, is oxidized to quinone imine **6** more rapidly than it can dimerize. At peak II_a potentials 4-chloro-5-hydroxytryptamine (**C**) is oxidized to dione **B**. It is probable that this process involves oxidation of **C** ($2e^-$, $2H^+$) to the 4-chloro quinone imine **7** which is attacked by water with loss of HCl to yield dione **B** (Figure 8).

At pH 2 and in very dilute solutions tryptamine-4,5-dione (**B**) slowly reacts to give dimer **E** (Figure 7) which consists of residues of dione **B** and 5-hydroxytryptamine-4,7-dione (**9**, Figure 9). Indeed, the cyclic voltammetry and UV-visible spectrum of **E** are very similar to those expected for an equimolar mixture of the latter two compounds. Formation of **E** has been rationalized by an initial reaction of dione **B** with water to give 4,5,7-tryhydroxytryptamine (**8**, Figure 9)[50]. The latter compound is very easily oxidizable by air to give 5-hydroxytryptamine-4,7-dione (**9**, Figure 9). Experiments have shown that addition of **9** to dilute solutions of **B** in 0.01 M HCl leads to formation of **E**[50]. Thus, dione **B** must attack **9** to yield **10** which is air-oxidized to the dimer **E** as shown in Figure 9. It should be noted that because of the instability of **E** ^1H-NMR experiments

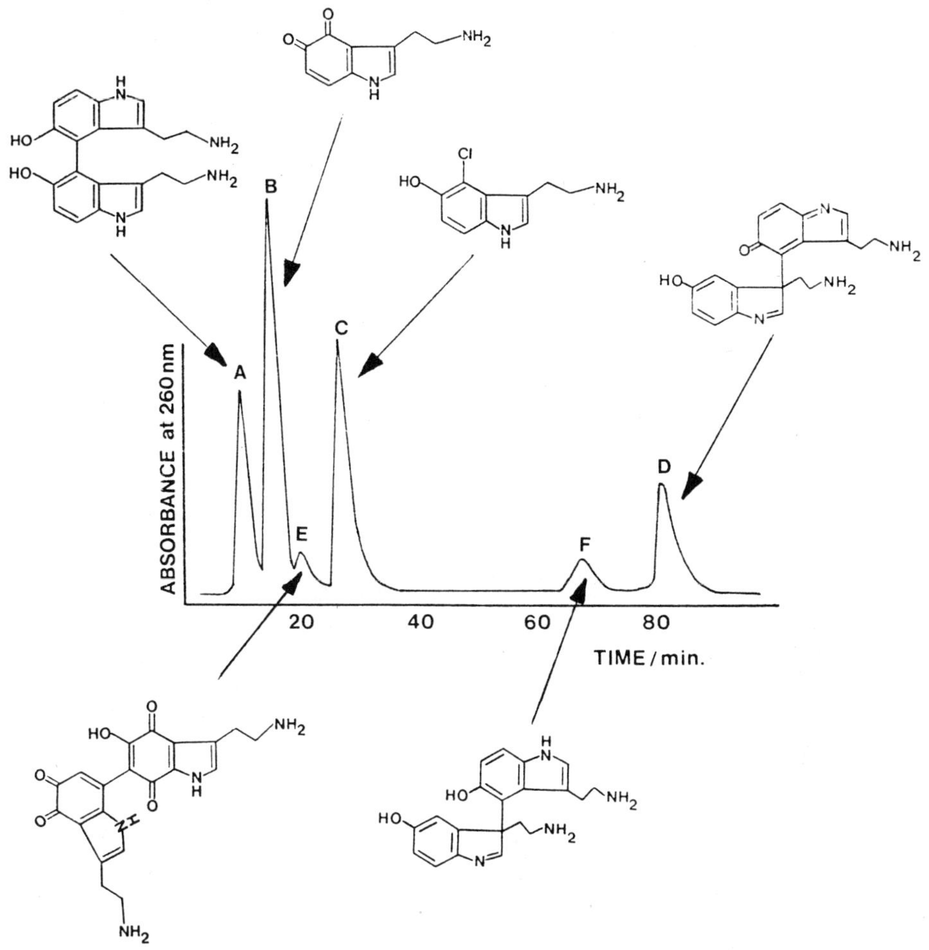

Figure 7. Structures of products of oxidation of 5-hydroxytryptamine (≤ 30 μM) in 0.01 M HCl. **A:** 4,4'-bi-5-hydroxytryptamine; **B:** tryptamine-4,5-dione; **C:** 4-chloro-5-hydroxytryptamine; **D:** 3-[4'-(tryptamine-5-one)]-5-hydroxytryptamine; **E:** 7-[6'-(5-hydroxytryptamine-4,7-dione)]-tryptamine-4,5-dione; **F:** 3,4'-bi-5-hydroxytryptamine.

could not be performed and, as a result, the exact sites at which the two indole residues are linked could be different to that shown in structure **11**.

2. 5-HYDROXYTRYPTOPHAN (5-HTPP)

Cyclic voltammograms of 5-HTPP at pH 2 (Figure 10) show that after having scanned through oxidation peak I_a a reversible couple (peak I_c/I_a') appears on the reverse sweep. At 5-HTPP concentrations above 0.1 mM and using very slow sweep rates (≤ 20 mVs^{-1}) a second oxidation peak appears at potentials slightly more positive than peak I_a. This peak grows, relative to peak I_a, with increasing concentration of 5-HTPP[51]. Above a sweep rate of 20 mVs^{-1} the more positive oxidation peak merges with peak I_a and cannot be observed. Thus, in Figure 10B, the rather broad shape of peak I_a is simply due to overlap with the more positive peak.

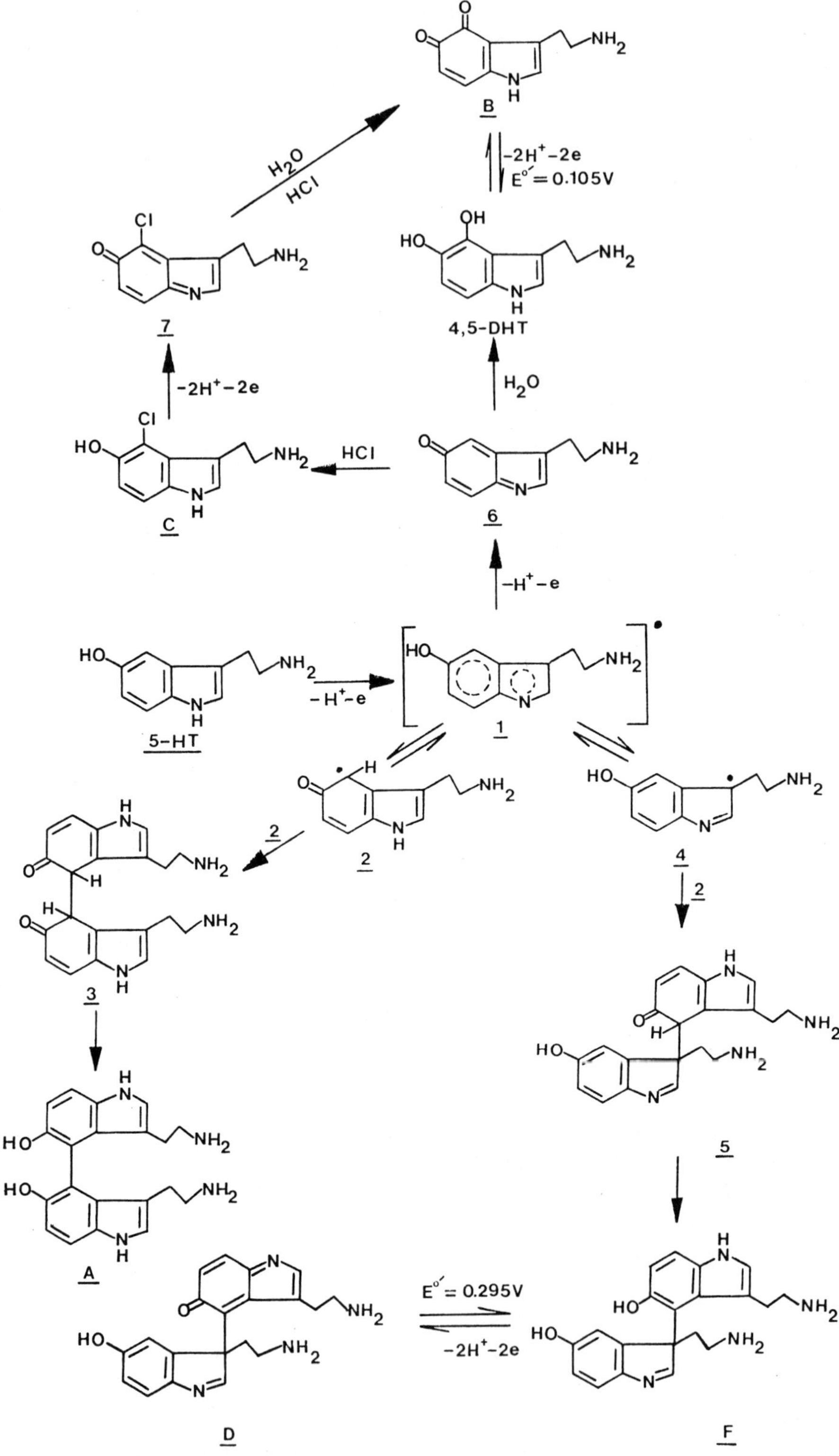

Figure 8. Mechanism proposed for the electrochemical oxidation of 5-hydroxytryptamine at low pH.

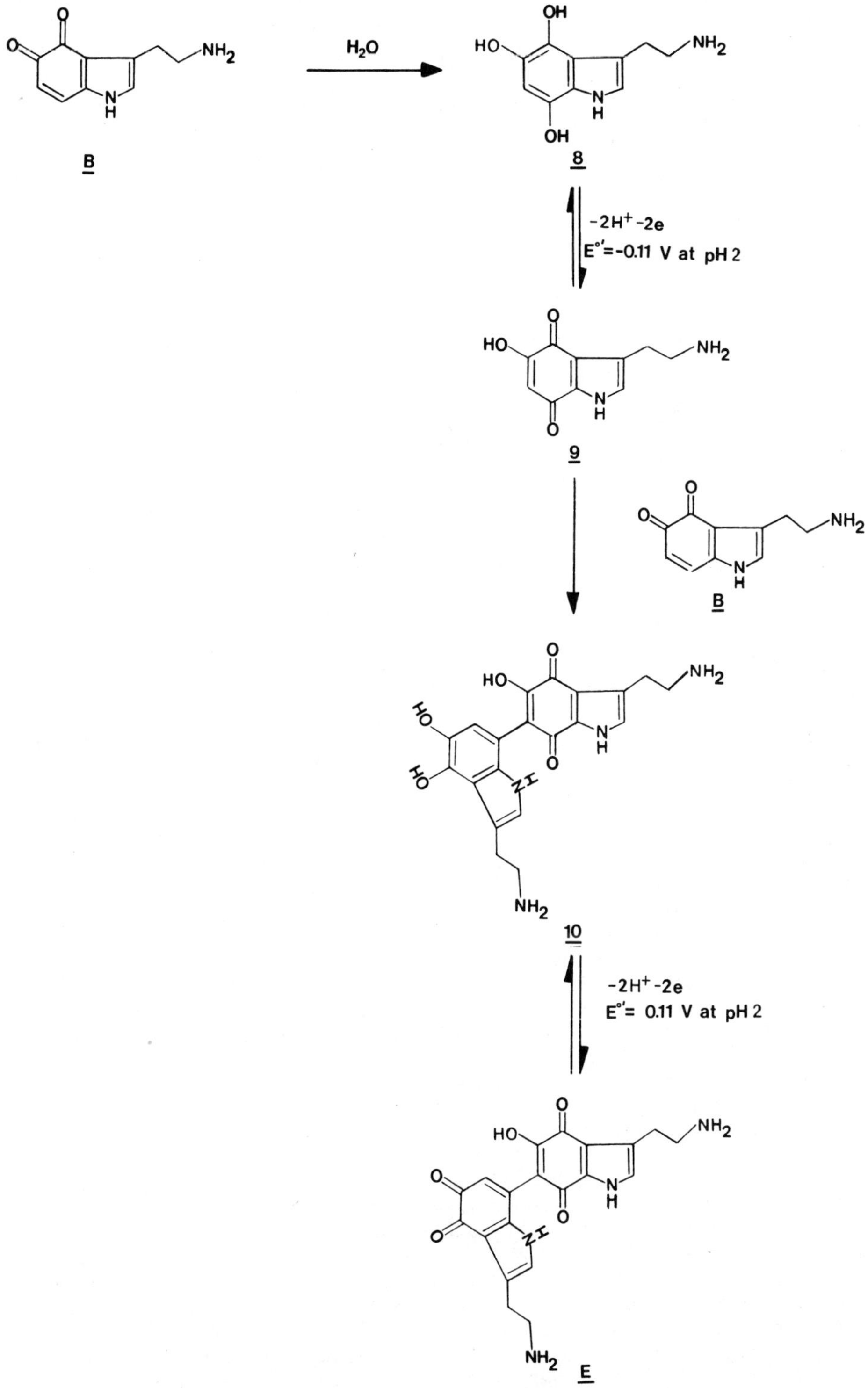

Figure 9. Reaction scheme proposed for formation of dimer **E** from tryptamine-4,5-dione (**B**). (Adapted by permission of the American Chemical Society from M.Z. Wrona and G. Dryhurst, J. Org. Chem., 52, 2817 (1987).

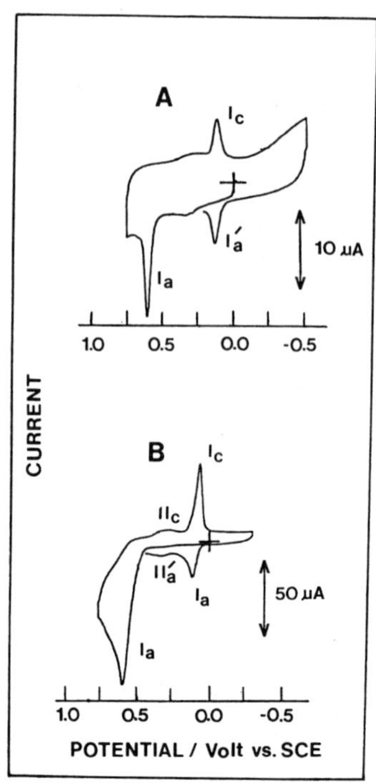

Figure 10. Cyclic voltammogram at the PGE of (A) 0.009 mM and (B) 0.83 mM 5-hydroxytryptophan in phosphate buffer (μ = 0.1) pH 2.14. Sweep rate: 200 mVs^{-1}. (Reprinted by permission of the American Pharmaceutical Association from K. Humphries and G. Dryhurst, J. Pharm. Sci., 76, 839 (1987).

Under such conditions a rather indistinct couple characterized by peaks II_c/II_a' appears at more positive potentials than peaks I_c/I_a' (Figure 10B). As was the case with 5-HT, 5-HTPP is also very strongly adsorbed at the PGE[51]. Coulometric experiments indicate that electrolyses of increasing concentrations of 5-HTPP at any given potential in the region of peak I_a causes a systematic decrease in the measured n-value. At any given initial concentration of 5-HTPP the n-value increases with increasingly positive applied potential. The maximum n-values were thus observed when very low concentrations of 5-HTPP were oxidized at relatively high potentials. Electrooxidation of 30 μM 5-HTPP at 0.70 V gave an n-value of 4 and a single product (*vide infra*).

Figure 11 shows a high performance liquid chromatogram of the product mixture obtained following controlled potential electrooxidation of 5-HTPP at $E_{p/2}$ for peak I_a. Liquid chromatographic peaks A and B were found[51] to be due to diastereomers of 4,4'-bi-5-hydroxytryptophan. In fact chromatographic peak B (Figure 11) could be separated into at least two isomers using a strong cation exchange column. Several minor unidentified products were eluted under chromatographic peak C. Peak D is due to tryptophan-4,5-dione, and cyclic voltammograms of this compound revealed it to be the species responsible for reversible peaks I_c/I_a' observed in cyclic voltammograms of 5-HTPP (Figure 10). Chromatographic peaks E and F were due to diastereomers of a dimeric compound containing residues of 5-HTPP and 6-hydroxyquinoline.

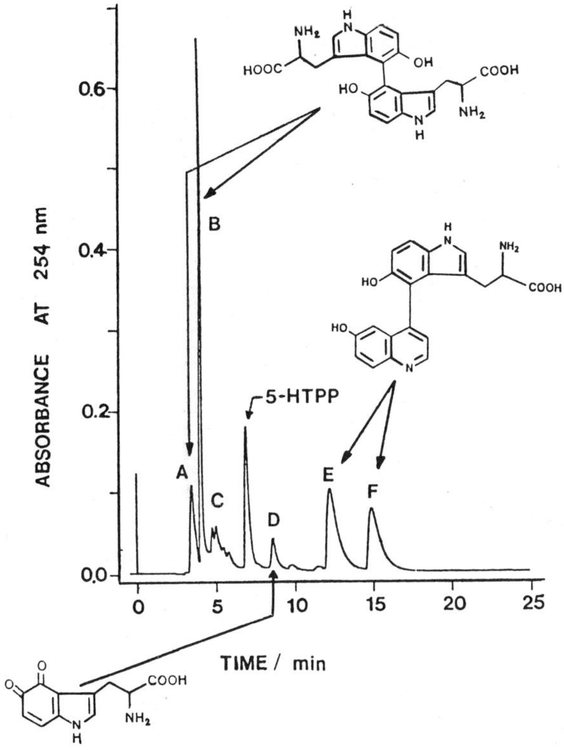

Figure 11. High performance liquid chromatogram of the product solution formed after controlled potential electrooxidation of 1.18 mM 5-hydroxytryptophan at 0.50 V in dilute HCl pH 2.5. Column: Brownlee RP-18, 250 x 0.7 mm. Mobile phase: 1.0 ml concentrated NH_3 dissolved in 500 ml of 4% MeCN in H_2O adjusted to pH 3.5 with formic acid. Flow rate: 2.5 ml min^{-1}. Volume injected: 650 µl. (Adapted by permission of the American Pharmaceutical Association from K. Humphries and G. Dryhurst, J. Pharm. Sci., 76, 839 (1987).

The yields of the various products shown in Figure 11 were very dependent upon the concentration of 5-HTPP oxidized and the applied potential. As noted earlier, oxidations of micromolar solutions of 5-HTPP at high potentials (e.g. 0.70 V) at pH 2 gave purple tryptophan-4,5-dione as the sole product in a $4e^-$ reaction. With increasing concentrations of 5-HTPP and lower applied potentials the dimeric products appear as major products.

Based upon such information it has been concluded[56] that the initial electrooxidation of 5-HTPP yields a very reactive radical intermediate. The oxidation of 5-HTPP yields only one simple dimer, 4,4'-bi-5-hydroxytryptophan (**14**, Figure 12). This behavior differs, therefore, from that reported earlier for 5-HT where two dimers are formed at low concentrations[50] and many more when high concentrations are oxidized[55]. Nevertheless, it appears reasonable to conclude that the initial $1e^-$, $1H^+$ oxidation of 5-HTPP yields radical **11** (Figure 12). This then dimerizes as conceptualized in Figure 12 to yield the diastereomers of dimer **14**. Dependent upon the applied potential and concentration of 5-HTPP oxidized part or all of radical **11** can be oxidized ($1e^-$, $1H^+$) to quinone imine **15** which is rapidly attacked by water to yield 4,5-dihydroxytryptophan (**16**).

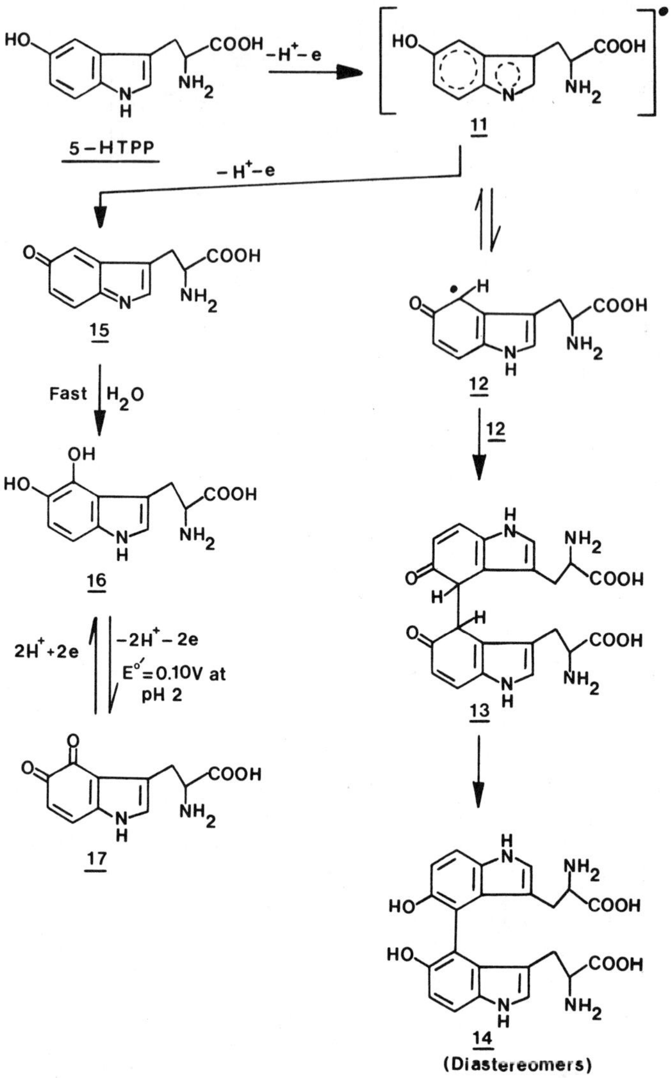

Figure 12. Reaction scheme proposed for the electrochemical oxidation of 5-hydroxytryptophan (5-HTPP) in acid solution to the diastereomers of 4,4'-bi-5-hydroxytryptophan (**14**) and tryptophan-4,5-dione (**17**).

Oxidation of the latter compound, which is much easier to oxidize than is 5-HTPP, yields dione **17** (Figure 12). The reversible peaks I_c/I_a' observed in cyclic voltammograms of 5-HTPP are due to the **17**/**16** couple.

The formation of diastereomers of 4-[4'-(6-hydroxyquinolyl)]-5-hydroxytryptophan (**27**) upon oxidation of 5-HTPP is a rather unusual reaction. It seems unlikely that these diastereomers would be formed from different intermediates than are the 4,4'-dimers (**14**) or tryptophan-4,5-dione (**17**). The structures of the indole-quinoline dimers suggest that the C(4)-position of one 5-HTPP residue must attack the C(3)-position of another residue. There are several ways that a coupling of 5-HTPP residues would occur. It was originally proposed[51] that quinone imine **15** is attacked by 5-HTPP to yield the indole-indolenine dimer **20** (Figure 13). However, it seems more probable that radical **11** in its resonance forms **18** and **19** (Figure 13) couples to give the same indole-indolenine dimer **20**. The 1,2-double bonds of indolenines are known to undergo additional reactions[56] and the addition of water to the indolenine residue of **20** would lead to **21** (Figure 13). Cleavage of the latter indoline across the 1,2-bond would then give the N-formyl-p-aminophenol derivative **22**. Electrochemical oxidation of this derivative should be very easy[57] giving in a $2e^{-}$-$2H^{+}$ reaction quinone imine **23** (Figure 13). Hydrolysis of **23** would give the p-quinone **24**. The expected cyclization reaction then occurs to give **25** which, following decarboxylation and further electrochemical oxidation, yields the indole-quinoline dimer **27** as shown in Figure 13.

3. SOME BIOCHEMICAL IMPLICATIONS

The electrochemical studies reported above and elsewhere[49-51,55] establish that the initial step in the oxidation of the neurotransmitter 5-HT and of its immediate precursor in the CNS, 5-HTPP, is a $1e^{-}$, $1H^{+}$ reaction yielding an extremely reactive radical intermediate. While radical species are thought to be formed upon autoxidation of 5-HT and related 5-hydroxyindoles at high pH[43] little attention seems to have been given to the possibility of forming radical species from such compounds in the CNS even though radicals are well known to be potentially harmful in living tisues[58]. As noted in the INTRODUCTION considerable speculation has centered about the possible anomalous formation of more highly hydroxylated indoles in the CNS and that it is such compounds which are the neurotoxins which might lead to certain mental illnesses. The work reported by this group provides the first unequivocal evidence that the primary radical formed upon oxidation of 5-HT is itself easily oxidized to a very reactive quinone imine. This is very rapidly attacked by water to yield 4,5-DHT which, being more easily oxidized than 5-HT, is further oxidized to tryptamine-4,5-dione. It is known that when injected intracranically in mice, 4,5-DHT is neurotoxic although the mechanism by which it exerts its neurotoxic effect is not known. It is worth noting that 4,5-DHT and tryptamine-4,5-dione constitute a reversible redox system. Furthermore, there is some evidence that the CNS is actively redox buffered*. Thus, regardless of the actual redox form of 4,5-DHT injected into brain the active redox buffer decides the proportion of reduced and oxidized species present.

It is probably fair to state that the mechanisms underlying the neurotoxicity of dihydroxytryptamines such as 4,5-, 5,6- and 5,7-DHT are not known. Nevertheless, there is much speculation about but very little experimental evidence to support such mechanisms in the literature. Most of these speculations implicate autoxidation of the dihydroxytryptamines to quinones or quinone imines. One theory suggests that these quinonoid compounds are attacked by nucleophiles such as thiol residues on nerve ending proteins leading the cross linking and irreversible polymerization of such proteins[60-62]. The neurotoxicity of quinonoid intermediates has also been proposed to be caused by their interference with the electron transport chain[63] or to formation of reduced oxygen species (H_2O_2, HO·, O_2^{-}) which are cytotoxic[64-68]. A major goal of the work underway in this laboratory is to understand the chemistry which underlies the neurotoxic properties of dihydroxytryptamines.

*This is probably an ascorbate-dehydro ascorbate redox buffer poised at -0.200 V vs SCE[59].

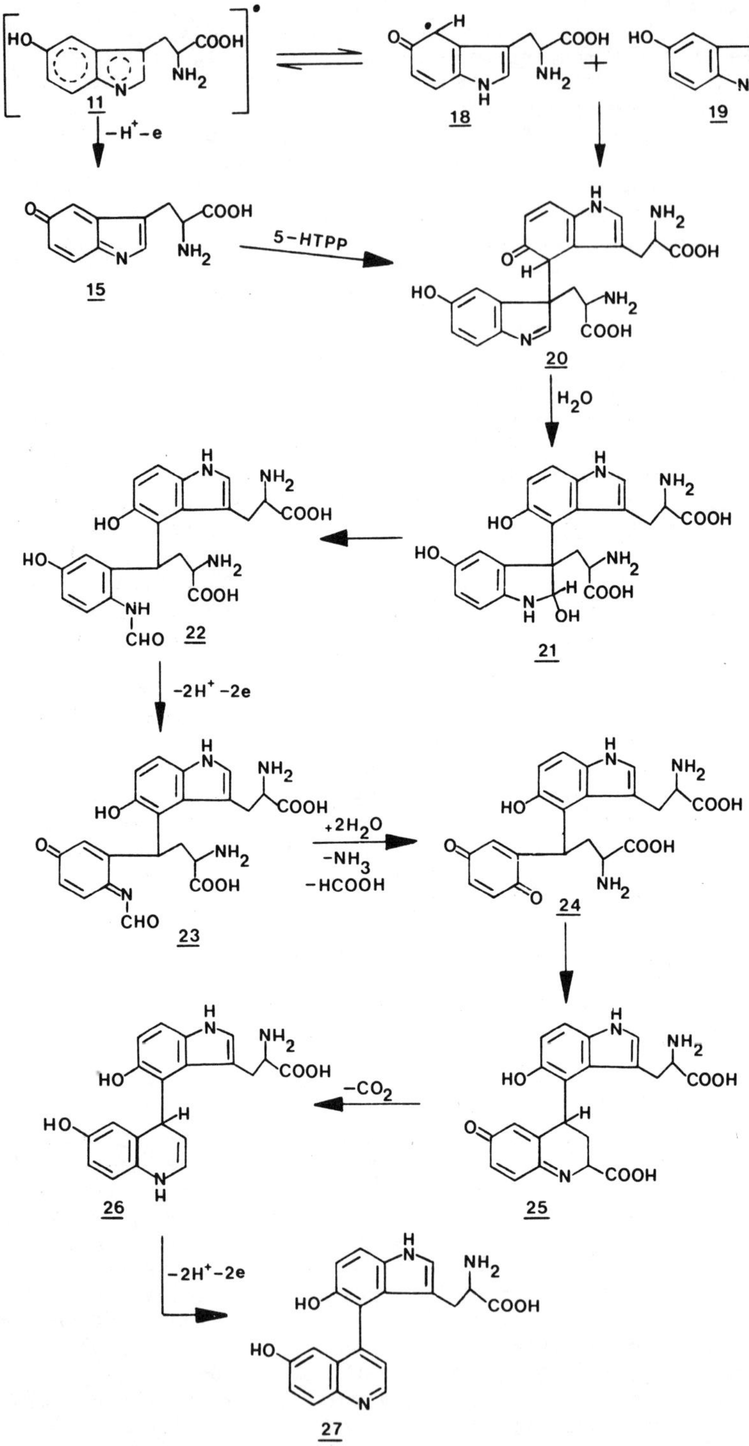

Figure 13. Reaction scheme proposed for the electrochemical and chemical reactions leading to formation of the diastereomers of 4-[4'-(6-hydroxyquinolyl)]-5-hydroxytryptophan (27).

The reaction scheme shown in Figure 9 indicates that tryptamine-4,5-dione (B) can be slowly (at pH 2) attacked by water to yield 4,5,7-trihydroxytryptamine (8) which is rapidly air-oxidized to 5-hydroxytryptamine-4,7-dione (9). When 5-HT is oxidized at pH 2 and the product solutions analyzed by HPLC without solutions being concentrated (e.g., by freeze-drying) then **9** is not observed as an isolatable product[50,55]. However, following such a concentration step **9** can be isolated as an orange product[49]. The reasons for these observations are complex and will be discussed elsewhere[69]. Nevertheless, there is substantial evidence that **9** is a minor intermediate (as in Figure 9) or product[49] of oxidation of 5-HT. This same compound is also a major product of electrochemical oxidation of 5,7-DHT at pH 2 and of both electrochemical and autoxidation at pH 7[49]. This observation prompted us to evaluate the neurotoxicity of **9**. When **9** is injected intracranially in mice it shows a greater general toxicity than all other previously studied neurotoxins which are believed to involve oxidation/reduction mechanisms in their mode of action. Table 1 presents some comparative data for various catecholamine and indoleamine neurotoxins which show that the LD_{50} for **9** (21 μg) is less than one-half that of, for example, 5,7-DHT (55 μg). A second interesting result concerns the depletion of endogenous neurotransmitters by **9**. Seven days

Table 1. LD_{50} values for some selected catecholamine and indoleamine neurotoxins.

Structure[a]	Name	$LD_{50}/\mu g^{b}$	Reference
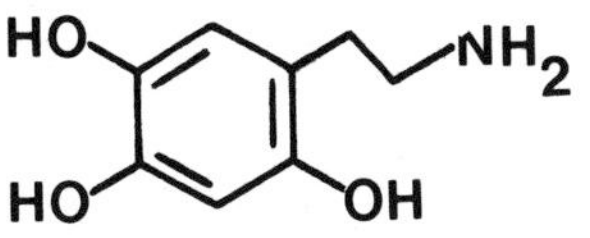 HO, HO, OH, NH_2	6-Hydroxydopamine	80	70
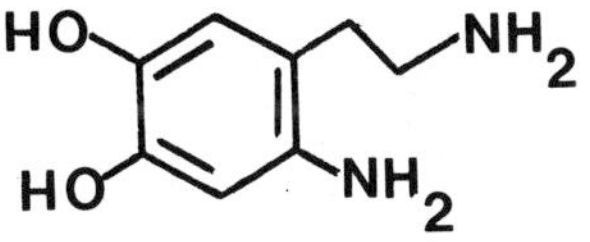 HO, HO, NH_2, NH_2	6-Aminodopamine	50	70
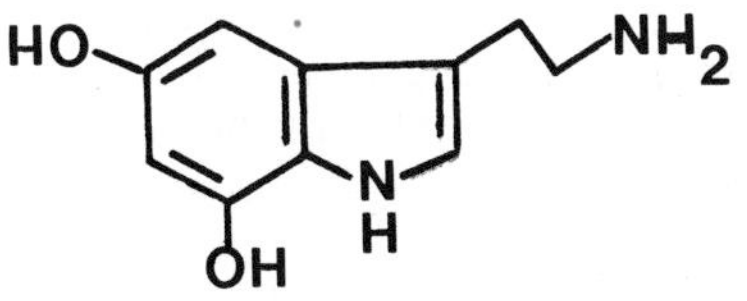 HO, OH, N H, NH_2	5,7-Dihydroxytry-ptamine	55	71
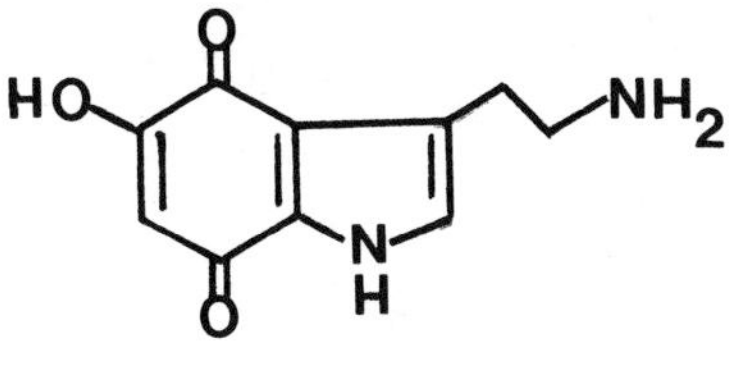 HO, O, O, N H, NH_2	5-hydroxytryptamine-4,7-dione	21	49

[a] A review of the ways in which the catecholamine neurotoxins exert their effect in the CNS has appeared[63].

[b] Compounds injected intracranially. Dose contained in 5 μl of 1 mg ml^{-1} ascorbic acid is isotonic saline (0.9% NaCl in water).

after an intracranial injection of 20 μg of **9**, whole mouse brain norepinephrine levels were lowered to 87 ± 1% of control values. This result was significant and, further, is reminiscent of the norepinephrine depletion of ca. 80% of controls 10 days after injection reported by Massotti et al.[71] using a 21 μg dose of 5,7-DHT. However, the 20 μg dose of **9** did not produce any significant decline of 5-HT levels in contrast to those observed with 5,7-DHT[71]. This latter result is presumed to be the direct result of the lack of selective uptake of **9** by serotonergic neurons. Thus, it has been assumed[49] that addition of the 4-oxo group of **9** compared to 5,7-DHT has effected change in its uptake properties with respect to 5-HT neurons. The fact that **9** is a major autoxidation product of 5,7-DHT[49] suggests that if the latter type of reaction is important in the CNS then it is probable that at least part of the neurotoxicity of 5,7-DHT is due to its partial conversion to **9**.

The reaction schemes shown in Figures 8, 9, 12 and 13 indicate that the oxidation chemistry of 5-HT and 5-HTPP is quite complex. Furthermore, radical intermediates and 4,5-dihydroxytryptamine derivatives are formed under very mild oxidizing conditions and such compounds are either expected to be or indeed are neurotoxic. Of all the other products of oxidation of 5-HT and 5-HTPP which have been thus far isolated only one, 5-hydroxytryptamine-4,7-dione (**9**), has been tested for its neurotoxic properties and it is clearly the most powerful indolic or catecholamine neurotoxin yet discovered. It does not, however, appear to exhibit any selective toxicity for particular neurons. Nevertheless, it seems probable that many other hydroxylated oxidation products of 5-HT, 5-HTPP and the other 5-hydroxyindoles found in the CNS might be neurotoxic.

While it is too early yet to speculate about the relationship of our electrochemical results to the etiology of mental illnesses, it is quite clear that CNS indoles are very easily oxidizable compounds. In addition their oxidation chemistry is extraordinarily rich and has already been shown to yield neurotoxic products. Thus, the rather wild suggestions that anomalous oxidation chemistry of 5-HT might be an underlying cause of mental illnesses certainly appears reasonable on the basis of our, admittedly, preliminary experimental results.

ACKNOWLEDGMENTS

The experimental work reported in this paper was supported by the NIH through Grant No. GM-32367. One of us (K.H.) would like to thank Conoco for the award of a fellowship during part of this work.

REFERENCES

1. O. Beck, Substance and Alcohol Actions / Misuse, 5, 211 (1985).

2. A.H. Amin, T.B.B. Crawford, J.H. Gaddum, J. Physiol. (London), 126, 596 (1954).

3. B.N. Twarog and I.H. Page, Am. J. Physiol., 175, 157 (1953).

4. I.H. Page, Physiol. Rev., 38, 277 (1958).

5. R. Novotna, Physiol. Bohemslov., 29, 243 (1980).

6. S. Udenfriend, E. Titus, H. Weissbach and R.E. Peterson, J. Biol. Chem., 219, 335 (1956).

7. A. Sjoerdma, T.E. Smith, T.D. Stevenson and S. Udenfriend, Proc. Exp. Biol. Med., 89, 36 (1955).

8. C. Mitoma, H.S. Posner, H.C. Reitz and S. Udenfriend, Arch. Biochem. Biophys., 61, 431 (1956).

9. V. Erspamer, J. Physiol. (London), 127, 118 (1955).

10. W.M. McIsaac and I.H. Page, J. Biol. Chem., 234, 858 (1959).

11. S. Kveder, S. Iskric and D. Keglivic, Biochem. J., 85, 447 (1962).

12. D. Eccleston, A.T.B. Moir, H.W. Reading and I.M. Ritchie, Br. J. Pharmacol. 28, 367 (1966).

13. D.W. Woolley and E. Shaw, Proc. Natl. Acad. Sci. USA, 40, 228 (1954).

14. J.H. Gaddum, J. Physiol. (London), 119, 363 (1953).

15. H.E. Himwich, S.S. Kety and J.R. Smythies, Amines and Schizophrenia, Pergamon Press, New York, 1967.

16. D.W. Woolley, The Biochemical Bases of Psychoses, Wiley, New York, 1962.

17. J.C. de la Torre, Dynamics of Brain Monoamines, Plenum Press, New York, 1972.

18. R.D. Kaplan and J.J. Mann, Life Sci., 31, 583 (1982).

19. C.E. Dalgleisch, Proceedings of the Symposium on 5-Hydroxytryptamines, Pergamon Press, New York, 1957, p. 60.

20. N. Eriksen, G.M. Martin and E.P. Benditt, J. Biol. Chem., 235, 1662 (1960).

21. H.G. Baumgarten, A. Björklund, L. Lachenmeyer, A. Nobin, U. Stenevi, Acta Physiol. Scand. Suppl., 373, 1 (1971).

22. H.G. Baumgarten, K.D. Evetts, R.B. Holmes, L.L. Iversen and G. Wilson, J. Neurochem., 19, 1587 (1972).

23. H.G. Baumgarten, M. Goethert, A.F. Holstein and H.G. Schlossberger, Z. Zellforsch., 128, 115 (1972).

24. A. Björklund, A. Nobin and U. Stenevi, Brain Res., 53, 117 (1973).

25. A. Björklund, A. Nobin and U. Stenevi, Z. Zellforsch., 145, 479 (1973).

26. A. Saner, L. Pieri, J. Moran, M. DaPrada and A. Pletscher, Brain Res., 76, 109 (1974).

27. H.G. Baumgarten and L. Lachenmeyer, Z. Zellforsch., 135, 399 (1972).

28. A. Björklund, A.S. Horn, H.G. Baumgarten, A. Nobin and H.G. Schlossberger, Acta Physiol. Suppl., 429, 29 (1975).

29. C.C. Porter, D.C. Titus, B.E. Sanders and E.V.C. Smith, Science, 126, 1014 (1957).

30. D. Sanker and V. Siva, Fed. Proc., Am. Soc. Exp. Biol., 18, 441 (1959).

31. G. Curzon and L. Vallet, Biochem. J., 74, 279 (1960).

32. B. Frieden and H.S. Hsieh, Exp. Med. Biol., 74, 505 (1976).

33. B.C. Barrass, D.B. Coult, R.M. Pindor and M. Skeels, Biochem. Pharmacol., 22, 2891 (1973).

34. B.C. Barrass and D.B. Coult, Biochem. Pharmacol., 21, 677 (1972).

35. G. Martin, N. Eriksen and E.P. Benditt, Fed. Proc., Am. Soc. Exp. Biol., 17, 477 (1958).

36. H. Blashko and A.S. Milton, Br. J. Pharmacol., 15, 42 (1960).

37. H. Blashko and W.G. Levine, Br. J. Pharmacol., 15, 625 (1960).

38. G.M. Martin, E.P. Benditt and N. Eriksen, Arch. Biochem. Biophys., 90, 208 (1960).

39. T. Uemura and T. Shimazu, Biochem. Biophys. Res. Commun., 93, 1074 (1980).

40. J.J. Blum and N.S. Ling, Biochem. J., 73, 530 (1959).

41. S.G.A. Alivisatos, H.G. Williams-Ashman, Biochim. Biophys. Acta, 86, 392 (1964).

42. D.C. Borg, Proc. Natl. Acad. Sci. USA, 53, 829 (1965).

43. E. Perez-Reyes and R.P. Mason, J. Biol. Chem., 256, 2427 (1981).

44. E. Walaas and O. Walaas, Arch. Biochem. Biophys., 95, 151 (1961).

45. T. Uemura, H. Matsushita, M. Ozawa, A. Fiori, E. Chiesara, FEBS Lett., 101, 59 (1979).

46. T. Uemura, T. Shimazu, R. Miura, T. Yamano, Biochem. Biophys. Res. Commun., 93, 1074 (1980).

47. G. Dryhurst, Electrochemistry of Biological Molecules, Academic Press, New York, 1977.

48. G. Dryhurst, K.M. Kadish, F. Scheller and R. Renneberg, Biological Electrochemistry, Vol. 1, Academic Press, New York, 1982.

49. M.Z. Wrona, D. Lemordant, L. Lin, C.L. Blank and G. Dryhurst, J. Med. Chem., 29, 499 (1986).

50. M.Z. Wrona and G. Dryhurst, J. Org. Chem., 52, 2817 (1987).

51. K. Humphries and G. Dryhurst, J. Pharm. Sci., 76, 839 (1987).

52. J.M. Saavendra, M. Brownstein and J. Axelrod, J. Pharmacol. Exp. Ther., 186, 508 (1973).

53. M.J. Brownstein in Basic Neurochemistry, G.J. Siegel, B.W. Albers, B.W. Angraff and R. Katzman, Eds., Little, Brown, Boston, 1981, p. 220.

54. M.D. Gershon and J. Tamir in Serotonin, Current Aspects of Neurochemistry and Function, B. Haber, S. Gabay, M.R. Issidorides and S.G.A. Alivisatos, Eds., Plenum Press, New York, 1981, p. 43.

55. M.Z. Wrona and G. Dryhurst, submitted to J. Org. Chem. (1988).

56. W.A. Remers in Indoles, W.J. Houlihan, Ed., Wiley-Interscience, New York, Part I, 1972, p. 191.

57. M. D. Hawley and R.N. Adams, J. Electroanal. Chem., 10, 376 (1965).

58. W.A. Pryor, Ed., Free Radicals in Biology, Academic Press, New York, Volumes 1- , 1976- .

59. R.L. McCreery, Ph.D. Dissertation, University of Kansas, 1974. Cited in C.L. Blank, R.L. McCreery, R.M. Wightman, W. Chey, R.N. Adams, J.R. Reid and E.E. Smissman, J. Med. Chem., 19, 178 (1976).

60. A. Rotman, J.W. Daly and R.C. Crevelling, Mol. Pharmacol., 12, 887 (1976).

61. H.P. Klemm, H.G. Baumgarten and H.G. Schlossberger, J. Neurochem., 35, 1400 (1980).

62. H.G. Schlossberger, Ann. N.Y. Acad. Sci., 305, 25 (1978).

63. G. Cohen and R.E. Heikkila, Ann. N.Y. Acad. Sci., 305, 74 (1978).

64. H.G. Baumgarten, H.P. Klemm, L. Lachenmeyer, A. Björklund, N. Lovenberg and H.G. Schlossberger, Ann. N.Y. Acad. Sci., 305, 3 (1978).

65. H.P. Klemm, H.G. Baumgarten and H.G. Schlossberger, J. Neurochem., 35, 1400 (1980).

66. H.G. Baumgarten, S. Jenner, A. Björklund, H.P. Klemm and H.G. Schlossberger in "Biology of Serotoninergic Transmission", N.N. Osborn, Ed., Wiley, New York, 1982, Chap. 10.

67. H.G. Baumgarten, H.P. Klemm, J. Sievers and H.G. Schlossberger, Brain Res. Bull., 9, 131 (1982).

68. C.R. Crevelling and A. Rotman, Ann. N.Y. Acad. Sci., 305, 57 (1978).

69. M.Z. Wrona and G. Dryhurst, J. Pharm. Sci., in press (1988).

70. T. Hsi, MS Thesis, University of Olkahoma, 1979, p. 51.

71. M. Massotti, A. Scotti de Carolis and V.G. Longo, Pharm. Biochem. Behav., 2, 796 (1974).

ORGANIC ELECTROCHEMISTRY, IV[1]:

ELECTROCHEMICAL DESULFURIZATION REACTIONS OF THIOLUMAZINES

Wolfgang Pfleiderer and Bernd S. Schulz

Department of Chemistry
University of Konstanz
P.O. Box 5560
D-7750 Konstanz / West Germany

INTRODUCTION

A great number of naturally occurring pteridines[2-5] have a substituent carbon side-chain in the 6-position of the heterocycle. This is because the biosynthetic pathway[6] of the pteridine nucleus involves in the first and most important step an interconversion of GTP into 7,8-dihydroneopterin-3'-triphosphate. Neopterin, biopterin, the drosophila eye pigments sepiapterin and deoxysepiapterin as well as leucettidine, a lumazine derivative, are typical representatives of more or less biological and biochemical importance in this series. The availability of this type of compound is based on chemical syntheses which afford a regioselective condensation of the appropriate components either in a Gabriel-Isay type reaction[3-5], starting from a 5,6-diaminopyrimidine, or in a Taylor synthesis[3-5,7], forming first the pyrazine moiety in the build-up of the pteridine nucleus.

A more modern approach for the direct introduction of a carbon side-chain into the pteridine nucleus has been developed by homolytic nucleophilic substitution reactions[8,9], especially using acyl radicals[10-15] as well as alkyl radicals[16] as reactive species. These reactions, however, take place regioselectively with 6,7-unsubstituted pteridine derivatives at the most electron-deficient 7-position leading to the unnatural isomers[11,12]. Direction of the incoming nucleophile towards the C-6 atom can only be achieved in the presence of a 7-substituent[13-16]. A reasonable "protecting" group for C-7 has been the alkylmercapto and thione function, since it was found that the difficulties encountered with the Raney-nickel desulfurization reaction[17] in the pteridine series could be overcome using Raney-cobalt and copper-aluminum alloy, respectively[13,14].

However, there are also reports in literature[18-20] about the successful removal of thio groups by electrochemical reduction which regenerate from 6-thiopurine, for example[18], rapidly the N-1=C-6 bond by loss of H_2S to give purine and on immediate further reduction 1,6-dihydropurine. These results prompted us to investigate the electrochemical behavior of various thiolumazines in respect to an elimination of H_2S as an alternative desulfurization reaction.

EXPERIMENTAL

Chemicals

The starting materials 1,3-dimethyl-7-thiolumazine[21], 1,3,6-trimethyl-7-thiolumazine[22], 1,3-dimethyl-6-thiolumazine[23], 6,7-diphenyl-4-thiolumazine[24] and 7-hydroxy-1,3-dimethyllumazine[25] were synthesized according to procedures in the literature. Buffer solutions were prepared from analytical reagent grade chemicals. Nitrogen (Linde) used for deoxygenating purposes was used with no further purification.

Materials

Tlc was performed on silica gel thin-layer sheets (F 1500 LS 254) and cellulose thin-layer sheets (F 1440) from Schleicher & Schüll. Preparative column chromogatography was carried out on silica gel Merck 60 (0.063 to 0.2 mesh). Separation of the electrolysis chambers was achieved by Nafion 125 membrane from E.I. Dupont de Nemours, Plastic Division, Wilmington, USA.

Apparatus

UV-spectra were measured with a Perkin-Elmer Lambda 5 spectrophotometer and NMR-spectra with a Bruker WN-250 spectrometer. Macroscale electrolyses were performed in a U-shaped cell[26] with a stirred mercury pool under N_2. The potential was controlled by a Jaissle potentiostat, Model 5000T-B, and the progress of the reaction followed by a coulometer developed and built by the electronic workshop of the University of Konstanz. A silver/silver chloride electrode (SSE) was used as a reference.

Reactions

7-Amino-1,3-dimethyllumazine (2)[27]

a) A mixture of 200 ml of 1 N ammonium formate and 20 ml of concentrated ammonia were placed in an electrolysis cell[26] and then 0.224 g (mmole) of 1,3-dimethyl-7-thiolumazine (1) dissolved in the cathodic compartment. Electrolysis was performed at -860 mV against SSE for 1 day. The reaction was stopped after uptake of 12.4 electron equivalents. The reaction solution was extracted continuously by $CHCl_3$ in a perforator, the extract evaporated and the residue recrystallized from $EtOH/H_2O$ to give 0.18 g (86%) of colorless crystals of m.p.> 360°C. Lit.[27] m.p. > 360°C. The product was identified by chromatographic and spectroscopic comparisons with an authentic sample.

b) Electrolysis of 0.224 g (1 mmole) of 1,3-dimethyl-7-thiolumazine (1) in 200 ml of 1 N ammonium formate at -500 mV for 3 days led to the separation of a precipitate, which was collected after concentration of the reaction solution to half of its volume. Recrystallization gave 0.165 g (80 %) of **2**.

7-Methylamino-1,3-dimethyllumazine (3)[27]

Electrolysis of 0.224 g (1 mmole) of 1 was performed in 200 ml of 1 N methylammonium formate at -500 mV for 4 hours leading to an uptake of 2.3 electron equivalents. Extraction of the reaction solution by $CHCl_3$, evaporation to dryness and recrystallization of the residue from MeOH gave 0.08 g (35%) of colorless crystals of m.p. 345°C. Lit.[27] m.p. 348°C.

7-Anilino-1,3-dimethyllumazine (**4**)

200 ml of a 1 N solution of aniline were acidified by HCl to pH 3.8 and then 0.224 g (1 mmole) of **1** was added to the cathodic chamber of the reaction vessel. Electrolysis took place at -500 mV for 1 days, during which 2.3 electron equivalents were used. A precipitate separated during the reaction which was filtered off by suction. Recrystallization from EtOH gave 0.175 g (62%) of colorless crystals of m.p. 320°C (decomp.).
UV (MeOH): λ_{max} 223, 292, 364 nm; lg ε 4.31, 4.04, 4.34.
^{1}H-NMR (D_6-DMSO): 8.50 (s, 1 H, 6-H); 7.4-7.8 (m, 5 H, C_6H_5); 3.82 (s, 3 H, N-1-CH_3); 3.61 (s, 3 H, N-3-CH_3).
$C_{14}H_{13}N_5O_2$ (283.3) Calc.: C 59.36, H 4.63, N 24.72;
Found: C 59.34, H 4.68, N 24.46.

7-Hydroxy-1,3-dimethyllumazine (**5**)[25]

To a mixture of 150 ml of 1 N HCl and 100 ml of EtOH in the electrolysis cell was added 0.224 g (1 mmole) of **1**. Electrolysis at -1400 mV was stopped after 20 hours and uptake of 2 electron equivalents. The reaction solution was evaporated to dryness, and then the residue was recrystallized from water to give 0.054 g (26%) of colorless crystals of m.p. 265°C. Lit.[25] m.p. 264°C.

1,3,8-Trimethylxanthine (**10**)

0.224 g (1 mmole) of **1** was dissolved in 170 ml of pH 6.5 phosphate buffer and electrolyzed at -1.4 V. The reaction is stopped after 1 day, the solution in the cathode compartment evaporated to dryness followed by extraction with hot EtOH. The extract was again evaporated and the residue purified by column chromatography on silica gel with $CHCl_3$ and $CHCl_3$/MeOH (19/1) to give 0.121 g (62 %) of colorless crystals of m.p. 330°C. Lit.[28] m.p. 330°C. The product was chromatographically and spectroscopically identical with an authentic sample.
UV (MEOH): λ_{max} 205, 271 nm; lg ε 4.30, 3.97.

1,3,8-Trimethylxanthine (**10**) and 1,3-Dimethyllumazine (**12**)

Analogous electrolysis like the preceding procedure at -1.6 V led to the formation of 0.06 g (30%) of **10** and 0.033 g (17%) of **12**, which were separated by silica gel column chromatography in $CHCl_3$/MeOH (49/1) and identified by spectroscopic and chromatographic comparisons with authentic samples.

8-Ethyl-1,3-dimethylxanthine (**11**) and 1,3,6-Trimethyllumazine (**13**)

Cathodic reduction of 0.238 g (1 mmole) of 1,3,6-trimethyl-7-thiolumazine (**9**)[22] was performed in 150 ml of phosphate buffer pH 6.3 at -1.4 V. After 24 hours the solution was evaporated to dryness, the residue extracted with hot EtOH and then again evaporated. The reaction mixture was separated and purified by silica gel chromatography (21 x 1.5 cm) with $CHCl_3$/MeOH (49/1) to give 0.014 g (7 %) of **13**; m.p. 202°C; Lit.[29] m.p. 202-203°C and 0.096 g (46%) of **11** as colorless crystals of m.p. 270°C; Lit.[30] m.p. 270°C.

5,6-Dihydro-7-hydroxy-1,3-dimethyllumazine (**14**)

a) Electrolysis of 0.208 g (1 mmole) of 7-hydroxy-1,3-dimethyllumazine (**5**) in 150 ml of phosphate buffer pH 6.4 led to an uptake of 2.5 electron equivalents within 4.5 h at -1.4 V. The reaction solution was then evaporated to dryness, the residue extracted with hot EtOH, from which on concentration to a smaller volume a precipitate separated to yield 0.14 g (67%) of colorless crystals of m.p. 204°C.
UV (MeOH): λ_{max} 206, [239], 328 nm; lg ε 4.34, [3.71], 3.79.
^{1}H-NMR (D_6-DMSO): 10.72 (s, O-H); 4.71 (s, N-H); 3.50 (s, 2 H, CH_2); 3.17 (s, N-H-CH_3).

b) In a mixture of 50 ml of water and 50 ml of ethanol were suspended 1.04 g (0.005 mole) of **5** and 0.1 g of PtO_2 and then hydrogenated in a shaking apparatus under H_2 and normal pressure. After 1 day the starting material had dissolved and H_2 uptake came to a stop. The catalyst was filtered off and the filtrate evaporated to a small volume until crystals separated. The precipitate was collected after cooling and yielded 0.85 g (81%) of colorless crystals of m.p. 204°C.

$C_8H_{10}N_4O_3$ (210.2)	Calc.:	C 45.71,	H 4.80,	N 26.66;
	Found:	C 45.48,	H 4.84,	N 26.45.

1,3,8-Trimethylxanthine (**10**) and 5-Acetylamino-6-amino-1,3-dimethyluracil (**15**)

An analogous reaction as described in the preceding procedure a) was performed with 0.208 g (1 mmole) of **5** at -1.8 V for 24 h. The reaction solution in the cathode compartment was evaporated to dryness, the residue extracted with hot EtOH and then again evaporated. This mixture was separated by column chromatography on silica gel in $CHCl_3$/MeOH (19/1) to give 0.088 g (45%) of **10** of m.p. 330°C and 0.014 g (6%) of 5-acetamino-6-amino-1,3-dimethyluracil (**15**) as a crystalline powder of m.p. 280°C; Lit.[28] m.p. 281°C.
UV (MeOH): λ_{max} 221, 267 nm; lg ε 3.55, 4.17.
^{1}H-NMR (D_6-DMSO): 8.35 s, 1-H, N-H); 6.57 (s, 2H, NH_2); 3.28 (s, 3 H, N-1-CH_3); 3.09 (s, 3 H, N-3-CH_3); 1.91 (s, 3 H, CH_3).

7,8-Dihydro-1,3-dimethyl-6-thiolumazine (**20**)

In a mixture of 20 ml of pyridine and 100 ml of phosphate buffer pH 7 0.224 g (1 mmole) of 1,3-dimethyl-6-thiolumazine (**19**) were dissolved and electrolyzed at -1.9 V. After 3 h the reaction solution was evaporated to dryness. From the residue was extracted, by hot ethanol, 0.126 g (55%) of yellowish crystals of m.p. 295°C.
UV (MeOH): λ_{max} 249, 291, 355 nm; lg ε 3.73, 4.00, 3.79
^{1}H-NMR (D_6-DMSO): 11.28 (s, 1H, S-H); 7.62 (s, 1 H, NH); 4.17 (s, 2 H, CH_2); 3.25 (s, 3 H, N-1-CH_3); 3.14 (s, 3 H, N-3-CH_3).

$C_8H_{10}N_4O_2S$ (226.2)	Calc.:	C 42.48,	H 4.46,	N 24.77;
	Found:	C 42.31,	H 4.38,	N 24.56.

6,7-Diphenyllumazine (**22**)

0.332 g (1 mmole) of 6,7-diphenyl-4-thiolumazine (**21**) was dissolved in 150 ml of dioxane[24], and then added to 60 ml of 0.015 M tetrabutylammonium hydrogen sulfat solution in the electrolysis cell. Electrolysis of the solution (pH 3.8) at -1.4 V for 4 days led according to tlc to a new product. Work-up was performed by evaporation, extraction of the residue by $CHCl_3$, concentration to a smaller volume and chromatography on a silica gel column (22 x 1.5 cm) with $CHCl_3$ and followed by $CHCl_3$/MeOH (19/1). Two fractions of 0.087 g (26%) of starting material **20** and 0.135 g (43%) of 6,7-diphenyllumazine (**22**) of m.p. 322-325°C; lit.[31] m.p. 320-325°C were obtained. The identity of the products was checked by chromatography and UV-spectroscopy with authentic materials.

RESULTS AND DISCUSSION

During the electrochemical reduction of 1,3-dimethyl-7-thiolumazine (**1**) in ammonium formate solution it was noticed, in contrast to the expected desulfurization of the thio group, that a displacement reaction with formation of 7-amino-1,3-dimethyllumazine (**2**) took place. This mild interconversion seems to be of general application and little dependent on the pH, since analogous reaction conditions using methylammonium formate and aqueous anilinium chloride solution, respectively, formed the 7-methyl-amino-(**3**) and 7-anilino-1,3-dimethyllumazine (**4**) in good yields. Secondary amines, on the other hand, did not react at all owing perhaps to steric reasons.

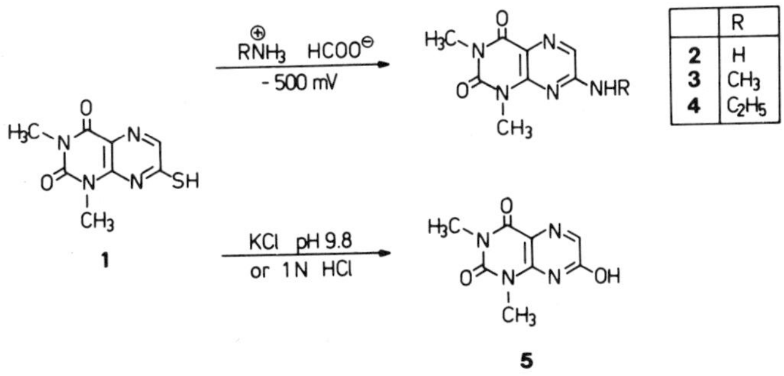

	R
2	H
3	CH_3
4	C_2H_5

In order to eliminate the nucleophilic component of the electrolyte the electrochemical behavior of **1** in potassium chloride and 1 N hydrochloric acid solution was investigated. It was found that the cathodic reduction is associated with a formal hydrolysis of the thio group leading to 7-hydroxy-1,3-dimethyllumazine (**5**). Since **1** is chemically stable at room temperature towards aminolysis and hydrolysis it is obvious that a reduced intermediate will be prone to nucleophilic displacement reactions. It is assumed that **1** is first reduced in a 2 electron/2H$^+$-step to 1,3-

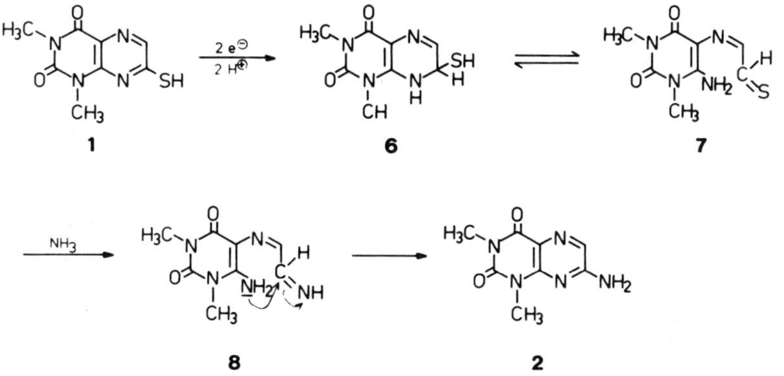

dimethyl-7-thio-7,8-dihydrolumazine (**6**), which equilibrates in a ring-chain-tautomerism with **7**. The generated thioaldehyde function possesses a high electrophilic potential and reacts easily with nucleophiles to form, with amines, the corresponding imines (**8**) or on hydrolysis the aldehyde, which leads on subsequent cyclization to the isolated reaction products **2-5**.

Modulation of the working potential to -1.4 V revealed, in phosphate buffer pH 7, a new result, since **1** reacted under these conditions with a ring contraction and formation of 1,3,8-trimethylxanthine (**10**). Further increase of the potential to -1.6 V led to a mixture of **10** and a small amount of 1,3-dimethyllumazine (**12**) showing that electrochemical desulfurizations of thiolumazines are in principle possible but cannot be considered as the main reaction pathway with preparative applications. Cathodic reduction of **1** at -1.9 V finally generated **10** in 90% yield. A first hint regarding the mechanism of the ring contraction was derived from the electroreduction of 1,3,6-trimethyl-7-thiolumazine (**9**), which was converted at -1.4 V in phosphate buffer into 8-ethyl-1,3-dimethylxanthine (**11**) and a smaller amount of 1,3,6-trimethyllumazine (**13**).

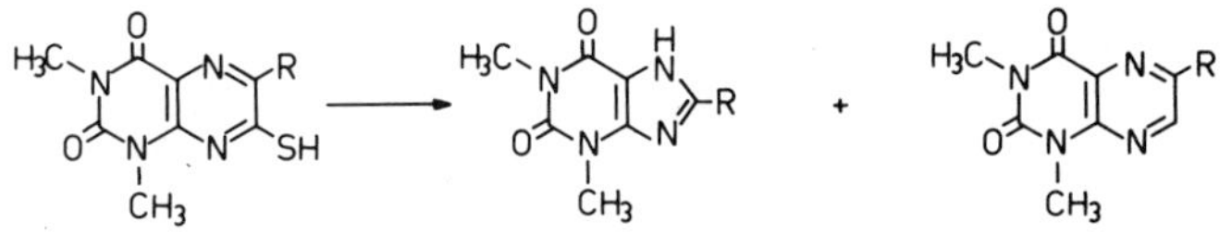

	R
1	H
9	CH_3

	R
10	CH_3
11	C_2H_5

	R
12	H
13	CH_3

This result indicates that the 5-6 bond has to be cleaved first as postulated also recently for the chemical reductive interconversions of 7-alkoxy-1,3-dimethyllumazines into the corresponding 8-alkyltheophyllines[32]. An indirect proof for the proposed mechanism of the pteridine-purine ring contraction (Scheme 5) was deduced from the electrochemical reductions of 7-hydroxy-1,3-dimethyllumazine (**5**). In phoshate buffer pH 6.6 a $2e^-/2H^+$-reduction took place at -1.4 V yielding 5,6-dihydro-7-hydroxy-1,3-dimethyllumazine (**14**), whereas at more negative potential (-1.8 V) a mixture of 1,3,8-trimethylxanthine (**10**) and 5-acetylamino-6-amino-1,3-dimethyluracil (**15**) was formed.

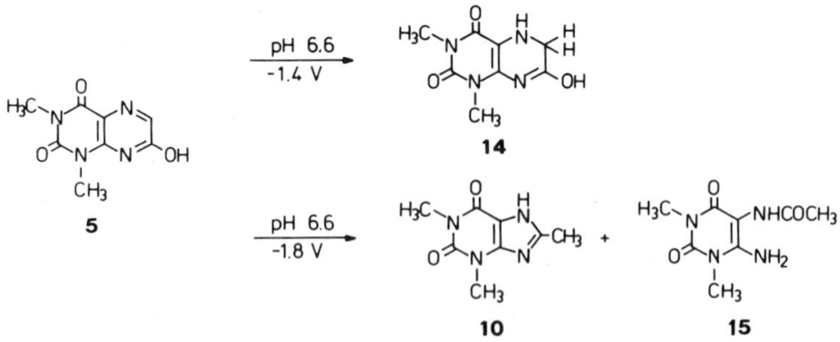

pH 6.6, -1.4 V → **14**

pH 6.6, -1.8 V → **10** + **15**

5

The latter compound (15) is derived from the isomeric labile 6-acetyl-amino-5-amino-1,3-dimethyluracil (18), which seems to be the initial cleavage product showing either cyclization to the purine derivative (10) or acyl migration to the thermodynamically more stable isomer **15**.

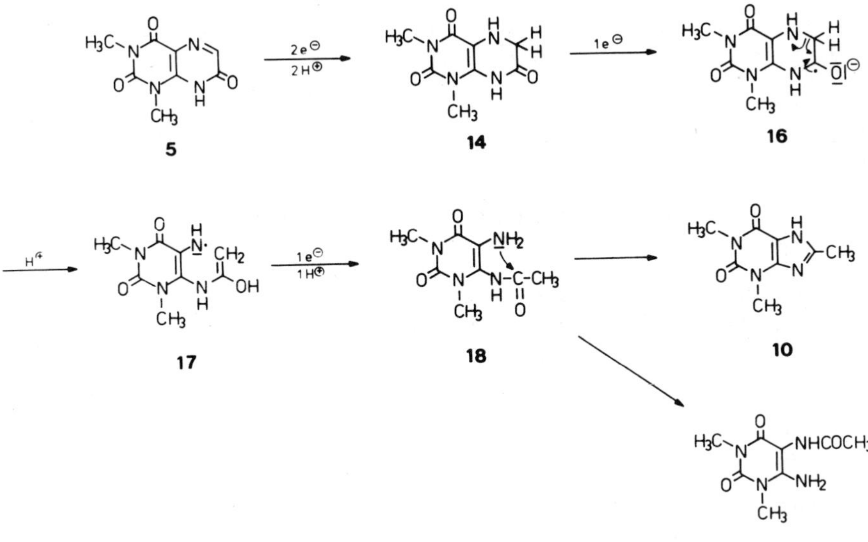

Analogous electrochemical reductions with 1,3-dimethyl-6-thiolumazine **19** did not reveal new facts, since under a broad variety of conditions only the corresponding 7,8-dihydro derivative **20** could be detected.

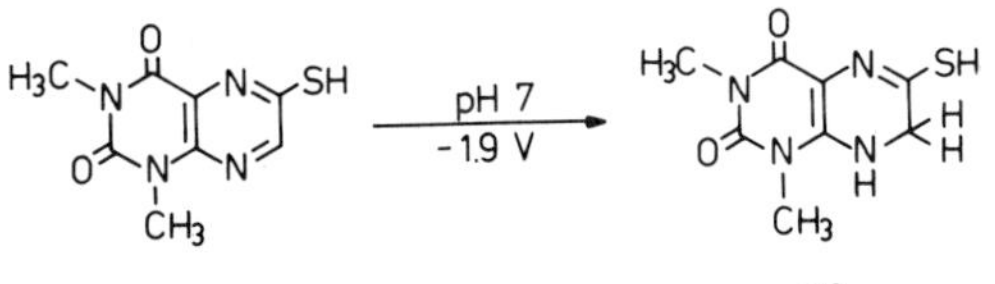

Also 6,7-diphenyl-4-thiolumazine (**21**) could not be desulfurized analogously to 6-thiopurine, but showed instead hydrolysis of the thio function at a relatively negative potential of -1.4 V and an acid pH of 3.8 to yield 6,7-diphenyllumazine (**22**).

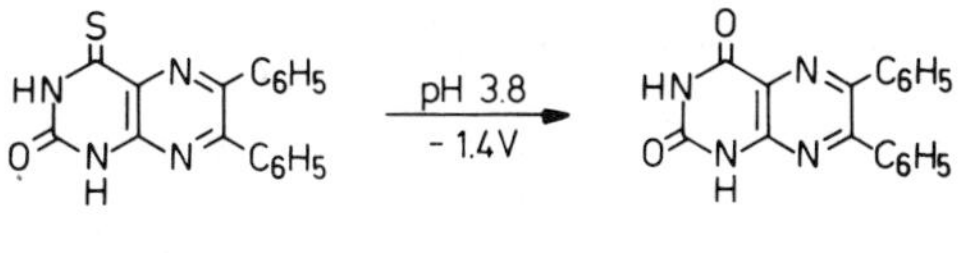

REFERENCES

1. Part III: R. Gottlieb and W. Pfleiderer, Liebigs Ann. Chem., 1451 (1981).

2. W. Pfleiderer, Angew. Chem., Int. Ed. Engl., 3, 114 (1964).

3. W. Pfleiderer in "Folates and Pterins", E.L. Blakley and S.J. Benkovic, Eds., J. Wiley & Sons, New York, Vol. 2, p. 43 (1985).

4. W. Pfleiderer in "Comprehensive Heterocyclic Chemistry", A.R. Katritzky and C.W. Rees, Eds., Pergamon Press, Oxford, Vol. 3, Part 2B, p. 262 (1984).

5. W. Pfleiderer in "Unconjugated Pterins in Neurobiology", W. Loevenberg and R.A. Levene, Eds., Taylor & Francis, London, p. 29 (1987).

6. G.M. Brown in "Folates and Pterins", R.L. Blakley and S.J. Benkovic, Eds., J. Wiley & Sons, New York, Vol. 2, p. 115 (1985).

7. E.C. Taylor and P.A. Jacobi, J. Am. Chem. Soc., 96, 6781 (1974); 98, 2301 (1976).

8. F. Minisci, Synthesis, 1 (1973)

9. F. Minisci and O. Porta, Advanc. Heterocycl. Chem., A.R. Katritzky and A.J. Boulton, Eds., Academic Press, New York, Vol. 16, p. 123 (1974).

10. W. Pfleiderer, R. Baur, M. Bartke and H. Lutz in "Chemistry and Biology of Pteridines", J.A. Blair, Ed., W. de Gruyter, Berlin, p. 93 (1983).

11. R. Baur, E. Kleiner and W. Pfleiderer, Liebigs Ann. Chem., 1798 (1984).

12. R.C. Boruah, R. Baur and W. Pfleiderer, Croat. Chem. Acta, 59, 183 (1986).

13. R. Baur, T. Sugimoto and W. Pfleiderer, Chem. Lett. 1025 (1984).

14. W. Pfleiderer, Tetrahedron Lett., 1031 (1984).

15. R. Baur and W. Pfleiderer, Israel J. Chem., 27, 81 (1986).

16. T. Sugimoto, S. Murata, S. Matsuura and W. Pfleiderer, Tetrahedron Lett., 4179 (1986).

17. E.C. Taylor and E. Wachsen, J. Org. Chem., 43, 4254 (1978).

18. G. Dryhurst, J. Electrochem. Soc., 116, 1357 (1969).

19. G. Dryhurst, J. Electrochem. Soc., 117, 1118 (1970).

20. G. Dryhurst, Electrochemistry of Biological Molecules, Academic Press, New York (1977).

21. Z. Kazimierczuk and W. Pfleiderer, Chem. Ber., 112, 1499 (1979).

22. A. Heckel and W. Pfleiderer, Helv. Chim. Acta, 69, 1095 (1986).

23. A. Heckel and W. Pfleiderer, Helv. Chim. Acta, 69, 708 (1986).

24. T.E. Gorizdra, Khim. Geterotsikl. Soedin., 5, 908 (1969); Chem. Abstr., 72, 111 423 (1970).

25. W. Pfleiderer, Chem. Ber., 90, 2588 (1957).

26. H. Lund in "Advances in Heterocyclic Chemistry", A.R. Katritzky and A.J. Boulton, Eds., Acad. Press, New York, Vol. 12, p. 233 (1970).

27. H. Steppan, J. Hammer, R. Baur, R. Gottlieb and W. Pfleiderer, Liebigs Ann. Chem., 2135 (1982).

28. H. Bredereck, I. Hennig, W. Pfleiderer and G. Weber, Chem. Ber., 86, 333 (1953).

29. Y. Kang, R. Soyka and W. Pfleiderer, J. Heterocycl. Chem., 24, 597 (1987).

30. J.H. Speer and A.L. Raymond, J. Am. Chem. Soc., 75, 114 (1953).

31. E.C. Taylor, J.A. Carbon and D.R. Hoff, J. Am. Chem. Soc., 75, 1904 (1953).

32. T. Sugimoto, N. Nishioka, S. Murata and S. Matsuura, Heterocycles, 24, 1565 (1986).

DETERMINATION OF CATIONIC DRUGS ON A NAFION-COATED ELECTRODE

Tomokazu Matsue*, Atsushi Aoki and Isamu Uchida

Department of Molecular Chemistry and Technology
Faculty of Engineering
Tohoku University
Aramaki Aoba, Sendai 980 / Japan

INTRODUCTION

There has been considerable interest in electrochemical determinations in both static and flow systems because of their high sensitivity. Among extensive studies, some efforts[1-3] have recently been focused on treatment of electrode surfaces to improve reproducibility or analyte selectivity in high performance liquid chromatography (HPLC) and flow injection analysis (FIA). We report here the electrochemical determination of several cationic drugs on Nafion-coated electrodes both in static and flow systems.

Nafion is a perfluorosulfonated ion-exchange polymer and has received a great deal of attention as a modifier for polymer-coated electrodes mainly because of its outstanding chemical stability and excellent ionic conductivity. Nafion-coated electrodes have been proved to be effective for the determination of electroactive compounds[4-6]. The attractive features of Nafion which are useful for electroanalytical purposes are its preconcentrating ability[7,8] of cations and its ionic selectivity.

Taking advantage of these unique properties, we have carried out the indirect determination of electroinactive cationic drugs with Nafion-coated electrodes both in static and flow systems. Indirect electrochemical determination[3,9] is one of the attractive methods since it removes the limitation of direct electrochemical determination. The present method is based on the competitive partitioning effect between an electroactive compound and cationic drugs into a Nafion layer, and has been applied to several ammonium drugs using (ferrocenylmethyl)trimethylammonium ion (FA^+) as an electroactive chemical agent. It has been demonstrated that low concentrations of electroinactive drugs can be determined by electrochemical methods since Nafion has a strong affinity for the drugs. The direct determinatin of catecholamines in the presence of their metabolites has also been carried out using a Nafion-coated electrode in flow systems. Since the metabolites are rejected by the surface Nafion, the determination of catecholamines can easily be performed in the present system.

EXPERIMENTAL

Chemicals

A Nafion solution (Nafion 117, 1100 EW, 5% solution) was obtained from Aldrich Co. and was diluted with methanol. (Ferrocenylmethyl)trimethylammonium (FA^+) perchlorate was prepared from the iodide salt of FA^+ (Tokyo Kasei Co.) by using an aqueous solution of $NaClO_4$, and was recrystallized three times from water. All drugs were obtained from commercial sources and used without further purification. The structures of these druges, including metabolic reactions of catecholamines, are shown in Figure 1.

Apparatus

Cyclic voltammetry was carried out with a PAR model 273. The cell was a water-jacket type and kept at 25°C. The working electrode was a 3 mm diameter glassy carbon (GC) disk (GC-20, Tokai Carbon Co.) mounted in a Teflon rod and polished finally with alumina powder (50 nm average particle size, Buehler Co.) to give a mirror-like surface. A Pt wire was used as a counter electrode and potentials were referred to a saturated calomel electrode (SCE). The apparatus used for HPLC and FIA was a JASCO high pressure pump model 880-PU equipped with an electrochemical flow cell (self-made, working electrode: 3 mm in diameter GC-20). The potential control and current monitor were made by a dual potentiostat (self-made). A 10 µl orl 50 µl sample loop was used for the injection port. The column used was a 120 mm x 4 mm SUS packed with Unisil C18 (5µm average size), Gaskuro Kogyo).

Surface coating

The Nafion coating for static systems was prepared by pipetting 2 µl of the 0.025% Nafion solution onto a clean GC surface and allowing the solvent to evaporate at room temperature. This process was repeated several times using methanol without Nafion to avoid island-like surface coatings. These procedures are important to obtain stable coatings and reproducible results. The Nafion coating on the GC surface of the flow cell was performed by the spin-coating method (4000 rpm) using a 3% Nafion solution.

RESULTS AND DISCUSSION

Determination of electroinactive drugs in a static system

Cyclic voltammetry of FA^+ with a Nafion-coated electrode showed near-ideal behavior for a thin film[10]. The peaks were almost symmetrical with very little diffusional tailing and the peak current was proportional to scan rate when the scan rate is less than 200 mV s^{-1}. The peak width at half-peak height was slightly larger than the theoretical value, 90 mV[10] predicted for a surface species, which may indicate that the entrapped FA^+ exists under various circumstances. The surface concentration of $-SO_3^-$ sites on the electrode was estimated to be 3.4×10^{-9} mol equivalent cm^{-2} from the loading amount of FA^+ entrapped in the Nafion, and the average thickness of the film was calculted to be ca. 24 nm using the density of wet Nafion (1.58 g cm^{-3}) in Na^+ form[15]. This value agrees reasonably well with the value (ca. 40 nm) estimated from the amount of the Nafion on the surface. Although the thickness of the Nafion film on the GC surface might

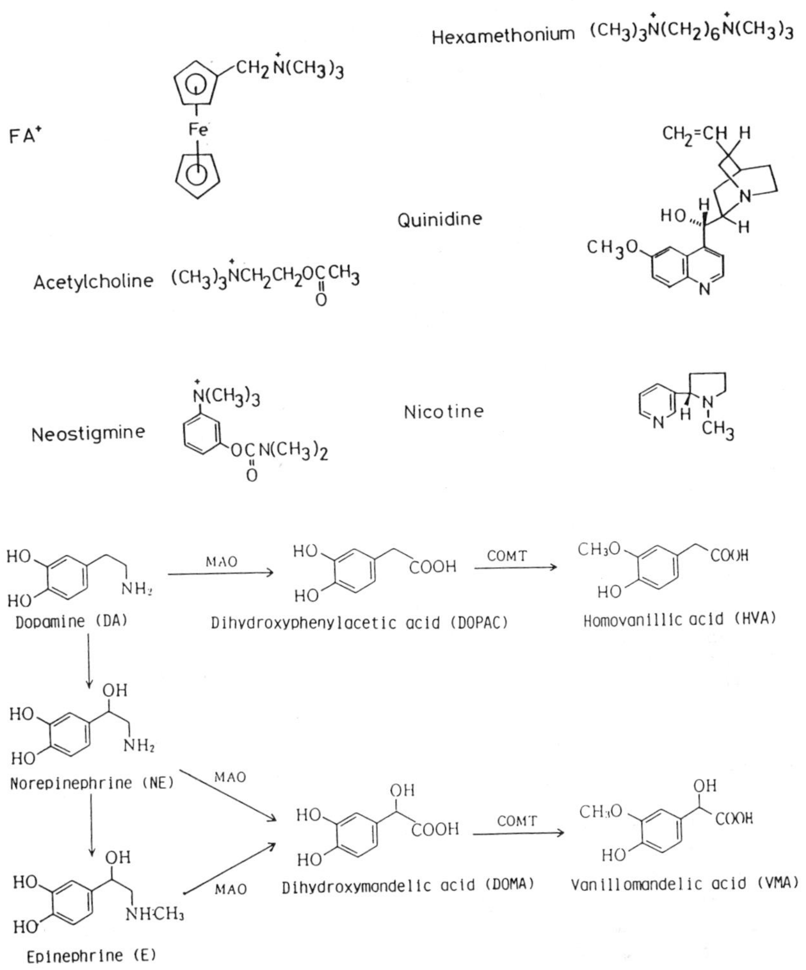

Figure 1. Structures of drugs of interest.

be uneven, FA^+ in the outside solution equilibrates fairly rapidly. Near equilibrium for the cation exchange between Na^+ and FA^+ was attained within 10 min after the Nafion-coated electrode was dipped into the solution containing FA^+. The apparent half-wave potential of FA^+ observed on the Nafion-coated electrode was 0.34 V vs SCE, which is ca. 50 mV negative of the value on a bare GC electrode. This difference in the half-wave potential indicates the stabilization[12-14] of the oxidized species by electrostatic interaction with the negatively charged sulfonato group. This phenomenon contrasts well with that observed on a perfluorinated carboxylic acid polymer[13].

The concentration of FA^+ inside the Nafion membrane is very sensitive to the presence of other cations. Figure 2 shows cyclic voltammograms observed with the Nafion-coated GC electrode with 7.2×10^{-6} M FA^+ and different concentrations of hexamethonium (autonomic ganglionic blocking drug) in the electrolyte solution. The addition of hexamethonium resulted in a decrease in the peak current for FA^+. This phenomenon clearly demonstrated that some FA^+ initially trapped inside the Nafion membrane is replaced by hexamethonium added to the electrolyte solution (Figure 3).

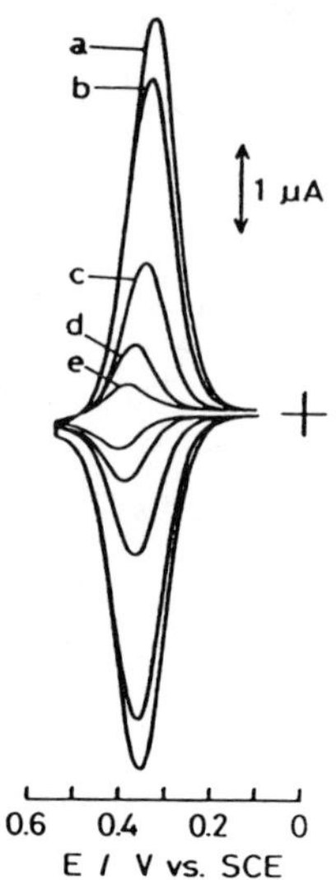

Figure 2. Cyclic voltammograms for FA^+ in the presence of hexamethonium on the Nafion-coated electrode. Concentration of hexamethonium: a) 0; b) 1.0 x 10^{-6}; c) 1.0 x 10^{-5}; d) 1.0 x 10^{-4}; e) 1.0 x 10^{-3} M. Scan rate: 0.05 V/s. Concentration of FA^+: 7.2 x 10^{-6} M.

The extent of the decrease in the peak current was dependent on the concentration of hexamethonium in the solution and thus the determination of concentration becomes possible by using this phenomenon. As shown in Figure 2, the presence of 1.0 x 10^{-6} M hexamethonium in solution induces a clear decrease in the peak current for FA^+. We have used several other cationic drugs (acetylcholine (cholinergic drug), neostigmine (cholinergic drug), nicotine (autonomic ganglionic blocking drug) and quinidine (anti-arrhythmic drug)) to clarify the influence of these drugs on the voltammetric response of FA^+ and to investigate the applicability of Nafion-

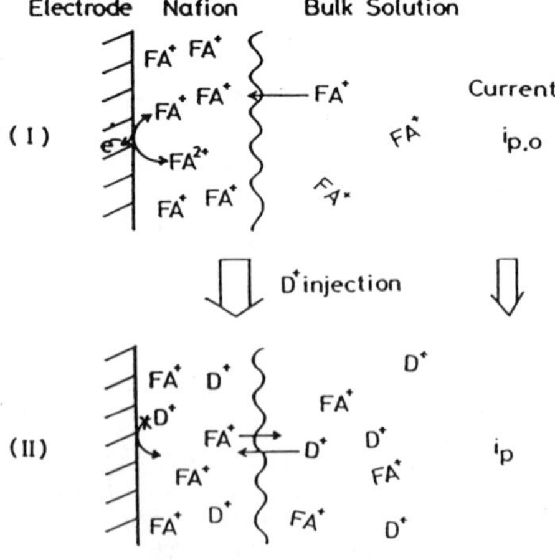

Figure 3. Principle of the electrochemical determination of drugs.

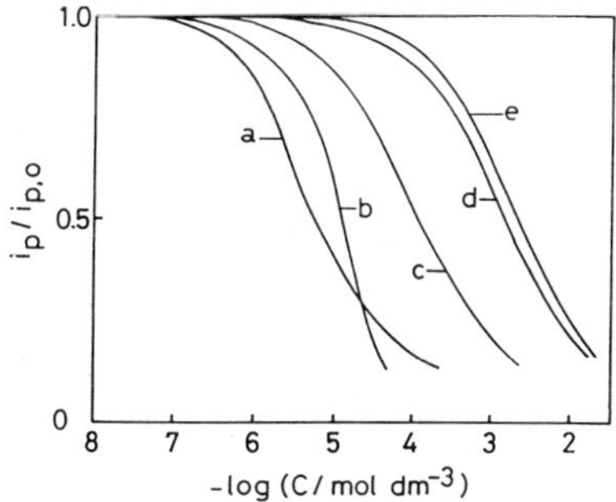

Figure 4. Relationship between $i_p/i_{p,o}$ and the concentration of drugs. a) Hexamethonium; b) Quinidine; c) Neostigmine; d) Nicotine; e) Acetylcholine.

coated electrodes for the determinatin of these drugs. The drugs induced a decrease in the oxidation peak current for FA^+ as was observed for hexamethonium. Figure 4 demonstrates the relationship between the $i_p/i_{p,o}$ value (where i_p and $i_{p,o}$ are the peak currents with and without drugs, respectively) and the concentration of several drugs in the electrolyte solution. All drugs investigated here show a clear decrease in the $i_p/i_{p,o}$ value in the fairly low concentration ranges of the drugs. Therefore, the concentration of a drug can be determined by using these curves. The detection limits for the drugs in the present procedure are as follows: hexamethonium, 2×10^{-7}; quinidine, 5×10^{-7}; neostigmine, 2×10^{-6}; nicotine, 1×10^{-5}; acetylcholine, 5×10^{-5} M.

In these measurements, cyclic voltammograms were recorded after 5 min immersion of the Nafion-coated electrode into the sample solution containing a specific amount of a drug. Hexamethonium, acetylcholine and neostigmine equilibrated fairly rapidly and a near equilibrium state was obtained after 5 min of immersion. Only small changes in the voltammetric peak current were observed on subsequent scans. However, ca. 30 min was necessary to reach a near equilibrium state for quinidine. These differences in the equilibration time seems to be mainly dependent on the size of the cation. For analytical purposes it is not necessary to obtain an equilibrium response. The same measurements were carried out using a number of Nafion-coated GC electrodes. The results were quite reproducible and the variation of the relative peak current under different runs were within 5%. When the Nafion-coated GC electrode used for these measurements

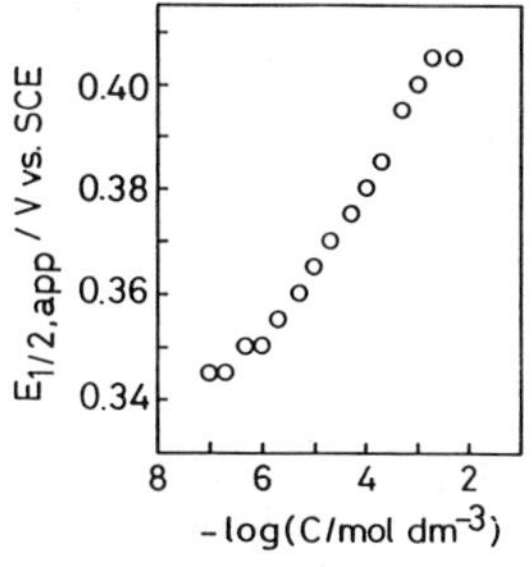

Figure 5. Relationship between $E_{1/2,app}$ and the concentration of hexamethonium.

was transferred to an electrolyte solution containing 7.2 x 10^{-6} M FA^+ without any drug, the anodic peak current recovered to its original value. The Nafion film on the GC surfce seems to be quite stable and the electrode can be used for at least several measurements.

The apparent half-wave potential ($E_{1/2,app}$) of FA^+ on the Nafion-coated GC is dependent on the loading level of hexamethonium. Figure 5 shows the relationship between the $E_{1/2,app}$ value and the concentration of hexamethonium in the electrolyte solution. The half-wave potential (0.34 V vs SCE) observed on the Nafion-coated GC without hexamethonium is negative compared to the value on a bare GC (0.392 V vs SCE). The $E_{1/2,app}$ value shifts to positive potentials with increasing concentration of hexamethonium and surpasses the value on a bare GC at hexamethonium concentration higher than 10^{-3} M. At the low concentrations of hexamethonium, the $E_{1/2,app}$ value is governed by the stabilization[11-14] of the oxidized FA^+ by the favorable electrostatic interaction of the divalent cation (FA^{2+}) with the sulfonate groups in the Nafion film. However, incorporation of hexamethonium into the hydrophilic phase in the Nafion film provides a more hydrophobic environment relative to the original state[14]. This environmental change obviously destabilizes the highly charged species, which implies a positive shift of the $E_{1/2,app}$ of FA^+ entrapped in the Nafion film. Therefore, the shift in $E_{1/2,app}$ is caused by the compensation of extra stability due to electrostatic interaction by the increasing hydrophobicity of the ionic cluster[14]. This trend was also observed for other drugs.

The incorporation of bulky drugs also caused changes in the cyclic voltammetric shapes of FA^+. Voltammograms in the presence of quinidine in solution showed obvious diffusional tails after the anodic and cathodic peaks. In addition, the peak separation in this case was ca. 50 mV at 100 mV s^{-1}, which can be compared to ca. 20 mV observed in the original voltammogram. The peak separation gradually increased with increasing concentration of quinidine. These changes in the voltammetric shape are probably caused by the decrease in the apparent diffusion coefficient of FA^+ in the Nafion film since the anodic peak current is no longer linearly dependent on scan rate. The incorporation of bulky quinidine into the ionic cluster will decrease the hydrodynamic mobility of FA^+ in the hydrophilic region[14]. On the other hand, such an effect on the cyclic voltammetric shape in the presence of hexamethonium in solution was relatively small.

The present procedure for determination of cationic drugs is based on the competitive partitioning effect between FA^+ and the drug into Nafion, and thus the detection limits are largely dependent on the strength of the interaction. This interaction is correlated with the ion-exchange selectivity coefficient, K^D_{Na} [4], which is the equilibrium constant for the following ion-exchange reaction:

$$D^{n+} + n(SO_3^-Na^+)_{Naf} \rightleftharpoons [(SO_3^-)_nD^{n+}]_{Naf} + nNa^+ \tag{1}$$

and given by

$$K^D_{Na} = x_D(C_{Na})^n/(x_{Na})^nC_D \tag{2}$$

where C_{Na} and C_D are the concentrations of Na^+ and a drug in an aqueous solution and x_{Na} and x_M are the equilibrium ionic fractions of $-SO_3^-$ sites occupied by each ion, respectively. Thus, for monocationic drugs the following simple equation is derived:

$$x_{FA,o}/x_{FA} = k^D_{Na}C_D/(C_{Na}+K^{FA}_{Na}C_{FA}) + 1 \tag{3}$$

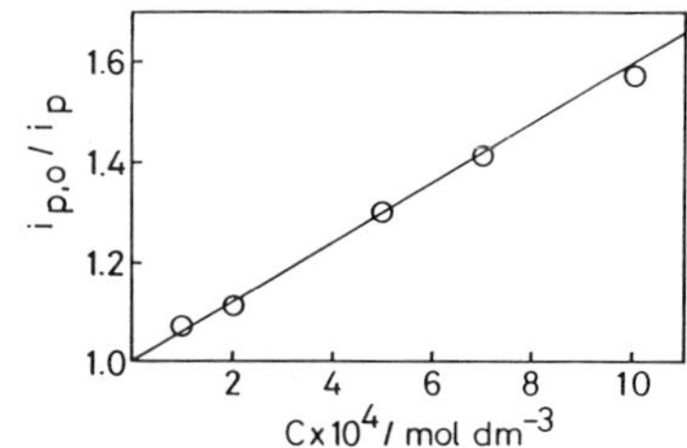

Figure 6. Relationship between $i_{p,o}/i_p$ and the concentration of acetylcholine.

where $x_{FA,o}$ and x_{FA} represent the x values for FA^+ without and with a drug, respectively. The $x_{FA,o}/x_{FA}$ value is directly proportional to the ratio of the peak current, $i_{p,o}/i_o$, observed at steady-state. Figure 6 shows plots for $i_{p,o}/i_p$ vs concentrations of quinidine and acetylcholine at low concentration ranges. The plots show quite good linearity and the K^D_{Na} values can be determined from the slopes of the plots by using the K^{FA}_{Na} value (7.3 x 10^4) reported by Szentirmay and Martin[4]. Since the experimental points deviate from the line at high concentrations, the measurements were restricted to relatively low concentration ranges. The deviation may be indicative of a structural change of Nafion. Hexamethonium is a dicationic drug and the plot does not show a linear relationship. The K^D_{Na} value in this case has been determined by curve fitting between the experimental and theoretical current ratios. Table 1 shows the ion-selectivity coefficients. It is obvious that these drugs have a strong affinity for Nafion. The K^D_{Na} value seems to be related to the hydrophobicity of the drugs. In a series of monocationic drugs, relatively hydrophobic quinidine shows a large value, whereas hydrophilic acetylcholine shows a small value.

Determination of electroinactive drugs in flow systems

The above phenomena have been applied to determinations by FIA and HPLC. The flow system used in HPLC measurements is shown in Figure 7. The detector electrode surface was covered with Nafion. The mobile phase contained FA^+ and a steady oxidation current was observed. When an injected cationic drug flows over the Nafion coated electrode maintained at an oxidizing potential for FA^+ some of the drug penetrates into the Nafion film. This invasion of the drug perturbs the equibrated state between FA^+ and oxidized FA^+ (FA^{2+}) in the Nafion film. This change in the concentra-

Table 1. Ion-selectivity coefficients for the drugs.

Cation	$K^{D^{n+}}_{Na^+}$
FA^+	7.3 x 10^4
Acetylcholine (n=1)	5.6 x 10^2
Hexamethonium (n=2)	2.0 x 10^5
Quinidine (n=1)	1.0 x 10^5
Neostigmine (n=1)	2.8 x 10^4
Nicotine (n=1)	3.7 x 10^3

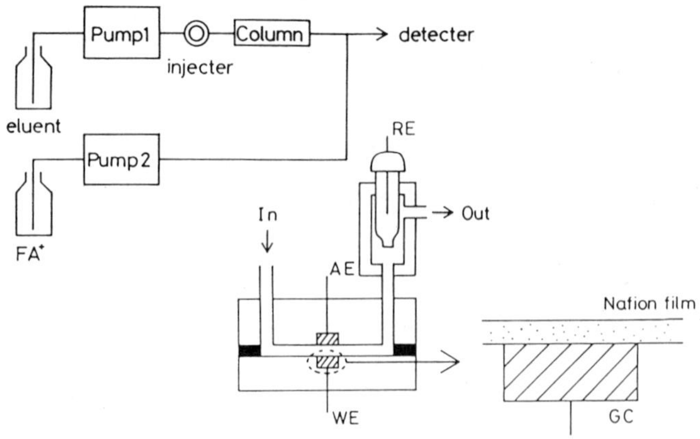

Figure 7. Schematic representation of the flow system.

tion induces a change in the oxidation current for FA^+. When the electrode potential was set at 0.70 V vs Ag/AgCl, at which FA^+ is fully oxidized, a positive peak was observed on passage of an electroinactive drug. The invasion of the drug causes a decrease in the concentration of the oxidized FA^+ in the Nafion layer and then FA^+ flows back into the Nafion film after passage of the drug. The observed positive peak is probably caused by oxidation of the inflowed FA^+. The response was greatly dependent on the potential applied to the detector electrode. At 0.40 V vs Ag/AgCl a negative peak appeared, corresponding to a cationic drug, and sensitivity was better than that observed at 0.70 V vs Ag/AgCl. Figure 8 shows the responses at various electrode potentials on injection of hexamethonium. The dependence of the peak current on the electrode potential is shown in Figure 9. It is obvious that the peak current depends on potential, and the maximum response for hexamethonium is observed at ca. 0.40 V vs Ag/AgCl, which is close to the apparent formal potential. As described in the above section, the apparent formal potential shifts to more positive values with increasing concentration of a cationic drug in the solution phase. Therefore, the peak observed at 0.40 V vs Ag/AgCl is probably correlated with the change in the apparent formal potential. When the apparent formal potential shifts to a positive value as a result of the

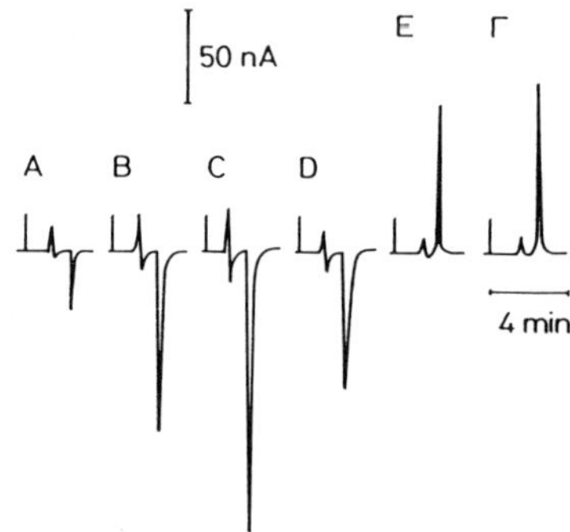

Figure 8. Chromatograms for 1.0×10^{-3} M hexamethonium. Applied potential: A) 0.31; B) 0.35; C) 0.40; D) 0.44; E) 0.50; F) 0.70 V vs Ag/AgCl. Solution A: Acetate buffer (pH 5.3), 1.0 ml/min. Solution B: 1.0×10^{-4} M FA^+, 0.1 ml/min.

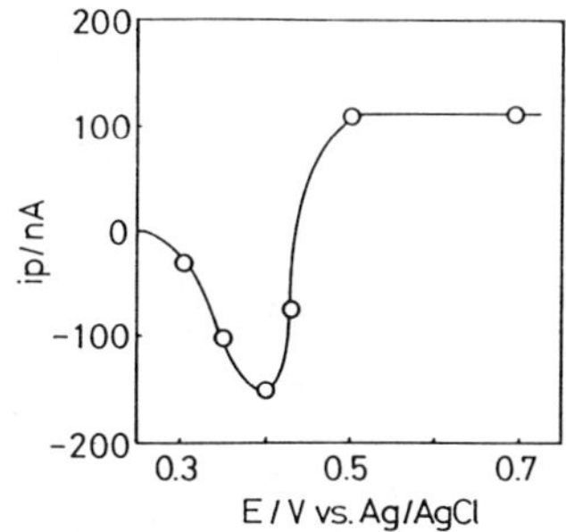

Figure 9. Dependence of the peak current on electrode potential.

effect of a cationic drug, the oxidized FA^+ present in the Nafion film is suddenly reduced back to FA^+, which causes a negative current peak. The present system has also been applied to an FIA system. Almost the same current response was observed and the peak current increased linearly with the concentration of hexamethonium (1×10^{-6} M - 1×10^{-4} M). The present system needs further investigation and improvement in various analysis conditions will undoubtedly bring the detection limit to lower values.

Since all of the drugs employed here were electroinactive in a practical potential range, determination by a direct electrochemical method is difficult. In addition, neither acetylcholine nor hexamethonium has a chromophore in the molecule and thus the determination by adsorption spectroscopy is also difficult. It should be emphasized that the simple method proposed here can be used for determination of these drugs and will be applicable to many cationic drugs.

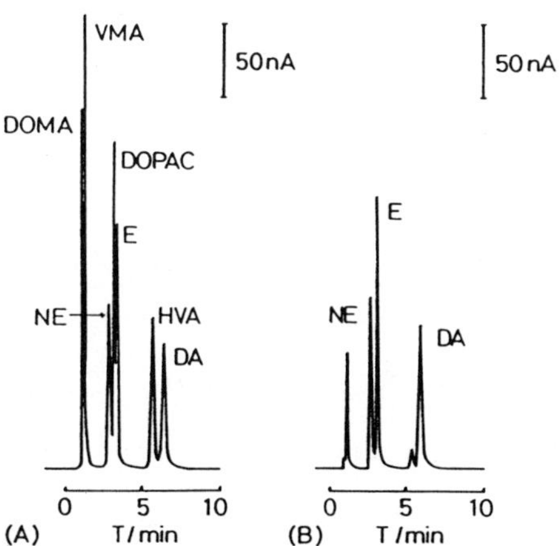

Figure 10. Chromatograms for 1.0×10^{-5} M catecholamines and their metabolites at the bare GC (A) and the Nafion-coated GC (B). Applied potential: 0.80 V vs Ag/AgCl. Mobile phase: 0.1 M acetate buffer (pH 5.3). Flow rate: 1.0 ml/min.

Determination of catecholamines in a flow system

Catecholamines are deeply involved in signal transmission through synaptic regions. Therefore, the determination of catecholamines is important in studying nervous systems and also for clinical purposes. Although various methods for the determination have been proposed, electrochemical methods seem to be the most popular at the present. Since the samples from biological sources (e.g., brain tissue, serum, urine etc) contain various metabolits, HPLC has usually been employed for analysis of catecholamines. Although various compositions for mobile phases have been proposed, a relatively long analysis time is necessary to obtain a good separation of catecholamines and their metabolites. It is desirable to detect catecholamines selectively to reduce the analysis time. The metabolites are formed by an enzymatic deamination reaction and are anions at neutral pH. However, catecholamines are organic cations in such a pH region due to the protonation of the amino group. This difference can be used for the selective determination of catecholamines on Nafion-coated electrodes since Nafion rejects anions[2]. We have investigated the selective determination of catecholamines in the presence of their metabolites by HPLC analysis. The flow electrochemical cell was the same as shown in Figure 7. The electrode was coated with Nafion by spin coating. Figure 10 shows chromatograms of a sample solution containing catecholamines and metabolites obtained at bare and Nafion-coated electrodes (film thickness: ca. 100 nm). It is obvious that the surface Nafion acts effectively as a permeation barrier for the metabolites. Under the present conditions, determination of epinephrine and norepinephrine on the bare electrode is very difficult because a peak for dihydroxyphenylacetic acid, a metabolite, appears between the two peaks for the catecholamines. However, the interfering peak disappears in the chromatogram obtained at a Nafion-coated electrode, which makes it possible to determine epinephrine and norepinephrine. Another important observation is that peaks for the catecholamines on the Nafion-coated electrode is slightly higher than those on the bare electrode. This phenomenon indicates accumulation of catecholamines in the Nafion film by electrostatic interaction during passage over the electrode surface. The thickness of the Nafion film affected the chromatographic response. With a thicker film, the peaks for catecholamines were relatively small and broad though the undesired peaks for metabolites are almost eliminated. With a thinner film, it is difficult to remove the peaks for metabolites. The thickness of 100-200 nm seems to be suitable for determination of catecholamines. It is emphasized that the analysis time can be effectively reduced by the present method. It takes more than 10 min to obtain complete separation of the catecholamines and metabolites on the bare electrode.

REFERENCES

1. J. Wang and L.D. Hutchins, Anal. Chem., 57, 1536 (1985).

2. J. Wang, P. Tuzhi and T. Golden, Anal. Chim. Acta, 194, 129 (1987).

3. Y. Ikariyama and W.R. Heineman, Anal. Chem., 58, 1803 (1986).

4. M.N. Szentirmay and C.R. Martin, Anal. Chem., 56, 1898 (1984).

5. G. Nagy, G.A. Gerhardt, A.F. Oke, M.E. Rice, R.N. Adams, R.B. Moore III, M.N. Szentirmay and C.R. Martin, J. Electroanal. Chem., 188, 85 (1985).

6. T. Matsue, U. Akiba and T. Osa, Anal.Chem., 58, 2096 (1986).

7. C.R. Martin, I. Rubinstein and A.J Bard, J. Am. Chem. Soc., 104, 4817 (1982).

8. N. Oyama and F.C. Anson, J. Electrochem. Soc., 127, 247 (1980).

9. J. Ye, R.P. Baldwin and K. Ravichandran, Anal. Chem., 58, 2340 (1986).

10. A.J. Bard and L.R. Faulkner, Electrochemical Methods, John Wiley & Sons, New York, 1980, Chapter 12.

11. C.R. Martin and K.A. Dollard, J. Electroanal. Chem., 159, 127 (1983).

12. H. Braun, W. Storck and K. Doblhofer, J. Electrochem. Soc., 130, 807 (1983).

13. Y.H. Tsou and F.C. Anson, J. Electrochem. Soc., 131, 595 (1984).

14. I. Rubinstein, J. Electroanal. Chem., 188, 227 (1985).

ADSORPTIVE STRIPPING VOLTAMMETRIC STUDIES OF BIOLOGICALLY-SIGNIFICANT COMPOUNDS

Joseph Wang

New Mexico State University
Las Cruces, New Mexico 88003

INTRODUCTION

Over the last five years adsorptive stripping voltammetry has evolved and grown into a very powerful and versatile electroanalytical technique[1]. In this version of stripping analysis the preconcentration step is based solely on the interfacial accumulation of the analyte at the electrode surface. The voltammetric response of the surface-confined species is directly related to the surface concentration, with the adsorption isotherm providing the relationship between the bulk and surface concentrations of the adsorbate. The surface-active characteristics of numerous compounds of biological significance (that commonly complicate their conventional voltammetric measurements) can be exploited for obtaining effective adsorptive accumulation. Extremely low levels of compounds such as riboflavin, tetracyclines, methotrexate, cardiac glycosides, mitomycin C, tricyclic antidepressants, cytochrome c, or insulin, can thus be determined at mercury and carbon electrodes. Besides the powerful bioanalytical utility, a useful knowledge of the interfacial and redox behaviors of these, and other, biological compounds is obtained, hence providing valuable insights into their biological function. The interfacial accumulation of macromolecules, such as cytochrome c or ferritin, results in a more favorable interaction between the electrode and the redox center of the molecule, thus allowing direct electrical communication. Adsorptive stripping voltammetric studies of biologically-significant compounds will be described in the following sections, together with the fundamental characteristics of the technique.

PRINCIPLES

Adsorptive stripping voltammetry is similar to conventional stripping voltammetry in that the analyte is preconcentrated onto the working electrode prior to the voltammetric measurement. However, it differs from the conventional scheme in the approach used to preconcentrate the analyte. Unlike the electrolytic process employed in conventional stripping voltammetry, adsorptive accumulation (that does not include any faradaic process) forms the basis for adsorptive stripping voltammetry.

To achieve maximum sensitivity in the subsequent voltammetric response, optimum conditions for maximum adsorption should be utilized during the preconcentration step. The amount of analyte accumulated on the electrode surface is a composite of many variables such as the solvent, electrode material, potential, time, pH, ionic strength, mass-transport, and temperature. Different analytes would respond differently to changes in these parameters, which could be useful analytically. In general, only strongly adsorbing species with relatively large ($>10^{-4}$ cm) adsorption coefficients are suitable for adsorptive stripping measurements.

The rate of forming the absorbed layer is affected by the rate of transporting the analyte from the bulk of the solution to the electrode surface and the rate of the actual adsorption. The slower of these two processes then becomes the rate controlling step in the accumulation process. For most of the biological compounds determined by the adsorptive approach, the adsorption at the surface is fast and the overall rate of accumulation is governed by mass-transport. Hence, as in conventional stripping voltammetry, forced convection is employed during the preconcentration step.

The voltammetric response of the surface-confined analyte is directly related to its surface concentration. Using linear scan measurements, under ideal Nernstian behavior, the faradaic peak current arising from the presence of Γ mol/cm^2 of an attached analyte is given by

$$i_p = \frac{n^2 F^2 \Gamma A v}{4RT} \qquad (1)$$

where v is the scan rate, and n, F, A, R and T have conventional symbology.

Adsorption isotherms provide the equilibrium relationship between concentration of the adsorbate on the surface and in the bulk of the solution. The most frequently used isotherm is that of Langmuir:

$$\Gamma = \Gamma_m \left(\frac{BC}{1 + BC} \right) \qquad (2)$$

where Γ_m is the surface concentration corresponding to a monolayer coverage, and B is the adsorption coefficient. At low adsorbate concentrations, $1 \gg BC$, and the surface concentration is directly proportional to the bulk concentration. Hence, a linear response is commonly observed at low analyte levels (10^{-7} - 10^{-10}M) for which the method is usually applied.

The surface coverage (Equation 2) and hence the resulting adsorptive stripping response (Equation 1), differ from compound to compound based upon its polarity, size or orientation. It is this dependence of the response upon the adsorbate properties that provides useful information in the studies of biologically-significant compounds.

BIOANALYSIS

The clinical need for measuring extremely low levels of compounds of pharmaceutical and biological significance was the main motivation for the development of adsorptive stripping voltammetry. In particular, the adsorptive approach has been shown to be very attractive as a highly sensitive, rapid, and inexpensive method for measuring potent drugs (anticancer, cardiovascular, etc.) with low and narrow therapeutic margins.

A great number of adsorptive stripping procedures for measuring low levels of important biological and pharmaceutical compounds have appeared during the last several years. These studies include trace measurements of reducible compounds, such as digoxin, tetracycline, diazepam, mitomycin C, streptomycin or folic acid, as well as quantitation of the oxidizable phenothioazine and tricyclic-antidepressant compounds. These and other schemes are summarized in Table 1. Measurements of reducible compounds are performed at the hanging mercury electrode, while carbon electrodes are usually employed when oxidizable compounds are concerned. The hanging mercury drop electrode (especially with the static mercury drop design) offers the advantages of reproducible surface area and self-cleaning properties. In addition, because of its low background current extremely low detection limited are obtained. For example, Figure 1 illustrates differential pulse voltammograms obtained after successive standard additions of riboflavin, each addition affecting a 5×10^{-9} M increase in concentration. Convenient quantitation at the nanomolar level is possible following one minute accumulation. Longer preconcentration periods yield a detection limit of 2.5×10^{-11} M ribofavin[14]. Sub-nanomolar detectability has been reported for many other reducible compounds (see Table 1). With carbon electrodes (used for measurements over the anodic potential range) a cleaning step may be required when the analyte is not desorbed during the voltammetric scan. In addition, higher detection limits (approx. 1×10^{-9} M) and lower reproducibility (5-10% RSD) are observed, compared to analogous measurements at the hanging mercury drop electrode. An interesting observation is the penetration of phenothiazine compounds into the interior of carbon paste electrodes, indicating a mixed adsorptive/extractive accumulation process. Surface-active biological compounds exhibiting both oxidation and reduction processes (such as the antineoplastic agents daunorubicin and methotrexate) can be studied at both carbon and mercury surfaces, with each electrode offering certain advantages.

In addition to procedures for low molecular weight compounds, recent studies have demonstrated that utility of adsorptive stripping voltammetry for measuring macromolecules of biological significance. For example, the adsorptive accumulation of DNA has allowed an enhancement of the sensitivity by two orders of magnitude compared to pulse polarographic procedures[18]. Other studies of biomacromolecules are described in the following sections.

The susceptibility of the method to the presence of surfactants could pose serious problems for direct assays of samples of biological origin. Various approaches to correct for these coadsorption effects have been suggested. A proper choice of the preconcentration potential or time and

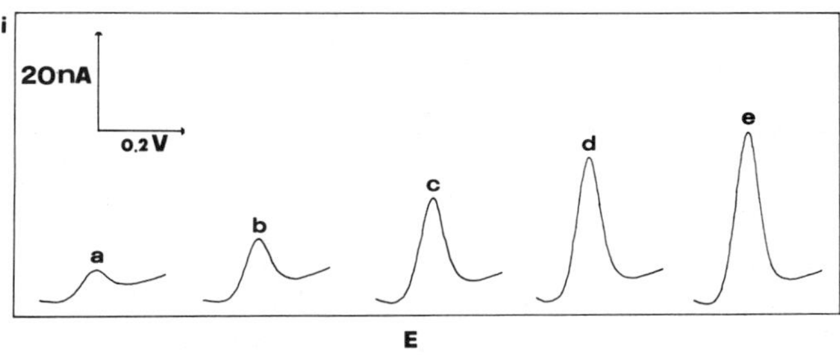

Figure 1. Voltammograms obtained for solutions of increasing riboflavin concentration in 5×10^{-9} M steps (a-e). Preconcentration for 60 s at -0.2 V.

Table 1. Examples of Adsorptive Stripping Procedures for Measuring Small Compounds of Biological and Pharmaceutical Significance.

Compound	Working Electrode	Supporting Electrolyte	Detection Limit	Ref.
Bilirubin	HMDE*	Sodium acetate	$5x10^{-10}M$	2
Chlorpromazine	Carbon paste	Phosphate buffer	$5x10^{-9}M$	3
Cimetidine	HMDE	Hydrochloric acid	$4x10^{-9}M$	4
Codeine	HMDE	Sodium hydroxide	$1x10^{-8}M$	5
Desipramine	Glassy carbon	Phosphate buffer	$1.7x10^{-8}M$	6
Diazepam	HMDE	Acetate buffer	$5x10^{-9}M$	7
Digoxin	HMDE	Sodium hydroxide	$2x10^{-10}M$	8
Folic acid	HMDE	Sulfuric acid	$1x10^{-10}M$	9
Methotrexate	HMDE	Phosphate buffer	$2x10^{-9}M$	10
Mytomycin C	HMDE	Borate buffer	$2x10^{-9}M$	11
Progesterone	HMDE	Sodium hydroxide	$2x10^{-10}M$	12
Reserpine	Carbon paste	Phosphate buffer	$2x10^{-9}M$	13
Riboflavin	HMDE	Sodium hydroxide	$2.5x10^{-11}M$	14
Streptomycin	HMDE	Sodium hydroxide	$7x10^{-10}M$	15
Tetracycline	HMDE	Borate buffer	$2x10^{-9}M$	16
Vitamin B_{12}	HMDE	Ammonium acetate	$2x10^{-9}M$	17

* HMDE = hanging mercury drop electrode

the standard additions method can minimize surfactant interferences when modest (10-50%) peak depressions are involved. More severe peak depressions usually require a separatory technique to isolate the analyte from coadsorbing interferences. For example, Hernandez et al.[19] recently reported the determination of bromazepam and camazepam in human serum following a solvent extraction separation. Alternatively, it is possible to perform the separation step in-situ, at the electrode surface, using a permselective polymeric coating such as cellulose acetate[20]. Interferences due to solution-phase electroactive species can be easily corrected by transferring the electrode from the complex sample to a blank solution before the measurement step. For example, such a medium-exchange procedure has allowed trace measurements of dopamine in the presence of a large concentration excess of ascorbic acid[21]. By using the medium-exchange procedure, it is also possible to evaluate the interfacial behaviour of biological molecules under more versatile conditions. The manifold of flow injection systems is particularly suitable for performing preconcentration/medium-exchange/voltammetric measurement, as it allows rapid assays of small-sample volumes, and convenient achievement of medium exchange[3].

ACHIEVEMENT OF ELECTRICAL COMMUNICATION

In addition to its quantitative aspects, adsorptive stripping voltammetry provides important contributions to our knowledge of biological compounds. In particular, considerable recent activity in our laboratory has focussed on the achievement of direct electron transfer for various biomacromolecules. The strategy here is to form the "protein/-electrode complex", essential for a facile redox process, using an unmodified electrode. Electron-transfer rates are known to decay rapidly (exponentially) upon increasing the distance between the electrode and the redox center of the biomacromolecule. Binding of such molecules to the surface may thus be effective for electron-transfer enhancement. Other laboratories have concentrated on the use of electrode modifiers (e.g., 4.4'-bipyridine) for facilitating the binding and hence the redox process. Our work indicates that similar advantages can be achieved following adsorptive accumulation at an ordinary mercury drop electrode. Using adsorptive stripping voltammetry, adsorption-related conformational changes occur during the preconcentration step. As a result of the preferred orientation, the overall free energy of activation for electron transfer (during the subsequent measurement step) is significantly lower.

For example, knowledge of the redox behavior of the iron storage protein ferritin can reveal important information regarding its biological function. Because the iron core of ferritin is located deep within the protein (surrounded by a 25 Å thick shell), a redox process cannot be observed using conventional voltammetric procedures. It is possible, however, to establish direct electrical communication between the iron core of native ferritin and the hanging mercury drop electrode following a controlled adsorptive accumulation. Figure 2 shows differential pulse voltammograms for 7×10^{-8} M ferritin following different preconcentration periods. While voltammetric response is not observed in solution-phase measurements (curve a - no accumulation), well-defined peaks are observed following the adsorptive accumulation (curves b and c). It appears that the adsorption process brings the iron core sufficiently close to the surface to allow electron transfer at practical rates. The data of Figure 2 also indicate that extremely low (nanomolar) levels of proteins can be measured as a result of the adsorptive accumulation.

In another study, submicromolar levels of the heme protein cytochrome C were measured following controlled adsorptive accumulation at the hanging mercury drop electrode[22]. The redox behavior of this protein has been the subject of intense studies. The controlled interfacial accumulation of cytochrome C brings the heme group close to the mercury surface. Hence, well-defined peaks are observed (Figure 3); analogous measurements without preconcentration (dotted lines) do not permit quantitation. The electron transfer of the adsorbed cytochrome C is not restricted to the first monolayer but can proceed, through this layer, to the layers formed on top. The consequence of this behavior is a favorable adsorptive stripping response, as saturation effects are minimized. The "complexes" of cytochrome c or ferritin with the mercury electrode are useful analogs of protein/protein complexes, which are the basis of many biological redox processes.

Adsorptive stripping procedures for measuring other biomacromolecules have been developed recently in our laboratory. As a result, submicromolar and nanomolar levels of ferredoxin, chlorophyll, insulin, trypsin and

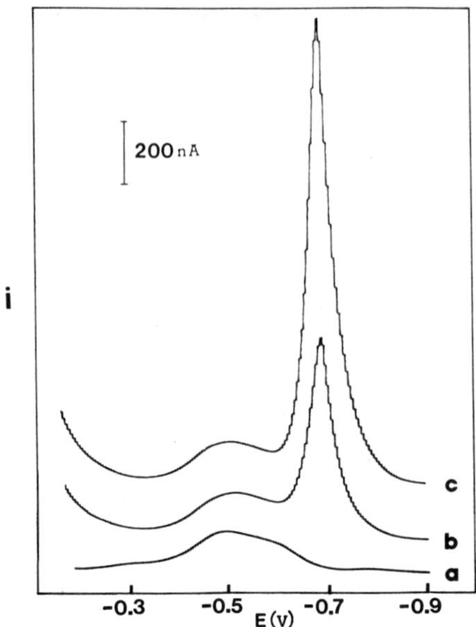

Figure 2. Voltammograms for 8 x 10^{-7} M ferritin after different preconcentration times: (a) 0; (b) 300; (c) 600 s. Preconcentration potential, +0.05 V. Electrolyte, phosphate buffer (pH 7.4).

chymotrypsin can be easily quantified. Because of the strong adsorption of these large compounds, the interference of co-existing surfactants is substantially lower (compared to that observed when small molecules are measured).

INTERFACIAL BEHAVIOR OF BIOLOGICALLY IMPORTANT COMPOUNDS

Voltammetry is one of the few techniques which can provide information on behavioral changes in biological molecules upon their interaction with a charged surface, which is of great significance in living organisms.

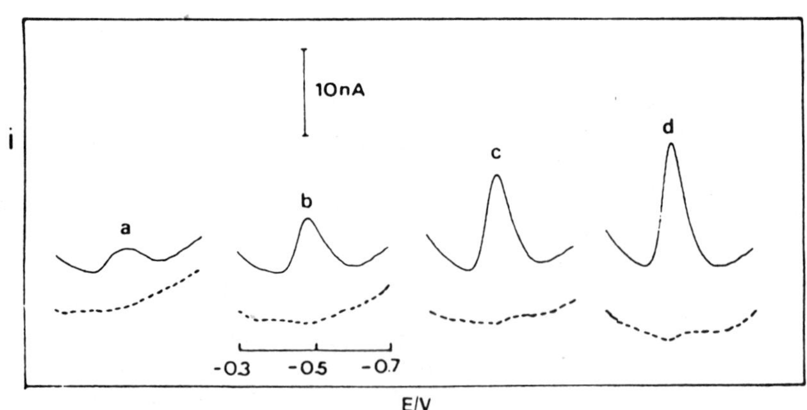

Figure 3. Voltammograms obtained for solutions of increasing cytochrome C concentration, 1.0 -4.0 μM (a-d). Preconcentration for 180 s at -0.10 V.

The basis for this information is the similarity between the charged electrode-solution interface and a cell surface-biological fluid interface. Conventional polarographic techniques and the dropping mercury electrode are not well suited for such interfacial studies. For example, in numerous instances the solution concentration of the bioactive substance is low, and complete adsorption equilibrium cannot be attained during the life of a drop. In contrast, controlled adsorptive accumulation at a stationary electrode (as in adsorptive stripping measurements) allows longer times for adsorption equilibrium to occur and produces a large subsequent faradaic response. Such attainment of adsorption equilibrium permits interpretation of adsorption phenomenon on the basis of thermodynamic arguments. In addition, interfacial data can be achieved at significantly lower concentrations (compared to those used in conventional polarographic experiments). This may be of interest when adsorbates exhibiting concentra-ation-induced orientational transitions are concerned, and is relevant to studies of the activity of drugs (with low therapeutic levels). This, and other properties, make adsorptive stripping voltrammetry particularly useful for obtaining new knowledge on the behavior of biologically important compounds.

Consider, for example, the structurally similar tricyclic antidepressants shown in Figure 4. These compounds exhibit a different adsorptive stripping response because of their different interfacial properties[6]. For example, Figure 5 shows adsorptive stripping calibration plots for the compounds shown in Figure 4. For trimipramine (c), the peak current increases linearly with concentration. In contrast, desipramine and imipramine exhibit a curvature in the response at high concentrations. Such in-situ generated adsorption isotherms indicate different extents of surface coverage. The latter was estimated from controlled-adsorption/-cyclic voltammetric experiments. Values of 3.9×10^{-10}, 5.9×10^{-10}, and 8.6×10^{-10} mol/cm^2 (corresponding to 43, 32, and 19 nm^2/adsorbed molecule) were calculated for desipramine, imipramine, and trimipramine, respectively. It appears that minor differences in the side chain of antidepressants have a profound effect upon their orientation; imipramine and desipramine yield the expected planar orientation of the tricyclic ring system and the carbon surface, while for trimipramine a vertically oriented adsorbed layer is indicated. Adsorption isotherms may be useful also for obtaining information regarding intermolecular interactions between absorbed molecules, as well as on interactions between adsorbed molecules and the surface. Concentration-induced orientational transitions may also be detected from the construction of adsorptive stripping calibration curves.

The behavior of the anticancer agent cis-dichlorodiammineplatinum (II), or cisplatin, is another example of the utility of the method[23]. The adsorptive-stripping/cyclic voltammetric (saturation) data yield a surface coverage of 5.1×10^{-10} mol/cm^2. Each adsorbed molecule thus occupies an area of 33 Å^2. This value is in excellent agreement with the area (34 Å^2) estimated from the crystal structure of cisplatin. Such agreement suggests the existence of a packed geometrically planar layer of adsorbed cisplatin at the surface.

Additional insights regarding the interaction of biological molecules at biosurfaces can be achieved from various adsorptive stripping parametric experiments involving systematic changes of variables such as preconcentration potential, solution pH, or scan rate. Adsorptive stripping competition experiments may also be very useful. In all cases, the new knowledge of the interfacial behavior is accompanied by simultaneous information regarding heterogeneous electron transfer of the biological compound. Utilizing the medium-exchange concept, the redox process may be evaluated independent of the conditions under which adsorption occurs.

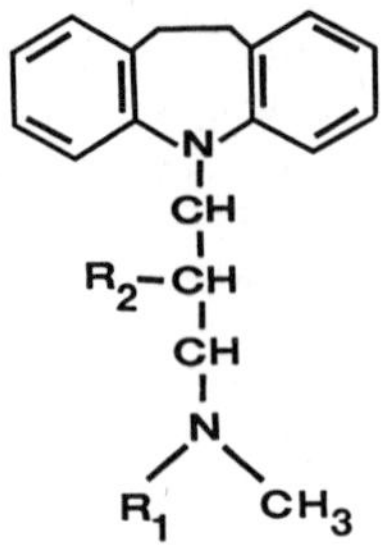

Desipramine	R_1=H	R_2=H
Trimipramine	R_1=CH_3	R_2=CH_3
Imipramine	R_1=CH_3	R_2=H

Figure 4. Structural formulas of tricyclic antidepressants.

Capacitance and optical observations can be very useful for supplementing the adsorptive stripping information.

Apart from small molecules, insights into the adsorption of biomacromolecules are very important for understanding their role in many biological processes. The adsorption of such molecules is considerably more complex than that of small ones, and adsorptive stripping voltammetry has been very useful for obtaining new knowledge of their interaction with electrodes[18,22,24].

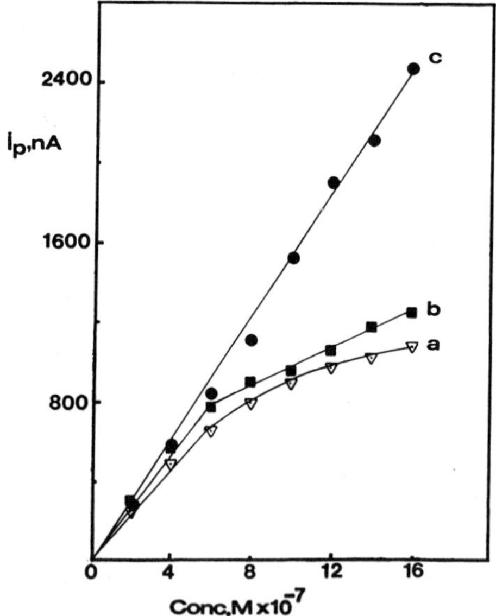

Figure 5. Adsorption stripping calibration plots for imipramine (a), desipramine (b), and trimipramine (c). One-min accumulation at the glassy carbon electrode.

CONCLUSIONS

The results described above show that adsorptive stripping voltammetry is an important tool in investigating the behavior of biological compounds and for detecting such compounds at extremely low levels. The adsorptive approach is thus very attractive from the perspective of both fundamental studies and bioanalytical applications. The range and capabilities of adsorptive stripping voltammetry continue to expand at a rapid rate. For example, it is possible to combine catalytic processes with a controlled interfacial accumulation of the catalyst for ultratrace measurements, with detection limits as low as 10^{-12} M[25]. Similarly, useful information regarding electroinactive biological compounds can be achieved in adsorptive stripping tensammetric studies, where the measured current change is proportional to the change in differential capacitance. Hence, an enormous potential remains for using adsorptive stripping voltammetry in the area of bioelectrochemistry.

ACKNOWLEDGEMENTS

The adsorptive stripping voltammetric studies described in this chapter were supported by the National Institutes of Health, Grant No. GM30913-04.

REFERENCES

1. J. Wang, Am. Lab., 17 (5), 41 (1985).
2. J. Wang, D.B. Lou and P.A.M. Farias, J. Electroanal. Chem. 185, 61 (1985).
3. J. Wang and B.A. Freiha, Anal. Chem., 55, 1285 (1983).
4. A. Weber, M. Shah and J. Osteryoung, Anal. Chim. Acta, 154, 105 (1983).
5. R. Kalvoda, Anal. Chim. Acta, 138, 11 (1982).
6. J. Wang, M. Bonakdar and C. Morgan, Anal. Chem., 58, 1024 (1986).
7. R. Kalvoda, Anal. Chim. Acta, 162, 197 (1984).
8. J. Wang, J.S. Mahmoud and P.A.M. Farias, Analyst, 110, 855 (1985).
9. D.B. Luo, Anal. Chim. Acta, 198, 277 (1986).
10. J. Wang, P. Tuzhi, M.S. Lin and T. Tapia, Talanta, 33, 707 (1986).
11. J. Wang, M.S. Lin and V. Villa, Anal. Lett., 19, 2293 (1983).
12. J. Wang, P.A.M. Farias and J.S. Mahmoud, Anal. Chim. Acta, 171, 195 (1985).
13. J. Wang, T. Tapia and M. Bonakdar, Analyst, 111, 1245 (1986).
14. J. Wang, D.B. Luo, P.A.M. Farias and J.S. Mahmoud, Anal. Chem., 57, 158 (1985).
15. J. Wang and J.S. Mahmoud, Anal. Chim. Acta, 186, 31 (1986).

16. J. Wang, T. Peng and M.S. Lin, Bioelectrochem. Bioenerg., 15, 147 (1986).

17. H. Sawamoto, J. Electroanal. Chem., 195, 395 (1985).

18. E. Palecek, P. Boublikova and F. Jelen, Anal. Chim. Acta, 187, 99 (1986).

19. L. Hernandez, A. Zapardiel, J.A.P. Lopez and E. Bermejo, Analyst, 112, 1149 (1987).

20. J. Wang, M. Bonakdar and M.M. Pack, Anal. Chim. Acta, 192, 215 (1987).

21. J. Wang and B.F. Freiha, J. Electroanal. Chem., 151, 273 (1983).

22. J. Wang and M.S. Lin, J. Electroanal. Chem., 221, 257 (1987).

23. J. Wang, T. Peng and M.S. Lin, Bioelectrochem. Bioenerg., 16, 395 (1986).

24. E. Palecek, Bioelectrochem. Bioenerg., 15, 275 (1986).

25. J. Wang, J. Zadeii and M.S. Lin, J. Electroanal. Chem., 237, 281 (1987).

MODIFIED ELECTRODES USED TO ELUCIDATE ELECTROCHEMICAL BEHAVIORS OF BIOLOGICALLY SIGNIFICANT MOLECULES

G.J. Patriarche*, J.-M. Kauffmann and J.-C. Viré

Free University of Brussels (U.L.B.)
Institute of Pharmacy
Campus Plaine, 205/6, Bd du Triomphe
B-1050 Brussels / Belgium

INTRODUCTION

Recent research efforts have been concentrated on new chemically and non-chemically modified electrode developments. This investigation field has stimulated a great interest to control more and more directly the characteristics of the electrode surface and research into the interactions of molecules of biological importance has brought progress in the understanding of certain electrode mechanisms and interfacial behaviors. Modified, as well as classical electrochemical sensors, are also developing exponentially to produce more and more small portable or disposable devices to detect biologically or pharmacologically important substances in vitro or in vivo.

Murray at the University of North Carolina and Anson at Caltech have been the pioneers in the field of designing molecular films onto electrodes to study electrocatalysis, microstructures, electroreleasing involving expulsion of reagents or species from the electrode surface, preconcentration processes or tentative manufacturing of solid-state electrochemical cells that requires no liquids to function. These subjects have been extensively reviewed[1-5]. This important field has also been covered during the 38th Meeting of the International Society of Electrochemistry[6].

The development of an electrode reaction is highly dependent on the nature of the electrode-solution interface. Most electrochemical processes are still performed using classical metallic or carbon electrodes. However, modifications brought to the surface of the working electrode may result in the enhancement of particular properties which can be exploited in electroanalytical chemistry. In such a way, many electrode materials, such as metals, metal oxides but also carbon based electrodes have been submitted to chemical or non-chemical modifications. Most of the immobilized compounds are electroactive, the electron transfer having to be reversible. They are mainly used as catalysts for electrochemical reactions which cannot be performed at conventional electrodes. In addition, a judicious choice of a catalyst exhibiting a particular structure may also provide more or less specificity towards certain molecules or ions. Electroinactive substances may also be immobilized and act as intermediates for subsequent immobilization. They may exhibit catalytic properties without being oxidized or reduced themselves, or improve selectivity.

Several immobilized species, depending on their structure, may also entrap, more or less selectively, different ions. Such a modified electrode is able to preconcentrate the substrate to be analyzed from the solution, giving rise to a marked increase of sensitivity.

Catalytic properties and also the structural configuration of the immobilized compound often depend on the applied potential. Thus, this parameter can be used to define the best operating conditions. Such a property may be employed to gradually release some compounds from an electrode. This may be of great interest for controlled release of drugs.

METHODS OF IMMOBILIZATION

Three main procedures are generally used to modify electrode surfaces.

The covalent bonding of the immobilized species implies that functional groups must be present on the electrode surface. Metal electrodes have to be treated by oxidation techniques to form metal oxide groups at their surface[7,8]. Metal oxide electrodes such as SnO_2, TiO_2 or RuO_2 can be modified without any pretreatment[8-10]. Carbon electrodes already possess surface groups but additional groups can be created using chemical, thermal or radio-frequency pretreatment[4]. Particular modifications can also be performed using a bridge molecule which links the functional group to the electrode surface. Silane derivatives[11-14] and cyanuric chloride[15-17] have been proposed as linking compounds.

Thermal or mechanical pretreatment in an oxygen-free atmosphere may also activate the electrode surface, allowing the covalent bonding of functional groups[18,19].

Immobilization can also be made through the adsorption process, which is easier to perform but generally gives rise to less stable sensors[20,21]. The resulting adsorbed layer is often thicker than that obtained by covalent bonding.

A third method to modify electrodes is the formation of a polymer layer at the surface of the electrode[14,22-24]. The electrode is dipped into a solution containing the polymer and the solvent is evaporated. Polymerization of the monomer can also be induced at the electrode surface by electrochemical or other physical means.

CHARACTERIZATION OF THE MODIFIED ELECTRODES

A new modified electrode has to be characterized in order to point out its physical, chemical and electrochemical properties. An important field in interfacial electrochemical problems is the determination of the structure of the electrode surface and by this way a great deal of information has been accumulated on adsorbed layers of molecules and ions. Electron microscopy and spectroscopic techniques are employed to identify the structural characteristics of the modified surface. Cyclic voltammetry is largely used to define the electrochemical behaviours of the sensor and to elucidate mechanistic aspects of the electrode reaction. A.C. voltammetry has also been applied[25]. As both spectroscopic and electrochemical results are often required to obtain valuable information, the use of spectroelectrochemical methods may be very useful[26-28].

It is generally accepted that the rate of charge transport through polymer films is basically determined by the rate of the electron exchange between redox centers and that the polymer morphology has also a large influence. This type of material also gives rise to kinetic studies related to the propagation of charges through the coating, diffusion of the substrate and rate of the chemical reaction between the substrate and the mediator[29-35]. A theoretical point of view has been recently described by Andrieux[36].

APPLICATIONS OF POLYMER COATED ELECTRODES

Among polymer coated electrodes which have been the object of active investigations during this last decade, particular attention has been paid to the conductive polypyrrole films, obtained by electrooxidation of pyrrole in acetonitrile. Such electrodes have been used to study the electrochemical behaviours of the quinone-hydroquinone redox couple[37,38] and tetrathiafulvalene[39]. The controlled release of ferrocyanide from polypyrrole by reduction of the polymer has been demonstrated[40]. As an application, electroinactive anions can be determined using a polypyrrole modified electrochemical detector in flow-injection analysis[41]. Pyrrole can be polymerized from aqueous solutions[42]. Enlarging the modification field, the polymerization step may be preceded by a chemical reaction between pyrrole and another substrate[43].

Polymerization of 4-vinylpyridine has also been performed, resulting in a non-electroactive and non-conducting coating[34,44]. The kinetics of charge transport has been measured when this polymer contains electrostatically trapped $Fe(CN)_6^{3-}$, $IrCl_6^{3-}$ or tris(2,2'-bipyridine)osmium-(III/II) complexes[33,35]. Incorporating $IrCl_6^{2-}$, the electrode is able to catalyze the oxidation of iron(II)[45]. A composite electrode coated with a mixture of cellulose acetate and poly(vinylpyridine) has been described[46]. This mixture allows the binding of counterionic reactants in acidic media, while maintaining the size exclusion discriminative properties of cellulose acetate.

A detailed study of the physical and electrochemical properties of electrodeposited poly(vinylferrocene) has been presented[32,47]. The effect of different solvents on the electrochemical response of this polymer, deposited from a radio-frequency plasma, has been reported[48]. The poly(vinylferrocene/ferricinium) system has also been proposed as a reference electrode for non-aqueous media[49].

Structurally similar to polypyrrole, pyrazoline derivatives can be polymerized[50,51]. Spectroelectrochemical experiments show that an optical change accompanies the charge transfer process[28].

Sulfonated films of polystyrene crosslinked with divinylbenzene are capable of ion exchange with $Ru(bpy)_3^{2+}$ or $Ru(NH_3)_6^{2+}$. They offer a marked stability in acetonitrile and water[44]. Tris(2,2'-bipyridine)osmium(III/II) complexes can also be used as electron transfer centers[52]. Poly(nitrostyrene) has also been proposed as a catalyst for the reduction of organic dihalides[53].

Nafion, a perfluorinated sulfonated polymer introduced by Rubinstein and Bard[54,55], which is able to immobilize electroactive cations by electrostatic binding, has been coated on gold electrodes in order to display the electrochemical behaviours of the ferrocene-ferricinium couple[56]. $Ru(NH_3)_6^{3+}$ and $Ru(bpy)_3^{2+}$ have also been used as redox centers[57]. Nafion has been proposed to modify microvoltammetric electrodes for in vivo use[58].

The electrocatalytic effect of viologen (4,4'-bipyridine) derivatives has been demonstrated for the reduction of denatured cytochrome c using different voltammetric techniques[59]. Polymerized as polycyanurviologen to modify glassy carbon electrodes, these sensors have been used to catalyze the reduction of oxygen and ferricyanide[60]. The incorporation of viologen in polypyrrole films may decrease the potential required to obtain the charge transfer. This fact has been exploited for the reduction of a dibromoalkane[61].

Other polymer films have also been proposed, such as amino-substituted polyphenylene oxide[62], alkylamine-siloxane[63], polyaniline[64] or zeolite in order to immobilize electroactive metallic tetramethylpyridinium porphyrins[65].

OTHER MODIFIED ELECTRODES

The most important problems encountered with modified solid electrodes are the inconvenience in the initial preparation and the poor reproducibility during extended use due to surface deactivation.

Another way to modify electrodes and which circumvents these problems is to incorporate the modifying compound into the carbon paste of a carbon paste electrode[66]. Such a senor is easy to construct, eliminates the immobilization procedure, offers a surface easy to renew and allows the incorporation of numerous compounds. Several electrodes have been prepared following this procedure by incorporation of quinones[66], aromatic diamines[67], metalloporphyrins[68] and metallic phthalocyanins[69]. These latter compounds have been selected as redox catalysts for electrochemical detection in HPLC[71,72].

Cobalt(II) phthalocyanin has been incorporated into a carbon paste electrode to determine several sulphur containing compounds[73,74].

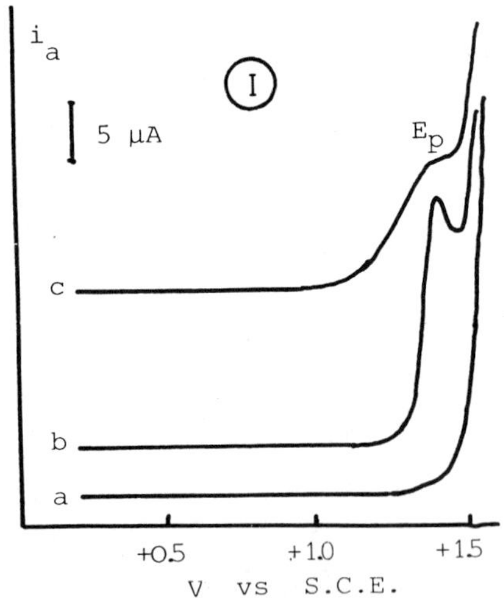

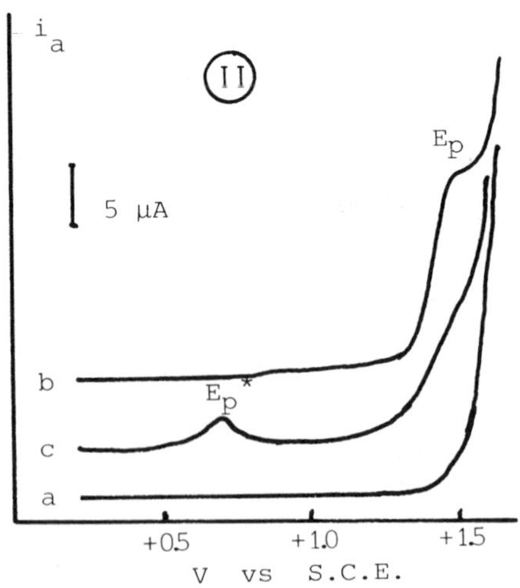

Figure 1. Voltammetric curves. Scan rate: 5 mVs^{-1}.
a) Supporting electrolyte: 0.05 M H_2SO_4;
b) cystine $2x10^{-4}$ M; c) cysteine $2x10^{-4}$ M.
I: Carbon paste electrode; II: modifed carbon paste electrode.

Cysteine and cystine

The electrochemical oxidation of cysteine and cystine is difficult to accomplish using solid state electrodes since the two molecules are electroinactive at the carbon electrode[75,76]. With the carbon paste electrode, a judicious selection of the supporting electrolyte permits the electroactivity of these two derivatives to be shown. In 0.05 M sulfuric acid, due to the large accessible anodic potentials, cysteine and cystine are oxidized at very positive potentials, close to the oxidation of the solvent (Figure 1, I, curves b, c). However, the study of these derivatives and especially of cysteine is difficult due to the poor resolution of the anodic peak. The oxidation of cysteine at the surface of the carbon paste corresponds to a slow electrochemical reaction as well as to a spreading of the anodic peak: $E_p - E_{p/2} = 170$ mV. On the contrary, the oxidation process of cystine, leading to the formation of cysteic acid[76,77], corresponds to a much faster reaction: $E_p - E_{p/2} = 50$ mV.

The catalytic activity of metallic phthalocyanins with respect to biochemical systems is well known. One can interpret the electrocatalytic process as being an oxidation on the surface of the electrode of the cobalt(II) phthalocyanin to cobalt(III) phthalocyanin, followed by a homogeneous oxidation in a solution of thiol according to a redox process which regenerates the cobalt(II) of the organometallic complex. This interpretation confirms the observations of Baldwin and coworkers[69].

A study of the response of the electrode as a function of the percentage of cobalt phthalocyanin ranging from 2 to 10% w/w indicated an optimum concentration of 2% of the redox mediator.

On the surface of the modified electrode (EPCM), the oxidation of cysteine gave a voltammetric curve consisting of a peak due to the electrocatalytic oxidation, E_p*, and, at more positive potentials, a shoulder E_p due to the oxidation of cysteine on the carbon paste (Figure 1, II). The peak of the electrochemical oxidation is analytically useful in the acid pH range (lower than 6), the peak being maximum in very acidic medium. A study of the height of peak E_p* as a function of the concentration was carried out by the method of standard addition using a differential pulse technique. Successive recordings were made without renewing the surface of the electrode. A mechanical cleaning by stirring the solution for 30 sec between each measurement suffices to restore the electrode surface. The voltammograms present an identical morphology in terms of concentration and the response of the electrode is perfectly linear between 4.0×10^{-6} M and 6.0×10^{-5} M. At concentrations higher than 6.0×10^{-5} M, a deviation from linearity is observed.

Concerning cystine, the oxidation of the molecule at the EPCM takes place at potentials too positive to benefit from the catalytic activity of the cobalt(II/III) phthalocyanin couple. The voltammetric curves comprise only the oxidation of the cystine on the surface of the carbon paste (Figure 1, II, curve c). Nevertheless, the amount of cystine could be determined, following electrolysis at a mercury pool, by cleavage of the disulfide bond, transforming it into two thiol molecules. This type of manipulation requires a fairly long time lapse, owing to the adsorption phenomenon revealed in the polarographic recording[69]. The adsorption of these compounds on the surface of the mercury film effects a slowdown of the reduction, which can take several hours to occur. Due to this fact, the electrolysis current decreases very slowly, until low residual values (tenths of microamperes) are observed. The liquid is then sampled under tension and directly transferred into the voltammetric cell.

Lipoic acid

Lipoic acid, or 6,8-dithiolane octanoic acid, is widely distributed in living organisms, intervening in hydrogen transport and acyl radicals by acting as a necesssary coenzyme in the oxidative decarboxylation of pyruvate. If one considers the electrochemical oxidation of the lipoic acid at the surface of the carbon paste electrode (EPC), in contrast to cystine, the molecule is electroactive at potentials less positive than +1.0 V vs SCE. The voltammetric recordings show a perfectly defined oxidation peak which indicates a fast kinetic reaction: $E_p - E_{p/2} = 50$ mV.

On the surface of the EPCM, the presence of the redox mediator causes a catalytic effect, negligible because of the elevated kinetics of the heterogeneous electrochemical reaction between lipoic acid and the EPC. One observes in addition a slight shift in the peak potential towards less positive values. The linearity of the ratio $i_{p/v^{1/2}}$ between 5 and 500 mVs^{-1} sweep rates implies an electrochemical process controlled by the diffusion of lipoic acid. A study of the shift of the anodic peak potential as a function of pH shows a zero slope of the E_p - pH relationship in the pH range of 1 to 11. The height of the oxidation peak is maximal at pH 4. Above this value, one observes a regular decrease of the oxidation current.

The quantitative determination of the lipoic acid was conducted in a comparative manner using both electrodes (EPC and EPCM). A restricted linear range is obtained on the surfce of the EPCM, resulting from the importance of the residual current at elevated sensitivities, the current originating from the electroactivity of the incorporated cobalt phthalocyanin.

Disulfiram

Disulfiram, (bis(diethyl thiocarbamyl)disulfide), is a compound currently used for the treatment of alcoholism. On the surface of the EPC in the pH 1 to 11 range, the oxidation of the molecule gave voltammetric curves exhibiting a single well defined oxidation peak. As in the case of lipoic acid, the oxidation takes place at lower potentials, +1.10 V vs SCE, in a rapid electrochemical process: $E_p - E_{p/2} = 60$ mV.

The presence of the redox mediator in the carbon paste causes a slight shift of the potential to less positive values, the reaction being controlled by the diffusion of the electroactive species. At the surface of the EPCM, the shift of the peak potential as a function of pH shows a zero slope in the 1 to 4 pH range. Above pH 4, the peak potential, E_p, undergoes a shift of 20 mV per pH unit to less positive values.

The quantitative determination of the molecule was effected at the surface of both EPC and EPCM electrodes in acidic medium in order to obtain a better resolution of the peak.

Other sulphur containing compounds

Figure 2 exhibits the use of this type of electrode for the determination of thiourea (Figure 2a), methylthiouracil (Figure 2b) and 6-mercaptopurine (Figure 2d). Oxidation of the compounds occurs at less positive potentials with a high rate of electron transfer. Thiamazole (Figure 2c) can also be analyzed using the EPCM but its electrochemical behaviors are very similar to those obtained with the conventional one.

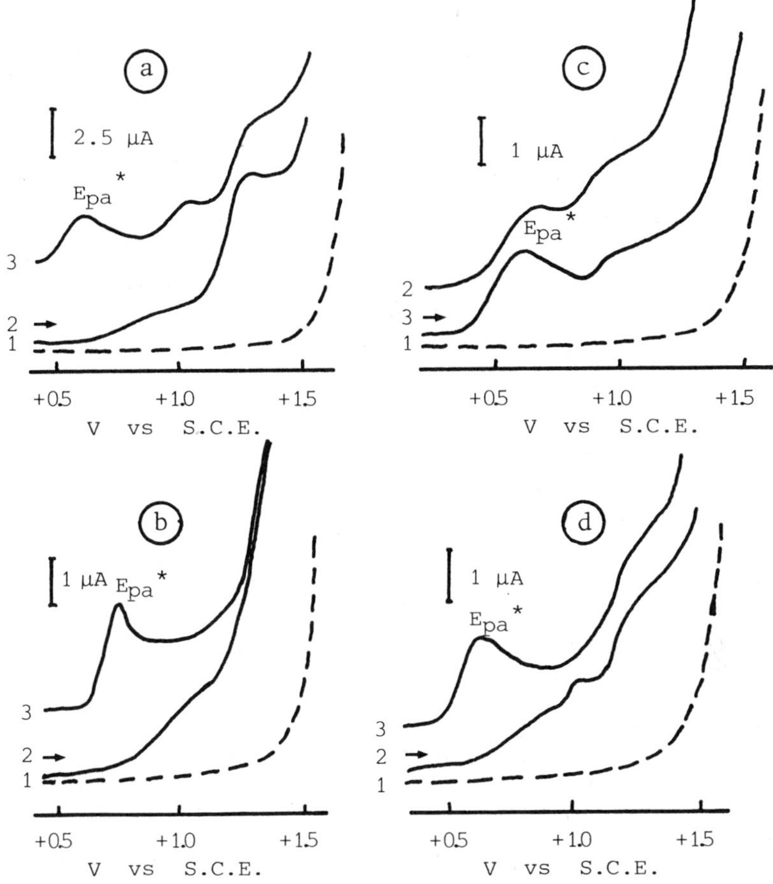

Figure 2. Voltammetric curves. Scan rates: 10 mVs^{-1}.
1: Supporting electrolyte: 0.05 M H_2SO_4
2: Carbon paste electrode
3: Modified carbon paste electrode
a) Thiourea; b) methylthiouracil; c) thiamazole;
d) 6-mercaptopurine (all at the $2x10^{-4}$ M concentration).

Hydrogen Peroxide

Iron phthalocyanin incorporated into a carbon paste electrode exhibits a catalase-like activity and has been successfully used for the electrochemical oxidation of hydrogen peroxide[7,8].

Figure 3 compares the oxidation of hydrogen peroxide at carbon paste and mixed paste containing iron phthalocyanin at pH 5.5. The electro-catalytic response of the modified electrode was reflected by the appearance of an oxidation wave with a potential peak of +0.59 V vs SCE. Under the same conditions, no corresponding oxidation wave was observed at the conventional CPE. The explanation of this catalytic effect is probably due to an interaction between iron and hydrogen peroxide. A characteristic shift of about 60 mV per pH unit was observed between pH 3 and 9. The pH of 5.5 was chosen to obtain a well-defined peak. Several types of buffers have been tested, but no particular influence on the catalytic activity of iron phthalocyanin was observed. The modified electrode employed in this

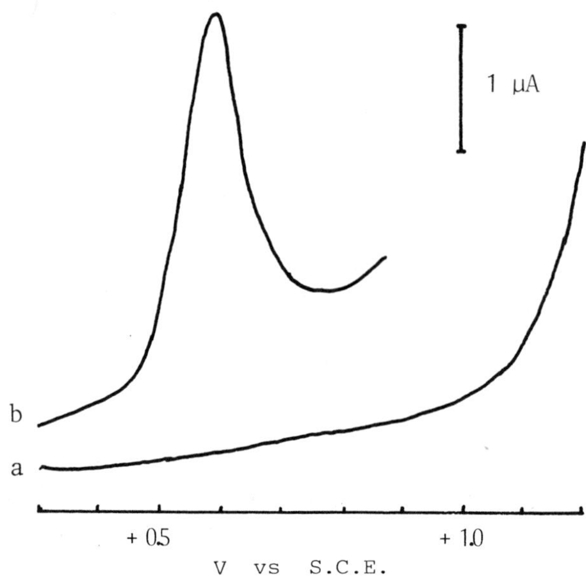

Figure 3. Voltammetric curves of $1x10^{-3}$ M H_2O_2 at carbon paste (curve a) and modified carbon paste (curve b) electrodes. Phosphate buffer 0.05 M, pH 5.5; sweep rates: 5 mVs^{-1}.

work represents an iron-phthalocyanin doping level of 2% w/w. Other catalyst levels of 1, 5 and 10% have also been tested. The 1% paste cannot be used because of low sensitivity. Electrodes containing higher concentrations of modifying agent (5-10%) were not found to yield significantly higher current levels and thus were not investigated further.

Cyclic voltammetry of hydrogen peroxide obtained on iron-phthalocyanin electrodes indicates an irreversible process, no corresponding peak being observed on the reverse scan under the employed conditions.

The current response of the modified electrode towards hydrogen peroxide was linear from 6 x 10^{-4} M to 5 x 10^{-5} M. The detection limit was 5 x 10^{-5} M.

Platinum(II) anti-tumor drugs

Cis-platin
(CDDP)

Carboplatin
(JM 8)

The investigation of the above represented platinum(II) derivatives at solid electrodes was undertaken in order to obtain further insights into the redox behavior and aqueous stability of these clinically interesting drugs and in order to elucidate the processes which govern the platinum particle electrodeposition step.

To achieve these purposes, the carbon paste electrode (CE) has been preferentially utilized since the electrode exhibits low residual currents within a large available potential range and is not subject to memory effects due to the surface renewing technique. Moreover, in contrast to platinum or glassy carbon electrodes, the CPE is not subject to surface evolution phenomena. Indeed, the formation of oxides, the rapid evolution of hydrogen at platinum electrodes or the appearance of surface functionalities at glassy carbon surfaces may interfere in the overall electrode process[79]. Finally, it was of interest to establish whether or not free chloride ions play a significant catalytic role at the electrode-solution interface at carbon electrodes as is the case at platinum electrodes (the halide bridged pathway)[80].

The measurements have been carried out by cyclic voltammetry in 0.1 M sodium perchlorate solutions in the absence and in the presence of chloride ions[81-83]. The latter play a significant role in the stability of the platinum complexes in aqueous solutions, as shown below for cis-platin, JM 8 being much more stable[84,85]:

$$cis\text{-}(H_3N)_2PtCl_2 \rightleftharpoons cis\text{-}(H_3N)_2Pt(Cl)(OH_2) + Cl^- \rightleftharpoons (H_3N)(H_2N)Pt(OH_2)_2 + Cl^-$$

cis-$PtCl_2$ cis-Pt(Cl)(H_2O)   cis-Pt$(H_2O)_2$

Thus, in order to minimize these hydrolysis phenomena, care was taken to freshly prepare the solutions, the measurements being performed as soon as the solutions were ready. The platinum microparticle deposition has been confirmed by SEM observations of CPE surfaces. Figure 4 shows several typical cyclic voltammetric (CV) curves of JM 8 as a function of free chloride ion concentration. CV and SEM studies have demonstrated that in the absence or in the presence of chloride ions, JM 8 is not reduced to Pt(O) in contrast to cis-platin. In the latter case, increasing the chloride ion concentration to 1 x 10^{-1} M produces just a slight increase in the reduction peak height but further increase of chloride ion concentration drastically inhibits Pt(O) formation. Considering the oxidation process at the CPE, both JM 8 and cis-platin are oxidizable in a diffusion controlled process. The oxidation of the former occurs at high positive potentials (+1.50 V vs SCE), while cis-platin is oxidized at less positive potentials giving two peaks (E_{paI} = +1.20 V, E_{paII} = +1.45 V). These two peaks have been characterized by studying cis-platin ageing in aqueous solution. The first-peak, corresponding to cis-platin oxidation, diminishes gradually while the second, corresponding to the oxidation of the cis-platin hydrated species, increases with time.

By reversing the scan direction after the oxidation peak of JM 8, several cathodic peaks are observed at very negative potentials illustrating the high degree of irreversibility of the process (Figure 4). The analysis of the appearance and disappearance of these peaks as a function of chloride ion concentration has permitted their nature to be elucidated. Considering the observations of Martin[86] concerning the mole fraction of Pt(II) versus chloride and those of Hubbard relative to the electro-oxidation of platinum(II) complexes at platinum electrodes[87]; moreover, taking into account that perchlorate ions are non-complexing agents[88] and

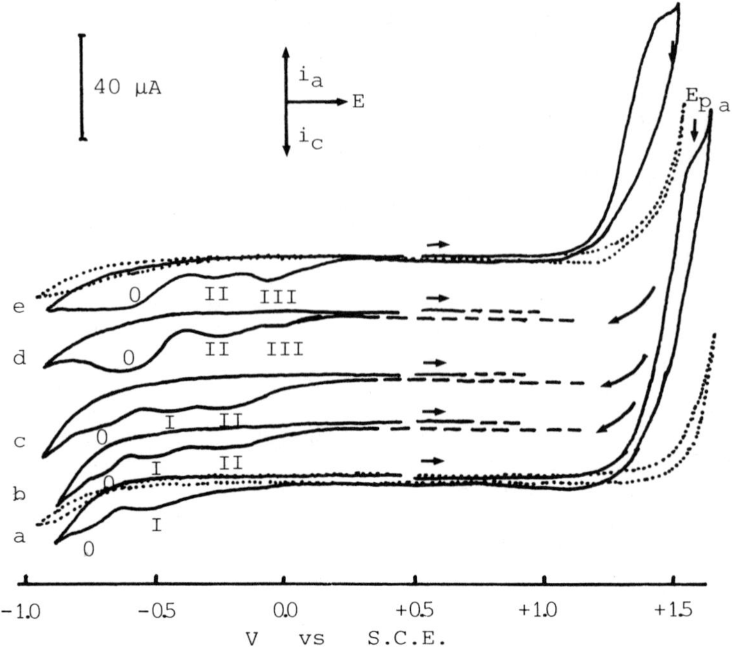

Figure 4. Cyclic voltammograms of 1×10^{-3} M JM 8 in 0.1 M $NaClO_4$ at the CPE. Scan rate: 20 mVs^{-1}. Starting potential: +0.5 V; initial anodic scan.
a) $Cl^- = 0$; b) $Cl^- = 10^{-3}$ M; c) $Cl^- = 10^{-2}$ M; d) $Cl^- = 10^{-1}$ M; e) $Cl^- = 1$ M. Surface renewed after each cyling.
Dotted line = supporting electrolyte.

assuming that the oxidation of the molecule involves a single two-electron step to platinum(IV), the following structures can be suggested:

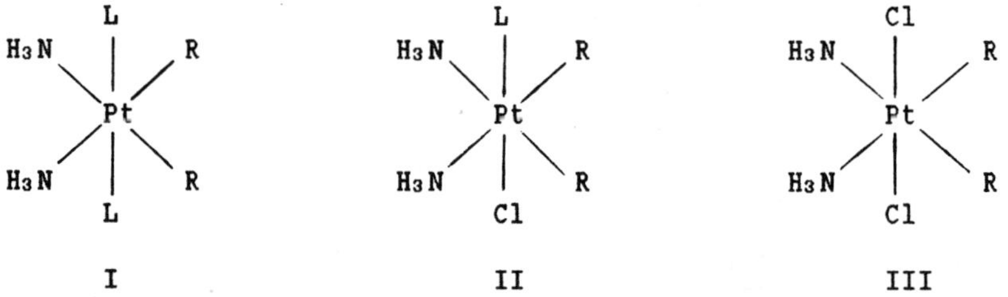

L = H_2O or OH; R = organic ligand.

The number and respective concentration ratio of these oxidized forms being directly related to the free chloride ion concentration, structure I appears in the absence of chloride ion, structures I and II between 1×10^{-3} and 1×10^{-2} M in chloride ion and structures II and III between 1×10^{-2} and 1 M in chloride ion. These oxidized forms of JM 8 are reduced in two steps, giving new platinum(II) structures which are further reduced in a common step (E_{pc_0}) to give Pt(0).

Due to the fact that JM 8 is not reducible and that organic ligands are facile leaving groups compared to ammonia[88,89], it may be assumed that the reduction in steps I, II and III gives, respectively, structures I', II' and III' with elimination of the organic ligand:

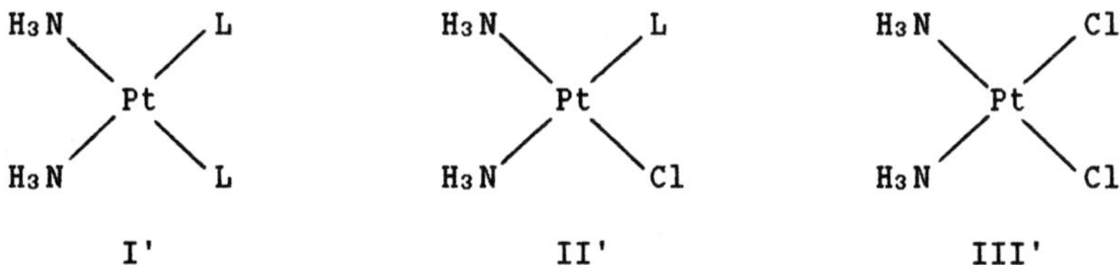

The intermediate reduction steps of cis-platin oxidized structures at the CPE are not so well defined; the processes rapidly gave important Pt(O) formation and concommitant hydrogen evolution at the electrode surface. The Pt(O) electrodeposition markedly affects further scans on the same surface as shown in Figure 5 in the case of JM 8. Indeed, we may see by reversing after cathodic steps II or III a new redox couple formation E_{pa2}* and E_{pc2}* which corresponds to the catalytic oxidation and reduction of species II' and III' at the platinum microparticles. Oxidation of species I', which possess no halide ligand in its structure, is not detected at the platinum particles. In the case of cis-platin, several oxidation-reduction cycles conducted on the same surface produce the same electrode modification process as observed for JM 8, but to a greater extent since cis-platin itself is reducible. The shape of the quasi-reversible couple E_{pa}*/E_{pc}* is particularly well marked in the presence of chloride ions. With continued cycling, the peaks increase and the reversibility of the reaction and available potential ranges become restricted.

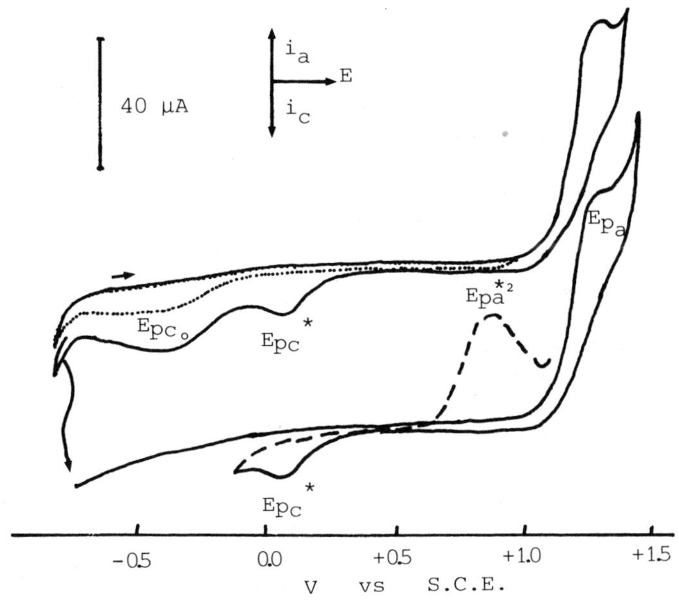

Figure 5. Cyclic voltammograms of 1×10^{-3} M JM 8 in 0.1 M $NaClO_4$ and 1 M NaCl at the platinized CPE as a function of inversion potential. Scan rate: 20 mVs^{-1}; starting potential: -0.8 V, initial anodic scan. Several cycles on the same surface.
(.......) first cycle, (———) second and third cycles, (-------) fourth anodic scan.

In conclusion, cyclic voltammetric experiments conducted with a carbon paste electrode have permitted the elucidation of the redox behavior of cis-platin and of a new platinum(II) derivative, JM 8. The extended potential range permits the study of Pt(II) complexes possessing no halide ligands which are difficult to study at platinum or glassy carbon electrodes. The case of JM 8 is of particular interest since the molecule is fairly stable and exhibits a slight platinum electrodeposit after a prior oxidation step. Moreover, the reduction peaks of the oxidized forms of JM 8 are sufficiently distinct to allow analysis of their evolution as a function of free chloride ion concentration. This indirectly permits the structures of the oxidized forms of JM 8 to be suggested. It appears that the reduction of the platinum(II) oxidized structures of JM 8 does not regenerate the starting molecule but gives rise to the appearance of cis-platin structures, and the greater the number of halide ligands in the oxidized products, the easier is their reduction. Furthermore, the use of the CPE has permitted the conclusion that non-chloride ligands, in this case aquo or organic, tend to stabilize the platinum(II) species as well towards oxidation as towards reduction. The positive effect of the in-plane chloride ligands on the ease of oxidation and reduction of the platinum(II) species at the CPE supports the hypothesis of an in-plane bridging effect as earlier reported at platinum electrodes by Hubbard[80,87]. The presence of free chloride ion in the solution exerts a negligible effect on the ease of platinum(II) oxidation at the CPE, contrary to the behaviors observed at platinum electrodes. Finally, the electrodeposition of platinum particles is only observed for platinum(II) species possessing halide(s) ligands in their structures; the process is only slightly influenced by chloride concentration lower than 1×10^{-1} M but is drastically inhibited at high concentrations. This seems to rule out the existence of an axial bridging effect at the CPE.

POTENTIOMETRIC SENSORS

Polymer film coated electrodes, mainly platinum or carbon electrodes, have also been prepared for potentiometric measurements. The first such electrodes have shown a response to pH variations, their selective activity being explained by the ability of protons, by virtue of their small size, to move through the polymer matrix to the platinum or carbon surface[90,91].

Polymer coated electrodes can be made selective to particular compounds by entrapping in the polymer salts, ion-exchangers or chelating ligands, but these active components tend to be leached from the matrix, limiting the sensitivity of the electrode. This inconvenience can be avoided by using components which are covalently attached to the polymer[92]. However, physical entrapment is often easier to perform and renewal of the surface can be made very rapidly.

A specific sensor for the determination of certain neuroleptics (dibenzodiazepines) has been developed by coating a graphite spectroscopic rod with a film of polyvinyl chloride (PVC) previously dissolved in cyclohexanone. The solvent mixture, dioctylphthalate-nitrobenzene, is used as plasticizer. Inside the film, dibenzodiazepine-tetraphenyborate is incorporated as an ion-pair. The graphite rod is immersed in the mixture for 30 min and dried at room temperature for one hour. This procedure is repeated each day and allows reproducible measurements.

The Nernstian response of the electrode was found to be between 10^{-2} and 10^{-5} M[93]. Interferences were studied. Organic substances containing the pyrrolidine, piperazine or piperidine group present a certain degree of interference. Other types of neuroleptics of the benzodiazepine family do not interfere. Methods for the accurate assay of dibenzodiazepines in tablets and injections have been described[93].

A potentiometric titration procedure using tetraphenylborate has been proposed and various determinations can be carried out on pharmaceuticals at low levels of concentration using the same electrode[94].

For drugs, liquid and polyvinyl chloride membrane electrodes which are sensitive and selective for atropine have been described. They are based on the inclusion of an atropine-reineckate ion-pair complex in the PVC matrix[95].

OTHER DEVELOPMENTS

Electroinactive polymers have also been utilized as supports for carbon-based electrodes assigned to voltammetric analysis. Such an electrode has been constructed by spraying a suspension of colloidal graphite in poly(methylmethacrylate) dissolved in butyl acetate onto a metallic support. A thin conductive film is obtained after drying[96-99]. Another carbon-polymer electrode has been developed by mixing and heating at 150°C a mixture of carbon black and polyethylene in a ratio of 30 to 45% w/w. Thin disks are obtained by pressing the paste at 170°C under 200 $kg \cdot cm^{-2}$ [100,101].

Figure 6 compares the electrochemical behaviors of both sensors with those recorded with conventional glassy carbon and carbon paste electrodes, using the representative hexacyanoferrate(II/III) couple in 0.1 M KCl. As expected, the kinetics of the electrochemical reaction is greatly influenced by the nature of the electrode material. At the graphite spray electrode (Figure 6a), the hexacyanoferrate couple behaves almost the same as at the glassy carbon electrode (Figure 6b), but at the former electrode, the reversibility is somewhat improved. The carbon-

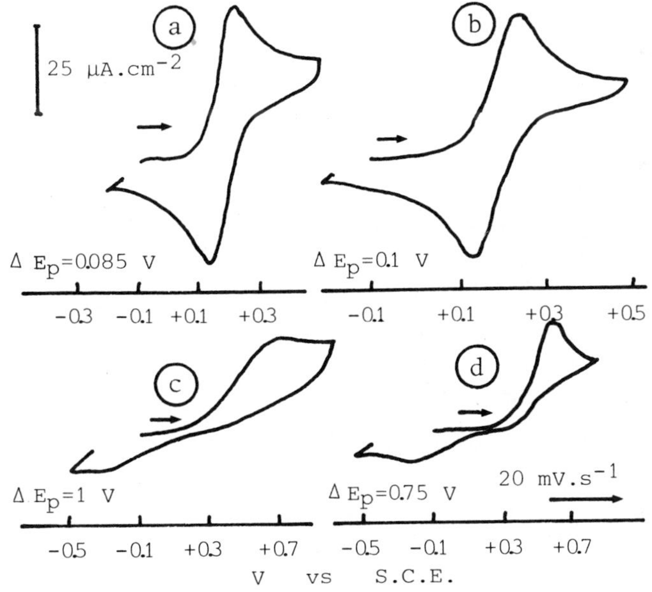

Figure 6. Cyclic voltammetry of the hexacyanoferrate(II/III) couple. Hexacyanoferrate(II) 5 x 10^{-4} M, 0.1 M KCl; scan rate: 20 mVs^{-1}. a) Graphite spray electrode; b) glassy carbon electrode; c) carbon-polyethylene electrode; d) carbon paste electrode.

polyethylene (Figure 6c) and carbon paste (Figure 6d) electrodes exhibit fairly large peak potential separation; currents are low and electrochemical activation by cycling between -0.7 V and +1.3 V vs SCE has no marked effect upon their respective response.

Again, the nature of the electrode material influences the peak separations of the reversible N-acetyl-p-quinoneimine couple (A/A') appearing when phenacetin is oxidized (Figure 7). However, the potential of the main oxidation peak is not markedly shifted. On the other hand, carbon-polymer and carbon paste electrodes exhibit a lower residual current than the graphite spray or the glassy carbon electrodes.

The graphite spray is easy to renew: after cleaning the surface with acetone to remove the graphite film, a new spray ensures a very reproducible surface without any memory effect. The restoration of the glassy carbon surface is more difficult and the reproducibility of the carbon paste is lower, owing to the heterogeneous structure of the paste. The carbon-polymer electrode offers the advantage of a very wide accessible potential range in the positive side, allowing the investigation of drugs not oxidizable at the other solid electrodes[102-105].

In respect to the shape of the voltammograms, it is actually difficult to predict the most satisfactory electrode for studying certain pharmacological compounds. It seems reasonable, therefore, before initiating any investigations, to test a large number of commercially available or laboratory-made electrodes in order to obtain the best results.

A carbon-polymer electrode, obtained by mixing 25% of carbon black to a copolymer of ethylenevinyl acetate with 9% vinyl acetate, has been used as a detector in HPLC for the investigation of epinephrine and norepine-

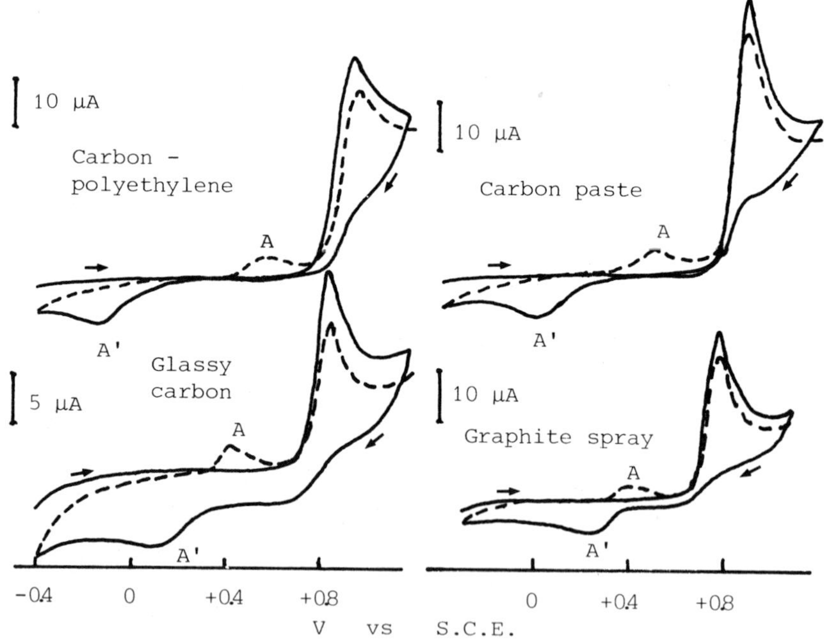

Figure 7. Cyclic voltammetry of 5 x 10^{-4} M phenacetin. Phosphate buffer pH 6.5 with 10% methanol. Scan rate: 20 mVs^{-1}. The dotted lines represent a second scan on the same surface.

phrine, since the determination of neurotransmitters is a major application of HPLC-ED[106]. Firstly the carbon-polymer electrode was compared with a conventional glassy carbon electrode (both materials having been included in identical cell designs) in terms of the baseline currents obtained at various applied potentials. The results of this study show that the currents obtained using the glassy carbon electrode were in all cases higher than those obtained using the carbon-polymer electrode. However, the potential of the latter reached any applied value about five times faster (on average) than the glassy carbon at identical applied potentials. This is of obvious importance when chromatovoltammetric curves are being constructed where potential values have to be changed following injection, and when the electrode material has to be cleaned or replaced.

The current response of the two electrode materials at potential values on the plateau of the chromatovoltammetric curves for both compounds at +0.65 V (for the glassy carbon electrode) and at +0.75 V (for the carbon-polymer electrode) was investigated. For both compounds, the currents obtained were similar using both electrode materials. In terms of noise, the carbon-polymer electrode gave rise to much steadier baselines, especially when the detectors were operated near the limit of detection for the two compounds. In terms of linearity of response, both electrode materials showed linearity between 5 pg and 20 ng injected for both compounds. The limit of detection in both cases was about 2 pg injected.

The reproducibility of both detectors was checked using a 200 pg injection of both compounds, and the coefficients of variation were found to be 3.4% and 3.0% for norepinephrine and epinephrine, respectively, for the glassy carbon electrode, and 2.9% and 2.8%, respectively, for the carbon-polymer electrode (based on the results obtained for 8 successive injections).

The results from this study show that the carbon-polymer electrode is comparable to the glassy carbon electrode with respect to current response (and hence sensitivity). It is superior to the glassy carbon electrode with respect to noise levels, stabilisation time and accessible potential range, but inferior in terms of requiring a higher potential for the oxidation of model compounds, and the kinetics of electron transfer. This new material, however, is much easier to handle than glassy carbon in terms of machining and is cheaper. It is anticipated that this new electrode will find many applications for the determination of oxidizable molecules of biological and environmental significance.

ACKNOWLEDGEMENTS

Thanks are expressed to the "Fonds National de la Recherche Scientifique" (F.N.R.S. Belgium) for help to one of us (G.J.P.), to the SPPS (Belgium Politic Research, ARC) contract No. 86/91-89, and to the Office of the Scientific Politic of Belgium, contract PREST No. 278/84.

REFERENCES

1. R.W. Murray, in: "Electroanalytical Chemistry", A.J. Bard, Ed., M. Dekker, New York, Vol. 13, p. 191 (1984).

2. L.R. Faulkner, Chem. Eng. News, 62, 28 (1984).

3. A.J. Bard, J. Chem. Educ., 60, 302 (1983)

4. J. Schreurs and E. Barendrecht, Recl. Trav. Chim. Pays-Bas, 103, 205 (1984).

5. G.J. Patriarche, J.-C. Viré and J.-M. Kauffmann, Chim. Nouv., 2, 91 (1984).

6. Extended Abstracts of the 38th Meeting of the International Society of Electrochemistry, Vol. and 2, Maastricht, The Netherlands, Sept. 13-18 (1987).

7. A.B. Fischer, M.S. Wrighton, M. Umana and R.W. Murray, J. Am. Chem. Soc., 101, 3442 (1979).

8. K.N. Kuo and R.W. Murray, J. Electrochem. Soc., 129, 756 (1982).

9. N.R. Armstrong, A.W. Lin, M. Fujihira and T. Kuwana, J. Electroanal. Chem., 48, 741 (1976).

10. J. Fujihira, T. Osam, D. Hursh and T. Kuwana, J. Electroanal. Chem., 88, 741 (1978).

11. P.R. Moses, L.M. Wier, J.C. Lennox, H.O. Finklea, J.R. Lenhard and R.W. Murray, Anal. Chem., 50, 576 (1978).

12. K.N. Kuo, P.R. Moses, J.R. Lenhard, D.C. Green and R.W. Murray, Anal. Chem., 51, 745 (1979).

13. J. Facci and R.W. Murray, J. Electroanal. Chem., 112, 221 (1980).

14. K. Itaya and A.J. Bard, Anal. Chem., 50, 1487 (1978).

15. A.W.C. Lin, P. Yeh, A.M. Yacynych and T. Kuwana, J. Electroanal. Chem., 84, 411 (1977).

16. M.F. Dantartas, J.F. Evans and T. Kuwana, Anal. Chem., 51, 104 (1979).

17. D.C.S. Tse, G.P. Royer and T. Kuwana, J. Electroanal. Chem., 98, 345 (1979).

18. S. Mazur, T. Matusinovic and K. Cammann, J. Am. Chem. Soc., 99, 3888 (1977).

19. R. Nowak, F.A. Schultz, M. Umana, H. Abruna and R.W. Murray, J. Electroanal. Chem., 94, 219 (1978).

20. M. Sharp, M. Petersson and K. Edström, J. Electroanal. Chem., 95, 123 (1979).

21. L.L. Miller and M.R. Van De Mark, J. Am. Chem. Soc., 100, 639 (1978).

22. M.F. Dantartas, K.R. Mann and J.F. Evans, J. Electroanal. Chem., 110, 379 (1980).

23. M.C. Pham, P.C. Lacaze, J.E. Dubois, J. Electroanal. Chem., 99, 331 (1979).

24. C. Degrand and L.L. Miller, J. Electroanal. Chem., 117, 267 (1981).

25. E. Laviron, J. Electroanal. Chem., 100, 263 (1979).

26. T. Kuwana and N. Winograd, in: "Electroanalytical Chemistry", A.J. Bard, Ed., M. Dekker, New York, Vol. 7, p. 1 (1974).

27. E.M. Genies, G. Bidan and A.F. Diaz, J. Electroanal. Chem., 149, 101 (1983).

28. F.B. Kaufman and E.M. Engler, J. Am. Chem. Soc., 101, 547 (1979).

29. K. Shigehara, N. Oyama and F.C. Anson, J. Am. Chem. Soc., 103, 2552 (1981).

30. D.A. Buttry and F.C. Anson, J. Electroanal. Chem., 130, 333 (1981).

31. A.F. Diaz, J.I. Castillo, J.A. Logan and W.Y. Lee, J. Electrochem. Chem., 129, 115 (1981).

32. P..J. Peerce and A.J. Bard, J. Electroanal. Chem., 112, 97 (1980).

33. J. Facci and R.W. Murray, J. Electroanal. Chem., 124, 339 (1981).

34. H.O. Finklea and R.S. Vithanage, J. Electroanal. Chem., 161, 283 (1984).

35. X. Chen, P. He and L.R. Faulkner, J. Electroanal. Chem., 222, 223 (1987).

36. C.P. Andrieux, in: "Electrochemistry, Sensors and Analysis", M.R. M.R. Smyth and J.G. Vos, Eds., Elsevier, Amsterdam, Vol. 25, p. 235 (1986).

37. R.C.M. Jakobs, L.J.J. Janssen and E. Barendrecht, Electrochim. Acta, 30, 1313 (1985).

38. A. Haimerl and A. Merz, J. Electroanal. Chem., 22, 55 (1987).

39. A.F. Diaz, W.Y. Lee, A. Logan and D.C. Green, J. Electroanal. Chem., 108, 377 (1980).

40. L.L. Miller and B. Zinger, Q.X. Zhou, J. Am. Chem. Soc., 109, 2267 (1987).

41. Y. Ikariyama and W.R. Heineman, Anal. Chem., 58, 1803 (1986).

42. T.F. Otero, R. Tejada and A.S. Elola, Polymer, 28, 651 (1987).

43. P. Audebert, G. Bidan and M. Lapkowski, J. Electroanal. Chem., 219, 165 (1987).

44. B.R. Shaw, G.P. Haight and L.R. Faulkner, J. Electroanal. Chem., 140, 147 (1982).

45. N. Oyama and F.C. Anson, Anal. Chem., 52, 1192 (1980).

46. J. Wang and P. Tuzhi, J. Electrochem. Soc., 134, 586 (1987).

47. A. Merz and A.J. Bard, J. Am. Chem. Soc., 100, 3222 (1978).

48. P. Daum and R.W. Murray, J. Electroanal. Chem., 103, 289 (1979).

49. R.M. Kannuck, J.M. Bellama, E.A. Blubaugh and R.A. Durst, Anal. Chem., 59, 1473 (1987).

50. A. Diaz, J. Am. Chem. Soc., 99, 5838 (1977.

51. F.B. Kaufman, A.H. Schroeder, V.V. patel and K.H. Nichols, J. Electroanal. Chem., 132, 151 (1982).

52. E.T. Turner Jones and L.R. Faulkner, J. Electroanal. Chem., 222, 201 (1987).

53. J.B. Kerr and L.L. Miller, J. Electroanal. Chem., 101, 262 (1979).

54. I. Rubenstein and A.J. Bard, J. Am. Chem. Soc., 102, 6641 (1980).

55. I. Rubenstein and A.J. Bard, J. Am. Chem. Soc., 103, 5007 (1981).

56. I. Rubenstein, J. Electroanal. Chem., 176, 359 (1984).

57. L.D. Whiteley and C.R. Martin, Anal. Chem., 59, 1746 (1987).

58. E.W. Kristensen, W.G. Kuhr and R.M. Wightman, Anal. Chem., 59, 1752 (1987).

59. J. Haladjian, R. Pilard and P. Bianco, J. Electroanal. Chem., 184, 391 (1985).

60. P. Janda, J. Wever and L. Kavan, J. Electroanal. Chem., 180, 109 (1984).

61. L. Coche, A. Deronzier and J.-C. Moutet, J. Electroanal. Chem., 198, 187 (1986).

62. J.E. Dubois, P.C. Lacaze and M.C. Pham, J. Electroanal. Chem., 117, 233 (1981).

63. K.N. Kuo and R.W. Murray, J. Electroanal. Chem., 131, 37 (1982).

64. E.M. Genies, A.A. Syed and C. Tsintairs, Mol. Cryst. Liq. Cryst., 121, 181 (1985).

65. B. De Vismes, F. Bediouni, J. Devynck and C. Bied Charreton, J. Electroanal. Chem., 187, 197 (1985).

66. K. Ravichandran and R.P. Baldwin, J. Electroanal. Chem., 126, 293 (1981).

67. K. Ravichandran and R.P. Baldwin, Anal. Chem., 55, 1586 (1983).

68. E.S. Takeuchi and R.W. Murray, J. Electroanal. Chem., 188, 49 (1985).

69. M.K. Halbert and R.P. Baldwin, Anal. Chem., 57, 591 (1985).

70. A.B.P. Lever, M.R. Hempstead, C.C. Leznoff, W. Lui, M. Melnick, W.A. Nevin and P. Seymour, Pure & Appl. Chem., 58, 1467 (1986).

71. M.K. Halbert and R.P. Baldwin, Anal. Chim. Acta., 187, 89 (1986).

72. L.M. Santos and R.P. Baldwin, Anal. Chem., 59, 1766 (1987).

73. C.R. Linders, G.J. Patriarche and J.-M. Kauffmann, Anal. Letters, 19, 193 (1986).

74. C.R. Linders, J.-M. Kauffmann and G.J. Patriarche, J. Pharm. Bel., 41, 373 (1986).

75. D.G. Davis and E. Bianco, J. Electroanal. Chem., 12, 254 (1966).

76. J. Zagal, C. Fierro and R. Rozas, J. Electroanal. Chem., 119, 403 (1981).

77. Z. Samec, Zh. Malysheva, J. Koryta and J. Pradac, J. Electroanal. Chem., 65, 573 (1975).

78. C.R. Linders, B.J. Vincke and G.J. Patriarche, Anal. Letters, 19, 1831 (1986).

79. K. Shimazu, D. Weisshar and T. Kuwana, J. Electroanal. Chem., 223, 223 (1987).

80. C.N. Lai and A.T. Hubbard, Inorg. Chem., 11, 2081 (1972).

81. F. Mebsout, J.-M. Kauffmann and G.J. Patriarche, J. Pharm. Biomed. Anal., 5, 223 (1987).

82. F. Mebsout, J.-M. Kauffmann and G.J. Patriarche, J. Pharm. Biomed. Anal., in press.

83. J.-M. Kauffmann, G.J. Patriarche, F. Mebsout, in: "Electrochemistry, Senors and Analysis", M.R. Smyth and J.G. Vos, Eds., Elsevier, Amsterdam, Vol. 25, p. 367 (1986).

84. S.J. Lippard, in: "Platinum, Gold and other Metal Chemotherapeutic Agents", S.J. Lippard, Ed., A.C.S. Symposia Series (1982).

85. I.S. Krull, X.D. Ding, S. Baverman, C. Selavka, F. Hochberg and L.A. Sternson, J. Chromotogr. Sci., 21, 166 (1983).

86. R.B. Martin, in: "Platinum, Gold and other Metal Chemotherapeutic Agents", S.J. Lippard, Ed., A.C.S. Symposia Series, p. 231 (1982).

87. J.P. Cushing and A.T. Hubbard, J. Electroanal. Chem., 23, 183 (1969).

88. M.E. Howe Grant and S.J. Lippard, in: "Metal Ions in Biological Molecules", H. Siegel, Ed., M. Dekker, New York, Vol. 11, p. 63 (1980).

89. F. Elferink and W.J.F. Van Der Vijgh, J. Chromatogr., 320, 379 (1985).

90. W.R. Heineman, H.J. Wieck and A.M. Yacynych, Anal. Chem., 52, 345 (1980).

91. G. Cheek, C.P. Wales and R.J. Nowak, Anal. Chem., 55, 380 (1983).

92. R.S. Lawton and A.M. Yacynych, Anal. Chim. Acta, 160, 149 (1984).

93. G.J. Patriarche and J.R. Sepulchre, Analusis, 14, 351 (1986).

94. G.J. Patriarche and J.R. Sepulchre, J. Pharm. Belg., 41, 293 (1986).

95. S.S.M. Hassan and F.S. Tadros, Anal. Chem., 56, 542 (1984).

96. J.-M. Kauffmann, A. Laudet and G.J. Patriarche, Anal. Letters, 15, 763 (1982).

97. J.-M. Kauffmann, A. Laudet and G.J. Patriarche, Talanta, 29, 1077 (1982).

98. G.J. Patriarche, J.-M. Kauffmann and J.-C. Vire, J. Pharm. Biomed. Anal., 1, 469 (1983).

99. J.-M. Kauffmann, A. Laudet, J.-C. Vire, G.J. Patriarche and G.D. Christian, Microchem., J., 28, 357 (1983).

100. M.P. Prete, J.-M. Kauffmann, J.-C. Vire, G.J. Patriarche, B. Debye and G. Geuskens, Anal. Letters, 17, 1391 (1984).

101. J.-M. Kauffmann, M.P. Prete, J.C. Vire and G.J. Patriarche, Fres. Z. Anal. Chem., 321, 172 (1985).

102. J.-M. Kauffmann, G.J. Patriarche, J.-C. Vire and W.R. Heineman, Analyst, 110, 349 (1985).

103. G.J. Patriarche, J.-C. Vire and J.-M. Kauffmann, Anal. Proc., 22, 202 (1985).

104. J.-M. Kauffmann, G.J. Patriarche, M. Chateau-Gosselin and B. Gallo Hermosa, Talanta, 33, 733 (1986).

105. G.J. Patriarche, J. Pharm. Biomed. Anal., 4, 789 (1986).

106. J.-M. Kauffmann, C.R. Linders, G.J. Patriarche and M.R. Smyth, Talanta, submitted for publication.

STUDIES OF THE KINETICS OF MEDIATED ELECTROCHEMICAL OXIDATION AT POLYMER-COATED ELECTRODES

Michael Sharp

Department of Analytical Chemistry
University of Umea
S-90187 Umea / Sweden

INTRODUCTION

During the past few years interest in the electrochemical behaviour of biological molecules has been further stimulated by the quest for practical biosensors[1]. These devices combine the inherent selectivity of a biochemical process, e.g., an enzyme reaction, with the advantages of measuring the analyte concentration in terms of a potential or current. Both potentiometric (equilibrium) and amperometric (dynamic) modes of operation, however, require facile electron transfer between the biomolecule and the surface of an electrode. For most large, electroactive biomolecules such direct heterogeneous electron transfer is extremely slow and even for the measurement of equilibrium potentials, which requires only very small currents, a suitable mediator must be added to the solution. Clearly, provision for charge transfer mediation is essential for the successful construction of sensors of both types. The selection of an appropriate mediator is usually based on the fulfillment of certain criteria. Its heterogeneous and homogeneous redox reactions must take place rapidly at a well-defined potential in the medium of interest and involve the transfer of a definite number of electrons to and from stable redox species. In addition the mediator should be adequately soluble, usually in aqueous media at pH ~ 7, and should not inhibit the reaction of the biomolecule with its substrate. Very few substances meet all these demands.

The possibility of permanently attaching molecules to the surfaces of common electrode materials without greatly affecting their electrochemical properties, however, widens the choice of mediator considerably and offers a number of advantages. Solubility requirements can be relaxed, separation problems are avoided and only very small total amounts are needed to provide a high concentration of mediator at an optimum location on the electrode surface. Several successful applications of this approach have been reported. The oxidation of ß-nicotinamide adenine dinucleotide (NADH) has attracted particular attention since it is produced from the co-factor, NAD^+, which is needed in over 250 different enzyme reactions. In principle the rate of the enzymatic process, and hence the concentration of a given substrate, may be easily monitored by measuring the current associated with the reconversion of NADH to NAD^+ in an amperometric device. At conventional electrode materials such as platinum, gold, and different form forms of carbon, however, this reaction is slow and complicated by associated chemical reactions and fouling by reaction products. Significantly higher reaction rates have been obtained at electrodes coated with monolayers of catechols[2], hydroquinones[3] and redox dyes[4,5] but poor stability arising from desorption of the mediator species or from side reactions has prevented the preparation of practical systems.

Improved durability can be obtained by coating the electrodes with thin polymer films in which the redox mediator sites are covalently- or electrostatically-bound. Such systems are much more complex than electrodes covered with monolayers and an adequate description of their properties must take into account the various charge-transfer and mass-transport steps which contribute to the overall electrochemical reaction. These are included in the theoretical treatments of polymer-modified electrodes which have been developed for steady-state techniques by Andrieux, Saveant and co-workers [6-8] in terms of characteristic current densities according to Figures 1 and 2, where D_E is the "electron" diffusion coefficient, D and D_s are substrate diffusion coefficients pertaining to the solution and polymer phases respectively, Γ is the total amount of mediator confined to the polymer coating, C_s is the concentration of substrate in the solution, κ is the substrate solution-coating partition coefficient, ϕ is the film thickness, δ is the diffusion-layer thickness at the coating-solution interface and n is the number of electrons transferred. Further details may be found in the original papers. The relative magnitudes of i_E, i_A and i_s, which are experimentally accessible by different electrochemical techniques, allow the rate determining processes to be identified and the system to be kinetically classified. Evaluation of i_k permits the second-order rate constant, k, for the mediator-substrate reaction to be calculated. Electron transfer between the electrode and the polymer-attached mediator, corresponding to i_H in Figure 1, is normally assumed to be rapid. High frequency ac-impedance studies have recently shown that this is the case for the systems examined below[9-11]. Complete kinetic analysis of this kind can then be used as a basis for optimizing the system.

The purpose of the present article is to provide, with the aid of selected examples, a brief survey of the electrochemical methods most commonly used in characterising the properties of polymer-modified electrodes and to illustrate the different types of behaviour which may be encountered when employing such systems to catalyse or mediate the electrochemical reactions of biomolecules. A comprehensive review on chemically-modified electrodes, which includes details of synthetic methods and appropriate electrochemical and auxiliary techniques for studying their properties, is available[12].

REDOX-POLYMER COATED ELECTRODE

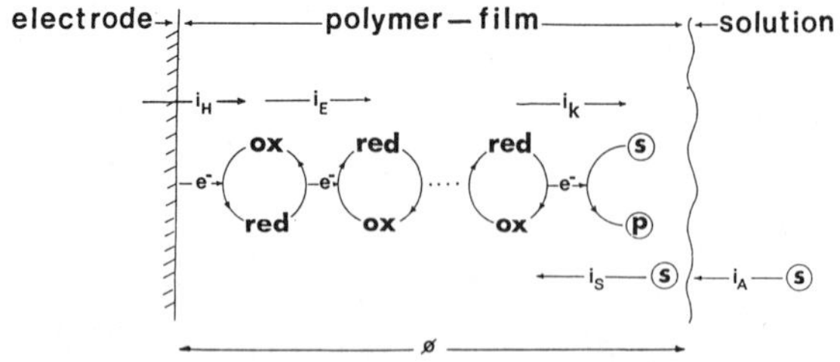

Figure 1. Schematic diagram of a polymer-coated electrode in a solution containing a substrate S. Current densities corresponding to the different contributing steps are defined in Figure 2.

Charge propagation through polymer film:

$$i_E = nF.D_E.\Gamma/\phi^2$$

Substrate solution-diffusion (Levich):

$$i_A = nF.D.C_S/\delta$$

Substrate film-diffusion:

$$i_S = nF.\kappa.D_S.C_S/\phi$$

Cross-exchange reaction:

$$i_k = nF.\kappa.k.\Gamma.C_S$$

Figure 2. Characteristic current densities defined according to Andrieux-Saveant theory.

EXPERIMENTAL

Chemicals

Potassium hexachloroiridate(IV) and potassium hexacyanoferrate(III) were obtained commercially and used directly. Tris-(2,2-bipyridyl)iron(II) perchlorate was synthesized according to Reference 13. Catechol(1,2-dihydroxygenzene), L-dopa(3-(3,4-dihydroxyphenyl)alanine), and tiron(1,2-dihydroxybenzene-3,5-disulphonic acid) were obtained commercially. Catechol was purified by sublimation; L-dopa and tiron were used as received. Structural formulae for the forms of these three substrates which predominate under the chosen experimental conditions are shown in Figure 3. Horse-heart cytochrome-c (Type VI) was purchased from the Sigma Chemical Co. in its oxidised form and was dialysed anaerobicaly at 5°C for 24 h following reduction with ascorbic acid. Final concentrations were determined from spectroscopic absorption ratios at 550 and 565 nm. Commercial samples of poly(4-vinylpyridine), (PVP, molecular masses 750,000 g and

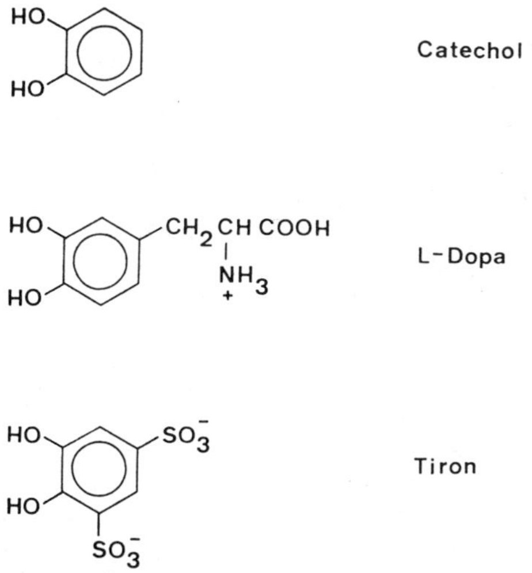

HO, HO (benzene ring) — Catechol

HO, HO (benzene ring) $CH_2CH(COOH)\overset{+}{N}H_3$ — L-Dopa

HO, HO (benzene ring) SO_3^-, SO_3^- — Tiron

Figure 3. Structural formulae of catechol substrates.

143,000 g) were dissolved in 2-propanol to give solutions with a polymer concentration of 0.5 wt%. A solution of Nafion R, perfluorosulphonic acid membrane, DuPont, equivalent mass 1100 g) was obtained from CG Processing Inc., Rockland, DE, and was diluted with 2-propanol to a final concentration of 0.5 wt%. Polymer films were prepared by micropipetting precise volumes of the above solutions onto the electrode surfaces and allowing the solvent to evaporate at room temperature. PVP coatings were stabilized either by cross-linking with 1,6-dibromohexane[14] or by addition of ternary co-polymer[15]. Quaternized poly(4-vinylpyridine) films (QPVP) were prepared by exposing pre-deposited and cross-linked coatings to a 5 % solution of methyl iodide in dimethylformamide overnight. Mediator species were incorporated electrostatically into the polymer films by immersing the coated electrodes in solutions of $K_3Fe(CN)_6$, K_2IrCl_6 and $Fe(bipy)_3(ClO_4)_2$ in the appropriate electrolyte. The supporting electrolyte for catechol, L-dopa and tiron was aqueous CF_3COONa (0.1 M) adjusted to pH = 2 with CF_3COOH. Cytochrome-c was dissolved in buffer solutions (pH = 7) containing Trizmabase (0.05 M) and Na_2SO_4 (0.05 M).

Apparatus

Working electrodes were constructed by pressure-fitting discs of glassy-carbon (active areas 0.20 and 0.34 cm^2), edge-plane pyrolytic graphite (active area 0.29 cm^2) and gold (active area 0.20 cm^2) into insulating sleeves which could be used in both stationary and rotating-disc electrode configurations. Pretreatment comprised polishing with alumina slurries (1.0, 0.3 and 0.05 μm grain-sizes) and ultrasonic cleaning in ethanol.

Bioanalytical Systems BAS 100 and BAS 100A Electrochemical Analyzers, a PAR 174 polarograph and a pulse generator-transient recorder combination[16] were used for performing cyclic voltammetry, linear-sweep voltammetry, and chronocoulometry. Rotating-disc current-potential curves were recorded with a Tacussel model EDI electrode operating at rates of 100 to 2500 rpm. All measurements were made at ambient temperature (21±2°C) in a single-compartment cell employing saturated calomel (SCE) or saturated sodium chloride calomel (SSCE) reference electrodes and platinum discs as counter electrodes.

RESULTS AND DISCUSSION

Since the same experimental and interpretive procedures were applied throughout, they are treated in some detail for the first of the systems which is described below and only the more important features are considered for the remainder.

Characterisation of redox-polymer modified electrodes

Cyclic voltammetry at different applied potential scan rates is normally the first technique employed to study the properties of polymer-modified electrodes. From current-potential curves recorded with solutions containing supporting electrolyte only and at scan rates sufficiently low for thin layer conditions to prevail both the formal potentials and the amounts of redox species incorporated in the surface coating can easily be measured. In addition, deviations from ideal behaviour provide information on interactions within the polymer phase and, when uncompensated resistance effects can be accounted for, on the reversibility of electron transfer at the electrode-coating interface. Figure 4 shows a typical cyclic voltammogram obtained from a glassy-carbon electrode covered with

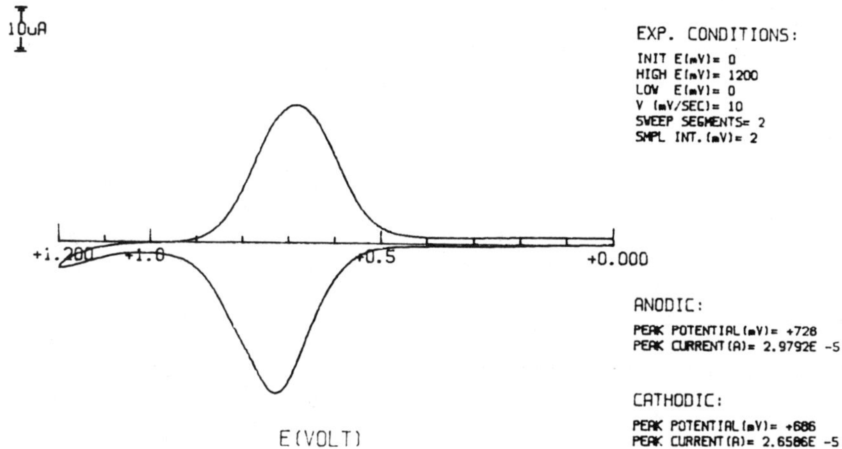

Figure 4. Cyclic voltammogram obtained for a glassy-carbon electrode coated with a PVP:$IrCl_6^{2-}$ film.

a PVP film containing the $IrCl_6^{3-/2-}$ redox couple. The formal potential, $E^{o'}$ = 0.71 V vs. SCE, estimated from the average of the peak potentials, was close to that observed for the same couple at a bare electrode in the same medium. Integration of the voltammogram showed that the film contained a total 1.88 x 10^{-8} mol$\cdot$cm^{-2} of $IrCl_6^{3-}$, which corresponds to compensation of approximately 25% of the charged sites available on the polymer. On the basis of a film density of 0.65 g$\cdot$cm^{-3}, the concentration of redox species thus approaches 0.5 M.

Charge propagation through the polymer phase may involve both electron-hopping between neighbouring redox sites and ionic motion and both mechanisms have been shown to conform with a diffusion-based model[17]. The maximum rate of charge transport can be estimated from the characteristic current density, i_E, which has frequently been calculated from the slopes, S, of charge vs. (time)$^{-1/2}$ plots derived from chronocoulometric experiments according to Equation 1[18].

$$I_E = \frac{\pi\, nS^2}{4F\Gamma} \qquad (1)$$

Such measurements are restricted to short times to ensure the validity of semi-infinite conditions. However, recent objections have been raised to this method of measuring i_E which may underestimate the true value by a factor of 10-100[19-21]. Undoubtedly, large structural changes and charge compensating ionic movements take place in the polymer coating when the oxidation state of the mediator is altered by an applied potential step and these may control the rate of the conduction process. Such effects would not be present during a mediated reaction at steady state when the coating contains the redox species in a single oxidation state. Nevertheless, chronocoulometry was employed in the present studies with due recognition of the fact that the values obtained may represent lower limits. For the PVP-$IrCl_6$ system, the charge propagation current density was found to be 3.4 mA$\cdot$cm^{-2}. Although much smaller values of i_E of the order of

0.3-0.4 mA·cm^{-2} were measured for the Nafion coating used, charge propagation currents were much larger than any of the limiting currents observed in rotating-disc electrode experiments. The transport of charge through both PVP and Nafion coatings was not considered to be rate determining in any of the mediated oxidation reactions to be described below.

Mediated oxidation of L-dopa at a PVP-$IrCl_6^{2-}$ coated electrode

Linear sweep voltammetry yields a current peak at a potential of 0.73 V vs. SCE for the oxidation of L-dopa in aqueous CF_3COONa (0.1 M, pH = 2) at a bare glassy-carbon electrode. This potential is close to that where $IrCl_6^{3-}$ is oxidized and only a single, composite wave comprising currents from both the mediated and direct oxidation of L-dopa was observed in rotating-disc experiments when the glassy-carbon electrode was coated with a PVP-film in which $IrCl_6^{3-}$ had been incorporated. At potentials more positive than that required to convert $IrCl_6^{3-}$ to $IrCl_6^{2-}$, current plateaus were recorded which were dependent on both electrode rotation rate, ω, and on the L-dopa substrate concentration, C_s. Levich plots of limiting current vs. $\omega^{1/2}$ are shown in Figure 5

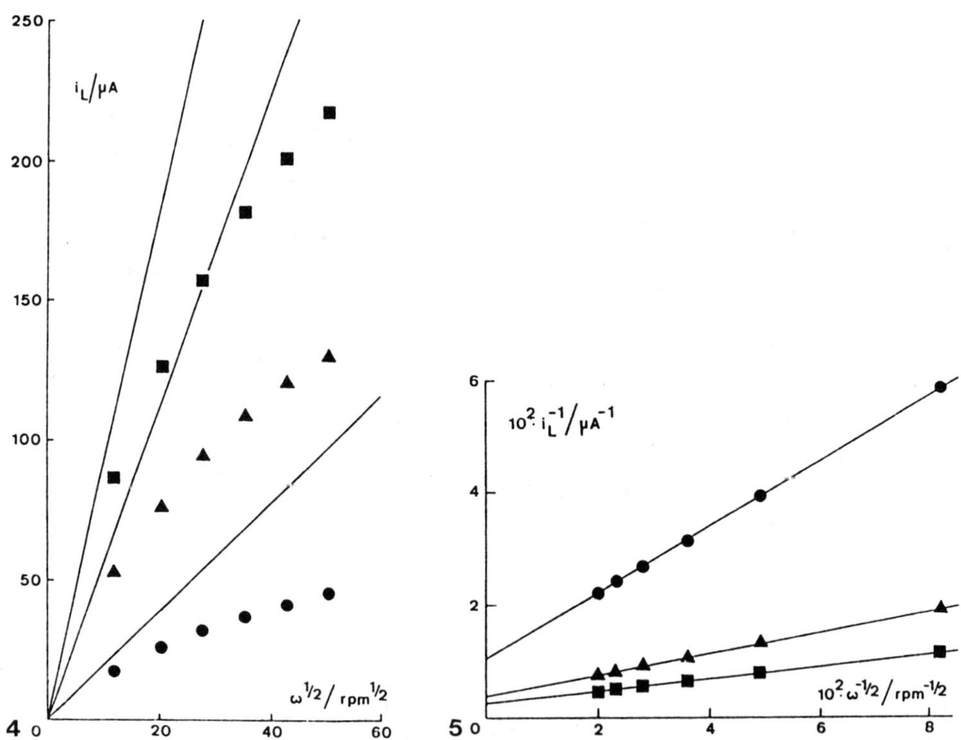

Figure 5. Levich plots of i_L vs. $\omega^{1/2}$ for the oxidation of L-Dopa at a GC-PVP:$IrCl_6^{2-}$ electrode. The experimental points correspond to the following L-Dopa concentrations; (●) 0.213, (▲) 0.614 and (■) 0.986 mM. The solid lines refer to responses at the bare electrode.

Figure 6. Koutecky-Levich plots of i_L^{-1} vs. $\omega^{-1/2}$ corresponding to the data shown in Figure 5. The solid lines refer to responses at the bare electrode.

together with the linear plots (solid lines) obtained for similar measurements at the bare electrode. The deviations of the data points from the lines indicate that a kinetic limitation is imposed by the surface coating, i.e., the rate of the $IrCl_6^{2-}$-L-dopa reaction is not sufficiently fast to consume all the substrate which reaches the coating-solution interface. Corresponding Koutecky-Levich reciprocal plots of i_L^{-1} vs. $\omega^{-1/2}$, shown in Figure 6, were strictly linear with slopes equal to those obtained for the oxidation of L-dopa at the bare electrode. The reciprocals of the intercepts at $\omega^{-1/2} = 0$ define the film currents, i_F. The remaining parameter needed to proceed further with the kinetic analysis is the characteristic substrate permeation current density, i_s. In the present case this was determined to be $i_s = 0.88$ mA$\cdot$cm^{-2} for a substrate concentration of 1 mM from the intercepts of Koutecky-Levich plots for the oxidation of L-dopa at electrodes coated with identical PVP films to those used for the mediation experiments but which contained the non-mediating surrogate anion $Fe(CN)_6^{3-}$ instead of $IrCl_6^{2-}$ in an effort to preserve morphology. The values of i_s for other substrate concentrations were calculated from the relevant equation in Figure 2. In favourable cases (see the mediated oxidation of catechol at a Nafion-$Fe(bipy)_3^{3+}$ coated electrode below), two distinct waves corresponding to direct and mediated reactions may be observed which greatly facilitates the determination of i_s.

Considered collectively, the experimental observations, i.e., the presence of a single wave in the rotating-disc voltammogram, the linearity of the Koutecky-Levich plots, and the relative magnitudes of the characteristic current densities, $i_E > i_s$ and i_A, at all values of ω and C_s employed, assign the system to the "SR" kinetic category in the diagnostic tables provided by Andrieux and Saveant[7]. This implies that the overall rate of oxidation of L-dopa is controlled jointly by substrate permeation within the polymer phase and by the kinetics of the bimolecular reaction itself and that the latter occupies a region of the coating close to the film-solution interface. The film currents, i_F, derived from the Koutecky-Levich plots are then defined by Equation 2 [7],

$$i_F = (i_K \cdot i_s)^{1/2} \tag{2}$$

which allows values for the kinetic current, i_K, to be calculated since i_s is known. The dependence of i_K on substrate concentration is shown in Figure 7, and conforms with the appropriate equation in Figure 2. The slope of this plot provides a value for the product of the second order rate constant for the mediator-substrate reaction and the substrate partition coefficient of $\kappa k = 4.9 \pm 0.46 \times 10^2$ $M^{-1}\cdot s^{-1}$. This agrees well with a value of $\kappa k = 5.3 \pm 0.3 \times 10^2$ $M^{-1}\cdot s^{-1}$ observed for the same reaction at a PVP coating stabilized by the addition of a ternary copolymer[15]. From a kinetic viewpoint these two types of coating thus appear to be quite similar. The homogeneous solution rate constant for reaction between $IrCl_6^{2-}$ and L-dopa under comparable conditions[22] is of the order of 9×10^3 $M^{-1}\cdot s^{-1}$ and the much lower value of κk pertaining to the polymer phase might indicate either diminished reactivity, corresponding to a lower k, or poor partitioning of the substrate into the coating for which $\kappa < 1$. Similar measurements with catechol as substrate[15] yielded values of $\kappa k = 3.5 \pm 0.5 \times 10^3$ $M^{-1}\cdot s^{-1}$ which are in much better accord with the homogeneous rate constant of $k_h = 4.3 \times 10^3$ $M^{-1}\cdot s^{-1}$ [22]. This result suggests that the reactivities of these two catechols are not greatly different in the polymer and solution phases and that the lower value of κk for L-dopa arises from a decreased concentration of this substrate inside the coating. The L-dopa molecule is positively charged under the chosen experimental conditions and would be expected to experience at least partial Donnan exclusion by the similarly charged PVP

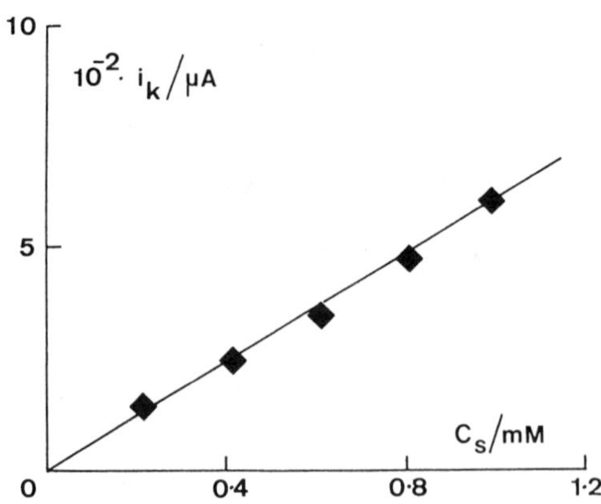

Figure 7. Plot of i_K vs. C_s for the oxidation of L-Dopa at a GC-PVP:$IrCl_6^{2-}$ electrode.

matrix. Such electrostatic repulsion would not occur for the non-charged catechol species and κ would not differ greatly from unity.

Estimates of the thickness of the zone, μ, within which the mediator-substrate reaction proceeds can be obtained from Equation 3 [6,8],

$$\mu = (i_s/i_K)^{1/2} \cdot \phi \tag{3}$$

From the measured values of i_s and i_K it is found that the reaction between $IrCl_6^{2-}$ and L-dopa occupies 65-70% of the cross-linked PVP coating whereas that between $IrCl_6^{2-}$ and catechol takes place within the outer 20% of PVP films stabilised by the ternary co-polymer.

Mediated oxidation of catechol at a Nafion-$Fe(bipy)_3^{3+}$ coated electrode

Equation 3 indicates that for redox-polymer mediated reactions that comply with "SR" kinetics, fast mediator-substrate cross-reaction rates (large i_K) and poor substrate permeation through the polymer phase (low i_s) will tend to decrease the thickness of the reaction zone. Situations may thus arise in which only the outermost layers of mediator sites attached to the coating participate in the reaction with substrate. Such a surface process is observed for the oxidation of catechol at a carbon electrode modified with a Nafion film containing $Fe(bipy)_3^{3+}$ as mediator. The homogeneous solution rate constant for the oxidation of catechol by $Fe(bipy)_3^{3+}$ is estimated[23] to be 2 x 10^6 $M^{-1} \cdot s^{-1}$ and is roughly 1000 times faster than the rate of oxidation of the same substrate by $IrCl_6^{2-}$. The rotating-disc current-potential curves recorded for this system (Figure 8) show two waves, the first of which at E ~ 0.45 V corresponds to the direct oxidation at the underlying carbon surface of catechol species which are able to penetrate the Nafion film. Substrate film-diffusion current densities, i_s, were determined from the rotation rate dependence of the limiting currents of this first wave and were found to be approximately ten times lower than those measured for the permeation of catechol through PVP films of comparable thickness. Limiting currents measured at potentials more positive than that required for converting $Fe(bipy)_3^{2+}$ to the mediating form $Fe(bipy)_3^{3+}$ were much larger than those of the first wave and show that mediated oxidation of catechol also takes place. Levich plots of these limiting currents, shown in Figure 9, contain two interesting features. At low concentrations of substrate, complete agreement is found

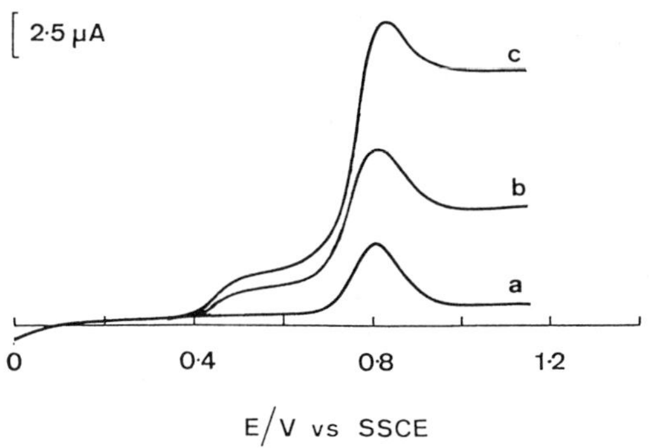

Figure 8. Rotating-disc current potential curves recorded at a scan-rate of 5 mV·s-1 and ω = 160 rpm for the oxidation of catechol at an EPG-Nafion:$Fe(bipy)_3^{3+}$ electrode. The catechol concentrations were (a) 0, (b) 0.0467 and (c) 0.116 mM.

between the currents measured (points) and those predicted for the oxidation of catechol at the bare electrode by the Levich equation (solid lines). Under these conditions mass-transport alone controls the rate of oxidation; no kinetic limitations are apparent within the range of electrode rotation-rates examined. Since $i_E > i_A$ and i_S at all the concentrations and rotation rates examined, this system also conforms with "SR" kinetics but the values of i_K derived from the Koutecky-Levich plots via Equation 2 are found to be very much larger than i_S. This result implies that the mediated oxidation of catechol is restricted to a very thin reaction zone at the coating-solution interface and may involve only the outermost monolayer of $Fe(bipy)_3^{3+}$ species. Figure 9 also shows that deviations from the Levich lines occur and an approach to a concentration- and rotation rate-independent current becomes apparent as the substrate concentration and rotation rate are increased. A kinetic limitation has thus been reached under these conditions. These observations can be interpreted in terms of the model for the surface reaction shown in Figure 10, which includes precursor complex formation prior to a rate determining intramolecular charge transfer. Analysis of this model shows that limiting current densities appropriate to rotating-disc electrode experiments can be expressed in Koutecky-Levich form as the sum of three terms according to Equation 4,

$$\frac{1}{i_L} = \frac{1}{nFk_2\Gamma_m} + \frac{1}{nFk_2K_e\Gamma_mC_s} + \frac{1}{nFk_1\Gamma_mC_s} \qquad (4)$$

where k_1 ($M^{-1}\cdot s^{-1}$) is the rotation rate-dependent pseudo rate constant for precursor formation, k_2 (s^{-1}) is the first-order rate constant for intramolecular electron-transfer, K_e (M^{-1}) is the precursor complex formation constant, and Γ_m ($mol\cdot cm^{-2}$) is the surface density of mediator species. Since the $Fe(bipy)_3^{3+}$-catechol reaction is mass-transport controlled at low C_s, its second-order rate constant, k_2K_s, must be large compared with k_1 and the second term on the right of Equation 4 can be neglected. At high substrate concentrations, the third term will also be small and the limiting current, $i_L = nFk_2\Gamma_m$, becomes independent of C_s and ω as required by experiment. When Γ_m is known, the rate of electron transfer within the precursor complex, k_2, can be estimated. For the present system, i_L = 0.13 $mA\cdot cm^{-2}$ and $\Gamma_m = 3.4 \times 10^{-11}$ $mol\cdot cm^{-2}$, which yield a value of

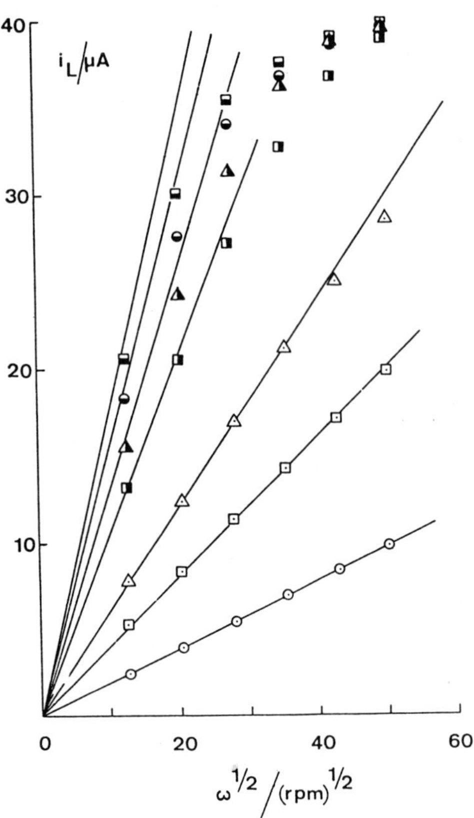

Figure 9. Levich plots of i_L vs. $\omega^{1/2}$ for the oxidation of catechol at EPG-Nafion:$Fe(bipy)_3^{3+}$ electrodes. The catechol concentrations were: (○) 0.0243, (□) 0.0467, (Δ) 0.0698, (◨) 0.116, (▲) 0.139 (◒) 0.162 and (⬓) 0.184 mM. The solid lines represent responses at the bare electrode.

$k_2 = 20\ s^{-1}$ for the rate of electron transfer within the $Fe(bipy)_3^{3+}$... catechol complex. No value for the formation constant of the latter can be obtained from the experimental data.

Mediated oxidation of tiron at a Nafion-$Fe(bipy)_3^{3+}$ coated electrode

The model shown in Figure 10 and its associated Equation 4 also provide an interpretation of the kinetics of oxidation of tiron at an electrode coated with a Nafion film containing $Fe(bipy)_3^{3+}$. In this system Donnan exclusion of the negatively-charged tiron species by the similarly charged Nafion matrix is expected and clear evidence of the inability of the substrate to penetrate the coating is given by a comparison of the rotating-disc current-potential curves obtained at coated and bare electrodes shown in Figure 11.

The substrate film-diffusion current density, i_s, is thus equal to zero in this case and the mediated oxidation reaction, the existence of which is confirmed by the limiting currents observed at potentials positive of that needed to generate $Fe(bipy)_3^{3+}$, is definitely of surface character. The limiting currents measured for this system were largely independent of

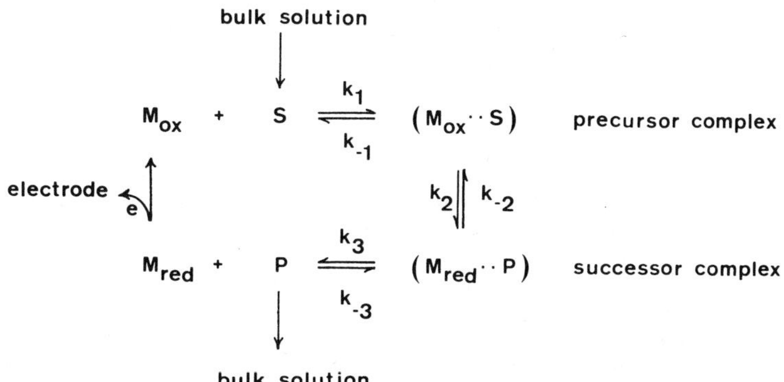

Figure 10. Model for the oxidation reactions of catechol and tiron at the surfaces of Nafion films containing $Fe(bipy)_3^{3+}$.

rotation rate and very much lower than those predicted by the Levich equation even at low substrate concentrations, as shown by the Levich plots in Figure 12. In addition the limiting currents appear to approach a concentration independent value as C_s is increased. This is more clearly seen in the plot of i_L vs. C_s shown in Figure 13. These observations are consistent with kinetic control of the mediated reaction at low substrate concentrations and a saturation in the rate of this reaction, similar to that found for the oxidation of catechol, as C_s becomes higher. Since $k_1 > k_2K_e$, the appropriate (Michaelis-Menten) form of Equation 4 to accommodate the results becomes,

$$\frac{1}{i_L} = \frac{1}{nFk_2\Gamma_m} + \frac{1}{nFk_2K_e\Gamma_mC_s} \tag{5}$$

Figure 14 shows that the plot of i_L^{-1} vs C_s^{-1}, prepared from the data in Figure 13, is linear in accordance with Equation 5 and values of $k_2 = 8s^{-1}$ and $K_e = 485\ M^{-1}$ were derived from its slope and intercept. The second-

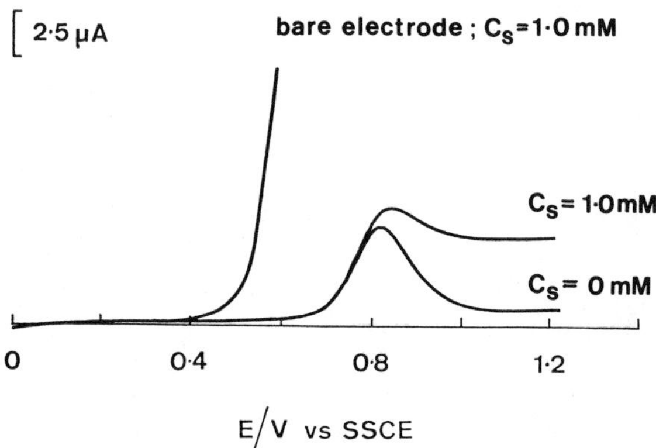

Figure 11. Rotating-disc current-potential curves recorded at a scan-rate of 5 $mV \cdot s^{-1}$ and ω = 160 rpm for the oxidation of tiron at an EPG-Nation:$Fe(bipy)_3^{3+}$ electrode. The limiting current for the oxidation of tiron (1.00 mM) at the uncoated electrode was 101 µA.

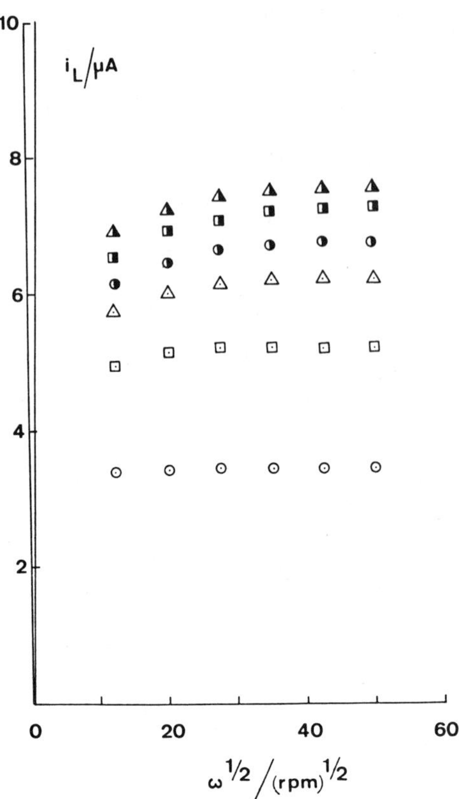

Figure 12. Levich plots of i_L vs. $\omega^{1/2}$ for the oxidation of tiron at an EPG-Nafion:$Fe(bipy)_3^{3+}$ electrode. The tiron concentrations (○) 1.00, (□) 1.96, (△) 2.89, (◑) 3.78, (◨) 4.64 and (▲) 5.45 mM. The solid line shows the Levich plot for response at the uncoated electrode for a tiron concentration of 1.00 mM.

order rate constant for the $Fe(bipy)_3^{3+}$-tiron reaction, $k_{surf} = k_2K_e = 4 \times 10^3\ M^{-1}\cdot s^{-1}$ is found to be in reasonable agreement with that observed for the oxidation of tiron by the structurally similar species $Fe(phen)_3^{3+}$ in homogeneous solution[23], $k_h = 3 \times 10^3\ M^{-1}\cdot s^{-1}$, and the kinetics do not appear to have been greatly affected by incorporating $Fe(bipy)_3^{3+}$ in the Nafion coating.

Mediated oxidation of ferrocytochrome-c at a QPVP-$Fe(CN)_6^{3-}$ coated electrode

The redox reactions of cytochrome-c proceed rapidly under physiological conditions and with a range of inorganic and organic complexes in homogeneous solution but only very slowly at solid electrodes. Significant binding between the dissolved reactants at specific sites on the cytochrome molecule is known to be necessary for facile electron transfer and the lack of similar interactions at an electrode surface is considered to be the cause of the slow heterogeneous reaction. Increased charge transfer rates are observed when precursor complex formation is enhanced by an adsorbed layer of 4,4'-bipyridine on a gold surface[24,25]. In this instance, however, the adsorbed species does not function as a true mediator. Since

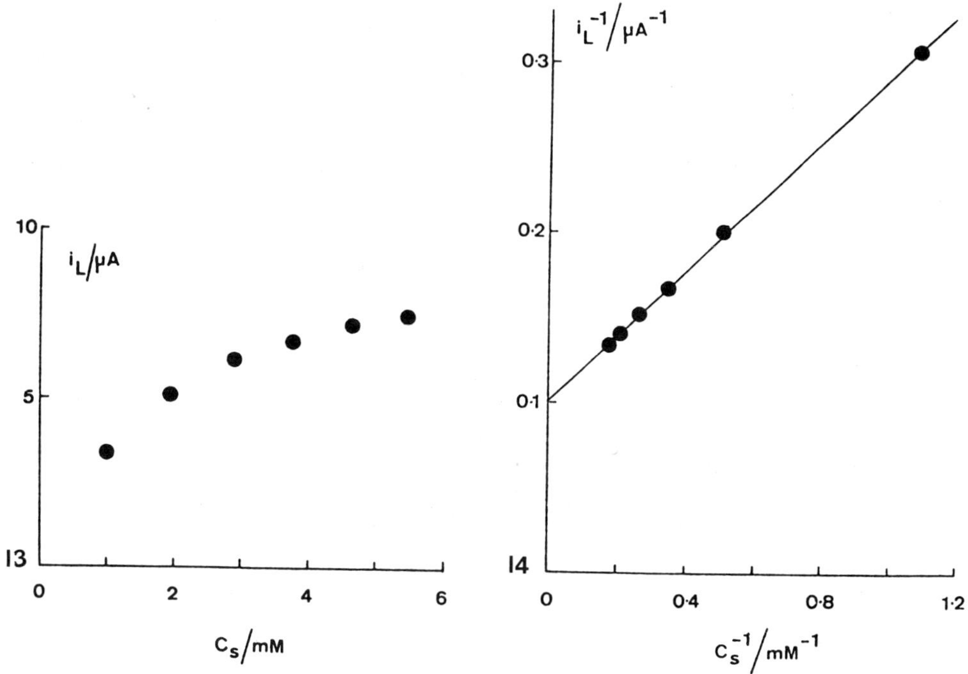

Figure 13. Plot of i_L vs. C_s for the oxidation of tiron at an EPG-Nafion:$Fe(bipy)_3^{3+}$ electrode.

Figure 14. Michaelis-Menten plot of i_L^{-1} vs. C_s^{-1} corresponding to the data shown in Figure 13.

$Fe(CN)_6^{3-}$ rapidly oxidises ferrocytochrome-c in solution at pH ~ 7 [26,27], and is itself readily regenerated by re-oxidation of the $Fe(CN)^{4-}$ produced at a metal or carbon electrode, this species is expected to behave as a charge mediator when contained within a polymer coating. To examine this possibility, $Fe(CN)_6^{3-}$ was incorporated into films of QPVP on a gold electrode. Limiting currents, which indicated that mediated oxidation occurred, were obtained from rotating-disc electrode experiments at different substrate concentrations and different surface coverages of the mediator and were used to determine the kinetics of ferrocytochrome-c oxidation by polymer-bound $Fe(CN)_6^{3-}$. Figure 15 shows examples of Levich plots of i_L vs. $\omega^{1/2}$ recorded at a given C_s for two different surface coverages. The deviations of the experimental points from the calculated Levich plot (solid line) are indicative of kinetic limitations. Since the ferrocytochrome-c molecule is highly positively charged at pH ~ 7, Donnan exclusion by the QPVP matrix should prevail and the mediated oxidation restricted to the coating-solution interface. This was confirmed by attempts to measure the substrate film-diffusion current density with QPVP films containing the non-mediating surrogate anion $IrCl_6^{3-}$. Limiting currents in such experiments differed insignificantly from background values indicating that cytochrome-c did not permeate the film. In these circumstances, the film currents derived from Koutecky-Levich plots are equal to kinetic curents, i_k, and the second-order rate constant for the mediated reaction, which was calculated from the relevant equation in Figure 2 where $\Gamma = \Gamma_m$ the surface density of mediator sites, was found to be 1.5×10^4 $M^{-1} \cdot s^{-1}$. This is very much lower than the corresponding rate constant for reaction between ferrocytochrome-c and $Fe(CN)_6^{3-}$ in

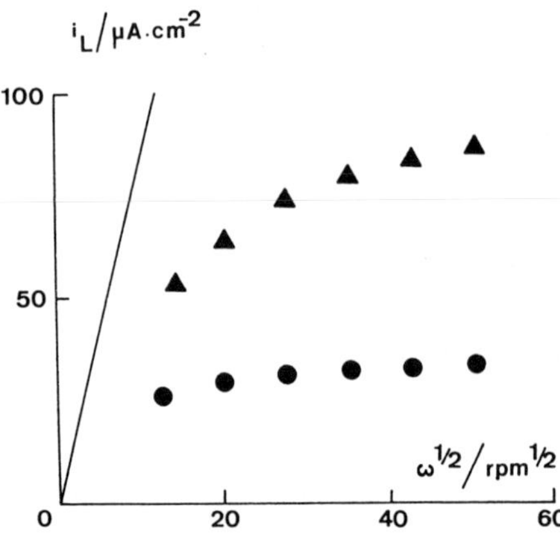

Figure 15. Levich plots of i_L vs $\omega^{1/2}$ for the oxidation of ferrocytochrome-c at a gold electrode coated with QPVP containing $Fe(CN)_6^{3-}$. The substrate concentration was C_s = 1.95 M and mediator coverages were (○) 1.9 x 10^{-11} and (Δ) 3.6 x 10^{-11} mol·cm^{-2}. The solid line represents the calculated Levich plot for response at the uncoated electrode.

homogeneous solution where values within the range 6-16 x 10^6 $M^{-1} \cdot s^{-1}$ have been reported[26]. Precursor complex formation, in which precise orientation of the exposed heme group in the cytochrome molecule is a prerequisite for electron transfer, may be more easily attained for the solution reaction than for the surface process where the negatively charged $Fe(CN)_6^{3-}$ species is at least partially buried in the positively charged QPVP coating. The slower rate for the surface reaction may result from either decreased complex stability (lower K_e), from retarded electron transfer within the complex (lower k_2), or from a combination of both. Intramolecular electron transfer rates, in particular, are highly dependent on distance and an increase in the separation of redox sites within the ferrocytochrome-$Fe(CN)_6^{3-}$ complex of the order of 0.4 nm would be sufficient to reduce k_2 by a factor of 300 [28]. Without further knowledge of the structural details of the precursor complex, such considerations must remain speculative.

CONCLUSIONS

The application of thin polymer films offers a versatile method for permanently attaching electron-transfer mediators to electrode surfaces in high local concentration. Such electrodes can be used to promote the electrochemical oxidation or reduction of dissolved species which normally react only very slowly at bare metal or carbon surfaces. Fairly detailed kinetic characterisations of these systems can be achieved using cyclic voltammetry, chronocoulometry and rotating-disc electrode measurements in conjunction with available theoretical treatments. Although mediator-substrate reaction rates measured in homogeneous solution may serve as a guide to the behaviour of the modified electrode, the above results suggest that it may be necessary to resolve overall second-order rate constants into precursor complex formation constants and intramolecular electron transfer rates to better estimate the maximum currents which can be sustained by the mediated process.

ACKNOWLEDGEMENTS

The author thanks Don Montgomery and Fred Anson, Caltech, who were involved in the studies of catechol oxidations at PVP-coatings stabilised by a ternary co-polymer, Britta Lindholm, University of Umea, who carried out most of the work with ferrocytochrome-c, and Elsevier Sequoia S.A. for permision to reproduce Figures 8, 9, 11-15. Financial support was provided by the Swedish Natural Science Research Council.

REFERENCES

1. "Biosensors: Fundamentals and Applications", A.P.F. Turner, I. Karube and G.S. Wilson, Eds., Oxford Scientific Publications, Oxford (1987).

2. H. Jaegfeldt, A. Torstensson, L. Gorton and G. Johansson, Anal. Chem., 53, 1972 (1981)

3. D. Tse and T. Kuwana, Anal. Chem., 49, 1589 (1977).

4. H. Huck, A. Schelter-Graf, J. Danzer, P. Kirch and H. Schmidt, Analyst, 109, 149 (1984).

5. L. Gorton, A. Torstensson, H. Jaegfeldt and G. Johansson, J. Electro-anal. Chem., 161, 103 (1984).

6. C.P. Andrieux, J.M. Duman-Bouchiat and J.M. Saveant, J. Electroanal. Chem., 131, 1 (1982).

7. C.P. Andrieux and J.M. Saveant, J. Electroanal. Chem., 134, 163 (1982).

8. C.P. Andrieux and J.M. Saveant, J. Electroanal. Chem., 142, 1 (1982).

9. R.D. Armstrong, B. Lindholm and M. Sharp, J. Electroanal. Chem., 202, 69 (1986).

10. B. Lindholm, R.D. Armstrong and M. Sharp, J. Electroanal. Chem., in press.

11. B. Lindholm, E.-L. Lind and M. Sharp, in preparation.

12. R.W. Murray in "Electroanalytical Chemistry", Vol. 13, A.J. Bard, Ed., Dekker, New York (1984), p. 191.

13. F.H. Burstall and R.S. Nyholm, J. Chem. Soc., 3570 (1952).

14. B. Lindholm and M. Sharp, J. Electroanal. Chem., 198, 37 (1986).

15. M. Sharp, D.D. Montgomery and F.C. Anson, J. Electroanal. Chem., 194, 247 (1985).

16. M. Sharp, Electrochimica Acta, 28, 301 (1983).

17. C.P. Andrieux and J.M. Saveant, J. Electroanal. Chem., 111, 377 (1980).

18. F.C. Anson, J.M. Saveant and K. Shigehara, J. Am. Chem. Soc., 105, 1096 (1983).

19. M. Majda and L.R. Faulkner, J. Electroanal. Chem., 137, 149 (1982).

20. X. Chen, P. He and L.R. Faulkner, J. Electroanal. Chem., 222, 223 (1987).

21. C.E. Chidsey, B.J. Feldman, C. Lundgren and R.W. Murray, Anal. Chem., 58, 601 (1986).

22. E. Mentasti, E. Pelizzetti and C. Baiocchi, J. Chem. Soc. Dalton Trans., 132 (1977.

23. E. Pelizzetti and E. Mentasti, Z. Phys. Chem., 105, 21 (1977).

24. M.J. Eddowes and H.A.O. Hill, J. Chem. Soc. Chem. Commun., 71 (1977).

25. W.J. Albery, M.J. Eddowes, H.A.O. Hill and A.R. Hillman, J. Am. Chem. Soc., 103, 3904 (1981).

26. A.J. Ahmed and F. Millett, J. Biol. Chem., 256, 1611 (1981).

27. Y. Ilan and A. Schafferman, Biochem. Biophys. Acta, 548, 161 (1979).

28. J.J. Hopfield, Proc. Natl. Acad. Sci. U.S.A., 71, 3640 (1974).

ELECTROCHEMICAL PREPARATION OF METALLOPHTHALOCYANINE FILMS

Sei-ichiro Iijima*, Fumio Mizutani,
Yoshio Tanaka and Kunihiro Ichimura

Research Institute for Polymers and Textiles
1-1-4 Yatabe-Higashi
Tsukuba, Ibaraki 305 / Japan

INTRODUCTION

Organic thin films have been extensively investigated in connection with the attempts to apply organic materials to electronic and electrochemical devices. Phthalocyanine and its metallo-derivatives are particularly of interest in this field; a wide variety of physical and physicochemical properties applicable to such devices, i.e., photoconductivity[1], photovoltaic effect[2,3], electrochromism[4], electrocatalysis[5-8], have been reported for their films.

Electrochemical deposition, including electropolymerization, of organic molecules has recently been shown to be a useful method for easily obtaining uniform and adhesive films on electrically conductive substrates[9-16]. Phthalocyanines, however, have received little attention in the field of electrochemical deposition on account of their very low solubility in most organic solvents as well as in aqueous solutions. On the other hand, several studies have been reported on the preparation of films of porphyrins, the analogous macrocyclic compounds, by oxidative electropolymerization[14,15].

We noted that octaalkoxyphthalocyanines show exceptionally high solubilities in various organic solvents, and have tried to prepare electrodeposited films from these derivatives. In this paper, we report about a method of film preparation using zinc 1,4,8,11,15,18,22,25-octaethoxyphthalocyanine ($(EtO)_8PcZn$, Figure 1) under potential sweep conditions. The films obtained were found to be electrochemically active and to catalyze the reduction of O_2 on Au electrodes.

EXPERIMENTAL

Materials

Zinc 1,4,8,11,15,18,22,25-octaethoxyphthalocyanine ($(EtO)_8PcZn$) was prepared by the reaction of 3,6-diethoxyphthalonitrile with $ZnCl_2$ in the presence of 1,8-diazabicyclo[5.4.0]-7-undecene[17,18], and purified by column chromatography on alumina. Found: C, 61.85; H, 5.37; N, 11.76%. Calcd for $C_{48}H_{48}N_8O_8Zn$: C, 61.97; H, 5.20; N, 12.04%. The field desorption mass spectrum of the sample showed the peaks which correspond to M^+ (m/e = 928 for the peak with the most comon isotopic abundances) and 1/2 M^+ only;

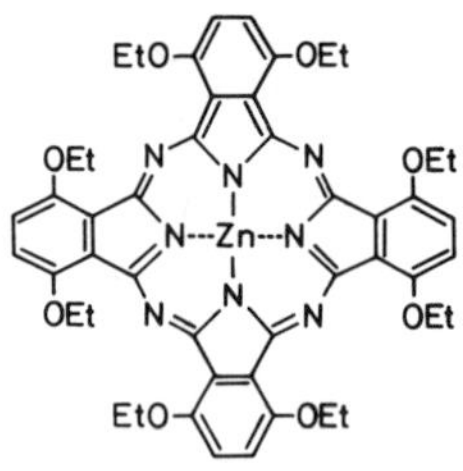

Figure 1. Zinc 1,4,8,11,15,18,22,25-octaethyoxyphthalocyanine ($(EtO)_8PcZn$).

this indicated the high purity of the sample. Tetra-n-butylammonium perchlorate ($n\text{-}Bu_4NClO_4$) was purchased from Nakarai Chemicals and recrystallized from water-methanol. Spectroquality dichloromethane was dried over 0.4 nm molecular sieves and then distilled.

Electrochemical experiments

A standard three-electrode configuration was employed. A platinum disk (TOA Electronics, area 0.20 cm^2) or gold disk (Bioanalytical Systems, area 0.03 cm^2) was used as the working electrode, and a platinum wire, as the counter electrode. The reference electrode for the experiments in dichloromethane solutions was an Ag/Ag^+ couple, which was made up by a silver wire and an acetonitrile solution containing 0.01 M silver perchlorate and 0.09 M $n\text{-}Bu_4NClO_4$. An Ag/AgCl electrode was used for the experiments in perchloric acid.

Spectroscopy

A Shimadzu UV-200 spectrophotometer equipped with an integrating sphere was used for the measurements of reflection spectra. The X-ray photoelectron spectra were obtained by using a Shimadzu ESCA 750, which was equipped with a Mg source ($Mg\alpha_{1,2}$) operated at a power of 240 W.

RESULTS AND DISCUSSION

$(EtO)_8PcZn$ is (unlike most phthalocyanines) sufficiently soluble in dichloromethane to permit electrochemical studies in this solvent. Figure 2a shows the cyclic voltammogram (CV) of 0.8 mM $(EtO)_8PcZn$ in an argon-saturated dichloromethane solution containing 0.1 M $n\text{-}Bu_4NClO_4$; the CV was recorded at a Pt electrode in the potential range of -0.5 to 1.0 V. Two pairs of redox peaks, which would correspond to two consecutive one-electron processes, were observed in the CV. The half-wave potentials, which were obtained from the average of the anodic and cathodic peak potentials, were 0.09 and 0.48 V for the first and second oxidation, respectively. The CV curve was virtually unchanged by repeated potential sweeps.

When the potential region was expanded to more positive values, an irreversible anodic peak appeared around at 1.6 V. The CV obtained through repeated sweeps between -1.0 and 1.7 V is shown in Figure 2b. In this case, increases in the oxidation and reduction current envelope between -0.5 and 1.2 V were observed with each successive potential sweep. This

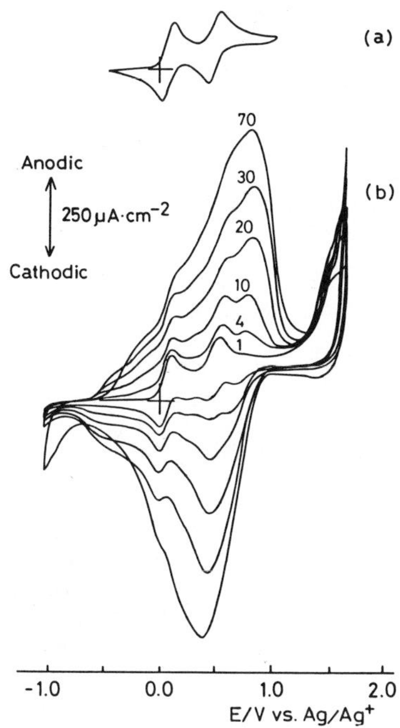

Figure 2. (a) Cyclic voltammogram of 0.8 mM $(EtO)_8PcZn$ in the potential range of -0.5 to 1.0 V. (b) Cyclic voltammograms of 0.8 mM $(EtO)_8PcZn$ obtained by repeated potential sweeps in the range of -1.0 to 1.7 V. Scan rate, 100 mVs^{-1}. Working electrode, Pt. Electrolyte, 0.1 M m-Bu_4NClO_4/dichloromethane. Numbers in (b) refer to those of the cyclic scans.

behavior is characteristic of deposition of a material on the electrode which is capable of being repetitively charged and discharged[15]. The redox waves finally settled to show a complicated shape with an anodic and a cathodic peak around 0.9 and 0.4 V, respectively. The height of the irreversible anodic peak around 1.6 V gradually decreased during the repetition of potential sweep.

Upon the oxidation of $(EtO)_8PcZn$ mentioned above, using a potential sweep technique, a uniform, blue film was formed on the Pt electrode. The deposited film showed total insolubility in organic solvents as well as aqueous solutions, while it is usually known that phthalocyanines are slightly soluble in certain organic solvents such as N,N-dimethylacetamide and N,N-dimethylformamide. After this coated electrode was placed in phthalocyanine-free 0.1 M n-Bu_4NClO_4/dichloromethane solution, the cycling of the electrode potential gave the CVs shown in Figure 3, owing to the oxidation and reduction of the deposited film itself. The area of the anodic or cathodic peak of the CVs was proportional to the scan rate of the electrode potential, as expected for a surface-modified electrode.

Both the visible and X-ray photoelectron spectra of the electro-deposited film indicated that the metal-ring structure of $(EtO)_8PcZn$ was maintained in the film. Before these spectral measurements, the film was kept at -0.5 V for 5 min in 0.1 M n-Bu_4NClO_4/dichloromethane solution, for neutralization. Although the peak in the visible spectrum of the film obtained by reflection method was rather broad, the wavelength of the absorption maximum was almost identical to that of an $(EtO)_8PcZn$ film

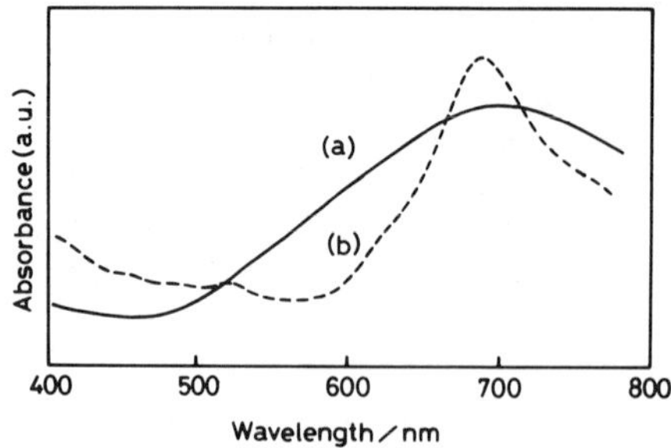

Figure 3. Cyclic voltammograms of the electrodeposited film prepared from $(EtO)_8PcZn$ on a Pt electrode. Scan rate, (a) 100 mVs^{-1}, (b) 50 mVs^{-1}, (c) 20 mVs^{-1}. Electrolyte, 0.1 M $n\text{-}Bu_4NClO_4$/dichloromethane.

deposited on Pt in vacuo (Figure 4). X-ray photoelectron spectra in the region of N $1s_{1/2}$ and Zn $2p_{3/2}$ were obtained for both the electrodeposited film and a powdered sample of $(EtO)_8PcZn$. The ratios of N to Zn intensities, corrected for the differences in photoionization cross section, were in good agreement for them both (9.2 for the depositied film and 9.4 for $(EtO)_8PcZn$ powder). The deviations of the values from 8, the ratio of N to Zn in the chemical composition of $(EtO)_8PcZn$, seem to be ascribable to the differences in mean free path between the N $1s_{1/2}$ and Zn $2p_{3/2}$ photoelectrons[19].

Although the chemical structure of the electrodeposited film is not clear at present, its insolubility and broad visible spectrum[14,15] mentioned above are suggestive of the electropolymerization of $(EtO)_8PcZn$. It could be pointed out that there is a similarity between the electrochemical deposition in this study and the oxidative electropolymerization of dibenzo-18-crown-6; the coupling of the aromatic rings, which are substituted by ether groups, occurs by the oxidation of the crown ether in dichloromethane[16]. From the area of CV peaks obtained for the films in 0.1 M $n\text{-}Bu_4NClO_4$/dichloromethane solution (see Figure 3), the maximum apparent coverage of the phthalocyanine sites were estimated to be 4×10^{-8} mol/cm^2 on the assumption that the number of electrons involved in the redox reaction of phthalocyanine unit in the deposited film was two.

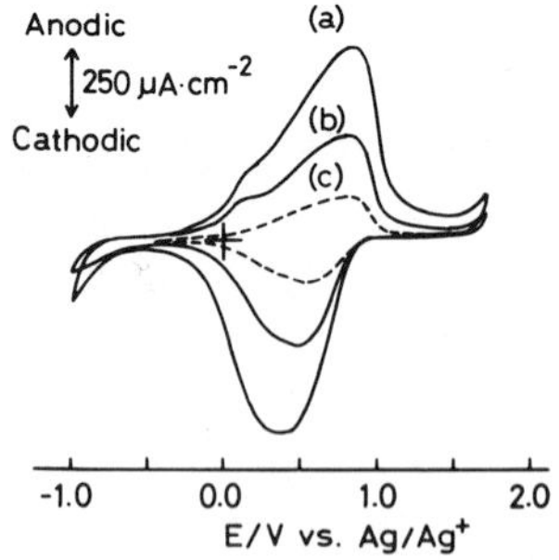

Figure 4. Visible spectra of (a) the electrodeposited film prepared from $(EtO)_8PcZn$ on a Pt electrode, and (b) a vacuum-deposited film of $(EtO)_8PcZn$ on a Pt plate, obtained by reflection method.

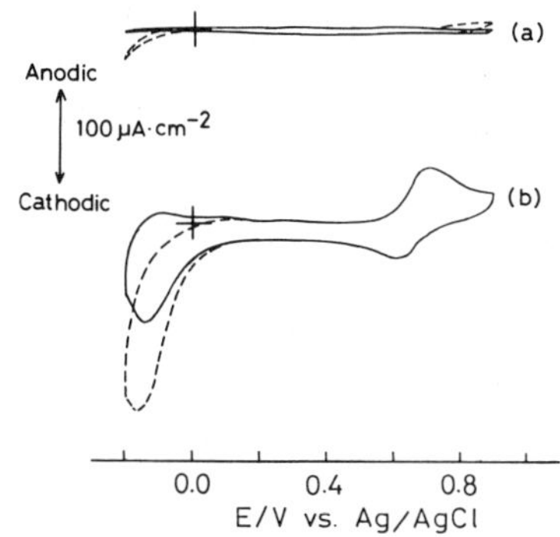

Figure 5. Cyclic voltammograms for oxygen reduction (———, Ar sat.; -------, air sat.) of (a) a bare Au electrode and (b) the Au electrode modified with the electrodeposited film prepared from $(EtO)_8PcZn$, in 0.1 M $HClO_4$. Scan rate, 50 mVs^{-1}.

The electrocatalytic reduction of O_2 on the electrodes modified with metallophthalocyanines has attracted much interest, in connection with the search for cathode materials for oxygen fuel cells[5-7]. In this study, the metallophthalocyanine film electrochemically prepared from $(EtO)_8PcZn$ on an Au electrode was tested for the reduction of O_2. A typical CV obtained for the modified electrode in an argon-saturated 0.1 M $HClO_4$ solution is shown in Figure 5b (solid line). In the CV, the redox waves probably due to the reduction of the film, were observed below 0 V in addition to the waves in the higher potential range. When the solution was saturated with air, the cathodic peak around -0.15 V increased in size owing to the reduction of O_2 (dashed line in Figure 5b). From a comparison of the CVs of the modified electrode with those of a bare Au electrode (Figure 5a), it is recognized that the deposited film on the Au electrode exhibited an electrocatalytic activity for the reduction. Reduction on the modified electrode was observed at a potential ca. 0.2 V higher than that on the bare Au electrode. As for metallophthalocyanines, the electrocatalytic activity for the O_2 reduction has been reported almost exclusively on compounds having a transition metal atom such as Fe and Co. It is noteworthy that the deposited film prepared from a zinc phthalocyanine derivative showed such an appreciable effect.

REFERENCES

1. C.F. Hackett, J. Chem. Phys., 55, 3178 (1971), and references quoted therein.

2. Z.D. Povovic and R.O. Loutfy, J. Appl. Phys., 52, 6190 (1981), and references quoted therein.

3. N. Minami, K. Sasaki and K. Tsuda, J. Appl. Phys., 54, 6764 (1983).

4. G.C.S. Collins and D.J. Schiffrin, J. Electroanal. Chem., 139, 335 (1982).

5. J. Zagal, P. Bindra and E. Yeager, J. Electroanal. Chem., 127, 1506 (1980), and references quoted therein.

6. R.A. Bull, F.-R. Fan and A.J. Bard, J. Electrochem. Soc., 131, 687 (1984).

7. F. Mizutani, S. Iijima, Y. Tanabe and K. Tsuda, Synth. Metals, 18, 111 (1987).

8. L.M Santos and R.P. Baldwin, Anal. Chem., 58, 848 (1986).

9. A.F. Diaz and J. Bargon, "Handbook of Conducting Polymers", T.A. Skotheim, Ed., Marcel Dekker, New York (1985), Vol. 1, Ch. 3.

10. P.J. Peerce and A.J. Bard, J. Electroanal. Chem., 114, 89 (1980).

11. P.K. Ghosh and T.G. Spiro, J. Electrochem. Soc., 128, 1281 (1981).

12. P. Denisevich, H.D. Abruna, C.R. Leidner, T.J. Meyer and R.W. Murray, Inorg. Chem., 21, 2153 (1982).

13. H. Nishihara and K. Aramaki, Chem. Lett., 1063 (1986).

14. K.A. Macor and T.G. Spiro, J. Am. Chem. Soc., 105, 5601 (1983).

15. B.A. White and R.W. Murray, J. Electroanal. Chem., 189, 345 (1985).

16. V. Le Berre, R. Carlier, A. Tallec and J. Simonet, J. Electroanal. Chem., 143, 425 (1982).

17. O. Ohno, K. Ichimura and T. Sasaki, 51st National Meeting of the Chemical Society of Japan, Kanazawa, October 1985, Abstr. No. 4A01.

18. Z. Witkiewicz, R. Dabrowski and W. Waclawek, Mater. Sci., 2, 39 (1976).

19. D. Briggs, "Handbook of X-Ray and Ultraviolet Photoelectron Spectroscopy", D. Briggs, Ed., Heyden, London (1978), Ch. 4.

SELECTIVE DETERMINATION OF PHOSPHATE WITH PhoE PORIN-LECITHIN MEMBRANE ELECTRODE

Tadashi Matsunaga, Akinori Shigematsu, and Yoshio Asakura

Department of Applied Chemistry for Resources
Tokyo University of Agriculture & Technology
Koganei Toyko 184 / Japan

Hideo Abe and Susumu Takezawa

Yamamoto Co. Ltd. R & D
Tokorozawa Saitama 359 / Japan

INTRODUCTION

The presence of phosphate in the environment originating from detergent and fertilizers causes water pollution. The determination of phosphate in water is, therefore, important in environmental analysis. The phosphorus-molybdate blue method can be employed for the determination of phosphate[1], but requires a complicated procedure and is not suitable for on-line measurements. An enzyme sensor based on the inhibition of alkaline phosphatase by phosphate has been developed[2]. However, this sensor requires additional glucose and glucose-6-phosphate. The outer membrane of gram-negative bacteria acts as a molecular sieve with defined exclusion limits. This property results largely from an outer membrane protein, called porin, which forms a large water-filled pore through the hydrophobic core of the outer membrane[3]. The PhoE porin protein is produced as a major outer membrane protein under condition of phosphate deprivation[4]. The PhoE porin has a specific permeability function in enhancing the uptake of phosphate into the periplasm. In this study, PhoE porin-lecithin membranes were prepared on a basal plane pyrolytic graphite electrode and an anion-selective polymer membrane electrode. The selectivity of the PhoE porin-lecithin membrane was studied by applying cyclic voltammetry to the porin graphite electrode. The PhoE porin-lecithin membrane electrodes were used for the selective determination of phosphate and phosphocompound.

EXPERIMENTAL

Materials

Peptone and yeast extract were purchased from Kyokuto Pharmaceutical Co., Ltd. (Tokyo, Japan). Riboflavin and riboflavin phosphate (flavin mononucleotide, FMN) were obtained from Kanto Chemical Co., Inc. (Tokyo, Japan). Lecithin from egg was purchased from Wako Pure Chemical Co. (Osaka, Japan). Other reagents were commercially available analytical reagents or laboratory grade materials. Deionized water was used in all procedures.

Microbial cells

Escherichia coli K-12 CE 1237 and CE 1241 were a gift from Ben Lugtenberg (State University of Utrecht) and were cultured aerobically at 37°C in yeast broth consisting of 1% peptone, 0.7% yeast extract, 0.5% NaCl and 0.1% $NaH_2PO_4 \cdot 2H_2O$.

Apparatus

The electrode system for the characterization of the PhoE porin-lecithin membrane is depicted in Figure 1. A basal plane pyrolytic graphite (BPG) electrode was coated with a PhoE porin-lecithin membrane. The electrode system consisted of the PhoE porin-lecithin membrane BPG electrode with a surface area of 0.19 cm^2, a counter electrode (platinum wire), and a reference electrode (saturated calomel electrode; S.C.E.). Cyclic voltammograms were obtained with a potentiostat (Hokuto Denko, Model HA301), a function generator (Hokuto Denko, Model HB104) and an X-Y recorder (Riken Denshi, F35). The measurement cell was of all-glass construction, approximately 10 ml in volume, incorporating a conventional three-electrode system. An anion-selective polymer membrane electrode was also coated with a PhoE porin-lecithin membrane. Potentiometric measurements were made with an electrometer (Hokuto Denko, Model HE-101A) in conjunction with a recorder (Riken Denshi, Model SP-J3C).

Preparation of PhoE porin

Late-exponential phase cells were centrifuged at 4°C and 10,000 xg for 6 min and suspended in 10 mM Tris-HCl buffer, pH 8.0. Preparation of

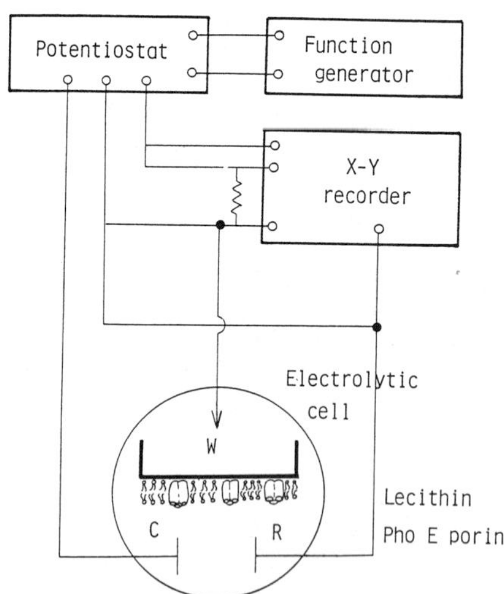

Figure 1. Schematic diagram of the PhoE porin-lecithin membrane-BPG electrode system for measurement of phosphocompounds.
W: Working electrode (basal plane pyrolytic graphite electrode)
C: Counter electrode (platinum wire)
R: Reference electrode (saturated calomel electrode)

porin from this cell suspension was performed by the modified procedure of Lakey et al.[5]. The cell suspension was passed four times through a French pressure cell at 800 Kg/cm^2. The extract was centrifuged at 3,000 xg for 20 min to remove unbroken cells. Then, the supernatant was centrifuged at 100,000 xg for 20 min. The pellet was resuspended in 2% SDS-10 mM Tris-HCl, pH 7.5; the suspension was kept at 37°C for 30 min and then centrifuged at 15°C and 100,000 xg for 30 min. The peptidoglycan-porin pellet was further purified by repeating this step another 4 times. The peptidoglycan-porin was resuspended in NaCl buffer (1% Triton X-100, 0.4 M NaCl, 3 mM NaN_3, 0.05% 2-mercaptoethanol, 5 mM EDTA, 50 mM Tris-HCl, pH 7.2) and incubated at 37°C for 2 h. The suspension was then centrifuged at 15°C and 100,000 xg for 30 min. The supernatant was concentrated to half of its original volume by ultrafiltration and applied to a Sepharose-4B column. The resultant PhoE porin was checked by SDS-polyacrylamide gel electrophoresis. The amount of PhoE porin was determined with the Bio-Rad Protein Assay (Bio-Rad, Richmond, CA). The extracted PhoE porin solution was dialyzed against 4L of 50 mM Tris-HCl buffer (pH 7.2) containing 3 mM NaN_3 before using.

Preparation of porin-lecithin membrane electrodes

A PhoE porin-lecithin membrane-BPG electrode was prepared as follows: n-decane containing 0.5% egg lecithin and 0.25% cholesterol was brushed on the BPG electrode and dried in air. The resulting lecithin membrane-BPG electrode was inserted in 10 ml, 50 mM Tris-HCL buffer (pH 7.0), and coated again with the n-decane solution containing lecithin and cholesterol. After the lecithin membrane turned black, extracted PhoE porin was added to the lecithin membrane-BPG electrode and Tris-HCl buffer solution system. The anion-selective polymer membrane electrode was an Ag/AgCl electrode (0.422 cm^2) coated with a PVC membrane containing 6% methyltridodecyl ammonium chloride and 30% nitrophenyloctyl ether. A PhoE porin-lecithin membrane-anion selective membrane electrode was prepared in the same way as the PhoE porin-lecithin membrane-BPG electrode described above.

Procedure

The PhoE porin-lecithin membrane electrodes were inserted in the sample solution. Cyclic voltammetry was run in the range -1.0 to 1.0 V vs SCE. The potential was measured in combination with the saturated calomel electrode with an electrometer and displayed on a recorder.

RESULTS AND DISCUSSION

Porin-Lecithin Membrane-BPG Electrode System

Cyclic voltammetry at the porin-lecithin membrane-BPG electrode

Figure 2 shows cyclic voltammograms of FMN at a lecithin membrane electrode (A) and a porin-lecithin membrane electrode (B). Cathodic waves appeared at -0.45 V and -0.57 V vs SCE at the porin-lecithin membrane electrode on the first scan in the negative direction. After scan reversal an oxidation peak was observed at -0.52 V. The peak potentials of FMN at the porin-lecithin membrane electrode were similar to those at a bare BPG electrode, although the peak currents were one-sixth to one-eighth their size at the latter electrode. No peak current was observed at the lecithin membrane electrode. The peak current of riboflavin was not obtained at either the lecithin or the porin-lecithin membrane electrode (Figure 3).

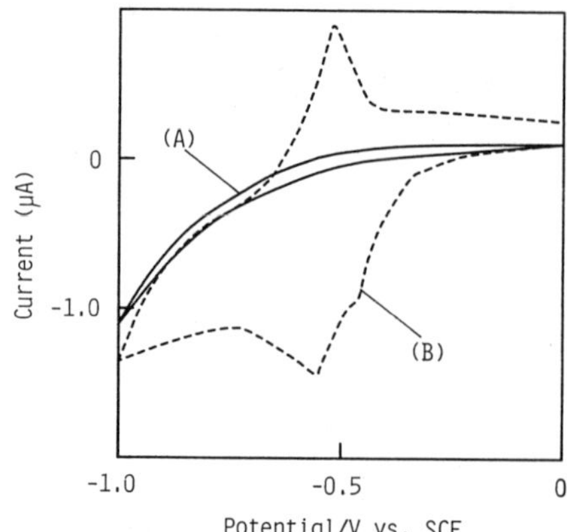

Figure 2. Cyclic voltammograms of FMN by (A) lecithin membrane BPG electrode and (B) PhoE porin-lecithin membrane BPG electrode. Scan rate was 10 mV/s. FMN concentration was 60 μM. The experiments were performed in 50 mM Tris-HCl buffer (pH 7.0).

Selectivity of PhoE porin-lecithin membrane-BPG electrode

The selectivity of the PhoE porin-lecithin membrane electrode is indicated in Table 1. The cyclic voltammograms were obtained at the PhoE membrane electrode for L-Cysteine, riboflavin, FMN, NADH and $FADH_2$. Peak currents were obtained for FMN. L-cysteine and riboflavin, which have no phosphate, did not show a peak current. Peak currents were not obtained for NADH and $FADH_2$. The molecular weights of NADH and $FADH_2$ were more than 700. The PhoE porin-lecithin membrane is apparently permeable to phospho-compounds having molecular weights less than 700 daltons.

Effect of amount of PhoE porin on peak current

Figure 4 shows the relationship between the peak current of FMN and the amount of PhoE porin in the lecithin membrane-BPG electrode. The peak

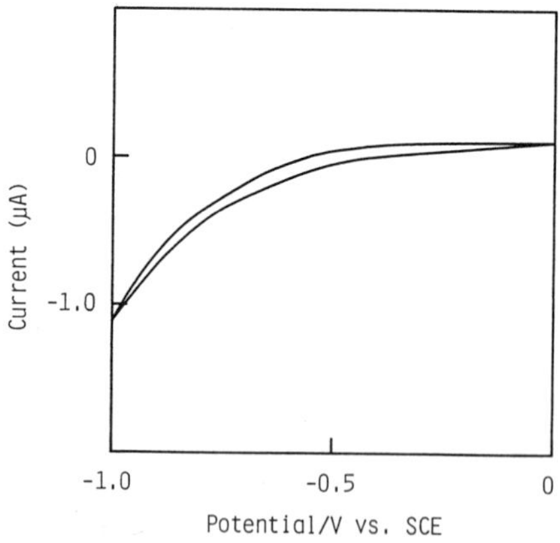

Figure 3. Cyclic voltammograms of riboflavin at a PhoE porin-lecithin membrane-BPG electrode and lecithin membrane electrode. Both voltammograms are similar. Scan rate was 10 mV/s. Riboflavin concentration was 60 μM. The experiments were performed in 50 mM Tris-HCl buffer (pH 7.0).

Table 1. Redox potentials of cyclic voltammograms of electroactive substrates observed at a PhoE porin-lecithin membrane BPG electrode.

Compound	M.W.	Number of phosphate group	Potential(V vs. SCE) Red. peak	Ox. peak
L-Cysteine	121.2	0	---	---
Riboflavin	376.4	0	---	---
FMN(H_2)	478.4	1	-0.48,-0.56	-0.52
NAD(H)	709.4	2	---	---
FAD(H_2)	829.6	2	---	---

current increased linearly with increasing amounts of PhoE porin from 6 µg protein/ml to 30 µg protein/ml. However, the peak current became constant above 30 µg protein/ml. The peak current was 40% of that observed at a bare BPG electrode when the porin content in the lecithin membrane was optimum (30 µg/ml). Therefore, the lecithin membrane contained PhoE porin at 30 µg/ml hereafter.

Relationship between peak current and FMN concentration

The relationship between the peak current and the FMN concentration was obtained. The electrode gave a linear current response over the range 0.3 - 1.4 mM FMN. The slope was 13.4 µA/mM.

Porin-Lecithin Membrane-Anion Selective Electrode System

Potential response at the PhoE porin-lecithin membrane-anion-selective electrode

Phosphate was measured with the anion-selective electrode coated with a PhoE porin-lecithin membrane. When phosphate solution was added to the

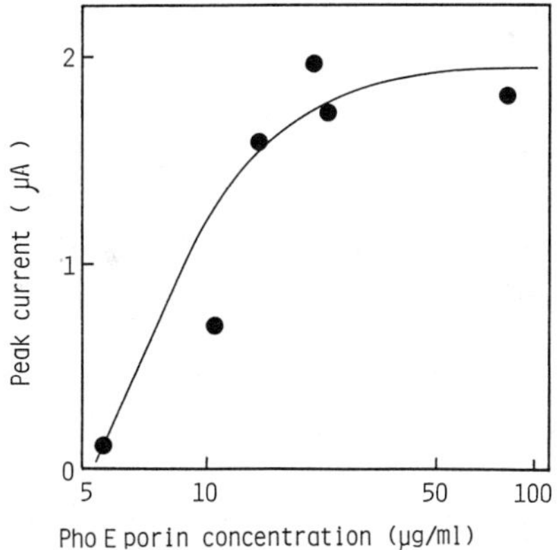

Figure 4. Relationship between peak current and PhoE porin concentration in buffer inserted PhoE porin-lecithin membrane-BPG electrode. FMN concentration was 0.44 mM and experiments were performed in 0.1 M Tricine buffer (pH 7.4), 25°C.

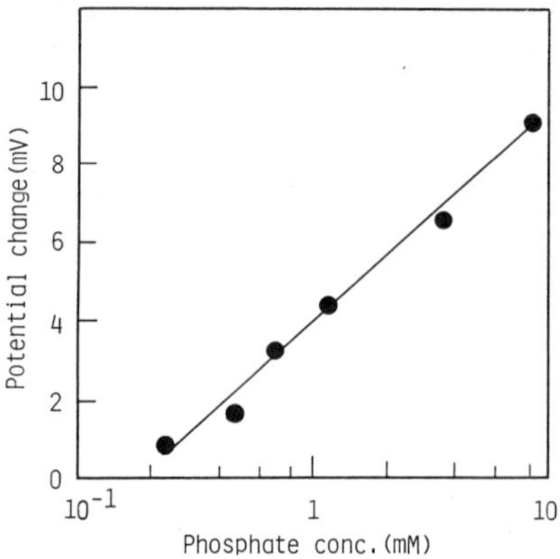

Figure 5. Calibration curve for phosphate using PhoE porin-lecithin membrane-anion-selective electrode. The experiments were performed in 0.1 M Tricine buffer (pH 7.4), 25°C.

PhoE-lecithin membrane electrode system the potential changed until it reached a maximum value. The maximum value was attained in all cases within 10 min, and the maximum potential was dependent on the phosphate concentration. When the PhoE porin-lecithin electrode system was removed from the sample solution and placed in Tricine buffer solution, the potential returned to its initial level.

Selectivity of the PhoE porin-lecithin membrane-anion-selective electrode

The selectivity of the PhoE porin-lecithin membrane electrode was studied for acetate, nitrate, nitrite, formate, citrate, succinate, propionate, chloride and phosphate. The potential changes were 6.5 mV for phosphate and 1.2 mV for chloride when the concentrations of ions were 3.5 mM. No potential change was obtained for acetate, nitrate, nitrite, formate, succinate, and propionate. Therefore, phosphate can be selectively determined in the absence of chloride.

Relationship between potential change and phosphate concentration

Figure 5 shows the relationship between the potential change in the PhoE porin-lecithin membrane electrode and the concentration of phosphate. The potential increased with increasing phosphate concentration. A linear relationship was obtained between the potential change and the logarithm of phosphate concentration in the range 0.2 - 9 mM.

Until now, ion-selective electrodes have been developed for a wide variety of ions and a phosphate ion electrode has been reported[6]. However, a practical phosphate ion electrode has not been reported because of serious interferences by common ions such as chloride, nitrate and sulfate. The principle of the enzyme sensor for phosphate was as follows. Glucose, formed from glucose-6-phosphate by alkaline phosphatase, was determined electrochemically with the glucose oxidase electrode. Although this method does not suffer from interference by ions coexisting in samples, the enzyme sensor is also sensitive to other oxyacids, such as arsenate, tungstate, molybdate and borate ions. Recently, a microbial sensor for phosphate consisting of immobilized *Chlorella vulgaris* and an oxygen electrode[7]. The PhoE porin-lecithin membrane electrode can detect phosphate in the range 0.2 - 9 mM whereas the detectable concentration at the microbial electrode was 8 - 70 mM.

REFERENCES

1. APHA AWWA WPCF: "Standard Methods for the Examination of Water and Waste Water", American Public Health Association (1971).

2. G.G. Guilbault and M. Nanjo, Anal. Chim. Acta, 78, 69 (1975).

3. H. Nikaido, in "Bacterial Outer Membranes", M. Inoue, Ed., Wiley-Interscience, New York (1980), p. 361.

4. N. Overbeeke and B. Lugtenberg, Eur. J. Biochem., 126, 113 (1982).

5. J.H. Lakey, J.P. Watts and E.J.A. Lea, Biochim. Biophy. Acta, 817, 208 (1985).

6. F.R. Shu and G.G. Guilbault, Anal. Lett., 5, 559 (1972).

7. T. Matsunaga, T. Suzuki and R. Tomoda, Enzyme Microb. Technol., 6, 355 (1984).

MEMBRANE BIOENERGETICS AS VIEWED FROM RECONSTITUTION EXPERIMENTS

H. Ti Tien

Membrane Biophysics Laboratory
Department of Physiology
Michigan State University
East Lansing, Michigan 48824-1101

INTRODUCTION

There is a plethora of literature on mitochondria and chloroplasts whose main task is energy transduction. That is, most of the energy for cellular processes in the form of ATP (Adenosine triphosphate) is produced by the organelle membranes of mitrochondria and chloroplasts[1-3]. The mechanism of energy transduction has been elegantly explained by Mitchell's chemiosmotic hypothesis[4]. Briefly, mitochondria contain the enzymes and coenzymes that mediate the electron transport from NADH (a reduced nucleotide) down the chain of redox proteins to the ultimate electron acceptor, oxygen. As electrons move down the redox protein chain, an electrochemical gradient of protons is generated. Free energy is conserved in the synthesis of ATP as a result of the discharge of the proton gradient across the cristae membrane of mitochondria. The essential feature of the Mitchell theory is "energy coupling" among the various membrane-bound processes: electron transfer, redox reaction, ion and proton gradient, membrane potential and ATP synthesis. Obviously, these membrane-based activities, many of which are electrical in nature, should be probed and investigated by electrochemical methods, as have been carried out on whole cells and nerve axons by electrophysiologists. Unfortunately, mitochondria, along with other organelles such as chloroplasts, are too tiny to stick microelectrodes into them for reliable experiments, although there have been several isolated reports[5,6]. In order to test the chemiosmotic hypothesis, many researchers have in the past resorted to reconstituted membrane studies using artificial bilayer lipid membranes (planar BLMs and liposomes)[7-10]. In this paper a brief description of the essential aspects of membrane bioenergetics will be given. This will be followed by the experimental techniques used. Finally, a review of reconstitution experiments with respect to membrane energetics will be presented.

BASICS OF MEMBRANE BIOENERGETICS

Sources of Bioenergy

The sources of life's energy are limited to two: (1) the energy of light in the range from 400 nm to 1000 nm, in the spectrum of solar radiation, and (2) the chemical energy stored in organic compounds by virtue of their constituent atomic and molecular configuration. In either case, we are speaking about electronic energy. In fact, as pointed out by Szent-Gyorgyi[11], the discoverer of vitamin C, life is nothing but a movement of electrons:

> "It is common knowledge that the ultimate source of all our energy and negative entropy is the radiation of the sun. When a photon interacts with a material particle on our globe it lifts one electron from an electron pair to a higher level. This excited state as a rule has but a short lifetime and the electron drops back within 10^{-7} to 10^{-8} seconds to the ground state giving off its excess energy in one way or another. Life has learned to catch the electron in the excited state, uncouple it from its partner and let it drop back to the ground state through its biological machinery utilizing its excess energy for life processes".

Cellular Energy Metabolism[2]

In a cell (or organism), the sum total of all chemical reactions that take place is defined as metabolism[2]. Through metabolism, cells make use of resources to obtain useful energy and building materials for their continued existence, growth, and replication. In order for these to happen, a large negative free energy change ($\Delta G^{o\prime} < 0$) is necessary as will be defined in the following section. For now, a reaction central to cellular metabolism is the oxidation of glucose to CO_2 and H_2O ($\Delta G^o =$ -686 Kcal/mol), the free energy of which is conserved in the phosphorylation of ADP to ATP. More specifically, the foodstuffs (carbohydrates, fats and proteins) one ingests are broken down enzymatically into simpler molecules in three states (catabolism) before the cells can use them. First, carbohydrates, fats, and proteins are broken down into sugars, fatty acids, glycerol, and amino acids, respectively. Second, sugars and fatty acids upon oxidation are converted into the acetyl groups of acetyl coenzyme A (acetyl CoA) in mitochondria. Third, the acetyl CoA is degraded to CO_2 and H_2O with the production of ATP. In this paper, we are mainly interested in the mechanisms of ATP synthesis and hydrolysis as viewed from membrane bioenergetics.

The Free Energy Concept

Life as far as we know in our universe depends upon spontaneous processes such as flowing of heat from a warmer body to a cooler one, water at the mountain top to its valley below and excited electrons at a higher energy level to the ground state in accordance with the second law of thermodynamics, which is expressed in the entropy function, $\Delta S = Q/T > 0$. To assess ΔS, which is the sum of $(Q/T)_{system}$ and $(Q'/T)_{surroundings}$, is a difficult task. Experimentally, the free energy function (ΔG) introduced by Gibbs is used, which is obtained by combining the first and second law of thermodynamics[12]. The Gibbs free energy ($\Delta G = \Delta H - T\Delta S$) is by far most useful in bioenergetics, for it can be related to the equilibrium constant (K) of a reaction ($\Delta G^o = -RT \ln K$) and to the standard reduction-oxidation (redox) potential (E^o) of an electrochemical cell ($\Delta G^o = -nF\Delta E^o$).

As will be mentioned presently, energy transductions in living cells are membrane-based redox reactions. It should be stated here that ΔG^o and E^o as expressed are in their standard states, usually at one atmosphere pressure, 25°C, with reactants and products at unit activity. ΔG and E expressed without superscript 'o' are for reactions other than the standard state conditions. In bioenergetics, $\Delta G^{o\prime}$ and $E^{o\prime}$ are used, with all components present at 1 M except $[H^+]$ at pH = 7. For all spontaneous processes at constant temperature (T) and pressure (P), $\Delta G<0$, which is the maximum work possible except PV work. For a given process involving one or more reactions, these are called coupled reactions which are crucial in explaining cellular energy transduction. Thus, in bioenergetics the free energy concept is important for it indicates the capacity of doing useful work for which ΔG is negative. This concept applies to a single reaction as well as to a group of reactions so long as the overall ΔG is less than zero. Finally, a remark should be made about useful work which is a product of two terms, intensity x capacity; examples are electrical work (electric potential x charge) and transport work (chemical potential x mass).

The Electrochemical Potential

In bioenergetics the basic unit of action is the cell and its organelles (e.g. chloroplast and mitochondria) of which the cell's plasma membrane along with its organelle membranes play the central role. For a cell to be viable, nutrients and toxic waste products must be transported across the membrane, for which ΔG must be negative otherwise metabolic energy must be supplied.

Since a compound is a chemical species, an intensive factor called the chemical potential ($\bar{\mu}$) can be assigned to it. If the chemical species also carried electric charges, then the term assigned to it is called the electrochemical potential (μ). By definition, $\bar{\mu} = \bar{\mu} + zF\psi = \mu^o +$ RT ln a + zF , where μ^o is a constant, R is a gas constant, T is absolute temperature, a = activity coefficient (γ) multiplied by concentration [ ], z is the valence, F is the Faraday constant and ψ is the electrical potential[12]. It can be shown that $\bar{\mu} = (\partial G/\partial n)_{P,T}$. For transmembrane processes of interest, the driving force is the electrochemical potential gradient in the direction (x) perpendicular to the membrane and is given by

$$\frac{d\bar{\mu}}{dx} = \frac{d\mu}{dx} + \frac{zFd\psi}{dx} \approx \frac{RT\, d\ln C}{dx} + \frac{zFd\psi}{dx} \qquad (1)$$

where $d\bar{\mu} = \bar{\mu}_i - \bar{\mu}_o$ and $d\mu = \mu_i - \mu_o$. When Equation (1) is combined with material transport across the cell membrane, the well known Nernst-Planck equation can be derived, upon which many important membrane transport equations are based[10]. It will be seen later that Equation (1) is of central importance in membrane bioenergetics.

Energy Flow and Coupled Processes in Bioenergetics

In thermodynamics, one deals with closed and open systems, the difference between the two being that the latter involves the exchange of matter in addition to energy (heat and work). Clearly, a cell is an open system. Similarly, organelles such as chloroplasts and mitochondria are also open systems. Other energy transducing systems of interest are found in bacteria, in visual receptors. We shall mainly focus out attention on the thylakoid membrane of chloroplasts and the cristae membrane of mitochondria. It is particularly noteworthy that energy transduction and material transport in these two systems are coupled; the products of photosynthesis are utilized as the reactants in respiration, and vice versa. This is illustrated in Figure 1. The interdependence of photosynthesis and respiration was dramatically demonstrated by Priestley in 1771 (see Figure

1). He observed that a green plant in a closed bell-jar withered after some time, in spite of sunlight, and a mouse in another closed bell-jar died. However, when the plant and mouse were placed in the same closed bell-jar in the presence of sunlight, both thrived.

The essential feature of Priestley's experiment is the coupled processes of photosynthesis and respiration in that it represents on a global basis, a steady-state operation brought about by the flow of energy from the ultimate high energy source of the sun to the low potential sink of the earth[13,14]. As pointed out by Morowitz[15], it is not the energy itself that makes life possible, but the flow of energy. Thus, from the observations of Priestley in the 1770's to the 1970's a clear picture has emerged after two centuries of research that energy flow and coupling are what bioenergetics is all about; they are accomplished through the energy transducing membranes of chloroplast thylakoids and mitochondrial cristaes[13,14], as shown in Figure 1.

Nature's Energy Transducers

The fundamental unit of vital cellular functions is the lipid bilayer which provides the structural framework for intercalating of other membrane constituents. For functions, the most important of which is energy transduction. Thus, a variety of energy transducing apparatuses made of lipids, proteins, pigments, and their complexes must reside in the membrane. Some of these energy transducing machineries may be termed as bioenergy transducers listed in Table 1.

Unsolved Problems in Membrane Bioenergetics

What are some of the unsolved problems in membrane bioenergetics in the 1980's and beyond? In photosynthesis, there are three, namely (i) the primary steps of photon conversion, (ii) electron transport and oxygen evolution, and (iii) ATP synthesis. In respiration, the main unsolved

Table 1. A Classification of Biotransducers*

Basic Unit	Membrane Involved	Energy Transduction
Chloroplast	Thylakoid Membrane	Light to Chemical
Mitochrondrion	Cristae Membrane	Food of Chemical
Individual cell	Plasma Membrane	Chemical to Mechanical
Nervous System	Nerve Membrane	Chemical to Electrical
Retina	Rod and Cone Segment Membranes	Light to Electrical
Inner Ear	Tectorial Membrane	Sound to Electrical
Muscle	Actomyosin	Chemical to Mechanical

*Adapted from References 17 and 18

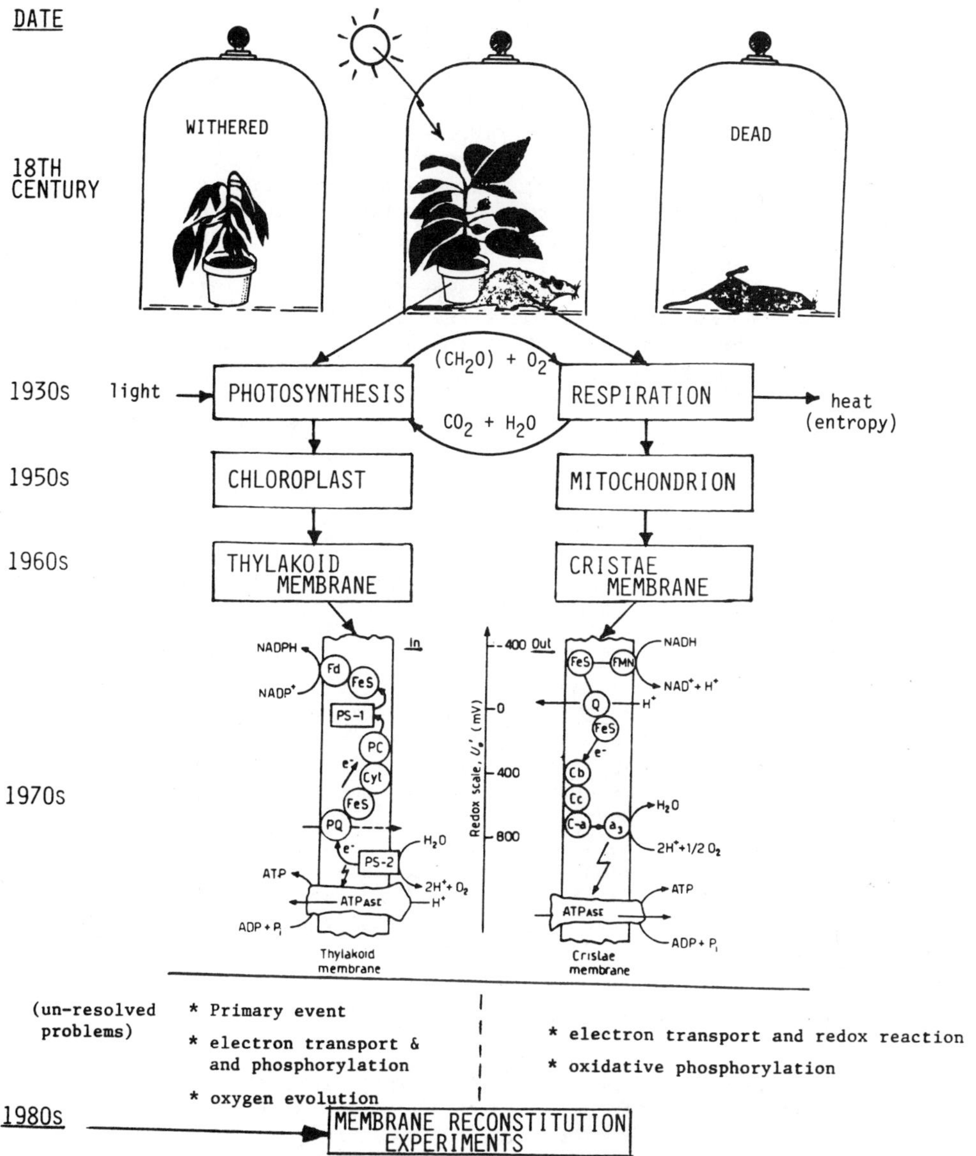

Figure 1. Major events in membrane bioenergetics research. Priestley's experiment demonstrating the interdependence of photosynthesis and respiration in 1771. In the 1930's, new insights were gained by comparative studies. In the 1950's, the ultrastructures of chloroplast and mitochondrion were revealed by electron microscopy. In the 1960's, advances in membrane biophysics and biochemistry together with the chemiosmotic hypothesis of Mitchell and model bilayer lipid membranes development resulting in evermore detailed pictures of the thylakoid membrane of chloroplasts and the cristae membrane of mitochondria in the 1970's. Unsolved problems for the 1980's and beyond are listed. One approach to solving these problems is by membrane reconstitution experiments.

problems appear to be quite similar; they are electron and proton transport, energy coupling, and ATP synthesis. All of these are membrane-bound problems. To date, in view of what is known about the thylakoid membrane, cristae membrane and other energy-transducing membranes, they are far too complicated for direct experimentation when the structural-functional relationships are considered at the molecular level (see Figure 1). For example, a number of hypotheses have been proposed for photon conversion in green plant photosynthesis, one of which is the mechanism analogous to that taking place in semiconductor silicon solar cells[13]. If so, the most direct kind of experiment would be to place electrodes across the membrane and measure its voltage under illumination. Unfortunately, such an approach is still very difficult owing to the microscopic size of the thylakoid. Similar experimental difficulties are encountered for mitochondria when one wishes to probe them electricaly. Therefore, an alternative approach is to resort to membrane reconstitution experiments using bilayer lipid membranes (planar BLMs and spherical liposomes)[7,10]. As will be seen in the following sections, these reconstituted membrane systems are very useful for investigating energy transduction and transport problems.

Mechanisms of Membrane Transport

Of central importance in membrane bioenergetics are primary particles such as electrons, holes (vacant sites left by electrons) and protons, and a vast variety of chemical compounds. All of these must be transported within as well as across the membrane. The driving force is due to the electrochemical gradient for passive and facilitated transport. Ion "pumps" driven by metabolic energy (ATP) for active transport are known, an outstanding example of which is the Na^+/K^+ ATPase pump. Four basic mechanisms have been proposed: carrier, channel, hopping and tunneling. For ions such as Na^+ and K^+, water-filled channels seem to be most efficient pathways for transport. For electrons and holes a tunneling mechanism is a plausible one. Protons may be transported by a hopping mechanism. It is beyond the scope of this paper to discuss any of these mechanisms here. The fact is that ion carriers, pigments and protein "channels" can be reconstituted into experimental bilayer lipid membranes displaying unique transport characteristics[16].

EXPERIMENTAL

Planar bilayer lipid membranes (BLM) with diameters up to 2 mm separating two aqueous solutions, which are especially suitable for electrical measurements, can be formed by a variety of techniques. Spherical bilayer lipid membranes (liposomes) ranging from 200 Å to 1 cm diameter can also be formed by numerous methods[7,8,17]. A brief description of the BLM system only will be given here along with experimental setups.

Planar Bilayer Lipid Membranes (BLM)[7,10]

To facilitate handling, a simple cell assembly was used. This consists of a 10 ml Telfon cup (commercially available), which is held in a Lucite holder, such as shown in Figure 2, having a second chamber of equal volume. A small aperture of about 0.5 - 2 mm diameter is punched in the side of the Teflon cup. Since electrical measurements can provide a simple means of monitoring the behavior of BLM and its interactions with different modifiers and agents, they are in common use in BLM investigations. Because of the very high resistance of unmodified BLM (10^9 ohm cm^2) and large capacitance (0.5 - 1 $\mu F cm^{-2}$), care is needed in the insulation

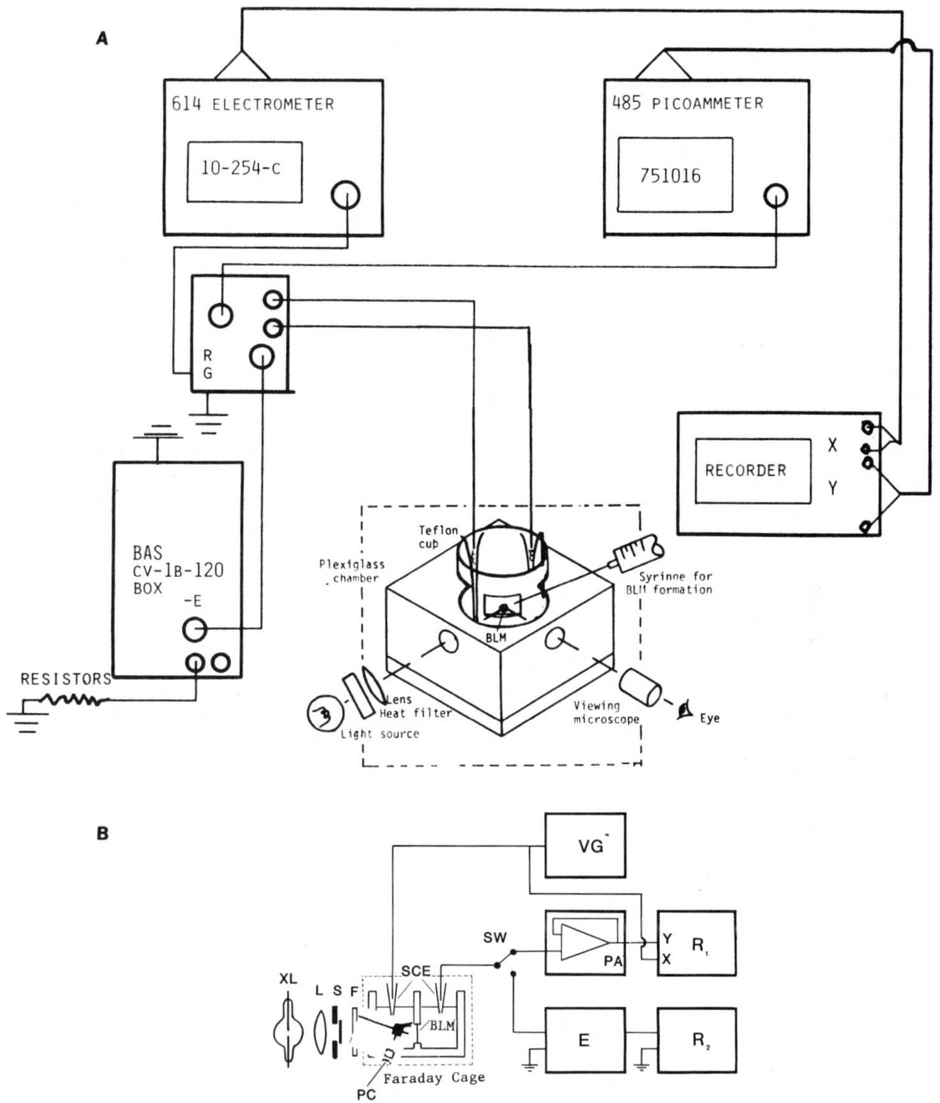

Figure 2. Experimental setups for investigating transport and energy transduction in reconstituted bilayer lipid membranes (planar BLM). (A) Schematic representation of the setup for electrical measurements. BAS CV-1B-120 Box, made by Bioanalytical Systems Inc., Ohio, is used for cyclic voltammetry studies (see Reference 34). (B) Block diagram for investigating photoelectric effects in pigmented BLM. XL= xenon lamp, L = lens, S = shutter, F = filter, PC = plexiglass chamber, SCE = saturated calomel electrode, SW = switch, VG = R_2 = x-t recorder (from J. Kutnik and H.T. Tien, Biophys. J., 49, 484a, 1986 and Photobiochem. Photobiophys., 7, 319, 1984).

and shielding of the membrane chamber, electrodes, cables, switches and any connections to avoid current leakage and antenna-like action of the setup. Therefore, it is generally advisable to enclose the whole membrane chamber in a Faraday cage. For d.c. measurements, different types of so-called nonpolarizable electrodes may be used, including saturated calomel and silver/silver chloride (Ag/AgCl) electrodes. However, for several kinds of experiments using pulse or a.c. techniques, bright platinum or platinum coated with platinum black, though polarizable, are preferable. The temperature of the cell assembly (Teflon cup and Lucite holder) may be maintained either by placing the whole assembly in a thermostatic chamber or by circulating water from a thermostat through a glass coil placed around the Teflon cup. For effective stirring, a magnetic stirrer with very small stirring bars may be used, if required, but should be switched off during measurements, particularly for cyclic voltammetry (CV) studies. During the experiments, if required, the solution in one or both chambers may be changed with the BLM in place, by using infusion-withdrawal pumps in which closely matched syringes should be used to prevent the membrane from bulging.

For light induced effects in pigmented BLMs, essentially the same set-up can be used except a strong light source is needed. A xenon arc lamp or a halogen projector lamp has been found suitable. For obtaining spectra of the BLM, a monochromator is inserted between the BLM and the light source (see Figure 2B).

Variable d.c. voltage sources may be constructed from a dry cell and precision potentiometer (e.g., 10-turn Helipot). The portion of applied voltage developed across the BLM (the applied voltage also drops on electrodes, electrolyte solutions on both sides of BLM, and input resistor R_i) is monitored with a high impedance electrometer. A low input impedance picoammeter may also be connected in series to the BLM. One side of the BLM (one calomel electrode) is at ground potential. The critical component of such a system is an electrometer, which must have a high input impedance, usually more than 10 times greater than that of BLM. Excellent instruments fulfilling all requirements are commercially available electrometers (Keithley Instruments, Inc., Cleveland, Ohio). For voltammetric studies of BLM such as CV, an IBM instrument (EC/225) has been found suitable (see Reference 7 for more details).

BLM Formation

The seminal method introduced in the early 1960's is still being practiced with some minor modifications. With the cell chambers filled with an appropriate bathing solution (usually 0.1 M KCl), a drop of lipid solution is introduced to the aperture in the side of the Teflon cup using a Hamilton syringe with a dispenser attachment (PB-600-1). Owing to the small diameter of the aperture, a low power (20-60x) microscope is used to observe the membrane. After formation of the BLM, the electrical properties can be investigated with the instruments described above. To exchange or add small amounts of solutions on either side of BLM, a disposable or an adjustable micropipette may be used. If a given amount of a new solution is added to one side, the same amount of bathing solution should be added to the other side of the BLM to maintain equal solution levels on both sides.

Methods of Study

If the BLM system to be investigated is symmetrical, i.e., a symmetrical BLM separates two identical electrolyte solutions of equal ionic strength, then in the absence of external voltage there is no residual transmembrane potential. A transmembrane potential may be generated, however, whenever there is a difference in ionic strength, ionic species or redox compounds across the BLM.

RESULTS AND DISCUSSION

Reconstituted Bilayer Lipid Membranes As A Model For the Cristae Membrane

Mitochondria are known as the "power plants" of aerobic cells, the primary function of which is fatty acid oxidation to CO_2 and H_2O, and ATP synthesis. As mentioned in the INTRODUCTION, the chemiosmotic hypothesis of Mitchell suggests that coupled electron and ion movements are crucial to redox protein chain energy transduction into ATP movement. The key questions to be answered are: (i) how are electrochemical potential gradients ($\Delta\mu/\Delta x$) of protons across the biomembranes generated, (ii) how are electrons, ions and chemical species transported across the membrane, and (iii) how are such potential gradients ($\Delta\mu/\Delta x$) used to drive the synthesis of ATP?

From a biophysical point of view, the use of whole mitochondrion (or chloroplast) to test the "molecular" theory of Mitchell does not seem satisfying. The translocation of ions, such as protons, across a complex system, such as a cristae, is in itself ill-defined. The interactions between various fluxes (ion and water movements, electron and hole transport, etc.) are far too complex to be amenable to a simple analysis. At the present time, direct tests with the mitochondrial membranes are difficult. Experimental testing of the chemiosmotic hypothesis, using simpler model systems such as planar BLM and spherical liposomes are, therefore, in order.

Among the first to employ a BLM system to test one aspect of Mitchell's hypothesis were Bielawski, Thompson and Lehninger[19] who studied the effect of 2,4-dinitrophenol (DNP) on the BLM conductivity. The BLMs were formed from a solution of purified lecithin, chlolesterol, and n-decane. They found that the uncoupling agent (DNP), at a concentration of 10^{-3} M, decreases electrical resistance of BLM to less than 0.4 % of the original values (i.e., from 1.5×10^{7} to 6.1×10^{4} Ω cm^{-2}). The effects of uncoupling agents of oxidative phosphorylation on BLM were examined systematically by Liberman et al.[20]. The major findings of their work are as follows:

(1) In general, the uncoupling agents (uncouplers) raise the conductivity of BLM by several orders of magnitude, and the effect depends strongly on the pH of the bathing solution, being most effective at the pK value of the uncoupler.

(2) The effectiveness of the uncouplers studied follows a series: tetrachlorotrifluoromethylbenzimidazole (TFB) > p-trifluoromethyloxycarbonylcyanide phenylhydrazone (FCCP) > DNP > salicylic acid > acetoacetic ester. The order of effectiveness is similar to that found in the mitochondria.

(3) The current/voltage curves of the BLM in the solution containing the uncoupler are nonlinear, and depend on pH and the buffer capacity of the solution.

(4) In the presence of TFB, the BLM behaves as a hydrogen electrode. The modified BLM is said to be more permeable to H^+.

(5) The uncoupler is considered a carrier of H^+. The operation is described as follows: on one side of the membrane, a proton reacts with the uncoupler in the BLM; this uncoupler-H^+ complex diffuses across the membrane, and to the other side of the membrane discharging the H^+; the uncoupler returns to the other side. Liberman and colleagues[20] suggested that their findings agree with the scheme of oxidative phosphorylation as envisioned by Mitchell, and conclude that BLM are a good model for the mitochondrial membrane.

The pH-dependence of the effect of TFB and other uncouplers on the electrical conductivity of the BLM described above has been studied in detail by Sotnikov and Melnik[21]. They were interested in establishing the relationship between the pK_a value of the modifier (e.g., DNP) and the conductivity of the BLM and found that the effectiveness of these compounds in lowering the conductivity follows their pK values. The conductivity for picric acid-modified BLM was about $10^4\ \Omega\ cm^{-2}$. More recently, Wardak[22] has investigated the effect of picric acid on the transport of H^+ and K^+ and found that the salt of picric acid is highly permeable in the BLM.

Other uncouplers such as carbonylcyanide phenylhydrazone (CCP) derivatives have also been reported by Ting, Wilson, Chance, and Koppelman[23]. Although they also observed a resistance decrease by a number of compounds, the conclusion they reached is that no simple correlation exists between the relative effectiveness of the uncouplers on the conductance of BLM. Since literally several hundreds of agents are known at the present time, a theory is needed to explain the observed phenomena. Bruner[24] has, in fact, attempted to established a correlation between the pK_a value and the membrane potential arising from ion concentration gradients. It has been found that substances which uncouple oxidative phosphorylation in mitochondrial membranes, usually increase the electrical conductance of phospholipid BLM. Among these uncouplers is a group chemically known as weak acids. For this group the conductance of BLM, when measured as a function of pH at fixed uncoupler concentration, shows a maximum at a pH approximately equal to the pK_a value of the uncoupler used. Corresponding maxima in electrical potential arising from ion concentration gradients are also observed. Bruner[24] has suggested that charge transport is by direct transfer of either protons or anions of the uncoupler. A fixed charge density at the membrane surface is required in Bruner's theory.

As pointed out by Mitchell, all uncouplers are weak acids (HW) with relatively high solubility in the lipid phase of the membrane. In BLMs, weak acids act as carriers for H^+ and the charged translocating species is either the anionic form of the weak acid, W^-, or the complex of the anion, HW_2^- and the undissociated acid, HW. Thus, the concentration of the undissociated HW and W^- should be governed by the pH of the outer and inner bathing solutions[25]. At a steady state in the membrane phase

$$J_{HW} - J_{W^-} = 0 \tag{2}$$

where the J's denote molecular fluxes. If two solutions of different pH are separated by a BLM formed from a neutral or zwitterionic lipid, a membrane potential (E_m) should be observed according to the Nernst equation:

$$E_m = \frac{RT}{F} \ln \frac{[W^-]_i}{[W^-]_o} = \frac{[H^+]_o}{[H^+]_i} \tag{3}$$

since the anion and its complex (HW_2^-) are the only charge translocating species. The behavior of uncouplers is readily understood by considering the following reactions as shown by Finkelstein[25]:

$$HW \rightleftharpoons H^+ + W^- \tag{4}$$

$$K_{a_1} = \frac{[H^+]\ [W^-]}{[HW]} \tag{5}$$

$$HW_2^- \rightleftharpoons HW + W^- \tag{6}$$

$$K_{a_2} = \frac{[HW]\ [W^-]}{[HW_2^-]} \tag{7}$$

Note that $[HW_2^-] \ll [HW] + [W^-]$. Therefore, $[W^-]_{total} = [HW] + [W^-]$. It is further assumed that the concentration $[HW_2^-]$ is directly proportional to [HW] at the interfaces. Experimentally, it has been found that DNP and other uncouplers exhibit a quadratic dependence of conductance on concentration and pH with a sharp maximum conductance at $pK_{a_1} = pH$. According to Finkelstein[25], it can be readily shown that

$$[HW_2^-] = \frac{K_{a_1}}{K_{a_2}}\ [W]^2_{total}\ \frac{[H^+]}{(K_{a_1} + [H^+])^2} \tag{8}$$

Letting the BLM conductance (G) be:

$$G = k[HW_2^-] \tag{9}$$

we then have, from Equation (8)

$$G = k\ \frac{K_{a1}}{K_{a2}}\ [W^-]^2_{total}\ \frac{[H^+]}{(K_{a1} + [H^+])^2} \tag{10}$$

Equation (10) indeed predicts that the membrane conductance goes through a maximum at $pK_{a1} = pH$ and is proportional to the square of the total anion concentration, $[W^-]$ total. The results of Wardak[22] are consistent with the analysis of Finkelstein[25]. In this connection, Dilger et al.[26] suggested that because of the ability of weak acids to transport protons across BLM and to uncouple oxidative phosphorylation in mitochondria, some fraction of the lipid bilayer component of the cristae membrane must have a dielectric constant greater than 2.2. This conclusion is based on the experiments of Dilger et al.[26], who used chlorodecane as the BLM ($\varepsilon = 4.5$) instead of n-decane ($\varepsilon = 2$). In the presence of thiocyanate, the conductance of chlorodecane-BLM is one thousand times larger than n-decane-BLM. The specific capacitances, which are proportional to ε/t_m, are 0.73 $\mu F/cm^2$ and o.39 $\mu F/cm^2$ for chlorodecane BLM and n-decane-BLM, respectively. Thus, the observed increase could be due to either an increase or a decrease in t_m (membrane thickness). Since Dilger and colleagues[26] have observed by X-ray diffraction experiments that there is little difference in thickness, the effect must be due to the increase in BLM dielectric constant.

A voltage-dependent anion channel material from Pramecium mitochondria has been incorporated into BLM. The transmembrane voltage is in the vicinity of zero and decreases sharply with both positive and negative voltage[27]. These mitochondrial outer membrane porin channels appear to be anion selective, in contrast to bacterial porin channels[28,29]. Mention should also be made about the incorporation of an invertebrate respiratory protein into BLM. Menestrina et al.[30] reported that keyhole limpet hemocyanin interacts with BLM increasing their conductance by the formation of cation-selective hydrated channels through the membrane. When the interaction occurs only on one side of the BLM, the membrane conductance is voltage dependent, showing a characteristic sigmoid shape similar to those obtained from the mollusc, Paludina vivipaa, showing it has the property of open channels in BLM. In oxidized cholesterol BLM, channel conductance depends on the type of electrolyte present (e.g., 0.2 M NaCl, 0.4 M $BaCl_2$ and 0.6 M $CaCl_2$), ranging from about 20 to 500 pS (picosiemens). Menestrina et al. discuss their interesting results in terms of the Gouy-Chapman electrical double-layer theory and suggest that molluscan hemocyanins are a class of channel-forming proteins.

One of the important and well-studied cristae membrane enzymes is cytochrome C oxidase[69], which catalyzes the final reaction of respiration by reducing molecular oxygen to water with the electrons from its substrate. Several investigators have incorporated cytochrome C oxidase into BLM (see review by Shamoo and Tivol[31]), and found that, in the presence of oxidized cytochrome c on one side, a potential difference is generated upon reduction of the substrate. This transmembrane potential could be reversed or prevented by cyanide. However, the BLM conductance was little affected by cyanide or ascorbate.

Turning now to the other electrical properties, typically a biomembrane has been represented by an equivalent circuit consisting of a combination of resistor (R_m) and capacitor (C_m) in parallel[7]. Thus, it is worth remembering that energy may be stored in the membrane capacitor as given by

$$dW_{el} = E_m C_m dE_m \qquad (11)$$

or

$$W_{el} = 1/2 C_m E_m^2 \qquad (12)$$

where W_{el} is the electrical work, C_m and E_m are, respectively, membrane capacitance and potential. In this connection two aspects of cristae membrane from an experimental point of view will be described: electron transport and ATP synthesis. As already mentioned, central to membrane bioenergetics are the transmembrane redox reactions. The process entails the generation, separation and transport of charges (electrons, holes and protons), leading eventually to redox reactions on opposite sides of the membrane. The energy stored in separated charges and redox compounds may be used to drive vital processes such as oxidative phosphorylation. In the cristae membrane of mitochondria[2], the redox protein components embedded in the lipid bilayer are poised to perform electron-transfer and redox reactions. The exact location and orientation of these membrane proteins and other components are not known (see Figure 1). The generally accepted arrangement, however, is that membrane proteins and other components are closely intercalated among lipid molecules which are organized in the form of a lipid bilayer[17,32,33,55]. It should be of interest to incorporate electron-transfer proteins and co-enzymes into BLMs and to investigate their redox properties (e.g. measurements of $E^{o\prime}$ values in the past have been obtained by the classical potentiometric technique without "membrane" participation. The potentiometric technique is frequently complicated by the low electron exchange currents at Pt electrodes. Since accurate E°' values are important in determining the sequence of electron transfers among redox enzymes in electron transport chains, their measurements in "membrane-bound" states are self-evident. Up to now the measurement of E°' values of electron transport chain components in association with lipid bilayers has not been possible, owing simply to the fact that no suitable technique existed. With the availability of electron-conducting BLMs coupled with the CV technique developed recently[34], an entirely new approach is open for unraveling the complicated electron-transfer and redox reactions in the process of oxidative phosphorylation.

The second aspect of the mitochondrial process is ATP synthesis. As proposed by Mitchell, it establishes a proton and/or potential gradient which is used to drive ATP synthesis from ADP and phosphate ions. Racker and Stoeckenius reported photophosphorylation in reconstituted proteoliposomes from a mixture of bacteriorhodopsin, beef-heart ATPase, and soybean phospholipids (see Oosawa and Imai[35]). This outstanding, often cited experiment does not appear to have been repeated by others. It would be, therefore, of interest to test the chemiosmotic hypothesis using planar BLM. In the BLM system, both pH and applied potential gradients can be easily created.

The reverse process of ATP synthesis is hydrolysis, which is used to provide energy for a host of cellular activities such as muscle contraction, biocompound synthesis and active transport[2,3,36]. The last mentioned process is of special significance since differential ion concentration gradients are crucial to the living state. For example, almost all cells maintain a high concentration of K^+ and a low concentration of Na^+ inside. The so-called Na^+/K^+ pump is believed to exist in the membrane, which actively transports Na^+ out and K^+ in to the cell against their respective electrochemical potential gradients. The Na^+/K^+ pump, made of protein (the Na^+/K^+ ATPase) is powered by the energy resulting from the hydrolysis of ATP. To test this hypothesis, a most interesting experiment, using the BLM system as a model for the plasma membrane, has been reported by Jain, Strickholm, and Cordes[37]. In their experiment, the BLM was formed in the usual manner. ATP was added to one side of the BLM, and this was followed by the addition of a preparation of membrane-bound ATPase (Na-K dependent fraction). After these additions, a decrease in the membrane resistance was observed. The current across the BLM and open-circuit membrane potential was monitored. Jain et al.[37] observed a positive current across the BLM. This was interpreted as being due to the transport of Na^+, which is analogous to that found in the natural membrane. The current in their reconstituted system was said to be dependent on the presence of Na^+ in the bathing solution, the magnitude of which depends on ionic strength. Further, the presence of ouabain (a cardiac glucoside), a compound known to block Na^+ transport in red cells, inhibited the current across the ATPase-modified BLM. It seems probable that, by introducing ATPase into the bathing solution, "molecular sodium pumps" could have been inserted into the BLM, which may account for the phenomena. In this connection, mention must be made of the paper by Fendler et al.[38], who investigated the transport activity of purified Na^+/K^+ ATPase incorporated into a BLM. Stationary pump currents were obtained after the addition of monensin and valinomycin (ionophores). In the absence of ADP and inorganic phosphate (P_i) the half saturation for the maximal effect was obtained at 6.5 μm released ATP. Further, pump activity was obtained only in the presence of Na^+, K^+ and Mg^{2+}. The electrical response was blocked completely by ouabain and vanadate (ATPase inhibitors). Fendler and colleagues[38] suggest that their method offers the possibility of studying by direct electrical measurements Na^+/K^+ ATPase in transport. In our laboratory, we have observed the effect of ATPase, followed by the addition of ATP, on on the BLM current[7]. The result is shown in Figure 3. Both in our experiments and those reported by Jain et al.[37], open-circuit membrane potentials as large as 130 mV were observed. Development of the membrane potential and short-circuit current is dependent upon the presence of all the substrates of Na^+/K^+ ATPase as well as upon the enzyme itself. Further, the short-circuit membrane current thus generated could be diminished by applying voltages of opposite polarity. At about 175 mV positive with respect to the ATPase-free side, the current, presumably due to the Na^+ flux, was zero.

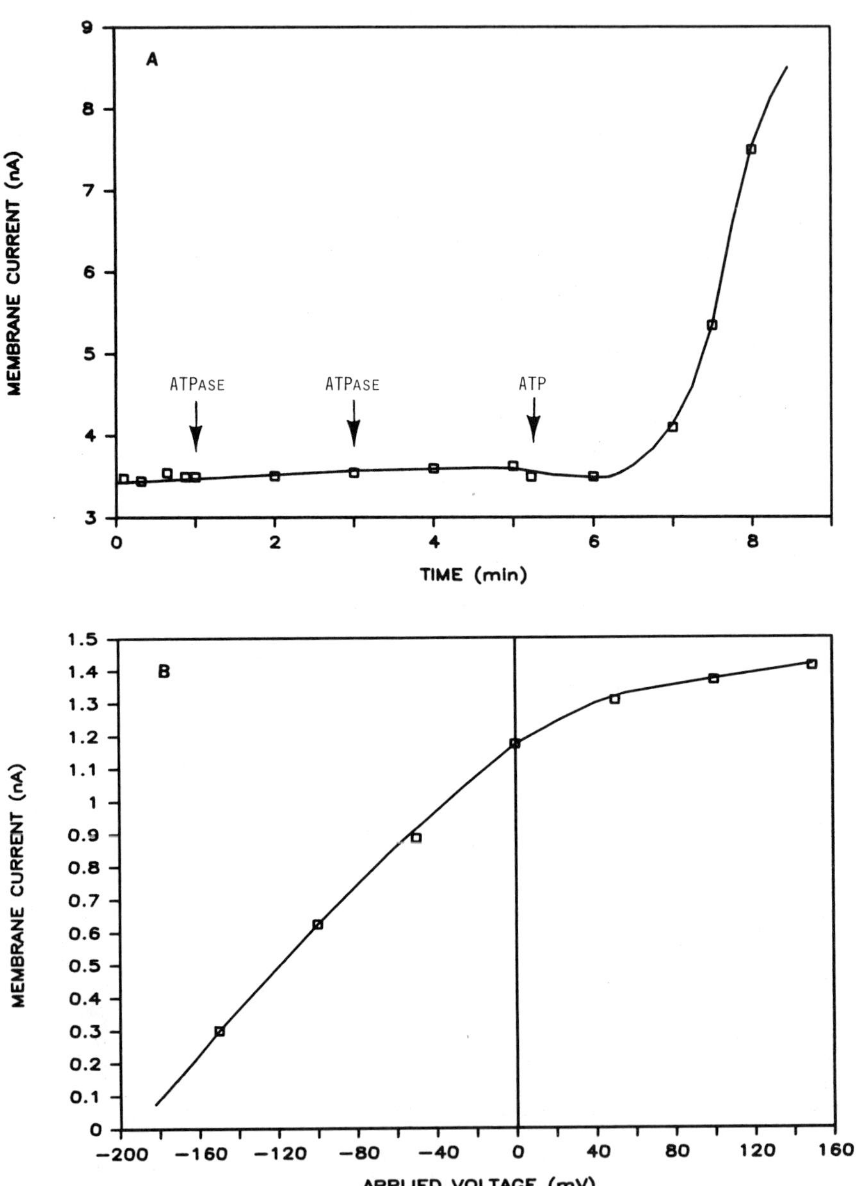

Figure 3. The effect of ATP and ATPase on the electrical properties of of reconstituted BLM. (A) Generation of membrane current upon additions of ATPase followed by ATP as a function of time. (B) Current/voltage curve of a reconstituted ATPase-BLM (see References 7, 13, 37 and 38).

Table 2. Major Landmarks In Photosynthesis Research.

Year	Investigator	Finding or Insight
1776	Joseph Priestley	Oxygen evolution
1782	Jean Senebier	Carbon dioxide reduction
1796	Jan Ingenhousz	Carbon dioxide reduction
1845	Julius R. Mayer	Energy "transduction"
1931	C.B. van Niels	Generalized equation for photosynthesis
1943	S.C.J. Ruben	ATP and NADPH for CO_2 reduction
1949	E. Katz	Semiconduction theory
1951	M. Calvin, A.A. Benson and J.A. Bassham	Dark reactions
1960	R. Hill and F. Bendall	Two light reactions
1961	P. Mitchell	Chemiosmotic hypothesis
1965	K. Muehlethaler, H. Moor and Szarkowskin	Bilayer leaflet model
1966	A.T. Jagendorf and E. Uribe	ATP synthesis by pH gradient
1968	W. Junge and H.T. Witt	Transmembrane electric field in thylakoids

Analysis of data obtained used the relationships $\Delta G^{o\prime} = -RT\ln K$, where K = [ADP] $[P_i]$/[ATP] = 10^{-4}, and $\Delta G^{o\prime} = -nFE_m$, where E_m is the membrane potential at $I_m = 0$, one can calculate the quantity, "n", the number of protons translocated across the membrane per mol of ATP hydrolyzed. This turns out to be about 3. For ATP synthesis, Lane et al.[39] have recently reported that the entire mitochondrial electron transport chain (from NADH dehydrogenase to cytochrome oxidase) requires the translocation of at least 12 protons per oxygen atom. How these results are correlated remains to be seen[69].

Pigmented Bilayer Lipid Membranes as a Model for the Thylakoid Membrane

Photosynthesis in green plants is by far the most crucial redox reaction on Earth, for life depends upon the transduction of solar energy into biochemical energy. In simplest terms, the overall process of photosynthesis can be considered as light-driven, coupled CO_2 reduction and H_2O oxidation. The important landmarks of photosynthesis research are summarized in Table 2. In spite of more than 200 years of efforts, however, we are still far from solving all the major problems of photosynthesis[40-42]. Among the outstanding problems that remain to be solved at the molecular level are: (i) the primary events of quantum conversion,

(ii) electron transport and photophosphorylation, and (iii) oxygen evolution. It appears that all major unsolved problems are intimately connected with the lipid bilayer structure of the thylakoid membranes. To facilitate an understanding and to maximize the effort, photosynthesis research has been operationally divided into two areas: the photophysicochemical domain and the dark biochemical domain. An overall view of photosynthesis by green plants may be seen in the flow diagram below.

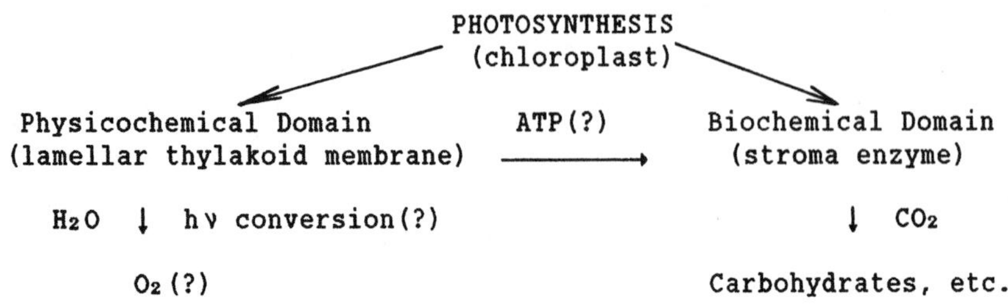

In the biochemical domain in the absence of light, carbohydrate is formed from CO_2 by a series of enzymatic reactions driven by ATP and NADPH. The process is well understood. In the photophysicochemical domain, the lipid bilayer structure or the thylakoid membrane is strongly implicated in the two different photoreactions. Therefore, the resolution of the three major problems (identified by question marks in the flow diagram) cannot be achieved without, first of all, having some appreciation of the role of the thylakoid membrane in terms of chemical composition and molecular organization.

Under a light microscope, a green leaf shows numerous cells containing disc-shaped bodies (about 7 μ in diameter, 1 μ in width and about 20 or so per cell). These bodies are the so-caled chloroplasts, where chlorophylls (Chl) and other pigments are concentrated. The chloroplast is the site of photosynthetic activity. Under an electron microscope, each chloroplast can be seen to be bound by a plasma membrane envelope in which thylakoids are located. Thylakoids are seen to be highly organized lamellar structures called grana. The thickness of these structures is on the order of 40-120 Å. Here we come to the limit of resolving power of the electron microscope. So, from here on, it is the end of the "facts" and the beginning of imagination or speculation. To Menke[41-43], these grana looked like a stack of disc-shaped vesicles, named thylakoids. Thylakoids are double membranes closed in themselves, which are embedded either singly or in stacks (grama) in the stroma. The thylakoids are the structural units of the photosynthetic lamellar systems. From the images of electron micrographs, the structure of the thylakoid membrane has been variously interpreted. According to Park and Pon[41-43], the membrane consists of an ultrathin lamella with attached globular units, termed quantasomes, which is considered as the morphological expression of the photosynthetic unit. A different interpretation has been given by Howell and Moudrianakis and they suggest that the entire membrane is involved in photoreactions[41-43]. Weier and Benson have interpreted their images in a somewhat unique fashion, and conjured up a highly intricate picture[41-43]. On the basis of X-ray diffraction studies, the ultrastructure of chloroplast membrane has been analyzed by Kreutz[41-43], and the subject reviewed by him. Using the freeze-etching technique, Muehlethaler[40] has suggested a picture for the thylakoid membrane based upon the bimolecular leaflet model. The salient feature in all these proposed models lies in their oriented bilayer lipid core, onto which other important cellular constituents, such as proteins and pigments, may interact through either ionic or van der Waals attraction, or both. Muehlethaler's interpretation is of special interest, in view of the experiments using the pigmented bilayer lipid membrane as a model for the thylakoid membrane to be discussed in a later section.

The Equations of Photosynthesis

Photosynthesis research may be roughly divided into three periods:

Period I:

Major advances were made in the 1930's after R. Hill showed that isolated chloroplasts can evolve O_2 (when a suitable electron acceptor A, or oxidant, was provided).

$$CO_2 + H_2 \xrightarrow{\text{light}} (CH_2O) + O_2 \tag{13}$$

Period II:

In the 1950's, Arnon and his colleagues[41-43] demonstrated that isolated chloroplasts can carry out not only photophosphorylation but also CO_2 reduction. Two types of phosphorylation are known:

(A) Cyclic - which involves no net electron transport from a donor to and acceptor, and

(B) Non-cyclic - where

$$ADP + P_i + NADP + 2H_2O \longrightarrow ATP + NADPH_2 + H_2O + 1/2\ O_2 \tag{14}$$

In this process, ATP formation is associated with NADP reduction (in contrast to NAD oxidation in mitochondria) and O_2 evolution.

Period III: (Early 1960's)

In order to explain these two types of photophosphorylation and the enhancement effect observed by Emerson in the 1940's and 1950's, the so-called "Z" scheme, or two photosystems, was proposed. It is generally accepted that Photosystem I (PS-I) and Photosystem II (PS-II) together provide reduced nicotinamide adenine dinucleotide phosphate (NADPH) and adenosine triphosphate (ATP) for the CO_2 reduction. However, the details of these systems remained obscure and they were the focal point of much research activity. Concurrently, with this development was the proposal of Mitchell[4], which has come to be known as the chemiosmotic hypothesis, to explain energy transduction. In Mitchell's formulation the membranes must be relatively impermeable to ions. The driving force for phosphorylation is due to the electrochemical gradient of protons, which could come about as a result of electron transport. The results obtained by Jagendorf, Neumann and Hind, with chloroplast suspensions subjected to pH "shock", are generally cited as evidence in support of the chemiosmetic hypothesis[41-43].

Equation (13) has been known for over 100 years; it tells us about the stoichiometry but nothing about the mechanism of photosynthesis. In articles and books dealing with photosynthesis published in the 1960's Equation (13) was frequently given, which we now know to be incorrect or at best misleading in at least two respects, from the well-known facts: (1) To transform 1 mole of CO_2 into 1 g-atom of carbon in the carbohydrate requires about 112 Kcal (5.2 ev). Since this quantity of free energy must come from sunlight, the question one usually asks is: "Does the sunlight reaching the earth contain energetic photons of this magnitude?" The answer is "NO", since we know that the sunlight which reaches the green plants consists of light predominantly in the range of 400-1300 nm (88%) which in terms of energy is between 1-3 ev or 20-70 Kcal/quantum. Thus, it is impossible to expect that the reaction as indicated in Equation (13) is achieved in a single step on the basis of elementary energetic considerations. (2) The work of van Niels and others has conclusively demonstrated that CO_2 does not take part at all in any photochemical reactions. CO_2 reduction involves a series of enzymatic reactions that can and do take place in the dark. Thus, van Niels, by a comparative approach, suggested a generalized equation

$$CO_2 + 2H_2A \longrightarrow (CH_2O) + H_2O + 2A + \Delta G^\circ \tag{15}$$

where H_2A = H_2O, H_2S, Isopropranol, etc. (all of these are hydrogen donors). Implicit in Equation (15) is that the O_2 comes from H_2O. Further, van Niels proposed that the essence of green plant photosynthesis is water splitting;

$$H_2O \longrightarrow [H] + [OH] \tag{16}$$

where [H] and [OH] are an unspecified reducing entity and oxidizing entity, respectively. Translating van Niels' Equation (16) into modern terminology, the scheme may be depicted as follows:

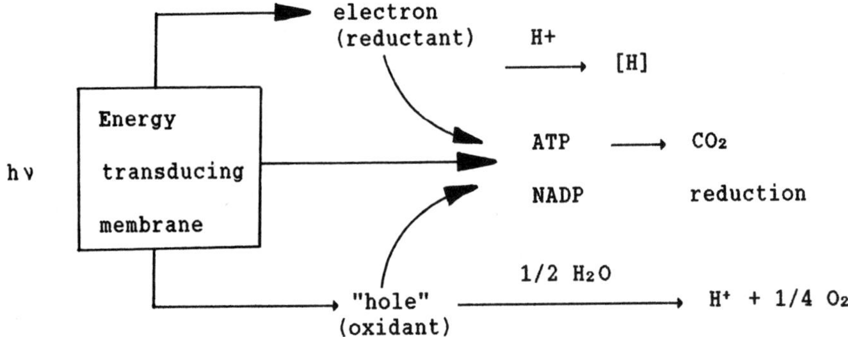

Redox Reactions

Photosynthesis may also be considered as a coupled redox reaction, in which it is involved in the gain or loss of electrons as follows:

$$A + D \rightleftharpoons A^- + D^+ \tag{17}$$

where A and D denote an electron receptor and donor, respectively, and A^- and D^+ are the species in their respective reduced and oxidized state.

In terms of membrane dynamics, the light-induced redox reactions can be simplified as shown in Figure 4. This membrane system consists of two interfaces with donor molecules on one side and acceptor molecules on the other. The pigment molecules, denoted by P_o and P_i, at each membrane/-solution interface are assumed to be capable of excitation by light, charge separation and electron transfer, thereby leading eventually to stabilized products (A^- and D^+). This simplified energy transducing scheme will be considered further in the section on pigmented bilayer lipid membranes.

Reconstitution Experiments

With the above background information, a series of experiments has been undertaken to investigate (i) whether a BLM could be constituted from photoactive pigments, and (ii) the possibility of using such a BLM system as an energy-transducer in the study of certain aspects of photophysical and photochemical processes. The results of these experiments are summarized in the following paragraphs.

BLM formed from photoactive compounds, such as chlorophylls, porphyrin, covalently-linked porphyrin-donor/acceptor complexes, etc., exhibit two interesting phenomena: (i) a photovoltaic effect, and (ii) photoconductivity. As a model for the thylakoid membrane of the chloroplast, BLM of this type are generated from various sources such as extracts of spinach leaves, chlorella, and chlorophylls obtained from commercial sources. In the absence of light excitation, the d.c. resistance of BLM is high ($>10^7\ \Omega\text{-cm}^2$) and the dielectric breakdown voltage is generally less than 200 mV. Further, current/voltage curves of Chl-BLM appear to be linear, up to 150 mV. The thickness of the membrane was measured. Taking the experimentally determined thickness at its face value, the BLM generated from chloroplast pigments is pictured as similar to those of liquid crystals in two dimensions (Figure 4). It is suggested that, in the case of chlorophyll BLM, the hydrophobic portion (phytyl group) of the molecule extends inward, while the hydrophilic head (porphyrin) is situated at the aqueous solution/membrane interface. The high d.c. resistance of the membrane is attributed to the liquid hydrocarbon-like

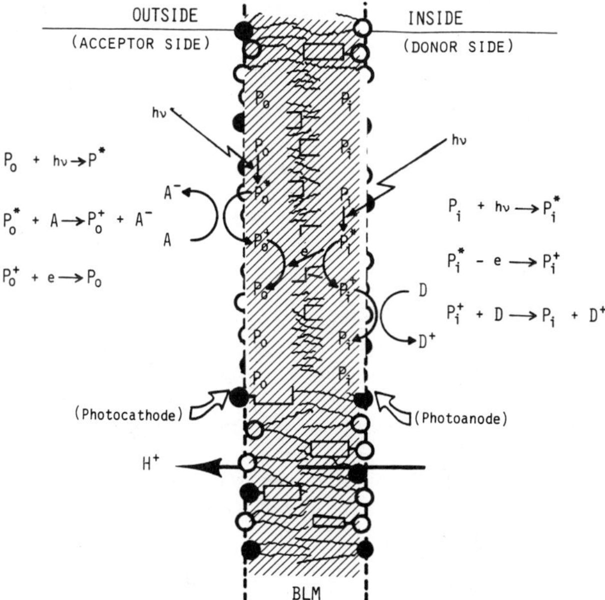

Figure 4. The redox electrode model of photosynthesis illustrating the mechanism of photon-induced charge separation and electron transfer processes in a pigmented BLM separating two aqueous solutions. P and P* denote the pigment in the ground and excited state, respectively. A = electron acceptor, D = electron donor, A^- and D^+ are reduced and oxidized forms. Subscripts 'o' and 'i' indicate the outer and inner side of the membrane, and P^+ is oxidized P. (Adapted from Nature, 219, 272, 1968; 227, 1232, 1970).

interior. Several orientations of the porphyrin group may be possible, i.e., the porphyrin plate may sometimes lie parallel to the biface, sometimes perpendicular, and at other times somewhere in between. A slant orientation is favored, which is based on the data of photovoltage polarization and conformation analysis obtained by Brasseur et al.[44], who found that the porphyrin ring has an orientation angle of 45° ± 5° to the plane of the BLM.

The kinetics of photopotential was studied by Liu and Mauzerall[45], who explained the decay by a rate constant that exponentially decreases with the distance between the pigment cation and ferricyanide. The other conclusion from their study is that the pigment cation does not transverse the BLM in less than 10 msec. To examine the 'action' spectra of pigmented BLM, a technique known as photoelectrospectrometry has been developed, which is based on the photoelectric effect and optical spectroscopy[46]. Thus far, only the range of UV and visible wavelengths has been reported. In this connection Koyama et al.[47] obtained a spectrum of carotenoid-BLM by resonance Raman spectroscopy, which should be considered as a major advance in BLM spectroscopy.

For electron transfer and charge separation studies, BLMs have been formed from lipid solutions containing Chl and porphyrins with and without added modifiers such as quinones and carotenes. The addition of quinones and/or carotenes is due to the well-documented evidence that quinones serve as the primary electron acceptor and carotenes as effective light-gathering accessory pigments. Rich and Brody[48] found that Chl-BLM in the presence of dehydroxy-carotenoids gives rise to much greater photocurrents than either the simple carotenes or the diketocarotenoids. BLM formed using pheophytin (a Mg-free chlorphyll) generate a very small photocurrent regardless of the carotenoids used. This is explained in terms of lack of interaction between pheophytin and carotenoids. The negligible photocurrent in pheophytin-BLM might also be due to the mismatching of redox potential between pheophytin and ferricyanide, thus preventing the latter to be reduced by the former. The interaction between chlorophyll and β-carotene in BLM has also been studied by Vacek et al.[49] who observed a decrease of photopotential upon repeated flash excitation, which might be owing to lipid excitation. Sielewiesiuk[50] reported the photooxidation of Chl in BLM on the basis of a decrease in photocurrent, whereas Kadoshnikov and Stolovitsky[51] described the results of spectral photoelectrochemical investigations of chlorophyll BLM and liposomes. Chl in these membranes has been shown to have monomeric (λ = 640 nm) and crystalline (λ = 740 nm) forms. The BLM, with incorporated Chl and pheophytin, generated photopotentials of opposite signs according to Kadoshnikov and Stolovitsky[51]. In this connection, it is interesting to note that Hattenbach et al.[52] raised the question, "Does phytochrome interact with lipid bilayers?". Phytochrome, a plant chromoprotein, which was claimed to respond reversibly to red light of different wavelengths (667 and 725 nm), was incorporated into a BLM. However, Hattenbach and colleagues[52] found no detectable change in membrane conductance during illumination with red light (660 nm). Their finding in BLM, however, does not rule out the regulatory role of phytochrome in plant membranes.

Towards a more complicated experiment, reconstitution with an entire reaction center (i.e., Photosystem I of green plants) has also yielded a photoelectric effect[53]. Upon illumination, the PS-I reaction center (RC) containing BLM generated transient and steady-state light-induced voltages and currents. The wavelength dependence of the photoresponse matched the absorption spectrum of RC. The main features of the photoresponse are explained by the redox membrane electrode model (see below). The observed photosignals are due to the transfer of electrons from a donor (D) to an acceptor (A) located on opposite sides of the BLM (Figure 4). This RC-containing BLM provides direct evidence that the RC is an intrinsic membrane protein that spans the lipid bilayer. In this connection the experiments reported by Seibert et al.[54] on an RC-doped semiconductor electrode and by Tabushi et al.[31] in liposomes are of interest.

The separation of charges in chloroplasts by light is believed to be among the first steps involved in the primary process (i.e., quantum conversion) of photosynthesis. Chemical compounds, such as 2,4-dinitrophenol (DNP), can uncouple electron transfer from ATP formation in the electron transport chain. The similar effect of DNP on oxidative phosphorylation in the mitochondria, as already mentioned, is well-known. The effect of DNP, ρ-trifluoromethoxy-carbonylcyanide-phenylhydrazone (FCCP), and dicoumarol on the electrical conductivity of Chl-BLM has been examined.[7] Below 10^{-5} M of FCCP little change in the conductivity was observed. Above this the conductivity increased rapidly, as a function of increasing FCCP concentration, until 10^{-4} M. Similar effects have been obtained with DNP and dicoumrol, although the effect took place at different concentrations. The order of effectiveness, when compared at the same concentrations, is FCCP > dicoumarol > DNP. The peak corresponds roughly to the pK value of the uncoupler, which suggest that H^+ is the charge transported across the membrane. The uncoupler, which has appreciable lipid solubility, serves as a carrier of hydrogen ions. It is interesting to note that the pH dependence effect of the uncoupler does not appear to depend on the composition of the lipid solution used for BLM formation, since the lecithin-BLM reported by Liberman and colleagues displayed similar characteristics[20]. The effect of these modifiers (FCCP and DNP) on nonlinear I/V curves may be explained in terms of ion association in the lipid moiety, and dissociation of the ionizable groups of the lipids which must be governed by the pH of the bulk solution, and possibly the electric field. Thus, the selectivity of the BLM would be controlled by the ionizable groups at the biface, whereas the ion specificity depends upon the solubility of lipid-soluble complex. The effect of DCMU and CMU (inhibitors) on Chl-BLM showed the concentration dependence of both photo emf and transmembrane dark voltage[7]. In both cases the photo emfs decreased with increasing concentration, while the dark transmembrane voltages increased linearly with concentration. Furthermore, the membrane resistance decreased by one order of magnitude in the presence of 4×10^{-3} M DCMU or 5×10^{-3} M CMU. Both DCMU and CMU are well-known poisons of photosynthesis, and in the Chl-BLM system, they may act in a manner which prevents the separation of charges in the membrane. Finally, varying the pH of the bathing solution changed the magnitude of the photo emf of Chl-BLM. However, the waveform of the action spectrum remained unchanged. A maximum photo emf at pH slightly above 5 was observed, which coincided with both of the two maxima, resistance and photo emf, that peak at pH around 5, indicating the photoresponse is closely related to BLM resistance.

Toward Molecular Photosynthesis

The key step in photosynthesis is believed to be photon-driven charge separation. It has been known for nearly two decades now that, indeed, the light induced charge separation demonstrated in artificial BLM systems is one of the most convincing evidences at a molecular level[57], which have been described at length in the preceding paragraphs. Some of the current efforts towards understanding of photon-induced charge separation and transfer in natural photosynthesis will now be presented. The central question is the molecular detail as to how the conversion of photons absorbed into a potential difference across the membrane takes place; it is this membrane potential difference (or voltage) which drives the subsequent biochemical stages of photosynthesis. There are at least two approaches to answering the question of photon-induced charge separation and transfer using the pigmented BLM systems.

As an experimental model, a unique feature of the BLM system is that coupled photon-driven redox reactions may be independently activated at the two interfaces of the BLM, thereby mimicking the Z-scheme of photosynthesis. One way to accomplish this is to use ZnTPP (or MgTPP), as the photoabsorber in the BLM, in conjunction with tris(2,2'-bipyridine) ruthenium ion, $Ru(bpy)_3^{2+}$, methyl-viologen (MV^{2+}, 1,1'-dihexadecyl-4,4'-bipyridium) as an electron acceptor. EDTA (ethylenediamine-N,N,N',N'-tetraacetate) as electron donor, and vitamin K (VK_1) as a hydrogen carrier. In one system, the BLM phase contained VK_1 and MV^{2+} with $Ru(bpy)_3^{2+}$ added to one side of the bathing solution and EDTA to the other side. The second system differed from the above by having only ZnTPP added to the BLM-forming solution. Photovoltages with a viologen-free side were positive 50 mV and 90 mV, respectively, for the first and second systems. The time course for the photocurrent is as follows: photocurrent rises rapidly, which is followed by a slow increase, then it decays to a steady value. When the light is switched off, the current goes back to the dark value. Since in both systems the photoresponses were observed, $Ru(byp)_3^{2+}$ must be bound to the BLM. The presence of ZnTPP in the BLM greatly enhanced the response, which implies facilitated electron transport. Photoexcitation of $Ru(bpy)_3^{2+}$ and/or ZnTPP leads to viologen reduction, as evidenced by the polarity of the photopotentials. Since acceptor (MV^{2+} or Fe^{3+}) and donor (EDTA) are physically immobilized, transmembrane redox reactions must take place. The most direct interpretation of the data is that reduction of $Ru(bpy)_3^{2+}$ occurs on the side containing EDTA with transmembrane electron tunneling facilitated by ZnTPP and VK_1. The drop in photocurrent may be caused by the diffusion of reduced MV^+, which could become more soluble in the membrane phase. As in the case of Chl-BLM, electron transfer is possible from a donor (D) to an acceptor (A), if the bands of respective D and A are overlapped. It should be noted that a photoactive pigment can serve as either a donor or acceptor in the process. When electrons and holes are generated in the BLM, they will polarize the surroundings and attract charges of opposite sign across the lipid bilayer. These charges will, in turn, attract the electrons (or holes) toward the oppositely charged interface, resulting in the reduction of the interfacial potential. Similarly, it can be shown that the interfacial barrier for charge transport can also be reduced by an externally applied voltage across the BLM[58]. The problem of light-induced electron transfer has been addressed by many investigators[1,14,75].

Instead of using chlorophylls and their derivatives such as porphyrins, special "molecularly" designed compounds have been incorporated into BLMs[59]. This brings us to the second approach. The working hypothesis is that photon-induced charge separation into electrons and holes on pigment molecules is the crucial first step but equally crucial is how to prevent the light-generated electrons and holes from recombination (or back reaction) which results in energy loss. For efficient charge separation and transfer, therefore, it seems highly probable that a close proximity between a donor species (e.g. chlorophyll or porphyrin) and an electron acceptor molecule (e.g. quinone) is a prerequisite. To test this hypothesis, several groups of workers have synthesized covalently linked porphyrin-quinone and porphyrin-carotene complexes[60-62]. Incorporation of these newly available compounds, as a model for the initial photophysicochemical event in the reaction center of photosynthesis, into reconstituted BLM is of evident interest and has been carried out in our laboratory[59]. The photopotentials obtained on some of these new BLM systems are indeed exciting in that they are larger by at least a factor of three (the highest values hitherto reported) than those of chlorophyll-containing BLMs (see Figure 4). If a pigmented BLM separating two different redox couples is illuminated, the basic experimental results are as follows: a change in membrane potential ($E_{op} = E_{h\nu} - E_{dark}$ occurs under open-circuit conditions, and a change in membrane current ($I_{sc} = I_{h\nu} - I_{dark}$) under short-circuit conditions. E_{op} or I_{sc} vs time curves can take several forms, depending on the experimental conditions. Three possible mechanisms have been proposed to explain the BLM photoelectric effect, one of which is the redox electrode mechanism[57,64]. Electronic charges generated as a result of light absorption cause one side of the BLM-solution interface to be reducing and the other side oxidizing. Before discussing the redox membrane electrode model, one may recall the usual picture of a BLM separating two electrolyte solutions consisting of a liquid hydrocarbon core (~ 50 Å) sandwiched between two polar regions which contain the sites for ion binding. These sites, if charged, determine the interfacial potentials, and constitute one of the two major factors governing charge transport, the other factor being the hydrophobic layer. The number of pigment molecules in a BLM is estimated to be about 3 x 10^{13} molecules per cm^2 with an average distance of about 10 ± 5 Å between photoactive molecules[40].

Evidence favors a redox electrode mechanism in which the excited pigment molecule is directly involved. It seems reasonable that the excited pigment molecule (P) ejects an electron to an acceptor (A). This is followed by accepting an electron from a donor (D) by the oxidized pigment molecule (P^+). The separated charges are prevented from recombining by the presence of the lipid bilayer core of the BLM and the energy barrier between pigment molecules. Thus, an electric potential should appear shortly after the absorption of photons by the pigments in the BLM. In the case of covalently linked porphyrin-quinone (P-Q) or porphyrin-carotene (P-C) complexes, the distance for photon-driven electron transfer is much less than 10 Å. Thus, for the P-Q containing BLM interposed between two redox solutions, we have intramolecular charge separation/-transfer as well as interfacial charge separation/transfer. It seems highly probable that the intramolecular process in the P-Q molecules is so rapid that only the singlet state is involved. On the other hand, electron transfer between excited P-Q molecules and electron acceptors and donors may involve the less energetic and less efficient triplet state. For photo-active excited molecules embedded in the bilayer lipid core of the BLM, the most likely process of charge transfer is by electron tunneling. In this connection, the work of Joran et al.[65] is important in that they report the effect of a 4 Å change in distance on the rates of light-induced electron transfer for P-Q complexes separated by rigid hydrocarbon spacers. According to Joran and his colleagues[65], a marked increase in the rate of charge separation, by more than an order of magnitude, is observed in the P-Q complex over the first 0.2 eV. The process takes about 100 ps. Above this, the rate of charge separation is almost constant; it is practically independent of whether a non-polar or highly polar solvent is used.

It is worth restating that one of the aims of pigmented BLM studies is to elucidate the mechanism of quantum conversion of photosynthetic thylakoid membranes[57,64]. In natural photosynthesis, the uniqueness of having an ultrathin lipid membrane, besides providing an appropriate environment for pigments and redox proteins, serves as a barrier to suppress the back reaction of the photoproducts. With the photosystems located in different parts of the membrane, these make possible the up-grading of the energy of photons as well. As shown in Figure 4, for every two photons absorbed, only one electron and one hole are involved in the ensuing redox reactions. According to the redox electrode model, two modes of operation are possible, i.e., via a one-photon or a two-photon process. In the former case, the ion-radicals (P^+ and P^-) generated are reduced and oxidized, respectively, by D and A. In the two-photon process, the reaction of P* with A on the acceptor side (left) produces A^- and P^+. Being in the membrane phase, this P^+ is highly likely to be reduced by an adjacent P* on the donor side (right) of the membrane. The resulting P^+ on the donor side is, in turn, reduced by D. The overall process is the sum of the equations given in Figure 4 and Equation 17.

In the two-photon scheme, it is envisioned that transmembrane diffusion of P^+ (or P^-) is unlikely. Rather, the electron translocation across the membrane is explained in terms of a tunneling mechanism between P^+ and P*. This suggestion may be checked experimentally by measuring the photoeffects of the system as a function of temperature[7,34,66].

Finally, mention should be made that a pigmented BLM has been likened to a double-Schottky barrier (or p-n type) cell, with the aqueous solution playing the role of metal[14,67]. Bearing this in mind, one side of the membrane acts as a photocathode (p-type junction) and the other acts as a photoanode (n-type junction) as illustrated in Figure 4. The photovoltages observed are influenced by the redox compounds (electron donors and acceptors) present in the bathing solution and/or externally applied voltage[58]. In the scheme shown, two types of processes can be distinguished: photosynthetic and photocatalytic. In the former, the overall free energy is positive ($\Delta G^o > 0$) in that external energy input is necessary to drive the uphill reaction. In photocatalytic processes which are thermodynamically favored ($\Delta G^o < 0$) the energy of light is not stored but merely used to overcome the energy barrier of the reaction[68]. In this connection, the BLM system appears to be of interest in areas of solar energy conversion[67,70-75] and molecular electronics[76] as well. It remains to be seen what additional physical-chemical developments are necessary to bring about some envisioned practical applications.

ACKNOWLEDGEMENTS

The author would like to acknowledge the generous support of the NIH (GM-14971) and ONR(N00014-85-K0399). Thanks are due to Theresa Hubbard for her excellent assistance in the preparation of the manuscript.

REFERENCES

1. G. Hauska and G. Orlich, J. Memb. Sci., 6, 7 (1980).

2. D.E. Metzler, "Biochemistry: The Chemical Reactions of Living Cells", Academic Press, Inc., New York (1977).

3. R.P.F. Gregory, "Biochemistry of Photosynthesis", Wiley, New York (1978).

4. P. Mitchell, FEBS Sym., 28, 353 (1972); J.M. Palmer and D.O. Hall, Prog. Biophys. Mol. Biol., 24, 125 (1972).

5. A.A. Bulychev, Andrianov, G.A. Kurella and F.F. Litvin, Biochim. Biophys. Acta, 420, 336 (1976).

6. K.W. Kinnally and H. Tedeschi, FEBS Lett., 62, 41 (1976); C. Bowman and Tedeschi, Nature, 280, 597 (1979).

7. H.T. Tien, "Bilayer Lipid Membranes (BLM): Theory and Practice", Marcel Dekker, Inc., New York (1974).

8. R. Antolini, A. Gliozzi and A. Gorio, Eds., "Transport in Membranes: Model Systems and Reconstitution", Raven Press, New York (1982), pp. 57-75.

9. R. Blumenthal and R.D. Klausner, in: "Membrane Reconsitution", G. Poste and G.L. Nicholson, Eds., Elsevier, New York (1982), p. 43.

10. S.G. Davison, Ed., "Progress in Surface Science", Vol. 19, No. 3, Pergamon Press, New York (1985), pp. 169-274.

11. A. Szent-Gyorgyi, in: "Light and Life", W.D. McElroy and B. Glass, Eds., Johns Hopkins, Baltimore (1961), p. 7.

12. E.A. Guggenheim, "Thermodynamics", North-Holland, Amsterdam (1950).

13. M.L. Hair, Ed., "Chemistry of Bio-Surfaces", Marcel Dekker, Inc., New York (1971), pp. 234-348.

14. H.T. Tien, Separation Science and Technology, 15, 1035 (1980).

15. H.J. Morowitz, Biosystems, 14, 14 (1981); Am. J. Physiol., 235, R99 (1978).

16. G. Benga, Ed., "Water Transport in Biomembranes", CRC Press, Boca Raton (1987).

17. H.D. Brown, Ed., "Chemistry of the Cell Interface", Part A, Academic Press, New York (1971), pp. 205-254.

18. P.F. Weller, Ed., "Solid State Chemistry and Physics", Vol. 2, Marcel Dekker, Inc., New York (1973), pp. 847-903.

19. J. Bielawski, T.E. Thompson and A.L. Lehninger, Biochem. Biophy. Res. Commun., 24, 948 (1966).

20. E.A. Liberman, V.P. Topaly, L.M. Tsofina, A.A. Jasaites and V.P. Skulachev, Nature, 222, 1076 (1969).

21. P.S. Sotnikov and E.I. Melnik, Biofizik, 13, 185 (1968); N. Takeguchi, T. Saitoh, M. Morii, K. Yoshikawa and H. Terad, J. Biol. Chem., 260, 9158 (1985).

22. A. Wardak, Stud. Biophys., 102, 155 (1984).

23. H.P. Ting, D.F. Wilson and B. Chance, Arch. Biochem. Biophys., 141, 141 (1970).

24. L.J. Bruner, Biophysik, 6, 242 (1970).

25. A. Finkelstein, Biochim. Biophys. Acta, 205, 1 (1970).

26. J.P. Dilger, L.R. Fischer and D.A. Haydon, Chem. Phys. Lipids, 30, 159 (1979).

27. S.J. Schein, M. Colombini and A. Finkelstein, J. Membrane Biol., 30, 99 (1976).

28. R. Benz, K. Janko, W. Boos and P. Läuger, Biochim. Biophys. Acta, 511, 305 (1978).

29. G. Xu, B. Shi, E.J. McGroarty and H.T. Tien, Biochim. Biophys. Acta, 862, 57 (1986).

30. G. Menestrina, D. Maniacco and R. Antolini, J. Memb. Biol., 71, 173 (1983).

31. I. Tabushi, T. Nishiya, M. Shimoura, T. Kunitake, H. Inokuchi and T. Yagi, J. Am. Chem. Soc., 106, 219 (1984); A.E. Shamoo and W.F. Tivol, Curr. T. Mem., 14, 57 (1980).

32. A.N. Martonosi, Ed., "Membranes and Transport", Plenum Press, New York (1982), pp. 165-171.

33. J.D. Robertson, J. Cell Biol., 91, 189 (1981).

34. H.T. Tien, J. Phys. Chem., 88, 3172 (1984); Bioelectrochem. Bioenerg., 15, 19 (1986).

35. F. Oosawa and N. Imai, Eds., "Dynamic Aspects of Biopolyelectolytes and Biomembranes", Kodansha-Elsevier, Tokyo and Amsterdam (1982), pp. 445-456.

36. D.F. Wilson, H.P. Ting and M.S. Koppelman, Biochemistry, 10, 2897 (1971).

37. M.K. Jain, A. Strickholm and E.H. Cordes, Nature, 222, 871 (1969).

38. K. Fendler, E. Grell, M. Haubs and E. Bamberg, EMO J., 4,3079 (1985).

39. M.D. Lane, P.L. Pedersen and A.S. Mildvan, Sci., 234 (1986).

40. J. Barber, Ed., "Photosynthesis in Relation to Model Systems, Topics in Photosynthesis", 3, Elsevier/North-Holland, New York (1979), pp. 115-173.

41. G. Akoyunoglou, Ed., "Fifth International Congress on Photosynthesis", Balaban International Science Services, Rehovot, Israel, 1, pp. 254-262 (1981).

42. J. Biggins, Ed., "Progress in Photosynthesis Research", Vol. 1, Martinus Nijhoff Publishers, Dordrecht, The Netherlands (1987), pp. 209-211.

43. J.M. Olson and G. Hind, Eds., "Chlorophyll-Proteins, Reaction Centers and Photosynthetic Membranes", Brookhaven Symposia Biology, Vol. 28, pp. 105-131 (1977).

44. R. Brasseur, J.D. Meutter and J.-M. Ruysschaert, Biochim. Biophy. Acta, 764, 295 (1984).

45. T.M. Liu and D. Mauzerall, Biophy. J., 48, 1 (1985).

46. J.R. Lopez and H.T. Tien, Biochim. Biophy. Acta, 597, 433 (1980).

47. Y. Koyama, M. Komori and K. Shiomi, J. Coll. Int. Sci., 90, 293 (1982).

48. M. Rich and S.S. Brody, FEBS Lett., 143(1), 45 (1982).

49. K. Vacek, O. Valent and A. Skuta, Gen. Physl. B., 1, 135 (1982).

79 (1982).

50. J. Sielewiesiuk, Stud. Biophys., 96, 117 (1983).

51. S.I. Kadoshnikov and Yu.M. Stolovisky, Bioelectrochem. Bioenerg., 9,

52. A. Hattenbach, J. Gundel, G. Hermann, D. Haroske and E. Müller, Biochem. Physiol. Pflanzen., 177, 611 (1982).

53. J.R. Lopez and H.T. Tien, Photobiochem. Photobiophy., 7, 25 (1984).

54. M. Seibert, A.F. Janzen and M. Kendall-Tobias, Photochem. Photobio., 35, 193 (1982).

55. G. Dryhurst, K.M. Kadish, F. Scheller and R. Renneberg, "Biological Electrochemistry", Vol. 1, Academic Press, New York (1982), pp. 497-515.

56. E.A. Liberman and V.P. Topaly, Biofizika, 14, 452 (1969).

57. H.T. Tien, Nature, 219, 272 (1968).

58. M. Blank and E. Findl, Eds., "Mechanistic Approaches to Interaction of Electric and Electromagnetic Fields with Living Systems", Plenum Press, New York (1987).

59. N.B. Joshi, J.R. Lopez and H.T. Tien, J. Photochem., 20, 139 (1982).

60. A.L. Moore, G. Dirks, D. Gust and T.Z. Moore, Photochem. Photobiol., 32, 691 (1980).

61. I. Tabushi, N. Koga and M. Yakagita, Tetrahedron Lett., 3, 257 (1979).

62. S. Nishitani, N. Kurata, Y. Sakata, S. Misumi, M. Migita, T. Okada and N. Nataga, Tetrahedron Lett., 22, 2099 (1981).

63. C.-B. Wang, H.T. Tien, J.R. Lopez, Q.-Y. Liu, N.B. Joshi, Photobiochem. Photobiophys., 4, 177 (1982).

64. H.T. Tien and S.P. Verma, Nature, 227, 1232 (1970).

65. A.D. Joran, B.A. Leland, P.M. Felknew, A.H. Zewail, J.J. Hopfield and P.B. Dervan, Nature, 327, 508 (1987).

66. H.D. Mettee, W.E. Ford, T. Sakai and M. Calvin, Photochem. Photobiol., 39, 679 (1984).

67. J.R. Bolton, Ed., "Solar Energy and Fuels", Academic Press, New York (1977), pp. 167-225.

68. P.K. Shieh and H.T. Tien, Bioenergetics, 6, 45 (1974).

69. M. Wikstrom, Nature, 308, 558 (1984).

70. G. Porter and M.D. Archer, Indisc. Sci. Rev., 1, 119 (1976).

71. M. Calvin, Photochem. Photobiol., 23, 425 (1976).

72. J. Kiwi and M. Gratzel, J. Am. Chem. Soc., 100, 6314 (1978).

73. M. Gratzel, Ber. Bun. Ges., 84, 981 (1980).

74. N.N. Lichtin, Chemtech., 10, 252 (1980).

75. S. Murermanzi and H.T. Tien, Int. J. Ambient Energy, 7(1), 3 (1986).

76. F.L. Carter and H. Wohltjen, Eds., "Proc. 3rd International Symposium on Molecular Electronic Devices, Arlington, VA / USA, 6-8 October 1986", North-Holland, Amsterdam.

SURFACE CHARGE DETERMINES THE AGGREGATION OF HEMOGLOBIN SUBUNITS AS PREDICTED BY THE SURFACE FREE ENERGY

Martin Blank

Physiology-Cell Biophysics
Columbia University
630 W. 168th Street
New York, N.Y. 10032

INTRODUCTION

There are many observations in the literature which suggest that the surface charge is an important factor in controlling the interactions between proteins. Thus, oligomeric proteins exist in many different shapes and sizes, but they all tend to disaggregate as the pH is further away from the isoelectric point, or the ionic strength increases. A good example is the tobacco mosaic virus (TMV), which has long been studied as a model for protein aggregation reactions. The virus, a hollow cylinder about 3000A long and 180A in outside diameter, is made up of 2130 copies of the same 18,000 molecular weight protein, arranged in a helix. An RNA chain containing 6400 residues is wound along the inner face of the helix. The TMV protein will "self-assemble" into aggregates containing different numbers of subunits, and at the isoelectric point, pH = 4.7, the predominant form is the fully assembled multi-subunit helix. As the protein is progressively charged, it exists in smaller and smaller aggregates of subunits until it becomes a monomer at pH = 9. However, even at pH = 9 there can still be appreciable aggregation at high ionic strength, presumably due to the charge screening effect of high salt. These observations are consistent with accepted ideas of electrostatic repulsion, but this explanation does not include other contributions to the interaction energies.

In the search for general principles governing assembly, Lauffer[1] has noted that oligomer aggregation reactions are "entropy driven". This means that a large positive entropy change must accompany these reactions because they proceed spontaneously (i.e., there is a net negative free energy change) with a positive change of enthalpy. The entropy probably arises from the release of bound water molecules when the protein oligomers aggregate. This description is compatible with the energetics of aggregation, but it does not indicate why the process only occurs in certain ranges of pH, i.e., when the surface charge has particular values.

An explanation that includes the pH dependence arises from a more detailed thermodynamic analysis based on Langmuir's ideas[2] about the surface free energy of molecules applied to proteins in solution. The equilibrium degree of aggregation is determined by the minimum in the total free energy, but the free energy of the system is approximated by the surface free energy. One can make this assumption because the internal interactions in the oligomer are not affected by aggregation, and the surface free energy change is the main contribution to the energetics of the process. The loss of bound waters upon oligomer interaction, cited earlier, is equivalent to the loss of protein/water interface. In both approaches, the changes in interfacial area or bound water are the major source of the free energy change that drives the process, but the surface free energy model has the advantage of bringing in the surface charge directly and explaining the effects of pH and ionic strength on aggregation phenomena.

Surface Free Energy Model

Figure 1 can be used as a model for a spherical protein molecule in an aqueous medium where the interfacial free energy is low. The surface free energy increases when the droplet splits, because of the positive ΔA, so this process will not occur spontaneously. However, an increase in the surface charge (σ) makes the splitting more favorable, i.e., a negative ΔF results when the charge is spread over a larger area. In this way, the surface free energy can be used to evaluate the energetics of aggregation/disaggregation reactions. The surface free energy of a protein can be calculated simply in terms of the area of the molecule in contact with the aqueous solution and the surface charge density[3]. Because dissolved proteins have low interfacial energies, the electric charge is an important contributing factor.

When these ideas were applied to hemoglobin, where changes in the interfacial molecular area can be estimated from X-ray data and the surface charge from titration data, it was possible to determine the pH's at which the tetramer will disaggregate into two dimers. The calculations agreed with observations regarding the magnitudes of the pH and the relative dissociation constant, as well as the effects of oxygenation and ionic strength[3]. The same model also accounted for the acid and the alkaline Bohr effects of hemoglobin oxygenation[4], as well as the variation of the viscosity at high concentrations of hemoglobin in aqueous solution[5]. All of these results suggest that the model has reasonable predictive value and is useful for describing oligomeric equilibria.

EXPERIMENTAL

To test the quanitative aspects of the model, we decided to vary the charge on the hemoglobin tetramer and accurately measure the extent to

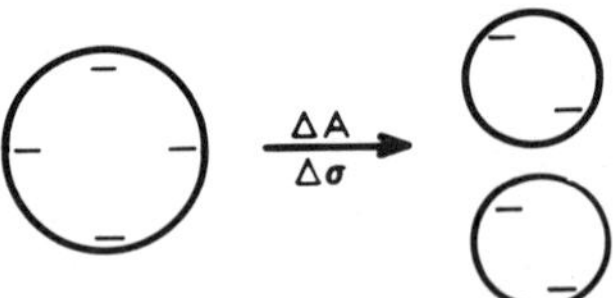

Figure 1. A model for calculating the energetics of disaggregation reactions where a spherical protein breaks into two spherical subunits. The change in area (ΔA) is positive and the change in charge density ($\Delta\sigma$) is negative. The net free energy change depends upon the magnitudes of the two terms and, in general, becomes negative at higher charge densities. (Reproduced with permission from Colloids Surf.).

which disaggregation occurs. Since disaggregation leads to an increase in osmotic pressure, we can measure the osmotic pressure[6] for hemoglobin solutions as a function of pH and compare it to the pressure observed in solutions of a protein of comparable molecular weight that does not disaggregate, i.e., bovine serum albumin (BSA). The difference in pressure is due to an increase in disaggregation. In the titration of both protein solutions, the miniumum in the pressure occurs at the isoelectric point (IEP), a fact that enables us to fix the IEP with great accuracy. Details of the technique and the measurements can be found in a recent publication[7].

RESULTS

The osmotic pressures obtained during alkaline titration are shown in Figure 2 for hemoglobin (Hb) and BSA as functions of the charge on the molecules. (The standard errors of the 10 sets of measurements are in the range of 1.0 - 2.2 mm water). The BSA curve represents the osmotic pressure generated by a charged biopolymer as a result of the complex Gibbs Donnan equilibrium that includes the contribution of the dissolved NaCl. In principle, the hemoglobin solution should give the same osmotic pressure at the same molar concentration and biopolymer charge, so the higher pressure observed there must represent an increase in concentration due to disaggregation. The difference between the two curves as a function of charge is shown separately as the dashed curve. The difference curve indicates that disaggregation occurs at all degrees of charge, and that above 28 charges per molecule or pH = 9.5, where the difference curve is higher than the BSA curve, disaggregation goes beyond the dimer to the monomer.

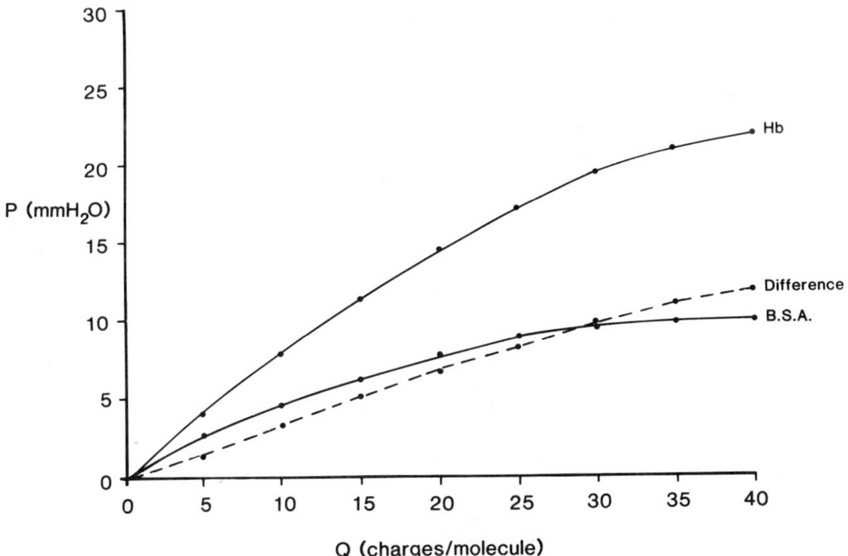

Figure 2. The colloid osmotic pressure (in mm H_2O) as a function of the number of charges per molecule in the alkaline region of hemoglobin (Hb) and BSA. The difference in pressure between the two protein solutions at comparable charges/molecule is indicated as the dashed line labeled difference. (Reproduced with permission from Bioelectrochem. Bioenerg.)

The excess concentration of osmoticaly active molecules, C, can be calculated from the dashed curve, and the ratio of C to the initial concentration, C_o, is a direct measure of the degree of disaggregation. C/Co is linear with charge until the average charge/molecule is about 28 and pH = 9.5, close to the point usually seen in plots of the average molecular weight where disaggregation becomes significant at higher concentations in the alkaline range[8].

Using these measures of disaggregation, it is possible to calculate the equilibrium constant, K_D. A plot of K_D vs Q, on semi-log coordinates is given in Figure 3. At low levels of charge, K_D appears to be an exponential function of Q, and the extrapolated value of K_D at the IEP is 2.0×10^{-7} moles/liter. Ackers[9] gives a value of 1.08×10^{-6} moles/liter at pH 7.4. At pH 7.4, our value of $K_D = 3 \times 10^{-6}$ moles/liter, is in excellent agreement with the earlier measurements. The difference between the two values may be due to the somewhat different ionic compositions in the experiments.

In our earlier paper on disaggregation equilibria in hemoglobin, we calculated values of K_D using our surface free energy model together with published titration data[3]. Based on the titration data reported here, the agreement shown in Figure 3, between our predictions and the measurements is excellent. There can be no doubt that the surface free energy changes are a good measure of the total free energy change occurring in hemoglobin on disaggregation. The surface free energy increases rather sharply as the molecule charges up, and disaggregation is apparently a mechanism for buffering the free energy change.

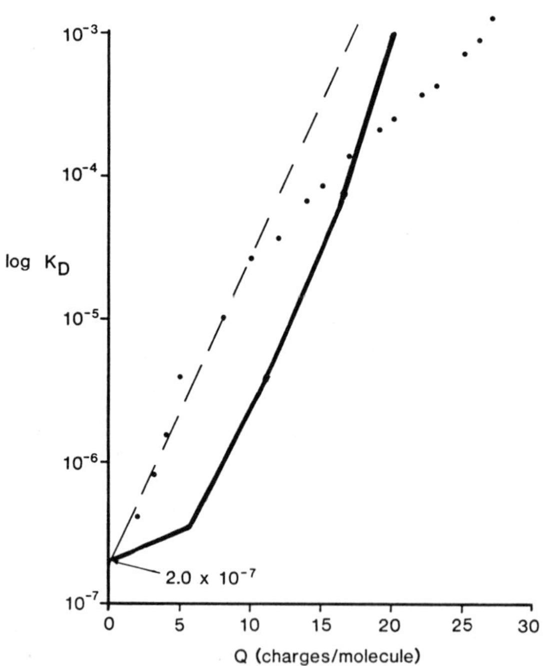

Figure 3. The logarithm of the disaggregation equilibrium constant, K_D, as a function of the charge, Q, on the hemoglobin molecule. The points and dashed line are experimental, while the solid line is based on the surface free energy calculation. (Reproduced with permission from Bioelectrochem. Bioenerg.)

If we assume that disaggregation leads to a 26% increase in interfacial area (as would be expected in the case of spheres), the charge density decreases. A comparison of the charge density of hemoglobin and BSA indicates that the effect is not very great. However, since the repulsive free energy is a steep function of charge density, the small difference is important. In other types of respiratory pigments that have more subunits, e.g., hemocyanins, the buffering capacity due to disaggregation is much greater.

DISCUSSION

Biopolymer Aggregation Equilibria

In dissolved and dispersed systems, as the surface-to-volume ratio increases, the surface energy becomes a larger fraction of the total energy. The surface energy should be especially important in the size range of many globular proteins (typical dimensions ~ 10^{-6} cm), which is smaller than the classic colloids (~ 10^{-4} cm) whose stability is determined by the surface free energy. In oligomeric proteins, it appears possible to account for the energetics of aggregation-disaggregation reactions on the basic of surface free energy alone.

Oligomer aggregation in proteins that form large globular or long rod-like assemblies is a process that appears to be governed by the same physical factors despite the differences in the compositions and structures of the proteins. Since assembly does not involve molecular rearrangements within the subunits, but rather interactions between the subunit surfaces, it is reasonable to approximate the total free energy changes in these processes by the surface free energy. Assuming that the surface free energy is the sole contribution to the total free energy, we have estimated changes in surface free energy in terms of molecular area and charge, and calculated the equilibrium constant of aggregation-disaggregation reactions. This approach leads to a surprisingly accurate quantitative account of the process.

Our measurements indicate that hemoglobin tetramers disaggregate into dimers at all degrees of charge, and above pH 9.5 there is further disaggregation into monomers. The process depends upon total hemoglobin concentration, and one would expect the values of K_D to change with both hemoglobin and salt concentrations, as well as with pH. At higher concentrations of hemoglobin, such as found in erythrocytes, there would be less of a tendency to disaggregate. The variation of K_D with concentration is undoubtedly due to the same forces that cause the "salting in" and "salting out" effects of proteins in solution.

Hemoglobin Properties

The surface free energy model (SFE) that has been used here to characterize the energetics of aggregation reactions involving hemoglobin subunits, can be used to describe any protein reaction where there are changes in the area or surface charge. The changes that occur as a result of the reactions of hemoglobin with oxygen (and other gases) can be used to demonstrate that the equilibrium constant for this reaction will be a function of the pH and ionic strength of the solution. For hemoglobin, the pH dependence is known as the Bohr effect, and the SFE model gives both the acid and alkaline Bohr effects with the maximum equilibrium constant below the iso-electric point[4]. Here again the calculated points are remarkably close to the observations indicating that the simple SFE model is quite powerful.

One consequence of this approach to hemoglobin reactions is the development of an alternative quantitative description for these reactions. The extra free energy driving oxygenation arises from the conformational change, i.e., the change in area and surface charge when the heme groups combine with oxygen. One can, therefore, describe the displacement of the equilibrium constant in terms of the surface free energy of the conformational change, rather than by introducing the Hill coefficient, which has been used as a measure of cooperativity but is poorly understood. Comparing the SFE approach with the conventional one, we see that the Hill coefficient is directly related to a Gibbs surface excess[10], and, therefore, not a constant as it is frequently described.

The general model shown in Figure 1 should describe oligomer association reactions under any conditions. We have used the model to calculate the distribution of molecular species when the tetrameric hemoglobin, $(\alpha\beta)_2$, continues to aggregate beyond the tetramer to the n-mer $(\alpha\beta)_n$ as the concentration of the solution increases[5]. The various species contribute to an increase in the viscosity of the solution as a result of the increased chain length and also due to the formation of a rigid structure when the chain becomes long enough to fold upon itself. The calculations assume the same loss of area when an αβ subunit adds at any stage, and the subsequent increase in charge density. From the distribution of molecular species, it is posible to estimate the viscosity of the solution and to show that it increases markedly at higher hemoglobin concentrations. This model indicates the possibility of high aggregates of hemoglobin and provides a rationale for the kinds of processes that do occur when the abnormal sickle hemoglobin aggregates.

Gating Processes in Oligomeric Channels

The integral membrane proteins associated with ion transport in excitable cells (e.g., the Na channel, the acetylcholine receptor, the Na, K-ATPase), appear to be generally composed of associated oligomers or the associated helices of a single chain. This form of organization is not necessarily limited to the transport function, but the need to form an aqueous pathway in a hydrophobic bilayer structure appears to require an assembly of proteins with specialized hydrophilic and hydrophobic regions organized in a cylindrical geometry. The mechanism of gating in such assemblies is still not clear, but the opening and closing of a channel would appear to depend upon the same physical principles that govern aggregation and disaggregation in oligomeric structures. Our model for a voltage-gated channel[11], based on aggregation phenomena in multi-subunit proteins, appears to account for a number of important properties of channels. (See Figure 4.)

Considering the sodium channel in the squid axon, the membrane has a much higher negative charge on the outside than on the inside, so we can consider the subunits disaggregated (i.e., open) on the outer surface, where the charge is high, and aggregated (i.e., closed) on the inner surface, where the charge is low. Upon depolarization there is a shift of some negative charge from the outer to the inner surface, and this causes the sodium channel to open. Qualitatively, the properties of this voltage sensitive ion channel are compatible with observed behavior in terms of:

1) the steady state distribution of charge;
2) the magnitude and direction of charge flow during depolarization;
3) the range of surface charge where opening (disaggregation) occurs;
4) the cation selectivity of the open channel; and
5) the cation binding.

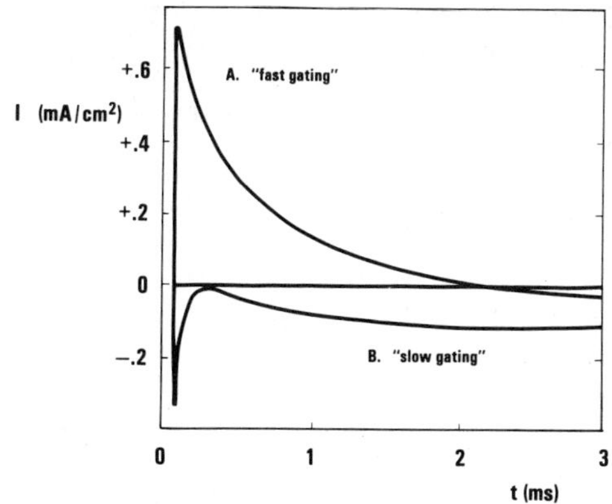

Figure 4. The ionic current in a voltage gated channel, I (in mA/cm^2) vs the time t (in milliseconds) for different rates of channel opening. Fast gating is characteristic of the sodium channels in excitable membranes and leads to a calculated "sodium" current. Slow gating, about an order of magnitude slower, leads to a calculated "potassium" current. The parameters used in calculating the two currents are identical except for the one that controls the gating current, but this leads to a great difference in the apparent specificity of the ionic current. (Reproduced with permission from Bioelectrochem. Bioenerg.)

This voltage sensitive channel model has recently been incorporated into the Surface Compartment Model (SCM) of ion transport that uses electrodiffusion equations and focuses on ionic processes in the electrical double layer regions at membrane surfaces. When a membrane is depolarized at constant voltage and the channels open as a result of the charge shifts known as gating channels, the SCM gives the ionic currents normally seen in the sodium and potassium channels of excitable membranes, i.e., an early inward current for fast gating sodium channels, followed by an outward current for slower gating potassium channels. The ion selectivity of a channel appears to be directly related to the gating currents following depolarization that determine the rate of channel opening. Depolarization not only causes the charge shifts in the channel proteins, but also starts the changes of ion concentration gradients between the electrical double layers on the two sides of a membrane. If channel opening is fast, one obtains ion fluxes that reflect the high sodium ion gradients present in the steady state. If the channel opens slowly, the gradients have changed and one obtains fluxes due to the other cation gradients, i.e., mainly potassium ions. This model of channel dynamics has provided an understanding of the role of kinetic factors in selectivity and has also shown that ion fluxes across biological membranes appear to obey the same electrodiffusion equations as ions in solution.

CONCLUSION

The surface free energy model appears to predict the disaggregation of hemoglobin tetramers into dimers at alkaline pH's with remarkable accuracy, considering the simplicity of the model and the approximations made in the calculations. The results of the model have already given insight into such hemoglobin properties as the Bohr effects and the Hill coefficient. Our present results provide additional quantitative evidence of the usefulness of the surface free energy model for aggregation-disaggregation equilibria. The surface free energy does appear to act as the potential in this equilibrium.

The results thus far suggest that the model would be useful in all equilibria involving changes in the area or charge of protein/solution interfaces where there are no internal rearrangements. The opening of membrane channels, which is probably accompanied by an increase in interfacial area, appears to follow the surface free energy model. The model may also be useful in determining the principles governing protein insertion in membranes, where interfacial interactions with the membrane matrix are involved.

ACKNOWLEDGEMENTS

Supported in part by Research Contract DAAL 03-86-K-0162 from the Army Research Office.

REFERENCES

1. M.A. Lauffer, in: "Entropy Driven Processes in Biology", Springer-Verlag, New York, 1975.

2. I. Langmuir, Chem. Rev., 13, 147 (1933).

3. M. Blank, Colloids and Surfaces, 1, 139 (1980).

4. M. Blank, J. theoret. Biol., 51, 127-134 (1975).

5. M. Blank, J. theoret. Biol., 108, 55-64 (1984).

6. Y. Kakiuchi, T. Arai, M. Horimoto, Y. Kikuchi and T. Koyama, Am. J. Physiol., 236(4), F419 (1979).

7. M. Blank and L. Soo, Bioelectrochem. Bioenerg., 17, 349 (1987).

8. E. Antonini and M. Brunori, in: "Hemoglobin and Myoglobin in their Reactions with Ligands", North-Holland, Amsterdam, 1971.

9. G.K. Ackers, Biophys. J., 32, 331 (1980).

10. M. Blank, J. Electrochem. Soc., 123, 1653 (1980).

11. M. Blank, J. Electrochem. Soc., 134, 343 (1987).

ELECTROCATALYSIS IN MEMBRANE MIMETIC MEDIA

James F. Rusling*, Chun-Nian Shi and Eric C. Couture

Department of Chemistry (U-60)
University of Connecticut
Storrs, Connecticut 06288

Thomas F. Kumosinski

Agricultural Research Service
U.S. Department of Agriculture
Philadelphia, Pennsylvania 19118

INTRODUCTION

Nature uses membrane-based microstructures which organize redox enzymes and reactants to provide living organisms with highly efficient redox processes. Micellar solutions and microemulsions contain surfactant aggregates which can be considered crude models for biological membranes. We have employed surfactant aggregates to bind redox catalysts which are cycled at an electrode to their reactive forms. Taking reductions as an example, the electrochemically generated reduced form of an aggregate-bound catalyst can react with a nonpolar, organic substrate bound to hydrophobic regions of the same surfactant aggregate. Such organized systems are potential models for dark biological redox processes at membranes. Other advantages include providing non-toxic, aqueous media for organic redox reactions of nonpolar substrates, and enhancing rates of reactions due to compartmentalization of reactants in the aggregates[1].

Recently, the study of chemical processes in surfactant assemblies has been termed "membrane mimetic chemistry"[1]. To orient readers to this interdisciplinary research area, a brief discussion of spontaneously organized surfactant assemblies is given first. Surfactants have positive, negative, or neutral hydrophilic head groups attached to hydrocarbon tails containing 6-20 carbons. Figure 1 shows oversimplified structures of surfactant aggregates useful in designing organized electrocatalytic reactions. Exact structural details are still a matter for controversy[1-4]. Micelles in water are dynamic aggregates of amphiphilic surfactant molecules (about 1.5-4.0 nm in diameter) with hydrophobic and hydrophilic regions. Typical ionic surfactants which form micelles are sodium dodecylsulfate (SDS) and hexadecyltrimethylammonium bromide, commonly known as cetyltrimethylammonium bromide (CTAB). Like other surfactant aggregates, ionic micelles have a surface potential at the so-called "Stern layer" containing the charged head groups, represented by the circles in Figure 1. The value and sign of this surface potential depend on the charge of the head group and the concentration of salt. Microemulsions are optically clear, microheterogeneous fluids consisting of surfactant, solvent, and one or more additional components, usually alcohol or hydrocarbon. Oil-in-water (o/w) microemulsions have water as the bulk phase and contain aggregates

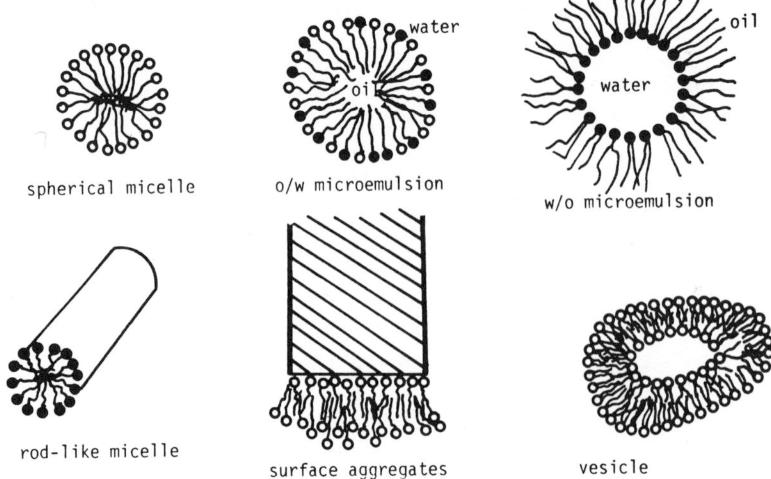

Figure 1. Oversimplified representations of structures of aggregates formed by surfactants.

that resemble swollen aqueous micelles[1]. Water-in-oil (w/o) microemulsions have hydrocarbon as the continuous phase and aggregates similar to reverse micelles[1,4]. Aggregates in microemulsions are usually larger (about 3 to 20 nm diameter) than micelles, and have a better capacity for dissolving solutes. Surfactant vesicles are cell-like compartments enclosed by bilayers of surfactants having two alkyl chains. These bilayers closely resemble natural membranes. Vesicles can be stable for weeks and are much less dynamic than micelles and microemulsions[1]. Surfactant assemblies have been used extensively in models for biological photo-redox events and other enzyme-catalyzed processes[1-4].

In electrochemical studies, surfactant will nearly always be adsorbed to the electrode-solution interface[5]. At extreme positive or negative applied potentials and surfactant concentrations well above the critical micelle concentration, surfactant molecules may be organized on the electrode surface in thick aggregates that profoundly affect the electrochemistry[6].

Solutes bound to micellar and microemulsion aggregates are in rapid equilibrium with free solute and aggregates. Rate constants for the exit of neutral solutes from ionic micelles are estimated[1] at 10^4-$10^6 s^{-1}$. Rates for recapture are near or at diffusion control. Rates of these processes are similar to equilibration rates between micelles and surfactant momomers. Solutes in micelles and microemulsions are most probably solubilized near the Stern layer or in the interfacial region. Hydrophilic or hydrophobic environments will be preferred depending on the properties of the particular solute. There is little evidence for deep solute penetration into the core of the aggregate[1]. Vesicles have much slower dynamics of solute equilibria.

The use of micelles to solubilize non-polar organic compounds in water for electrochemical measurements was first described by Proske[7] in 1952. Recent work has shown that ion radicals produced at electrodes can be stabilized[8-13] by coulombic and hydrophobic interactions with micelles. Such stabilization in aqueous media can protect ion radicals in a catalytic cycle from undesirable side reactions, including protonation.

Mediating electron transfer in micelles seems to have been reported first by Kuwana and co-workers[14,15], who used electrogenerated ferricinium ion in non-ionic micelles for redox titrations of cytochrome C and cytochrome C oxidase. Neutral micelles have been used to solubilize cytochrome C oxidase in spectroelectrochemical titrations[16], but examples of other applications are few.

EXPERIMENTAL

Chemicals and Solutions

All catalysts and substrates were of the highest purity available, and were recrystallized when necessary. CTAB was Fisher certified grade, 99.8 %, or from Eastman of about the same purity as shown by identical or higher m.p. (238-240°C, lit.[6] 237-239°C). CTAB recrystallized from water gave identical voltammetric results. SDS from Aldrich was washed four times with diethyl ether, then crystallized twice from 95 % ethanol. Water with a specific resistance >12 megaohm-cm prepared with a Sybron/Barnstead Nanopure water purification system was used for ultrasonic cleaning of electrodes, and to prepare all aqueous solutions. Ultrasonication was used to aid dissolution in micellar solutions. All micelle solutions were equilibrated with stirring and ultrasonication at the temperature of the experiment for one or more days prior to use. Some of the more water-insoluble solutes, such as 9-phenylanthracene and ferrocene required up to a week for solubilization, which was considered complete when time-independent cyclic voltammetric peaks were obtained. All other chemicals and solvents were reagent grade or better.

Apparatus and Procedures

Conventional 3-electrode cells, commercial potentiostats with external feedback IR compensation, and procedures similar to those described previously[6,17a] were used for electrochemical experiments. Working electrodes were a PARC Model 9323 hanging-drop-mercury electrode (HDME, A = 0.019 cm^2) or a highly polished glassy carbon disk electrode (GCE, A = 0.071 cm^2). A Pt wire served as the counter electrode and the reference was a saturated calomel electrode (SCE). For the HDME, a fresh Hg drop was used for each experiment. A polishing method described previously was used for GCE[17a] and was repeated prior to each voltammetric scan. Area of the GCE was estimated electrochemicaly by using the Randles-Sevcik equation and CV peak for the oxidation of ferrocene in acetonitrile ($D = 2.4 \times 10^{-5}$ $cm^2 s^{-1}$). Experiments in all CTAB solutions were thermostatted at 30.0±0.1°C; for SDS the temperature was usually 25.0±0.1°C.

RESULTS AND DISCUSSION

Comparative Kinetics In Micelles

Reduction of Halogenated Biphenyls

Electroreductions of organic catalysts 9-phenylanthracene[6] (9-PA) and 1,2-dicyanobenzene[17b] (1,2-DCB or o-phthalonitrile) were chemically reversible and diffusion controlled at cyclic voltammetric (CV) scan rates between about 0.2 and 50 V/s in aqueous 0.1 M CTAB containing 0.1 M tetraethylammonium bromide (TEAB). The anion radicals of 9-PA and 1,2-DCB were stabilized against reaction with proton donors by the cationic micellar aggregates. The anion radical of 1,2-DCB had a lifetime on the order of a second. The anion radical of 9-PA appears to be more stable and is formed in aggregates at the surface of the electrode, as discussed below.

9-PA anion radical rapidly reduced 4-bromobiphenyl (4-BB) to biphenyl in 0.1 M CTAB with an enhanced rate compared to isotropic solvent (Table 1). Quantitative bulk electrolytic reduction of 0.02 mmol of 4-BB in 25 mL 0.1 M CTAB was effected on stirred mercury pool electrodes in 2.5 h with 20 % decomposition of the catalyst. Time for complete conversion to biphenyl and amount of catalyst decomposed were significantly smaller compared to similar experiments in surfactant-free N,N-dimethylformamide (DMF)[18]. Diffusion controlled CV and chronocoulometric data[6] for 0.2 mM 9-PA in 0.1 M CTAB were used to obtain an apparent diffusion coefficient (D') of 10^{-3} $cm^2 s^{-1}$. This is much too large to attribute to a diffusing micelle-bound species. Furthermore, at scan rates (v) below 5 mV s^{-1}, CV's for the 9-PA anion radical were not diffusion controlled as at higher v, but had a symmetric peak shape attributable to a thick surfactant layer at the surface of the electrode. Thus, at the potential required (-2.2 V vs SCE) to reduce 9-PA in 0.1 M CTAB, the catalytic reduction of 4-BB takes place in a thick, spontaneously organized surfactant film on the electrode surface. In addition to voltammetric results[6], support for existence of a thick film comes from differential capacitance, ellipsometry[19], and reflectance infrared spectroscopy[20].

Electrocatalytic dehalogenation of halobiphenyls in dry DMF proceeds by the pathway in Scheme I[21,22]. The first step is fast transfer of an electron from electrode to catalyst A (Equation 1), and the rate-determining step (rds) in DMF is homogeneous electron transfer from the anion radical of the catalyst to halobiphenyl ArX (Equation 2). This rds is thermodynamically unfavorable, and the overall reaction is driven by rapid cleavage of the anion radical of the halobiphenyl in Equation 3.

Scheme I

$$A + e = A^{\overline{\cdot}} \qquad (1)$$

$$ArX + A^{\overline{\cdot}} \underset{k_2}{\overset{k_1}{=}} ArX^{\overline{\cdot}} + A \qquad (2)$$

$$ArX^{\overline{\cdot}} \overset{k}{\rightarrow} Ar\cdot + X^- \qquad (3)$$

$$Ar\cdot + A^{\overline{\cdot}} \rightarrow A + Ar^- \qquad (4)$$

$$Ar^- + (H^+) \rightarrow ArH \qquad (5)$$

Subsequently, fast electron transfer and/or chemical steps, represented by Equations 4 and 5, yield biphenyl as the ultimate product. In reactions following Scheme I, an increase in cathodic current for the reduction of the catalyst occurs upon addition of substrate because of recycling of A at the electrode (Equations 2 and 4). Rate constants for the rds can be obtained from the so-called catalytic efficiency, i.e., the ratio of catalytic (i_c) to diffusion (i_d) peak currents, or by nonlinear regression analysis of voltammetric curves[6,21,22].

Table 1. Observed Rates of Reduction of Organohalides by Reduced Catalysts

Catalyst	Substrate	Homogeneous System: Solvent	k_{obs} 1/(Ms)	Micellar System: [CTAB] (M)	k_{obs} 1/(Ms)
9-PA	4-BB	DMF	310	0.1	10^7
1,2-DCB	PCB[a]	DMF	115	0.1	$3x10^3$
Tris(bpy)Co(II)	allylCl	MeCN	100	0.1	220

[a] 2,2',5,5'-tetrachlorobiphenyl

The kinetics of catalytic reduction of 4-BB in CTAB solutions involves micellar catalysis of the electron transfer (ET) between 9-PA anion radical and 4-BB (Equation 2) in a thick layer of surfactant at the electrode. The effective rate constant for this ET in 0.1 M CTAB increased more than three orders of magnitude compared (Table 1) to the same reaction in surfactant-free DMF. The rate-determining step in CTAB was not Equation 2 as in DMF. In CTAB decomposition of the 4-BB anion radical (Equation 3) became kinetically important. The major cause of the kinetic alterations was compartmentalization of the reactants in high concentrations in surfactant aggregates at the surface of the electrode. The same catalytic reaction was not as successful in non-ionic igepal micelles, which did not provide good stabilization for 9-PA anion radicals.

Reduction of 2,2',5,5'-tetrachlorobiphenyl (2,2',5,5'-PCB) with electrogenerated 1,2-DCB anion radical could not be followed quantitatively with cyclic voltammetry because of severe overlap of the two one-electron reduction peaks of the catalyst. However, square-wave voltammetry (SWV) allowed a clear demonstration of a catalytic increase in peak current (Figure 2) for 1,2-DCB upon addition of 2,2',5,5'-PCB. Fitting catalytic current ratios (Figure 3) with the appropriate working curves[23] indicated a 25-fold increase in observed rate in 0.1 M CTAB as compared to DMF (Table 1). Occurring at about -1.6 V vs SCE, this reaction did not show the obvious surface character found in the 9-PA/4-BB system. An apparent diffusion coefficient (D') of 3×10^{-6} $cm^2 s^{-1}$ for 2 mM 1,2-DCB in 0.1 M CTAB and the lack of symmetric CV's at any scan rate suggests that 1,2-DCB diffuses to the electrode partially bound to CTAB micelles. A measured distribution ratio of 3.8 for 1 mM DCB in equal volumes of water/hexadecane further suggests that considerable 1,2-DCB resides in the water phase in solutions of 0.1 M CTAB. The dehalogenation reaction probably involves diffusion of both the micelle-bound and free catalyst, which accept electrons upon arrival at the electrode, and transfer these electrons to the PCB residing in the hydrophobic region of the micelle. Preparative electrolysis with analysis of products by HPLC confirmed stepwise removal of chloride from the substrate but proceeded with 94% decomposition of the catalyst in 3.5 h. These results suggest that the anion radical of the weakly bound 1,2-DCB is not sufficiently protected from side reactions to be practical as a preparative-scale catalyst.

Adsorption of 1,2-DCB can be induced at Hg electrodes in a microemulsion prepared by adding 1-butanol to aqueous CTAB micelles. This is illustrated by decreased peak separation and increased peak symmetry in microemulsions of butanol/CTAB/water/KBr as the amount of butanol is increased (Figure 4). Preliminary work on the reduction of 2,2'5,5'-PCB by SWV showed 15-30% increases in catalytic efficiency at low frequency (f) when 0.1 M 1-butanol was added to 0.1 M CTAB. These results suggest

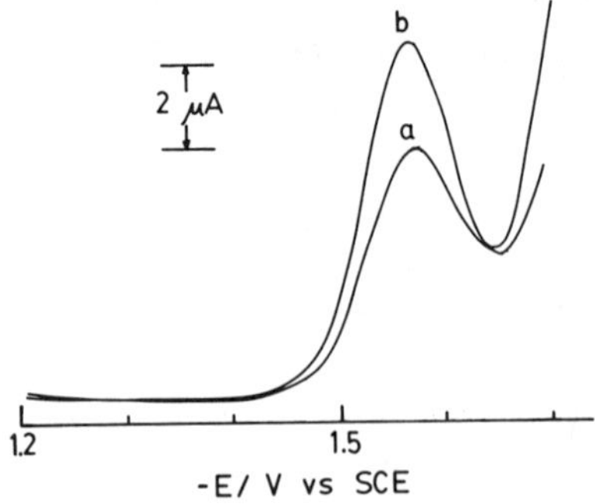

Figure 2. Square-wave voltammograms of 1 mM 1,2-DCB in 0.1 M CTAB: (a) alone; (b) with 5 mM 2,2'5,5'-tetrachlorobiphenyl present. f = 30 Hz, amplitude = 25 mV, step = 2 mV.

that control of microemulsion composition may be important for modifying reaction environments. We are presently beginning to explore other catalytic reactions in microemulsions.

Reduction of Allyl Halides

We have also used dicationic tris(2,2'-bipyridyl)cobalt(II) and derivatives in 0.1 M solutions of SDS and CTAB as catalysts for reducing allyl halides[24]. Electrochemical, UV and NMR spectroscopic results showed that the complexes were bound to both cationic and anionic micellar aggregates. The Co(II) complexes were bound close to the charged Stern

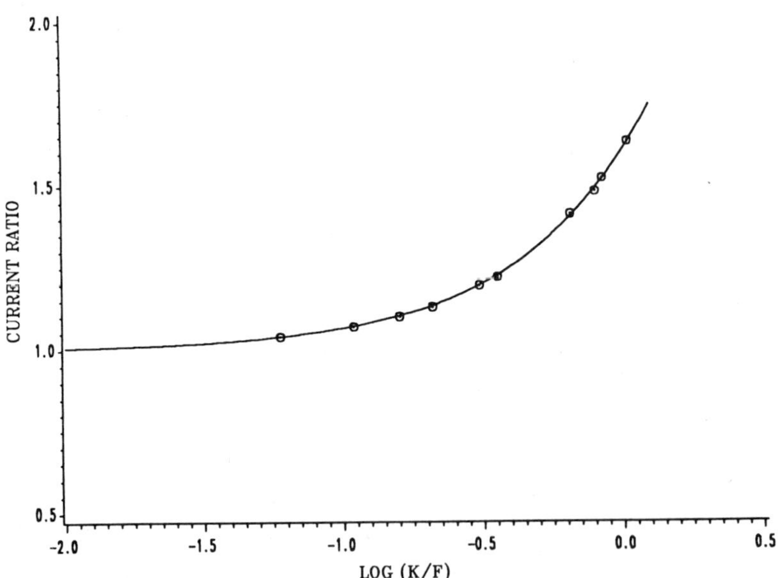

Figure 3. Catalytic efficiency ratio (i_c/i_d) plotted vs log(k/f) as working curve for amplitude = 50 mV and [substrate]/[catalyst] = 5. Circles are experimental ratios for k = 3 x 10^3 $M^{-1}s^{-1}$.

layer of SDS micelles, and to hydrophobic regions of CTAB micelles. In CTAB, hydrophobic interactions between the dicationic complexes and the micelles overcome coulombic repulsion with the positive surface potential of the micelles. Rates of electron transfer at glassy carbon electrodes were slower in surfactant solutions than in acetonitrile (MeCN), following the order MeCN > CTAB > SDS[24b].

The bipyridyl complexes of Co(II) showed electrocatalytic activity for reduction of allyl halides to 1,5-hexadiene in micellar media. Catalysis lowered the overpotential for reduction of allyl chloride in 0.1 M SDS and CTAB by 1.4 V compared to direct reduction. Yields of about 60% of 1,5-hexadiene were obtained from electrolyses at carbon felt electrodes. Small micellar enhancements of reaction rates were found for tris(2,2'-bipyridyl)cobalt(II) (Table 1). Catalytic efficiency followed the order CTAB > SDS = acetonitrile. Preliminary work with the long chain derivative bis(2,2'-bipyridyl)(4,4'-hexadecyl-2,2'-bipyridyl)cobalt(II)

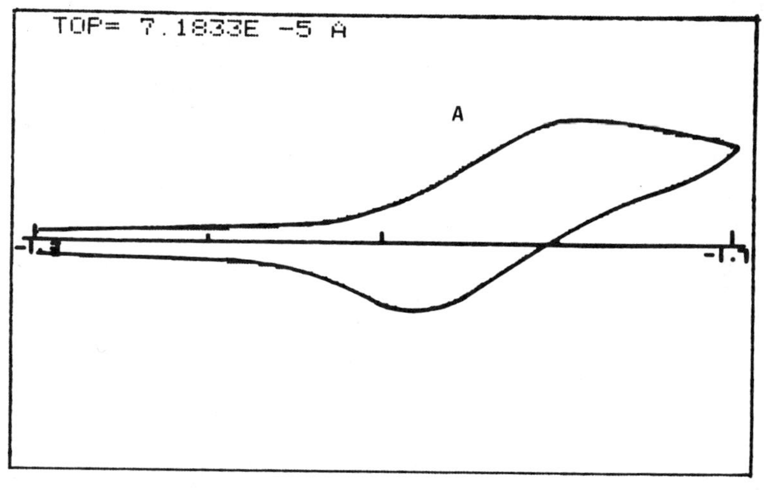

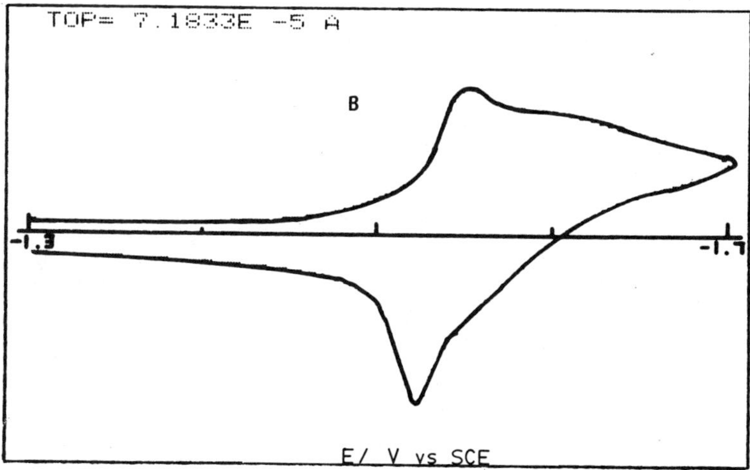

Figure 4. Cyclic voltammograms at 10 Vs^{-1} of 1 mM 1,2-DCB at an HDME in: (a) 0.1 M aqueous CTAB, 0.1 M KBr; (b) microemulsion containing 0.2 M 1-butanol, 0.1 M CTAB, 0.1 M KBr and water.

showed that its catalytic efficiency in 0.1M SDS was equivalent to that of the fastest system studied, i.e., tris(2,2'-bipyridyl)cobalt(II) in CTAB.

Reduction of Alkyl Dihalides with Metallophthalocyanines in CTAB

Metal phthalocyanines (MPc) and other macrocyclic complexes are highly efficient catalysts for a wide variety of reductions and oxidations[25]. Water soluble MPc tetrasulfonates have been extensively studied as catalysts for reducing oxygen[25d-i] for possible fuel cell applications. MPc's also have been used as chemical catalysts for a variety of organic redox reactions. An attractive advantage of the iron and cobalt macrocycles is their high catalytic efficiency and the nearly reversible electrochemical accessibility of M(III), M(II), and M(I) oxidation states[26].

FePc is a model for iron porphyrins, which are of central importance in binding oxygen to hemoglobin in blood, and as prosthetic groups in cytochromes. Iron porphyrins have been proposed as mediators in biodehalogenation[27] and in metabolic activation of toxic chemicals in humans[28]. By a related process, reduction of alkyl vicinal dibromides to alkenes can be accomplished by generating[29] the Co(I) form of the macrocyclic vitamin B_{12} complex at an electrode. The electrode serves a function similar to a redox enzyme by cycling Co(II) to Co(I). Such reactions are also catalyzed by CoPc and FePc.

The following electrode reactions occur in aprotic organic solvents at relatively fast rates on glassy carbon, pyrolytic graphite[30], mercury and platinum electrodes[26]:

$$M(II)Pc + e = [M(I)Pc]^-$$

$$[M(I)Pc]^- + e = [M(I)Pc]^{2-} \qquad (6)$$

The first one-electron reductions of Fe(II)Pc and Co(II)Pc are metal centered. Reduction of the M(I)Pc- species involves the phthalocyanine ring and leads to an $[M(I)Pc]^{2-}$ radical. Adsorption of FePc and CoPc was found on the carbon electrodes, and was strongest for CoPc[30].

Underivatized MPc's are notoriously water-insoluble, but can be solubilized sufficiently in CTAB micelles and CTAB/butanol/water microemulsions to observe well-defined square-wave voltammograms (SWV) at mercury and glassy carbon electrodes. CoPc and FePc concentrations on the

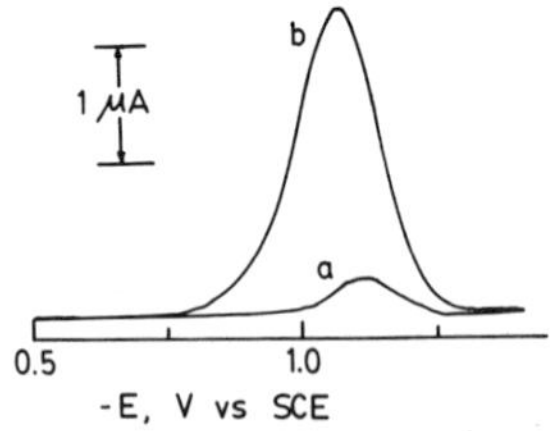

Figure 5. Square wave voltammograms of saturated solutions of FePc in 0.1 M CTAB, 0.1 M TEAB at an HDME: (a) FePc alone, (b) FePc and 5 mM 1,2-dibromobutane. f=15 Hz, amplitude = 25 mV, step = 2 mV.

order of 30 μm can be attained in 0.1M CTAB at 30°C. We observed the two reductions in Equation 6 for both Co(II)Pc and Fe(II)Pc in 0.1 M CTAB at potentials similar to those found in nonaqueous solvents. 1,2-Dibromobutane (DBB) can be catalytically reduced by Fe(I)Pc- or Co(I)Pc-electrogenerated at mercury in 0.1 M CTAB (Figure 5). SWV's of the catalytic reactions in the presence of 5 mM DBB show peak currents about 10-fold greater and peak potentials shifted positive compared to the catalyst alone. Both these observations are characteristic of good catalytic efficiency. The product of the reduction of DBB is presumably butene as found when Vitamin B_{12} is the catalyst[29]. Catalytic efficiency is increased upon addition of 1-butanol to the 0.1 M CTAB.

Diffusion of Micelle-Bound Catalysts

Models for Diffusion to Electrodes

As a first step toward a quantitative description of electrocatalysis in micelles, we developed models for diffusion of micelle-bound electroactive molecules. Measurement of apparent diffusion coefficients (D') of electroactive solutes in micellar solutions by voltammetry or chronocoulometry has been used to estimate the fraction of total electroactive solute "probe" bound to micelles[10-13,24b,31-33], as well as micellar size. If a single size distribution of micelles (i.e., a monodisperse system) exists, D' can be expressed by a two-state model:

$$D' = f_a D_o + f_b D_1 \tag{7}$$

where f_b is the fraction of electroactive solute bound to micelles, and D_o and D_1 are the diffusion coefficients of the free solute in water and the micelle, respectively. The fraction of free solute in water is $f_a = 1 - f_b$. If D_o and D_1 are known, Equation 7 provides a means to estimate f_b. Equilibrium of the probe (X) between micelle (M) and water can be used to expand Equation 7 in terms of a distribution or formation equilibrium constant.

We found reproducible decreases in D' with increasing total concentration of X (C_X) for a series of reversible redox couples in solutions of cationic and anionic micelles. Our approach to modeling these data follows from that used for multisite binding equilibria of small ligands to macromolecules, as described by Tanford[34] and Wyman[35]. Consider a micelle capable of binding n solute molecules by the equilibria:

$$M + i\,X = MX_i \qquad K'_i = [MX_i]/\{[M][X]^i\} \tag{8}$$

The concentration of bound X is given[35] by:

$$X_b = [M] \sum_{i=1}^{n} i K'_i x^i \tag{9}$$

where K'_i is the apparent equilibrium constant in Equation 8 and x is the concentration of unbound X. The fraction of bound X is then given by:

$$f_b = X_b/C_X; \quad C_X = x + X_b \tag{10}$$

Assuming $x \ll C_X$ and considering only the overall equilibrium with i = n, Equations 7-10 can be used to obtain the approximate relation:

$$D' = D_o/(1 + C_M K^n C_X^{n-1}) + D_1 C_M K^n C_X^{n-1}/(1 + C_M K^n C_X^{n-1}) \tag{11}$$

where C_M is the total concentration of micelles and nK'_n has been replaced by K^n for ease of computation. K is an apparent formation constant per binding site[34]; the binding sites are considered equivalent. For Equation 11 to explain the observed dependence of D' on C_X, there must be more than one binding site for X per micelle, i.e., n>1. Consideration of all the intermediate stepwise equilibria for i = 0 to n leads to complicated expressions for D', the use of which rapidly becomes impractical as n increases[36].

When two types or sizes of micelles exist in the medium, D' depends on equilibria of X with two micelles (M_1 and M_2), as follows:

$$M_1 + nX = M_1X_n \qquad K'_1 = [M_1X_n]/\{[M_1][X]^n\} \tag{12a}$$

$$M_2 + mX = M_2X_m \qquad K'_2 = [M_2X_m]/\{[M_2][X]^m\} \tag{12b}$$

The required three-state expression for D' is:

$$D' = f_aD_0 + f_{b1}D_1 + f_{b2}D_2 \tag{13}$$

where f_{b1} and f_{b2} are the fractions of X bound to micelles M_1 and M_2 with diffusion coefficients D_1 and D_2, respectively. If $x \ll C_X$, using the relevant expressions for f_a, f_{b1} and f_{b2} in Equation 13 yields the approximate expression:

$$\begin{aligned} D' = {} & D_0/(1 + C_{M1}K_1^nC_X^{n-1} + C_{M2}K_2^mC_X^{m-1}) + \\ & D_1C_{M1}K_1^nC_X^{n-1}/(1 + C_{M1}K_1^nC_X^{n-1} + C_{M2}K_2^mC_X^{m-1}) + \\ & D_2C_{M2}K_2^mC_X^{m-1}/(1 + C_{M1}K_1^nC_X^{n-1} + C_{M2}K_2^mC_X^{m-1}) \end{aligned} \tag{14}$$

where C_{M1} and C_{M2} are the total concentrations of M_1 and M_2, respectively. As in Equation 11, binding constants are expressed as $nK'_1 = K_1^n$ and $mK'_2 = K_2^m$. The K's are again apparent values only.

Testing the Diffusion Models

Methyl viologen dication (MV^{2+}) and ferrocene (Fc) in 0.1 M SDS/0.1 M NaCl, and ferrocene in 0.1 M CTAB containing either 0.1 M NaCl or KBr were used to obtain D' at a series of concentrations of the electroactive probe. Standard CV criteria confirmed reversible electrochemistry for MV^{2+} and Fc in the ionic micellar systems chosen[12,31]. Plots of forward peak current (i_p) vs $V^{1/2}$ showed excellent linearity for each electroactive compound. The Randles-Sevcik equation[37] was used to compute the apparent diffusion coefficient D' from the slopes of i_p vs $V^{1/2}$ plots at each concentration C_X.

The Monodisperse Model

Nonlinear regression analysis onto the simplest model, Equation 11, was first used to analyze D' vs C_X data. This was done with a general program for nonlinear regression using Equation 11 in the subroutine which computes the dependent variable[36]. The regression prameters were D_0, D_1, and $C_M^{1/n}K$. The value of the integer n was fixed in each regression analysis. A series of regressions was done, beginning with n = 2 and increasing n in successive regressions until the best fit to the data was obtained. The best fit was chosen as that for the n-value giving the smallest minimum sum of squares of residuals consistent with low correlation for all pairs of parameters.

Plots of D' vs C_X compared to the computed regression curves show (Figure 6) that the general trend in D' is explained by Equation 11. However, quantitative agreement of computed and experimental data is poor and deviation or residual plots of $(D'_{measd}-D'_{calcd})/SD$ vs C_X [SD = standard deviation of the regression] revealed systematic errors in the model. These results led us to test more complex models. Nevertheless, D_0 and D_1 computed from regression using Equation 11 agreed within about 20-30% with reported values for MV^{2+} and Fc (SDS data). Values for Fc from the 0.1 M CTAB data were smaller than expected for diffusion of the free molecule.

The Two-micelle Model

Nonlinear regression analysis of D' vs C_X data onto Equation 14 was done by using fixed values of n and m. A series of regressions was done for each set of data to find the combination of n/m values giving the minimum SD consistent with low parameter correlation. The parameters were D_0, D_1, D_2, $C_{M1}{}^{1/n}K_1$, $C_{M2}{}^{1/m}K_2$. Excellent fits of Equation 14 to experimental data were obtained for the optimum n/m combination in all cases. In addition, an alternative five parameter sequential binding model[36] gave results nearly identical to Equation 14. For the best values of n/m, random residuals were obtained and SD values were significantly lower than those using Equation 11 in all cases. F-tests accounting for the difference in degrees of freedom in the two regression analyses being compared yielded high probabilities that the sum of squares of residuals (RSS) for fits with Equation 14 was significantly smaller than RSS for fits with Equation 11 (Table 2). These results show that models assuming two types of micelles were more realistic than the model assuming a monodisperse system. This is visually evident by comparing Figure 6, representative of fits with Equation 11, with Figure 7 representing a fit of the same data to Equation 14.

There is considerable evidence for polydispersity in micellar systems of CTAB and SDS at surfactant and salt concentrations of 0.1 M or above[1a,38-40]. For example, for CTAB in 0.1 M alkali bromide solutions, both spherical and larger rod-like aggregates were proposed to coexist on the basis of light scattering and NMR results[38,40]. Extrapolating from these studies, M_1 is probably spherical or globular and M_2 larger and rod-like.

Comparisons of D_0 values for the probes computed from regressions (Table 2) on the SDS data are in good agreement with reported values, but D_0 values are again low for Fc in CTAB. The latter result could be rationalized by assuming strong association of Fc with submicellar aggregates, recently proposed as a vehicle for reactant exchange in micellar systems[41]. If this is the case, very little free Fc would diffuse to the electrode, and D_0 would reflect Fc bound to submicellar surfactant aggregates.

Computed values of D_1 for the smaller SDS and CTAB micelles (M_1) show excellent agreement with reported values (Table 2). Differences in D_1 for CTAB/KBr and CTAB/NaCl are greater than can be explained by experimental error. This is not surprising, since counter ion effects on the size of ionic micelles are well-known. For example, NMR tracer diffusion coefficient measurements[40] in salt-free systems gave 1 x 10^{-6} cm^2s^{-1} (33°C) for 0.15 M CTACl and 0.65 x 10^{-6} cm^2s^{-1} for 0.14 M CTAB. Similarly, average aggregation numbers in salt-free systems estimated by fluorescence quenching were 104 in 0.03 M CTAB and 89 in 0.03 M CTACl[44]. Thus, bromide ion promotes the formation of larger micelles. This is consistent with our smaller values of D_1 and D_2 for the CTAB/KBr system when compared to CTAB/NaCl.

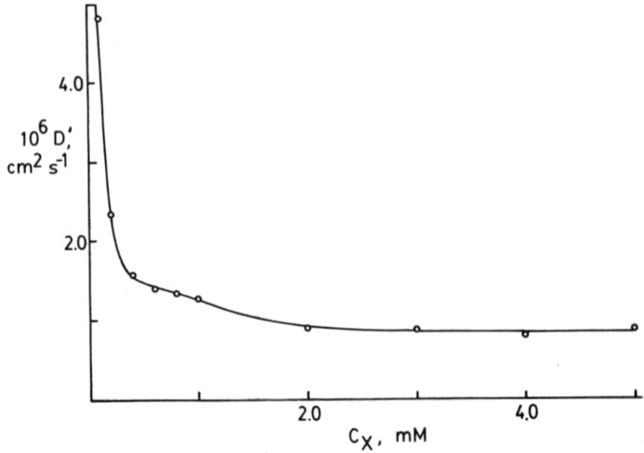

Figure 6. Influence of concentration of methyl viologen on D' in 0.1 M SDS/0.1 M NaCl. Circles are experimental data; line is best fit from nonlinear regression onto Equation 11 with n = 3. Parameters were $10^6 D_0$ = 9.1 $cm^2 s^{-1}$, $10^6 D_1$ = 1.01 $cm^2 s^{-1}$, $C_M^{1/n}K$ = 4.8 $mM^{1/n-1}$. Reprinted with permission from Reference 36. Copyright 1988 American Chemical Society.

Computed values of D_2 are in reasonable agreement for the two SDS systems. A significant difference in D_2 for Fc/CTAB when the salt is changed is consistent with the counter ion effect.

Binding constant parameters (Table 3) are in good agreement for the two SDS and the two CTAB systems, respectively. Binding constants for the smaller micelles are large. Their values can be estimated by neglecting M_2 and using reported average aggregation numbers to obtain C_{M1}. An aggregation number of about 100 for 0.1M CTAB/KBr[24b] gives an estimated C_{M1} = 1 mM. Using this C_{M1}, the parameter containing K_1 in Table 3, and the relation $K'_1 = K_1^n/n$, yields 7.5 x 10^{13} M^{-4} for the apparent binding constant of ferrocene to CTAB micelles. Similarly, using an aggregation

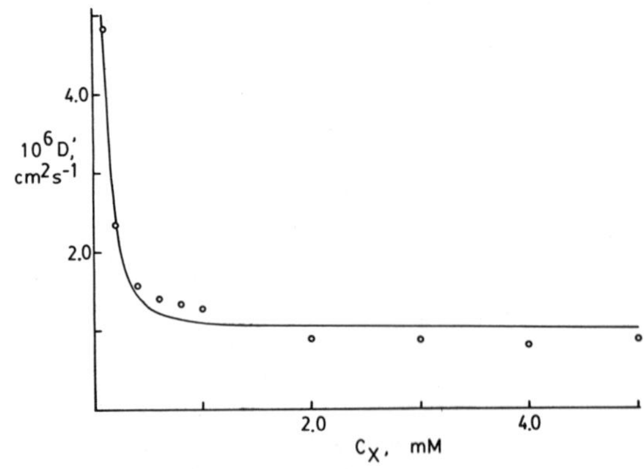

Figure 7. Comparison of data in Figure 6 with best fit (line) from nonlinear regression onto Equation 14. Parameters as in Tables 2 and 3. Reprinted with permission from Reference 36. Copyright 1988 American Chemical Society.

Table 2. Diffusion Parameters[a] from Nonlinear Regression Analysis of D' vs C_X Data onto Equation 14 for Best Values of n and m

System	Solute	n/m	$10^6 D_0$ found	lit(ref)	$10^6 D_1$ found	lit(ref)	$10^6 D_2$ found	%P[b]
SDS/NaCl	MV^{2+}	4/8	6.89±0.21	6.56(12)	1.41±0.03	1.45(42)	0.84±0.02	>99
SDS/NaCl	Fc	4/14	5.19±0.04	6.7[c]	1.45±0.01	1.45	0.99±0.01	>99
CTAB/NaCl	Fc	6/7	1.41±0.01	6.7[c]	1.01±0.04	0.5-1.0 (40,43)	0.77±0.04	>99
CTAB/KBr	Fc	4/7	1.25±0.09	6.7[c]	0.73±0.15	0.5-0.83 (43)	0.38±0.02	>70

[a]Units of $cm^2 s^{-1}$; all surfactant and salt concentrations at 0.1 M.
[b]Probability confidence level with which Equation 14 can be considered a significantly better model than Equation 11 after correction for degrees of freedom in each analysis.
[c]Estimated from Stokes-Einstein equation in Reference 31.

number 74 reported for 0.07 M SDS in 0.1 M NaCl[45] yields 1.2 x 10^{14} M^{-4} as an estimated binding constant for both methyl viologen and ferrocene to SDS micelles. These values are only rough estimates of binding strength, since M_2 has been neglected and K_1 is not a true equilibrium constant. The sequence of binding to the two types of micelles or the role of solute-induced aggregation of micelles could not be determined from the data.

Thus, models based on multisite binding equilibria successfully describe the dependence of D' on C_X at a fixed micelle concentration when the electroactive probe is almost totally bound to the micelles. The electrode reaction of the probe must not be accompanied by adsorption or chemical reaction. Nonlinear regression analysis of the data enables an assessment of the importance of polydispersity and an accurate estimate of diffusion coefficients of the micelles. Parameters proportional to binding constants are also obtained, and these can be converted to apparent binding constants if the micelle concentrations are known. The models also quantitatively predict the observed[32] decrease in measured diffusion coefficient with concentration of micelles. This work shows that the micellar system we have used for electrocatalysis are probably polydisperse.

Table 3. Binding Parameters from Nonlinear Regression Analysis of D' vs C_X Data[a] onto Equation 14

System	Solute	$C_{M1}{}^{1/n}K_1$ $mM^{1/n-1}$	$C_{M2}{}^{1/m}K_2$ $mM^{1/m-1}$	SD
SDS/NaCl	MV^{2+}	4.96±0.12	2.00±0.09	0.032
SDS/NaCl	Fc	5.04±0.03	1.85±0.01	0.0023
CTAB/NaCl	Fc	3.96±0.08	3.26±0.32	0.0091
CTAB/KBr	Fc	4.16±1.08	3.04±0.79	0.0198

[a]From same analyses and data as in Table 2.

CONCLUSIONS

We have shown that membrane mimetic media such as micelles, microemulsions, and surface aggregates can be used to compartmentalize reactants in second-order electrocatalytic reactions to enhance overall reaction rates. With hydrophobic catalysts and substrates, high concentrations of reactants are realized in the aggregates. Modes for enhancement of reaction rates include (i) binding of reactants to aggregates at the electrode surface, observed so far only at extreme negative potentials with positively-charged CTAB aggregates, and (ii) binding of reactants to aggregates in the organized bulk medium which then diffuse to an electrode to initiate the catalytic cycle. Similar reactant binding may be important for the high efficiency of membrane-bound redox reactions in living systems. Further, surfactant systems enable the reduction of nonpolar organic substrates in largely aqueous media, thus avoiding toxic and expensive organic solvents in synthetic and analytical applications.

The framework of a quantitative description of electrochemistry and electrocatalysis in micelles and microemulsions is illustrated by regression models for diffusion of aggregate-bound catalysts. Besides providing an excellent method for characterizing catalyst-aggregate systems, multisite binding holds promise as the basis for development of kinetic models from which actual rate constants in the aggregates might be derived.

This paper represents a progress report on our beginning research in the exciting area of electrocatalysis in membrane mimetic media. Ongoing work is aimed at exploring the use of oil-in-water and water-in-oil microemulsions, and vesicles, and at controlling the formation of surface aggregates.

ACKNOWLEDGEMENT

The authors are grateful to coworkers named in references of joint publications who were instrumental in beginning much of the research described herein. We are grateful for financial support from U.S. PHS Grant ES03154 awarded by the National Institute of Environmental Health Sciences, and from the Petroleum Research Fund administered by the American Chemical Society.

REFERENCES

1. (a) J.H. Fendler, "Membrane Mimetic Chemistry", Wiley: New York, 1982.
   (b) J.H. Fendler, Ann. Rev. Phys. Chem., 35, 137 (1984).
   (c) J.H. Fendler, J. Phys. Chem., 89, 2730 (1985).
   (d) J.H. Fendler, J. Phys. Chem., 84, 1485 (1980).

2. (a) M. Gratzel in "Energy Resources through Photochemistry and Catalysis", M. Gratzel (Ed.), Academic: New York, 1983, pp. 71-98.
   (b) M. Gratzel, Modern Aspects Electrochem., 15, 83-165 (1983).

3. C. Tanford, "The Hydrophobic Effect", 2nd Ed., Wiley: N.Y., 1980.

4. (a) P.L. Luisi and B.E. Straub, "Reverse Micelles", Plenum: N.Y., 1984.
   (b) P.L. Luisi and L.J. Magid, CRC Critical Reviews in Biochemistry, 20, 409 (1986).

5. N. Shinozuka and S. Hayano in "Solution Chemistry of Surfactants", Vol. 2, K.L. Mitall (Ed.), Plenum, N.Y., 1979, pp. 599-623.

6. J.F. Rusling, C.N. Shi, S.S. Shukla and D.K. Gosser, J. Electroanal. Chem., 240, 201 (1988).

7. G.E.O. Proske, Anal. Chem. 24, 1834 (1952).

8. G.L. McIntire and H.N. Blount, J. Am. Chem. Soc., 101, 7720 (1979).

9. G.L. McIntire and H.N. Blount in "Solution Behavior of Surfactants", Vol. 2, K.L. Mitall and E.J. Fendler (Eds.), Plenum, N.Y., 1982, pp. 1101-1123.

10. G.L. McIntire, D.M. Chiappardi, R.L. Casselberry and H.N. Blount, J. Phys. Chem., 86, 2632 (1982).

11. G. Meyer, L. Nadjo, J.M. Saveant, J. Electroanal. Chem., 119, 417 (1981).

12. A.E. Kaifer and A.J. Bard, J. Phys. Chem., 89, 4876 (1985).

13. M.J. Eddowes and M. Gratzel, J. Electroanal. Chem., 152, 143 (1983).

14. Y. Fujihira, T. Kuwana and C.R. Hartzel, Biochem. Biophys. Res. Commun., 61, 488 (1974).

15. P. Yeh and T. Kuwana, J. Electrochem. Soc., 123, 1334 (1976).

16. F. Hawkridge, Virginia Commonwealth University, personal communication.

17. (a) G.N. Kamau, W.S. Willis and J.F. Rusling, Anal. Chem., 57, 545 (1985). (b) J.F. Rusling, C.-N. Shi and E. Couture, University of Connecticut, 1987, unpublished results.

18. T.F. Connors and J.F. Rusling, J. Electrochem. Soc., 130, 1120 (1983).

19. (a) M. Humphries, Ph.D. Thesis, University of Bristol, U.K., 1975. (b) J.B. Hayter and R.J. Hunter, J. Electroanal. Chem., 37, 7181 (1972). (c) J.B. Hayter, M.W. Humphreys, R.J. Hunter and R.J. Parsons, J. Electroanal. Chem., 56, 160 (1974).

20. R. Nichols, J.F. Rusling and A. Bewick, Southampton University, 1986, unpublished results.

21. R. Nichols and T.F. Connors, Anal. Chem., 55, 776 (1983).

22. (a) J.V. Arena and J.F. Rusling, Anal. Chem., 58, 1481 (1986). (b) J.V. Arena and J.F. Rusling, J. Phys. Chem., 91, 3368 (1987).

23. J. Zeng and R.A. Osteryoung, Anal. Chem., 58, 2766 (1986).

24. (a) J.F. Rusling and G.N. Kamau, J. Electroanal. Chem., 187, 355 (1985). (b) G.N. Kamau, T. Leipert, S.S. Shukla and J.F. Rusling, J. Electroanal. Chem., 233, 173 (1987). (c) G.N. Kamau and J.F. Rusling, J. Electroanal. Chem., 240, 217 (1987).

25. (a) R. Scheffold, G. Rytz and L. Walder in "Modern Synthetic Methods", Vol. 3, R. Scheffold (Ed.), Wiley, New York: 1983, pp. 355-439. (b) H. Eckert and I. Ugi, Angew. Chem., 87, 847 (1975). (c) H. Eckert and Y. Kiesel, Angew. Chem. (Int. Ed.), 29, 473 (1981). (d) J.P. Randin, Electrochim. Acta, 19, 83 (1974).

(e) J.A.R. van Veen and C. Visser, Electrochim. Acta, 24, 921 (1979). (f) J.F. Van Baar, J.A.R. van Veen, J.M. van der Eijk, T.J. Peters and N. de Wit, Electrochim. Acta, 27, 1315 (1982). (g) J. Zagal, E. Munoz and S. Ureta-Zamartu, Electrochim. Acta, 27, 1373 (1982). (h) J. Zagal, P. Bindra and E. Yeager, J. Electrochem. Soc., 127, 1506 (1980) and refs. therein. (i) S. Zecevic, B. Simic, E. Yeager, A.B.P. Lever and P.C. Minor, J. Electroanal. Chem., 196, 339 (1985) and refs. therein.

26. (a) D.W. Clack, N.S. Hush and I.S. Woolsey, Inorganica Chim. Acta, 19, 129 (1976). (b) A.B.P. Lever and J.P. Wilshire, Can. J. Chem., 54, 2514 (1976). (c) K.M. Kadish, L.A. Bottomley and J.S. Cheng, J. Am. Chem. Soc., 100, 2731 (1978). (d) A.B.P. Lever and J.P. Wilshire, Inorg. Chem., 17, 1145 (1978).

27. (a) W.B. Jacoby, Ed., "Enzymatic Basis of Detoxification", Vol. I and II, Academic: New York, 1980. (b) R.S. Wade, N.O. Belser and C.E. Castro, Biochemistry, 24, 204 (1985) and refs. therein. (c) R.S. Wade and C.E. Castro, Inorg. Chem., 24, 2862 (1985).

28. (a) B. Halliwell and J.M.C. Gutteridge, "Free Radicals in Biology and Medicine", Clarendon: Oxford (U.K.), 1985. (b) P.J. van Bladeren, J. Am. Coll. Toxicol., 2, 73 (1983). (c) R.P. Mason in "Free Radicals in Biology", Vol. V, W.A. Pryor (Ed.), Academic: New York, 1982, pp. 161-222. (d) S.D. Aust and B.A. Svingen in "Free Radicals in Biology", Vol. V, W.A. Pryor (Ed.), Academic: New York, 1982, pp. 1-28.

29. J.F. Rusling, T.F. Connors and A. Owlia, Anal. Chem., 59, 2123 (1987).

30. A. Owlia and J.F. Rusling, J. Electroanal. Chem., 234, 297 (1987).

31. J. Georges and S. Desmettre, Electrochim. Acta, 29, 521 (1984).

32. R. Zana and R.A. Mackay, Langmuir, 2, 109 (1986).

33. Y. Ohsawa and S. Aoyagui, J. Electroanal. Chem., 136, 353 (1982).

34. C. Tanford, "Physical Chemistry of Macromolecules", Wiley: N.Y., 1961, pp. 526-586.

35. J. Wyman, Adv. Protein Chem., 19, 223-286 (1964).

36. J.F. Rusling, C.-N. Shi and T.F. Kumosinski, Anal. Chem., in press (1988).

37. A.J. Bard and L.R. Faulkner, "Electrochemical Methods", Wiley: New York, 1980.

38. T. Imae, R. Kamiya and S. Ikeda, J. Coll. Interface Sci., 108, 215 (1985).

39. S. Hayashi and S. Ikeda, J. Phys. Chem., 84, 744 (1980).

40. H. Fabre, N. Kamenka, A. Khan, G. Lindblom, B. Lindman and G.J.T. Tiddy, J. Phys. Chem., 84, 3428 (1980).

41. A. Malliaris, J. Lang, J. Sturm and R. Zana, J. Phys. Chem., 91, 1475 (1987).

42. (a) D.F. Evans, S. Mukherjee, D.J. Mitchell and B.W. Ninham, J. Colloid Interf. Sci., 93, 184 (1983). (b) R.M. Weinheimer, D.F. Evans and E.L. Cussler, J. Colloid Interf. Sci., 80, 357 (1931).

43. (a) R.B. Dorshow, J. Briggs, C.A. Bunton and D.F. Nicoli, J. Phys. Chem., 86, 2388 (1982). (b) J. Briggs, R.B. Dorshow, C.A. Bunton, and D.F. Nicoli, J. Chem. Phys., 76, 775 (1982). (c) B. Lindman, M.-C. Puyal, N. Kamenka, R. Rymden and P. Stilbs, J. Phys. Chem., 88, 5048 (1984).

44. E. Roelants and F.C. DeSchryver, Langmuir, 3, 209 (1987).

45. N.J. Turro and A. Yekta, J. Am. Chem. Soc., 100, 5951 (1978).

SOME TECHNIQUES FOR INVESTIGATION OF ELECTROCHEMICAL AND ION TRANSPORT PROPERTIES OF BIOLOGICAL CELL MEMBRANES: THE CASE OF CHARA CORALLINA

F. Homble, A. Jenard and H.D. Hurwitz*

Laboratoire de Thermodynamique Electrochimique
Universite Libre de Bruxelles
Faculty of Sciences,
C.P. 160, Ave. F.D. Roosevelt 50
B-1050 Brussels / Belgium

INTRODUCTION

In the case of living cells and cellular organites, many specific problems arise in the determination of electrochemical properties such as membrane potentials and conductances under stationary or transient conditions, surface charge densities, surface potentials and ionic specific conductivities. These problems prevent a direct application of methods used in artificial systems like model membranes which are made of lipidic bilayers with protein inclusions. In living systems, one must take into account:

(a) Stringent stability conditions and important time effects which limit experimental in vivo procedures and reproducibility of measurements.

(b) Physiological and morphological reactions to imposed chemical and electrical conditions, e.g., the strong external pH dependence, action potential, etc...

(c) Multiplicity and diversity of transport mechanisms and of the ionic constituents generally present in the normal life conditions.

(d) Transversal dynamic and lateral fluidity related to the surface phase configuration of the biochemical membrane constituents.

(e) Relative undetermination of the surface topology conferring to the interphase three-dimensional properties.

(f) Variety of surface sites characterized by rather undefined surface charge densities.

There are essentially three types of transferable chemical species across cell membranes, governing the electrochemical properties:

(1) Species which are neither actively transferred by themselves, nor involved in chemical reactions (at equilibrium) with others which would be themselves actively transferred.

The condition of stationarity and the finiteness of the interior cell volume impose the equilibrium distribution of these species. Only their passive exchange fluxes across the mebrane. This can be measured by using isotopic labeling.

(2) Species which are actively transferred and not involved in chemical equilibrium with other transferable species.
Their overall flux is zero by mutual compensation of active and passive transfers. The two classes so far defined will be called unreactive species.

(3) Species involved in at least one chemical equilibrium with other transferable species (reactive species). Their flux balance must take into account the chemical reactions in the lateral phases.

We endeavor in this work to review some theoretical and experimental treatments concerning the transport of all these ions across plant cell membranes. Which role can be assigned to each of these ions in the determination of the electrical properties of cells? Most of these ions are simultaneously actively pumped and passively transported through channels. Can be identify among them some prevailing electrochemical actors (e.g., the K^+ or H^+ ions) responsible for the values of the resting membrane potential? We shall tackle first this problem of the origin of these overall macroscopic properties and try to understand the complex way by which the ionic pump activities adjust to the membrane permeabilities and external conditions in order to maintain the living conditions in the cell. The general criteria which will be given in terms of fluxes and permeabilities rely on the nature and specificity of chemical pumps and ionic channels.

Many attempts have been made to model the electrical properties of cell membranes with equivalent electrical circuits made of resistors and capacitors connected in parallel. The membrane impedance can then be analyzed in terms of conductance and capacitance of these hypothetical circuits which should reflect some properties of the ionic processes in membranes. Different methods of impedance measurements will be described. Current clamp and especially voltage clamp techniques are very efficient for the investigation of the transient electric responses of membranes. These techniques allow to study the ion transport kinetics through the membrane. Thus, we dispose of several means of investigation which provide information on the whole cell membrane electrical properties in steady state and dynamic conditions. We shall explore next these phenomena at a microscopic level and, therefore, describe more recent methods permitting to perform the electrical measurements on single channels. The principles of the patch-clamp techniques will be presented and results for the single K^+ channel will be discussed. Hence, the fascinating stage is reached at which the elucidation of macroscopic electrical effects from the knowledge of elementary ionic processes becomes possible. We expect now that the synergy which all these processes need to organize life may soon be investigated and that an answer may be found to our initial questions about the intricacy of regulating ionic processes with regard to the resting membrane potential.

The experimental material which is used in this article consists of membranes of the isolated internodal cells of the freshwater green plants Characeae and mainly of Chara corallina. The cytoplasmic ion concentration of these cells can be determined directly because of their size. These giant cylindrical cells of corallina have a regular size, a diameter of about 0.7 to 1.2 mm and a length which can reach 20 cm. These cells easily survive after being isolated from the mother plant. The cell is composed

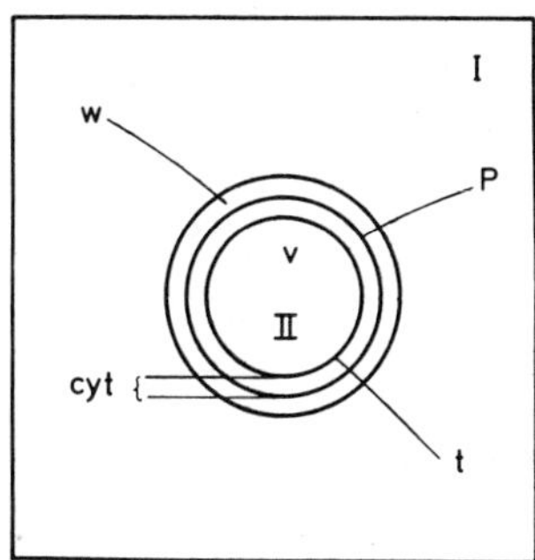

Figure 1. Schematic representation of the internodal Chara cell. V = vacuole; t = tonoplast; cyt = cytoplasm; P = plasmalemma; w = cell wall; I refers to the outside medium and II to the inside medium.

of a large internal vacuole which occupies about 90% of the intracellular medium surrounded by a thin cytoplasmic layer. This layer is bounded on the vacuole side by the tonoplast membrane and on the external side by the plasmalemma membrane (Figure 1) which is itself protected by the pecto-cellulosic cell-wall against direct contact with the external solution.

ION TRANSPORT AND MEMBRANE POTENTIAL

Ion transfer across cell membranes results from simultaneous active pumping and passive diffusion. It is basically the operation of the ion pumps which produces the electrochemical potential differences for the ions across the membrane, $\Delta\tilde{\mu}_i$. These quantities in turn become the driving forces for the passive ionic fluxes whose values determine the resting membrane potential E_M. The same arguments as developed in animal cells have been used for many years in plant cells in order to relate E_M with the ionic compositions inside and outside the cell. It was suggested without much experimental proof, that in response to treatments where the pumps are stopped by means of suitable metabolic inhibitors or blocking agents, the internal ionic activities would drop slowly with time in such a way as to keep instantaneous quasi-stationary values of E_M. According to this view, E_M could only be accounted for by strictly passive ion transport and ion pumps ought to be neutral. Consequently, the ion pumps do not transfer net electrical current across the membrane and the total diffusive or passive ion transport across the membrane should correspond to a total zero electrical current, thus

$$\sum_i Z_i J_i^A = \sum_i Z_i J_i^P = 0 \tag{1}$$

where J_i is the flux of species i and superscripts A and P refer to the active and passive transport, respectively.

The Goldman equation

$$E_M = E_M^G = \frac{RT}{F} \ln \frac{P_{K^+} C_{K^+}^{I} + P_{Na^+} C_{Na^+}^{I} + P_{Cl^-} C_{Cl^-}^{II}}{P_{K^+} C_{K^+}^{II} + P_{Na^+} C_{Na^+}^{II} + P_{Cl^-} C_{Cl^-}^{I}} \tag{2}$$

becomes valid with the conditions state above, applied to the major univalent cations and anions. Its derivation relies moreover on the assumption of a constant electric field in the membrane interior. In Equation (2), the P_i's are the permeability coefficients and C_i^I and C_i^{II} are respectively the external and internal concentrations. We deduce immediately from Equation (2) that the value of E_M^G can only vary within extreme values set by the Nernst potential of the ions that exhibit the highest permeability P_i. Let us note that the true significance of the P_i coefficients still remains controversial, although one agrees to relate proportionally their values to the initial transfer rate of labelled ions in radiotracer measurements performed in the same experimental conditions as the determination of E_M. Hodgkin and Katz[2] have assumed that $P_i = RT\ u_i k/d$, where u_i is the ionic mobility, d the thickness of the membrane and k the Donan partition coefficient of i at the membrane boundaries. In connection with the determination of E_M, it must be stressed that the internal activity of any diffusible ion which does not participate to any active transfer is determined by the equality of its Nernst potential E_i with E_M. The transport rate of such ion is given by its exchange transfer rate across the membrane. For nonmarine systems, like characean cells, where E_{K+} is often close to E_M[3], one would infer that the K^+ ion pump rate is much lower than K^+ exchange rate.

The Na^+ ion active effusion was first considered with the intention of explaining the quite general inversion of the sodium to potassium concentration ratio between the inside and outside of the cell and the negative sign of the rest potential. With fresh water plants, there is also some evidence of anion pumping since a higher concentration of diffusible anion is found inside the cell than in the bathing medium[4], although one should expect some Donan exclusion of these ions because of the presence of negatively charged macromolecules in the cell.

For most plant cells placed in ordinary life conditions, we cannot accept any more today the validity of the Goldman equation for E_M. Many investigations made obvious that the model of neutral pumps and related passive diffusion are inadequate[6-8]. Among numerous experimental facts that bear evidence as to the transfer by the pumps of a net electric charge across the membrane, the so-called electrogenic process, let us mention the phenomena of hyperpolarization where the potential E_M is displaced towards a more negative value than E_M^G, the strong correlation between inhibition of pumps and membrane conductance and the important temperature effects on E_M and on the membrane conductivities[7,9] The recognition of an electrogenic transfer across plant membranes has led to several theoretical treatments with a view to correct the Goldman equation. Unfortunately, these methods lack theoretical consistency. They generally admit that one type of ionic pump is prevailingly responsible for the overall electrogenic effect in each type of membrane system. For example, in the case of marine algae like Acetabularia[10], the influx generated by an electrogenic Cl^- pump is considered as the main electrogenic source. In non-marine plants like in characean cells, the strong external and/or internal pH dependence of both membrane potential and conductance have let to assess the role of an electrogenic proton pump[9,11]. A perfect selective proton transfer mechanism is easily conceivable and its role appears particularly suggestive in relation to the paradigmatic frame of reference of the chemiosmotic theories[12,13]. However, we may fairly ask the question whether experimental proofs and theoretical argumentation are nowadays sufficiently conclusive in order to retain the predominant proton pump as the most plausible model for steady state electrogenic transfer in these cells.

Before reconsidering this problem, we wish first to describe shortly the treatments used with respect to this model.

Electrogenic proton pumps in characean cells

The cell is represented by a single compartment of finite volume II in contact across a living membrane with an external medium I of fixed composition and pH and of unlimited dimension. The whole layer which lies between the vacuole of an aquatic plant cell and the aqueous solution in which it lives (Figure 1) is treated globally as a single transversally homogeneous membrane. In principle it is assumed that an increased sophistication of the model, e.g., distinction between tonoplast and plasmalemma membranes, sets of ionic conductance of separate specific channels should not affect the overall validity of the phenomenological relations. Thus, the results are also valid for the plasmalemma alone.

Working on Nitella clavata, Kitasato[11] put forward the hypothesis that a large passive influx of protons had to be correlated to an electrogenic efflux of protons. Therefore, he introduced C^{I}_{H+} and C^{II}_{H+}, respectively, in the numerator and denominator of Equation (2), both terms being affected by a permeability factor P_{H+} of high value. He assumed further that the electrogenic pump works as a current source, i.e., as though it has infinite impedance, and he described the membrane potential by the equation

$$E_M = E^G_M - \frac{FJ^A_{H+}}{g^m} \approx E_{H+} - \frac{FJ^A_{H+}}{g^m} \tag{3}$$

where g^m is the chord conductance of the membrane and E_{H+} the Nernst potential of H^+. A theoretical justification of this expression cannot be readily found unless one refers to the equation given by Rapoport[14]

$$E_M = \frac{\sum_i g^m_i E_i}{\sum_i g^m_i} - \frac{\sum_i Z_i F J^A_i}{\sum_i g^m_i} \tag{4}$$

which is only valid in the case of a linear relationship between the passive flux and the membrane potential. According to the result obtained by integration of the Nernst-Planck equation including all the model specifications given above and the constant field approximation, such linear relationship is correct only at vanishing values of both E_M and E_{H+}. Otherwise, one gets a more general expression of Equation (2) encountering the condition

$$\sum_i Z_i \; J^P_i = - \sum_i Z_i \; F \; J^A_i = -I^A \tag{5}$$

which tacitly implies that the balance of fluxes is realized, thus

$$J^P_i = -J^A_i. \tag{6}$$

This expression can be written

$$E_M = (RT/F) \ln \frac{\sum_{i+} P_i C^I_i + \sum_{i-} P_i C^{II}_i - (RT/F^2)(I^A/E_M)}{\sum_{i+} P_i C^{II}_i + \sum_{i-} P_i C^I_i - (RT/F^2)(I^A/E_M)} \tag{7}$$

where (I^A/E_M) can be put equal to g^m and the summations are made over i^+ and i^- corresponding respectively to monovalent cations and anions. Furthermore, it is found experimentally that E_M tends towards E_{K+} when the pump is inhibited and not towards E_{H+} as given by Equation (3). This

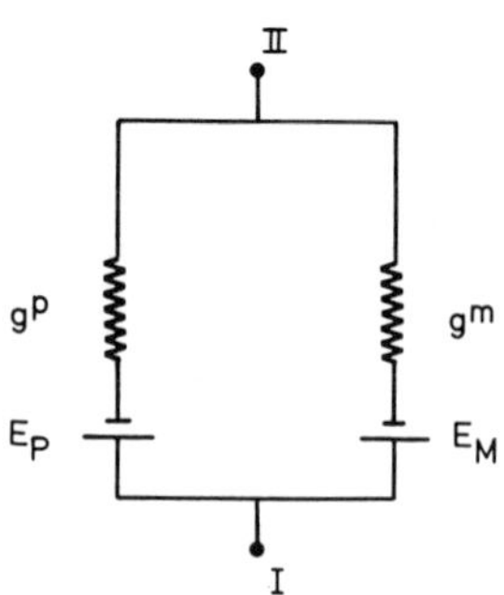

Figure 2. Equivalent circuit suggested in Reference 9 for the cell membrane in the case of an electrogenic pump.

remark has led Spanswick[9,15] to reject the hypothesis of a large passive H^+ influx and to propose an alternative explanation in terms of an electric analog for the membrane which consists of two branches connected in parallel (Figure 2). One branch represents the passive transport (E_M^G and g^m). In the other channel, an E.M.F. equal to E_p and a conductance g^p have been assigned to the pump; hence, the pump is postulated to have the properties of electric conductance, i.e., the flux through the pump would vary with the potential. Thus, according to the circuit of Figure 2:

$$E_M = (E_p\ g^p + E_M^G\ g^m)/(g^p + g^m) \tag{8}$$

Again, the value of E_p has been obtained from Rapoports's[14] thermodynamic treatment under the condition of the proton pump at equilibrium[9] ($J_{H^+}^A = 0$), a condition which does not seem to be realized in the case of Characean cells.

Indeed, the evaluation of the proton fluxes is subjected to large uncertainties. However, some experiments on perfused characean cells deprived from their tonoplast have led to the following results[16-18]:

1. The pump conductance is less than the passive conductance.
2. The permeability of H^+ is high ($P_{H^+} > 100\ P_{K^+}$).
3. The active efflux of H^+ is less than half of the overall flux.

Besides the remarks made above, which might notably weaken one's confidence in the electrogenic proton pump theory, let us stress some further complications. First, it is impossible to perform direct proton permeability measurements by the use of tritium in radiotracer methods. Second, several processes might be correlated to the external pH change, like effects caused by CO_2 transport, OH^- efflux, HCO_3^- uptake, the permeability of H^+ and ions via pores, and kinetic control of all ionic pumps.

Such considerations have led us to adopt a less restrictive modellistic approach[19]. For the diagnostic of steady state electrogenic ion transfer, we have used therefore a computation which starts from the Nernst-Planck equation, avoids a separation between diffusion and ion pumping, both appearing at once as antagonistic and complementary processes (see Equation 6), and which handles all transferable chemical species possibly involved in ion transfer across membrane without selecting *a priori* a main transport process.

Treatment of electrogenic transfer

The membrane is treated as the same simple system as described above (Figure 1). We assume, besides the electroneutrality inside the membrane (constant field hypothesis), a continuity of the electrical potential ψ from the bulk of the surrounding medium to the center of the cell. We are compelled, therefore, to assume that the potential at the outer Helmholtz plane, which is induced by the effective surface charge (dissociated functional groups or strongly adsorbed ions) is screened to a major extent by counterion-site pairing in the Nernst layer. We consider the three classes of transferable chemical species quoted in the introduction. Among the species not directly involved in chemical reactions, the calculation takes into account an anion (Cl^-), two monovalent cations (K^+, Na^+) and two divalent cations (Mg^{2+}, Ca^{2+}). As for the species involved in chemical reactions, we assume that they do not interact inside the membrane (independent channels). However, their flux balance must take into account the chemical reactions in both lateral phases. We handle only here two types of equilibria: the water dissociation and the first dissociation of the carbonic acid. The three flux balances of the five chemical species which are interconnected by the conditions for the equilibria are as follows:

$$J^A_{H^+}+J^P_{H^+}+J^A_{H_2O}+J^P_{H_2O}+J^A_{H_2CO_3}+J^P_{H_2CO_3} = 0 \qquad (9a)$$

$$J^A_{OH^-} + J^P_{OH^-} + J^A_{H_2O} + J^P_{H_2O} = 0 \qquad (9b)$$

$$J^A_{H_2CO_3} + J^P_{H_2CO_3} + J^A_{HCO_3^-} + J^P_{HCO_3^-} = 0 \qquad (9c)$$

As a matter of fact, the water molecules are not supposed to be actively transferred ($J^A_{H_2O} = 0$). Let us notive that in characean cell membranes the water permeability is very high[20-22] (of the order of 10^{-6} to 10^{-7} s^{-1} atm^{-1}) if compared to the ionic permeability.

Finite space charge densities occur in two diffuse regions in contact with both sides of the membrane. Their presence results from the disturbance of ionic distribution due to the action of the specific pumpings of ions. The thickness of these regions is supposed to be small in comparison with the cell dimensions. The Poisson-Boltzmann equations are written after linearization as follows, the axes x being perpendicular to the membranes

$$\frac{d^2\psi^I}{dx^2} = \alpha_F^2 \, \psi^I \, ; \qquad \frac{d^2\psi^{II}}{dx^2} = \alpha_{II}^2 (\psi^{II} - E_M) \qquad (10a,b)$$

The potential is taken as zero in the bulk of the external medium. In the interior of the membrane, we have

$$\frac{d^2\psi}{dx^2} = 0 \qquad (10c)$$

and

$$\alpha^2 = (F^2/\varepsilon RT) \, (\sum_i Z_i^2 \, C_i) \qquad (11)$$

The dielectric constant ε is taken as equal to 80 for both the external and internal media and to 6 inside the membrane (ε_M). The solution of Equation (10a,b,c), taking into consideration the continuity conditions

$$\psi^I(x \to -\infty) = 0; \quad \psi^I(x=-d/2) = \psi_s^I = \psi(x=-d/2);$$

$$\psi^{II}(x=d/2) = \psi_s^{II} = \psi(x=d/2); \quad \psi^{II}(x=\infty) = E_M \tag{12}$$

leads to:

$$\psi_s^I = \frac{\alpha^{II}\ \varepsilon_M\ E_M}{\alpha_I\ \alpha_{II}\ d\ \varepsilon_I + \varepsilon_M\ (\alpha_I + \alpha_{II})} \tag{13a}$$

$$\psi_s^{II} = \frac{\alpha_{II}\ (\alpha_I\ d\ \varepsilon_I + \varepsilon_M)\ E_M}{\alpha_I\ \alpha_{II}\ d\ \varepsilon_I + \varepsilon_M\ (\alpha_I + \alpha_{II})} \tag{13b}$$

at the two membrane-aqueous solution contact interfaces and d is the membrane thickness which has been taken equal to 10^{-8} m. Integration of the Nernst-Planck equation, including all the model specifications and Equation 6, gives

$$C_i^{II} = C_i^I \left\{1 + \frac{RT}{Z_i F\ C_i^I\ (\psi_s^{II} - \psi_s^I)} \cdot \left(\frac{J_i^A}{P_i}\right) \cdot \left[\exp.\left(\frac{Z_i F \psi_s^{II}}{RT}\right) - \exp.\left(\frac{Z_i F \psi_s^I}{RT}\right)\right]\right\} \exp.\left(\frac{-Z_i F E_M}{RT}\right) \tag{14}$$

The factor $J_i^A/P_i \equiv (J^A/P)_i$ represents a parameter which emphasizes the fact that the efficiency of a particular active ion flux in generating composition and potential difference is not accounted for by its magnitude only but also by the ability of the species to flow back passively through the membrane.

The results reported here were obtained with the following assumption for the relative permeability of the reactive species: $P_{H^+} = 10\ P_{OH^-} = 100\ P_{HCO_3^-}$. Finally, the system of equations (9, 13, 15) will be completed with the expressions of the chemical equilibria in both solutions (I and II); thus

$$K_W = C_{OH^-} C_{H^+} \tag{15a}$$

$$K_a = C_{H^+} C_{HCO_3^-} / C_{H_2CO_3} \tag{15b}$$

Discussion of numerical results

There have been few published studies about the cytoplasmic and vacuolar ionic contents of fresh water algae. Results obtained in our laboratory of vacuolar sap composition of Chara corallina as a function of the external pH are recorded in Table 1[23]. One observes a general trend corresponding to an increase of monovalent ion concentration and a decrease of the bivalent ion concentration with increasing pH, even though the vacuolar pH remains relatively constant. The corresponding

Table 1. Ionic vacuolar concentrations of Chara corallina cells treated 48 hours in a buffered solution in artificial pond water at different pH (MES-NaOH for pH 4.6 and 6.0; HEPES-NaOH for pH 7.5 and 8.5). Mean values of concentration ± s.e.m. and number of cells used are recorded as well as the theoretical Nernst potential E_i for K^+, Na^+ and Cl^-. The pH determination has been performed on 250 µl extracted from several cells. (Adopted from Reference 23.)

pH^{out}	K^+	Na^+	Ca^{++}	Mg^{++}	Cl^-	Vacuolar pH
	10^{-3} M	10^{-3} M	10^{-3} M	10^{-3} M	10^{-3} M	
	mV	mV	mV	mV	mV	
4.60	106 (7) ± 10.3	13 (7) ± 3.1	7 (7) ± 0.9	1.8 (7) ± 0.1	99 (6) ± 17.9	5.00
	- 179	- 126			+ 97	
6.00	147 (18) ± 4.7	16 (18) ± 3	5.8 ± 0.4	2.1 (18) ± 0.1	104 (7) ± 6.4	4.96
	- 187	- 130			+ 99	
7.50	169 (9) ± 7.5	7 (9) ± 1.7	4.1 (9) ± 0.3	1.36 (9) ± 0.11	146 (11) ± 6.4	5.00
	- 190	-109			+ 108	
8.50	174 (10) ± 6.3	18 (10) ± 4.8	3.9 (10) ± 0.3	1.87 (10) ± 0.04	139 (11) ± 4.5	4.78
	- 191	- 133			+ 106	

theoretical Nernst potentials E_i of each ionic species are also shown in the table. In the vacuolar sap, few ions are bound to cellular organites or macromolecules. Accordingly, the values of Table 1 are principally referring to free diffusible ions. In the cytoplasm[24-29], the Ca^{++} concentration is lower than 10^{-6} M; the Cl^- concentration is about 20.10^{-3} M, the pH is about 7.5 and the potential is nearly 20 mV more negative than in the vacuole. In Figures 3 and 4 are depicted the outer pH dependence of the resting membrane potential and of the membrane conductance of Chara cells. In Tables 2 and 3 are displayed values of the parametric quantities $(J^A/P)_i$ relevant to the different chemical species. These sets of numerical

Table 2. Parameters $(J^A/P)_i$ and $(J^A/P{\cdot}C^I)_i$ calculated for E_M = -0.120 mV; pH^I = 4.60; pH^{II} = 5.00 and for selected values of the concentrations C_i^I and C_i^{II} (T = 25°).

H^+	Cl^-	Na^+	K^+	Ca^{++} + Mg^{++}	HCO_3^-	i = ions
0.025	2.98	0.94	1.04	0.96	0.02	C_i^I (10^{-3}m/l)
0.01	100	18.70	105.96	10.21	0.32	C_i^{II} (10^{-3}m/l)
-7 10^{-2}	464	-4.0	4.7	-12.0	7.8	$(J^A/P)_i$ (10^{-3}m/l)
-2.8	156	-4.3	4.5	-12.5	391	$(J^A/P.C^I)_i$
E_M = -120 mV						

Table 3. Parameters $(J^A/P)_i$ and $(J^A/P \cdot C^I)_i$ calculated for E_M = -0.140 mV; pH^I = 6.00; pH^{II} = 5 and 7 and for selected values of the concentrations of C_i^I and C_i^{II} (T = 25°C).

pH	H^+	Cl^-	Na^+	K^+	Ca^{++}	Mg^{++}	HCO_3^-	i = ions
pH^I= 6	0.001	1.70	0.90	0.10	0.40	0.10	0.30	C_i^I $(10^{-3}$m/l)
pH^{II}= 5 pH^{II}= 7	0.01 0.0001	289	27.0	243	0.15	14.8	6.0	C_i^{II} $(10^{-3}$m/l)
pH^{II}= 5	1.4	1567	-5.2	6.3	-6.7	-1.7	171	$(J^A/P)_i$ $(10^{-3}$m/l)
	1380	922	-5.8	62	-17	-17	570	$(J^A/P.C.^I)_i$
pH^{II}= 7	$2.72\ 10^{-4}$	1567	-5.2	6.2	-6.7	-1.7	32	$(J^A/P)_i$ $(10^{-3}$m/l)
	0.272	922	-5.8	63	-17	-17	109	$(J^A/P.C^I)_i$
	E_M = -140 mV							

computed values are the solutions of the set of equations (9, 13, 14, 15) for all i's. The calculation was performed for external conditions which are close to natural living conditions and for internal ionic composition, pH and E_M which represent approximately the cytoplasmic and vacuolar states of Chara corallina cells. All these selected values of these external and internal variables are reported in Tables 2 and 3. A systematic analysis of these data leads to the following comments. (The use of the dimensionless flux parameters (ratio of two fluxes) recorded in Table 2 and 3 can sometimes make the discussion more intelligible.)

(1) Consistent values of E_M and of internal concentrations are generated if values of $J^A/P)_{Cl^-}$ and $(J^A/P)_{HCO_3^-}$ are both positive (inward flow)

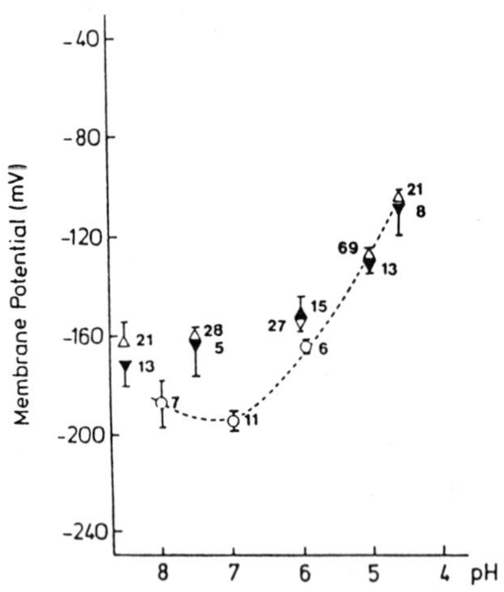

Figure 3. pH Dependence of the rest membrane potential of Chara cells treated 48 hours a) in buffered solution Δ, or b) in unbuffered solutions O. Open symbols are used for data obtained in light conditions and close symbols are used for dark conditions. The number of cells used is indicated in the graph (from Reference 29).

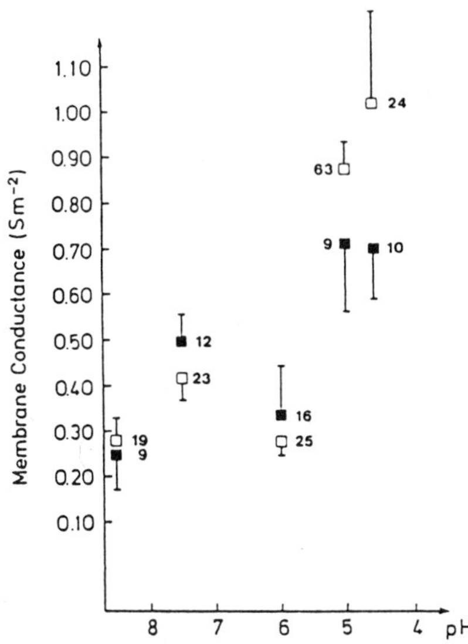

Figure 4. pH Dependence of the membrane conductance in the light (open symbols) and in the dark (close symbols) of Chara cells treated 48 hours in buffered solutions. The number of cells used is indicated in the graph (from Reference 29).

and are, at least, of one order of magnitude larger than values of all other $(J^A/P)_i$. This result expresses the fact of a high inward anion pumping rate and/or of a low membrane permeability with regard to anions. Diffusible anion concentrations inside the cell, notably that of Cl^-, are of the same order of magnitude as that of diffusible monovalent cations. The existence of actively transferred anions appears thus of prime importance for the generation of both E_M and the overall internal ionic concentration.

(2) Regarding the effect of the other pumps, their roles appear essential to fulfill the requirements of the internal ionic, particularly cationic, specific balance, meaning the respective proportion of Na^+ to K^+, of Mg^{++} to Ca^{++} and of monovalent to divalent cations. This acceptation of the ionic transfer process has been deduced by splitting $(J^A/P)_i$ in various components. From this computation, it is found that the ratio $[(J^A/P)_{Na+} / (J^A/P)_{K+}]$, respectively, for conditions of Table 2 and Table 3 becomes equal to -0.85 and -0.082. These negative values fulfill the requirement of the Na^+-K^+ ratio inversion (from 9 external to 0.11 internal). The combination of these negative values with the positive values of $[(J^A/P)_{Na+} + (J^A/P)_{K+}]$ reflects the existence of Na^+ efflux ($J^A_{Na+} < 0$) and K^+ influx ($J^A_{K+} > 0$). Since P_{K+} is usually assumed to be much larger than P_{Na+} [30,31], these results also involve the condition $|J^A_{Ka}| > |J^A_{Na+}|$, thus a higher pump efficiency for K^+. In all possible real life conditions and for E_M varying between -80 and -150 mV, we obtain $[(J^A/P)_{Na+}/(J^A/P)_{K+}] < 0$, but concomitantly the individual values of $(J^A/P)_{Na+}$ and $(J^A/P)_{K+}$ undergo considerable variations, even if the ratio (C^I_{Na+}/C^I_{K+}) is kept constant. These results suggest that the pumping of Na^+ and K^+ adjusts to the external medium in a delicate and complex way in order to maintain the internal Na^+/K^+ balance. This kind of regulation involves two types of effects: the pumping rate, of course, but also the permeability of the membrane to both monovalent ions.

(3) With regard to the divalent cations, it is obvious from Tables 2 and 3 that they are actively expelled from the cell. A closer insight of the dependence of the ratio $[(J^A/P)_{Mg^{++}}/(J^A/P)_{Ca^{++}}]$ with the external conditions indicates that it is nearly equal to the ratio $(C^I_{Mg^{++}}/C^I_{Ca^{++}})$. We might infer from this result, since the internal ionic concentrations are kept constant, that both bivalent cation expelling pumps although differing in efficiency are subordinated to an identical linear relationship with the external bivalent ionic concentrations.

(4) We now wish to focus our attention on the parameter $(J^A/P)_{H^+}$. From Tables 2 and 3, it is apparent that its value remains smaller than that of the parameter of all other ions. Some salient features of $(J^A/P)_{H^+}$ can be derived from some of our earlier calculations[19,23]. They show that at fixed external conditions and internal pH, there are no marked effects of the internal variables and of E_M on $(J^A/P)_{H^+}$, although the other parameters $(J^A/P)^i$ required to find back the internal conditions are changing. Conversely, at fixed internal conditions and external pH, the value of $(J^A/P)_{H^+}$ is also only weakly affected by changes of the external conditions. Finally, if all external and internal conditions are kept constant $(J^A/P)_{H^+}$ varies appreciably with the external pH in acid solutions (pH $<$ 7). It is reasonable to conclude from all these facts that the role of the proton transfer in the course of the ionic transfer process between the external and internal phases is restricted to the cytoplasmic and vacuolar pH regulation. The proton transfer exerts a minor effect on the other internal parameters.

At this state of our investigations, we can draw the following conclusions:

(a) Under physiological conditions all ions seem to be more or less pumped. Two different hypotheses might be suggested to explain this: Either every ion has its own specific pump mechanism, or alternatively there is a small number of pumps, but they are not limited to a single ion specificity.

(b) The cation balances $(C^{II}_{Na^+}/C^{II}_{K^+};\ Ca^{II}_{Ca^{++}}/C^{II}_{Mg^{++}}\ldots)$ probably depend on various cation pumps.

(c) In Chara, the existance of H^+/Cl^- cotransport is well established[32-36]. However, it has not been possible to prove whether the proton transport produces only the energy for the active Cl^- transport or both the energy for the transport and the resting membrane potential. Our approach helps to elucidate this problem. From our data, it seems that the membrane potential results essentially from the active anionic inward transport (Cl^- and HCO_3^- at alkaline pH) which itself might be driven by a proton-motive force. The weak selectivity displayed by the anionic pumps might account for the small differences of electrophysiological effects between the anionic species.

(d) Global proton pumping, from bulk to bulk, acts primarily as an internal pH regulator. Experimental evidence of the pH influence on (1) vacuolar composition, (2) resting potential and (3) isotopic flux, must not be confused with evidence of proton pumping, but rather should be considered as indicating that ion pump activities are pH dependent, which is often the rule for enzyme activities. Nevertheless, the quasi-independence of the internal pH to the external one (see Table 1) evidently proves the occurrence of H^+ pumping. A tempting hypothesis consists in assuming that the active pumping of H^+, combined with a very high passive permeability, constitutes the primary energetic driver of other ion pumpings, which derive their energy from the local passive retrodiffusion of H^+. What is known, for instance, about the first stages of photosynthesis in the thylakoid membrane contributes some arguments to this assumption? In this case, the proton circulation would not be apparent at the level of the bulk media. Only its indirect influence on other ion pumpings would be apparent in our results, because this proton circulation acts at the level of the internal mechanism of the pumps.

MEMBRANE IMPEDANCES

Many attempts made by biophysicists to understand the origin of electrical properties of the biological membranes are based on electrical equivalent circuits. The equivalent circuit most generaly proposed for the representation of membrane properties consists of a capacitor and a resistor in parallel (Figure 5a). Intuitively, this arrangement makes sense for a membrane considered as a plane, bearing electric charges and penetrated by purely conductive elements like ion channels. In presence of time- and voltage-dependent channels, a pseudo-inductive behavior will be observed and inductive elements should be added in series with the resistor (Figure 5b)[37]. In Chara corallina cells, the pseudo-inductive behavior may be ascribed to an increasing specific conductance of membranes for K^+ ions[46].

The electrical methods for measuring properties of ion transport through membranes are very attractive in comparison with tracer fluxes methods because of their accuracy and rapidity.

The large size of the Chara corallina cell makes it a convenient material for electrical studies. In one type of experimental set up which we have adopted in several occasions, the cell can be mounted in a Perspex chamber divided into two compartments. A small part of the cell protrudes outside the chamber (see Figure 9). Both compartments are filled with an isotonic solution, but a continuous flow of solution is maintained in the small compartment which is kept at a negative pressure. When the cell has lost its turgor, the externally protruding part is cut and an axial electrode (Pt/Ir wire) is introduced. In these conditions, the cell remains excitable and exhibits normal electrical responses. The external electrodes are made of Ag/AgCl.

Membrane impedance measurements can be achieved by using various methods. One possible approach consists to apply a sinusoidal current or voltage signal of a given frequency and to determine the corresponding voltage or current response by means of a lock-in amplifier or a vector impedance meter. This procedure must be repeated for every frequency of interest, which makes the measurements quite time consuming. Two

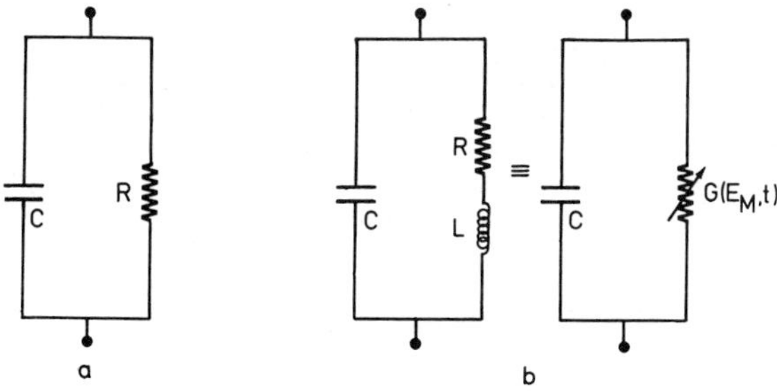

Figure 5. Equivalent circuits for the ion transport in membranes.

quantities are measured at each frequency: the phase angle between voltage and current signals and the ratio of the amplitudes of the voltage to the amplitude of the current.

It is, however, possible to have access to the frequency dependent impedance Z(f) using a Fourier analysis technique. In order to obtain the impedance Z(f) over a range of frequencies, we apply to the membrane a voltage or a current signal having measurable power at each frequency of interest. The Fourier transforms of the recorded respectively current response $F_i(f)$ or voltage response $F_v(f)$ as well as the Fourier transforms of the input signal are used to calculate Z(f) from the relation $Z(f) = F_v(f)/F_i(f)$. A white noise signal is generally injected through the membrane because it is a random signal having equal power spectral density at all frequencies[39]. An alternative procedure is given by transient analysis. In this case, the voltage response to a square current step is recorded. The ratio of the Laplace transform of the voltage to the Laplace transform of the current yields the operational impedance. With this technique, data gathering becomes rapid since all frequencies are recorded simultaneously rather than sequentially[40-42]. It should be stressed, however, that application of this method is restricted to simple systems because 1) the frequency spectrum of a step function is such that the amplitude increases exponentially with frequencies and 2) the determination of time constants becomes difficult if not impossible when the different time constants characterizing the voltage response are not sufficiently distant one from each other.

The results of impedance investigation are displayed either in a Nyquist plot where the imaginary part of the impedance is reported versus the real part or in a Bode plot which shows the frequency dependence of the phase shift and of the logarithm of the impedance magnitude. As pointed out by Diamond and Machen[43], the Bode plot has three advantages: First, phase angle is a much more sensitive indicator of changes in electrical circuit values; second, the linear magnitude scale of the Nyquist plot weights low frequency data because of their high impedance and third, the Nyquist plot appears to be poorly sensitive to phase shift effects at high frequencies. In Figure 6 is shown the Nyquist plot of the impedance spectra of Chara corallina membranes measured by means of sinusoidal alternating current method.

The operational impedance of the plant cell, as determined by the transient analysis, is displayed in Figure 7. In this case, the data are fitted by using the electrical equivalent circuit shown in the inlet of the figure. This circuit accounts for the plasmalemma membrane in series with the tonoplast membrane. The series resistance R_i takes into account the ohmic drop in the external solution, in the cell wal, in the cytoplasm and vacuolar solutions. The resistance R_{TO} and R_{PL} expresses the effects of electrogenic pumps and of passive ionic permeabilities respectively in the tonoplast and the plasmalemma membrane, the latter appearing five times more resistive. The calculated values of the capacitances which can be associated to the phenomena of interfacial charge distribution (capacitance and pseudocapacitance effects) are also quite different for both membranes. The capacitance is nearly twice as large for the plasmalemma as for the tonoplast membrane. This results confirms the well-known fact of the predominant role played by the electrochemical processes occurring at the plasmalemma interface. It is noteworthy that no pseudo-inductive effects are produced by the use of a low intensity current stimulus as in the case of the results reported in Figure 7. Bode plots presented by S. Ross[44] are reproduced in Figure 8. They result from impedance measurements on Chara corallina cells by means of the Fourier transform method. In Figure 8 are also depicted the overall conductances (Figure 8c) and

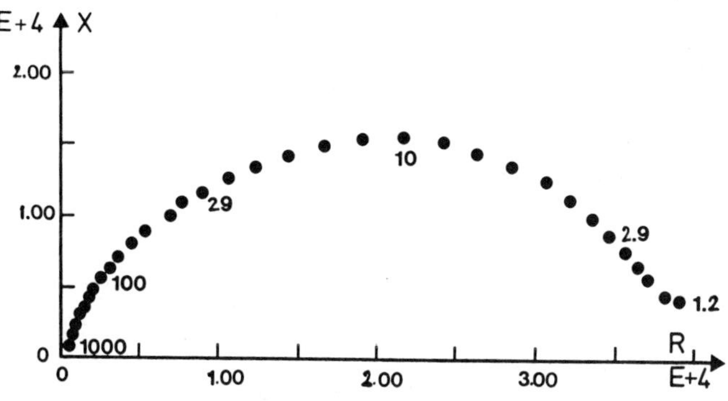

Figure 6. Nyquist plot of the impedance spectrum of Chara corallina under saturating light in artificial pond water at pH 6 (APW 6). The values indicate frequencies in Hz.

capacitances (Figure 8d) over the frequency range 3.125 - 100 Hz and the capacitance over the low frequency range of 0.0781 - 5 Hz (Figure 8e) of the system of both membranes in series. One should take notice of the large negative capacitance or pseudo-inductance which occurs at low frequencies (Figure 8e). This extreme low-frequency pseudo-inductance can be ascribed to a diffusion potential which develops in the unstirred layer located at the outer boundary of the cell membrane[45]. As a matter of fact, the charged species like the proton and the K^+ ion, which display the greatest membrane conductance, carry most of the current I through the membrane. It happens that these species have diffusion coefficients which, at least for H^+, are significantly greater than the average in the external bathing solution. Therefore, I becomes large than the purely diffusive driven current which is formally ensured by all charged species in solution. This difference will give rise to a diffusion potential gradient in the unstirred layer which will be out-of-phase with the current. As a result, the potential gradient will lead the current. This mechanism can produce a very negative capacitance in the appropriate frequency range.

CURRENT CLAMP AND VOLTAGE CLAMP TECHNIQUES USED FOR THE IONIC CONDUCTANCE DETERMINATION IN MEMBRANES

Both time- and voltage-dependent ionic conductances and voltage-independent ionic conductances are found in biological membranes. Two experimental methods have been developed in order to determine the relationship between the membrane potential and the ionic conductance: these are the constant current or current clamp and the constant voltage or voltage clamp techniques. In the case of the current clamp, one applies a constant current step to the membrane and records the resulting membrane potential changes[46].

In the case of the voltage clamp, the magnitude of the membrane potential is imposed and one monitors the resulting current which flows through the membrane. The current clamp method is of limited use when the time constants of ionic conductances are larger than the membrane time constant or when ionic conductances change at a threshold potential because in this case unstable potentials are arising which are not attainable by a constant current.

A current clamp system is easier to set up than a voltage clamp system. A constant current source is obtained with a power supply V_s in

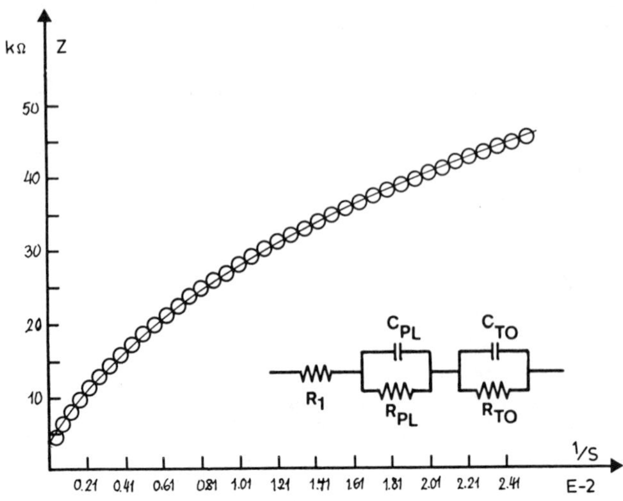

Figure 7. Frequency dependence of membrane impedance measured in an internode of Chara corallina under saturating light in APW 6 (Laplace transform method applied to transient response) ($1/s(sec^{-1}) = (j2\mu\nu)_{-1}$ where $j = \sqrt{-1}$).
O: experimental points
-: calculated membrane impedance by use of the equivalent circuit represented in the inlet, with the following values: for the phase and cell resistance R_1 = 1.0 kΩ cm²; for the tonoplast R_{TO} = 4.4 kΩ cm² and C_{TO} = 1.1 μF cm⁻²; for the plasmalemma R_{PL} = 26.2 kΩ cm² and C_{PL} = 1.9 μF cm⁻².

series with a high value resistor R_S[37]. The series resistor must be much larger than the membrane resistance in order to avoid any current change in response to a change in membrane conductance. This system is depicted in Figure 9. A voltage clamp system involves the use of a negative feedback circuit. In Figure 10 is displayed a simple and general scheme of a voltage clamp set-up. In this diagram, the properties of the electrometer which feels the membrane potential and the current to voltage converter I/V are both assumed to be ideal (no phase shift and potential perturbation). The membrane potential difference (E_M) is compared to a command potential (E) at the input of a large gain amplifier (G). The polarities are so arranged that the output voltage of G (Δ V) forces the membrane potential to be equal to the command potential. The access resistance R_A arises from the electrodes and R_S is the serial resistance which accounts for the solution conductivity. Assuming that the values of the circuit elements are at steady state, the operation of the voltage clamp circuit can be readily understood. The output voltage of the feedback amplifier G is equal to

$$\Delta V = (E - E_M)G \tag{16}$$

where G is the amplifier gain. Moreover, this output voltage is distributed between the voltage drop across the access resistance and the voltage drop across the membrane, so that

$$\Delta V = IR_A + E_M \tag{17}$$

Combining these two equations, we find

$$E_M = EG/(1 + G) - IR_A/(1 + G) \tag{18}$$

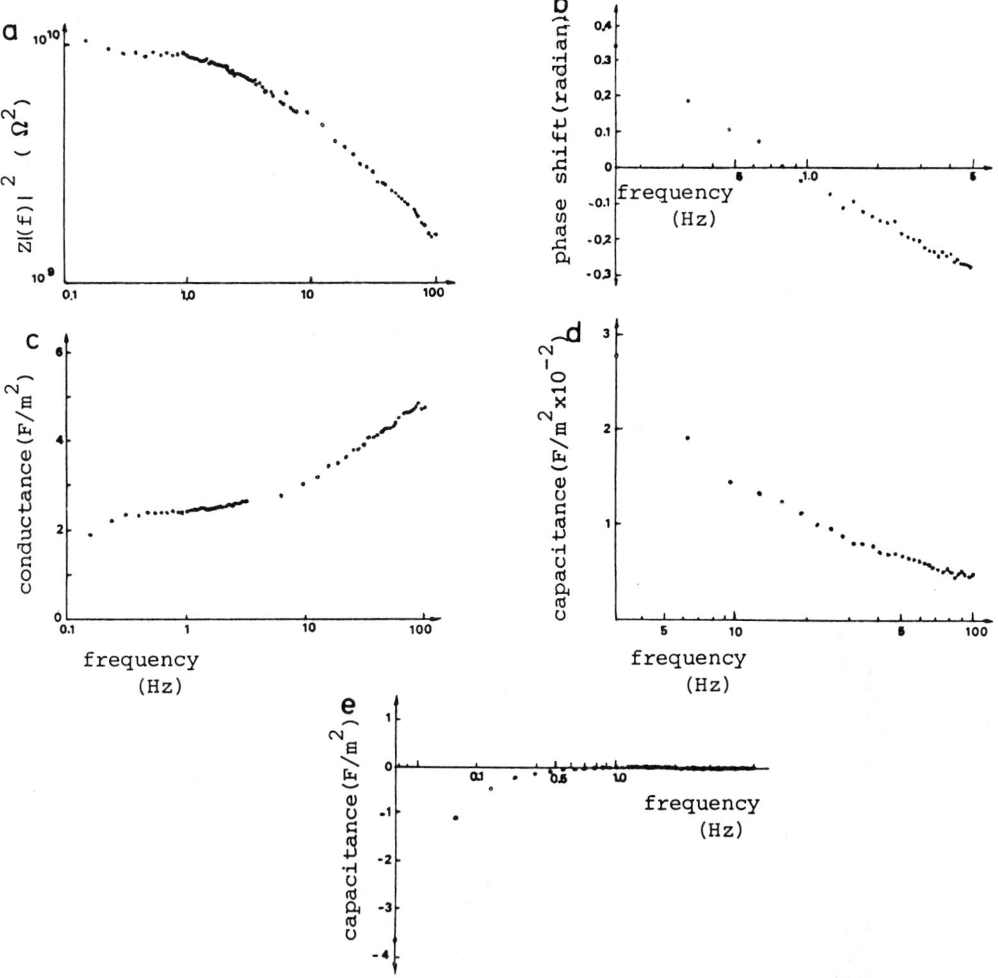

Figure 8. (a) Impedance magnitude measured in an internode Chara corallina in the light in APW at pH 7.
(b) Phase shift. (c) Conductance.
(d) Capacitance over the frequency range 3.125-100 Hz.
(e) Capacitance over the low frequency range 0.0781-5 Hz (from Reference 44).

Hence, we observe that a large gain for the feedback amplifier minimizes the effects of the access resistance and forces the membrane potential to be equal to the command potential. It is worth stressing the fact that E_M should be close to E. Because of the presence of R_S between the membrane and the voltage recording electrodes, E_M will differ from E by the voltage drop produced by the membrane current flowing across the series resistance R_S.

The time course of the inward current flowing through the membrane of a protoplasmic droplet formed from internodal cells of Chara corallina voltage clampled at different hyperpolarizing potentials is shown in Figure 11a[48]. Let us first point out that a protoplasmic droplet is obtained by gently cutting the lower end of a turgorless internodal cell

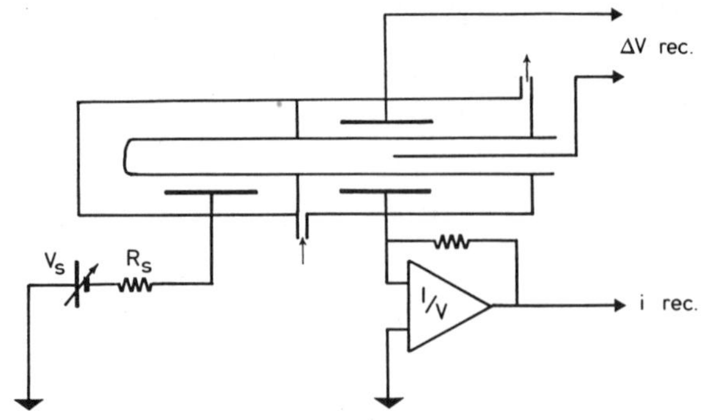

Figure 9. Schematic diagram of the experimental set-up used for current clamp techniques.

maintained vertically. The endoplasm which flows down from the open end is collected in an artificial vacuolar solution. In this solution, protoplasmic droplets form spontaneously. We used protoplasmic droplets of 40 to 80 μm diameter containing no chloroplasts. Cell attached patches are usually obtained about 30 min. after collecting the endoplasm. This time is required for the formation of membrane on the surface of the droplet[47]. It can be proved that the protoplasmic droplets are surrounded by a true biological membrane containing proteins. Obviously, the absence of active ATPase prevents any pump process, but, as Homble has experimentally observed, the K^+ channels have the same macroscopic electrical properties as those of the plasmalemma of intact cells[48]. The experimental set-up used is of the tight-seal whole cell made as described later in this work. One sees in Figure 11a that near the resting state the membrane of the protoplasmic droplet seems to behave in a purely passive manner, i.e., the current flowing through the membrane is proportional to the voltage difference applied across it and is independent of time. At potentials around 50 mV, more negative than the resting potential, the current flowing through the membrane decreases with time indicating that the membrane conductance is decreasing. At about -80 mV, after the first decrease in current is completed, an increase in current is observed, indicating that the membrane conductance is now increasing. For practical purposes, the first transient current will be called the early inward current and the

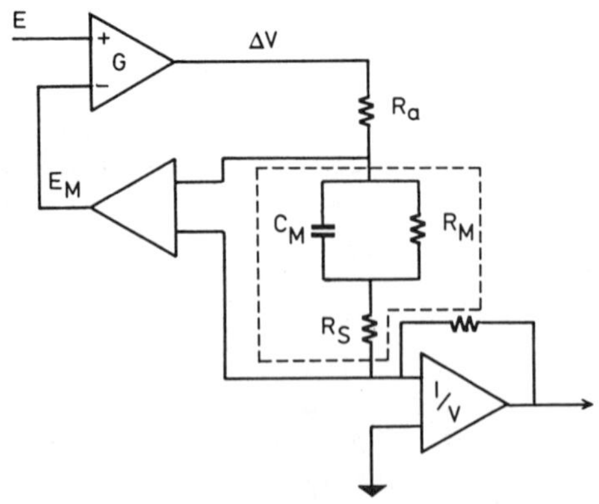

Figure 10. Schematic diagram of a voltage clamp set-up.

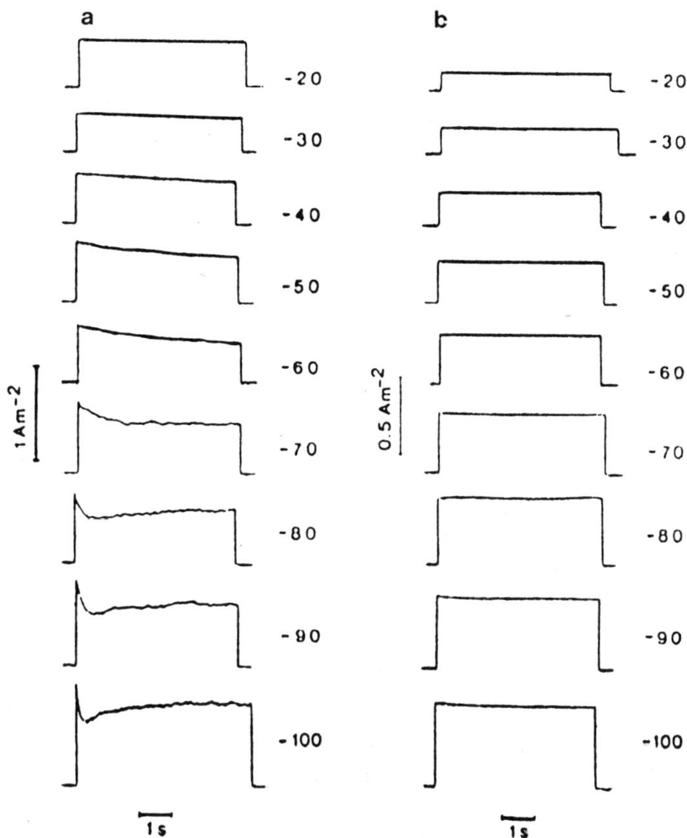

Figure 11. Time course of the current flowing through the membrane of protoplasmic droplets of Chara clamped at different values of membrane potential (mV) as indicated on the right of each curve, (a) in the absence, (b) in the presence of 20 mM TEA (from Reference 48).

second one will be called the delayed inward current. Over the range of potential investigated, the delayed inward curent displays large fluctuations and always appears after the time course of the early inward current had been completed. The time constant of the time course of both currents decreases with the applied voltage difference[48]. When tetraethylammonium (TEA) is present in the external solution, both early and delayed inward currents are blocked and the membrane current is no longer time-dependent (Figure 11, 13). This result suggests that both early and delayed inward currents are carried by K^+ ions since TEA is known to block K^+ channels of Chara[50,51]. In the presence of TEA and at membrane potentials for which the delayed inward current usually occurs, the current does not show large fluctuations, suggesting that the fluctuations observed in the absence of TEA were related to K^+ transport[48].

Figure 12 displays the response of the membrane potential to inward current pulses of different magnitudes. At low current density a purely passive response is observed. At 0.25 A m^{-2} and above, a spontaneous increase in voltage occurs which is followed by a slight voltage decrease in which large fluctuations can be observed. These results are similar to those observed on intact cells of Chara corallina[23], Nitella[52] and on

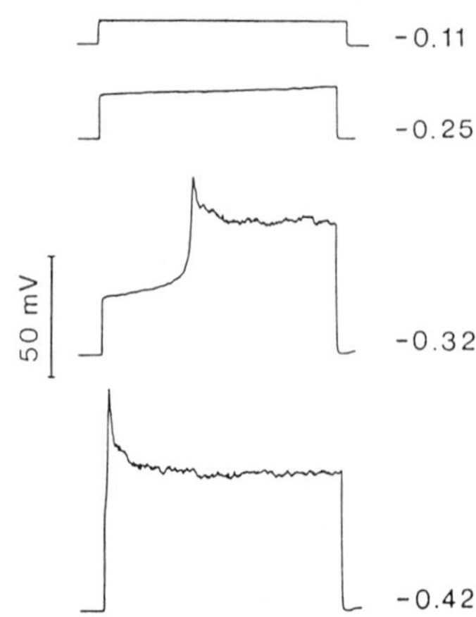

Figure 12. Time-dependence response of the membrane potential of protoplasmic droplets of Chara clamped at different values of the current density. The values to the right of each curve indicates the magnitude of the current density in $A.m^{-2}$ (from Reference 48).

protoplasmic droplets formed from internodal cells other than those of C. corallina[53]. The voltage dependence of the conductance associated with the early inward current and with the delayed inward current are shown in Figure 13. Over the range of voltage investigated in this study, the early inward current displays a sigmoidal voltage-dependent conductance. The delayed inward current also displays a non-linear voltage-dependent conductance. However, in this case, no saturation of the voltage conductance curve is obtained because the limit to which the membrane could be changed was -130 mV. (Above this value the tight seal in the patch clamp technique usually broke (see paragraph on patch clamp technique).) The minimum membrane conductance of the early inward current is significantly lower than the membrane conductance measured when K^+ channels are blocked by TEA which suggests that TEA does not fully block the K^+ channels.

In order to derive a mathematical expression which describes quantitatively the voltage-dependent conductance of K^+ channels associated with the early inward current in the protoplasmic droplets of Chara corallina, it has been assumed that each K^+ channel behaves like a dipole which can exist in two states: open and closed[48]. In this case, the K^+ conductance is given by

$$g_K = g_K^{max} \frac{\exp.-\alpha(E_M - E_{1/2})}{1 + \exp.-\alpha(E_M - E_{1/2})} \qquad (19)$$

where $E_{1/2}$ is the potential at which half of the channels are open, E_M is the membrane potential, g_K^{max} the maximum K^+ conductance and α is written for

$$\alpha \equiv z\ e\ (\cos\theta_{closed} - \cos\theta_{open})/kT \qquad (20)$$

where θ is the angle between the direction of the dipoles and the electric field, e the electronic charge and z the number of the electronic charges of the dipole.

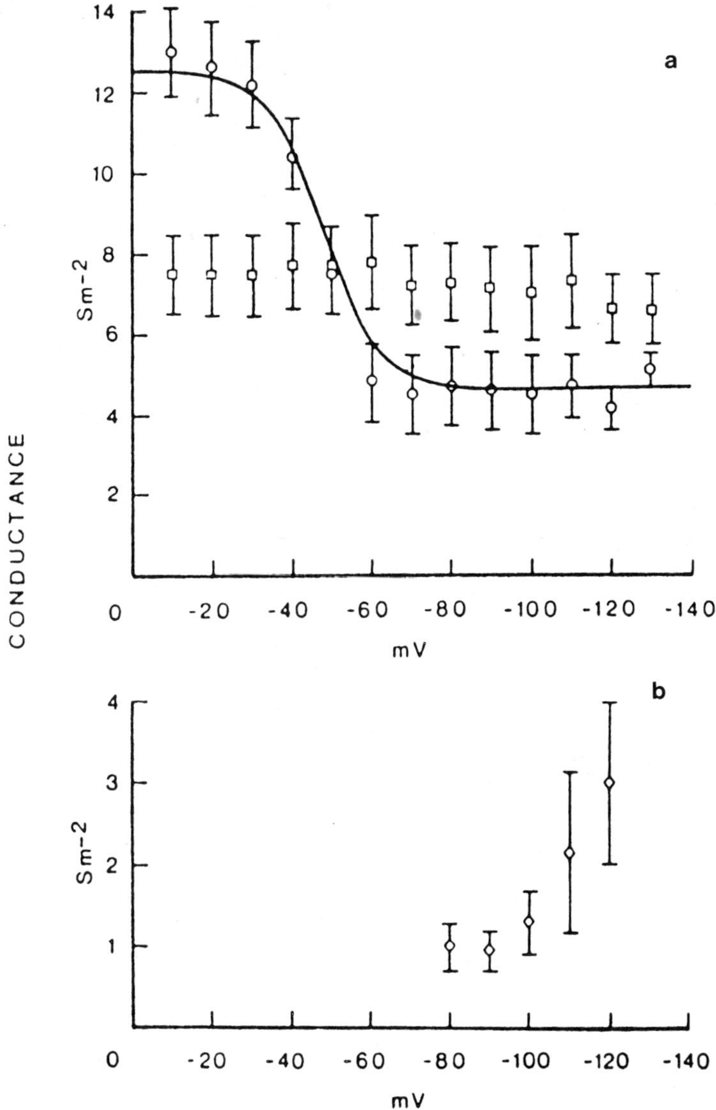

Figure 13. Voltage dependent conductance: (a) of the early inward K^+ current in the absence (O) or presence (□) of TEA, and (b) of the delayed inward K^+ current in the absence of TEA. The continuous line drawn in (a) has been calculated assuming a two state constant dipole model (from Reference 48).

According to the fact that the maximum value of ($\cos\theta_{open}-\cos\theta_{closed}$) is equal to one and that the maximum value of dipole length is the membrane thickness Homble[48] has estimated the minimum equivalent gating charge involved in the gating process of the K^+ channels to be equal to $z = 6$. The identification of the two K^+ voltage dependent conductances remains still very speculative.

THE PATCH CLAMP TECHNIQUE USED FOR SINGLE CHANNEL CONDUCTANCE DETERMINATION

In vivo, passive transport of ions occurs through selective channels which open and close randomly. These last ten years a number of research groups have tried to find a way which allows the determination of the electrical properties of a small number of channels. In 1981, a German research group at Göttingen succeeded in measuring the current flowing through a single ion channel with a very high accuracy[54]. This technique has been called patch clamp and is now applied in both animals and plant cells. The four different types of experimental configuration that can be achieved with the patch clamp techniques are described here after and in Figure 14.

The cell attached configuration

The initial step consists of pressing a clean fine tip (about 1 μm) glass pipette against the cellular membrane free of connective tisues. Upon application of suction a tight seal of extremely high resistance (about 10 G Ω) forms between the membrane and the inner wall of the glass pipette. This is called the cell-attached patch (Figure 14a). It permits the single channel activity recording in vivo. This configuration is the easiest to obtain. It disturbs the membrane and its physiological environment to a lesser extent than the other types of experimental configuration.

The inside-out configuration

Because of the high mechanical stability of the glass-membrane seal, further manipulation is possible starting from the cell-attached con-

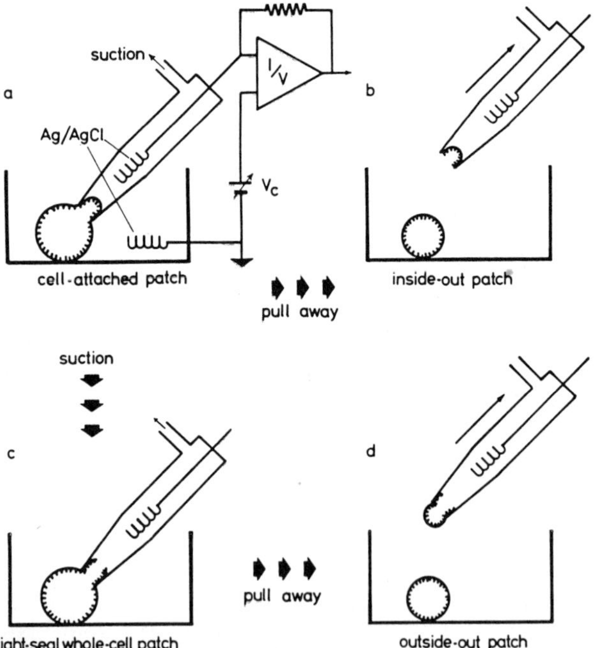

Figure 14. Schematic diagram of the procedure used for patch-clamp techniques. I/V high gain low noise current voltage converter; V_c voltage command; Ag/AgCl: silver chlorided electrodes.

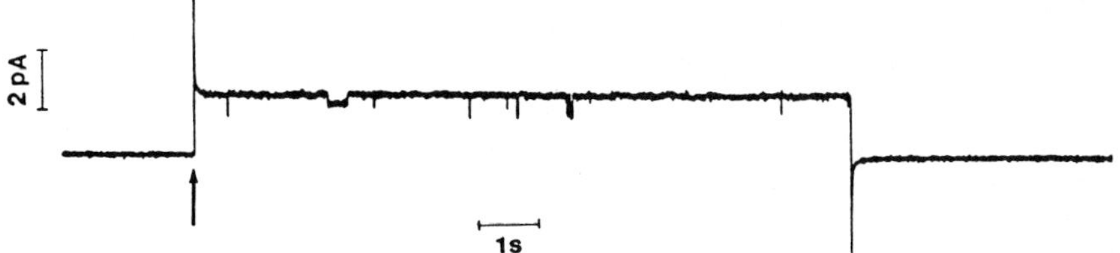

Figure 15. Time course of the current flow through the patch. The arrow indicates the moment at which the membrane is clamped at +20 mV. Channel activity can be seen after the fast initial transient capacitive current. Downwards deflection indicates channel closing (from Reference 49).

figuration. When the cell under investigation is attached to its support, a quick withdrawal will break the sealed patch membrane from the remaining cell membrane and form an inside-out patch configuration (Figure 14b).

The tight-seal whole cell configuration

The tight-seal whole cell configuration is created by the rupture of the patch membrane at the stage of cell-attached configuration (Figure 14c). It is obtained by increasing the suction in the pipette until the patch membrane breaks. This configuration allows to investigate small sized cells under voltage and current clamp techniques and to record overall current and voltage response of the total amount of channels present in the membrane.

The outside-out configuration

The outside-out configuration (Figure 14d) differs from the inside-out configuration by presenting the membrane intracellular face towards the pipette solution instead of the outer cellular face. Such configuration can be achieved by forcing the rupture of the sealed membrane from the cell fixed in its tight-seal whole cell configuration.

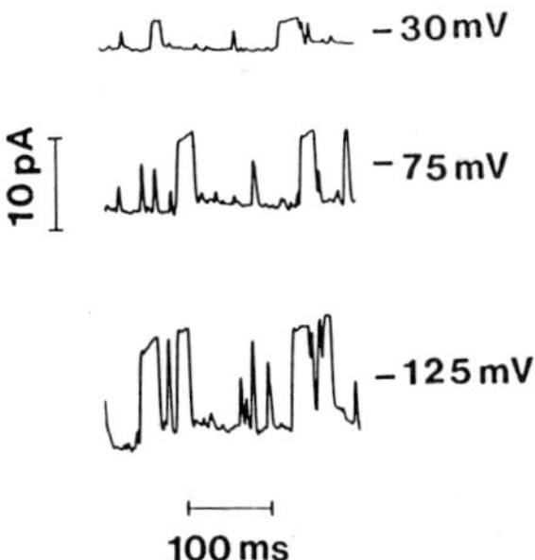

Figure 16. Current fluctuations of a single voltage-dependent K^+ channel at different hyperpolarizing potentials, indicated at the right. Downwards deflection indicates channel opening (from Reference 49).

The patch clamp techniques require a direct access to the membrane under investigation. This means that we must first "clean" the membrane from any connective tissue. In the case of plant cells for which the plasmalemma is surrounded by a cell wall two different methods can be used. The cell wall can be enzymatically digested in order to obtain protoplasts or, in the case of larges cells of characeae, a protoplasmic droplet can be formed from the cellular endoplasm, as described earlier.

The patch clamp electronic circuit consists of a current to voltage converter provided with a feedback resistor of high value which sets the amplification factor and with a low leakage and low noise input current in order to allow a resolution of a few picoamperes. A high order active filter and a capacitance compensating circuit are usually added in order to improve the noise and time constant performances of the current to voltage converter.

We describe hereafter some results of single channel analysis obtained[49] on protoplasmic droplets formed from Chara chorallina with the cell-attached configuration.

The time course of the electrical current flowing through the patched membrane when clamping the membrane potential from its resting value to a more positive one (membrane depolarization) is shown in Figure 15. At the onset of the voltage clamp (indicated by the arrow) there is a transient current pulse (capacitive current) followed by a steady level in which opening and closing events of membrane channels can be seen. The same kind of response is observed when the membrane potential is clamped at a value more negative than the resting value (membrane hyperpolarization) except that the current is in the opposite direction. High conductance single-channels with inward currents were observed at negative potentials. Figure 16 shows records for different voltage-clamp pulses as obtained during high channel activity. The current through the channel disappears at the resting membrane potential (-1.6 ± 0.8 mV, means ±s.e., n = 17) indicating that the ions flowing through the channel are at electrochemical equilibrium at the resting membrane potential. This suggests that the channel is selective to K^+ ions since Reeves et al.[55] have shown that K^+ is in electrical equilibrium, at the resting membrane potential, across the membrane of a protoplasmic droplet in the experimental solution used in the present study. Further evidence for the identification of the channel as a K' selective channel comes from the fact that if 20 mM TEA is added to the pipette solution, the current through the channel decreases by about 90 %. These results are consistent with the fact that TEA blocks the K^+ transport across the plasmalemma of internodal cells of Chara[46,50,51], of Nitella, of the Hydrodictyon[56], of Eremosphaera[57] and with the inhibition of TEA of the ionic current through single K^+ channels in animal cells[58].

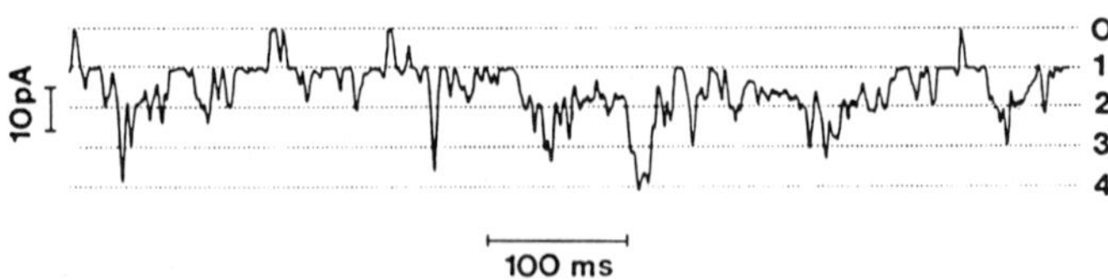

Figure 17. Multi-channel activity in a patched membrane clamped at -150 mV. Four different levels of a single channel current have been identified. Downwards deflection indicates channel opening (from Reference 49).

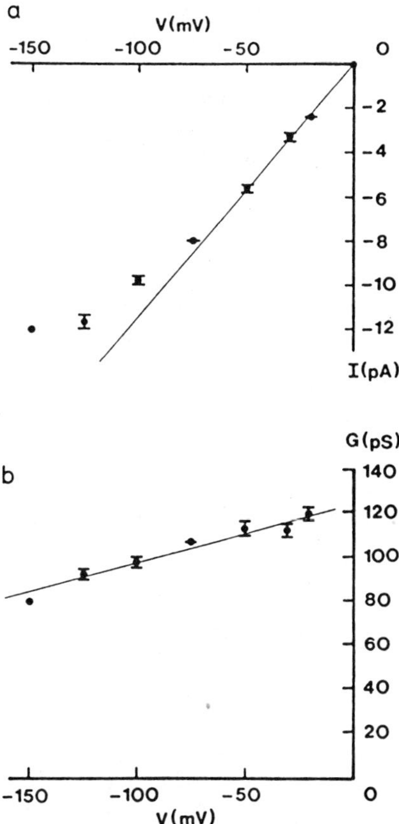

Figure 18. (a) Current-voltage relation for a single open K^+ channel recorded when the membrane is hyperpolarized. A straight line was drawn through the first four points to indicate the non-linear tendency of the curve (from Reference 49).
(b) Linear voltage-dependence of the conductance of a single open K^+ channel. Vertical bars indicate the magnitude of the standard errors (from Reference 49).

The kinetics of opening and closing of the K^+ selective channel were analyzed in patches in which the activity of the only one channel could be detected. The mean dwelling time of the open channel state is 196 ± 39 ms (n = 44) at -20 mV and 100 + 16 ms (n = 71) at -50 mV. The mean closed time of the channel at -20 mV is 47 ± 6 ms (n = 61) and is 44 ± 6 ms (n = 60) at -50 mV. These results show that the mechanism responsible for the closing of the K^+ selective channel is voltage dependent and that the channel spends less time open when the membrane is hyperpolarized. No data about the lifetime of single channel activity were obtained beyond -50 mV since further hyperpolarization of the membrane induced multichannel activity in the patch.

A typical example of such multichannel patch membrane clamped at a membrane potential of -150 mV is shown in Figure 17. In this example, at least four channels can be identified. Returning to a more depolarized level does not inhibit the activity of these channels. Thus, even when single channel activity is recorded, it is clear that the patch contains more channels that are inactive.

The current-voltage curve of a single channel current (Figure 18a) indicates that this K^+ channel has also a voltage-dependent conductance. The conductance of single K^+ channels increases linearly as the membrane potential becomes less negative (Figure 18b). The shape of the voltage dependence of a single K^+ channel is different from that measured on a whole cell (Figure 14). The voltage conductance curve of a whole cell has a sigmoidal shape (see Figure 13).

One can derive an explanation for this apparent discrepancy from the fact that the macroscopic voltage dependence of the conductance of a whole cell is determined mainly by the opening and closing frequency of single channels rather than by the change of the intrinsic conductance of an isolated single channel. At -75 mV, the single channel conductance is 106 ± 1 pS (n = 7) which is about three times the conductance of the voltage independent K^+ channel as obtained from guard cells[59]. (In plant cells, two guard cells surrounding the stomatal pores modulate its aperture, allowing the exchange of CO_2 and O_2 for photosynthesis and respiration, while minimizing water loss to the atmosphere.)

Our measurements for the specific membrane conductance give a value of about 10 $S.m^{-2}$ which is in agreement with published values of protoplasmicdroplet conductance published previously[60]. We can thus estimate that the channel density is about 0.1 channel per square micrometer of membrane. This is consistent with the observation noted above that even when single channel activity is recorded, the pipette (which has a tip diameter of about 5 μm) contains more channels that are not active. At the same potential (-75 mV) but in presence of 20 mM TEA, the measured single channel conductance drops to 11 ± 2 pS (n = 3). This is in agreement with the measurements involving TEA in animal cells[58] and indicates that the lifetime of the interaction between K^+ channel and the TEA molecule is so short that individual blocking events cannot be resolved by our measuring system. This life time would thus be less than 1 ms. At-75 mV, the opening of a single channel carries an equivalent electrical current of 8 pA. Therefrom, one can calculate[49] that, at this value of the membrane potential, the ions are transported by the K^+ channel at a rate of 5.10^7 ions/s. This is about four times that of the voltage-independent K^+ channel of guard cells.

CONCLUSION AND PERSPECTIVES

The numerical solution of a set of Nernst-Planck equations leads to new insights into the ionic transport processes which contribute to the general membrane function in the case of Chara corallina under natural life conditions. If one accounts not only for the proton, but also for the other ionic species of physiological importance, it can be shown that, at steady state, only a major transport of anions directed inwardly can give rise to both the membrane potential and the intracellular ionic composition, proton transport being essential for the intracellular ionic composition, proton transport being essential for the intracellular pH regulation and playing the role supposedly of an energy source for cotransport. There are no reasons to believe that the proton pumps are mainly responsible for the electrogenic effects in plants. The treatment also draws our attention to the large K^+ ion inward pumping rate as compared to the Na^+ ion outward pumping rate. The active K^+ flux is attributed to a high outward passive flux necessary to maintain the internal Na^+/K^+ balance. The method yields values of ratios of the active flux to the passive permeability of all ionic species actually required to fulfill the normal life conditions.

Electrical impedance measurements applied to living cells confirm the main role played by the plasmalemma interface. Presently, it is, however, well proved that the impedance of the plasmalemma membrane of Characean cells are frequency dependent. By applying Laplace analysis to transient signals, we were able to show that the membrane impedance spectrum obtained with cells stimualted successively by a hyperpolarizing and depolarizing current is neither synmetrical nor linear with respect to the magnitude of the current pulses around the resting state, even with signals generally considered as small[40]. In Chara, we ascribe these effects to the properties of potassium channels. Thus, when K^+ channels are operating in this way, the classical alternating current method cannot be applied to study the membrane impedance, since the use of sinusoidal signals implies that electrical properties are symmetrical. On the other hand, this lack of symmetry calls for a more complicated and time-consuming computer technique than simple Fourier analysis[61].

Protoplasmic droplets formed from the endoplasm of Characean cells could be used to study the distinct electrical behavior of K^+ channels in absence of pumps. Current clamp and voltage clamp techniques applied to the whole droplets lead to identificate two voltage dependent overall membrane conductances: the first related to an early inward K^+ curent, the second to a delayed inward K^+ current. It is obvious, however, that whole cell investigations by themselves cannot provide crucial tests which should bring out a decisive choice between various possible models used to elucidate a rather complex electrochemical behavior. In Chara such an answer can only be expected as a result of direct electrical investigation on a single K^+ channel in an experimental system which combines the protoplasmic droplet formation with the patch clamp technique recently developed. Voltage-dependence of single K^+ channel conductances recorded with this method are essentially different in shape and substance from that obtained on whole cells.

The knowledge acquired from our work on the microscopic level of a single channel compells us to reassess the meaning of the electrical phenomena observed at the level of the whole cell. In order to reproduce the macroscopic behavior of cell membranes, we shall endeavor to conceive a statistical analysis of opening and closing frequencies in a large ensemble of single channels and maybe, if the systen is non-ideal, to consider cumulative and cooperative interactions emerging from the assemblage of channels in the membrane. This approach is somewhat tantamount to passing from statistical mechanics to local thermodynamics. In the future, having thus recognized the distinctive electrical properties of single K^+ channels, one would, of course, be led to formulate some ideas about the possible conformation of the proteins responsible for the K^+ ion transport.

ACKNOWLEDGMENTS

This work was supported by a material grant of the Belgian Ministry of Scientific Policy (Action de Recherche Concertee) and of the Fonds National de la Recherche Collective (F.R.F.C.).

REFERENCES

1. J. Dainty, Ann. Rev. Plant Physiol., 13, 379 (1962).

2. A.L. Hodgkin and B. Katz, J. Physiol., 37, 208 (1949).

3. A.B. Hope and N.A. Walker, Aust. J. Biol. Sci., 14, 26 (1961).

4. E.A.C. MacRobbie, Quart. Rev. of Biophys., 3, 251 (1970).

5. U. Kishimoto, Ann. Rep. Sci. Works Fac. Sci. Osaka Univ., 7, 115 (1959).

6. N. Higinbotham, B. Etherton and R.J. Foster, Plant Physiol., 39, 196 (1964).

7. C.L. Slayman, J. Gen. Physiol., 49, 64 (1965).

8. R.M. Spanswick, J. Stolarek and E.J. Williams, J. Exp. Bot., 17, 1 (1967).

9. R.M. Spanwick, Biochim. Biophys. Acta, 288, 73 (1972).

10. D. Gradmann, J. Memb. Biol., 25, 183 (1975).

11. H. Kitasato, J. Gen. Physiol., 52, 60 (1968).

12. P. Mitchell, Adv. Enzymol., 29, 33 (1967).

13. C.L. Slayman, in: "Membrane Transport in Plants", U. Zimmermann and J. Dainty, Ed., Springer-Verlag, Berlin (1974), p. 107.

14. S.I. Rapoport, Biphys. J., 10, 246 (1976).

15. R.M. Spanswick, Ann. Rev. Plant Physiol., 32, 267 (1981).

16. M. Tazawa and T. Shimmen, in: "Current Topics in Membranes and Transport", Vol. 16, Slayman, Ed., Academic Press, New York (1982), p. 49.

17. G. Kawamura and M. Tazawa, Plant and Cell Physiol., 21, 547 (1980).

18. T. Shimmen and M. Tazawa, Plant Cell Physiol., 21, 1007 (1980).

19. A. Jenard, H.D. Hurwitz and F. Homble, Biophys. Chem., 25, 287 (1986).

20. N. Kamiya and M. Tazawa, Protoplasma, 46, 423 (1956).

21. J. Dainty and B.Z. Ginsburg, Biochim. Biophys. Acta, 79, 102 (1964).

22. E. Steudle and U. Zimmermann, Biochim. Biophys. Acta, 332, 339 (1974).

23. F. Homble, Ph.D. Thesis, Universite Libre, Bruxelles, Belgium (1984).

24. R.E. Williamson and C.C. Ashley, Nature, 296, 647 (1982).

25. R.M. Spanswick and A.G. Miller, Plant Physiol., 56, 664 (1977).

26. J.A. Raven and F.A. Smith, Can. J. Bot., 52, 1035 (1974).

27. E.A.C. MacRobbie, J. Exp. Bot., 20, 236 (1969).

28. J.L. Richards and A.B. Hope, J. Memb. Biol., 16, 121 (1974).

29. F. Homble, Bull. Classe des Sciences, Acad. Roy. Belgique, 5e serie, tome LXXI, 230 (1985).

30. A.B. Hope and N.A. Walker, "The Physiology of Giant Algae Cells", Cambridge Univ. Press, Cambridge (1975).

31. M.A. Bisson and D. Bartolomev, Plant Physiol., 74, 252 (1984).

32. F.A. Smith, J. Exp. Bot., 35, 43 (1984).

33. D. Sanders and H.P. Hansen, J. Memb. Biol., 58,139 (1981).

34. M.J. Beilby and N.A. Walker, J. Exp. Bot., 32, 43 (1981).

35. R.J. Reid and N.A. Walker, J. Memb. Biol., 78, 35 (1984).

36. R.J. Reid and N.A. Walker, J. Memb. Biol., 78, 157 (1984).

37. F. Homble and A. Jenard, J. Exp. Bot., 35, 1309 (1984).

38. D. Poussart, L.E. Moore and H.M. Fischman, Ann. N.Y. Acad. Sci., 303, 355 (1977).

39. S.M. Ross, Comp. Prog. Biomedicine, 15, 217 (1982).

40. F. Homble and A. Jenard, Plant Physiol., 919, 81 (1986).

41. F. Homble and A. Jenard, Bioelectrochem. and Bioenergetics, 17, 131 (1987).

42. A.A. Pilla, J. Electrochem. Soc., 117, 467 (1970).

43. J.M. Diamond and T.E. Machen, J. Memb. Biol., 72, 17 (1983).

44. S.M. Ross, in: "Electrical Double Layers in Biology", M. Blank, Ed., Plenum Press, New York (1986), p. 211.

45. J.M. Ferrier, J. Dainty and S.M. Ross, J. Memb. Biol., 85, 245 (1985).

46. F. Homble, J. Exp. Bot., 36, 1603 (1985).

47. T. Takaneka, T. Yoshida and H. Horic, Proc. Japan Acad., 49, 286 (1973).

48. F. Homble, Plant Physiol., 84, 433 (1987).

49. F. Homble, J.M. Ferrier and J. Dainty, Plant Physiol., 3, 53 (1987).

50. H.A. Coleman and G.P. Findlay, J. Memb. Biol., 83, 109 (1985).

51. M. Tazawa and T. Shimmen, Plant Cell Physiol., 21, 1535 (1980).

52. U. Kishimoto, Plant Cell Physiol., 7, 429 (1966).

53. I. Inoue, N. Ishida and Y. Kabatake, Biochim. Biophys. Acta., 230, 27 (1973).

54. O.P. Hamill, A. Marty, E. Neher, B. Sakmann and F.J. Sigworth, Plügers Arch., 391, 85 (1981).

55. M. Reeves, T. Shimmen and M. Razawa, Plant Cell Physiol., 26, 1185 (1985).

56. G.P. Findlay and H.A. Coleman, J. Memb. Biol., 75, 241 (1985).

57. K. Kohler, W. Steigner, J. Kolbowski, U.P. Hansen, W. Simonis and W. Urbach, Planta, 167, 66 (1986).

58. R. Grygorczyk and W. Schwarz, Eur. Biophys. J., 12, 57 (1985).

59. J.I. Schroeder, R. Hedrich and J.N. Fernandez, Nature, 312, 361 (1984).

60. E.D. Prishchepov, V.K. Andrianov, Yu.M. Abramenko, G.A. Kurella and Va. Scintitskikh, Sov. Plant Physiol., 28, 63 (1981).

61. P.Z. Marmarelis and V.Z. Marmarelis: "Analysis of Physiological Systems: The White-Noise Approach", Plenum Press, New York (1978).

VECTORIAL CONTROL OF CELL GROWTH WITH MICROELECTRODE

Hideaki Matsuoka*, Satoshi Matsumoto and
Madoka Kinoshita

Department of Industrial Chemistry
Faculty of Technology
Tokyo University of Agriculture and Technology
2-24-16 Nakamachi, Koganei, Tokyo 184 / Japan

INTRODUCTION

Quite a few investigations have been reported on the effects of electrical field on biological phenomena. These phenomena are usually composed of so many reactions that it is difficult to identify the net effects of electric field. In comparison with chemical reagents, an electric field has characteristic properties such as strength, exposure time, and exposure site that are easily controlled. Therefore, vectorial phenomena such as cell growth direction and morphological control associated with cell differentiation are interesting subjects with biological significance. Until now, however, only a few papers have appeared which demonstrate such vectorial control and elucidate the effects quantitatively[1-5].

In the present study, our attention has been focused on cell growth direction and its vectorial control by electric fields. The effects were investigated on a single cell with a microelectrode. The use of microelectrodes may be advantageous to focus the electric field on a cell and to emphasize the vectorial property.

Another important factor of electric stimulus is the mode and strength of the electric field. Considering the effects on cell growth, they would vary from no appreciable effect to positive (if any), negative, and finally to lethal effects, as the electric field strength increases.

However, Frohlich[6,7] proposed the concept of coherent excitation. According to this concept, a pulsing electric field may force the dipolar components of the cell membrane, which are otherwise vibrating randomly, into coherent vibration[8]. Such a coherent state should become a trigger of a critical step in the cell cycle. In such a case, a low energy stimulus causes a drastic phenomena. Thus, it is called a non-thermal effect.

If the coherent vibration of these dipoles is essential, the pulsing electric field is expected to be much more effective than the direct current electric field (DC field), as long as the coherent frequency components are contained. Therefore, the present study has been performed mainly with a pulsing electric field.

EXPERIMENTAL

Materials

Rizoctonia solani (a fungi which causes rice sheath blight) was used as an example of the cell growth type I-A (Figure 1), since it is well known to elongate its hypha at a comparatively high rate (10 to 40 $mm \cdot day^{-1}$). *R. solani* was cultured in a potato-dextrose broth at 25°C for 2-3 days. A flock of hypha was mechanically dissociated and a single hypha was taken as a sample.

A wild type strain of *Saccharomyces cerevisiae* and its cell division cycle (*cdc*) mutants were used as examples of the cell growth type II. The type strain was cultured in YPD medium (yeast extract 1%, polypeptone 2%, glucose 2%) at 30°C for 15 h. Among *cdc* mutants, *cdc* 35 and *cdc* 28 strain were used in the present study. These strains can be arrested by incubation at 38°C just before spindle pole body satellite formation (SPBSF) and spindle pole body emergence (SPBE), respectively[9-12] (Figure 2). The *cdc* strains cultured in YPD medium at 25°C for 15 h were collected and resuspended in a fresh YPD medium, and then incubated at 38°C for 2 h to arrest them at respective points. Then, the arrested cells were cooled to 25°C and the budding direction was measured with or without electric field.

PC12h-AG8 (a clonal variant of PC12 rat adrenal pheochromocytoma nerve cell line[13]) was used as an example of the cell growth type III. This cell was cultured in a medium composed of Dulbecco's modified Eagle medium (DMEM) 90%, precolostrum newborn calf serum (PNCS) 5%, and horse serum (HS) 5%. The medium for the cell diferentiation was prepared by

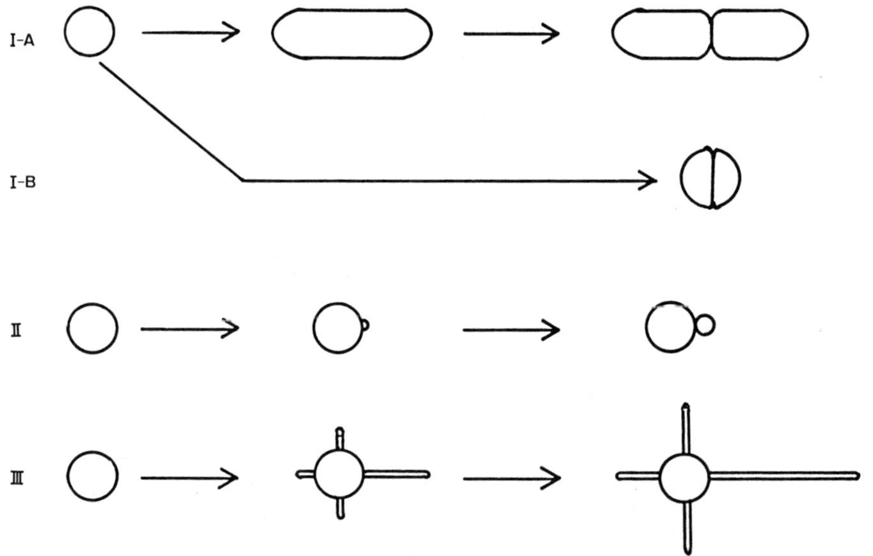

Figure 1. Growth types of cell.
I-A: A cell divides into daughter cells after elongation (ex. hypha of fungi).
I-B: A cell divides into daughter cells without changing the initial cell size (ex. egg of sea urchin).
II: A small bud grows into a daughter cell (ex. yeast).
III: A part of cell elongates without cell division (ex. neurite of nerve cell).

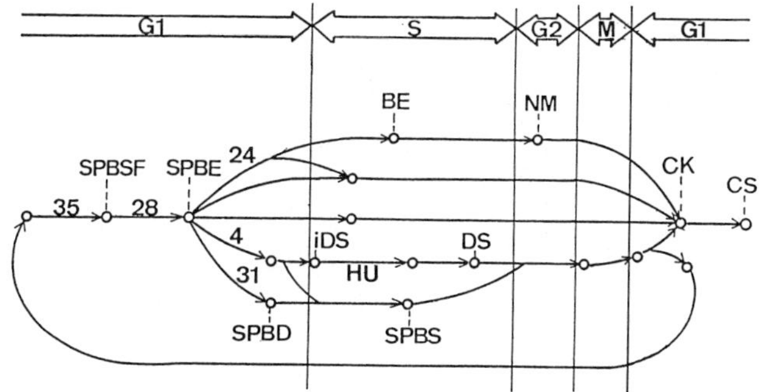

Figure 2. Arrested point of *cdc* mutants of *S. cerevisiae*.
SPBSF: Spindle-pole-body satellite formation
SPBE: Spindle-pole-body emergence
SPBD: Spindle-pole-body duplication
SPBS: Spindle-pole-body separation
BE: Bud emergence
NM: Nuclear migration
iDS: Initiation of chromosomal DNA synthesis
DS: Chromosomal DNA synthesis
CK: Cytokinesis
CS: Cell separation

modifying the DMEM-PNCS-HS medium: DMEM 45%, PNCS 5%, HS 5%, and Ham's F12 (F12) 45%. PC12h-AG8 cells cultured in the DMEM-PNCS-HS medium were transferred into a culture dish containing the cell differentiation medium. After incubation for 1 day, nerve growth factor (NGF) was added into the dish to initiate the neurite elongation.

Preparation of microelectrode

A glass tube (1.5 mm dia) was pulled with a capillary puller. The capillary was filled with an electrolyte (3.0 M KCl), into which platinum wire was inserted.

Application of electric field

Electric field was applied with microelectrodes as illustrated in Figures 3 and 4. In the case of fungi and yeast, a hypha or a cell was fixed at the tip of a glass capillary (2-3 μm dia) with slight suction. In the case of nerve cell, cells were spread in a culture dish and stood still for 1 day. After the adhesion of the cells on the bottom of the culture dish, two microelectrodes were set near particular cell(s) from both sides. The distance between the tips of the microelecrodes was 60 μm, 40 μm, 50μm for *R. solani, S. cerevisiae*, PC12h-AG8, respectively.

An alternating pulsing electric field (10 KHz) was applied on respect-tive cells. In the case of fungi, a DC field was also applied. Pulse height was estimated by the combined analysis of impedance measurement, analytical calculation, and computer analysis.

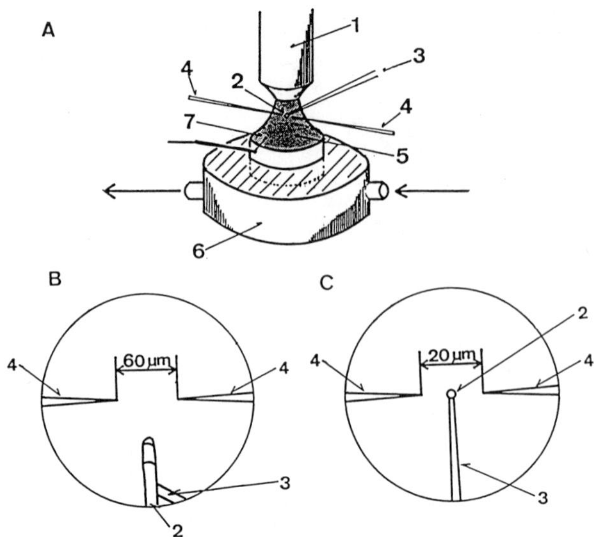

Figure 3. Experimental set-up for electric stimulation of fungi and yeast. 1. Microscope; 2. Cell; 3. Supporting capillary tube; 4. Microelectrode; 5. Medium; 6. Jacketed vessel; 7. Thermistor; B: Fungi; C: Yeast.

RESULTS

Estimation of electric field actually applied on a cell

For the particular case of yeast cell, the electric field actually applied on a cell (Vs) was estimated as follows. The ohmic resistance of

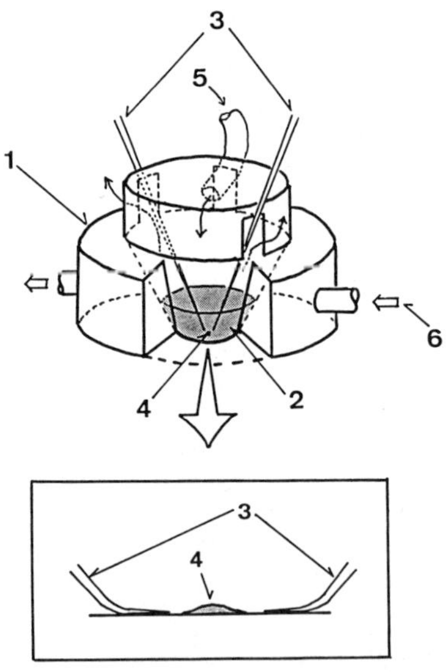

Figure 4. Experimental set-up for electric stimulation of nerve cell. 1. Thermostated culture dish; 2. Medium; 3. Microelectrode; 4. Cell; 5. Supply of covering gas (CO_2 5%-Air 95%, humidity 100%, 37°C); 6. Water (38°C).

a pair of microelectrodes were measured in a 3 M KCl solution and YPD medium, respectively. The resistance ranged from 10-40 MΩ. The resistance depended greatly on the diameter around the tip. A microelectrode suitable for the nerve cell experiment was more slender in shape and greater in resistance than for other experiments. The resistance of a typical microelectrode for the yeast cell experiment was about 25 MΩ. Using this value, the peak height of Vs was estimated at 9.2 V, when 15 V was applied on the electrode terminals. Under this condition, the potential gradient was analyzed by a computer program and estimated at 1.61 $KV \cdot cm^{-1}$ at the midpoint between the microelectrode tips, where the yeast cell was set. The cross-membrane potential at the cell membrane varied along the cell surface, as determined by the Pauly-Schwan's equation[14]. The maximum value was obtained at the cell membrane loci in the direction parallel with the electric field, where the peak height was 480 mV. At the loci in the direction perpendicular to the electric field, however, no cross membrane potential appeared. In a similar manner, Vs could be estimated for each case of fungi and nerve cell, respectively.

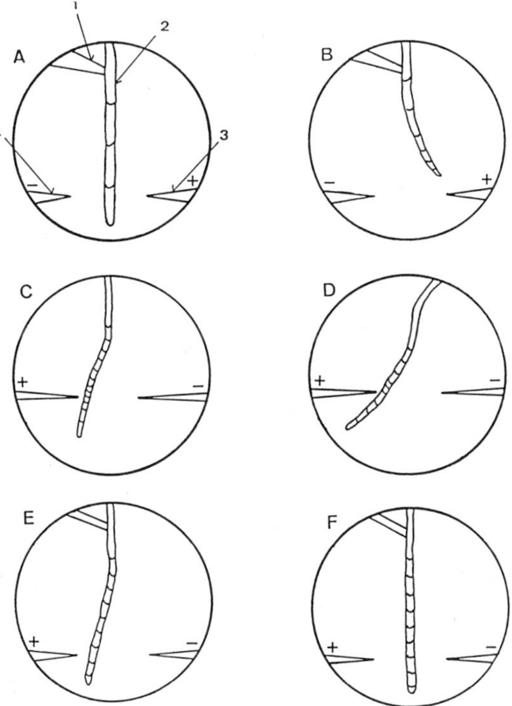

Figure 5. Elongation of *R. solami* hypha under electric field.

Electric Field: A: Monopolar pulse, V = 900 $V \cdot cm^{-1}$, 10 KHz; B: DC 90 $V \cdot cm^{-1}$; C: DC 360 $V \cdot cm^{-1}$; D: DC 360 $V \cdot cm^{-1}$; E: DC 1100 $V \cdot cm^{-1}$; F: DC 1100 $V \cdot cm^{-1}$. Each segment means elongation during 5-min interval (for A, B, E, F) and 3-min interval (for C, D), respectively.
1: Supporting capillary tube; 2. Hypha; 3: Anode; 4: Cathode. Distance between the tips of microelectrode is 60 μm.

Electric field effects on the elongation of *R. solani* hypha

Initially, an alternating pulsing electric field was applied, but no appreciable effect was observed on the hypha elongation. The pulse mode was then changed to a one directional pulsing field. Still no appreciable effect was obtained (Figure 5-A). In sharp contrast, the hypha bent towards the anode (positive electrode), when the DC field (90 $V \cdot cm^{-1}$) was applied (Figure 5-B). At the same time, cell growth ceased at the loci about 15 μm apart from the anode. Figure 5-C and D show similar results that hypha elongates towards the anode. In these cases, the growing rate was retarded near the electrode, but it recovered after the top of the hypha passed away from the anode. Confusing the matter, a hypha that elongated initially towards the anode under DC field (Figure 5-E) did not bend toward the anode any more the second time, though the electric field condition was the same as the initial time (Figure 5-F). These results indicate that the elongation of hypha is not influenced by the present condition of alternating electric field. It is also suggested that fungi would become more insensitive to electric field upon repeated exposure to an electric stimulus.

Electric field effects on the budding direction of yeast

The cell budding direction was determined for each cell of a wild type strain of *S. cerevisiae* as defined in Figure 6. Without electric field, cell budding occurred in every direction. The average of θ ($\bar{\theta}$) for 12 samples was 47° and its standard deviation (σ) was 32° as listed in Table 1. Thus, immobilization of a cell at the tip of a glass capillary caused no appreciable effect on the budding direction. In sharp contrast, θ and σ were 45° and 26° under an alternating electric field. The next problem was to determine which stage of a cell cycle is most critically influenced by an electric field. Hydroxyurea (HU) is well known to inhibit DNA synthesis of yeast cells and to arrest them around the starting point of the S-stage. Removal of HU restarts the cell cycle. Therefore, the cells treated with HU become a uniform population around the start point of the S-stage. As shown in Table 1, HU-treated cells also budded in the direction of electric field. This suggests that the critical point for the determination of budding direction should exist in the period after the initiation of S-stage.

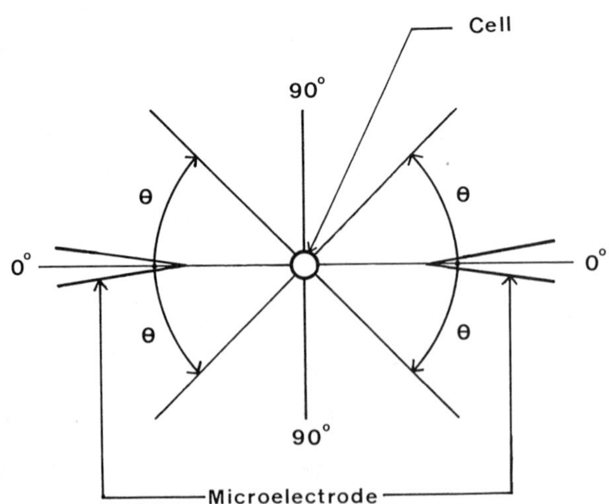

Figure 6. Definition of cell growth direction.

Table 1. Budding Direction of Yeast Cell Under Alternating Pulsing Electric Field.

Strain		EF(-)[1]			EF(+)[1]		
		N	θ[4] (°)	σ[5] (°)	N	θ(°)	σ(°)
Wild type	NT[2]	12	47	32	13	15	16
	HU[3]	12	45	26	7	17	18
cdc	*cdc*35	15	46	26	15	16	16
	*cdc*28	15	40	11	15	21	15
Theoretical value		-	45	26	-	-	-

1) EF(-), EF(+) means "without electric field" and "under electric field", respectively.

2) No treatment.

3) Treatment with hydroxyurea.

4) The average of θ defined as follows: $\bar{\theta} = \frac{1}{N}\sum_{i=1}^{N} \theta_i,\ 0 \le \theta_i \le 90$

5) Standard deviation:

$$\sigma = \sqrt{\frac{1}{N}\sum_{i=1}^{N} (\theta_i - \theta_i)^2}$$

From the standpoint of synchronization, *cdc* mutants strains are much more useful than HU treated cells for the assignment of critical point when budding direction should be determined. At first we used *cdc* 35 which can be arrested at the farthest point from the budding. From the result as shown in Table 1, it is concluded that the budding direction may be controlled by electric field after SPBSF, which suggests the critical point should exist after SPBSF.

As the next sample, *cdc* 28 was used in a similar manner as above. The results are included in Table 1. Directional dependency was also observed with the electric field, though $\bar{\theta}$ was comparatively larger than $\bar{\theta}$ obtained with *cdc* 35. These results demonstrate that the electric field is effective to cells existing in the period between SPBE and BE. More precise analysis is under investigation using other *cdc* mutants.

Electric field effects on the direction of neurite elongation

Initially, neurite elongation of a sample cell was checked as a control without electric stimulus. At first the sample cell stretched four short neurites, two of which became longer remarkably for 4 h. Then one of these two neurites further stretched towards another cell about 50 μm apart from the sample cell. After that, elongation and shrinking of neurites, movement of the sample cell, adhesion and dissociation of cells were repeatedly observed until after 50 h. Figure 7 shows the direction and length of each neurite which was recognized at respective times. The direction of elongation was defined as shown in Figure 6. From these results, no noticeable dependence was observed in the direction of neurite elongation.

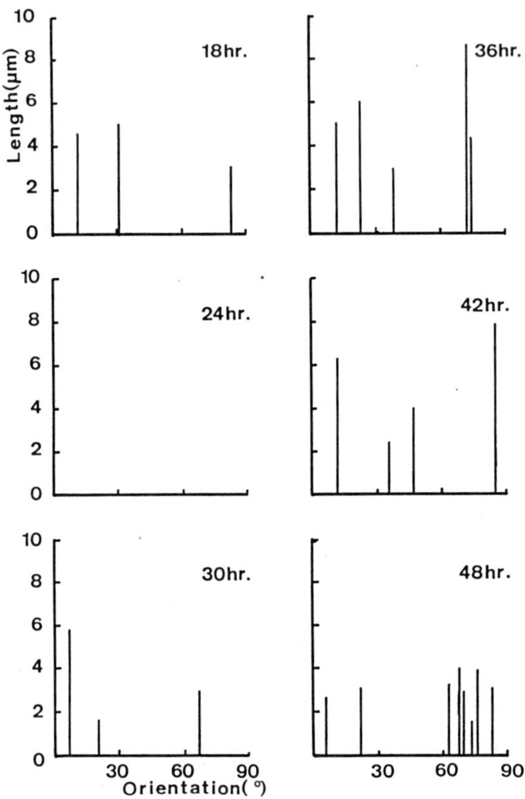

Figure 7. Neurite elongation of PC12h-AG8 without electric stimulus.

In the same manner, observation was performed under the alternating pulsing electric field. The peak height of the electric field was approximated as 470 $V \cdot cm^{-1}$. The cross membrane potential was estimated at about 0.7 V by assuming the cell diameter to be 20 μm. Until after 9 h, several very short neurites appeared. Then, one of them stretched noticeably in parallel with the electric field and became about 20 μm long at 12 h. This neutrite, however, shrank and another neurite stretched in the opposite direction at 18.5 h. The latter neurite held its length (40 μm) until after 21 h. After that the cell moved along this neurite. As demonstrated in Figure 8, the neurite showed a tendency that the direction of elongation was in parallel with that of the electric field. On the other hand, among the neurites which have already stretched, those in parallel with the electric field remained unchanged or further elongated,

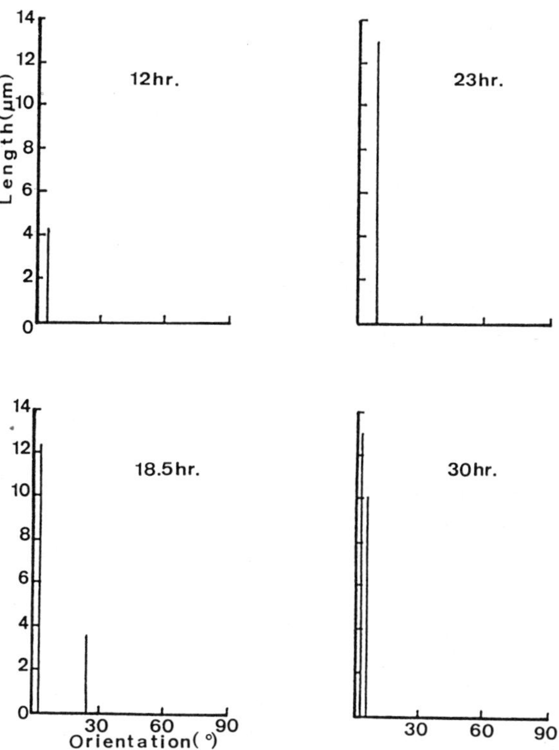

Figure 8. Neurite elongation of PC12h-AG8 under electric stimulus.

while those in other directions tended to shrink or disappear. Therefore, the neurite elongation is appreciably influenced by the alternating pulsing electric field.

DISCUSSION

Under the electric field, the growth direction of three types of cell were investigated. Yeast cell and nerve cell showed a tendency for their growth direction to be parallel to the electric field. Since an alternating electric field was applied, an electrophoretic effect should not occur to any significant extent. Rather, dielectric polarization should cause governing effects. Under the alternating electric field, the cell membrane sphere would be polarized and consequently deformed. Such a deformation might be a trigger for cell budding or for the initiation of neurite elongation in the case of yeast and nerve cell, respectively.

However, in the case of hypha, the electrophoretic effect might be important. It is supposed that some vacuoles containing cell wall hydrolyzing enzymes gather near the growing end of hypha. If these vacuoles are forced to move to one side, this side would become a growth initiating point. Since such vacuoles exist also in yeast, the electrophoretic effect should occur if a DC field was applied to yeast[15]. As regards neurite elongation, it was also reported earlier that the growing direction tended towards the cathode[4]. These papers support the opinion that vectorial biological phenomena could be controlled by DC field. In contrast, however, alternating electric fields are also effective for vectorial control in a different manner, as demonstrated in the present study.

ACKNOWLEDGMENTS

The authors are grateful to the following persons for kind advice and valuable discussions: As regards *R. solani* to Dr. D. Hosokawa and Dr. T. Teraoka, as regards *cdc* mutants to Dr. T. Ishikawa, Dr. Y. Ohsumi and Dr. Y. Ohya, as regards PC12h-AG8 to Dr. H. Hatanaka, and as regards computer analysis to Dr. S. Yamada.

REFERENCES

1. G. March and H.W. Beams, J. Cell. Comp. Physiol., 27, 139 (1946).

2. P. Vassilev, R. Dronzin, M. Kanazirska and G. Georgiev, Stud. Biophys., 90, 111 (1982).

3. N. Patel and M-m. Poo, J. Neurosci., 2, 483 (1982).

4. N.B. Patel, Z-P. Xie, S.H. Young and M-M. Poo, J. Neurosci. Res., 13, 245 (1985).

5. H. Matsuoka, S. Matsumoto, Y. Takekawa and N. Ai, Bioelectrochem. Bioenerg., 16, 235 (1986).

6. H. Fröhlich, Adv. Electron. Electron Phys., 50, 85 (1980).

7. H. Fröhlich and F. Kremer, Eds., "Coherent Excitations in Biological Systems", Springer-Verlag, Berlin (1983).

8. H.A. Pohl, in: "Bioelectrochemistry", H. Keyzer and F. Gutmann, Eds., Plenum Press, New York (1980), p. 273.

9. J.R. Pringle and L.H. Hartwell, in: "The Molecular Biology of Yeast Saccharomyces Life Cycle and Inheritance", J.N. Strathern, E.W. Jones and J.R. Broach, Eds., Cold Spring Harbor Laboratory, New York (1981), p. 97.

10. F. Boutelet, A. Petitjean and F. Hilger, EMBO J., 4, 2635 (1985).

11. J. Sy and Y. Tamai, Biochem. Biophys. Res. Commun., 140, 723 (1984).

12. S.A. Moore, Exp. Cell Res., 151, 542 (1984).

13. H. Hatanaka and H. Tamagawa, Proc. 3rd Int. Cell. Biol., Tokyo (1984).

14. H. Pauly, H.P. Schwan, Zeitsch. f. Naturforschung, 14b, 125 (1959).

15. A. Frey-Wyssling, Proc. Natl. Acad. Sci. U.S.A., 68, 636 (1971).

ELECTROCHEMICAL MANIPULATION OF SINGLE CELL WITH A MICROELECTRODE

Masashi Yaoita, Junichiro Kojima,
Hiroaki Shinohara and Masuo Aizawa*

Department of Bioengineering
Tokyo Institute of Technology
Ookayama, Meguro-ku
Toyko 152 / Japan

Yoshihito Ikariyama

Research Institute
National Rehabilitation Center for Disabled
Namiki, Tokorozawa
Saitama 359 / Japan

INTRODUCTION

The development of cell manipulation methods is of considerable current interest. There have been intensive investigations of dielectric effects on living cells, which has led to the commercialization of an apparatus for electric cell fusion[1-3]. Exposure of a cell suspension to an electric field or to an electric pulse on the order of 10^3 $V \cdot cm^{-1}$ results in the breakdown of the cellular membrane[4].

In a different line of research, we have proposed electrochemical manipulation of a single cell using either a bare or a conducting polymer film-coated microelectrode. In a previous report, we have shown with a three electrode system that erythrocytes burst on the surface of electrodes at an applied potential much lower than ca. 1.5 V vs. Ag/AgCl[5]. The cause of erythrocyte breakdown remains unsolved. Electrical as well as physico-chemical effects, caused by potential application, may induce erythrocyte lysis due to their susceptibility to breakdown by changes in pH, osmotic pressure, and so on.

Electrochemical cell destruction has also been observed with cultured human tumor cells, HeLa cells, adsorbed onto an In_2O_3 optically transparent electrode (OTE) whose potential range is from -0.8 V to 0 V and from 0.7 V to 2.0 V vs Ag/AgCl. It was noted that HeLa cells were deformed in the same potential range. We observed significant morphological changes of HeLa cells spread in the positive region of applied potential[6]. These cells swelled and burst in the negative potential region. A polypyrrole film-coated electrode was found to induce drastic effects on the destruction of erythrocytes and HeLa cells under moderate conditions, which might result from potential-controlled dynamic ion transfer on the polypyrrole surface.

In this report, we describe electrochemical effects on HeLa cells in the vicinity of a microelectrode. The feasibility of electrochemical manipulation of an individual cell is discussed.

EXPERIMENTAL

Cell Cultivation[7]

HeLa cells were cultured in a CO_2 incubator by controlling the CO_2 concentration at 5% at a constant temperature of 35°C. The culture medium was McCOY's 5A mixed with 10% PBS. Cultured HeLa cells were treated with trypsin, washed, and collected by centrifugation.

Electrode System

A microelectrode was made of a Teflon-coated platinum wire having a diameter of 50 μm. The microelectrode was installed with a micromanipulator. An Ag/AgCl reference electrode (1 mm (dia) x 10 mm) and

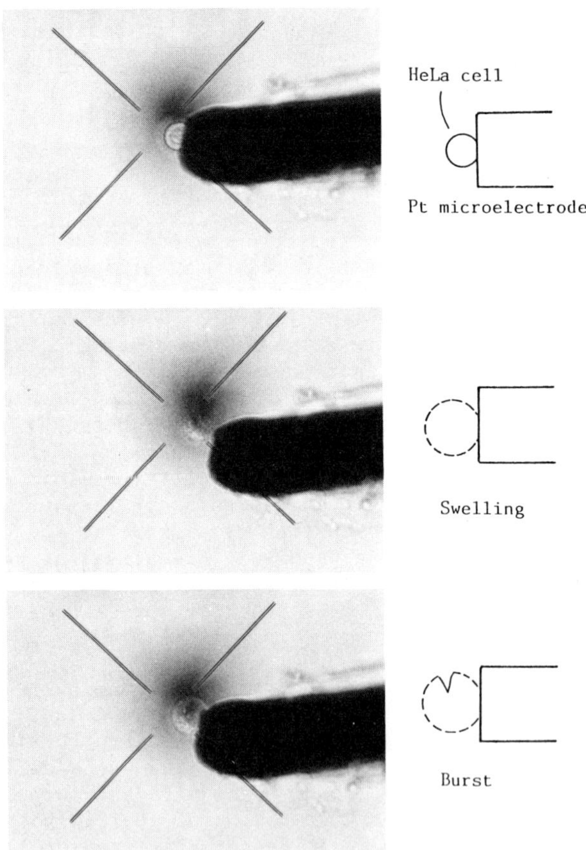

Figure 1. Electrochemical effect on the target cell in contact with a platinum microelectrode at a constant potential of -0.6 V vs Ag/AgCl.

platinum plate counter electrode (10 mm x 10 mm) were positioned in a Petri dish. The microelectrode potential was controlled with reference to an Ag/AgCl electrode with a potentiostat (Hokuto Denko Co., HA-301) and a function generator (Hokuto Denko Co., HB-104).

Polypyrrole Film Synthesis[8]

A polypyrrole thin membrane was synthesized and deposited on a platinum microelectrode by the electrochemical polymerization of pyrrole in a physiological salt solution. During this procedure oxygen was excluded by bubbling N_2 gas. The charge during polymerization was 5 μC which corresponded to a polymerization charge of a 500 nm-thick membrane. After polymerization, the polypyrrole film-coated microelectrode was assayed by cyclic voltammetry in a physiological salt solution.

Cell Observation

A microelectrode was placed in contact with a target HeLa cell suspended in physiological saline using a micromanipulator under an inverted microscope. Morphological changes of the HeLa cell were observed with a video camera and a video monitor, videotaped, and, if necessary, photographed. The diameter of the HeLa cell was measured on the video monitor.

RESULTS AND DISCUSSION

Cellular Morphological Change under the Potential Application

Using the micromanipulator, the microelectrode was placed in contact with the target HeLa cell. At a constant potential of -0.6 V vs Ag/AgCl the target cell first swelled and then burst. Photographs exhibiting these effects are shown in Figure 1. Figure 2 shows the change in the diameter of the HeLa cell as a function of time with the microelectrode maintained

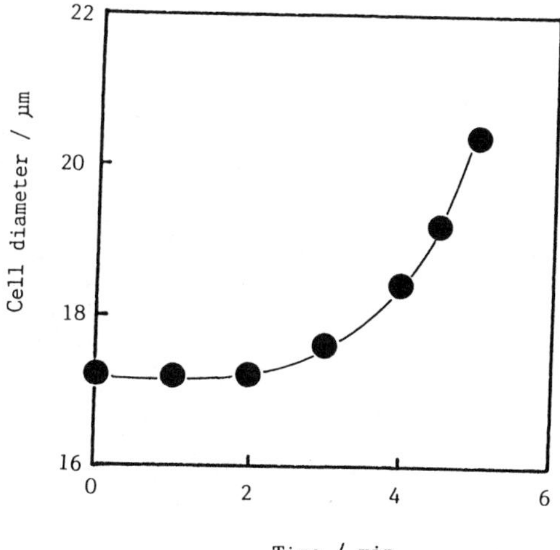

Figure 2. Time course of HeLa cell diameter change at a constant potential of -0.6 V vs Ag/AgCl.

at -0.6 V. The cell began to increase in diameter after 2 min. The diameter of the cell continued to increase for about 5 min until it burst. The diameter of the cell increased by about 18 % before it finally burst.

Electrochemical Effect on a HeLa Cell at Constant Potential

The electrochemical effect was evaluated by measuring the burst time of the cell. Electrochemical swelling and burst of the cells occurred in the potential range below 0.0 V. In contrast, no appreciable swelling or bursting of the cells was observed in the potential range above 0.0 V. It took 24 min until the cell burst at -0.1 V. At -0.8 V, HeLa cells burst after approximately 2 min. Thus, the elapsed time until cell burst depended markedly on the applied potential. Based on a platinum macroelectrode study, the pH at the interface of a platinum electrode, pH 5, became very alkaline (pH 10) at a potential below 0.0 V.

Electrochemical effects on HeLa cells were observed in the vicinity of the microelectrode surface. At a distance of 80 μm from the microelectrode surface, the cell burst within 20 min when the potential was controlled at -0.8 V. The duration time of the cell-burst depended sharply on the distance from the microelectrode. At 20 μm from the electrode, the cell burst within 2 min. Swelling and burst of the cells at a distance of 80 μm in a physiological salt solution suggest that these phenomena are caused by an electrochemically induced effect such as pH change.

Evaluation of Electrochemical Effect with Applied Rectangular Potential

To promote the electrochemical burst of cells, a rectangular wave potential was applied to the microelecrode. If, as was assumed, cell burst was caused by the local pH change at the electrode interface, a rectangular wave potential seemed to repeatedly cause the effect.

When the rectangular potential was modulated between -0.6 V and 0.6 V at a frequency of 5 Hz, swelling and bursting of the target cells were

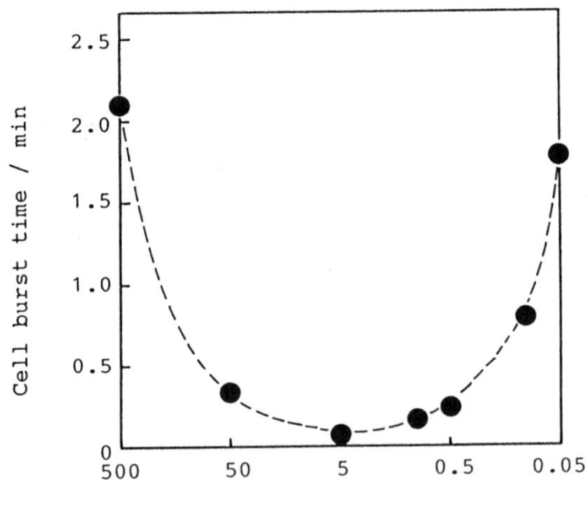

Figure 3. Frequency dependence of the target cell-burst time under a +0.8 V to -0.8 V peak to peak rectangular wave potential.

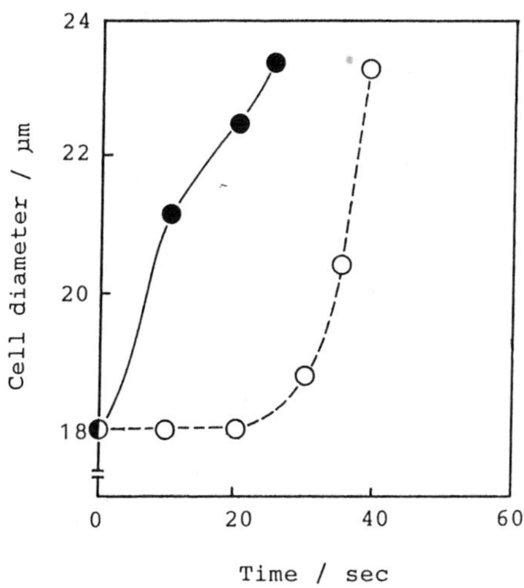

Figure 4. Time course of HeLa cell diameter change using a polypyrrole film-coated and a bare Pt microelectrode under a rectangular wave potential at 5 Hz.
-●-; Polypyrrole film-coated electrode at +0.4 V to -0.6 V peak to peak potential.
-○-; Bare Pt electrode at +0.6 V to -0.6 V peak to peak potential.

observed within only 39 s. Since HeLa cells required about 5 min to burst at a constant potential of -0.6 V, the rectangular wave potential seems to have amplified the electrochemical effect.

Figure 3 shows the relation between frequency and cell burst time, which was shortest with a rectangular wave potential between -0.8 V and 0.8 V. At 500 Hz and at 0.05 Hz, target cell-burst was observed within about 2 min. At a frequency of 5 Hz, the HeLa cell-burst time was the shortest and required only a few seconds.

At a frequency of 5 Hz the burst-time depended on the potential of the rectangular wave. When the largest potential was set at +0.8 V, the negative potential was closely related to the time for cell-burst. For example, when the modulation was set at 0.8 V and -0.4 V, a very long time (144 s) was required for the cell-burst. However, when the two potentials were 0.8 V and -0.8 V, the cell-burst occurred within 4 s. On the other hand, a less remarkable effect of the potential application on HeLa cells was observed when the positive potential was changed. By keeping the negative potential at -0.8 V, the burst-time was 45 s at a positive potential of 0.4 V. The increase of the positive potential caused the decrease in the burst time.

Amplification of Electrochemical Effect Using Polypyrrole Film-Coated Microelecrode

When a rectangular potential waveform from -0.6 V to 0.4 V at a frequency of 5 Hz was employed using a polypyrrole film-coated micro-electrode prepared by following the conditions described above, the target HeLa cell swelling and burst were induced within 20 s. In the latter potential range polypyrrole is reported to be stable. Figure 4 shows that

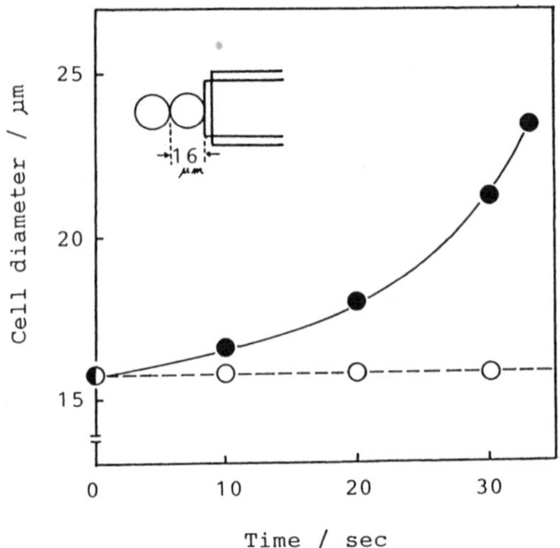

Figure 5. Selective electrochemical cell swelling and burst under +0.4 V to -0.6 V peak to peak rectangular wave potential at 500 Hz using a polypyrrole film-coated microelectrode.
-●-; Target cell in contact with the microelectrode.
-○-; Non-target cell in contact with the target cell at a distance of 16 μm from the electrode.

the electrochemical effect was amplified by a polypyrrole film-coated microelectrode. By the application of a rectangular potential from 0.6 V to -0.6 V at a frequency of 5 Hz with a bare platinum microelectrode the cells began to swell after 20 s and burst within 40 s. With the polypyrrole film-coated electrode the target HeLa cells were made to swell and burst more rapidly than with a bare platinum electrode. This amplification effect was enhanced at a higher frequency range. Because of this amplification at 500 Hz, the cells in contact with the polypyrrole-coated microelectrode burst at 33 s after application of the rectangular potential. The time was shorter than that (126 s) observed with a bare platinum electrode.

Using a high frequency potential we observed that the electrochemical effect became more significant than that with a low frequency. The further the position between the target cell and the tip of a microelectrode was, the more remarkable the effect that was seen. This observation suggests that cellular morphological change and burst were induced selectively to the target cell in contact with the electrode under these conditions. Figure 5 shows the selective effect when a rectangular wave potential from -0.6 V and 0.4 V at 500 Hz was employed. Under these conditions, the target HeLa cell on the surface of a polypyrrole film-coated microelectrode swelled and burst after 33 s. A non-target cell directly in contact with the target cell and at a position of 16 μm from the microelectrode showed no morphological changes for 10 min.

CONCLUSION

By miniaturizing electrodes, electrochemical effects noted in macroelectrode studies can be studied at single cells. Cell-burst could be controlled by the application of a potential through a microelectrode. Applied potentials from -0.1 V to -0.8 V induced cell-burst in the vicinity of the electrode. Using a square wave applied potential, the electrochemically-induced bursting of HeLa cells was caused more rapidly than that using a constant potential.

Employment of a polypyrrole-coated microelectrode brought about the amplification of the electrochemical effect. By the adoption of a high frequency square wave potential application, the susceptibility of the target cell at the interface of the microelectrode to bursting was remarkably enhanced. It should be noted that the polypyrrole film-coated electrode demonstrated interesting functions even under the conditions needed for a bare platinum electrode. This might be interpreted by the electrochemically induced influx of dopant anions onto the polypyrrole film matrix, which seems to result in drastic pH changes in the localized area, i.e., the interface between an electrode and bulk solution.

Using an electrochemical microsensor, substances from a target cell can be measured. Therefore, by the combination of the electrochemical manipulation of a single cell with a sensing electrode, we expect to apply the manipulation technique for extracting and measuring intercellular substances of a target cell which may lead to a new diagnostic method.

REFERENCES

1. M. Aizawa, T. Toyoshima, M. Yaoita and Y. Ikariyama, J. Chem. Soc. Japan, 1985, 1302 (1985).

2. H. Berg, K. Augsten, E. Bauer, W. Förster, H.-E. Jacob, P. Mühlig and H. Weber, Bioelectrochem. Bioenerg., 12, 119 (1984).

3. A. Kurischiki and H. Berg, Bioelectrochem. Bioenerg., 15, 513 (1986).

4. U. Zimmer, G. Pilwat and F. Riemann, Biophys. J., 1974, 881 (1974).

5. A.J.H. Sale and W.A. Hamilton, Biochim. Biophys. Acta, 163, 37 (1968).

6. M. Yaoita, H. Shinohara, Y. Hayakawa, T. Yamashita, Y. Ikariyama and M. Aizawa, Bioelectrochem. Bioenerg. (in press).

7. T.A. McCoy, M. Maxwell and P.F. Kruse, Proc. Soc. Exp. Biol. Med., 100, 115 (1959).

8. A.F. Diaz, K. Kanazawa and G.P. Gardini, J. Chem. Soc., Chem. Commun., 635 (1979).

INTRODUCTION OF NEW ALCOHOL BINDING SITES AT THE LIPID/WATER INTERFACE BY THE INCORPORATION OF MONOSIALOGANGLIOSIDES INTO DPPC LIPOSOMES

George P. Kreishman*, Cindy Graham-Brittain
and Harold E. Schueler

Department of Chemistry
University of Cincinnati
Cincinnati, Ohio 45221

Robert J. Hitzemann

Department of Psychiatry and Behavioral Sciences
S.U.N.Y.-Stony Brook
Stony Brook, New York 11794

INTRODUCTION

Following Overton's early observation of the direct relationship between the efficacy of an alcohol and oil/water partition coefficient of that alcohol[1], most research concerning the mechanism of action of ethanol and similar anesthetics has focused on the interior of neuronal membranes and model membrane systems. The lipid perturbation hypothesis simply states that the effects of alcohols result from changes in the fluidity of the interior of the membrane. The observed changes in membrane order, however, are usually small or occur at nonphysiological concentrations of alcohol[2].

If perturbation of the membrane interior is indeed the mechanism of anesthetic action, then the differences in alcohol sensitivity observed in test animals could be explained by differences in the animals' membrane-lipid composition. In the last decade, mice have been genetically bred solely on the basis of their ethanol sensitivity. Two genetically different strains of mice, the long sleep mice (LS, ethanol sensitive mice) and the short sleep mice (SS, ethanol resistant mice), now exist. The difference in sensitivity is so large that a lethal dose for the LS mice has no effect on the SS mice[3]. Analyses of total protein, types of fatty acids, lipid head groups, and all other membrane components have determined that the single major difference in the neuronal membrane composition of these mice is the monosialogangioside (GM_1) content[4,5]. The GM_1 concentration in membranes from the cerebellum tissue of LS mice is nearly three times greater than that in the tissue of SS mice. GM_1 has been shown to sensitize model membrane systems to the effects of ethanol; whereas other structurally similar lipids, such as sphingomyelin (SM), found in membranes of other cell types, do not[6]. The primary difference between these two lipids is in their head groups: GM_1 has a bulky, negatively-charged head, SM has the smaller, neutral phosphatidylcholine.

Comparative fluorescence studies by Harris show that while there is no measurable difference in the mobility of a fluorescent probe in the interior of ethanol-treated membranes of the LS and SS mice, there is a difference in the mobility of a surface probe, with the greater mobility at the surface of the LS mice membranes[7]. While the strains show a difference in their sensitivity to ethanol, they show the same sensitivity to the more hydrophobic alcohol butanol, even though both alcohols elicit the same disordering effects at the interior of the bilayer, albeit at different concentrations[3].

Recent research in our laboratory has opted to take an alternate approach in considering the interactions of alcohols with both the hydrophobic interior and the previously overlooked lipid-water interface. Using ^{2}H-NMR, it has been demonstrated that ethanol not only partitions to the interior of a membrane, it also binds in an apparently cooperative manner to the surface of model dipalmitoylphosphatidylcholine (DPPC) liposomes and synaptic plasma membranes[8,9]. The surface binding of ethanol has an ordering effect at the lipid/water interface that opposes the previously observed disordering induced by the ethanol at the interior as determined by delayed Fourier transform spectroscopy (DFT ^{1}H-NMR)[9]. Subsequently, Evers et al.[10] have discovered a similar immobilized site for the more lipophilic anesthetic, halothane.

Preliminary results are discussed which may provide a better understanding of the factors that are important in defining the properties of the lipid/water interface and their relationship to the overall goal of understanding the mechanism of anesthetic action.

EXPERIMENTAL

Chemicals

Both DPPC and GM_1 were purchased from Sigma Chemical Company, St. Louis, MO, and were used with no additional purification. The perdeuterated ethanol was obtained from MSD isotopes, and the D_2O and perdeuterated butanol were obtained from Aldrich.

Apparatus

All NMR spectra were acquired on a Nicolet NMC 300 MHz FT-NMR spectro meter with a magnetic field strength of 7.05 Tesla. The parameters for the DFT ^{1}H-NMR experiments are described in detail in Reference 9. Baselines were drawn between the minima of each resonance. The spectral intensities were obtained by the cut-and-weigh method and are expressed in arbitrary units. The ^{2}H-NMR spectra were obtained using the lock coil of the 5 mm ^{1}H probe at a spectral frequency of 46.066 MHz. The temperature of both the ^{1}H and ^{2}H experiments was maintained at 46°C using an NTC temperature control unit. Sonication was performed using a Cole-Palmer 8851 ultrasonic water bath.

Liposome Preparation

Stock solutions of DPPC and GM_1 in chloroform, and GM_1 in chloroform/methanol (1:0.1 v/v) were prepared prior to vesicle formation. To produce the desired DPPC to GM_1 ratio, which is expressed as a mole to mole ratio, appropriate volumes of these stock solutions were mixed in a round bottom flask, and the solvent was removed by aspiration using a rotary evaporator so that the lipids evenly coated the walls. After the apparent removal of the solvent, the samples were lyophilized at ambient temperature for several hours to ensure that any residual solvent was removed. The temperature of the dried lipid was equilibrated above the phase transition temperature (T_m) at 52°C in a constant temperature bath. H_2O or D_2O-phosphate buffered saline (pD = 7.6, 10 mM phosphate, 125 mM saline), also equilibrated to 52°C, was added to the sample, followed by frequent vortexing. The sample was sonicated for approximately 1-2 min. Deuterated alcohol was added to give the desired concentration, and the final volume was adjusted to 0.4 ml and a DPPC concentratiion of 5 mg/ml.

RESULTS

The effect of increasing GM_1 content on the DFT ^1H-NMR spectrum of DPPC liposomes at 46°C in the absence of alcohol is shown in Figure 1. At low concentrations of GM_1 (1.0 M:0.1 M and 1.0 M:0.2 M DPPC to GM_1), the intensities of the choline methyl resonance, the methylene resonance and the terminal methyl resonance all increase as compared to pure DPPC. At higher GM_1 concentrations (1.0 M:0.3 M), decreases in the spectral intensities of the choline methyl, methylene, and terminal methyl resonances are observed. As described elsewhere[9], changes in the observed spectral intensity of a given resonance in DFT ^1H-NMR are the result of changes in the membrane order around the reporting group. Any increase in the intensity results from an increase in disorder, while any decrease in intensity results from an increase in order. Therefore, GM_1 has a disordering effect at low concentrations but an ordering effect at high concentrations. Previous fluorescence polarization studies have shown that at high GM_1 concentrations, there is an increase in the melting temperature of the liposomes and a more ordered interior at 46°C[11]. One possible explanation of this behavior is that the electrostatic repulsion of the negatively charged sialic acid residues of the GM_1 polar head groups cause the gangliosides to separate via lateral diffusion of the lipids and induce an interdigitated state. This is consistent with the increase in the T_m upon formation of this state[12]. At the lower GM_1 concentrations, the observed disordering effect could be either a small change in order over the entire membrane or a large change in order for a localized region. Since the increase in the spectral intensities is dependent on the GM_1 concentration, the membrane disordering is probably localized around the GM_1 molecules.

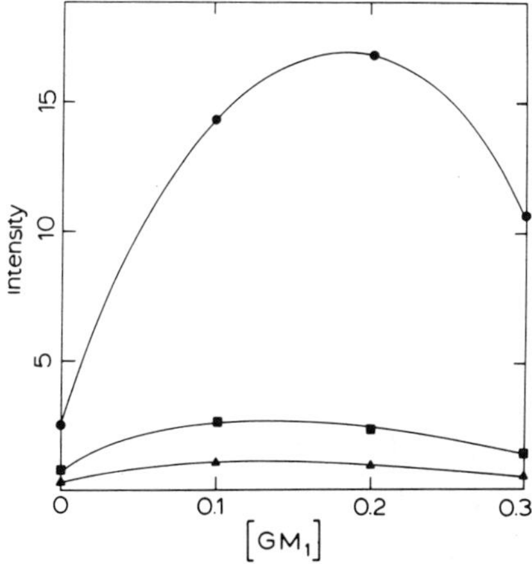

Figure 1. The spectral intensities of the methylene resonance (●), the choline methyl resonance (■) and the terminal methyl resonance (▲) of DPPC liposomes as a function of added GM_1 at 46°C. Concentrations are molar ratios of GM_1/DPPC.

Disruption of the close packing arrangement of a small number of neighboring lipids could occur due to the steric constraints of the bulky head group of GM_1. This creates "packing defects" where spaces between the lipid head groups are created on the surface of the bilayer.

^{2}H-NMR spectra of 1.0% (v/v) ethanol-d_6 in the presence of DPPC liposomes containing GM_1 concentrations of 0, 0.1, 0.2, and 0.3 M GM_1: 1.0 M DPPC at 46°C are shown in Figure 2. In the spectrum of ethanol with pure DPPC liposomes, a singlet for the CD_2 resonance at 3.7 ppm is observed. Upon addition of GM_1, this resonance is split into a doublet and reflects the binding of ethanol to the GM_1 induced "packing defects" at the lipid/water interface. The doublet is then the result of the rapid exchange of ethanol between the bulk water (a singlet) and the new site at the lipid/water interface (a doublet from quadrupolar splitting). A quadrupolar nucleus (such as ^{2}H) exhibits splitting (ΔV) when the molecule containing that nucleus is motionally restricted[13]. The magnitude of the splitting is determined by the attenuation of the quadrupolar coupling

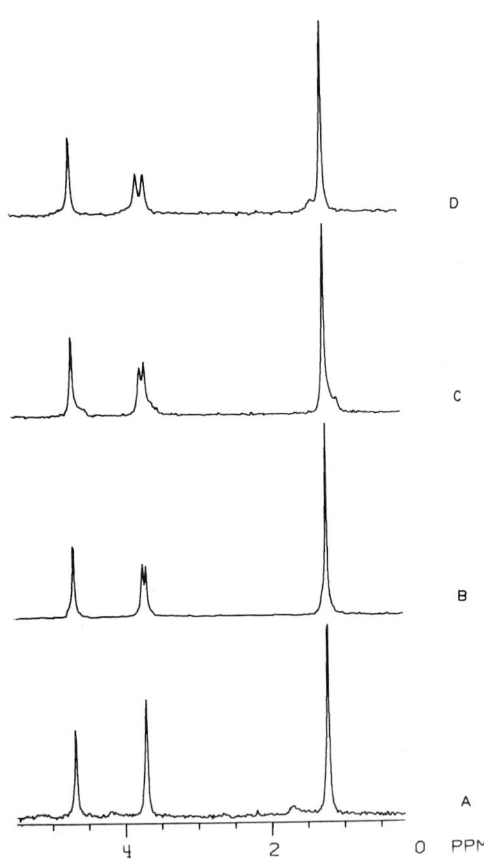

Figure 2. The ^{2}H spectrum of 1.0% v/v ethanol-d_6 in the presence of a) DPPC liposomes; b) DPPC/GM_1 (1.0 M:0.1 M) liposomes; c) DPPC/GM_1 (1.0 M:0.2 M) liposomes, and d) DPPC/GM_1 (1.0 M:0.3 M) liposomes, at 46°C.

constant (e^2qQ/h) by the restricted molecular motion, and is given by:

$$\Delta V = 3/2\ (e^2qQ/h)S$$

where S is the order parameter and ranges from 1 in a crystalline state to 0 in a rapidly tumbling state. Intermediate values occur for a pseudo-liquid such as the liquid-crystalline state of a lipid bilayer. S decreases toward the methyl end of an acyl chain, where molecular motion is less restricted. For a molecule undergoing fast exchange, the observed splitting would be given by:

$$\Delta V_{obs} = P_F\ \Delta V_F + P_B\ \Delta V_B$$

where P_F and P_B are the mole fractions of ethanol in the free and bound states, respectively. Since $\Delta V_F = 0$ for free ethanol, the observed splitting is simply:

$$\Delta V_{obs} = P_B\ \Delta V_B$$

While the ΔV_B for the methylene nuclei is large enough so that the resonance is split, the ΔV_B for the methyl nuclei is not; therefore, a singlet for the methyl resonance is observed. The splitting of the methylene resonance increases with increasing GM_1 concentration because the fraction of ethanol bound to this site increases with the formation of additional sites.

The DFT ^1H-NMR spectra of DPPC/GM_1 (1.0 M:0.1 M) liposomes in the presence of ethanol-d_6 are shown in Figure 3. At low ethanol concentrations, a small decrease in the choline methyl resonance is detected, indicative of surface ordering. The methylene resonance decreases markedly above ~ 1% ethanol. Above 2% ethanol, the choline methyl begins to increase. At 3% ethanol, there is ordering of the interior and disordering at the surface. This type of spectral behavior is consistent with the formation of an interdigitated state. As the distance between the head groups increases to accommodate the packing of the fatty acid chains, less restricted motion occurs at the surface while a more efficient packing of

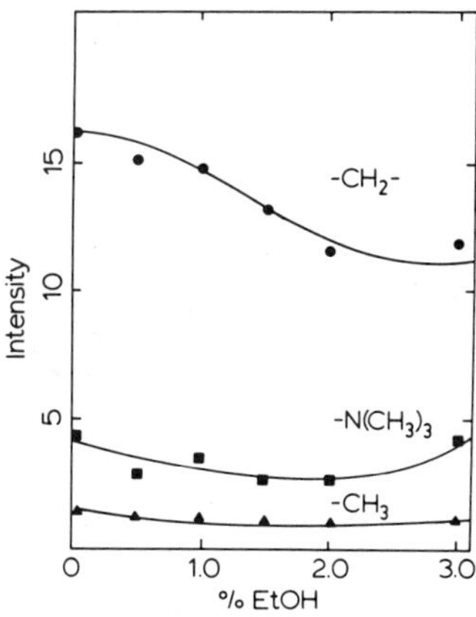

Figure 3. The spectral intensities of the methylene resonance (●), the choline methyl resonance (■) and the terminal methyl resonance (▲) of DPPC/GM_1 (1.0 M:0.1 M) liposomes as a function of added ethanol-d_6 at 46°C.

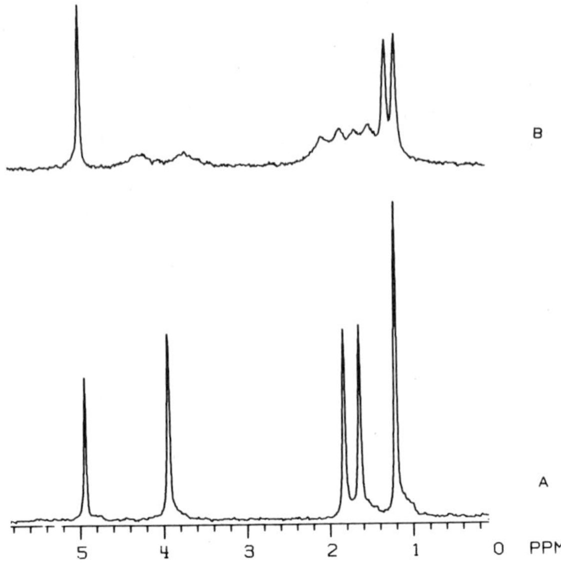

Figure 4. The ^2H spectra of 1.5% butanol-d_{10} in the presence of a) DPPC liposomes, and b) DPPC/GM_1 (1.0 M:0.1 M) liposomes, at 46°C.

the fatty acid chains restricts motion in the interior. Interdigitation need not occur over the entire bilayer but may be present only in small areas of the bilayer. A completely interdigitated state is induced by ethanol in pure DPPC liposomes but at much higher ethanol concentrations (>7%)[12,14]. The presence of GM_1 apparently promotes the formation of this state.

For the more lipophilic butanol, somewhat contrasting results are obtained. The ^2H spectra of butanol-d_{10} are shown in Figures 4 and 5. At least two types of binding can be discerned. From the decrease in overall intensities of the methylene and methyl resonances, slow exchange between a rigid binding site and the free state occurs in all cases. This site is tentatively assigned to interior binding as was shown for ethanol. The decrease is larger than for ethanol at equivalent concentrations, reflecting the larger partition coefficient for butanol[15]. In the presence of GM_1, splitting of the resonances is again observed and is attributed to binding to GM_1-induced "packing defects". With increasing butanol concentration, the splitting increases. The presence of butanol, therefore, must induce the formation of more of the new surface sites. In the DFT ^1H-NMR spectra of DPPC/GM_1 liposomes (1.0 M:0.1 M) with increasing butanol concentration, increases in the intensities of the choline, methylene, and methyl resonances are detected at all concentrations of butanol studied (Figure 6). Since the more hydrophobic butanol partitions to a greater extent in the interior of the bilayer[15], it has greater disruptive effects than ethanol. These effects may be so large that they mask more subtle surface effects which might be present at the concentrations used.

DISCUSSION

Lipids can adopt various forms of aggregates in solution: lamellar, interdigitated, hexagonal close pack and micellar. In some cases, the energy differences between these states is small, and conversion from one

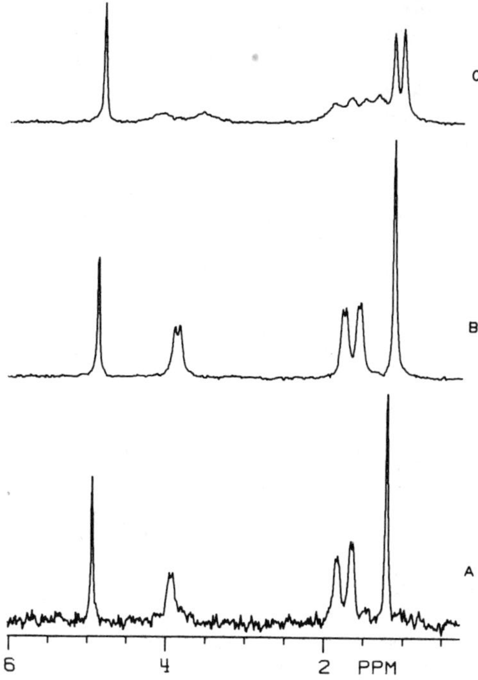

Figure 5. The concentration dependence of the ^{2}H spectrum of butanol-d_{10} in the presence of DPPC/GM_1 (1.0 M:0.1 M) liposomes at 46°C with a) 0.5% butanol-d_{10}, b) 1.0% butanol-d_{10} and c) 1.5% butanol-d_{10}.

form to another can be achieved by small changes in the environment or sample conditions. These can be changes in lipid composition, temperature, ionic strength, pH, etc.[16] or as in this study, the addition of alcohol.

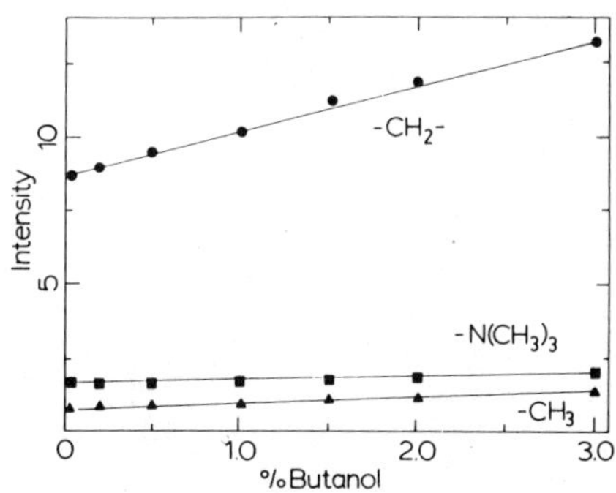

Figure 6. The dependence of the spectral intensities of the methylene resonance (●), the choline methyl resonance (■) and the terminal methyl resonance (▲) on butanol concentration for DPPC/GM_1 (1.0 M:0.1 M) liposomes at 46°C.

The presence of gangliosides at high concentrations in PC membranes increases the membrane order and increases the gel to liquid-crystalline phase transition temperature as detected in this study and by fluorescence probes[11]. The stabilized bilayer is possibly the result of the conversion of the bilayer to the interdigitated state. Interdigitation increases the average area occupied by the negatively charged sialo groups, thus lessening charge repulsion. This is similar to what is found for the negatively charged lipids such as PG, where repulsion between the head groups is sufficient to drive interdigitation[17]. At lower GM_1 concentrations, increased mobility of some of the lipids is observed. It is proposed that at temperatures just above the phase transition, the lipids near the GM_1 are mobilized; these regions bind alcohol and can be structurally altered by the presence of alcohol at physiologically relevant concentrations. It is the surface binding of alcohol to "packing defects" around the GM_1 molecules that makes the interdigitated state energetically favored in these localized areas. The hydrophilic hydroxyl group of the alcohol interacts favorably with the water at the lipid-water interface. In addition, the alkyl chains can undergo favorable hydrophobic interactions with the fatty acyl chains of neighboring lipids on the same side and the chain from a lipid on the opposing side of the membrane (Figure 7).

It has been shown that the area of the bilayer that is affected in this manner increases with increasing alcohol concentration. Although the concentrations of alcohol used in this study are above those encountered physiologically, it can be argued that small interdigitated areas exist at the lower concentrations, and it is only the sensitivity of the experimental techniques used that precludes the detection of these areas.

Even though the disrupted areas may be small, their physiological manifestations could be significant. Normal functioning of the neuronal membranes requires that the correct ionic gradients be maintained. In the RC description of gated ionic transport, there is a capacitive component, the lipid bilayer, and a switchable resistive component, the specific ionophore. One such RC system is the Cl^- specific ionophore in

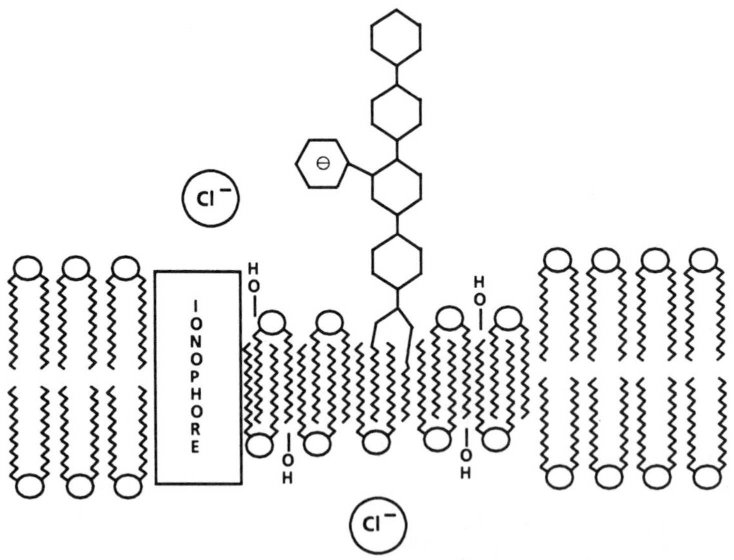

Figure 7. Proposed model for the GM_1 and alcohol-induced formation of regions of interdigitated lipids.

the postsynaptic membrane which is controlled by the neurotransmitter gammaaminobutyric acid (GABA)[18]. Binding of GABA to its receptor opens the ionophore, and Cl^- ions flow into the cells. The increased Cl^- ion concentration hyperpolarizes the membrane to such an extent that the threshold potential required to initiate the action potential cannot be achieved when acetylcholine binds to its receptor. The result is inhibition of neuronal transmission. One of the known actions of ethanol is the increased intake of Cl^- ions into nerve cells; however, this action is reversed by a known antagonist of ethanol action, Ro15-4513[19].

It is proposed that the alcohol-induced formation of small areas of the interdigitated state near the GM_1 can cause the hyperpolarization of the neuronal membrane by infusion of Cl^- ions, which elicits the behavioral effects of alcohols. Interdigitation could promote dielectric breakdown of the membrane or the induction of a conformational change in the ionophore protein such that ions can permeate the membrane. In either case, only a few ion-conducting sites need be activated for hyperpolarization to occur. Induction of increased permeability of metabolites across membranes has been demonstrated for the antibiotic polymyxin[20]. It has also been shown that polymyxin induces interdigitation[21].

In contrast to the reversible intoxication induced by low concentrations of alcohol, death results at some higher concentration. The large disruptive effects on membrane structure induced by butanol in this study, and by ethanol at higher concentrations in previous studies, could be the cause of the lethal effect of alcohol. In animals, the behavioral effects of ethanol can be blocked by pretreating the animal with the ethanol antagonist. But even with pretreatment, the animal cannot be protected from the lethal effect of ethanol. The behavioral effects and the lethal effect of alcohols must then be the result of two different mechanisms.

CONCLUSIONS

The disordering effects of alcohol and other anesthetics on the interior of membranes have been clearly demonstrated but cannot completely explain the mechanism of action of alcohol[22]. Our investigations have shown that alcohol-induced changes at the lipid/water interface are an integral part of this mechanism and warrant further study.

ACKNOWLEDGMENT

The authors wish to thank R. Adron Harris for helpful discussions on the content of this manuscript. Partial funding for purchase of the NMR was obtained from the National Science Foundation (CHE-8102974).

REFERENCES

1. E. Overton, Jena, Gustav Fischer, 101 (1901).

2. E. Rubin, Ed., Anal. N.Y. Acad. Sci., 492, 1 (1987).

3. R.A. Harris, personal communciation.

4. D. Ullman, R.C. Baker and R.A. Deitrich, Alcoholism: Clin. and Expt. Res., 11, 158-162 (1987).

5. R.C. Baker, Alcohol and Drug Research, 7, 291-299 (1987).

6. R.A. Harris, G.I. Groh, D.M. Baxter and R.J. Hitzemann, Mol. Pharmacol., 25, 410 (1984).

7. R.A. Harris, R. Burnett, S. McQuilkin, A. McClard and F.R. Simon, Anal. N.Y. Acad. Sci., 492, 125 (1987).

8. G.P. Kreishman, C. Graham-Brittain and R.J. Hitzemann, Biochem. Biophys. Res. Commun., 130, 301 (1985).

9. R.J. Hitzemann, H.E. Schueler, C. Graham-Brittain and G.P. Kreishman, Biochim. Biophys. Acta, 859, 189 (1986).

10. A.S. Evers, B.A. Berkowitz and D.A. d'Avignon, Nature, 328, 157-160 (1987).

11. R.J. Hitzemann, Chem. Phys. Lipids, 43, 25 (1987).

12. E. Rowe, Biochemistry, 22, 3299 (1983).

13. P.R. Allegrini, G. Van Scharrenburg, H.H. Mantsch and J. Seelig, Biochim. Biophys. Acta, 731, 448 (1983).

14. S.A. Simon and T.J. McIntosh, Biochim. Biophys. Acta, 773, 169 (1984).

15. Y. Katz and J.M. Diamond, J. Membrane Biol., 17, 101-120 (1974).

16. C.P.S. Tilcock and P.R. Cullis, Anal. N.Y. Acad.Sci., 492, 88 (1987) and references therein.

17. J.L. Ranck, T. Keira and V. Luzzati, Biochim. Biophys. Acta, 488, 432-441 (1977).

18. A. White, P. Handler, E.L. Smith, R.L. Hill and I.R. Lehman, "Principles of Biochemistry", McGraw-Hill, Inc., New York (1978), pp. 1115-1118.

19. P.D. Suzdak, J.R. Glowa, J.N. Crawley, R.D. Schwartz, P. Skolnick and S.M. Paul, Science, 234, 1243-1247 (1986).

20. D.R. Storm, K.S. Rosenthal and P.E. Swanson, Ann. Rev. Biochem., 46, 723-763 (1977).

21. J.L. Ranck and J.F. Tocanne, FEBS Lett., 143, 175 (1982).

22. C.D. Richard, K. Martin, S. Gregory, C.A. Keightly, T.R. Hesketh, G.A. Smith, G.B. Warren and J.C. Metcalfe, Nature, 276, 775 (1978).

INDEX